D1679120

S*pringer* **M***onographs in* **M***athematics*

Pekka Neittaanmaki Jürgen Sprekels
Dan Tiba

Optimization of Elliptic Systems

Theory and Applications

Springer

Pekka Neittaanmaki
Department of Mathematical
Information Technology
and
Department of Mathematics and Statistics
University of Jyvaskyla
Jyvaskyla, 40014
Finland
pn@mit.jyu.fi

Jürgen Sprekels
Angew. Analysis und Stochastik Abt. Part.
Differentialgleichg.
Weierstrass Institute for Applied Analysis
and Stochastics
Berlin, 10117
Germany
sprekels@wias-berlin.de

Dan Tiba
Institute of Mathematics
Romanian Academy
Bucuresti, 07070
Dan.Tiba@imar.ro

Mathematics Subject Classification (2000): 49-02, 49Q10, 35J85

ISBN-10: 0-387-27235-6 e-ISBN 0-387-27236-4
ISBN-13: 978-0387-27235-1

Printed on acid-free paper.

© 2006 Springer Science+Business Media, Inc.
All rights reserved. This work may not be translated or copied in whole or in part without the written permission of the publisher (Springer Science+Business Media, Inc., 233 Spring Street, New York, NY 10013, USA), except for brief excerpts in connection with reviews or scholarly analysis. Use in connection with any form of information storage and retrieval, electronic adaptation, computer software, or by similar or dissimilar methodology now known or hereafter developed is forbidden.
The use in this publication of trade names, trademarks, service marks, and similar terms, even if they are not identified as such, is not to be taken as an expression of opinion as to whether or not they are subject to proprietary rights.

Printed in the United States of America. (SHER)

9 8 7 6 5 4 3 2 1

springeronline.com

Preface

The present monograph is intended to provide a comprehensive and accessible introduction to the optimization of elliptic systems. This area of mathematical research, which has many important applications in science and technology, has experienced an impressive development during the past two decades. There are already many good textbooks dealing with various aspects of optimal design problems. In this regard, we refer to the works of Pironneau [1984], Haslinger and Neittaanmäki [1988], [1996], Sokołowski and Zolésio [1992], Litvinov [2000], Allaire [2001], Mohammadi and Pironneau [2001], Delfour and Zolésio [2001], and Mäkinen and Haslinger [2003]. Already Lions [1968] devoted a major part of his classical monograph on the optimal control of partial differential equations to the optimization of elliptic systems. Let us also mention that even the very first known problem of the calculus of variations, the *brachistochrone* studied by Bernoulli back in 1696, is in fact a shape optimization problem.

The natural richness of this mathematical research subject, as well as the extremely large field of possible applications, has created the unusual situation that although many important results and methods have already been established, there are still pressing unsolved questions. In this monograph, we aim to address some of these open problems; as a consequence, there is only a minor overlap with the textbooks already existing in the field.

The exposition concentrates along two main directions:

- the optimal control of linear and nonlinear elliptic equations, including *variational inequalities* and *control into coefficients problems*,
- problems involving unknown and/or variable domains, like general *shape optimization problems* defined on various classes of bounded domains in Euclidean space, or *free boundary problems* arising in various physical processes.

It should be noted that many shape optimization problems occur naturally as control into coefficients problems. A large and interesting class of examples of this type, to which the whole of Chapter 6 is devoted, concerns the optimization of basic mechanical structures like beams, plates, arches, curved rods, and shells.

There are strong connections between all these seemingly different types of problems. This fact has for the first time been illustrated in the so-called *map-*

ping method introduced by Murat and Simon [1976], which makes it possible to transform domain optimization problems into control into coefficients problems. Throughout this monograph, we will try to elucidate such connections. Another classical contribution to the solution of shape optimization problems is the *speed method*, which was introduced by Zolésio [1979] and thoroughly discussed in the above-mentioned publications.

One basic feature of this textbook is the endeavor to relax the needed regularity assumptions as much as possible in order to include large classes of possible applications. We have succeeded in this aim for several fundamental questions:

- The existence theory for general domain optimization problems presented in Chapter 2 requires just the uniform continuity of the domain boundaries.

- The existence theory and the sensitivity analysis for plates and for curved mechanical structures, mainly performed in Chapter 6, is established under regularity hypotheses that are one or two degrees (depending on the case) lower than those usually postulated in the scientific literature.

Another characteristic of this book is that we have tried to stress the application of optimal control methods even in the case of problems involving variable/unknown domains. In this respect, it should be mentioned that our techniques are close to the works of Lions [1968], [1983], Cesari [1983], Barbu [1984], [1993], and Barbu and Precupanu [1986]. We are thoroughly convinced that optimal control theory may provide a rather complete and reliable approach to the challenging problems involving the optimization of systems defined on variable domains. Many of the presented results in this direction, mostly in Chapter 5, are original contributions of the authors.

In order to give the reader a comprehensive overview of the subject, we also report on other important results from the existing literature. Whenever certain theoretical developments are already available in textbook form, our discussion will be limited to the shortest possible presentation.

The book is organized in six chapters that give a gradual and accessible presentation of the material, where we have made a special effort to present numerous examples, both at the theoretical and at the numerical level. The material covers

- motivating examples of "purely" mathematical nature or originating from various applications (in Chapter 1),

- general existence results for control and shape optimization problems (in Chapter 2),

- a sensitivity analysis of linear and nonlinear control problems in the absence of differentiability assumptions, based on various penalization methods (in Chapter 3),

- the presentation of the a priori estimates technique for the numerical approximation of control problems governed by linear or nonlinear elliptic equations (in Chapter 4),
- optimal control and other approaches in unknown domain problems including free boundaries and optimal design (in Chapter 5),
- a fairly complete optimization theory of curved mechanical structures like arches, curved rods, and shells (in Chapter 6).

The three appendices collect important notions and results from the theory of function spaces and elliptic equations, from convex and nonlinear analysis, and from functional analysis, which are frequently used throughout this monograph.

In Chapters 5 and 6, several rather complex geometric optimization problems are studied in detail and are completely solved, including numerical results. We do not discuss the questions that arise from the practical implementation of the presented methods on a computer or from the solving of the associated finite-dimensional problems, as they do not enter into the objective of this book.

Let us also mention at this place that in order to keep the exposition at a reasonable length and due to other reasons, several directions of active research, such as second-order optimality conditions, a posteriori error estimates, homogenization methods, and applications of shape optimization in fluid mechanics, could not be covered in this textbook. However, we have tried to provide the reader with the corresponding relevant references in some of these subjects.

Now we comment briefly on some examples and applications, and we make a more detailed presentation of the text. The aim is to give the reader, from the very beginning, a clear image about the problems and the questions that are studied in this book, and about their motivation and difficulties.

We consider first the simplest case of an elastic shell of constant thickness that admits a general cylindrical surface as its midsurface. We assume that the shell is clamped along two of its generators and the forces acting on it are constant along the generators and perpendicular to them. Consequently, it is clear that the resulting deformation of the shell is also constant along the generators.

It is enough to investigate a two-dimensional section perpendicular to the generators. The obtained structure in $\mathbf{R}^2$ is called an arch, and its deformation is described by the so-called Kirchhoff–Love model. We mention bridges, roads, industrial tubes, windows, roofs, among others, as real-life examples entering this description. The design of such structures puts several important questions to the engineer or the architect: maximize the mechanical resistence of the structure, minimize the total cost, fulfill all the (technological) constraints that are imposed, etc. In general, a "compromise" among the sometimes conflicting aims has to be found.

We indicate now the mathematical formulation of the Kirchhoff–Love model. If $\varphi = (\varphi_1, \varphi_2) : [0,1] \to \mathbf{R}^2$ is the parametrization of the arch with respect to its arc length and $c : [0,1] \to \mathbf{R}$ denotes its curvature, then the deformation vector $\bar{v} = (v_1, v_2) \in H_0^1(0,1) \times H_0^2(0,1)$ is the solution of

$$\int_0^1 \left[\frac{1}{\varepsilon}(v_1' - c\,v_2)(s)(u_1' - c\,u_2)(s) + (v_2' + c\,v_1)'(s)(u_2' + c_1\,u_1)'(s)\right] ds$$

$$= \int_0^1 (f_1\,u_1 + f_2\,u_2)(s)\,ds\,, \qquad \forall\, u_1 \in H_0^1(0,1)\,,\ \forall\, u_2 \in H_0^2(0,1)\,.$$

Here, $\sqrt{\varepsilon}$ represents the constant thickness of the arch and $[f_1, f_2] \in L^2(0,1)^2$ are, respectively, the tangential and normal components of the forces loading the clamped arch (assumed to act in its plane), while the tangential component v_1 and the normal component v_2 perform a similar representation for the deformation. The arbitrary functions $u_1 \in H_0^1(0,1)$ and $u_2 \in H_0^2(0,1)$ are test functions specific to the weak (variational) formulation of differential equations. Let us also mention that a complete study of this problem may be found in Ciarlet [1978, p. 432].

As the shape of the arch is completely characterized by its curvature c, the corresponding geometric optimization problems may be formulated as the minimization of some functional subject to the Kirchhoff–Love model as a side constraint and with the function c as the minimization parameter (control). For instance, one integral cost functional of interest is

$$\int_0^1 [v_2(s)]^2\,ds\,.$$

This means to find the form of the arch that has a minimal normal displacement in the sense of the above norm under the action of some known load (f_1, f_2). This is a natural safety requirement in many applications. Further (technological) constraints may be imposed directly on the admissible controls c or on the corresponding state (v_1, v_2).

We notice that the mere formulation of these problems requires the curvature c and its derivative (in the second term on the left side of the above equation). To ensure the integrability of such expressions one needs $\varphi \in W^{3,\infty}(0,1)^2$ or $\varphi \in C^3[0,1]^2$ for the corresponding parametrization. It is obvious that such requirements are inappropriate to the potential applications (see Figure 1.1 in Chapter 6, the Gothic arch). Moreover, some of the simplest and most popular discretization approaches (see Chapter 4) introduce nonsmooth approximations of φ in a natural way, and again the Kirchhoff–Love model cannot be applied. Such examples show that new mathematical methods have to be developed in order to relax the regularity hypotheses and to ensure a broad class of applications. In this book, a more sophisticated variational technique called the control variational method, based on control theory, is discussed. It is due to the authors and represents an alternative to the classical Dirichlet principle in the theory of elliptic equations. It is used for the analysis and optimization of Lipschitzian arches in Section 6.1 and of a simplified model of plates with discontinuous thickness in §3.4.2. More geometric optimization problems with mechanical background, such as optimal design of three-dimensional elastic curved rods and of general elastic shells, are studied by other methods in Sections 6.2 and 6.3. Thickness optimization problems for plates are investigated in §2.2.2 and Section 3.4. They are highly nonconvex optimization problems, but they still enjoy the property that they are defined in some known domain in the Euclidean space $\mathbf{R}^d$, $d \in \mathbf{N}$. In the above example, $d = 1$ and the domain is $]0,1[$.

We now present another example that involves unknown/variable domains. The application is related to the confinement of plasma in a tokamak machine. We denote by $\Omega \subset \mathbf{R}^2$ the smooth and bounded domain representing the cross section of the void chamber and by $D \subset \Omega$ its (unknown) subdomain occupied by the confined plasma (see Figure 2.1 in Chapter 1). Within the void region $\Omega\setminus\bar{D}$, the poloidal flux ψ satisfies (cf. Blum [1989, Ch. V]) the elliptic equation

$$-\frac{\partial}{\partial x}\left(\frac{1}{x}\frac{\partial\psi}{\partial x}\right) - \frac{\partial}{\partial y}\left(\frac{1}{x}\frac{\partial\psi}{\partial y}\right) = 0 \quad \text{in } \Omega\setminus\bar{D},$$

which is nonsingular $(x > c > 0)$ due to the natural choice of coordinates, based on the symmetry of the tokamak in $\mathbf{R}^3$. The boundary ∂D of the plasma is one of the unknowns of the problem, and this is an example of a free boundary problem. In order to identify it, one uses supplementary measurements on the outer boundary $\partial\Omega$:

$$\psi = f, \quad \frac{1}{x}\frac{\partial\psi}{\partial n} = g \quad \text{on } \partial\Omega.$$

One can introduce a shape optimization problem with minimization parameter given by the unknown domain $D \subset \Omega$, with performance index

$$\int_{\partial\Omega}\left|\frac{1}{x}\frac{\partial\psi}{\partial n} - g\right|^2 d\tau$$

obtained by the penalization of the second boundary condition and with side conditions given by the first boundary condition and the elliptic equation for ψ in $\Omega\setminus\bar{D}$. This formulation can be further refined by introducing a fictitious control variable and a Tikhonov regularization as in Example 1.2.6 in Chapter 1. Other simple examples of variable domain optimization problems may be found in §2.3.1. In Section 5.1, the relationship between free boundary problems and shape optimization problems is further explored, while §5.3.1 presents the connection between variable domain problems and control into coefficients problems via the classical mapping and speed methods. Since such a procedure demands high regularity properties for the unknown domains, we introduce in Section 5.2 several alternative approaches, based on control theory, which may be applied in more general situations. Moreover, in Section 2.3 a rather complete existence theory for variable domains optimization problems is developed under the mere (uniform) continuity assumption for the unknown boundaries. In Sections 2.1, 2.2 (existence), and Chapter 3 (optimality conditions), a rather complete presentation of control problems for linear and nonlinear elliptic equations, including variational inequalities, is given.

Although all of us have been actively involved in the study of optimization problems in infinite-dimensional spaces for many years, the origin of this book can be traced back to the lectures delivered by one of us in 1995 during the summer school that is organized annually by the University of Jyväskylä. These lectures have been published in the form of the report Tiba [1995b]. The following ten years were marked by an intensive cooperation between us that is witnessed by the publication of numerous papers in all of the research directions forming the subject of this monograph.

Much of the material covered in this volume is original and resulted from our studies when we were affiliated with the University of Jyväskylä, the Humboldt University Berlin, the Institute for Mathematics of the Romanian Academy of Sciences in Bucharest, and the Weierstrass Institute in Berlin. The financial support of these institutions, of the Academy of Finland, of the Alexander-von-Humboldt Foundation, and of the DFG Research Center MATHEON in Berlin, is gratefully acknowledged.

This monograph is addressed to a large readership, primarily to master's or doctoral students and researchers working in this field of mathematics. Much of this material will prove useful also to scientists from other fields where the optimization of elliptic systems occurs, such as physics, mechanics, and engineering.

During the preparation of this monograph, we obtained much encouragement and many helpful hints from a number of colleagues who cannot be named here. We are also indebted to Springer-Verlag, especially to Achi Dosanjh (New York), for their continuing encouragement.

Finally, we would like to thank Marja-Leena Rantalainen (Jyväskylä) and Jutta Lohse (WIAS Berlin) for their efforts in the excellent LaTeX setting of this text. We are also indebted to Dipl.-Math. Gerd Reinhardt (WIAS Berlin) for his help in solving the problems arising from the inclusion of the figures in the text. Of course, the authors carry the full responsibility for each occasional misprint or other possible mistake in this monograph.

Jyväskylä, Berlin, and Bucharest, March 2005

P. Neittaanmäki, J. Sprekels, and D. Tiba

A Brief Reader's Guide

The authors are fully aware of the fact that the reader of this volume will usually be interested in only a certain part of it. Therefore, we give some hints in order to facilitate the reader's orientation within the text.

The book is divided into six chapters, referred to as Chapter 1 to Chapter 6, and three appendices, referred to as Appendix 1 to Appendix 3. Each of the chapters consists of several "sections," called Section 1.1, Section 6.1, and so on. The sections themselves may be divided into several subsections, called "paragraphs" and referred to, for example, as §3.1.3. Also, these paragraphs may have subparagraphs denoted, for instance, by §3.1.3.1. Clearly, the latter refers to the first subparagraph of the third paragraph in the first section of Chapter 3.

Let us also comment on the numbering used in this textbook. Equations are numbered by three integers that refer to the corresponding chapter, section, and equation, in that order. If, for example, we refer to equation (4.2.6), then we mean the sixth equation in the second section of Chapter 4. Definitions, Theorems, Lemmas, Propositions, Corollaries, and Examples, are also numbered sectionwise within each chapter; typical examples are Theorem 5.2.1, Lemma 6.2.4, Definition 2.2.1, and so on. An exception to this rule is the numbering within the three appendices, where references are made in the form Proposition A1.1, Theorem A2.3, Definition A3.1, and the like, with obvious meaning. Remarks are not numbered. Finally, figures are numbered sectionwise within each chapter.

Contents

Preface v

A Brief Reader's Guide xi

Chapter 1. Introductory Topics 1

1.1 Some General Notions 1

1.2 Motivating Examples 3

1.2.1 Cost Functionals 3

1.2.2 Partial Differential Equations Setting 6

1.2.3 Applications 14

1.2.4 Variable Domains 20

Chapter 2. Existence **25**

2.1 A General Situation 25

2.2 Special Existence and Uniqueness Results 33

2.2.1 Second-Order Problems 33

2.2.2 Fourth-Order Problems 39

2.3 Variable Domains 44

2.3.1 Some Examples 45

2.3.2 General Dirichlet Problems 53

2.3.3 Neumann and Mixed Boundary Conditions 64

2.3.4 Partial Extensions 75

Chapter 3. Optimality Conditions **83**

3.1 Abstract Approaches 83

3.1.1 The Convex Case. Subdifferential Calculus 83

3.1.2 The Convex Case. Mathematical Programming 88

3.1.3 The Differentiable Case 92

3.1.3.1 Singular Control Problems 96

3.2 Penalization 102
3.2.1 The Standard Approach 103
3.2.2 Penalization of the State Equation 113
3.2.3 Semilinear Equations and Exact Penalization 127
3.3 Control of Variational Inequalities 143
3.3.1 An Abstract Result 143
3.3.2 Semilinear Variational Inequalities 148
3.3.3 State Constraints and Penalization of the Equation . . . 159
3.4 Thickness Optimization for Plates 168
3.4.1 Simply Supported Plates 168
3.4.2 Clamped Plates and Control Variational Methods 175
3.4.2.1 Penalization and Variational Inequalities . . . 183

Chapter 4. Discretization **193**

4.1 Finite Element Approximation of Elliptic Equations 193
4.1.1 The Finite Element Method 194
4.1.2 Error Estimates for the FE Equations 198
4.2 Error Estimates in the Finite Element Discretization of Control Problems 206
4.3 Semidiscretization 220
4.4 Optimal Control Problems Governed by Elliptic Variational Inequalities 226
4.4.1 The Ritz–Galerkin Approximation 228
4.5 Error Estimates in the Discretization of Control Problems with Nonlinear State Equation 238

Chapter 5. Unknown Domains **249**

5.1 Free Boundary Problems 249
5.1.1 The Dam Problem 250
5.1.2 Free Boundary Problems and Optimal Design 254
5.1.3 Shape Optimization of Systems with Free Boundaries . . 258
5.1.4 Controllability of the Coincidence Set 267
5.2 Direct Approaches 278
5.2.1 An Algorithm of Céa, Gioan, and Michel 278
5.2.2 Characteristic Functions 284
5.2.2.1 Structural Material Optimization 285

5.2.2.2 Bang-Bang Controls and Characteristic Functions 294
5.2.3 Controllability and Fictitious Domains Approaches . . . 297
5.2.3.1 Boundary Controllability 297
5.2.3.2 Distributed Controls 310
5.3 Domain Variations 323
5.3.1 Classical Approaches 323
5.3.2 Topological Asymptotics 334

Chapter 6. Optimization of Curved Mechanical Systems 341

6.1 Kirchhoff–Love Arches 342
6.1.1 Application of the Control Variational Method 343
6.1.2 Optimization of Nonsmooth Arches 351
6.1.3 Variational Inequalities for Arches 369
6.2 General Three-Dimensional Curved Rods 371
6.2.1 A Local Frame Under Low Differentiability Assumptions 372
6.2.2 A Generalized Naghdi-Type Model 374
6.2.3 Shape Optimization 389
6.3 Applications to Shells 406
6.3.1 A Generalized Naghdi Shell Model 406
6.3.2 Proof of Coercivity 413
6.3.3 Shell Optimal Design 422

Appendix 1. Convex Mappings and Monotone Operators 437

Appendix 2. Elliptic Equations and Variational Inequalities . . 451

Appendix 3. Domain Convergence 459

Bibliography 475

Index 495

Notation 505

Chapter 1

Introductory Topics

This first chapter brings a brief introduction to the problems to be studied in the following chapters. We present a large variety of examples involving different types of controls (distributed, boundary, pointwise, by the coefficients, linear, nonlinear, ...). All of them are governed by elliptic differential equations that are either defined in a given (fixed) spatial domain or in an a priori unknown domain. We also consider cases in which the domain itself is the minimization parameter (so-called *shape optimization*). For some of the examples, physical origin and practical relevance will be pointed out.

To avoid any unnecessary technicalities, we introduce the mathematical terminology mainly in the examples, in an informal manner. A brief rigorous account of the basic mathematical notions and results used throughout this monograph is contained in the three appendices at the end of the book, where relevant references are also given. It is, however, assumed that the reader has a working knowledge of the fundamental elements of analysis and functional analysis as presented, for instance, in the standard monographs by Rudin [1987] and Yosida [1980].

1.1 Some General Notions

We now discuss several definitions that are related to general optimal control problems. The setting adopted in this section simplifies the presentation and the systematization of the fundamental notions and is also motivated by a large class of examples and applications that will be described below in the next sections.

To begin with, let us consider three reflexive Banach spaces U, V, Z together with their respective dual spaces U^*, V^*, Z^*. By renorming, if necessary, we may assume without loss of generality that U, V, Z and their duals are strictly convex spaces. Moreover, let a Hilbert space H be given that is identified with its dual space and satisfies $V \subset H$ with continuous embedding. The scalar product in H and the pairing between V and its dual are denoted by $(\cdot\,,\cdot)_H$

and $(\cdot,\cdot)_{V^*\times V}$, respectively. The corresponding norms are denoted by $|\cdot|_H$, $|\cdot|_V$, $|\cdot|_U$, and so on; by $[\cdot,\cdot]$ we denote ordered pairs in product spaces.

Let $B : U \to Z$ be a linear and bounded operator, and let $A : V \to Z$ denote some (possibly nonlinear) operator. In many examples, we will have $Z = V^*$. We assume that for any fixed $f \in Z$ and any $u \in U$ (called *control*), the equation

$$Ay = Bu + f \tag{1.1.1}$$

has a unique solution $y \in V$ in a sense to be made precise (which is called the *state*). Consequently, (1.1.1) is sometimes named the *state equation.* In the applications to follow, y will be a weak solution to an elliptic problem. It may be defined in various ways, as one can see in Appendix 2 and in the subsequent examples. Later, we will also consider operators A depending directly on u, $Ay = A(u)y$, nonlinear operators B, and further generalizations.

Let a proper, convex, and lower semicontinuous mapping $L : V \times U \to]-\infty, +\infty]$ be given. We then introduce the abstract *control problem* (P) by

$$\inf\{L(y,u)\} \tag{1.1.2}$$

over all the pairs $[y,u]$ satisfying the state equation (1.1.1).

If $E = \operatorname{dom}(L) \subset V \times U$ denotes the closed convex set given by the effective domain of L (cf. Appendix 1), then we see that not all of the pairs $[y,u]$ satisfying (1.1.1) are meaningful for (1.1.2); indeed, some may give $L(y,u) = +\infty$. Consequently, the minimization in (1.1.2) is in fact considered only over all pairs $[y,u] \in E$ that satisfy (1.1.1). Such pairs are called *admissible* for the *cost functional* L or for the optimal control problem (P). We call E the *constraints set*, and we say in this case that the constraints are *mixed* since they involve both the state y and the control u.

It is quite standard in control theory to formulate the constraints explicitly, since they have their own motivation in the underlying applications. In general, $u = 0$ should be allowed as admissible control, corresponding to the case that no external influence is acting on the system.

Suppose now that some nonempty, closed, and convex sets $C \subset H$, $U_{ad} \subset U$ are given. We may then consider the separate constraints

$$\textit{control constraints} \quad u \in U_{ad}, \tag{1.1.3}$$

$$\textit{state constraints} \quad y \in C. \tag{1.1.4}$$

We get an equivalent formulation of the control problem (P) by including the constraints in the cost functional with the help of the *indicator function* $I_{C\times U_{ad}}$ of the set $C \times U_{ad} \subset H \times U$. To this end, we replace L by a new cost functional, namely by

$$L(y,u) + I_{C\times U_{ad}}(y,u). \tag{1.1.5}$$

In the following, the new cost functional (1.1.5) will again be denoted by L; this will not lead to any confusion. Now recall the definition of the indicator function (cf. Appendix 1) to see that $L(y,u) < +\infty$ only if $(y,u) \in C \times U_{ad}$, which means that any solution of the control problem (1.1.2) with the cost functional (1.1.5) automatically satisfies the constraints (1.1.3), (1.1.4). Of course, using the new cost functional (1.1.5) does not exclude the possibility that within the definition of the set E further (implicit) constraints are hidden.

In many cases it is advantageous to include only a part of the constraints in the cost functional while preserving the others in explicit form. If, for instance, only the control constraints are to be included, one considers the cost functional

$$L(y,u) + I_{V\times U_{ad}}(y,u). \qquad (1.1.5)'$$

Also for this cost functional the generic notation L may be preserved with no danger of confusion.

Let us summarize: a general formulation of the optimal control problem (P) consists of the following ingredients:

- a *cost functional* to be minimized ((1.1.2)),
- a *state system* ((1.1.1)),
- various *constraints* ((1.1.3), (1.1.4)).

A fundamental hypothesis for the control problem (P) is that of *admissibility*. It can be stated in the following form:

$$\exists\,[\overline{y},\overline{u}] \in C \times U_{ad} \text{ such that } L(\overline{y},\overline{u}) < \infty \text{ and } A\overline{y} = B\overline{u} + f. \qquad (1.1.6)$$

Without this assumption, the problem (P) may have an empty *admissible set* and be meaningless. For mathematical reasons, the case in which the admissible set of (P) is "rich" in some sense (typically, it has to be an open or a dense set with respect to some topology) is more interesting. Under such assumptions, we say that (P) is *nontrivial.* On the other hand, if (P) is "trivial," then its solution may be simple and thus not of mathematical interest.

Finally, let us mention that all the assumptions mentioned here can be relaxed in various ways; some of them may even be omitted. For instance, there is a rich literature on control problems without convexity hypotheses on L, E, C, U_{ad}, or allowing (1.1.1) not to be well-posed, and so on. One well-known alternative approach is to require various differentiability or generalized differentiability assumptions instead. In this connection, we refer to the monographs by Lions [1983] and Clarke [1983], where some extensions of this type are thoroughly examined. We shall study such topics in later sections of this monograph.

1.2 Motivating Examples

1.2.1 Cost Functionals

The cost functionals studied in this monograph will generally be of the form

$$L(y,u) = \theta(y) + \psi(u), \qquad (1.2.1)$$

where $\theta : V \to]-\infty, +\infty]$, $\psi : U \to]-\infty, +\infty]$ denote some proper, convex, and lower semicontinuous functions. A standard instance of this type is the quadratic functional

$$L(y,u) = \frac{\alpha}{2}\,|y - y_d|_V^2 + \frac{\beta}{2}\,|u|_U^2, \quad \alpha, \beta \geq 0, \tag{1.2.2}$$

where $y_d \in V$ is given.

The interpretation of (1.2.2) in connection with the control problem (P) is the following: we seek an admissible control $u \in U_{ad}$ such that the associated state $y \in C$ given by (1.1.1) is as close as possible to the "desired state" y_d. In addition, this control has to obey a minimal expenditure of energy condition (or minimal expenses condition, in general) reflected by the second term in (1.2.2). In fact, a compromise between the two (usually conflicting) aims "y close to y_d" and "minimal expenses" has to be found, and the relative importance of the criteria with respect to each other is expressed by the choice of the *weight coefficients* $\alpha, \beta \geq 0$.

As an anecdotal observation, we remark that the coefficients in (1.2.2) are chosen in this special form (as very frequently in the scientific literature) just because this "simplifies" the writing of the gradient of L, which plays a central role and is frequently used.

Notice that while (1.2.1), (1.2.2), and (1.1.2) define convex or even strictly convex functionals, the composed functional characterizing the control problem (P),

$$J(u) = L(y(u), u), \tag{1.2.3}$$

may be nonconvex. In fact, the state $y = y(u)$ defined by (1.1.1) may depend nonlinearly on u. If the operator A is linear, then J remains convex (or strictly convex), and any *optimal control* u^* is *global* (unique) if it exists. That is, the minimization property is valid with respect to the whole admissible set. The set of the global optimal controls is then convex. Otherwise, J may admit many *local* minimum points, in general. The existence of *optimal pairs* $[y^*, u^*]$ will be discussed in the next chapter. Their characterization, the development of methods to recover additional information on them, and their numerical approximation are among our basic objectives in this monograph.

Another fundamental example for a quadratic cost functional is obtained in the following way: Suppose that another Banach space W is given, and let $D : V \to W$ denote a linear and bounded operator (which in this connection is usually called an *observation operator*). We then consider the cost functional

$$L(y,u) = \frac{\alpha}{2}\,|Dy - \overline{y}_d|_W^2 + \frac{\beta}{2}\,|u|_U^2, \tag{1.2.4}$$

where $\overline{y}_d \in W$ has the same significance as y_d above. This setting is of particular practical importance and typically arises in situations in which the state y cannot be directly or fully observed, but only indirectly or in parts through the observation Dy. Typically, if (1.1.1) is a partial differential equation in a smooth domain, the operator D may be some trace operator on the boundary of the domain, a restriction operator to some subdomain, a partial differential operator of lower order, or the like.

A general form for the mappings θ, ψ occurring in (1.2.1) is obtained using integral functionals having *convex integrands*. To introduce such functionals, let $\Omega \subset \mathbf{R}^d$, $d \in \mathbf{N}$, be (Lebesgue) measurable, and suppose that $g : \Omega \times \mathbf{R}^m \to]-\infty, +\infty]$, $m \in \mathbf{N}$, satisfies the following conditions:

(i) $g(x, \cdot)$ is proper, convex, and lower semicontinuous for a.e. $x \in \Omega$.

(ii) g is measurable with respect to the σ-field of $\Omega \times \mathbf{R}^m$ generated by the product of the Lebesgue σ-field in Ω and the Borel σ-field in $\mathbf{R}^m$.

Such mappings g are called *normal convex integrands* (see Rockafellar [1970], Ioffe and Tikhomirov [1974], Levin [1985]). They have the basic property that the function $x \mapsto g(x, y(x))$ is measurable on Ω for any measurable function $y : \Omega \to \mathbf{R}^m$ (cf. Appendix 1, Proposition A1.1). Conditions (i), (ii) generalize the classical Carathéodory condition that $g(\cdot\,,\cdot)$ be finite and measurable in the first variable and continuous in the second.

For $y \in L^p(\Omega)^m$, $p \geq 1$, we then define the integral cost functional θ on $V = L^p(\Omega)^m$ by

$$\theta(y) = \begin{cases} \displaystyle\int_\Omega g(x, y(x))\, dx\,, & \text{if } g(\cdot, y(\cdot)) \in L^1(\Omega), \\ +\infty\,, & \text{otherwise.} \end{cases} \tag{1.2.1$'$}$$

Under appropriate conditions, θ turns out to be proper, convex, and lower semicontinuous (cf. Appendix 1). For the mapping ψ occurring in (1.2.1), we can proceed in a similar way. Also, one may consider the case that y is replaced by some Dy in (1.2.1)′.

Finally, let us point out a simple trick that is very useful in the numerical solution of optimal control problems. Suppose that $[y_0, u_0]$ is an admissible pair for (P), i.e., satisfies (1.1.1), (1.1.3), (1.1.4). Then, we may slightly modify the form of (1.2.4) by setting

$$\widetilde{L}(y,u) = \frac{\alpha}{2}\,|Dy - Dy_0|_W^2 + \frac{\beta}{2}\,|u - u_0|_U^2. \tag{1.2.4$'$}$$

The advantage of this form is that $[y_0, u_0]$ is obviously a global minimum (even when A is nonlinear and the corresponding $\widetilde{J}(u) = \widetilde{L}(y(u), u)$ is nonconvex) for the control problem ($\widetilde{\mathrm{P}}$) defined by (1.1.1), (1.1.4), (1.1.3), (1.2.4)′, with the *optimal value* equal to zero. Moreover, ($\widetilde{\mathrm{P}}$) has a structure that is very similar to that of (P). This a priori knowledge is helpful if one wants to test numerical code for the solution of (P). In particular, this idea is simple to apply when no state constraint (1.1.4) is imposed ($C = H$). Otherwise, even the question of finding an admissible pair $[y_0, u_0]$ may be very difficult due to the implicit character of (1.1.4).

1.2.2 Partial Differential Equations Setting

Here, we formulate several examples of elliptic state systems and related optimization problems that are among the objectives of this monograph. To this end, let a bounded domain $\Omega \subset \mathbf{R}^d$ with smooth boundary $\Gamma = \partial\Omega$ be given, and let $a_{ij} \in L^\infty(\Omega)$, $i,j = 1,\ldots,d$, define a (possibly nonsymmetric) coefficients matrix that satisfies with some fixed $a > 0$ the ellipticity condition

$$\sum_{i,j=1}^{d} a_{ij}(x)\,\xi_i\,\xi_j \geq a \sum_{i=1}^{d} \xi_i^2 \quad \text{for all } \xi \in \mathbf{R}^d \text{ and a.e. } x \in \Omega. \tag{1.2.5}$$

Example 1.2.1 Define the linear and bounded operator $A : V = H_0^1(\Omega) \to V^* = H^{-1}(\Omega)$ by

$$Ay = -\sum_{i,j=1}^{d} \frac{\partial}{\partial x_i}\Big(a_{ij}\frac{\partial y}{\partial x_j}\Big) + a_0\,y, \tag{1.2.6}$$

where $a_0 \in L^\infty(\Omega)$ with $a_0 \geq 0$ a.e. in Ω is given, and where the derivatives are understood in the sense of distributions. Let $U = L^2(\Omega)$, and let $B : L^2(\Omega) \to H^{-1}(\Omega)$ be the canonical injection operator, $Bu = iu = u$, for any $u \in L^2(\Omega)$. Then the state system (1.1.1) becomes a boundary value problem of Dirichlet type:

$$-\sum_{i,j=1}^{d} \frac{\partial}{\partial x_i}\Big(a_{ij}\frac{\partial y}{\partial x_j}\Big) + a_0\,y \;=\; u + f \quad \text{in } \Omega, \tag{1.2.7}$$

$$y \;=\; 0 \quad \text{on } \partial\Omega, \tag{1.2.8}$$

where $f \in L^2(\Omega)$ is fixed, and where (1.2.7), (1.2.8) have to be understood in the weak sense (see Appendix 2, Example A2.6), i.e.,

$$\sum_{i,j=1}^{d} \int_\Omega a_{ij}\,\frac{\partial y}{\partial x_j}\,\frac{\partial v}{\partial x_i}\,dx + \int_\Omega a_0\,y\,v\,dx = \int_\Omega (u+f)\,v\,dx \quad \forall\, v \in H_0^1(\Omega). \tag{1.2.9}$$

We say that we have a *distributed control* (or action) since u is defined in the domain Ω. A related situation is obtained if ω is a measurable subset of Ω and $B : L^2(\Omega) \to H^{-1}(\Omega)$ is given by $Bu = u\chi_\omega$, with χ_ω being the characteristic function of ω in Ω. Then the control action is again distributed, namely in ω, and (1.2.7) becomes

$$-\sum_{i,j=1}^{d} \frac{\partial}{\partial x_i}\Big(a_{ij}\,\frac{\partial y}{\partial x_j}\Big) + a_0\,y = u\,\chi_\omega + f \quad \text{in } \Omega. \tag{1.2.7$'$}$$

We indicate some possible choices for the cost functional (1.2.4) that are appropriate in this situation. If $W = L^2(\Omega)$ and $D : H_0^1(\Omega) \to L^2(\Omega)$ is the canonical injection, then

$$L(y,u) = \frac{\alpha}{2}\int_\Omega |y(x) - y_d(x)|^2\,dx + \frac{\beta}{2}\int_\Omega u^2(x)\,dx, \quad y_d \in L^2(\Omega), \tag{1.2.10}$$

and we have a *distributed observation*. If $Dy = \nabla y$, we have a distributed observation of the gradient of the solution:

$$L(y,u) = \frac{\alpha}{2}\int_\Omega |\nabla y(x) - \tilde{y}_d(x)|^2_{\mathbf{R}^d}\,dx + \frac{\beta}{2}\int_\Omega u^2(x)\,dx, \quad \tilde{y}_d \in L^2(\Omega)^d. \tag{1.2.11}$$

The domain Ω in (1.2.10), (1.2.11) may be replaced by some measurable subsets of Ω, at least in one of the integrals.

Let us assume now that the coefficients a_{ij}, a_0, $i,j = 1,\dots,d$, are sufficiently regular to guarantee that the solution y of (1.2.7), (1.2.8) belongs to $H^2(\Omega)$ (i.e., is a strong solution, cf. Appendix 2). Then, by virtue of the trace theorem (Appendix 2, Theorem A2.1), the *outer conormal derivative*

$$\frac{\partial y}{\partial n_A} = \sum_{i,j=1}^{d} a_{ij}\,\frac{\partial y}{\partial x_j}\,\cos(n, x_i) \tag{1.2.12}$$

on Γ (n is the outer unit normal to Γ) satisfies $\frac{\partial y}{\partial n_A} \in H^{1/2}(\Gamma)$. Taking some (relatively) open part $\Gamma_0 \subset \Gamma$, we may then choose as cost functional

$$L(y,u) = \frac{\alpha}{2}\int_{\Gamma_0} \left|\frac{\partial y}{\partial n_A} - \hat{y}_d\right|^2(\sigma)\,d\sigma + \frac{\beta}{2}\int_\Omega u^2(x)\,dx, \quad \hat{y}_d \in L^2(\Gamma_0). \tag{1.2.13}$$

In this case, we say that we have a *boundary observation* (while the control remains distributed in Ω).

To complete the definition of the control problem (P) for this example, let us discuss some instances of possible constraints. The simplest case is of course the *unconstrained* one when $U_{ad} = U = L^2(\Omega)$, $C = H = L^2(\Omega)$. One rough classification of the constraints is to distinguish between *local* and *global* ones. *Pointwise* constraints like

$$U_{ad} = \left\{u \in L^2(\Omega) : -1 \le u(x) \le 1 \text{ for a.e. } x \in \Omega\right\}, \tag{1.2.14}$$

$$U_{ad} = \left\{u \in L^2(\Omega) : u(x) \ge \ell(x) \text{ for a.e. } x \in \Omega,\ \ \ell \in L^2(\Omega) \text{ given}\right\}, \tag{1.2.15}$$

$$C = \left\{y \in H^1(\Omega) : |\nabla y(x)|_{\mathbf{R}^d} \le 1 \text{ for a.e. } x \in \Omega\right\}, \tag{1.2.16}$$

are of local type. Standard examples for constraints of global type are *integral* constraints like

$$U_{ad} = \left\{u \in L^2(\Omega) : |u|_{L^2(\Omega)} \le \mu\right\}, \quad \mu > 0, \tag{1.2.17}$$

$$C = \left\{y \in L^2(\Omega) : \int_\Omega y(x)\,dx \le 0\right\}. \tag{1.2.18}$$

A simple example of *mixed pointwise* constraints is given by

$$E = \left\{[y,u] \in L^2(\Omega) \times L^2(\Omega) : y(x) \le u(x) \text{ a.e. in } \Omega\right\}. \tag{1.2.19}$$

Let us briefly return to the control constraint (1.2.14). In this case, it is possible to introduce a new control $w \in L^2(\Omega)$ satisfying

$$u(x) = \sin(w(x)) \quad \text{a.e. in } \Omega.$$

Making corresponding substitutions, the optimal control problem (P) can be transformed into a control problem without constraints for w. The price to be paid for this simplification is that in (1.1.1) the dependence of the state on the new control variable w (more precisely, the operator corresponding to B) becomes nonlinear and that the convexity is lost. However, such simple tricks may be very effective in applications. For further details, we refer to Banichuk [1983, Chapter I].

We conclude this example with the remark that the above discussion of cost functionals and of constraints applies to any type of elliptic control problem. In the subsequent examples we will therefore focus our attention on the analysis of the state equation and control action.

Example 1.2.2 Let us now concentrate on *boundary control* problems. We begin with control action via Neumann boundary conditions, by considering the state system

$$Ay = f \quad \text{in } \Omega, \tag{1.2.20}$$

$$\frac{\partial y}{\partial n_A} = u \quad \text{on } \partial\Omega, \tag{1.2.21}$$

where A is given by (1.2.6), and where we assume that $a_0(x) \geq \mu > 0$ a.e. in Ω. The variational (weak) formulation of (1.2.20), (1.2.21) is obtained using Green's formula:

$$\int_\Omega \sum_{i,j=1}^{d} a_{ij} \frac{\partial y}{\partial x_i} \frac{\partial v}{\partial x_j}\, dx + \int_\Omega a_0\, y\, v\, dx = \int_\Omega f v\, dx + \int_{\partial\Omega} u\, v\, d\sigma \quad \forall v \in H^1(\Omega). \tag{1.2.22}$$

To recover the abstract setting (1.1.1), we fix some $f \in L^2(\Omega)$ and put $V = H^1(\Omega)$ and $U = H^{-1/2}(\partial\Omega)$. Moreover, $A : V \to V^*$ is generated by the left-hand side of (1.2.22) (cf. Appendix 2, Theorem A2.3), while $B : U \to V^*$ is defined by

$$(Bu, v)_{H^1(\Omega)^* \times H^1(\Omega)} = \int_{\partial\Omega} u\, v\, d\sigma \quad \forall v \in H^1(\Omega). \tag{1.2.23}$$

Obviously, the restriction of A to $H_0^1(\Omega)$ coincides with (1.2.6). Notice that also the choice $U = L^2(\partial\Omega)$ is possible with the same definition (1.2.23) of B.

Next, we turn our attention to control action via Dirichlet boundary conditions. It is known that the inhomogeneous Dirichlet boundary value problem does not admit a purely variational (weak) formulation and that a suitable translation has to be employed first in order to reduce the problem to the homogeneous case (Křížek and Neittaanmäki [1990]). In the setting of control problems the corresponding translation operator may be, roughly speaking, interpreted as the operator B. If the state system is described by (1.2.20) and

$$y = u \quad \text{on } \partial\Omega, \tag{1.2.24}$$

then we may fix $B : H^{-1/2}(\partial\Omega) \to L^2(\Omega)$ by $Bu = y_u$, where y_u satisfies (1.2.24) and

$$Ay_u = 0 \quad \text{in } \Omega. \tag{1.2.25}$$

We refer at this place to Appendix 2, Example A2.7, for the definition of a very weak solution of (1.2.24), (1.2.25), using the transposition method. We choose $V = V^* = L^2(\Omega)$, $U = H^{-1/2}(\partial\Omega)$, define a new operator $\widetilde{A} : V \to V^*$, $\widetilde{A}y = y$, and a new $\tilde{f} \in L^2(\Omega)$, given by

$$A\tilde{f} = f \quad \text{in } \Omega, \qquad \tilde{f} = 0 \quad \text{on } \partial\Omega. \tag{1.2.26}$$

If we write the abstract equation (1.1.1) in the form $\widetilde{A}y = Bu + \tilde{f}$, then it is equivalent to (1.2.20), (1.2.24).

The operator B is called the *Dirichlet mapping* and plays an essential role in this formulation.

Example 1.2.3 Let us also address the *pointwise control* of linear systems. We take $V = H_0^k(\Omega)$, $k > \frac{d}{2}$, with d being the dimension of Ω. By virtue of the Sobolev embedding theorem (Appendix 2, Theorem A2.2), we have $V \subset C(\overline{\Omega})$, and the Dirac functional $\delta_{x_0} : V \to \mathbf{R}$, $\delta_{x_0}(v) = v(x_0)$, with some given $x_0 \in \Omega$, is linear and continuous on V, that is, $\delta_{x_0} \in V^*$.

Let us put $U = \mathbf{R}$, and let $B : U \to V^*$ be given by $Bu = u\,\delta_{x_0}$, which is a linear and bounded operator. We assume $A : V \to V^*$ in the form

$$Ay = \sum_{|\alpha| \le k} (-1)^{|\alpha|} D^\alpha (a_\alpha(x)\, D^\alpha y), \quad a_\alpha \in L^\infty(\Omega), \tag{1.2.6$'$}$$

where the multi-index $\alpha = (\alpha_1, \ldots, \alpha_d) \in \mathbf{N}_0^d$, $\mathbf{N}_0 = \mathbf{N} \cup \{0\}$, and $|\alpha| = \alpha_1 + \ldots + \alpha_d$ is its length, the derivatives are taken in the distributional sense, and the coercivity condition

$$(Ay, y)_{H^{-k}(\Omega) \times H_0^k(\Omega)} \ge c\, |y|^2_{H_0^k(\Omega)} \quad \forall\, y \in H_0^k(\Omega), \tag{1.2.5$'$}$$

with some $c > 0$, is assumed to hold. Then the state equation (1.1.1) becomes

$$\sum_{|\alpha| \le k} (-1)^{|\alpha|} D^\alpha (a_\alpha(x)\, D^\alpha y) = Bu + f \quad \text{in } \Omega, \tag{1.2.27}$$

$$y = 0, \quad \frac{\partial y}{\partial n} = 0, \quad \ldots, \quad \frac{\partial^{k-1} y}{\partial n^{k-1}} = 0 \quad \text{on } \partial\Omega. \tag{1.2.28}$$

According to Appendix 2, this system admits a unique weak solution. Owing to the definition of B, the control u is concentrated in the point $x_0 \in \Omega$.

Example 1.2.4 We now examine *nonlinear* elliptic boundary value problems as state equations. We start with the semilinear case. Let A be defined as in (1.2.6) and consider a continuous mapping $\varphi : \Omega \times \mathbf{R} \times \mathbf{R} \to \mathbf{R}$ having a continuous derivative φ_y with respect to its second argument variable, and the property that for any $u \in L^s(\Omega)$, $s \geq \max\{2, \frac{d}{2}\}$ ($d =$ dimension of Ω), the mapping $\varphi(\cdot\,,\cdot\,,u(\cdot))$ is of Carathéodory type. Moreover, the following conditions are assumed:

$$|\varphi(x,0,v)| \leq M(x) + \hat{C}\,|v|, \quad \text{a.e. in } \Omega,\ v \in \mathbf{R}, \tag{1.2.29}$$

$$0 \leq \varphi_y(x,y,v) \leq (M(x) + \hat{C}\,|v|)\,\eta(|y|), \quad \text{a.e. in } \Omega,\ v, y \in \mathbf{R}, \tag{1.2.30}$$

with a function $M \in L^s(\Omega)$, a constant $\hat{C} > 0$, and a nondecreasing function $\eta : \mathbf{R}_+ \to \mathbf{R}_+$. The state equation has the form

$$Ay + \varphi(x, y(x), u(x)) = 0 \quad \text{in } \Omega, \tag{1.2.31}$$

$$y = 0 \quad \text{on } \partial\Omega. \tag{1.2.32}$$

If $a_{ij} \in C^1(\overline{\Omega})$, then (1.2.31), (1.2.32) has a unique strong solution $y \in W^{2,s}(\Omega) \cap H_0^1(\Omega) \cap L^\infty(\Omega)$; see Theorem A2.10 in Appendix 2 and the remark following it. In (1.2.31), (1.2.32), the control variable u appears implicitly. In order to fit this system into the formalism from Section 1.1, we put $B = 0$, and we allow $A = A(u)$, $u \in U = L^s(\Omega)$, to depend directly on the control parameter. One possible way to achieve this is to include the semilinear term $\varphi(x, \cdot, u)$ in the definition of $A(u)$ as a superposition (Nemytskii) operator (cf. Pascali and Sburlan [1978]).

One particular situation of interest is the *control in the coefficients* case. For instance, for $\varphi(x,y,u) = |u|y$ all the above assumptions are obviously fulfilled. The partial differential equation (1.2.31) then becomes linear with respect to y, but the dependence $u \mapsto y$, induced by it, is highly nonlinear. As a consequence, the associated optimization problems are nonconvex, and since they may have many local minima, they are stiff and hence difficult to solve numerically. An important application of this type arises in optimal shape design theory in connection with the so-called *mapping method*. For details, we refer to Pironneau [1984], Haslinger and Neittaanmäki [1988], as well as to the problems studied below in (1.2.51) and in §5.3.1.

Another important class of applications that may be described by control in the coefficients problems is given by the so-called *identification problems* to be discussed in Example 1.2.6 below.

Example 1.2.5 Let us assume for the moment that the symmetry condition $a_{ij} = a_{ji}$, $i, j = 1, 2, \ldots, d$, is fulfilled. Then the Dirichlet principle shows that the solution $y \in V = H_0^1(\Omega)$ to (1.2.7), (1.2.8) (or, equivalently, to the weak formulation (1.2.9)) admits the alternative variational characterization

$$\sum_{i,j=1}^{d} \int_\Omega a_{ij} \frac{\partial y}{\partial x_j} \frac{\partial y}{\partial x_i}\, dx + \int_\Omega a_0\, y^2\, dx - 2 \int_\Omega (u+f)\, y\, dx$$

$$= \operatorname*{Min}_{z\in V} \Big\{ \sum_{i,j=1}^{d} \int_\Omega a_{ij} \frac{\partial z}{\partial x_j} \frac{\partial z}{\partial x_i}\, dx + \int_\Omega a_0\, z^2\, dx - 2 \int_\Omega (u+f)\, z\, dx \Big\}. \tag{1.2.33}$$

Now let us consider the minimization problem when in (1.2.33) the full space V is replaced by a (nonempty) convex and closed set $S \subset V$. Again, there exists a unique minimizer $y_S \in S$ since the quadratic form in (1.2.33) is coercive and strictly convex (see Appendix 1). A straightforward computation shows that y_S is the unique solution to

$$\sum_{i,j=1}^{d} \int_\Omega a_{ij} \frac{\partial y_S}{\partial x_j} \Big(\frac{\partial y_S}{\partial x_i} - \frac{\partial z}{\partial x_i} \Big) dx + \int_\Omega a_0\, y_S\, (y_S - z)\, dx \le \int_\Omega (u+f)\, (y_S - z)\, dx, \tag{1.2.34}$$

for all $z \in S$. Since, in turn, any solution to (1.2.34) is also a solution to the minimization problem, then (1.2.33) (with V replaced by S) and (1.2.34) are in fact equivalent problems. Relation (1.2.34) is called a *variational inequality* associated with the closed and convex set S. Notice that the symmetry condition is not essential for the existence of a unique solution to the variational inequality (1.2.34), as follows from the Lions–Stampacchia theorem (see Appendix 2, Theorem A2.3), which is a generalization of the classical Lax–Milgram lemma.

Now let $I_S : V \to]-\infty, +\infty]$ denote the (proper, convex, and lower semicontinuous) indicator function of S in V. Then (1.2.34) may be reformulated in the form

$$\sum_{i,j=1}^{d} \int_\Omega a_{ij} \frac{\partial y_S}{\partial x_j} \Big(\frac{\partial y_S}{\partial x_i} - \frac{\partial z}{\partial x_i} \Big) dx + \int_\Omega a_0\, y_S\, (y_S - z)\, dx + I_S(y_S) - I_S(z)$$

$$\le \int_\Omega (u+f)\, (y_S - z)\, dx \quad \forall z \in V. \tag{1.2.34$'$}$$

More generally, let us consider for any proper, convex, and lower semicontinuous mapping $\Lambda : V \to]-\infty, +\infty]$ the variational inequality

$$\sum_{i,j=1}^{d} \int_\Omega a_{ij} \frac{\partial y}{\partial x_j} \Big(\frac{\partial y}{\partial x_i} - \frac{\partial z}{\partial x_i} \Big) dx + \int_\Omega a_0\, y\, (y - z)\, dx + \Lambda(y) - \Lambda(z)$$

$$\le \int_\Omega (u+f)\, (y - z)\, dx \quad \forall z \in V. \tag{1.2.35}$$

Then it follows directly from the theory of maximal monotone operators (cf. Appendix 1, Theorem A1.7) that (1.2.35) admits a unique solution $y \in \operatorname{dom}(\Lambda)$.

Moreover, using the subdifferential $\partial\Lambda$ of Λ, we may rewrite (1.2.35) as a semilinear elliptic *inclusion*, namely as

$$-\sum_{i,j=1}^{d}\frac{\partial}{\partial x_i}\Big(a_{ij}\frac{\partial y}{\partial x_j}\Big)+a_0\,y+\partial\Lambda(y)\ni u+f \text{ in } \Omega, \quad y=0 \text{ on } \partial\Omega. \qquad (1.2.35)'$$

Generally speaking, (1.2.35) or (1.2.35)′ may be viewed as extensions of the semilinear problem (1.2.31), (1.2.32) in the sense that the mapping φ is replaced by the nonsmooth and discontinuous (multivalued) subdifferential mapping $\partial\Lambda$.

In what follows, we give some important examples for possible sets S. We begin with the so-called *obstacle problem*:

$$S=\Big\{y\in H_0^1(\Omega)=V: y(x)\geq\mu(x) \text{ a.e. in } \Omega\Big\}, \qquad (1.2.36)$$

where $\mu\in H^2(\Omega)$ is a given function (called the *obstacle*) having the property that $\mu|_{\partial\Omega}\leq 0$, which ensures that S is nonempty.

Formally, the solution y of the *obstacle problem* (1.2.34), (1.2.36) will satisfy

$$\begin{aligned}
-\sum_{i,j=1}^{d}\frac{\partial}{\partial x_i}\Big(a_{ij}\frac{\partial y_S}{\partial x_j}\Big)+a_0\,y_S &= u+f \quad \text{a.e. in } \Omega^+=\{x\in\Omega: y_S(x)>\mu(x)\},\\
y_S &= \mu \quad \text{a.e. in } \Omega\setminus\Omega^+,\\
\sum_{i,j=1}^{d}\frac{\partial}{\partial x_i}\Big(a_{ij}\frac{\partial y_S}{\partial x_j}\Big)+a_0\,y_S &\geq u+f, \quad y_S\geq\mu, \quad \text{a.e. in } \Omega.
\end{aligned}$$

The "surface" $\partial\Omega^+\setminus\partial\Omega$ separating Ω^+ from $\Omega\setminus\overline{\Omega}^+$ is a priori unknown and is called the *free boundary* of the obstacle problem. The region $\Omega\setminus\Omega^+$, where y_S is equal to the obstacle, is called the *coincidence set*.

Next, we consider the set S in (1.2.34) that characterizes the so-called *elasto-plastic torsion problem*,

$$S=\Big\{y\in H_0^1(\Omega)=V: |\nabla y(x)|\leq 1 \text{ a.e. in } \Omega\Big\}. \qquad (1.2.37)$$

Again, we may (formally) define two subregions of Ω,

$$\text{the } \textit{plastic} \text{ region} \qquad \Omega_1=\Big\{x\in\Omega: |\nabla y_S(x)|=1\Big\},$$

$$\text{the } \textit{elastic} \text{ region} \qquad \Omega_2=\Big\{x\in\Omega: |\nabla y_S(x)|<1\Big\},$$

such that (1.2.34) becomes an equality in one of the subregions (namely in Ω_2).

Let us mention that for choice $\mu(x)=d(x,\partial\Omega)$ the two problems (1.2.36), (1.2.37) are in fact equivalent (cf. Brézis and Sibony [1971]).

We also notice that the solution y_S of the variational inequality (1.2.34) satisfies $y_S\in S$, obviously. But this should be distinguished from a state constraint (although here the form is similar to (1.2.16)), since it is automatically fulfilled.

Indeed, it follows from the Lions–Stampacchia theorem mentioned above that a unique solution $y_S \in S \subset V$ exists for any $u \in U = L^2(\Omega)$.

Unilateral problems, that is, problems involving inequalities in place of equations, may also be formulated on $\partial\Omega$. For instance, consider the set

$$S = \left\{ y \in V = H^1(\Omega) : y|_{\partial\Omega} \geq 0 \right\}. \tag{1.2.38}$$

In this case, the (formal) interpretation of (1.2.34) can be deduced from the following chain of formal calculations: first, we insert $z = y_S + v \in S$ for all $v \in \mathcal{D}(\Omega)$ in (1.2.34). Then we obtain

$$-\sum_{i,j=1}^{d} \frac{\partial}{\partial x_i}\Big(a_{ij}\,\frac{\partial y_S}{\partial x_j}\Big) + a_0\, y_S = u + f \quad \text{in } \mathcal{D}'(\Omega). \tag{1.2.39}$$

Next, multiplying (1.2.39) by any $z \in S$ and applying (formally) Green's formula, we find that

$$-\int_{\partial\Omega} \frac{\partial y_S}{\partial n_A}\, z\, d\sigma + \int_\Omega \sum_{i,j=1}^{d} a_{ij}\, \frac{\partial y_S}{\partial x_i}\,\frac{\partial z}{\partial x_j}\, dx + \int_\Omega a_0\, y_S\, z\, dx = \int_\Omega (u+f)\, z\, dx. \tag{1.2.40}$$

Then, we replace z in (1.2.34) by $z + y_S$, which is possible in view of (1.2.38), and use (1.2.40), to find that

$$\int_{\partial\Omega} \frac{\partial y_S}{\partial n_A}\, z\, d\sigma \geq 0 \quad \forall z \in S.$$

That is, we have

$$\frac{\partial y_S}{\partial n_A} \geq 0, \quad y_S \geq 0, \qquad \text{on } \partial\Omega.$$

Moreover,

$$\int_{\partial\Omega} y_S\, \frac{\partial y_S}{\partial n_A}\, d\sigma = 0,$$

which follows by using $z = y_S$ as test function in (1.2.39), (1.2.40), and by comparing with (1.2.34), where we put $z = 0$. Such boundary conditions are known as the *Signorini problem* and describe an elastic body Ω subject to volume forces $u + f$ and in contact with a rigid support body. This is an example of unilateral conditions on the boundary.

More generally, let $\Lambda : V = H^1(\Omega) \to]-\infty, +\infty]$ be defined by

$$\Lambda(y) = \begin{cases} \int_{\partial\Omega} j(y)\, d\sigma & \forall\, y \in V \text{ with } j(y) \in L^1(\partial\Omega), \\ +\infty & \text{otherwise}, \end{cases}$$

where $j : \mathbf{R} \to]-\infty, +\infty]$ is a proper, convex, and lower semicontinuous mapping. Then the variational inequality (1.2.35) (or, equivalently, the elliptic problem (1.2.35)′) has a unique solution $y \in V$ (see Barbu [1984]) that (formally) satisfies

$$-\sum_{i,j=1}^{d} \frac{\partial}{\partial x_i}\Big(a_{ij}(x)\,\frac{\partial y}{\partial x_j}\Big) + a_0\,y \;=\; u + f \quad \text{in } \mathcal{D}'(\Omega),$$

$$\frac{\partial y}{\partial n_A} + \partial j(y) \;\ni\; 0 \quad \text{on } \partial\Omega.$$

We remark that all the above formal arguments can be made rigorous provided that the solution y_S belongs to $H^2(\Omega)$ (strong solution). Boundary control action u may be studied as well.

1.2.3 Applications

We devote this paragraph to a first examination of some physically oriented applications. Further details and solutions of the problems will be provided later.

Example 1.2.6 We begin with a problem arising in the confinement of plasma in a *tokamak machine.* Let Ω be a smooth and bounded domain in $\mathbf{R}^2$ representing the cross section of the void chamber of a tokamak machine, and let $D \subset \Omega$ denote its (unknown) subdomain occupied by the confined plasma. Within the void region $\Omega \setminus \overline{D}$ the (unknown) poloidal flux ψ satisfies (cf. Blum [1989, Chapter V])

$$-\frac{\partial}{\partial x}\Big(\frac{1}{x}\frac{\partial \psi}{\partial x}\Big) - \frac{\partial}{\partial y}\Big(\frac{1}{x}\frac{\partial \psi}{\partial y}\Big) = 0 \quad \text{in } \Omega \setminus \overline{D}, \tag{1.2.41}$$

which is a nonsingular second-order linear elliptic equation since the natural choice of coordinates, based on the symmetry of the torus representing the tokamak in $\mathbf{R}^3$, yields $x > c > 0$ in Ω for some constant c (see Figure 2.1). The boundary ∂D of the plasma region is an unknown of the problem and represents a *free boundary.* It is characterized as a level set by the relation

$$M \in \partial D \text{ if and only if } \psi(M) = \sup_{x\in F} \psi(x),$$

where F (see Figure 2.1) represents physical devices called *limitators* that may have various shapes.

The only available data are the measurements on the outer boundary $\partial\Omega$:

$$\psi \;=\; f \quad \text{on } \partial\Omega, \tag{1.2.42}$$

$$\frac{1}{x}\frac{\partial \psi}{\partial n} \;=\; g \quad \text{on } \partial\Omega. \tag{1.2.43}$$

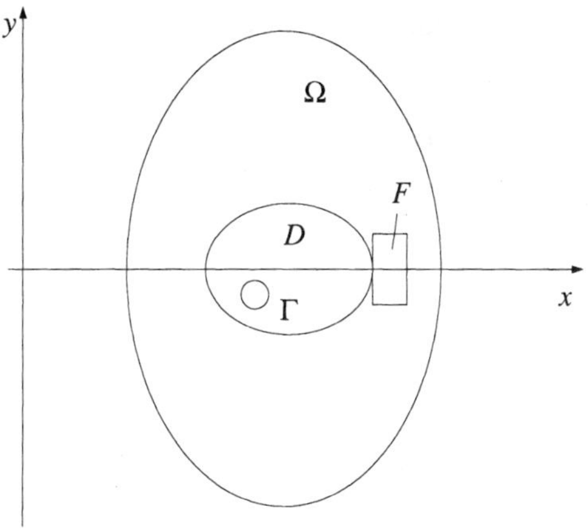

Figure 2.1. Schematic representation of the void chamber.

Thus, the problem to identify the subdomain D occupied by the plasma leads to an elliptic Cauchy problem ((1.2.41)–(1.2.43)) and is as such *ill-posed.* A *fictitious domain* approach to this problem consists in fixing some (artificial) smooth closed curve $\Gamma \subset D$ (see Figure 2.1), and defining the *least squares* boundary control problem in the domain Ω_0 limited by Γ and $\partial\Omega$,

$$\operatorname*{Min}_{u\in L^2(\Gamma)} \left\{ J(u) = \frac{1}{2} \left| \frac{1}{x}\frac{\partial\psi}{\partial n} - g \right|^2_{L^2(\partial\Omega)} \right\} \tag{1.2.44}$$

subject to

$$\begin{aligned} -\frac{\partial}{\partial x}\Big(\frac{1}{x}\frac{\partial\psi}{\partial x}\Big) - \frac{\partial}{\partial y}\Big(\frac{1}{x}\frac{\partial\psi}{\partial y}\Big) &= 0 \quad \text{in } \Omega_0, && (1.2.41)' \\ \psi &= f \quad \text{on } \partial\Omega, && (1.2.42)' \\ \psi &= u \quad \text{on } \Gamma. && (1.2.45) \end{aligned}$$

In view of the lack of coercivity in (1.2.44), a *Tikhonov regularization* technique may be used. We choose some *regularization parameter* $\varepsilon > 0$ and replace the minimization problem (1.2.44) by

$$\operatorname*{Min}_{u\in L^2(\Gamma)} \left\{ J_\varepsilon(u) = \frac{1}{2} \left| \frac{1}{x}\frac{\partial\psi}{\partial n} - g \right|^2_{L^2(\partial\Omega)} + \frac{\varepsilon}{2} |u|^2_{L^2(\Gamma)} \right\}, \tag{1.2.46}$$

subject to (1.2.41)′, (1.2.42)′, and (1.2.45). This results in a standard boundary control problem with boundary observation and a linear state system. The convergence analysis for $\varepsilon \searrow 0$ was performed in Neittaanmäki and Tiba [1995], Neittaanmäki, Räisänen, and Tiba [1994]; see §5.2.3.1.

While the regularized problem (1.2.46) appears to be easy to solve, the sensitivity to measurement errors, which is intrinsic to all ill-posed problems, remains an important problem, and the interpretation of the results in terms of the original problem turns out to be a difficult task (see Falk [1990]).

Another category of problems that may be handled via control methods are the so-called *identification problems*. Suppose that some physical system (for instance, the equilibrium position of a clamped membrane) is described by the following mathematical model:

$$-\sum_{i,j=1}^{d} \frac{\partial}{\partial x_i}\Big(a_{ij}(x)\frac{\partial y}{\partial x_j}\Big) + a_0(x)\,y \;=\; f(x) \quad \text{in } \Omega, \tag{1.2.7$''$}$$

$$y(x) \;=\; 0 \quad \text{on } \partial\Omega, \tag{1.2.8$''$}$$

where $\Omega \subset \mathbf{R}^d$ is a smooth bounded domain and where the assumption (1.2.5) is fulfilled. In many applications it turns out to be difficult to measure or to have precise a priori knowledge of all the coefficient functions, which usually depend on the physical properties of the membrane or other parameters. On the other hand, it is natural to assume that some observation of the real state (the deflection of the membrane) $\tilde{y}$ of the system, denoted by $D\tilde{y}$, is available via measurements.
Suppose that $a_0 \in L^\infty(\Omega)$, $a_0 \geq 0$, is the unknown coefficient. Then the least squares approach to its determination leads to the problem

$$\operatorname*{Min}_{a_0 \in L^\infty(\Omega)_+} \Big\{ J(y) = \frac{1}{2}\,|Dy - D\tilde{y}|_W^2 \Big\} \tag{1.2.47}$$

subject to (1.2.7)$'$, (1.2.8)$'$, and where W is the associated *observation space*, i.e., $D : H_0^1(\Omega) \to W$ is linear and continuous. We obtain a control in the coefficients problem (compare with Example 1.2.4), and one clear difficulty is its nonconvexity; in addition, it is also noncoercive. Therefore, a Tikhonov regularization technique is indicated also in this situation.

The above examples are special cases of *inverse problems*, an area of applications in which the optimal control approach is a standard method.

Example 1.2.7 Next, we describe some optimization problems involving geometric parameters, generally called *optimal shape design* problems. One such case, called the *optimal layout of materials*, is introduced as follows, starting from (1.2.7)$''$, (1.2.8)$''$ (in Example 1.2.6) and (1.2.47). We assume that $a_{ij}(x) = \delta_{ij}a(x)$ (δ_{ij} is the Kronecker symbol) and $a_0(x) = 1$ in Ω. The coefficient a can be interpreted as the thermal conductivity of the body given by Ω. We assume that the body consists of different materials having the thermal conductivities k_i, $i = \overline{1,m}$, that is,

$$a(x) = \sum_{i=1}^{m} \chi_i(x)\,k_i,$$

where χ_i is the characteristic function in Ω of the region occupied by the material indexed by i.

We then may ask the following question: If a fixed heat source f is given, what is the *optimal distribution* of the materials that maximizes the temperature y in a given subdomain $\omega \subset \Omega$ (or on some open part $\Gamma_0 \subset \partial\Omega$, etc.)?

To solve this problem, we may take one of the cost functionals

$$\operatorname*{Min}_{a}\left\{-\int_{\omega} y(x)\,dx\right\}, \tag{1.2.47$'$}$$

$$\operatorname*{Min}_{a}\left\{-\int_{\Gamma_0} y(\sigma)\,d\sigma\right\}. \tag{1.2.47$''$}$$

The minimization parameters are the subsets of Ω occupied by the various materials. Equivalently, one can use the characteristic functions $\chi_i,\ i=\overline{1,m}$, as control unknowns. Apparently, we can interpret the problem as a control into coefficients problem, where 0 and 1 are the only admissible values for the controls. For a detailed discussion, we refer the reader to Tartar [1975], Pironneau [1984, §8.4], and §2.3.4, §5.2.2.1.

Let us now briefly comment on a stationary variant of the so-called *electrochemical machining process*. To this end, we consider the bounded domains $C \subset E \subset D \subset \Omega$ in $\mathbf{R}^3$ where D is variable (see Figure 2.2). In $D \setminus C$, we consider the obstacle problem (compare with (1.2.34), (1.2.36))

$$\int_{D\setminus C} \nabla y(x)\cdot\nabla(y-z)(x)\,dx \le \int_{D\setminus C} f(x)\,(y(x)-z(x))\,dx, \tag{1.2.48}$$

$$\forall\, z \in S = \left\{w \in H^1(D\setminus C) : w|_{\partial C}=0,\ w|_{\partial D}=1,\ w \ge 0 \text{ a.e. in } D\setminus C\right\},$$

$$y \in S. \tag{1.2.49}$$

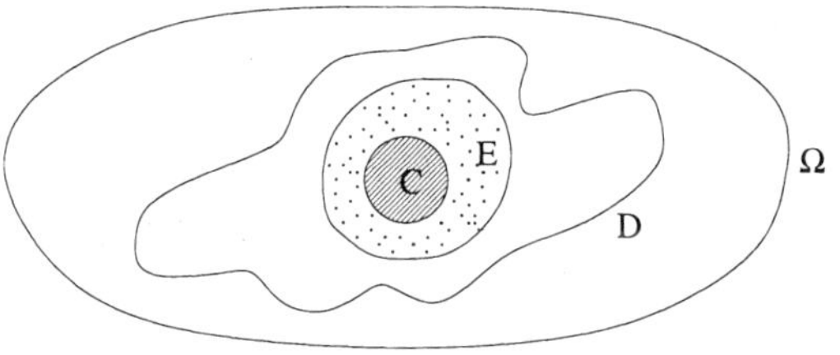

Figure 2.2. The electrochemical machining process.

The connection with the electrochemical machining process is the following: $D \subset \Omega$ represents the machine that contains a given core C (a hole, for instance) that cannot be influenced by the process. The sets ∂C and ∂D represent the electrodes, and the boundary condition $y=1$ on the boundary ∂D indicates that some fixed constant voltage is applied. If $\tilde{y}$ is the extension of y by 0 inside C, then the condition on ∂D in (1.2.49) should be understood in the sense $\tilde{y}-1 \in H_0^1(D)$. The desired shape for the metallic workpiece to be shaped is given by $E \setminus C$ and prescribed by the choice of E. The coincidence set

$$E_y = \{x \in D \setminus C : y(x) = 0\}$$

gives the final shape of the metallic workpiece obtained around C. Its boundary is a free boundary. This is another example of a free boundary problem expressed via the variational inequality (1.2.48), (1.2.49) (compare Elliott and Ockendon [1982]). We now formulate an optimal design problem related to it. Since we cannot expect to realize the desired shape $E \setminus C$ exactly, we want to find at least a domain D such that the associated coincidence set E_y satisfies

$$E_y \supset E \setminus C. \tag{1.2.50}$$

That is, we want to design the shape of the *machine itself* (by determining D) such that the metallic workpieces to be produced satisfy (1.2.50). Taking into account the definition of E_y, the least squares approach leads to the following optimization problem:

$$\operatorname*{Min}_{E\subset D\subset\Omega} \Big\{ J(D) = \frac{1}{2}\int_{E\setminus C} y^2(x)\,dx \Big\}, \tag{1.2.51}$$

subject to (1.2.48), (1.2.49). This is a quadratic control problem governed by a variational inequality. It is nonconvex, and the minimization parameter is just the subdomain D where the variational inequality is defined.

Let us discuss a mathematical approach in a very simple particular case of (1.2.48)–(1.2.51). Namely, we take $C = \emptyset$, $E =]0, \frac{1}{2}[\times]0, 1[$, $\Omega =]0, 1[\times]0, 1[\subset \mathbf{R}^2$, and $D = \{(x_1, x_2) \in \Omega : x_1 < \varphi(x_2)\}$ with some smooth function $\varphi : [0, 1] \to [\frac{1}{2}, 1]$. The minimization parameter remains the variable domain D, which is completely described by the mapping φ.

We now employ the so-called *mapping method* studied by Murat and Simon [1976] and Pironneau [1984] and mentioned above in Example 1.2.4. To this end, we transform the variable subdomain D through the coordinate transformation $x_1 \mapsto x_1/\varphi(x_2)$, $x_2 \mapsto x_2$, onto the fixed domain Ω. This has the advantage that we can work on a fixed domain now; however, one has to pay a price for this "simplification": as a consequence of the transformation, the mapping φ, together with its derivatives, will now enter into the coefficients of the state system (1.2.48), (1.2.49). Hence, we end up with a control in the coefficients problem, which again shows that this class of problems is closely related to optimal shape design problems.

In general, the control in the coefficients problem is difficult to treat since supplementary requirements have to be satisfied for it: φ appears in the coefficients together with its derivatives, or the control coefficients may take only certain values, and so on. Moreover, high regularity conditions must be imposed on the unknown boundary of the domain D for the transformation to be applicable, which is unnatural in many examples.

We continue with a fundamental application from *mechanics* that arises naturally as a control into coefficients problem. In Chapter 6 an investigation of general curved mechanical structures will be performed. Here, we consider systems governed by fourth-order equations of the form

$$\Delta(bu^3 \Delta y) = f \quad \text{in } \Omega, \tag{1.2.52}$$

where $f \in L^2(\Omega)$, $u \in L^\infty(\Omega)$, and $b > 0$ is a constant. If Ω is one- or two-dimensional, such models are used in the literature for the computation of the deflection y of beams or of plates having thickness $u > 0$ and subject to the transverse load f. The coefficient b is a material constant, which we assume normalized to unity, $b = 1$. Equation (1.2.52) may be complemented by various boundary conditions:

$$y = \Delta y = 0 \quad \text{on } \partial\Omega \quad \text{(simply supported plates)}, \tag{1.2.53}$$

$$y = \frac{\partial y}{\partial n} = 0 \quad \text{on } \partial\Omega \quad \text{(clamped plate)}. \tag{1.2.54}$$

In the one-dimensional case, with $\Omega =]0, 1[$, the boundary conditions

$$y(0) = y'(0) = 0, \tag{1.2.55}$$

$$y''(1) = (u^3 y'')'(1) = 0, \tag{1.2.56}$$

correspond to a cantilevered beam, that is, clamped at the left end and free at the right, see Figure 2.3(a) and (b).

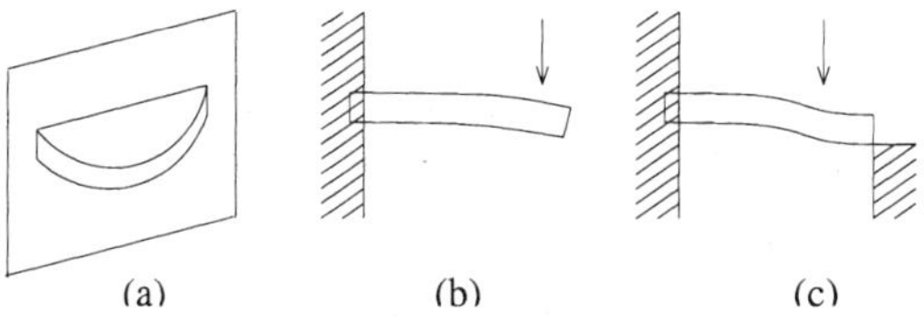

Figure 2.3. (a) A partially clamped plate. (b) A cantilevered beam. (c) A unilaterally supported beam.

We will also recall the model of a beam that is fixed at $x = 0$ and unilaterally supported at $x = 1$. To this end, we put $V = \{y \in H^2(0, 1) : y(0) = y'(0) = 0\}$. The state equation is the following: Find some $y \in \tilde{S} = \{z \in V : z(1) \geq 0\}$ such that

$$(A(u)y, z - y)_{V^* \times V} \geq \int_0^1 f(x)\,(z(x) - y(x))\,dx \quad \forall\, z \in \tilde{S},$$

where $A(u)y = (bu^3 y'')''$ and $f \in V^*$; see Figure 2.3(c). We remark here that in §6.1.3 a similar variational inequality for Kirchhoff-Love arches will be studied.

If $u(x) \geq m > 0$ a.e. in Ω, the bilinear form corresponding to (1.2.52) is coercive in $H^2(\Omega)$, and we easily deduce the existence of a unique weak solution in $H^2(\Omega)$ (compare Example 1.2.3). A classical optimization problem associated with each of the boundary value problems (1.2.52)–(1.2.56) is the minimization of the weight (or, equivalently, of the volume, since the density is assumed constant):

$$\text{Min}\left\{\int_\Omega u(x)\,dx\right\}. \tag{1.2.57}$$

The load f is supposed to be fixed, and the minimization parameter is the thickness u. Natural control and state constraints have to be imposed on u and y:

$$0 < m \leq u(x) \leq M \quad \text{a.e. in } \Omega, \tag{1.2.58}$$

$$y(x) \geq -\tau \quad \text{a.e. in } \Omega, \tag{1.2.59}$$

where m, M, τ are given positive constants. Relation (1.2.59) signifies that the deflection y should be "small," which is a typical safety requirement.

Another significant minimization problem is of identification type:

$$\text{Min}\left\{\int_\Omega (y(x) - y_d(x))^2\,dx\right\}, \tag{1.2.60}$$

i.e., we want to find the thickness u such that a "desired" (or observed) deflection $y_d \in L^2(\Omega)$ results. In this case the following control constraints may be of interest:

$$|u'(x)| \leq \gamma \quad \text{for a.e. } x \in [0,1], \qquad \int_0^1 u(x)\,dx = \alpha, \tag{1.2.58$'$}$$

with given positive constants γ and α.

We conclude by noticing that the above examples are further indications for the relationship between optimal shape design problems (involving geometric optimization parameters) and control into coefficients problems. In Chapter 5, we will introduce methods based on the use of characteristic functions and penalization, or on certain approximate controllability-type properties of elliptic equations, which also reduce geometric optimization problems to optimal control problems, that is, to analytic ones. At the same time, they avoid the use of control into coefficients formulations and are advantageous for numerical purposes.

1.2.4 Variable Domains

In Example 1.2.7 above, we have introduced some optimization problems that involve unknown geometric parameters and enter the field of optimal shape design problems. The main point was that using various techniques (characteristic functions, mapping method), or by their genuine formulations, such problems may be expressed as control into coefficients problems. We have also noticed that such problems are difficult to handle: the necessary transformations are not always possible or require high regularity hypotheses, the associated numerical procedures are stiff, and so on.

In this section, we briefly discuss *direct* formulations of variable domain problems and their optimization. We do this for a very general class of elliptic problems that will be studied in greater detail in Chapters 2 and 5.

Example 1.2.8 Let $\mathcal{O}$ denote a prescribed family of open subsets of a given bounded domain $D \subset \mathbf{R}^d$, and let $\Omega \in \mathcal{O}$. We define in D the nonlinear partial differential operator

$$Ay = \sum_{|\alpha| \le l} (-1)^{|\alpha|} D^\alpha A_\alpha(x, y, \dots, D^l y), \quad x \in D, \tag{1.2.61}$$

for $y \in W^{l,p}(D)$, $l \in \mathbf{N}$, and $p \ge 2$. The coefficient functions $A_\alpha : D \times \mathbf{R}^T \to \mathbf{R}$, $(x, \xi) \mapsto A_\alpha(x, \xi)$, where T denotes the total number of partial derivatives in $\mathbf{R}^d$ from order 0 up to order l (which are collected in the vector $\xi \in \mathbf{R}^T$), are assumed to satisfy the following conditions:

$A_\alpha(\cdot\,, \xi)$ is measurable in D for all $\xi \in \mathbf{R}^T$, and $A_\alpha(x, \cdot)$ is continuous on $\mathbf{R}^T$ for a.e. $x \in D$. (1.2.62)

$$|A_\alpha(x, \xi)| \le C\left(|\xi|_{\mathbf{R}^T}^{p-1} + \mu(x)\right), \text{ where } \mu \in L^q(D) \text{ and } q^{-1} + p^{-1} = 1. \tag{1.2.63}$$

$$\sum_{|\alpha| \le l} (A_\alpha(x, \xi) - A_\alpha(x, \eta))(\xi_\alpha - \eta_\alpha) \ge 0 \quad \forall \xi, \eta \in \mathbf{R}^T \text{ and a.e. } x \in D. \tag{1.2.64}$$

The nonlinear operator A in (1.2.61) is called the *generalized divergence operator* or the *Leray–Lions operator*. Linear elliptic equations of order $2l$ (see (1.2.27)) form a special subclass corresponding to the case $p = 2$. The operator $A : W_0^{l,p}(D) \to W^{-l,q}(D)$ is monotone and hemicontinuous, hence maximal monotone (cf. Appendix 1, Proposition A1.4). Moreover, if the coercivity condition

$$(Ay, y)_{W^{-l,q}(D) \times W_0^{l,p}(D)} \ge c_1 \, |y|^p_{W_0^{l,p}(D)} + c_2 \quad \forall y \in W_0^{l,p}(D), \tag{1.2.65}$$

with constants $c_1 > 0$ and $c_2 \in \mathbf{R}$, is satisfied, then A is surjective, and the realization of A in $L^2(D)$, denoted by $A_{L^2(D)}$, is maximal monotone and surjective (see Example A1.14).

The assumptions and the definitions (1.2.61)–(1.2.65) are directly inherited by any $\Omega \in \mathcal{O}$, since functions from $W_0^{l,p}(\Omega)$ may be trivially extended by zero onto the whole domain D.

Let us now consider a bounded family of functions $\{f_\Omega\} \subset L^2(\Omega)$, indexed by $\Omega \in \mathcal{O}$. Then, according to (1.2.62)–(1.2.65) (cf. Appendix 1), the state equation

$$Ay_\Omega = f_\Omega \quad \text{in } \Omega \tag{1.2.66}$$

has at least one solution in

$$\operatorname{dom}(A_{L^2(\Omega)}) = \left\{y \in W_0^{l,p}(\Omega) : Ay \in L^2(\Omega)\right\}. \tag{1.2.67}$$

If the inequality in (1.2.64) is strict, then the solution is unique. Relation (1.2.67) also expresses the regularity properties of y_Ω. A weak formulation of (1.2.66) is

$$\sum_{|\alpha|\le l} \int_\Omega A_\alpha(x, y_\Omega, \dots, D^l y_\Omega)\, D^\alpha v(x)\, dx = \int_\Omega f_\Omega(x)\, v(x)\, dx \quad \forall v \in W_0^{l,p}(\Omega). \tag{1.2.68}$$

If a family $\{h_\Omega\}$ of functions such that $h_\Omega \in W^{l,p}(\Omega)$ for any $\Omega \in \mathcal{O}$ is given, then we may as well impose inhomogeneous boundary conditions in the form

$$y_\Omega - h_\Omega \in W_0^{l,p}(\Omega), \tag{1.2.69}$$

according to §2.3.2.

Next, we introduce a general cost functional,

$$\operatorname*{Min}_{\Omega\in\mathcal{O}} \left\{\int_\Omega L(x, y_\Omega, \dots, D^l y_\Omega)\, dx\right\}, \tag{1.2.70}$$

where L satisfies the same measurability and continuity conditions as the functions A_α in (1.2.62), and the growth condition (compare (1.2.63))

$$0 \le L(x,\xi) \le C\left(|\xi|^t_{\mathbf{R}^T} + \eta(x)\right), \tag{1.2.63$'$}$$

with some function $\eta \in L^1(D)$ and some $1 \le t < p$.

The problem (1.2.66), (1.2.70) is a very general shape optimization problem in arbitrary dimension, with nonlinear elliptic equations of arbitrary order and homogeneous or (in the case of (1.2.69)) inhomogeneous boundary conditions. For instance, the simple case with the Laplace operator,

$$-\Delta y_\Omega = 1 \quad \text{in } \Omega, \qquad y_\Omega = 0 \quad \text{on } \partial\Omega,$$

is obtained with $l = 1$, $p = 2$, and arbitrary $d \in \mathbf{N}$.

Another important example arises if different boundary conditions are imposed:

$$\operatorname*{Min}_{\Omega\in\mathcal{O}} \left\{\int_\Omega l(x, y(x), \nabla y(x))\, dx\right\} \tag{1.2.71}$$

subject to

$$\int_\Omega \nabla y(x) \cdot \nabla v(x)\, dx + \int_\Omega y(x)\, v(x)\, dx = \int_\Omega f_\Omega(x)\, v(x)\, dx \quad \forall\, v \in H^1(\Omega). \tag{1.2.72}$$

Obviously, the relations (1.2.71), (1.2.72) represent an optimal design problem governed by a homogeneous Neumann boundary value problem. To this problem (as in all the other cases presented in this paragraph) we may add various constraints on the state y (for instance, positivity) and/or on the domains from $\mathcal{O}$ (for instance, a prescribed measure).

Chapter 2

Existence

This chapter is devoted to a thorough presentation of the existence theory for many of the problems introduced in Chapter 1. We begin with abstract results, continue with several specialized existence and uniqueness theorems, highlighting various difficulties and ideas to overcome them, and close the chapter with a general theory for variable domain problems that has recently been developed by the authors. The text also includes numerous comments and remarks on other approaches and research directions, providing a survey of the available mathematical literature. For general existence results in the calculus of variations and in the optimal control of ordinary differential equations we recommend to the reader the works of Buttazo [1989] and Cesari [1983], [1990].

2.1 A General Situation

In order to prove the existence of minimum points in optimization problems, one usually postulates compactness properties with respect to some topology for the (nonempty) set on which the problem is defined, as well as the lower semicontinuity of the cost functional with respect to the same topology. Under these premises, the Weierstrass theorem can be applied, which, if the problem is also convex, takes the particularly satisfactory form of Theorem A1.6 in Appendix 1.

These assumptions are quite restrictive, and in many problems some or all of them fail to be valid, so that existence or uniqueness of solutions cannot be shown. However, the existence of so-called *δ-solutions*, that is, of parameters at which the cost functional attains a value within an interval of radius $\delta > 0$ around the optimum value, is always guaranteed and suffices from the viewpoint of many practical applications. Moreover, even in the case that the problem admits minimum points, the necessary approximations in the numerical solution of nontrivial examples eventually yield only δ-solutions for the original problem. Equivalently, efficient minimizing sequences are "necessary and sufficient" for the solution of minimization problems.

It should be noted that in optimization problems a δ-solution may be "far away" from the solution itself (for instance, $x^* = 0$ is the minimizer of the function $f(x) = \delta|x|$ on $\mathbf{R}$, while $x = \pm 1$ are δ-solutions). This situation clearly differs from looking for the solution of a (differential) equation where that solution or a close approximation thereof with respect to a convenient topology has to be determined.

We conclude that δ-solutions are very important in optimization problems, but we also stress the fact that the stability properties of numerical algorithms for the determination of minimum points are improved in the case that existence (and uniqueness) are valid.

We begin our investigation by briefly discussing the fundamental question of admissibility, i.e., that the set of admissible controls should be nonempty. First, we study the control problem (P) given by (1.1.1), (1.1.2), (1.1.3), (1.1.4), where the spaces U, V, the sets C, U_{ad}, the cost functional L and the linear operators A, B fulfill the conditions of Chapter 1 with $Z = V^*$. Later on, in Theorem 2.1.3, we will also examine the situation in which the state operator A depends directly on the control u, as in Examples 1.2.4 and 1.2.7.

The pairs $[y, u] \in V \times U$ satisfying (1.1.1), (1.1.3), (1.1.4) form the *admissible set* for the control problem (P), which is a closed and convex subset of $V \times U$. Sometimes, the condition $L(y, u) < +\infty$ may be also included in the concept of admissibility. Since U_{ad} and C are closed and convex, they may be very "thin," and the admissible set may be empty. Therefore, the basic assumption that (P) has at least one admissible pair may be a stringent condition and difficult to verify. In fact, in many engineering applications it even suffices to find just one admissible pair. Of course, the difficulty originates from the state constraint (1.1.4), since without it ($C = H$) any $u \in U_{ad}$ is admissible.

One way to relax this hypothesis, inspired by the concept of δ-solutions, is to replace the constraints by

$$y \in C_\delta, \quad \text{where } C_\delta = \{y \in H : \exists\, z \in C \text{ with } |y - z|_H \leq \delta\}, \tag{2.1.1}$$

$$u \in (U_{ad})_\delta, \quad \text{where } (U_{ad})_\delta = \{u \in U : \exists\, w \in U_{ad} \text{ with } |w - u|_U \leq \delta\}. \tag{2.1.2}$$

Obviously, C_δ and $(U_{ad})_\delta$ are closed and convex subsets of H and U, respectively, and have a nonempty interior. The pairs $[y, u] \in V \times U$ satisfying (1.1.1), (2.1.1), (2.1.2) are called *δ-admissible* for (P). Sometimes also small perturbations of the state equation (1.1.1), in a sense to be made precise, may be taken into account (see Proposition 2.1.4 and (2.1.15) below).

Next we show that the question of admissibility (or δ-admissibility) may be equivalently formulated as an optimal control problem by adding a nonlinear term to the state equation (1.1.1). To this end, we denote by $\psi : U \to]-\infty, +\infty]$ and $\alpha : H \to]-\infty, +\infty]$ the indicator functions of $(U_{ad})_\delta$ and C_δ, respectively, where $\delta > 0$ is fixed. Then it is easily seen that a pair $[y, u]$

satisfies the constraints (1.1.1), (2.1.1), (2.1.2) if and only if it solves the minimization problem

$$\text{Min}\left\{\psi(u)\right\} \quad \text{subject to (1.1.1) and (2.1.1).} \tag{2.1.3}$$

We now associate to the minimization problem (1.1.1), (2.1.1), (2.1.3) the following optimal control problem governed by variational inequalities (see Appendix 2) without any explicit constraints:

$$\text{Min}\left\{\psi(u) + \frac{1}{2}\,|w|_{V^*}^2\right\}, \tag{2.1.4}$$

subject to

$$Ay + w = Bu + f, \quad w \in \partial\alpha(y). \tag{2.1.5}$$

The hypotheses are the same as in Section 1.1 of Chapter 1, where we assume that $A : V \to V^*$ is coercive. The idea to deal with the state constraints via the state system (2.1.5) seems more natural than to penalize them in the cost functional, which is a standard procedure in the literature. We refer to Chapter 3 in connection with this latter method. We have the following result.

Proposition 2.1.1 *The set of* δ-admissible *pairs for (P) coincides with the set of optimal pairs to the problem (2.1.4), (2.1.5).*

Proof. First notice that it follows from the general results of Appendix 2 that for any $u \in U$ there exist uniquely determined elements $y \in V$, $w \in V^*$, such that y is the solution of (2.1.5) corresponding to u. Any δ-admissible pair for (P) obviously satisfies (2.1.5) with associated $w = 0$ and thus is optimal for (2.1.4), (2.1.5) with optimal value zero. Conversely, any optimal pair $[\overline{y}, \overline{u}]$ for (2.1.4), (2.1.5) satisfies $\psi(\overline{u}) = 0$ and $\overline{w} = 0$, that is, $\overline{u} \in (U_{ad})_\delta$, and $[\overline{y}, \overline{u}]$ satisfies (1.1.1). Since clearly $\overline{y} \in C_\delta$ by the definition of variational inequalities, it follows that $[\overline{y}, \overline{u}]$ is δ-admissible for (P), which finishes the proof. □

Remark. The above equivalence is an example for the so-called *variational inequality method* for state constrained control problems (see (2.1.14), (2.1.15) below). Notice that we can further relax the constraints by replacing the indicator function α by a sufficiently smooth regularization α^ε, where $\varepsilon > 0$ is some regularization parameter. It may be obtained by the Yosida–Moreau regularization or as in Example A1.13 in Appendix 1. We then have $w = (\alpha^\varepsilon)'(y)$, where $(\alpha^\varepsilon)'$ is Lipschitzian and smooth, and (2.1.4) attains the standard form (1.2.1) without convexity. By solving this last problem, one may construct δ'-admissible pairs for (P) such that $\delta' > \delta$.

Remark. Finally, let us notice that the admissibility property is in fact a *controllability* property (with constraints) for the equation (1.1.1): when u ranges in U_{ad}, is it possible that the associated solution y "hits" a "target" in C? The difficulty of this type of question is well known. For a comprehensive investigation, we refer the reader to Henry [1978] and Cârja [1988], [1991].

We now give an existence result for the problem (P) in the general setting of Chapter 1 when $Z = V^*$.

Theorem 2.1.2 *Assume that* $A : V \to V^*$ *is a linear and bounded operator having a linear and bounded inverse. Moreover, let* U_{ad} *be bounded or* L *be coercive in* u *uniformly with respect to* y*, that is,*

$$\lim_{|u|_U \to \infty} L(y,u) = +\infty \text{ uniformly in } y. \tag{2.1.6}$$

Then (P) has at least one optimal pair $[y^*, u^*]$ *provided the admissibility condition (1.1.6) is fulfilled.*

Proof. If U_{ad} is bounded, we may redefine $L(\cdot,\cdot)$ by $+\infty$ outside $V \times U_{ad}$ as in (1.1.5)′. Then (2.1.6) is fulfilled. Hence, we have to examine only this case. The admissibility condition (1.1.6) ensures the existence of a minimizing sequence $\{[y_n, u_n]\}$ such that

$$\lim_{n\to\infty} L(y_n, u_n) = \inf((\mathrm{P})) < +\infty. \tag{2.1.7}$$

In view of (2.1.6), $\{u_n\}$ is bounded in U. By (1.1.1), and since $A^{-1} : V^* \to V$ is bounded, $\{y_n\}$ is bounded in V. Therefore, for a subsequence, which is again indexed by n, $[y_n, u_n] \to [\overline{y}, \overline{u}]$ weakly in $V \times U$ for some $[\overline{y}, \overline{u}] \in U \times V$. Passing to the limit as $n \to \infty$ in (1.1.1), we conclude that $\overline{y}$ is the state corresponding to $\overline{u}$. Obviously, $[\overline{y}, \overline{u}]$ belongs to $C \times U_{ad}$, which is weakly closed in $V \times U$. Since L is weakly lower semicontinuous, we finally obtain that

$$L(\overline{y}, \overline{u}) \le \liminf_{n\to\infty} L(y_n, u_n) = \inf((\mathrm{P})).$$

Therefore, $[\overline{y}, \overline{u}]$ is an optimal pair for (P), which we redenote by $[y^*, u^*]$. □

Remark. If strict convexity is assumed for L, then the optimal pair $[y^*, u^*]$ is unique. The above approach is related to the "direct method" in the calculus of variations, which is based on the construction of minimizing sequences. In the next chapter the "indirect approach" (via optimality conditions) will be thoroughly discussed.

Let us now prove a general existence result for (P) when (1.1.1) is nonlinear and A may depend directly on u as in Example 1.2.4 and Example 1.2.7. We make the following assumptions:

$$A(u) : V \to V^* \text{ is linear and bounded for every } u \in U_{ad}. \tag{2.1.8}$$

$$\varphi : V \to]-\infty, +\infty] \text{ is proper, convex, and lower semicontinuous.} \tag{2.1.9}$$

We replace the operator A from (1.1.1) by $A(u) + \partial\varphi$ and the state equation (1.1.1) by

$$A(u)y + \partial\varphi(y) \ni Bu + f. \tag{1.1.1$'$}$$

We then denote by (P)$'$ the optimal control problem defined by (1.1.1)$'$, (1.1.2), (1.1.3), (1.1.4), under the conditions (2.1.8), (2.1.9) and the standard hypotheses.

Finally, we assume that $A(u) : V \to V^*$ is coercive and depends continuously on u, that is, that

$$(A(u)y, y)_{V^* \times V} \geq m|y|_V^2 \quad \forall\, y \in V, \quad \text{for some } m > 0, \tag{2.1.10}$$

$$u_n \to u \text{ in } U \quad \Rightarrow \quad A(u_n) \to A(u) \text{ in } L(V, V^*). \tag{2.1.11}$$

The operator $B : U \to V^*$, $f \in V^*$, and the sets C and U_{ad} are as in Theorem 2.1.2 with the modifications specified below.

Theorem 2.1.3 *Suppose that the admissibility condition (1.1.6) and (2.1.8)–(2.1.11) are fulfilled. Then (P)$'$ has at least one optimal solution $[y^*, u^*]$ provided that one of the following conditions is satisfied:*

(a) *$U_{ad} \subset U$ is compact (not necessarily convex).*

(b) *$C \subset V$ is compact (not necessarily convex), $U_{ad} \subset U$ is bounded, and $A(\cdot)$ is linear and bounded in u.*

Proof. (a) Let $\{u_n\} \subset U$ be a minimizing sequence. Since U_{ad} is compact, we may without loss of generality assume that $u_n \to u$ strongly in U for some $u \in U_{ad}$. Then $Bu_n \to Bu$ strongly in V^* and $A(u_n) \to A(u)$ strongly in $L(V, V^*)$.

Now take the dual pairing in $V^* \times V$ of the terms in (1.1.1)$'$ with $y_n - y_1$. Invoking the monotonicity of $\partial\varphi$, we find that with some $v_n \in \partial\varphi(y_n)$,

$$\begin{aligned}
&(Bu_n + f, y_n - y_1)_{V^* \times V} \\
&\quad = (A(u_n)y_n, y_n - y_1)_{V^* \times V} + (v_n - v_1, y_n - y_1)_{V^* \times V} + (v_1, y_n - y_1)_{V^* \times V} \\
&\quad \geq (A(u_n)(y_n - y_1), y_n - y_1)_{V^* \times V} + (A(u_n)y_1, y_n - y_1)_{V^* \times V} - |v_1|_{V^*}|y_n - y_1|_V.
\end{aligned}$$

From (2.1.10) and (2.1.11) it then follows that there is a constant $C_1 > 0$ such that

$$m\,|y_n - y_1|_V^2 \leq C_1\,|y_n - y_1|_V \quad \forall\, n \in \mathbf{N}.$$

Therefore, $\{y_n\}$ is bounded in V, and there is some subsequence, again indexed by n, such that $y_n \to y$ weakly in V, where $y \in C$. Using the (linear and continuous) dual operator $A(u)^*$ associated with $A(u)$, we find that for any $v \in V$,

$$\begin{aligned}
&|(A(u_n)y_n, v)_{V^*\times V} - (A(u)y, v)_{V^*\times V}| \\
&\quad \le |(A(u_n)y_n, v)_{V^*\times V} - (A(u)y_n, v)_{V^*\times V}| \\
&\qquad + |(A(u)y_n, v)_{V^*\times V} - (A(u)y, v)_{V^*\times V}| \\
&\quad \le |v|_V\, |A(u_n)y_n - A(u)y_n|_{V^*} + |(y_n - y, A(u)^*v)_{V\times V^*}| \\
&\quad \le |v|_V\, |A(u_n) - A(u)|_{L(V,V^*)}\, |y_n|_V + |(y_n - y, A(u)^*v)_{V\times V^*}| \to 0
\end{aligned}$$

as $n \to \infty$. Moreover, (2.1.10) yields

$$\begin{aligned}
\liminf_{n\to\infty}(A(u_n)y_n, y_n - y)_{V^*\times V} &\ge \liminf_{n\to\infty}(A(u_n)y, y_n - y)_{V^*\times V} \\
&= \lim_{n\to\infty}(A(u_n)y, y_n)_{V^*\times V} - \lim_{n\to\infty}(A(u_n)y, y)_{V^*\times V} = 0. \qquad (2.1.12)
\end{aligned}$$

Multiplying (1.1.1)′ by $y_n - y$, and invoking the definition of $\partial\varphi$ (cf. Appendix 1), we find that

$$(A(u_n)y_n, y_n - y)_{V^*\times V} + \varphi(y_n) \le (Bu_n + f, y_n - y)_{V^*\times V} + \varphi(y).$$

But then, owing to the lower semicontinuity of φ,

$$\limsup_{n\to\infty}(A(u_n)y_n, y_n - y)_{V^*\times V} \le \varphi(y) - \liminf_{n\to\infty}\varphi(y_n) \le 0. \qquad (2.1.13)$$

From (2.1.12), (2.1.13), we conclude that

$$\lim_{n\to\infty}(A(u_n)y_n, y_n - y)_{V^*\times V} = 0.$$

On the other hand,

$$(A(u_n)y_n - A(u_n)y, y_n - y)_{V^*\times V} \ge m|y_n - y|_V^2,$$

that is, $y_n \to y$ strongly in V, for a subsequence again indexed by n. The demiclosedness of $\partial\varphi$ (compare Appendix 1) then shows that y is the solution to (1.1.1)′ corresponding to u, that is, the pair $[y, u]$ is admissible for (P)′. In view of the weak lower semicontinuity of L, it is also optimal (and denoted by $[y^*, u^*]$).

(b) If $\{u_n\}$ is a minimizing sequence for (P)′, then it follows from (b) that $\{u_n\}$ is bounded, therefore $u_n \to u \in U_{ad}$ weakly in U, and that $y_n \to y \in C$ strongly in V, on a subsequence. Moreover, under our assumptions, we have for any $v \in V$,

$$\begin{aligned}
&|(A(u_n)y_n - A(u)y, v)_{V^*\times V}| \\
&\quad \le |(A(u_n)y_n - A(u_n)y, v)_{V^*\times V}| + |(A(u_n)y - A(u)y, v)_{V^*\times V}| \\
&\quad \le |A(u_n)|_{L(V,V^*)}|y_n - y|_V|v|_V + |(A(u_n)y - A(u)y, v)_{V^*\times V}| \to 0, \quad n \to \infty.
\end{aligned}$$

From this point, the rest of the proof proceeds as for (a). □

Remark. The above existence and admissibility results are rather standard. Similar statements may be found, for instance, in Haslinger, Neittaanmäki, and Tiba [1987], Haslinger and Neittaanmäki [1988], and Tiba [1990].

Remark. Since equation (1.1.1)′ is nonlinear, the problem (P)′ is, in general, nonconvex even though L is convex. Thus, local minima are to be expected for (P)′, the set of global minima is not necessarily convex, and the optimal pair may not be unique. A general situation to which Theorem 2.1.3 applies is described in Appendix 2, Theorem A2.9. A completely different framework will be studied in the next paragraph.

Remark. One can use Theorem 2.1.2 and the subsequent remark to derive existence and uniqueness results in Examples 1.2.1–1.2.3, as well as in the plasma identification problem considered in Example 1.2.6 of Chapter 1. Theorem 2.1.3 may be applied in certain cases discussed in Examples 1.2.4 and 1.2.5, as well as in the identification problem from Example 1.2.6.

We close this section by briefly presenting an approximation technique that generalizes the ideas used in the proofs of Theorem 2.1.3 and of Proposition 2.1.1.

For the treatment of the state constraint (1.1.4), we indicate an approximation result for the problem (P)′ that gives a better error estimate than the usual penalization method (compare Chapter 3 below). This idea was originally introduced in Tiba [1986] as the *variational inequality method*, and it has a wide range of applications (see Tiba [1990, Chapter III], Haslinger and Neittaanmäki [1988, Chapter 10], and the references therein).

To this end, let $I_C : V \to]-\infty, +\infty]$ be the indicator function of $C \subset V$. We associate to (P)′ the approximating problem

$$\text{Min}\left\{L(y,u) + \frac{1}{2}\,|w|^2_{V^*}\right\} \tag{2.1.14}$$

subject to (1.1.3) and to the state system

$$A(u)y + \partial\varphi(y) + \varepsilon w \ni Bu + f, \quad w \in \partial I_C(y), \tag{2.1.15}$$

for some parameter $\varepsilon > 0$.

From Appendix 1 it follows that if φ has a continuity point $x_0 \in C$ or if C has interior points in $\text{dom}(\varphi)$, then $\partial\varphi + \varepsilon\,\partial I_C = \partial\varphi + \partial I_C = \partial(\varphi + I_C)$. Thus, equation (2.1.15) has a similar structure to that of (1.1.1)′ in the problem (P)′, and existence and uniqueness of the solution of (2.1.15) follow in a standard way under the coercivity assumption (2.1.10). The advantage is now that the state constraints (1.1.4) are no longer explicit but penalized in the state equation instead.

Under appropriate compactness assumptions, the existence of at least one optimal pair $[y_\varepsilon, u_\varepsilon]$ for the above control problem follows again from Theorem 2.1.3. We denote by $w_\varepsilon \in \partial I_C(y_\varepsilon)$ the selection of ∂I_C occurring in (2.1.14), and by $L^\varepsilon(y,u)$ the approximating cost functional (2.1.14). Obviously, if $[y,u]$ is admissible for (P)′, then $[y,u]$ is admissible for (2.1.14), (2.1.15), (1.1.3), and the associated $w \in \partial I_C(y)$ is given by $w = 0$, that is, $L^\varepsilon(y,u) = L(y,u)$, and we have the same cost value. This also holds for $[y^*, u^*]$ and $w^* = 0$, and we see that $\{u_\varepsilon\}$ is bounded in U and $\{w_\varepsilon\}$ is bounded in V^*, since

$$L^\varepsilon(y_\varepsilon, u_\varepsilon) = L(y_\varepsilon, u_\varepsilon) + \frac{1}{2}\,|w_\varepsilon|^2_{V^*} \le L(y^*, u^*), \tag{2.1.16}$$

and L is assumed to be coercive in u uniformly with respect to y (recall (2.1.6)). Finally, multiplying by y_ε in (2.1.15), and invoking (2.1.10), we find after a short calculation that $\{y_\varepsilon\}$ is bounded in V.

We have the following convergence result for the approximating sequences $\{u_\varepsilon\}$, $\{w_\varepsilon\}$, $\{y_\varepsilon\}$.

Proposition 2.1.4 *Assume that $A(\cdot)$ satisfies (2.1.11) and that $U_{ad} \subset U$ is compact. Then there is a sequence $\varepsilon_n \searrow 0$ such that $u_{\varepsilon_n} \to u^*$ strongly in U, $y_{\varepsilon_n} \to y^*$ strongly in V, $w_{\varepsilon_n} \to 0$ strongly in V^*, and $L^{\varepsilon_n}(y_{\varepsilon_n}, u_{\varepsilon_n}) \to L(y^*, u^*)$, as $n \to \infty$.*

Proof. Since the proof closely resembles that of Theorem 2.1.3, we only sketch the argument. By virtue of the boundedness results established above, we have, for a suitable subsequence $\varepsilon_n \searrow 0$, $u_{\varepsilon_n} \to \bar{u}$ strongly in U (since U_{ad} is compact), $y_{\varepsilon_n} \to \bar{y}$ weakly in V, $\varepsilon_n w_{\varepsilon_n} \to 0$ strongly in V^*. In view of the continuity assumption (2.1.11), a calculation similar to (2.1.12), (2.1.13), shows that $y_{\varepsilon_n} \to \bar{y}$ strongly in V, and $\bar{y}$ is the solution to (1.1.1)′ corresponding to $\bar{u}$. Moreover, by the demiclosedness of maximal monotone operators, $\partial\varphi(y_\varepsilon) \to \ell \in \partial\varphi(\bar{y})$. We thus can pass to the limit in (2.1.15) to conclude that $[\bar{y}, \bar{u}]$ is an admissible pair for the initial optimal control problem (P)′.

By virtue of the lower semicontinuity of L and (2.1.16), we see that $[\bar{y}, \bar{u}]$ is optimal for the problem (P)′, and we redenote it by $[y^*, u^*]$. Then, clearly $L^{\varepsilon_n}(y_{\varepsilon_n}, u_{\varepsilon_n}) \to L(y^*, u^*)$, and $w_{\varepsilon_n} \to 0$ strongly in V^*, which ends the proof of the assertion. □

Next, we show that the control u_ε is *suboptimal* for the initial problem (P)′. Let $y^\varepsilon \in V$ be the unique solution of (1.1.1)′ corresponding to u_ε. Subtraction of the inclusions (1.1.1)′ and (2.1.15) leads to

$$A(u_\varepsilon)y^\varepsilon - A(u_\varepsilon)y_\varepsilon + v^\varepsilon - v_\varepsilon = \varepsilon w_\varepsilon\,, \tag{2.1.17}$$

with suitable $v^\varepsilon \in \partial\varphi(y^\varepsilon)$, $v_\varepsilon \in \partial\varphi(y_\varepsilon)$. Taking the dual pairing of (2.1.17) with $y^\varepsilon - y_\varepsilon$, and using (2.1.10) and the monotonicity of $\partial\varphi$, we find that

$$m|y^\varepsilon - y_\varepsilon|_V \le \varepsilon |w_\varepsilon|_{V^*}. \tag{2.1.18}$$

But $y_\varepsilon \in C$, by the definition of the variational inequality (2.1.15). Therefore, (2.1.18) and Proposition 2.1.4 establish the following result.

Corollary 2.1.5 *Assume that* $L(y,u) = \theta(y) + \psi(u)$, *and suppose that* θ *is a continuous convex function on* V *and* ψ *is proper, convex, and lower semicontinuous. Then we have:*

(i) $\lim_{\varepsilon \to 0} L(y^\varepsilon, u_\varepsilon) = L(y^*, u^*)$.

(ii) $[y^\varepsilon, u_\varepsilon]$ *satisfies (1.1.1)'.*

(iii) $u_\varepsilon \in U_{ad}$, *and* $\inf\{|y^\varepsilon - v|_V : v \in C\} \le c\varepsilon$, *with a constant* $c > 0$ *that is independent of* ε.

Proof. We know that $y_\varepsilon \to y^*$ strongly in V, and (2.1.18) implies that also $y^\varepsilon \to y^*$ strongly in V. Then $\theta(y^\varepsilon) \to \theta(y^*)$, and (i) follows. The other two statements are clear from the previous argument; in fact, they do not need the special structure of L. □

Remark. It is in the above sense that we call u_ε *suboptimal* for the problem (P)'. The evaluation (iii) of the violation of state constraints is obtained in a stronger norm here than in the general penalization approach that will be discussed in full detail in Chapter 3. In examples involving function spaces and partial differential equations, (iii) yields uniform error estimates (compare Tiba [1990, Chapter III, 4]). In the case of optimal shape design (variable domain problems), an application of the variational inequality method is discussed in Tiba [1990, Chapter III, 5.1].

2.2 Special Existence and Uniqueness Results

It is clear that in nonconvex optimization problems more complex situations are possible, and many types of arguments can be applied. In the following, we indicate several such cases in which special direct arguments lead to positive answers to the question of the existence of optimal pairs or even to their uniqueness.

2.2.1 Second-Order Problems

We begin with an example in which, in comparison with the preceding section, the state equation has the "wrong" monotonicity behavior, that is, the increasing nonlinearity occurs on the right-hand side of the state equation and not as, for example, in Theorem 2.1.3.

Let Ω be a smooth domain in $\mathbf{R}^d$, and let $\alpha, \beta, \gamma \in L^\infty(\Omega)$ satisfy the conditions

$$0 < \alpha(x) < \beta(x) \quad \text{a.e. in } \Omega, \quad 0 < \gamma(x) \quad \text{a.e. in } \Omega.$$

We consider a set of admissible controls given by

$$U_{ad} = \{u \in L^\infty(\Omega) : \alpha(x) \leq u(x) \leq \beta(x) \;\; \text{a.e. in } \Omega\}. \tag{2.2.1}$$

For any $u \in U_{ad}$, the state of the system is given by the semilinear equation

$$-\Delta y = a(x, u, y) \quad \text{in } \Omega, \quad y = 0 \quad \text{on } \partial\Omega, \tag{2.2.2}$$

where $a : \Omega \times \mathbf{R}_+ \times \mathbf{R}$ is a Carathéodory mapping, i.e., measurable in x and continuous in u and y, which is bounded from below,

$$a(x, u, y) \geq a_0 > 0, \tag{2.2.3}$$

and sublinear,

$$|a(x, u, y)| \leq c_1 |y| + c_2(x), \tag{2.2.4}$$

for all admissible values for the variables, where $a_0 \in L^\infty(\Omega)$, $c_2 \in L^2(\Omega)$, and $0 < c_1 < c(\Omega)$. Here, $c(\Omega)$ denotes the Poincaré constant of Ω, that is, $c(\Omega)$ is the largest constant satisfying

$$\int_\Omega |y(x)|^2 \, dx \leq c(\Omega)^{-1} \int_\Omega |\nabla y(x)|^2 \, dx \quad \forall \, y \in H_0^1(\Omega).$$

Remark. The existence of a (not necessarily unique) solution $y \in H^2(\Omega) \cap H_0^1(\Omega)$ to (2.2.2) for any $u \in U_{ad}$ can be proved in a standard way using Schauder's fixed point theorem, even without any monotonicity condition on $a(x, u, \cdot)$. We briefly indicate the argument. To this end, set

$$\ell = |c_2|_{L^2(\Omega)} (c(\Omega) - c_1)^{-1} > 0,$$

and consider the ball

$$K = \Big\{ \tilde{y} \in L^2(\Omega) \, : \, |\tilde{y}|_{L^2(\Omega)} \leq \ell \Big\}.$$

We define the operator $S : K \to H^2(\Omega) \cap H_0^1(\Omega)$, $\tilde{y} \mapsto y$, where y is the (unique) solution to the linear elliptic problem

$$-\Delta y = a(x, u(x), \tilde{y}) \;\; \text{in } \Omega, \qquad y = 0 \;\; \text{on } \partial\Omega.$$

Then $S(K) \subset K$. Indeed, testing by y, we have for any $\tilde{y} \in K$ the chain of inequalities

$$c(\Omega) \, |y|^2_{L^2(\Omega)} \leq |y|^2_{H_0^1(\Omega)} = -\int_\Omega y \, \Delta y \, dx \leq \Big(c_1 \, |\tilde{y}|_{L^2(\Omega)} + |c_2|_{L^2(\Omega)} \Big) \, |y|_{L^2(\Omega)} \, ,$$

whence

$$|y|_{L^2(\Omega)} \leq c(\Omega)^{-1}(c_1\ell + |c_2|_{L^2(\Omega)}) = \ell$$

follows.

Next, we may employ Lebesgue's theorem and the compactness of the embedding of $H^2(\Omega) \cap H_0^1(\Omega)$ in $L^2(\Omega)$ to conclude that S is continuous on K. Since $S(K)$ is obviously relatively compact in $L^2(\Omega)$, Schauder's theorem (cf., for instance, Pascali and Sburlan [1978]) yields the existence of a fixed point of S in K, which then is a solution to (2.2.2).

From the maximum principle for the Laplacian (cf. Appendix 2, Theorem A2.8) it follows that any solution of (2.2.2) satisfies $y(x) \geq 0$ a.e. in Ω for any $u \in U_{ad}$. We also impose an explicit state constraint:

$$y(x) \leq \gamma(x) \quad \text{a.e. in } \Omega. \tag{2.2.5}$$

Finally, we introduce the cost functional

$$J(y,u) = \int_\Omega l(u(x))\,dx + \int_\Omega b(x, y(x))\,dx, \tag{2.2.6}$$

with a Carathéodory mapping b on $\Omega \times \mathbf{R}_+$, a continuous mapping l on $\mathbf{R}_+$, and with $[y,u]$ satisfying (2.2.1), (2.2.2), and (2.2.5). We denote by (P) the problem of minimizing J subject to all of these constraints. Notice that although (2.2.2) is in general not well-posed (there may be no uniqueness), the problem (P) has a clear meaning as the minimization of J over all pairs $[y,u]$ satisfying the given conditions.

We have the following existence result for the problem (P) in the case that $a(x,u,\cdot)$ has the "wrong" monotonicity behavior:

Theorem 2.2.1 *Assume that $a(x,u,\cdot)$ is increasing for a.e. $x \in \Omega$ and all $u \in \mathbf{R}_+$, $a(x,\cdot\,,y)$ is convex for a.e. $x \in \Omega$ and all $y \in \mathbf{R}$, l is convex and positive, and $b(x,\cdot)$ is increasing for a.e. $x \in \Omega$. Then (P) has at least one optimal pair $[y^*,u^*]$ provided there exists at least one admissible pair for (P).*

Proof. Let $[y_n,u_n]$ be a minimizing sequence for (P), that is,

$$J(y_n,u_n) = \int_\Omega l(u_n(x))\,dx + \int_\Omega b(x,y_n(x))\,dx \to \inf(\text{P}) \quad \text{as } n \to \infty, \tag{2.2.7}$$

where $[y_n,u_n]$ satisfy (2.2.1), (2.2.2), and (2.2.5). Obviously, $\{u_n\}$ is bounded in $L^\infty(\Omega)$. Multiplying (2.2.2) by y_n, integrating by parts, and using (2.2.4), we obtain (compare the above remark) that

$$|y_n|^2_{H_0^1(\Omega)} = \int_\Omega a(x,u_n(x),y_n(x))y_n(x)\,dx \leq c_1|y_n|^2_{L^2(\Omega)} + |c_2|_{L^2(\Omega)}\,|y_n|_{L^2(\Omega)}.$$

Since $c_1 < c(\Omega)$, the Poincaré inequality shows that $\{y_n\}$ is bounded in $L^2(\Omega)$. Invoking (2.2.4) again, we conclude that $\{a(\cdot\,,u_n,y_n)\}$ is bounded in $L^2(\Omega)$.

Hence, by the standard regularity theory for linear elliptic equations (cf. Appendix 2), $\{y_n\}$ is bounded in $H^2(\Omega) \cap H_0^1(\Omega)$. Therefore, we have for a subsequence, which is again indexed by n,

$$u_n \to \overline{u} \quad \text{weakly* in } L^\infty(\Omega), \quad y_n \to \tilde{y} \quad \text{strongly in } H_0^1(\Omega).$$

Notice that $\tilde{y}$ and $\overline{u}$ satisfy (2.2.5) and (2.2.1), respectively. Moreover, using the weak lower semicontinuity of convex functionals, Fatou's lemma, and the Carathéodory hypotheses on $b(\cdot\,,\cdot)$, we conclude that

$$\inf(\mathrm{P}) = \lim_{n\to\infty} J(y_n, u_n) \ge \int_\Omega l(\overline{u}(x))\,dx + \int_\Omega b(x, \tilde{y}(x))\,dx. \tag{2.2.8}$$

However, $[\tilde{y}, \overline{u}]$ is not necessarily an admissible pair for (P), since it may not satisfy (2.2.2). Now denote by $\tilde{w}$ the weak limit in $L^2(\Omega)$ (on a subsequence again indexed by n) of $w_n(x) = a(x, u_n(x), y_n(x))$. Passing to the limit as $n \to \infty$ in (2.2.2), we find that

$$-\Delta\tilde{y} = \tilde{w} \quad \text{in } \Omega. \tag{2.2.9}$$

The subsequent Lemma 2.2.2 then yields that

$$\tilde{w}(x) \ge w(x) = a(x, \overline{u}(x), \tilde{y}(x)) \quad \text{a.e. in } \Omega. \tag{2.2.10}$$

To verify this, we apply Lemma 2.2.2 to the sequences $u_n(x)$ and $v_n(x) = a(x, u_n(x), \tilde{y}(x))$, which, owing to (2.2.4), are bounded in $L^2(\Omega)$. We then obtain that

$$(\text{w-lim}_{n\to\infty} v_n)(x) \ge a(x, \bar{u}(x), \tilde{y}(x))\,, \text{ for a.e. } x \in \Omega.$$

Moreover, we have $v_n(x) - w_n(x) \to 0$ a.e. in Ω, by the uniform continuity of $a(x,\cdot,\cdot)$ in any rectangle in $\mathbf{R}_+ \times \mathbf{R}_+$ and by the boundedness of $\{u_n(x)\}$ and $\{y_n(x)\}$ for a.e. $x \in \Omega$, which then shows the validity of (2.2.10).

We now consider the sequence $\{z_n\}$, which is defined by the following recursion: $z_0 = \tilde{y}$ a.e. in Ω,

$$-\Delta z_n = a(x, \overline{u}, z_{n-1}) \quad \text{in } \Omega, \quad z_n = 0 \quad \text{on } \partial\Omega. \tag{2.2.11}$$

Next, we denote by $y_0 \in H_0^1(\Omega)$ the solution of $-\Delta y_0 = a_0$ in Ω, with the constant a_0 from (2.2.3). By virtue of (2.2.10) and (2.2.11), the maximum principle then implies that

$$\tilde{y} \ge z_1 \ge z_2 \ge \cdots \ge z_n \ge \cdots \ge y_0 \quad \text{a.e. in } \Omega. \tag{2.2.12}$$

Inequalities (2.2.12) and (2.2.4) imply that the sequence $\{a(x, \overline{u}, z_n)\}$ is bounded in $L^2(\Omega)$, that is, $\{z_n\}$ is bounded in $H^2(\Omega) \cap H_0^1(\Omega)$, by (2.2.11). Hence, by compact embedding, we have $z_n \to \overline{y}$ strongly in $H_0^1(\Omega)$ on a subsequence again indexed by n. Passing to the limit in (2.2.11), we see that $\overline{y}$ is a solution to (2.2.2) associated with $\overline{u}$:

$$-\Delta \overline{y}(x) = a(x, \overline{u}(x), \overline{y}(x)) \quad \text{a.e. in } \Omega, \qquad \overline{y} = 0 \quad \text{on } \partial\Omega,$$

and $\overline{y}(x) \leq \tilde{y}(x)$ due to (2.2.12). Clearly, $[\overline{y}, \overline{u}]$ is admissible for (P), and the above inequality, together with the monotonicity of $b(x, \cdot)$ and with (2.2.8), gives

$$\begin{aligned} J(\overline{y}, \overline{u}) &= \int_\Omega l(\overline{u}(x))\, dx + \int_\Omega b(x, \overline{y}(x))\, dx \\ &\leq \int_\Omega l(\overline{u}(x))\, dx + \int_\Omega b(x, \tilde{y}(x))\, dx \leq \inf(\mathrm{P}). \end{aligned}$$

Consequently, the pair $[\overline{y}, \overline{u}]$ is optimal for (P) (and redenoted by $[y^*, u^*]$), which finishes the proof. □

Lemma 2.2.2 *Let $\varphi : \mathbf{R} \to \mathbf{R}$ be a proper, convex, and lower semicontinuous mapping, and let $\{w_n\}$ and $\{\varphi \circ w_n\}$ be bounded sequences in $L^2(\Omega)$. Then there is a subsequence, which is again indexed by n, such that*

$$\text{w-lim}_{n\to\infty}(\varphi \circ w_n) \geq \varphi \circ (\text{w-lim}_{n\to\infty} w_n) \quad \text{a.e. in } \Omega.$$

Proof. We have $\{[\varphi \circ w_n, w_n]\} \subset \operatorname{Epi}\varphi$, $n \in \mathbf{N}$, where

$$\operatorname{Epi}\varphi = \{[z, q] \in L^2(\Omega) \times L^2(\Omega) : z(x) \geq \varphi(q(x)) \text{ a.e. in } \Omega\}.$$

$\operatorname{Epi}\varphi$ is obviously convex, and it is closed in $L^2(\Omega) \times L^2(\Omega)$, since, owing to the lower semicontinuity of φ, the strong limit of a sequence $\{[z_n, w_n]\} \subset \operatorname{Epi}\varphi$ satisfies the same pointwise inequality. But then $\operatorname{Epi}\varphi$ is also weakly closed. By the boundedness assumption, $\{[\varphi \circ w_n, w_n]\}$ contains a weakly convergent subsequence whose limit then belongs to $\operatorname{Epi}\varphi$. The assertion is proved. □

Remark. Theorem 2.2.1 is a variant of an existence result due to Tahraoui [1986], [1992], while the technique based on the use of the maximum principle is due to Lions [1968]. The uniqueness of the solution to problem (P) has been established in Tahraoui [1986] for some special cases.

We now turn our attention to a class of problems in which state equation and cost functional have a very close correspondence. In this special situation it will be possible to derive a general existence result.

To fix things, consider two reflexive Banach spaces V and U, and let for any fixed $u \in U$ the bilinear form $\tilde{a}(u, \cdot, \cdot) : V \times V \to \mathbf{R}$ be symmetric, bounded, and coercive; moreover, we assume that $\tilde{a}$ is linear with respect to u.

Take any fixed $f \in V^*$. Then for every fixed $u \in U$ there is a unique minimizer $y_u = y(u)$ for the "energy" $\frac{1}{2}\tilde{a}(u,y,y) - (f,y)_{V^* \times V}$ associated with u. We can therefore define the functional

$$J(u) = \operatorname*{Min}_{y \in V} \left\{ \frac{1}{2}\, \tilde{a}(u,y,y) - (f,y)_{V^* \times V} \right\} = \frac{1}{2}\, \tilde{a}(u,y(u),y(u)) - (f,y(u))_{V^* \times V}.$$

Proposition 2.2.3 *$J : U \to \mathbf{R}$ is concave with respect to u. If $\tilde{a}(\cdot\,,y,v)$ is continuous in u, then J is weakly upper semicontinuous.*

Proof. The mapping $u \mapsto \frac{1}{2}\tilde{a}(u,y,y) - (f,y)_{V^* \times V}$ is affine for any fixed $y \in V$, and it is well known that the minimization over a family of affine functionals generates a concave functional. Notice that under the given assumptions, the minimizer in the above definition of J exists for any u; that is, J attains finite values. Since $\tilde{a}(\cdot\,,y,v)$ is continuous, the lower envelope J is upper semicontinuous. Owing to its concavity and Mazur's theorem (see Yosida [1980]), it is also weakly upper semicontinuous. □

Example 2.2.4 It is known that the unique minimizer $y(u)$ of the energy functional is a weak solution to the variational equation (see Example A2.6 in Appendix 2)

$$\tilde{a}(u,y(u),v) = (f,v)_{V^* \times V} \quad \forall v \in V. \tag{1.1.1)$''$}$$

This shows that $J(u)$ can be equivalently expressed as

$$J(u) = -\frac{1}{2}\, \tilde{a}(u,y(u),y(u)) = -\frac{1}{2}\,(f,y(u))_{V^* \times V}.$$

If $U_{ad} \subset U$ is a nonempty, closed, convex, and bounded set, then Proposition 2.2.3 yields the existence of a maximizer in U_{ad} for J. The functional

$$-2\,J(u) = (f,y(u))_{V^* \times V},$$

known in the literature as the *compliance functional*, provides one example of a control in the coefficients problem for which the existence of a minimizer u^*, subject to the state equation (1.1.1)$''$, can be established without supplementary compactness assumptions. The control variable u is interpreted as the vector of all the coefficients (including the boundary conditions) that define the bilinear form $\tilde{a}$ and the associated differential operator. These coefficients may also appear in the higher-order terms, in contrast to the situation in Theorem 2.2.1. The differential operator may be of arbitrary order. For the case of fourth-order elliptic equations, an existence result of this type was established with other methods by Céa and Malanowski [1970]. Other problems related to fourth-order partial differential operators will be studied in the next paragraph.

If the possible minimizers of the energy functional associated with a fixed u have to respect a constraint, that is, have to belong to some nonempty, closed, and convex set $S \subset V$, then one may define a new functional J_S by replacing in the definition of J the minimization over the whole space V by a minimization over S. This corresponds to the case of variational inequalities (see Example 1.2.5 in Chapter 1). Proposition 2.2.3 then remains valid, but not equation (1.1.1)″. There is a severe limitation in applications: the functional to be optimized is strongly related to the energy of the system, more precisely, to the state equation. Another situation of this type is examined below in §2.3.1. In the paper by Tiihonen and Gonzalez de Paz [1994], one example with a strictly concave energy functional is investigated.

Remark. In the case that control coefficients appear in the leading terms of the differential operator, existence may in general be obtained via Theorem 2.1.3, provided that the set of admissible controls is assumed strongly compact in some appropriate space. For instance, in the problem of the optimal layout of materials (compare Example 1.2.7 in Chapter 1), which is very much discussed in the literature, the respective strong compactness hypothesis would be that the set of admissible coefficients a is bounded in $L^\infty(\Omega)$ with bounded gradients in $L^2(\Omega)$. However, since a is a linear combination of characteristic mappings in this problem, such an assumption is too strong and cannot be fulfilled.

An alternative approach is to define an extension (relaxation) of the optimization problem that has better lower semicontinuity properties with respect to some very weak convergence valid for the minimizing sequence in the original optimization problem. Such generalized convergence properties are known in the literature under various names: G-convergence, H-convergence, epiconvergence, variational convergence, and so on. For detailed expositions on these subjects, we refer to Attouch [1984], Buttazzo [1989], Zhikov, Kozlov, and Oleinik [1994], and Raitums [1997]. Lou [2005] uses relaxation theory to prove general existence results for semilinear elliptic control problems. In Chapter 5, we shall indicate some relaxation approaches in domain optimization. The general relaxation theory does not form the object of the present book; the interested reader may consult the monographs of Young [1969], Warga [1972], Gamkrelidze [1975], Lurie [1975], Raitums [1989], [1997], Allaire [2001]. However, we demonstrate in the next paragraph that for higher-order elliptic operators the situation is entirely different and classical existence properties are valid.

2.2.2 Fourth-Order Problems

In this section, we return to Example 1.2.7 in Chapter 1 and examine the questions of existence and uniqueness of optimal pairs for the cost functionals (1.2.57) and (1.2.60) under the constraints (1.2.52)–(1.2.54). We start with the case of the optimization of a simply supported plate (minimization of weight/volume):

$$\text{Min}\left\{\int_\Omega u(x)\,dx\right\}, \tag{2.2.13}$$

$$\Delta(u^3\Delta y) = f \quad \text{in } \Omega, \tag{2.2.14}$$

$$y = \Delta y = 0 \quad \text{on } \partial\Omega, \tag{2.2.15}$$

$$0 < m \le u(x) \le M \quad \text{a.e. in } \Omega, \tag{2.2.16}$$

$$y \in C, \tag{2.2.17}$$

where $C \subset L^2(\Omega)$ is a nonempty and closed, but not necessarily convex, set. The dimension of Ω is arbitrary, with the plate model corresponding to the case $\Omega \subset \mathbf{R}^2$. The state constraint (2.2.17) is very general; for example, (1.2.59) is of this type. In Section 3.4, we shall also examine the following general type of cost functionals,

$$\text{Min}\left\{\int_\Omega [\varphi(x, u(x)) + \psi(x, y(x))]\,dx\right\}, \tag{2.2.18}$$

which, for instance, include (1.2.60) and (2.2.13) as special cases.

Due to the structure of the boundary conditions (2.2.15), the state equation (2.2.14) can be equivalently rewritten as

$$\Delta z = f \quad \text{in } \Omega, \tag{2.2.19}$$
$$z = 0 \quad \text{on } \partial\Omega, \tag{2.2.20}$$
$$\Delta y = z\,l \quad \text{in } \Omega, \tag{2.2.21}$$
$$y = 0 \quad \text{on } \partial\Omega, \tag{2.2.22}$$

where $l = u^{-3} \in L^\infty(\Omega)$.

We notice that $z \in H^2(\Omega) \cap H_0^1(\Omega)$ is completely determined by $f \in L^2(\Omega)$ and can be viewed as a datum of the problem. The control system of interest is just (2.2.21), (2.2.22), which is linear with respect to the new control parameter l. Under the transformation $l = u^{-3}$, the constraint (2.2.17) remains unchanged, while (2.2.13), (2.2.16) become

$$\text{Min}\left\{\int_\Omega l^{-\frac{1}{3}}(x)\,dx\right\}, \tag{2.2.13$'$}$$

$$0 < M^{-3} \le l(x) \le m^{-3} \quad \text{a.e. in } \Omega. \tag{2.2.16$'$}$$

Apparently, the transformed cost functional (2.2.13)$'$ is strictly convex for l positive. Hence, Theorem 2.1.2 yields the following result.

Theorem 2.2.5 *If C is convex, then the problem (2.2.13)–(2.2.17) has a unique global optimal pair $[y^*, u^*]$ in $H^2(\Omega) \times L^\infty(\Omega)$ provided that the admissibility condition is fulfilled.*

Since the transformation $l = u^{-3}$ is nonlinear, the problem (P) given by (2.2.13)–(2.2.17) may be nonconvex even if C is convex. It may, therefore, have many local minimum pairs, but the global minimum is unique.

Remark. In Kawohl and Lang [1997] it is demonstrated that under supplementary hypotheses even the original problem (2.2.13)–(2.2.17) may be convex although the dependence $u \mapsto y$ remains nonlinear. In fact, if we assume $f \le 0$, then $z \ge 0$ a.e. in Ω by the strong maximum principle, applied to (2.2.19), (2.2.20). We have the representation

$$y(x) = -\int_\Omega G(x,y)\,\frac{z(y)}{u^3(y)}\,dy, \tag{2.2.23}$$

where G is the Green function corresponding to (2.2.21), (2.2.22), which is known to be nonnegative. Now let u_1 and u_2 satisfy (2.2.16), and let y_1, y_2 be the associated states according to (2.2.14), (2.2.15). Then $u_\lambda = \lambda u_1 + (1-\lambda)u_2$, $\lambda \in [0,1]$, satisfies (2.2.16). Since the real function $u \mapsto -u^{-3}$ is concave for $u \ge m$ and G and z are nonnegative, we deduce from (2.2.23) that for the associated state y_λ,

$$y_\lambda(x) \ge \lambda y_1(x) + (1-\lambda)y_2(x) \quad \text{a.e. in } \Omega. \tag{2.2.24}$$

In the case that C is given by (1.2.59), this implies that the set of admissible controls for (P) is convex (while the set of admissible pairs may remain nonconvex, since (2.2.24) is not an equality). Therefore, we see that for the cost functional (2.2.13), which depends only on u, (2.2.13)–(2.2.17) is in fact a convex optimization problem. We refer to Section 4.5 for further examples of this type and a complete discussion.

Remark. In order that the existence or uniqueness results remain valid for the case of the general cost functional (2.2.18), one needs that $\psi(x,y)$ and $\theta(x,l) = \varphi(x, l^{-1/3})$ are (strictly) convex integrands (cf. Appendix 1, Proposition A1.1). For instance, if $\varphi(x,u) = \lambda(x)\,u^{-3}$, then $\theta(x,l) = \lambda(x)\,l$ is even linear in l.

Let us now consider the case of clamped plates, i.e., if (2.2.15) is replaced by (see Example 1.2.7)

$$y = \frac{\partial y}{\partial n} = 0. \tag{2.2.25}$$

Theorem 2.2.6 *Let z satisfy (2.2.19) and (2.2.20), and let $l = u^{-3}$. Then $y \in H_0^2(\Omega)$ is a weak solution to the boundary value problem (2.2.14), (2.2.25) if and only if there is a function $h \in L^2(\Omega)$ that is harmonic in Ω in the sense of distributions such that y satisfies*

$$\Delta y = z\,l + h\,l \quad \text{in } \Omega. \tag{2.2.26}$$

Proof. Let y be a weak solution to (2.2.14), (2.2.25). Then

$$\int_\Omega (u^3 \Delta y - z)\, \Delta\phi \, dx = 0 \quad \forall \phi \in H_0^2(\Omega). \tag{2.2.27}$$

We put $h = u^3 \Delta y - z \in L^2(\Omega)$. Then (2.2.27) implies that $\Delta h = 0$ in the sense of distributions, and (2.2.26) is obtained just by dividing by u^3 in the definition of h. The converse direction of the assertion is obvious. □

Remark. Although l appears linearly in (2.2.26), the dependence $l \mapsto y$ is nonlinear since h itself also depends on y.

In the equivalent formulation (2.2.13)′, (2.2.26), (2.2.25), (2.2.16)′, (2.2.17) of the clamped plate optimization problem, one can interpret h as an additional control variable and the Neumann condition $\frac{\partial y}{\partial n} = 0$ as a new state constraint. The problem remains nonconvex even after the transformation, and the minimum is not unique in general. Concerning existence, we have the following result.

Theorem 2.2.7 *There exists at least one solution $u \in L^\infty(\Omega)$ for the problem (P_1) given by (2.2.13), (2.2.14), (2.2.25), (2.2.16), (2.2.17), provided that (P_1) has at least one admissible pair.*

Proof. Let $\{u_n\} \subset L^\infty(\Omega)$ be a minimizing sequence, that is, assume that

$$\int_\Omega u_n(x)\, dx \to \inf(P_1).$$

We put $l_n = u_n^{-3}$, and denote by $y_n \in H_0^2(\Omega)$ the corresponding solution of (2.2.14), (2.2.25) or, equivalently, of (2.2.25) and (2.2.26) with the corresponding harmonic functions $h_n \in L^2(\Omega)$. In view of (2.2.16), the sequences $\{u_n\}$ and $\{l_n\}$ are bounded in $L^\infty(\Omega)$, and we may assume that

$$u_n \to \hat{u}, l_n \to \hat{l}, \text{ both weakly* in } L^\infty(\Omega)\,,$$

for suitable subsequences again indexed by n. In general, we may have $\hat{l} \neq \hat{u}^{-3}$. We also notice that $\{y_n\}$ is bounded in $H_0^2(\Omega)$. Indeed, we have, inserting $\phi = y_n$ in (2.2.27) for $n \in \mathbf{N}$,

$$m \int_\Omega [\Delta y_n(x)]^2\, dx \le \int_\Omega u_n^3(x)[\Delta y_n(x)]^2\, dx = \int_\Omega f(x) y_n(x)\, dx \le |f|_{L^2(\Omega)} |y_n|_{L^2(\Omega)}.$$

Hence, for a suitable subsequence again indexed by n, $y_n \to \tilde{y}$ weakly in $H_0^2(\Omega)$. Since C is closed in $L^2(\Omega)$, we have $\tilde{y} \in C$. Moreover, the sequence $\{h_n = u_n^3 \Delta y_n - z\}$ remains bounded in $L^2(\Omega)$, and on a subsequence, $h_n \to \tilde{h}$ weakly in $L^2(\Omega)$, where $\tilde{h}$ is again harmonic in Ω in the sense of distributions.

From Lemma 2.2.8 below we conclude that $h_n(x) \to \tilde{h}(x)$ for every $x \in \Omega$ and $h_n \to \tilde{h}$ strongly in $L^s(\Omega)$, for any $s < 2$. Hence, we may pass to the limit in the right-hand side of (2.2.26) to obtain that $z\,l_n + h_n\,l_n \to z\,\hat{l} + \tilde{h}\,\hat{l}$ weakly in $L^2(\Omega)$ (where we also use that this sum is bounded in $L^2(\Omega)$). Then, $\tilde{y}$ satisfies

$$\Delta\tilde{y} = z\,\hat{l} + \tilde{h}\,\hat{l},$$

and by Theorem 2.2.6, the pair $[\tilde{y}, \tilde{u}]$, where $\tilde{u} = \hat{l}^{-1/3}$, is admissible for (P_1).

The weak lower semicontinuity of convex functionals finally yields that

$$\begin{aligned}\inf(P_1) &= \lim_{n\to\infty} \int_\Omega u_n(x)\,dx = \lim_{n\to\infty} \int_\Omega l_n^{-1/3}(x)\,dx \\ &\geq \liminf_{n\to\infty} \int_\Omega l_n^{-1/3}(x)\,dx \geq \int_\Omega \hat{l}^{-1/3}(x)\,dx = \int_\Omega \tilde{u}(x)\,dx \geq \inf(P_1).\end{aligned}$$

This concludes the proof of the assertion. □

Lemma 2.2.8 *Let $\Omega \subset \mathbf{R}^d$ be a bounded domain. Then the following hold:*

(a) *Suppose that $h_n \in L^2(\Omega)$, $n \in \mathbf{N}$, and $\tilde{h} \in L^2(\Omega)$ are harmonic in Ω in the sense of distributions and satisfy $h_n \to \tilde{h}$ weakly in $L^1(\Omega)$. Then $h_n(x) \to \tilde{h}(x)$ for all $x \in \Omega$.*

(b) *If $h_n(x) \to \tilde{h}(x)$ a.e. in Ω and $h_n \to \tilde{h}$ weakly in $L^p(\Omega)$ for some $p > 1$, then $h_n \to \tilde{h}$ strongly in $L^s(\Omega)$, for all $1 \leq s < p$.*

Proof. (a) The classical Weyl lemma (cf. Hörmander [1964]) yields that $h_n, \tilde{h} \in C^\infty(\Omega)$. Then, for any $x \in \Omega$, and any ball $B_\rho(x) \subset \Omega$ of radius ρ centered at x, we can apply the solid mean property to obtain that

$$h_n(x) = \frac{d}{w_d\rho^d} \int_{B_\rho(x)} h_n(y)\,dy \to \frac{d}{w_d\rho^d} \int_{B_\rho(x)} \tilde{h}(y)\,dy = \tilde{h}(x).$$

Here, d is the dimension of Ω and w_d denotes the "area" of the unit ball in $\mathbf{R}^d$.

(b) By Egorov's theorem, for any $\varepsilon > 0$ there is some measurable set $\Omega_\varepsilon \subset \Omega$ with $\text{meas}(\Omega \setminus \Omega_\varepsilon) < \varepsilon$ such that $h_n \to \tilde{h}$ uniformly in Ω_ε. Using Hölder's inequality and the boundedness of $\{|h_n|_{L^p(\Omega)}\}$, we have, with some $M > 0$,

$$\begin{aligned}&\int_\Omega |h_n(x) - \tilde{h}(x)|^s\,dx \\ &\quad= \int_{\Omega_\varepsilon} |h_n(x) - \tilde{h}(x)|^s\,dx + \int_{\Omega\setminus\Omega_\varepsilon} |h_n(x) - \tilde{h}(x)|^s\,dx \\ &\quad\leq \int_{\Omega_\varepsilon} |h_n(x) - \tilde{h}(x)|^s\,dx + \left(\int_{\Omega\setminus\Omega_\varepsilon} |h_n(x) - \tilde{h}(x)|^p\,dx\right)^{\frac{s}{p}} \text{meas}(\Omega \setminus \Omega_\varepsilon)^{\frac{p-s}{p}} \\ &\quad\leq \int_{\Omega_\varepsilon} |h_n(x) - \tilde{h}(x)|^s\,dx + M\varepsilon^{\frac{p-s}{p}} \ \leq\ c(\varepsilon)\end{aligned}$$

if $n \geq N(\varepsilon)$, where $c(\varepsilon) \to 0$ for $\varepsilon \to 0$. □

Remark. Lemma 2.2.8, point (b), is an extension of Lemma 1.3 in Lions [1969]. Point (a) seems to be new and is due to Sprekels and Tiba [1998/1999].

Remark. In the one-dimensional case a similar argument works for the boundary conditions (1.2.55), (1.2.56) (cantilevered beams). One essential feature is to construct a decomposition of the fourth-order equation into a system of two coupled second-order boundary value problems. In the case of (1.2.55), (1.2.56), the construction leads to two Cauchy problems for ordinary differential equations, and it seems difficult to extend the technique to higher dimensions (for instance, to partially clamped plates). We refer the reader to the discussion in Chapter 3, §3.4.2.1, for a special case of a partially clamped plate and its optimization.

Remark. From Theorem 2.2.7 and its proof we see that the "optimal" thickness $\tilde{u}$ is obtained by twice inverting the minimizing sequence $\{u_n\}$. If $\{u_n\}$ is pointwise convergent, then $\tilde{u} = \hat{u} = \lim_{n\to\infty} u_n$. This is the case usually appearing in the literature; see Haslinger and Neittaanmäki [1988], Casas [1990], Hlaváček, Bock, and Lovíšek [1985], Kirjner-Neto and Polak [1996], and Bendsøe [1984]. The results of this paragraph are due to Sprekels and Tiba [1998/99]; they demonstrate that the standard strong compactness assumption (namely, the boundedness of ∇u in some space of integrable functions) is not necessary for proving the existence of an optimal pair for fourth-order differential operators. They may also be interpreted in terms of the homogenization of fourth-order operators, especially Theorem 2.2.7. It should be noted that for second-order operators of the form $\operatorname{div}(\lambda \operatorname{grad}(\cdot))$, the counterexample of Murat [1971] shows that the boundedness of the gradient of the coefficients $\{\lambda(\cdot)\}$ is needed in order to pass to the limit.

2.3 Variable Domains

In this section we establish existence results for general shape optimization problems like those introduced in Example 1.2.8 in Chapter 1, where the assumptions on the unknown domains are very weak. Basically, we will require only that the boundaries be uniformly continuous in a sense to be made precise. We analyze Dirichlet, Neumann, and mixed boundary value problems associated with arbitrary-order elliptic equations or variational inequalities. It is known (see, for instance, Buttazzo and Dal Maso [1991, §4]) that an optimal domain may not exist if no additional assumptions are imposed on the boundaries of the open sets. While the general theory is covered in paragraphs two and three of this section, we discuss some "simpler" examples in the first paragraph, with the aim to introduce the reader to the main difficulties encountered in variable domain optimization.

2.3.1 Some Examples

We first discuss the following capacity (energy) optimization problem: Let $\Omega \subset \mathbf{R}^d$, $d \in \mathbf{N}$, be a given open, bounded domain. We denote by $A \subset \Omega$ some variable measurable set and put $D = \Omega \setminus A$. We then define $y_A \in H_0^1(\Omega)$ as the (unique) minimizer of the energy functional

$$J(y) = \int_\Omega |\nabla y(x)|^2 \, dx$$

over the nonempty, closed, and convex set

$$C = \left\{ y \in H_0^1(\Omega) : \int_\Omega \chi_A(x)(y(x) - 1)^2 \, dx = 0 \right\},$$

where χ_A is the characteristic function of A in Ω. If Ω and A are smooth domains and $A \subset \Omega$ is compact, then y_A is just the solution to the inhomogeneous Dirichlet boundary value problem

$$\Delta y_A = 0 \quad \text{in } D, \quad y_A = 0 \quad \text{on } \partial\Omega, \quad y_A = 1 \quad \text{on } \partial A,$$

extended to the whole domain Ω by putting $y_A(x) = 1$ for $x \in A$. The optimal value $E_d(A) = J(y_A)$ represents the potential energy of the system and is known in elementary potential theory as the *capacity* of the set A with respect to Ω.

If $C \neq \emptyset$, existence and uniqueness of the minimizer $y_A \in C$ are clear, since J is strictly convex, and owing also to Poincaré's inequality. The shape optimization problem (which we denote by (R)) is then to minimize $E_d(A) = J(y_A)$ subject to all the admissible choices of measurable sets $A \subset \Omega$ with $\text{meas}(A) = v$ (i.e., having a prescribed "volume"). In other words, we want to find a set $A \subset \Omega$ of given measure and with minimal capacity. Another possible physical interpretation results from multiplying the equation by y_A and formally integrating by parts. We obtain

$$0 = -\int_D \Delta y_A(x)\, y_A(x) \, dx = \int_D |\nabla y_A(x)|^2 \, dx - \int_{\partial\Omega \cup \partial A} \frac{\partial y_A}{\partial n} \, y_A \, d\sigma.$$

Taking the boundary conditions into account, we get

$$J(y_A) = \int_D |\nabla y_A(x)|^2 \, dx = \int_{\partial A} \frac{\partial y_A}{\partial n} \, d\sigma.$$

Hence, $J(y_A)$ represents the total heat flux lost through ∂A if we interpret y_A as a stationary heat distribution in D where ∂A and $\partial\Omega$ are maintained at the constant (relative) temperatures $y_A = 1$ and $y_A = 0$, respectively. The optimization problem (R) amounts to finding the shape of A minimizing this loss. It may be compared with the compliance minimization problem from Example 1.2.4.

We present an existence result for (R) obtained by a direct argument that is originally due to F. Murat and reported by Gonzalez de Paz [1982].

Theorem 2.3.1 *For any* $v \in]0, \text{meas}(\Omega)[$ *there is some measurable set* $A_0 \subset \Omega$ *with* $\text{meas}(A_0) = v$ *such that* $E_d(A_0) \leq E_d(A)$ *for all measurable sets* $A \subset \Omega$ *satisfying* $\text{meas}(A) = v$.

Proof. For any measurable set $A \subset \Omega$ with $E_d(A) < +\infty$ there is some $y_A \in H_0^1(\Omega)$ such that $E_d(A) = \int_\Omega |\nabla y_A(x)|^2\, dx$. Indeed, if $\{y_n\} \subset C$ is a minimizing sequence, then $\{y_n\}$ is bounded in $H_0^1(\Omega)$, and hence we may without loss of generality assume that $y_n \to y$ weakly in $H_0^1(\Omega)$ and pointwise a.e. in Ω. Thus, $y \in C$, and it follows from the weak lower semicontinuity of J that y is a minimizer (which we redenote by y_A).

Now let $v \in]0, \text{meas}(\Omega)[$ be fixed. Then there is some measurable set A_1 such that $\text{meas}(A_1) = v$ and $\overline{A}_1 \subset \Omega$. A_1 is an admissible set for the capacity optimization problem (that is, $C \neq \emptyset$), which therefore has a finite optimal value.

Let $\{A_n\}$ be a minimizing sequence of measurable subsets of Ω. We denote by χ_n the characteristic function of A_n in Ω and by $y_n = y_{A_n} \in H_0^1(\Omega)$ the function yielding $E_d(A_n) = \int_\Omega |\nabla y_n|^2\, dx$, $n \in \mathbf{N}$. Then $\{y_n\}$ is bounded in $H_0^1(\Omega)$, and on a subsequence again indexed by n, we may assume that

$$\begin{aligned} &y_n \to y^* \quad \text{weakly in } H_0^1(\Omega), \text{ strongly in } L^2(\Omega), \text{ and pointwise a.e. in } \Omega, \\ &\chi_n \to g^* \quad \text{weakly* in } L^\infty(\Omega), \end{aligned}$$

as well as $0 \leq g^*(x) \leq 1$ a.e. in Ω, and $\int_\Omega g^*(x)\, dx = v$. Set

$$\mathcal{A} = \{x \in \Omega : g^*(x) > 0\},$$

which is a measurable set in Ω, defined up to a set of zero measure. We have

$$v = \int_\Omega g^*(x)\, dx = \int_{\mathcal{A}} g^*(x)\, dx \leq \int_{\mathcal{A}} dx = \text{meas}(\mathcal{A}).$$

On the other hand,

$$\begin{aligned} 0 &= \lim_{n\to\infty} \int_\Omega \chi_n(x)(y_n(x) - 1)^2\, dx = \int_\Omega g^*(x)(y^*(x) - 1)^2\, dx \\ &= \int_{\mathcal{A}} g^*(x)(y^*(x) - 1)^2\, dx, \end{aligned}$$

since $\{y_n\}$ is strongly convergent in $L^2(\Omega)$. The last relation shows that $y^*(x) = 1$ a.e. in $\mathcal{A}$ and consequently, $E_d(\mathcal{A}) \leq \int_\Omega |\nabla y^*(x)|^2\, dx$.

By the weak lower semicontinuity of the norm, we also have

$$\int_\Omega |\nabla y^*(x)|^2\, dx \leq \liminf_{n\to\infty} \int_\Omega |\nabla y_n(x)|^2\, dx = \liminf_{n\to\infty} E_d(A_n) = \inf(R).$$

Hence, $E_d(\mathcal{A}) \leq \inf(R)$. Thus, it suffices to choose any measurable subset $A_0 \subset \mathcal{A}$ with $\text{meas}(A_0) = v$. Since $E_d(\cdot)$ increases with respect to set inclusion, we infer that A_0 respects all the constraints of (R) and is thus the sought minimizer. □

Remark. This result enjoys the remarkable property that except for measurability, no assumptions have to be imposed on the variable set A. However, there is a strong limitation on the possible cost criterion that "should" be given by the "energy" of the system. In this regard, it resembles the situation studied in Proposition 2.2.3. In Chapter 5, we will indicate a relaxation procedure and further properties for the problem (R).

We now continue with two of the simplest variable domain optimization problems. Further examples of this type may be found in the monographs by Neittaanmäki and Haslinger [1996] and Mäkinen and Haslinger [2003].

Example 2.3.2 Let $0 < c_0 < c_1$ and $c_2 > 0$ be fixed given constants. We put

$$E = \Big\{(x_1, x_2) \in \mathbf{R}^2 : 0 < x_1 < c_1,\ 0 < x_2 < 1\Big\},$$

$$D = \Big\{(x_1, x_2) \in E : 0 < x_1 < c_0\Big\},$$

and we consider the set of admissible control curves

$$\begin{aligned} U_{ad} \quad = \quad & \Big\{\alpha \in W^{1,\infty}(0,1) : 0 < c_0 \leq \alpha(x) \leq c_1,\ |\alpha(x) - \alpha(\bar{x})| \leq c_2|x - \bar{x}| \\ & \text{for all } x, \bar{x} \in [0,1]\Big\}. \end{aligned}$$

Obviously, U_{ad} is nonempty, closed in $W^{1,\infty}(0,1)$, and, by the Arzelà–Ascoli theorem, compact in $C[0,1]$. For any $\alpha \in U_{ad}$, we consider the set

$$\Omega(\alpha) = \Big\{(x_1, x_2) \in \mathbf{R}^2 : 0 < x_1 < \alpha(x_2),\ 0 < x_2 < 1\Big\}, \tag{2.3.1}$$

and we put $\mathcal{O} = \{\Omega(\alpha) : \alpha \in U_{ad}\}$. Obviously, every admissible domain $\Omega(\alpha) \in \mathcal{O}$ has a Lipschitz boundary $\partial\Omega(\alpha)$. Now let $f \in L^2(E)$ be given. Then we denote for any $\alpha \in U_{ad}$ the (unique) solution to the boundary value problem

$$-\Delta y = f \quad \text{in } \Omega(\alpha), \qquad y = 0 \quad \text{on } \partial\Omega(\alpha), \tag{2.3.2}$$

by $y(\alpha) \in H_0^1(\Omega(\alpha))$. Its extension by 0 to the larger domain E is denoted by $\tilde{y}(\alpha)$ and belongs to $H_0^1(E)$. With these denotations, the shape optimization problem to be studied then reads

$$\operatorname*{Min}_{\alpha \in U_{ad}} \left\{ \int_D |\nabla y(\alpha)(x) - y_d(x)|^2\, dx \right\}, \tag{2.3.3}$$

where $y_d \in L^2(D)^2$ is some given target function. We have the following continuity result.

Lemma 2.3.3 *Let $\{\alpha_n\} \subset U_{ad}$ satisfy $\alpha_n \to \alpha$ strongly in $C[0,1]$. Then $\tilde{y}(\alpha_n) \to \tilde{y}(\alpha)$ weakly in $H_0^1(E)$.*

Proof. Multiplying (2.3.2) for $n \in \mathbf{N}$ by $y(\alpha_n)$, and integrating by parts, we find that

$$|\tilde{y}(\alpha_n)|_{H_0^1(E)} = |y(\alpha_n)|_{H_0^1(\Omega(\alpha_n))} \leq |f|_{L^2(\Omega(\alpha_n))} \leq |f|_{L^2(E)}.$$

Therefore, we can assume for a subsequence, which is again indexed by n, that $\tilde{y}(\alpha_n) \to \tilde{y}$ weakly in $H_0^1(E)$. Let $y \in H^1(\Omega(\alpha))$ denote the restriction of $\tilde{y}$ to $\Omega(\alpha)$. We have to show that $y = y(\alpha)$, i.e., y is the solution of (2.3.2) corresponding to $\Omega(\alpha)$.

Since $\alpha_n \to \alpha$ strongly in $C[0,1]$, we have $\operatorname{supp}\varphi \subset \Omega(\alpha_n)$ for sufficiently large n, for any $\varphi \in \mathcal{D}(\Omega(\alpha))$. Consequently,

$$\int_{\operatorname{supp}\varphi} \nabla y(\alpha_n) \cdot \nabla\varphi\, dx = \int_{\Omega(\alpha_n)} \nabla y(\alpha_n) \cdot \nabla\varphi\, dx = \int_{\Omega(\alpha_n)} f\varphi\, dx = \int_{\operatorname{supp}\varphi} f\varphi\, dx,$$

whence, passing to the limit as $n \to \infty$,

$$\int_{\Omega(\alpha)} \nabla y \cdot \nabla\varphi\, dx = \int_{\Omega(\alpha)} f\,\varphi\, dx \quad \forall\varphi \in \mathcal{D}(\Omega(\alpha)).$$

Finally, let us verify that $\tilde{y}|_{E\setminus\Omega(\alpha)} = 0$, which, by virtue of the trace theorem (see Appendix 2, Theorem A2.1), will imply that $y = 0$ on $\partial\Omega(\alpha)$.

Since $\alpha_n \to \alpha$ strongly in $C[0,1]$, we have for any $\psi \in \mathcal{D}(E \setminus \overline{\Omega(\alpha)})$ that $\operatorname{supp}\psi \subset E \setminus \Omega(\alpha_n)$, for sufficiently large $n \in \mathbf{N}$. Thus,

$$0 = \int_{E\setminus\Omega(\alpha_n)} \tilde{y}(\alpha_n)\,\psi\, dx = \int_{\operatorname{supp}\psi} \tilde{y}(\alpha_n)\,\psi\, dx = \int_{E\setminus\Omega(\alpha)} \tilde{y}(\alpha_n)\,\psi\, dx,$$

and we can pass to the limit as $n \to \infty$ to infer that

$$0 = \int_{E\setminus\Omega(\alpha)} \tilde{y}\,\psi\, dx \quad \forall\psi \in \mathcal{D}(E \setminus \Omega(\alpha)).$$

Hence, $\tilde{y} = 0$ a.e. in $E \setminus \Omega(\alpha)$, whence, as explained above, $y = y(\alpha)$. Since the limit y does not depend on the choice of the subsequence, the assertion is proved. □

Remark. Since the last argument will be frequently used in the sequel, we indicate some details. It is known (cf. Yosida [1980, Ch. V.1]) that a sequence $x_n \to x$ weakly in the normed space X if and only if (i) $\{x_n\}$ is bounded, and (ii) $(\mu, x_n)_{X^*\times X} \to (\mu, x)_{X^*\times X}$ for every μ in a dense subset Λ in the dual space X^*. If Λ may be assumed countable, we set $\Lambda = \{\mu_l : l \in \mathbf{N}\}$, and we can construct

$$F(x) = \sum_{l=1}^{\infty} 2^{-l} \frac{|\mu_l(x)|}{1 + |\mu_l(x)|}.$$

Then, again by Yosida [1980, Ch. I.2], the condition (ii) is equivalent to $F(x_n - x) \to 0$ in $\mathbf{R}$. To the sequence $\{F(x_n - x)\}$ we can apply the argument from

Vulikh [1976, p. 171], that its convergence is equivalent to the fact that any subsequence admits a subsubsequence with the given limit. If strong convergence properties are valid for $\{x_n\}$ and the limit of any subsequence is the same, one can use directly the previous argument to conclude the convergence of the whole sequence.

Corollary 2.3.4 *The optimal shape design problem (2.3.1), (2.3.2), (2.3.3) admits at least one solution.*

Proof. Let $\{\alpha_n\} \subset U_{ad}$ be a minimizing sequence with the corresponding domains $\{\Omega(\alpha_n)\}$ and solutions $\{y(\alpha_n)\}$ of (2.3.2). Owing to the compactness of U_{ad} in $C[0,1]$, and invoking Lemma 2.3.3, we may without loss of generality assume that $\alpha_n \to \alpha$ strongly in $C[0,1]$ with some $\alpha \in U_{ad}$ and $\tilde{y}(\alpha_n) \to \tilde{y}(\alpha)$ weakly in $H_0^1(E)$. Since the cost functional (2.3.3) is weakly lower semicontinuous on $H^1(D)$, it follows that α is the desired minimizer and $\Omega(\alpha)$ the associated optimal domain. □

Remark. The Lipschitz property of the domain $\Omega(\alpha)$ plays an important role in the proof of Lemma 2.3.3 since it guarantees the applicability of the trace theorem. In §2.3.2 below this assumption will be removed by using another technique, which applies to general elliptic equations.

Remark. In Example 2.3.2 we have assumed that the performance index (2.3.3) is defined on a fixed domain D. The next example presents a case in which the cost functional depends on the variable part of the boundary of $\Omega(\alpha)$ itself. In this case stronger compactness assumptions have to be imposed in order to prove the existence of a minimizer.

Example 2.3.5 We maintain the notation from Example 2.3.2, imposing further restrictions on the admissible controls by putting, with fixed positive constants $c_3 > 0,\ c_4 > 0$,

$$\begin{aligned}\hat{U}_{ad} = \Big\{\alpha \in C^1[0,1] : c_0 \le \alpha(x) \le c_1,\ \ |\alpha'(x)| \le c_3,\\ |\alpha'(x) - \alpha'(\bar{x})| \le c_4\,|x - \bar{x}| \text{ for all } x, \bar{x} \in [0,1]\Big\}.\end{aligned}$$

By the Arzelà–Ascoli theorem, $\hat{U}_{ad}$ is a compact subset of $C^1[0,1]$. The shape optimization problem to be studied is given by

$$\underset{\alpha\in\hat{U}_{ad}}{\mathrm{Min}} \left\{ \int_{\Gamma(\alpha)} (y(\alpha)(\sigma) - y_d(\sigma))^2\, d\sigma \right\}, \tag{2.3.4}$$

subject to

$$-\Delta y(\alpha) + y(\alpha) = f \qquad \text{in } \Omega(\alpha), \tag{2.3.5}$$

$$\frac{\partial y(\alpha)}{\partial n} = 0 \qquad \text{on } \partial\Omega(\alpha), \tag{2.3.6}$$

where $y_d \in C(\overline{E})$ is a given target function, $\Omega(\alpha)$ is defined as in (2.3.1) for $\alpha \in \hat{U}_{ad}$, and $\Gamma(\alpha) = \{(x_1, x_2) \in \mathbf{R}^2 : x_1 = \alpha(x_2)\}$. It should be clear that the solution $y(\alpha)$ defined by (2.3.5), (2.3.6) differs from that considered in the previous Example 2.3.2. We have the following continuity result.

Lemma 2.3.6 *Let $\{\alpha_n\} \subset \hat{U}_{ad}$ satisfy $\alpha_n \to \alpha$ strongly in $C[0,1]$. Then, for some subsequence again indexed by n, $\tilde{y}(\alpha_n) \to y$ weakly in $H^1(E)$, with $y \in H^1(E)$ denoting some extension of $y(\alpha)$ to the whole set E.*

Remark. In the above Lemma 2.3.6, $\tilde{y}(\alpha_n)$ is defined as $\tilde{y}(\alpha_n) = T_n y(\alpha_n)$ with $T_n : H^1(\Omega(\alpha_n)) \to H^1(E)$ being some linear and bounded extension operator. Examples of such operators are the *Calderon extension* or the *method of extension by reflections*, cf. Adams [1975]. Chenais [1975] has shown that the boundedness constant of such operators depends in a bounded way on the Lipschitz constant of the domain boundary. In the next paragraphs of this chapter, we will introduce another technique that avoids the use of extension operators and allows us to relax the regularity hypotheses on the geometries.

Proof of Lemma 2.3.6. By the definition of the generalized solution to (2.3.5), (2.3.6), we have

$$\int_{\Omega(\alpha_n)} \nabla y(\alpha_n) \cdot \nabla \varphi \, dx + \int_{\Omega(\alpha_n)} y(\alpha_n)\varphi \, dx = \int_{\Omega(\alpha_n)} f\varphi \, dx \quad \forall \varphi \in C^\infty(\overline{E}). \quad (2.3.7)$$

By taking φ as an approximation to $\tilde{y}(\alpha_n)$ in $H^1(E)$, we can easily derive from (2.3.7) that $\{|y(\alpha_n)|_{H^1(\Omega(\alpha_n))}\}$ and $\{|\tilde{y}(\alpha_n)|_{H^1(E)}\}$ are bounded sequences, since the extension operators are uniformly bounded in uniformly Lipschitz domains. Possibly taking a subsequence, we may therefore assume that $\tilde{y}(\alpha_n) \to y$ weakly in $H^1(E)$. We have, for any $\varphi \in C^\infty(\overline{E})$,

$$\begin{aligned} \int_{\Omega(\alpha_n)} \nabla y(\alpha_n) \cdot \nabla \varphi \, dx &= \int_{\Omega(\alpha)} \nabla \tilde{y}(\alpha_n) \cdot \nabla \varphi \, dx + \int_{\Omega(\alpha_n)\setminus\Omega(\alpha)} \nabla y(\alpha_n) \cdot \nabla \varphi \, dx \\ &\quad - \int_{\Omega(\alpha)\setminus\Omega(\alpha_n)} \nabla \tilde{y}(\alpha_n) \cdot \nabla \varphi \, dx = I_1^n + I_2^n + I_3^n. \end{aligned}$$

We can estimate I_2^n as follows:

$$\begin{aligned} \left| \int_{\Omega(\alpha_n)\setminus\Omega(\alpha)} \nabla y(\alpha_n) \cdot \nabla \varphi \, dx \right| &\leq |\tilde{y}(\alpha_n)|_{H^1(E)} \left(\int_{\Omega(\alpha_n)\setminus\Omega(\alpha)} |\nabla \varphi|^2 \, dx \right)^{1/2} \\ &\leq c_\varphi \, |\tilde{y}(\alpha_n)|_{H^1(E)} \left[\operatorname{meas}(\Omega(\alpha_n) \setminus \Omega(\alpha))\right]^{1/2}, \end{aligned}$$

where $c_\varphi > 0$ is a constant depending only on φ. A similar inequality holds for I_3^n. Since $\alpha_n \to \alpha$ strongly in $C[0,1]$, we have $\operatorname{meas}(\Omega(\alpha_n) \setminus \Omega(\alpha)) \to 0$

and $\text{meas}(\Omega(\alpha) \setminus \Omega(\alpha_n)) \to 0$, as $n \to \infty$. Thus,

$$\lim_{n\to\infty} \int_{\Omega(\alpha_n)} \nabla y(\alpha_n) \cdot \nabla \varphi \, dx = \int_{\Omega(\alpha)} \nabla y \cdot \nabla \varphi \, dx \quad \forall \varphi \in C^\infty(\overline{E}).$$

For the other two terms in (2.3.7), the passage to the limit is of the same type, and we can infer that

$$\int_{\Omega(\alpha)} \nabla y \cdot \nabla \varphi \, dx + \int_{\Omega(\alpha)} y\varphi \, dx = \int_{\Omega(\alpha)} f\varphi \, dx \quad \forall \varphi \in C^\infty(\overline{E}).$$

Since the set of restrictions to $\Omega(\alpha)$ of functions in $C^\infty(\overline{E})$ forms a dense subset of $H^1(\Omega(\alpha))$ (cf. Adams [1975]), it follows that $y|_{\Omega(\alpha)} = y(\alpha)$, which ends the proof. □

Theorem 2.3.7 *Let $\{\alpha_n\} \subset \hat{U}_{ad}$ satisfy $\alpha_n \to \alpha$ strongly in $C^1[0,1]$. Then*

$$\lim_{n\to\infty} \int_{\Gamma(\alpha_n)} (y(\alpha_n) - y_d)^2(\sigma)\, d\sigma = \int_{\Gamma(\alpha)} (y(\alpha) - y_d)^2(\sigma)\, d\sigma.$$

Proof. Let $\alpha_n \to \alpha$ strongly in $C^1[0,1]$. By Lemma 2.3.6, we may assume that $\tilde{y}(\alpha_n) \to \tilde{y}(\alpha)$ weakly in $H^1(E)$ on a subsequence, which is again indexed by n.

Let us use the abbreviations $v_n = \tilde{y}(\alpha_n) - y_d$, $y_n = \tilde{y}(\alpha_n)$, and $v = \tilde{y}(\alpha) - y_d$. We have

$$\begin{aligned}
&\left| \int_{\Gamma(\alpha_n)} v_n^2 \, d\sigma - \int_{\Gamma(\alpha)} v^2 \, d\sigma \right| \\
&= \left| \int_0^1 v_n^2(\alpha_n(x_2), x_2) \sqrt{1 + (\alpha_n'(x_2))^2}\, dx_2 \right. \\
&\qquad \left. - \int_0^1 v^2(\alpha(x_2), x_2) \sqrt{1 + (\alpha'(x_2))^2}\, dx_2 \right| \\
&\le \left| \int_0^1 \left(v_n^2(\alpha_n(x_2), x_2) - v^2(\alpha(x_2), x_2) \right) \sqrt{1 + (\alpha_n'(x_2))^2}\, dx_2 \right| \\
&\qquad + \left| \int_0^1 v^2(\alpha(x_2), x_2) \left(\sqrt{1 + (\alpha_n'(x_2))^2} - \sqrt{1 + (\alpha'(x_2))^2} \right) dx_2 \right| \\
&= I_1^n + I_2^n.
\end{aligned}$$

Clearly, $I_2^n \to 0$, since $\alpha_n \to \alpha$ in $C^1[0,1]$. Denoting by C_i, $i \in \mathbf{N}$, positive constants that do not depend on n, we can estimate the first term by

$$\begin{aligned}
I_1^n &\le C_1 \int_0^1 |v_n^2(\alpha_n(x_2), x_2) - v^2(\alpha(x_2), x_2)|\, dx_2 \\
&\le C_2 \int_0^1 |v_n^2(\alpha_n(x_2), x_2) - v_n^2(\alpha(x_2), x_2)|\, dx_2 \\
&\qquad + C_2 \int_0^1 |v_n^2(\alpha(x_2), x_2) - v^2(\alpha(x_2), x_2)|\, dx_2 \\
&= C_2\, (I_{11}^n + I_{12}^n).
\end{aligned}$$

We have

$$\begin{aligned}
I_{12}^n &= \int_0^1 |v_n^2(\alpha(x_2),x_2) - v^2(\alpha(x_2),x_2)|\,dx_2 \\
&\le \int_0^1 |v_n^2(\alpha(x_2),x_2) - v^2(\alpha(x_2),x_2)|\sqrt{1+(\alpha'(x_2))^2}\,dx_2 \\
&= \int_{\Gamma(\alpha)} |v_n^2 - v^2|\,d\sigma = \int_{\Gamma(\alpha)} |\tilde{y}(\alpha_n) - \tilde{y}(\alpha)|\,|\tilde{y}(\alpha_n) + \tilde{y}(\alpha) - 2y_d|\,d\sigma \\
&\le C_3\left(\int_{\Gamma(\alpha)} |\tilde{y}(\alpha_n) - \tilde{y}(\alpha)|^2 d\sigma\right)^{\frac{1}{2}} \to 0 \quad \text{as } n \to \infty,
\end{aligned}$$

since $\tilde{y}(\alpha_n) \to \tilde{y}(\alpha)$ weakly in $H^1(E)$, and since the trace operator $H^1(E) \to L^2(\Gamma(\alpha))$ is compact (cf. Appendix 2). We also have

$$\begin{aligned}
I_{11}^n &= \int_0^1 |v_n(\alpha_n(x_2),x_2) - v_n(\alpha(x_2),x_2)|\,|v_n(\alpha_n(x_2),x_2) + v_n(\alpha(x_2),x_2)|\,dx_2 \\
&\le \int_0^1 |y_n(\alpha_n(x_2),x_2) - y_n(\alpha(x_2),x_2)|\,|y_n(\alpha_n(x_2),x_2) - y_d(\alpha_n(x_2),x_2) \\
&\qquad\qquad + y_n(\alpha(x_2),x_2) - y_d(\alpha(x_2),x_2)|\,dx_2 \\
&\quad + \int_0^1 |y_d(\alpha_n(x_2),x_2) - y_d(\alpha(x_2),x_2)|\,|y_n(\alpha_n(x_2),x_2) - y_d(\alpha_n(x_2),x_2) \\
&\qquad\qquad + y_n(\alpha(x_2),x_2) - y_d(\alpha(x_2),x_2)|\,dx_2 \\
&= I_a^n + I_b^n.
\end{aligned}$$

From the continuity of y_d in E, and since $\alpha_n \to \alpha$ strongly in $C^1[0,1]$, we can infer that $I_b^n \to 0$ as $n \to \infty$ (as in the estimate for I_{12}^n). For I_a^n, we have

$$\begin{aligned}
I_a^n &\le |y_n(\alpha_n) - y_n(\alpha)|_{L^2(0,1)}\,|y_n(\alpha_n) - y_d(\alpha_n) + y_n(\alpha) - y_d(\alpha)|_{L^2(0,1)} \\
&= |y_n(\alpha_n) - y_n(\alpha)|_{L^2(0,1)}\,\tilde{I}_a^n,
\end{aligned}$$

where

$$\begin{aligned}
\tilde{I}_a^n &\le |y_n(\alpha_n)|_{L^2(0,1)} + |y_n(\alpha)|_{L^2(0,1)} + |y_d(\alpha_n) + y_d(\alpha)|_{L^2(0,1)} \\
&\le \left(\int_0^1 |y_n(\alpha_n(x_2),x_2)|^2\sqrt{1+(\alpha_n'(x_2))^2}\,dx_2\right)^{1/2} \\
&\quad + \left(\int_0^1 |y_n(\alpha(x_2),x_2)|^2\sqrt{1+(\alpha'(x_2))^2}\,dx_2\right)^{1/2} + C_4 \\
&\le |y_n|_{L^2(\Gamma(\alpha_n))} + |\tilde{y}_n|_{L^2(\Gamma(\alpha))} + C_4 \\
&\le 2\,|\tilde{y}_n|_{H^1(E)} + C_4 \le C_5.
\end{aligned}$$

Finally, we have, for any $x_2 \in [0,1]$,

$$
\begin{aligned}
&|y_n(\alpha_n(x_2), x_2) - y_n(\alpha(x_2), x_2)|^2 \\
&= \left| \int_{\alpha(x_2)}^{\alpha_n(x_2)} \partial_1 y_n(x_1, x_2)\, dx_1 \right|^2 \le \left| \int_{\alpha(x_2)}^{\alpha_n(x_2)} |\partial_1 y_n(x_1, x_2)|^2\, dx_1 \right| \left| \int_{\alpha(x_2)}^{\alpha_n(x_2)} dx_1 \right| \\
&= |\alpha_n(x_2) - \alpha(x_2)| \left| \int_{\alpha(x_2)}^{\alpha_n(x_2)} |\partial_1 y_n(x_1, x_2)|^2\, dx_1 \right|.
\end{aligned}
$$

Therefore,

$$
\begin{aligned}
(I_a^n)^2 &\le C_5 \int_0^1 |y_n(\alpha_n) - y_n(\alpha)|^2\, dx_2 \\
&\le C_5 \int_0^1 \left| \int_{\alpha(x_2)}^{\alpha_n(x_2)} |\partial_1 y_n(x_1, x_2)|^2\, dx_1 \right| dx_2 \cdot \max_{x_2 \in [0,1]} |\alpha_n(x_2) - \alpha(x_2)| \\
&\le C_5 \int_E |\nabla \tilde{y}_n(x_1, x_2)|^2\, dx_1\, dx_2 \cdot \max_{x_2 \in [0,1]} |\alpha_n(x_2) - \alpha(x_2)| \to 0,
\end{aligned}
$$

as $n \to \infty$. Combining the above inequalities, we obtain the desired convergence result for the subsequence defined at the beginning of the proof. Since the limit is uniquely determined, the convergence holds for the entire sequence. □

Corollary 2.3.8 *The optimal shape design problem (2.3.4), (2.3.5), (2.3.6) associated with $\hat{U}_{ad}$ has at least one solution.*

Proof. The proof follows the same lines as that of Corollary 2.3.4, using Lemma 2.3.6, Theorem 2.3.7, and the compactness of $\hat{U}_{ad}$ in $C^1[0,1]$. We leave the details to the reader. □

Remark. The proof of Theorem 2.3.7 is due to Bedivan [1996]. It should be clear that the argument extends to other boundary conditions or more general elliptic operators and cost functionals, under appropriate regularity assumptions for the geometry.

2.3.2 General Dirichlet Problems

In this paragraph, we prove existence in shape optimization problems governed by nonlinear and inhomogeneous Dirichlet boundary value problems of arbitrary order with general cost functionals, as briefly introduced in Example 1.2.8 in Chapter 1.

We consider families of domains of class C that are all contained in a given open and bounded set $E \subset \mathbf{R}^d$ (where $d \in \mathbf{N}$ is arbitrary) and have further special properties. To this end, recall the notion of domains of class C given in Definition A3.1 in Appendix 3, and recall the notation introduced there, in particular, the meaning of the families $\mathcal{F}_\Omega$ and of the positive constants k_Ω, r_Ω, and a_Ω.

Now let $k > 0$ and $a > 0$ be given fixed constants. We then define the set

$$\begin{aligned} \widehat{\mathcal{O}} \;=\; & \{\Omega \subset E: \ \Omega \text{ is an open set of class } C \text{ with } k_\Omega \geq k > 0, \\ & r_\Omega \leq r < k, \ a_\Omega \geq a > 0\}. \end{aligned} \tag{2.3.8}$$

The (nonempty) set $\mathcal{O}$ of admissible controls Ω then consists of the set of all elements of $\widehat{\mathcal{O}}$ that have the additional property that they are connected (i.e., *domains*) and that the families $\mathcal{F}_\Omega$ of corresponding local charts can be chosen in such a way that the set $\mathcal{F} = \bigcup_{\Omega\in\mathcal{O}} \mathcal{F}_\Omega$ is equicontinuous and equibounded on $\overline{\tilde{B}(0,k)}$.

We claim that $\mathcal{O}$ is sequentially compact with respect to the Hausdorff–Pompeiu metric. Indeed, whenever $\{\Omega_n\} \subset \mathcal{O}$ is given, then Proposition A3.2(i) in Appendix 3 implies that there is an open set $\Omega \subset E$ such that for a subsequence that is again indexed by n, we have

$$\lim_{n\to\infty} \tilde{d}_H(\Omega_n, \Omega) = 0.$$

But then we can infer from Theorem A3.9 that Ω belongs to $\widehat{\mathcal{O}}$, and the Arzelà–Ascoli theorem implies that the family $\mathcal{F}_\Omega$ of corresponding local charts is again equicontinuous and equibounded on $\overline{\tilde{B}(0,k)}$. Moreover, Proposition A3.10 in Appendix 3 shows that also

$$\lim_{n\to\infty} d_H(\overline{\Omega}_n, \overline{\Omega}) = 0\,,$$

so that in view of Proposition A3.2(ii), the set $\overline{\Omega}$ is connected. Since Ω is of class C, then also Ω is connected, i.e., a domain. In conclusion, we have $\Omega \in \mathcal{O}$, and the claim is proved.

To each $\Omega \in \mathcal{O}$ we now associate given functions $f_\Omega \in L^2(\Omega)$ and $h_\Omega \in W^{l,p}(E)$, $l \in \mathbf{N}$, $p \geq 2$. The functions f_Ω may be extended (by zero) to the whole set E or to $\mathbf{R}^d$, where the respective extensions are again denoted by f_Ω and the family $\{f_\Omega\}_{\Omega\in\mathcal{O}}$ remains in $L^2(\mathbf{R}^d)$.

In E, we consider the partial differential operator (in the sense of distributions)

$$Ay = \sum_{|\alpha|\leq l} (-1)^{|\alpha|}\, D^\alpha A_\alpha(x, y, \ldots, D^l y), \quad x \in E, \tag{2.3.9}$$

for $y \in W^{l,p}(E)$, where α is a multi-index of length $|\alpha| \leq l$, and where the coefficients $A_\alpha : E \times \mathbf{R}^T \to \mathbf{R}$ (T denotes the number of partial derivatives in $\mathbf{R}^d$ from order 0 up to order l) satisfy the following conditions:

$A_\alpha(\cdot\,, \xi)$ is measurable in $x \in E$ for all $\xi \in \mathbf{R}^T$, and $A_\alpha(x, \cdot)$ is continuous on $\mathbf{R}^T$ for a.e. $x \in E$. (2.3.10)

There exist some $\hat{C}_1 > 0$ and some $\mu \in L^q(E)$ such that with $\frac{1}{p} + \frac{1}{q} = 1$,

$$|A_\alpha(x,\xi)| \le \hat{C}_1 \left(|\xi|_{\mathbf{R}^T}^{p-1} + \mu(x) \right) \quad \text{for all } (x,\xi) \in E \times \mathbf{R}^T. \tag{2.3.11}$$

$$\sum_{|\alpha|\le l} (A_\alpha(x,\xi) - A_\alpha(x,\eta))(\xi_\alpha - \eta_\alpha) \ge 0 \quad \forall \xi = (\xi_\alpha),\ \eta = (\eta_\alpha) \in \mathbf{R}^T,$$
$$\text{for a.e. } x \in E. \tag{2.3.12}$$

The nonlinear operator A defined in (2.3.9) is called the *generalized divergence operator* or the *Leray–Lions operator.* Linear elliptic operators of order $2l$ are special cases for $p = 2$. From Example A1.14 in Appendix 1 we know that $A : W_0^{l,p}(E) \to W^{-l,q}(E)$ is maximal monotone. If in addition, there exist constants $\hat{c}_1 > 0$ and $\hat{c}_2 \in \mathbf{R}$ such that the coercivity condition

$$\sum_{|\alpha|\le l} A_\alpha(x,\xi)\,\xi_\alpha \ \ge\ \hat{c}_1\,|\xi'|_{\mathbf{R}^{T'}}^p + \hat{c}_2, \quad \forall\,(x,\xi) \in E \times \mathbf{R}^T \tag{2.3.13}$$

is fulfilled (with ξ' denoting the vector made up of the components of ξ that correspond to the highest-order derivatives (see (2.3.9)), and T' being their total number), then A is coercive in $W_0^{l,p}(E)$ and surjective, and its realization $A_{L^2(E)}$ in $L^2(E)$ with the domain

$$\operatorname{dom}\left(A_{L^2(E)}\right) = \left\{ y \in W_0^{l,p}(E) : Ay \in L^2(E) \right\} \tag{2.3.14}$$

is maximal monotone and surjective.

The definitions (2.3.9), (2.3.14) and the properties following from the assumptions (2.3.10) to (2.3.13) are inherited by each admissible domain $\Omega \in \mathcal{O}$ for mappings in $W_0^{l,p}(\Omega)$. Consequently, for any $\Omega \in \mathcal{O}$ the nonlinear homogeneous Dirichlet boundary value problem

$$A\hat{y}_\Omega = f_\Omega \quad \text{in } \Omega, \qquad D^\alpha y_\Omega = 0 \quad \text{on } \partial\Omega, \quad \forall \text{ multi-indices } \alpha \text{ with } |\alpha| \le l-1, \tag{2.3.15}$$

has at least one solution $\hat{y}_\Omega \in \operatorname{dom}\left(A_{L^2(\Omega)}\right) \subset W_0^{l,p}(\Omega)$. Uniqueness follows if the inequality in (2.3.12) is strict for $\xi \ne \eta$.

In the inhomogeneous case we consider boundary conditions of the form $\hat{y}_\Omega = h_\Omega$, on $\partial\Omega$, with some given function $h_\Omega \in W^{l,p}(E)$. In this case a weak solution $y_\Omega \in W^{l,p}(\Omega)$ is defined by

$$\sum_{|\alpha|\le l} \int_\Omega A_\alpha(x, y_\Omega, \ldots, D^l y_\Omega)\, D^\alpha v\, dx \ = \ \int_\Omega f_\Omega\, v\, dx \quad \forall\, v \in W_0^{l,p}(\Omega), \tag{2.3.16}$$

$$y_\Omega - h_\Omega \in W_0^{l,p}(\Omega). \tag{2.3.17}$$

The existence of a solution to (2.3.16), (2.3.17) follows by considering the shifted mappings

$$\widetilde{A}_\alpha(x,\xi) = A_\alpha\left(x, \xi + \left[h_\Omega(x), \ldots, D^l h_\Omega(x)\right]\right) \quad \text{for } (x,\xi) \in E \times \mathbf{R}^T.$$

We associate with them the differential operator $\widetilde{A} : W_0^{l,p}(\Omega) \to W^{-l,q}(\Omega)$ that results if in (2.3.9) the coefficients A_α are replaced by $\widetilde{A}_\alpha$. Notice that there are constants $\hat{C}_2 > 0$, $\hat{c}_3 > 0$, $\hat{c}_4 \in \mathbf{R}$, such that for any $y \in W_0^{l,p}(\Omega)$,

$$\begin{aligned} \int_\Omega |\widetilde{A}_\alpha(x, y(x), \ldots, D^l y(x))|^q \, dx &\leq \hat{C}_2 \left(|y|_{W_0^{l,p}(\Omega)} + 1 \right), \\ \sum_{|\alpha| \leq l} \int_\Omega \widetilde{A}_\alpha(x, y(x), \ldots, D^l y(x)) \, D^\alpha y(x) \, dx &\geq \hat{c}_3 \left| D^l y \right|_{L^p(\Omega)^{T'}} + \hat{c}_4. \end{aligned}$$

This follows from (2.3.11), (2.3.13), and Clarkson's inequalities (see Hewitt and Stromberg [1965]).

We conclude that $\widetilde{A}$ is a well-defined operator in $W_0^{l,p}(\Omega)$ that is maximal monotone and surjective onto $W^{-l,q}(\Omega)$. The corresponding boundary value problem (2.3.15) admits at least one solution $\tilde{y}_\Omega \in W_0^{l,p}(\Omega)$, and it is easily verified that $y_\Omega = \tilde{y}_\Omega + h_\Omega$ satisfies (2.3.16), (2.3.17).

We assume now the boundedness of the data, namely, that there are constants $\hat{C}_3 > 0$ and $\hat{C}_4 > 0$ such that

$$|f_\Omega|_{L^2(\Omega)} \leq \hat{C}_3, \quad |h_\Omega|_{W^{l,p}(E)} \leq \hat{C}_4, \quad \forall \Omega \in \mathcal{O}, \tag{2.3.18}$$

and the strong monotonicity of A, namely, that there is some constant $\hat{c}_5 > 0$ such that

$$\sum_{|\alpha| \leq l} (A_\alpha(x,\xi) - A_\alpha(x,\eta))(\xi_\alpha - \eta_\alpha) \geq \hat{c}_5 \, |\xi - \eta|^2_{\mathbf{R}^T} \quad \forall \xi, \eta \in \mathbf{R}^T, \quad \text{and a.e. } x \in E. \tag{2.3.19}$$

We then consider the shape optimization problem

$$\operatorname*{Min}_{\Omega \in \mathcal{O}} \left\{ \int_\Omega L(x, y_\Omega(x), \ldots, D^l y_\Omega(x)) \, dx \right\}, \tag{2.3.20}$$

subject to (2.3.16) and (2.3.17). Here, $L : E \times \mathbf{R}^T \to \mathbf{R}$ is a function that satisfies condition (2.3.10) and obeys the polynomial growth condition

$$0 \leq L(x, \xi) \leq \hat{C}_5 \left(|\xi|^t_{\mathbf{R}^T} + \eta(x) \right), \tag{2.3.21}$$

with some $\hat{C}_5 > 0$, $\eta \in L^1(E)$, and $1 \leq t < p$.

In order to deal with the shape optimization problem (2.3.20), (2.3.16), (2.3.17), we will need a property of Sobolev spaces, which, thanks to the trace theorems (compare Theorem A2.1 in Appendix 2), is well known for Lipschitz domains. We are going to prove now that it remains true for domains of class C, for which trace theorems are not available.

Theorem 2.3.9 *Let Ω be an open and bounded set of class C. If $z \in H^1(\mathbf{R}^d)$ and $z = 0$ almost everywhere in $\mathbf{R}^d \setminus \Omega$, then $z \in H_0^1(\Omega)$.*

Proof. We first construct a finite number of particular open sets $\mathcal{O}_j$, $1 \le j \le m$, that have the property that

$$\partial\Omega \subset \bigcup_{j=1}^{m} \mathcal{O}_j. \tag{2.3.22}$$

Recalling Definition A3.1 in Appendix 3 and the notation introduced there, and since $\partial\Omega$ is compact, we can find finitely many continuous functions

$$g_j : \overline{\tilde{B}(0,k_\Omega)} \subset \mathbf{R}^{d-1} \to \mathbf{R}, \quad 1 \le j \le m,$$

such that

$$\partial\Omega = \bigcup_{j=1}^{m} \left\{ R_{g_j}(\tilde{s},0) + o_{g_j} + g_j(\tilde{s}) y_{g_j} : \ \tilde{s} \in \tilde{B}(0,k_\Omega) \right\}.$$

Let $B_j(0,k_\Omega)$ denote the $(d-1)$-dimensional ball in $\mathbf{R}^d$ centered at o_{g_j}, which is the image of $\tilde{B}(0,k_\Omega)$ under the transformation $(\tilde{s},0) \mapsto R_{g_j}(\tilde{s},0)+o_{g_j}$, $1 \le j \le m$.

Following the convention introduced in Appendix 3, we regard the points on the part of $\partial\Omega$ parametrized by the local chart g_j in the local coordinate system in the form $(s, g_j(s))$, where $s \in B_j(0,k_\Omega)$, for $1 \le j \le m$.

Now let $\lambda > 0$ denote the smallest of the lengths of all the interior and exterior segments (given by the segment property, see Appendix 3) that are associated with the local charts g_j, $1 \le j \le m$. We then choose $\mathcal{O}_j$, $1 \le j \le m$, as the union of the respective interior and exterior segments, which we may assume to generate a neighborhood of $\partial\Omega \cap \mathcal{O}_j$, $1 \le j \le m$. In view of the continuity of g_j on $\overline{B_j(0,k_\Omega)}$, we may also assume that

$$\max_{s \in \overline{B_j(0,k_\Omega)}} g_j(s) - \min_{s \in \overline{B_j(0,k_\Omega)}} g_j(s) \le \frac{\lambda}{4}. \tag{2.3.23}$$

Therefore, we can shift the system of local axes (more precisely, its origin o_{g_j}) along the "vertical" axis y_{g_j} in such a way that in the resulting new local coordinate system the open cylinder

$$\begin{aligned} V_j \;=\; \Bigg\{ (s,y) \in B_j(0,k_\Omega) \times \mathbf{R}^d : 0 = \min_{s \in \overline{B_j(0,k_\Omega)}} g_j(s) - \frac{\lambda}{4} \\ < y < \max_{s \in \overline{B_j(0,k_\Omega)}} g_j(s) + \frac{\lambda}{4} \Bigg\} \end{aligned}$$

satisfies $V_j \subset \mathcal{O}_j$ and $\cup_{j=1}^{m} V_j \supset \partial\Omega$.

There is an open set V_0 with $\overline{V}_0 \subset \Omega$ such that $\cup_{j=0}^{m} V_j \supset \overline{\Omega}$. We now choose a partition of unity $\{\psi_j\}_{j=\overline{0,m}}$ subordinate to the open covering $\{V_j\}_{j=\overline{0,m}}$, such that $\psi_j \in C_0^\infty(V_j)$, $\psi_j \ge 0$, $0 \le j \le m$, and

$$\sum_{j=0}^{m} \psi_j(x) = 1, \quad x \in \overline{\Omega}. \tag{2.3.24}$$

Let $z_j = z\,\psi_j$, $0 \le j \le m$. Then $z_j \in H_0^1(V_j)$, and (2.3.24) gives

$$z(x) = \sum_{j=0}^{m} \tilde{z}_j(x) \quad \forall\, z \in \mathbf{R}^d, \tag{2.3.25}$$

where $\tilde{z}_j$ is the extension by zero of z_j to $\mathbf{R}^d$, $0 \le j \le m$, and where we have used the fact that z vanishes almost everywhere in $\mathbf{R}^d \setminus \Omega$.

Clearly, $\tilde{z}_0 \in H_0^1(\Omega)$. From the subsequent Lemma 2.3.10 we can infer that also $\tilde{z}_j \in H_0^1(\Omega)$, $1 \le j \le m$. This, together with (2.3.25), concludes the proof of the assertion. □

Lemma 2.3.10 *Let $U \subset \mathbf{R}^{d-1}$ be an open and bounded set and $\tilde{E} = U \times \,]0, b[$ with some $b > 0$, and suppose that $g : U \to \mathbf{R}_+$ is a continuous mapping such that with some $c > 0$, $b \ge g(s) \ge c$ for all $s \in U$. Then $z \in H_0^1(\tilde{\Omega})$ for any function $z \in H_0^1(\tilde{E})$ that has compact support in $\tilde{E}$ and satisfies $z = 0$ almost everywhere in $\tilde{E} \setminus \tilde{\Omega}$, where $\tilde{\Omega} = \{(s, y) \in \tilde{E} : s \in U,\ y < g(s)\}$.*

Proof. We denote by Γ the part of $\partial\tilde{\Omega}$ represented by the graph of g. Obviously, we have $(s, g(s)-t) \in \tilde{\Omega}$ and $(s, g(s)+t) \in \mathbf{R}^d \setminus \tilde{\Omega}$ for all $(s,t) \in U \times (0, c)$; that is, the segment property is valid on Γ with "vertical" segments having a length of at least $c > 0$, both inside and outside $\tilde{\Omega}$. We define the "translated" mappings

$$z_t(s, y) = \tilde{z}(s, y + t), \quad y \in \,]0, b[\,, \quad t > 0, \quad s \in B, \tag{2.3.26}$$

where $\tilde{z}$ denotes the extension by zero of z to $\mathbf{R}^d$.

If $t < \min\{\frac{1}{2}\, d(\operatorname{supp} z, \partial\tilde{E}),\, c\}$, then $z_t \in H_0^1(\tilde{E})$ with $\operatorname{supp} z_t \subset \tilde{\Omega}$. This follows from the observation that $z_t = 0$ a.e. in the "interior band"

$$\{(s, y) \in \tilde{\Omega} : s \in U,\ g(s) - t < y < g(s)\}$$

of $\tilde{\Omega}$. Moreover, the interior band is a neighborhood of $\Gamma_t = \{w \in \Gamma : d(w, \partial\tilde{E}) > t\}$ in $\overline{\tilde{\Omega}}$, again by the segment property. In addition, there exists a neighborhood of $\partial\tilde{\Omega} \setminus \Gamma$ in which z_t vanishes almost everywhere, since $t < \frac{1}{2}\, d(\operatorname{supp} z, \partial\tilde{E})$. In conclusion, we have $z_t \in H_0^1(\tilde{\Omega})$ for sufficiently small $t > 0$.

Hence, in view of the continuity of the norm of $H^1(\tilde{\Omega})$ with respect to translations (cf. Hewitt and Stromberg [1965]), we can infer that $\lim_{t\to 0} z_t = z$ strongly in $H^1(\tilde{\Omega})$. Since z vanishes almost everywhere in $\tilde{E} \setminus \tilde{\Omega}$, we find that $z \in H_0^1(\tilde{\Omega})$, and the assertion is proved. □

The following result can be proved similarly as Theorem 2.3.9.

Corollary 2.3.11 *Let Ω be a bounded open set of class C in $\mathbf{R}^d$. If $z \in W^{l,p}(\mathbf{R}^d)$, $l \in \mathbf{N}$, $1 \le p < \infty$, and $z = 0$ almost everywhere in $\mathbf{R}^d \setminus \Omega$, then $z \in W_0^{l,p}(\Omega)$.*

Remark. Similar results are known for the case that $z = 0$ in $\mathbf{R}^d \setminus \Omega$ quasi-everywhere (that is, in the sense of capacity). A recent survey in this respect, with applications, is Henrot [1994]. We also refer to Theorem 4.5 (the Havin–Bagby theorem) in Heinonen, Kilpeläinen, and Martio [1993].

In this setting, the property of Ω expressed by Theorem 2.3.9 is called *stability* in the sense of Hedberg–Keldys. The existence result for the problem (2.3.20) reads as follows.

Theorem 2.3.12 *Suppose that the conditions (2.3.10)–(2.3.13), (2.3.18), (2.3.19), (2.3.21) are fulfilled, and suppose that the set $\mathcal{O}$ of admissible control domains satisfies (2.3.8) and has the property that $\mathcal{F} = \bigcup_{\Omega \in \mathcal{O}} \mathcal{F}_\Omega$ is equicontinuous and equibounded on $\overline{\tilde{B}(0,k)}$. Then the shape optimization problem (2.3.20), (2.3.16), (2.3.17) has at least one optimal domain $\Omega^* \in \mathcal{O}$.*

Proof. Let $\{\Omega_n\} \subset \mathcal{O}$ be a minimizing sequence for the problem (2.3.20). In view of the compactness properties of the set $\mathcal{O}$, we may assume that there is some $\Omega^* \in \mathcal{O}$ such that $\lim_{n\to\infty} \tilde{d}_H(\Omega_n, \Omega^*) = 0$, and that the corresponding characteristic functions satisfy $\chi_n \to \chi^*$ (with obvious new notation) a.e. in E, as well as $\chi_n \to \chi^*$ strongly in $L^r(E)$ for all $r \geq 1$.

Notice also that the Γ property (see Proposition A3.8 in Appendix 3) entails that for any compact set $K \subset \Omega^*$ there is some $n(K) \in \mathbf{N}$ such that $K \subset \Omega_n$ for $n \geq n(K)$.

We use the abbreviations h_n, f_n, y_n for the corresponding data and solutions to (2.3.16), (2.3.17) in Ω_n. Hypothesis (2.3.18) implies that $h_n \to h^*$ weakly in $W^{l,p}(E)$ and $f_n \to f^*$ weakly in $L^2(E)$, on a subsequence again indexed by n.

Taking $v = y_n - h_n$ in (2.3.16), $\Omega = \Omega_n$, invoking (2.3.13), and recalling that $p \geq 2$, we find that

$$\begin{aligned} \hat{c}_1 \sum_{|\alpha|=l} |D^\alpha y_n|^p_{L^p(\Omega_n)} + \hat{c}_2 \;&\leq\; \sum_{|\alpha|\leq l} \int_{\Omega_n} A_\alpha(x, y_n, \ldots, D^l y_n)\, D^\alpha y_n \, dx \\ &\leq C_1 \, |f_n|_{L^2(\Omega_n)} \, |y_n - h_n|_{L^p(\Omega_n)} + \sum_{|\alpha|\leq l} \int_{\Omega_n} A_\alpha(x, y_n, \ldots, D^l y_n)\, D^\alpha h_n \, dx, \end{aligned}$$

where we denote by C_j, $j \in \mathbf{N}$, generic positive constants that may depend on the given data, but not on n.

By virtue of (2.3.18), $\left\{|h_n|_{W^{l,p}(\Omega_n)}\right\}$ and $\left\{|f_n|_{L^2(\Omega_n)}\right\}$ are bounded sequences. Hence, invoking (2.3.11), Young's inequality, and the equivalence of norms in $\mathbf{R}^T$, we can easily show that the right-hand side of the above inequality is bounded by an expression of the form

$$\delta \, |y_n|^p_{W^{l,p}(\Omega_n)} + \frac{C_2}{\delta},$$

where $\delta > 0$ may be chosen arbitrarily small. Thus, we may invoke Poincaré's and Friedrichs' inequalities to conclude that the sequence $\left\{|y_n|_{W^{l,p}(\Omega_n)}\right\}$ is bounded. Therefore, we see that for any domain K such that $\overline{K} \subset \Omega^*$, we have $y_n \to y^*$ weakly in $W^{l,p}(K)$ for a suitable subsequence depending on K.

Now choose an increasing sequence $\{G_k\}_{k\in\mathbf{N}}$ of open and bounded sets such that

$$\overline{G}_k \subset \Omega^*, \quad k \in \mathbf{N}, \qquad \bigcup_{k\in\mathbf{N}} \overline{G}_k = \Omega^*.$$

Applying the above argument successively with increasing k to the compact sets $\overline{G}_k$, $k \in \mathbf{N}$, we deduce that for any multi-index α of length $|\alpha| \le l$, the function $D^\alpha y^*$ can be extended onto the whole set Ω^* to a well-defined mapping y^α.

We claim that $y^* = y^{(0,\dots,0)} \in W^{l,p}(\Omega^*)$ with $D^\alpha y^* = y^\alpha$, $|\alpha| \le l$. To verify this, let us consider any multi-index α with $1 \le |\alpha| \le l$. Suppose now that any $\varphi \in \mathcal{D}(\Omega)$ with $\operatorname{int}(\operatorname{supp}\varphi) \neq \emptyset$ is given. Then $y^* \in W^{l,p}(\operatorname{int}(\operatorname{supp}\varphi))$, and thus

$$\int_{\Omega^*} y^\alpha(x)\,\varphi(x)\,dx = \int_{\operatorname{supp}\varphi} D^\alpha y^*(x)\,\varphi(x)\,dx = (-1)^{|\alpha|}\int_{\Omega^*} D^\alpha\varphi(x)\,y^*(x)\,dx,$$

which shows that $y^\alpha = D^\alpha y^*$ in $\mathcal{D}'(\Omega^*)$.

It remains to prove that $D^\alpha y^* \in L^p(\Omega^*)$. To this end, let $k \in \mathbf{N}$ be fixed. By Proposition A3.8 in Appendix 3, we have $\overline{G}_k \subset \Omega_n$ for $n \ge n(G_k)$, and there is a suitable subsequence, again indexed by n, such that, by (2.3.18) and the weak lower semicontinuity of norms,

$$|y^*|_{W^{l,p}(G_k)} \le \liminf_{n\to\infty} |y_n|_{W^{l,p}(G_k)} \le \sup\left\{|y_n|_{W^{l,p}(\Omega_n)} : n \ge n(G_k)\right\} \le \hat{C}_4.$$

Now observe that the sequence $\{\chi_{G_k} D^\alpha y^*\}_{k\in\mathbf{N}}$ is bounded in $L^p(\Omega^*)$. Hence, on a suitable subsequence again indexed by k, we have $\chi_{G_k} D^\alpha y^* \to z_\alpha$ weakly in $L^p(\Omega^*)$ for some $z_\alpha \in L^p(\Omega^*)$. Clearly, $z_\alpha = D^\alpha y^*$, so that $D^\alpha y^* \in L^p(\Omega^*)$, as claimed.

Next, we show that y^* is the solution to (2.3.16), (2.3.17) associated with h^*, f^*, in Ω^*.

We put $z_n = y_n - h_n \in W_0^{l,p}(\Omega_n)$, and extend z_n by zero to a function $\tilde{z}_n$ defined on the whole space $\mathbf{R}^d$. Then $\{\tilde{z}_n\}$ is bounded in $W_0^{l,p}(E)$. Thus, we have $\tilde{z}_n \to \tilde{z}$ weakly in $W_0^{l,p}(E)$, for a subsequence again indexed by n. By compact embedding (cf. Theorem A2.2 in Appendix 2), we may therefore assume, possibly taking yet another subsequence indexed by n, that $\tilde{z}_n \to \tilde{z}$ strongly in $L^p(\Omega)$. Since, as noted above, $\chi_n \to \chi^*$ strongly in $L^r(\Omega)$ for any $r \ge 1$, we find that

$$\int_{E\setminus\Omega^*} |\tilde{z}(x)|\,dx = \int_E (1-\chi^*(x))|\tilde{z}(x)|\,dx = \lim_{n\to\infty}\int_E (1-\chi_n(x))|\tilde{z}_n(x)|\,dx = 0;$$

that is, $\tilde{z}$ vanishes almost everywhere in $E \setminus \Omega^*$. Corollary 2.3.11 then implies that $z = \tilde{z}|_{\Omega^*} \in W_0^{l,p}(\Omega^*)$. On the other hand, $\tilde{z} = y^* - h^*$ in Ω^*, since this is true on every compact subset of Ω^*, and we see that y^*, h^* satisfy (2.3.17).

To pass to the limit in (2.3.16), we use the fact that thanks to the above-mentioned boundedness of $\{|y_n|_{W^{l,p}(\Omega_n)}\}$ and to hypothesis (2.3.11), the sequences $\{|A_\alpha(\cdot, y_n, \ldots, D^l y_n)|_{L^q(\Omega_n)}\}$ are bounded.

Applying the same procedure as above in the construction of y^* on Ω^*, we may construct functions $a_\alpha \in L^q(\Omega^*)$, $0 \le |\alpha| \le l$, such that for any compact set $K \subset \Omega^*$ there exists a subsequence, again indexed by n, with

$$A_\alpha(\cdot\,, y_n(\cdot\,), \ldots, D^l y_n(\cdot\,)) \to a_\alpha \quad \text{weakly in } L^q(K),$$

for any multi-index of length $|\alpha| \le l$.

Now let $K \subset \Omega^*$ be any arbitrary, but fixed, compact set, and pick any nonnegative test function $\varphi \in C_0^\infty(\Omega^*)$ satisfying $\varphi(x) = 1$ for all $x \in K$. We estimate the expression

$$I_n = \int_E \varphi \sum_{|\alpha| \le l} \Big(A_\alpha(x, y_n, \ldots, D^l y_n) - A_\alpha(x, y^*, \ldots, D^l y^*) \Big) (D^\alpha y_n - D^\alpha y^*)\, dx.$$

Here, and in the following, we assume that n is so large that $\operatorname{supp}\varphi \subset \Omega_n$. Then I_n is meaningful by simply setting the integrand equal to zero in $E \setminus \operatorname{supp}\varphi$. We have

$$\begin{aligned} I_n \;&=\; \int_E \sum_{|\alpha| \le l} A_\alpha(x, y_n, \ldots, D^l y_n)\, D^\alpha(\varphi(y_n - y^*))\, dx \\ &\quad - \int_E \varphi \sum_{|\alpha| \le l} A_\alpha(x, y^*, \ldots, D^l y^*)\, (D^\alpha y_n - D^\alpha y^*)\, dx \\ &\quad - \int_E \sum_{|\alpha| \le l} A_\alpha(x, y_n, \ldots, D^l y_n)\, Z_n^{\alpha-1}\, dx \;=\; I_{1n} + I_{2n} + I_{3n}. \end{aligned}$$

Here, we have used the abbreviation

$$Z_n^{\alpha-1} = D^\alpha(\varphi(y_n - y^*)) - \varphi(D^\alpha y_n - D^\alpha y^*).$$

Notice that this expression does not contain derivatives of order l of y_n, for any α. Invoking the compactness of the embedding (compare Theorem A2.2(i) in Appendix 2) $W_0^{l,p}(E) \subset W_0^{l-1,p}(E)$, and selecting another subsequence, which is again indexed by n, we can infer that $Z_n^{\alpha-1} \to 0$ strongly in $L^p(E)$ for any $|\alpha| \le l$. Consequently, $\lim_{n\to\infty} I_{3n} = 0$, on this subsequence.

Next, we select yet another subsequence indexed by n from it such that $y_n \to y^*$ weakly in $W^{l,p}(\operatorname{supp}\varphi)$. On this subsequence, we also have $\lim_{n\to\infty} I_{2n} = 0$, and owing to the compactness of the embedding $W_0^{l,p}(E) \subset L^p(E)$, we may select another subsequence from it such that also

$$\lim_{n\to\infty} I_{1n} = \lim_{n\to\infty} \int_{\operatorname{supp}\varphi} f_n\, \varphi\, (y_n - y^*)\, dx \to 0,$$

where we have used (2.3.16).

Summarizing the above estimates, and invoking hypothesis (2.3.19), we conclude that for the subsequence just selected,

$$\lim_{n\to\infty}\left(\hat{c}_5\int_E \varphi \sum_{|\alpha|\le l} |D^\alpha y_n - D^\alpha y^*|^2\,dx\right) \le \lim_{n\to\infty} I_n = 0.$$

But since the test function φ is nonnegative in Ω^* with $\varphi|_K \equiv 1$, this implies that we can select yet another subsequence from the subsequence constructed above such that

$$D^\alpha y_n(x) \to D^\alpha y^*(x) \text{ as } n\to\infty, \quad \text{for a.e. } x\in K, \quad \text{for all } |\alpha|\le l.$$

The Carathéodory assumption (2.3.10) then yields that for this subsequence,

$$A_\alpha(x, y_n(x), \ldots, D^l y_n(x)) \to A_\alpha(x, y^*(x), \ldots, D^l y^*(x)) \quad \text{a.e. in } K.$$

Consequently, by applying Egorov's theorem, we can identify a_α a.e. in K. Since the compact set $K\subset\Omega^*$ was arbitrarily chosen, we conclude that

$$a_\alpha(x) = A_\alpha(x, y^*(x), \ldots, D^l y^*(x)) \quad \text{for a.e. } x\in\Omega^*,$$

which identifies the limits of the nonlinear terms in (2.3.16). Therefore, we may pass to the limit in (2.3.16) on a suitable subsequence to arrive at the conclusion that y^* is the (uniquely determined) solution to (2.3.16), (2.3.17) that corresponds to the domain $\Omega^*\in\mathcal{O}$ and the data h^*, f^*.

As the final step of the proof, we now show that Ω^* is an optimal domain that minimizes the cost functional in (2.3.20). Since L satisfies (2.3.10), we may carry out the same chain of arguments as above to arrive at the conclusion that for any domain K with $\overline{K}\subset\Omega^*$ (initially only for some subsequence depending on K, but by the uniqueness of the limit y^* eventually for the entire sequence), $y_n \to y^*$ weakly in $W^{l,p}(K)$ and

$$L(x, y_n(x), \ldots, D^l y_n(x)) \to L(x, y^*(x), \ldots, D^l y^*(x)) \quad \text{for a.e. } x\in K. \quad (2.3.27)$$

Assumption (2.3.21) and Vitali's theorem (cf. Vulikh [1976, p. 210]) imply that the convergence in (2.3.27) is even strong in $L^1(K)$, since it follows from Lemma 2.2.8(b) that $\{y_n\}$ converges strongly in $W^{l,t}(K)$ for any $1\le t<p$.

Now recall the increasing sequence $\{G_k\}_{k\in\mathbf{N}}$ of open and bounded sets introduced above that satisfy $\overline{G}_k\subset\Omega^*$, $k\in\mathbf{N}$, and $\bigcup_{k\in\mathbf{N}}\overline{G}_k = \Omega^*$. We have

$$\begin{aligned}
\inf_{\Omega\in\mathcal{O}}\left\{\int_\Omega L(x, y_\Omega, \ldots, D^l y_\Omega)\,dx\right\} &= \lim_{n\to\infty}\int_{\Omega_n} L(x, y_n, \ldots, D^l y_n)\,dx \\
\ge \liminf_{k\to\infty}\lim_{n\to\infty}\int_{G_k} L(x, y_n, \ldots, D^l y_n)dx &= \liminf_{k\to\infty}\int_{\Omega^*}\chi_{G_k} L(x, y^*, \ldots, D^l y^*)\,dx \\
\ge \int_{\Omega^*}\liminf_{k\to\infty}\left(\chi_{G_k} L(x, y^*, \ldots, D^l y^*)\right)dx &= \int_{\Omega^*} L(x, y^*, \ldots, D^l y^*)\,dx, \quad (2.3.28)
\end{aligned}$$

where the nonnegativity of L makes the use of Fatou's lemma possible. This concludes the proof of the assertion. □

Remark. An essential step in the proof of the continuity of the mapping $\Omega \mapsto y_\Omega$ is to show the almost everywhere convergence of the associated characteristic functions. We close this paragraph with a result that demonstrates that in certain situations this property is also necessary in order to guarantee the continuity of the mapping $\Omega \mapsto y_\Omega$.

We consider the simple case of the Laplace operator,

$$-\Delta y_\Omega = 1 \quad \text{in } \Omega, \qquad y_\Omega = 0 \quad \text{on } \partial\Omega, \tag{2.3.29}$$

where $\Omega \in \mathcal{O}$. The assumptions (2.3.10)–(2.3.13) are fulfilled with $p = 2$ and $l = 1$ for arbitrary space dimension d. Let us define

$$\hat{y}_\Omega(x) = \begin{cases} y_\Omega(x), & \text{for } x \in \Omega, \\ -y_{E\setminus\overline{\Omega}}(x), & \text{for } x \in E \setminus \overline{\Omega}, \end{cases} \tag{2.3.30}$$

with $y_{E\setminus\overline{\Omega}}$ given by (2.3.29) applied in $E\setminus\overline{\Omega}$. Then $\hat{y}_\Omega \in H_0^1(E)$, since (2.3.30) may be rewritten as

$$\hat{y}_\Omega(x) = \tilde{y}_\Omega(x) - \tilde{y}_{E\setminus\overline{\Omega}}(x) \quad \forall\, x \in E, \tag{2.3.31}$$

where $\tilde{y}_\Omega \in H_0^1(E)$ and $\tilde{y}_{E\setminus\overline{\Omega}} \in H_0^1(E)$ denote the extensions by zero to the whole set E of y_Ω and of $y_{E\setminus\overline{\Omega}}$, respectively. We have the following result.

Proposition 2.3.13 *Let $\{\Omega_n\} \subset \mathcal{O}$ and $\Omega \in \mathcal{O}$ be domains such that the strong maximum principle for the Laplacian is valid in Ω and in $E \setminus \overline{\Omega}$, as well as in Ω_n and in $E \setminus \overline{\Omega}_n$, for all $n \in \mathbf{N}$. If $\hat{y}_{\Omega_n} \to \hat{y}_\Omega$ a.e. in E, then $\chi_{\Omega_n} \to \chi_\Omega$ a.e. in E.*

Proof. First, we note that $\text{meas}(\partial\Omega) = 0$, since the compactness of $\partial\Omega$ implies that it can be covered by a finite union of graphs of continuous functions. By assumption, there is some set M with $\text{meas}(M) = 0$ such that

$$\lim_{n\to\infty} \hat{y}_{\Omega_n}(x) = \hat{y}_\Omega(x) \quad \text{for all } x \in E \setminus M. \tag{2.3.32}$$

Now let $x \in E \setminus (M \cup \partial\Omega)$ be given with $\hat{y}_\Omega(x) > 0$. Then, by the strong maximum principle, $x \in \Omega$. In addition, we must have $\hat{y}_{\Omega_n}(x) > 0$ for $n \geq n_1(x)$, that is, $x \in \Omega_n$, again by the strong maximum principle. Consequently, $\chi_{\Omega_n}(x) = \chi_\Omega(x) = 1$ for $n \geq n_1(x)$. By the same token, we conclude that for any $x \in E\setminus(M \cup \partial\Omega)$ with $\hat{y}_\Omega(x) < 0$, $x \in E\setminus\overline{\Omega}$, that is, $\chi_{\Omega_n}(x) = \chi_\Omega(x) = 0$, for $n \geq n_2(x)$. Since $M \cup \partial\Omega$ has zero measure, the assertion is proved. □

Remark. The general results of this paragraph are due to Tiba [1999], [2003]. We also notice that they cover Example 2.3.2, showing that the family of admissible domains defined in (2.3.1) can be enlarged considerably. Example 2.3.5, however, does not enter the present setting due to the boundary cost functional (2.3.4). It is clear that for domains of class C this type of cost criterion cannot be used in view of the absence of a corresponding trace theory. It is possible to study other restrictions like $\Omega \supset \omega$ and other cost functionals. This remark aims at pointing out the large diversity of shape optimization problems (which is also underlined by the examples given in Chapter 1) and the difficulties in developing a systematic theoretical approach to them.

2.3.3 Neumann and Mixed Boundary Conditions

We consider the model problem (cf. Chapter 1, (1.2.71), (1.2.72))

$$\operatorname*{Min}_{\Omega\in\mathcal{O}} \left\{ \int_\Omega l(x, y(x), \nabla y(x))\, dx \right\} \tag{2.3.33}$$

subject to the homogeneous Neumann boundary value problem (in weak form)

$$\int_\Omega \nabla y(x)\cdot\nabla v(x)\, dx + \int_\Omega y(x)\, v(x)\, dx = \int_\Omega f(x)\, v(x)\, dx \quad \forall\, v \in H^1(\Omega). \tag{2.3.34}$$

The family $\mathcal{O}$ of admissible domains is the same as in the previous paragraph, and we assume that $f \in L^2(E)$ is prescribed. Furthermore, the mapping $l : E\times \mathbf{R}\times\mathbf{R}^d \to \mathbf{R}$ is assumed to be nonnegative and measurable in $E\times\mathbf{R}\times\mathbf{R}^d$, where $l(x,\cdot,\cdot)$ is continuous on $\mathbf{R}\times\mathbf{R}^d$ for a.e. $x\in E$ and $l(x,s,\cdot)$ is convex on $\mathbf{R}^d$ for all $(x,s)\in E\times\mathbf{R}$.

Now let $\{\Omega_n\}$ be a minimizing sequence for the problem (2.3.34), (2.3.33). By the compactness properties of $\mathcal{O}$, we may again assume that $\lim_{n\to\infty} \tilde{d}_H(\Omega_n, \Omega^*) = 0$ for some $\Omega^*\in\mathcal{O}$, where the corresponding characteristic functions satisfy $\chi_{\Omega_n} \to \chi_{\Omega^*}$ pointwise a.e. in E and strongly in $L^r(E)$ for all $r\ge 1$. Let $y_n \in H^1(\Omega_n)$ and $y^*\in H^1(\Omega^*)$ denote the solutions of (2.3.34) corresponding to Ω_n and Ω^*, respectively.

Theorem 2.3.14 *For every domain K with $\overline{K}\subset\Omega^*$,*

$$y_n|_K \to y^*|_K \quad \textit{weakly in } H^1(K). \tag{2.3.35}$$

Proof. By the Γ property (cf. Proposition A3.8 in Appendix 3), we have $K\subset \Omega_n$ for $n\ge n(K)$ (which will be assumed in the following). Inserting $v = y_n \in H^1(\Omega_n)$ in (2.3.34), we readily verify that $|y_n|_{H^1(\Omega_n)} \le M$ for some $M>0$ that does not depend on n. Thus, on a subsequence again indexed by n, we have $y_n|_K \to \tilde{y}|_K$ weakly in $H^1(K)$ for some $\tilde{y}\in H^1(K)$. We rewrite (2.3.34) in the form

$$\int_K (\nabla y_n\cdot\nabla v + y_n v)\, dx - \int_E \chi_{\Omega_n} f v\, dx = -\int_{\Omega_n\setminus K} (\nabla y_n\cdot\nabla v + y_n v)\, dx, \quad \forall v \in C^1(\overline{E}). \tag{2.3.36}$$

We have

$$\left| \int_{\Omega_n \setminus K} (\nabla y_n \cdot \nabla v + y_n v)\, dx \right| \le |v|_{C^1(\overline{E})}\, |y_n|_{H^1(\Omega_n)}\, (\text{meas}(\Omega_n \setminus K))^{\frac{1}{2}}. \quad (2.3.37)$$

Passage to the limit as $n \to \infty$ in (2.3.36), (2.3.37), using the fact that $\chi_{\Omega_n} \to \chi_{\Omega^*}$ a.e. in E, yields that

$$\left| \int_K \nabla \tilde{y} \cdot \nabla v\, dx + \int_K \tilde{y} v\, dx - \int_{\Omega^*} f v\, dx \right| \le M\, |v|_{C^1(\overline{E})}\, (\text{meas}(\Omega^* \setminus K))^{\frac{1}{2}}. \quad (2.3.38)$$

Now consider an increasing sequence of open bounded sets G_k such that $\overline{G}_k \subset \Omega^*$ for all $k \in \mathbf{N}$ and $\bigcup_{k \in \mathbf{N}} \overline{G}_k = \Omega^*$. Applying the above arguments iteratively to the compact sets $\overline{G}_k$ with increasing k, we can extend the mapping $\tilde{y}$ to the whole set Ω^*, and arguing as in the proof of Theorem 2.3.12, we conclude that $\tilde{y} \in H^1(\Omega^*)$. Passing to the limit as $k \to \infty$ in (2.3.38), where K is replaced by G_k, we obtain that

$$\int_{\Omega^*} \nabla \tilde{y}(x) \cdot \nabla v(x)\, dx + \int_{\Omega^*} \tilde{y}(x)\, v(x)\, dx - \int_{\Omega^*} f(x)\, v(x)\, dx = 0, \quad \forall v \in C^1(\overline{E}). \quad (2.3.39)$$

Since Ω^* has the segment property, the restrictions of functions in $C^1(\overline{E})$ to Ω^* form a dense subset of $H^1(\Omega^*)$ (see Adams [1975]). Hence (2.3.39) implies that $\tilde{y} = y^*$. Thus, (2.3.35) is valid at least for the constructed subsequence. Since the limit point y^* is uniquely determined, it follows that (2.3.35) holds for the entire sequence (see the remark before Corollary 2.3.4). This ends the proof. □

Having proved Theorem 2.3.14, we are in a position to apply Theorem A3.15 in Appendix 3 to arrive at the conclusion that the limit domain Ω^*, together with the associated solution y^* of (2.3.34), is a minimizer of the cost functional in (2.3.33). We therefore have the following result.

Corollary 2.3.15 *Under the assumptions made above, the shape optimization problem (2.3.33), (2.3.34) has at least one optimal domain $\Omega^* \in \mathcal{O}$.*

Remark. It ought to be clear that more general differential operators can be treated similarly. However, the extension to inhomogeneous boundary data seems impossible for domains of class C, since then the weak formulation of the equation would involve boundary integrals. This constitutes an important difference in comparison with the case of Dirichlet boundary conditions.

Remark. The convexity assumption for $l(x, s, \cdot)$ is dispensable if the polynomial growth condition (compare (2.3.21))

$$0 \le l(x, s, \xi) \le \hat{C} \left(|(s, \xi)|^t_{\mathbf{R}^{d+1}} + \eta(x) \right),$$

with some $\hat{C} > 0$, $\eta \in L^1(E)$, and $t < 2$, is satisfied. Indeed, in this case we are (up to the different boundary conditions) in the same situation as in Theorem 2.3.12 (with $p = 2$), and we may proceed as in the final step in the proof of Theorem 2.3.12 to see that Ω^* minimizes the cost functional.

In the second part of this paragraph, we will discuss the case of *mixed boundary conditions*. The argument to be presented is based on a geometric compatibility condition (see (2.3.49)–(2.3.51) below) at those points where a Dirichlet condition changes into a Neumann condition and vice versa. We will demonstrate this in the case of a variational inequality of obstacle type as discussed in Example 1.2.5 in Chapter 1. With this, we aim to indicate that the theory developed in this section also applies to variational inequalities and free boundary problems.

We consider two fixed Lipschitz domains $C \subset \mathbf{R}^d$ and $E \subset \mathbf{R}^d$ with $\overline{C} \subset E$, and variable domains $\Omega \subset E$ of class C with $\overline{C} \subset \Omega$. To each Ω we associate a fixed (not necessarily connected) open set $D_\Omega \subset E$, $D_\Omega \not\subset \Omega$, having the property that $\Omega \cup D_\Omega$ is connected (i.e., a domain) and that $\Omega \cap D_\Omega$ consists of a finite number of connected components. We define the set

$$\begin{aligned}\mathcal{M} \;=\; \Big\{v:\; v = w|_{\Omega\setminus\overline{C}} \text{ for some } w \in C_0^\infty(\mathbf{R}^d), \text{ and there are an open}\\ \text{set } Q_w \text{ and a compact set } K_w \text{ with } \overline{D}_\Omega \subset Q_w,\; K_w \subset \Omega\setminus\overline{C},\\ \text{such that } w|_{Q_w\setminus K_w} = 0\Big\}.\end{aligned}$$

Notice that $\mathcal{M}$ forms a linear space: indeed, if $w \in \mathcal{M}$ and $\lambda \in \mathbf{R}$, then clearly $\lambda w \in \mathcal{M}$, and if $w_1, w_2 \in \mathcal{M}$ with associated sets Q_{w_i} and K_{w_i}, $i = 1, 2$, then $w_1 + w_2 \in \mathcal{M}$ with $Q_{w_1+w_2} = Q_{w_1} \cap Q_{w_2}$ and $K_{w_1+w_2} = K_{w_1} \cup K_{w_2}$. Therefore, the closure of $\mathcal{M}$ in the space $H^1(\Omega\setminus\overline{C})$,

$$V(\Omega) = \overline{\mathcal{M}}, \tag{2.3.40}$$

forms a closed subspace of $H^1(\Omega\setminus\overline{C})$, and is thus a Hilbert space with respect to the inner product $(\cdot\,,\,\cdot)_{H^1(\Omega\setminus\overline{C})}$. Let

$$V(\Omega)_+ = \{v \in V(\Omega):\; v \geq 0 \text{ a.e. in } \Omega\setminus\overline{C}\}$$

denote its positive cone. Then the set

$$\Lambda(\Omega) = \{v \in V(\Omega)_+ : v = 1 \text{ on } \partial C\} \tag{2.3.41}$$

is nonempty, closed, and convex, where the boundary condition in (2.3.41) has to be understood in the sense of traces (recall that ∂C is Lipschitz). A related definition of $V(\Omega)$ (under different geometric assumptions) in the framework of mixed boundary conditions for partial differential equations can be found in Rehberg and Gröger [1989].

In $\Omega \setminus \overline{C}$, we consider the variational inequality

$$\int_{\Omega\setminus C} \{\nabla y \cdot \nabla (y-v) + y\,(y-v)\}\,dx \le \int_{\Omega\setminus C} f\,(y-v)\,dx \quad \forall\, v \in \Lambda(\Omega). \quad (2.3.42)$$

If Ω is a Lipschitz domain, then the definition of $\Lambda(\Omega)$ and the weak formulation (2.3.42) correspond to the mixed boundary conditions

$$y = 1 \quad \text{on } \partial C, \qquad y = 0 \quad \text{on } \partial\Omega \cap D_\Omega, \qquad \frac{\partial y}{\partial n} = 0 \quad \text{on } \partial\Omega \setminus D_\Omega. \qquad (2.3.43)$$

The existence of a unique solution $y \in \Lambda(\Omega)$ to (2.3.42) follows immediately from the Lions–Stampacchia theorem (Theorem A2.3 in Appendix 2). We ought to remark at this place that the notation $V(\Omega)$ and $\Lambda(\Omega)$ is slightly misleading, since (2.3.40) and (2.3.41) clearly depend on the choice of the associated sets D_Ω. If D_Ω changes, then also $V(\Omega)$ and $\Lambda(\Omega)$ may do so. We will use the above notation for the sake of brevity in the following; possible confusion will be avoided by the context.

A physical example modeled by (2.3.42) is the electrochemical machining process considered in Example 1.2.7 of Chapter 1, with different notation. One difference to (1.2.48) is that (2.3.42), (2.3.43) account for the possibility that some part of the boundary, namely $\partial\Omega \setminus D_\Omega$, may be insulated. The metallic workpiece is formed around $\partial\Omega \cap D_\Omega$. The variational inequality (2.3.42) provides the shape of the workpiece via the *coincidence set* (see Elliott and Ockendon [1982], Barbu [1984, p. 165])

$$I_y = \{x \in \Omega \setminus \overline{C} : y(x) = 0\}. \qquad (2.3.44)$$

Here, we study the shape optimization problem corresponding to (2.3.42), which consists in finding the domain Ω (i.e., the machine) of minimal area such that the obtained metallic workpiece (2.3.44) satisfies certain requirements. A typical example is that I_y is required to contain a prescribed subset $D \setminus C$, with $D \subset \Omega \subset E$ being the desired shape of the workpiece. Mathematically, we may phrase this in the form

$$\underset{\Omega\in\widetilde{\mathcal{O}}}{\text{Min}} \left\{ \int_{\Omega\setminus C} dx \right\}, \qquad (2.3.45)$$

subject to (2.3.42) and to

$$y \le 0 \quad \text{a.e. in } D \setminus C, \qquad (2.3.46)$$

where the set $\widetilde{\mathcal{O}}$ of admissible domains has yet to be specified. Notice that (2.3.41), (2.3.46), and $y \in \Lambda(\Omega)$, imply that $y = 0$ a.e. in $D \setminus \overline{C}$; that is, we have $D \setminus \overline{C} \subset I_y$.

Let us now specify the assumptions on the set $\widetilde{\mathcal{O}}$ of admissible domains. First, we postulate that any $\Omega \in \widetilde{\mathcal{O}}$, together with its associated open set D_Ω, is contained in E, and that there is some open set $\tilde{C}$ with $\overline{C} \subset \tilde{C} \subset E$ such that $\tilde{C} \subset \Omega$ for every $\Omega \in \tilde{\mathcal{O}}$. The latter condition means that the "machine" Ω cannot degenerate to C in any part of its boundary. Also, we require that any $\Omega \in \tilde{\mathcal{O}}$ and all the (finitely many) connected components D_Ω^i, $1 \leq i \leq i_\Omega$, of D_Ω belong to the family $\widehat{\mathcal{O}}$ introduced in (2.3.8) (where we also refer to Definition A3.1 in Appendix 3). We thus have, in the terminology of (2.3.8) with obvious notation,

$$\begin{aligned}
&a_\Omega \geq a > 0, \quad a_\Omega^i \geq a > 0, \quad k_\Omega \geq k > 0, \quad k_\Omega^i \geq k > 0, \\
&r_\Omega \leq r, \quad r_\Omega^i \leq r, \quad 0 < r < k. \qquad (2.3.47)
\end{aligned}$$

We denote the corresponding families of local charts by $\mathcal{F}_\Omega$ and $\mathcal{F}_\Omega^i$, $i = \overline{1, i_\Omega}$, respectively, and we postulate that the family of local charts

$$\mathcal{F} = \bigcup_{\Omega \in \tilde{\mathcal{O}}} \left[\mathcal{F}_\Omega \cup \left(\bigcup_{i=1}^{i_\Omega} \mathcal{F}_\Omega^i \right) \right] \qquad (2.3.48)$$

is equicontinuous and equibounded on $\overline{\tilde{B}(0,k)}$ (in the notation of Definition A3.1 and its comments).

In addition to these standard assumptions, we need to impose a geometric compatibility condition at the intersection points between Ω and D_Ω^i, $i = \overline{1, i_\Omega}$. We postulate:

Whenever $x \in \partial\Omega \cap \partial D_\Omega^{i_0}$ for some $i_0 \in \{1, \ldots, i_\Omega\}$, then there is some neighborhood V_x of x such that both $\partial \left(\Omega \cup (E \setminus \overline{D_\Omega^{i_0}}) \right) \cap V_x$ and $\partial \left(\Omega \cup D_\Omega^{i_0} \right) \cap V_x$ can be represented in the same local system of axes by continuous functions that "extend" the representation of $\partial D_\Omega^{i_0}$ around x from $\partial D_\Omega^{i_0} \setminus \Omega$ to the above-mentioned sets. More precisely, we have

$$\begin{aligned}
\partial(\Omega \cup D_\Omega^{i_0}) \cap V_x &= \{(s, \tilde{g}_x(s)) : s \in B(0, k_\Omega^{i_0})\}, \qquad (2.3.49) \\
\partial \left(\Omega \cup (E \setminus \overline{D_\Omega^{i_0}}) \right) \cap V_x &= \{(s, \hat{g}_x(s)) : s \in B(0, k_\Omega^{i_0})\}, \qquad (2.3.50)
\end{aligned}$$

as well as $\hat{g}_x(s) \leq \tilde{g}_x(s)$ for all $s \in B(0, k_\Omega^{i_0})$. These two mappings are equal on the subset of $B(0, k_\Omega^{i_0})$ that corresponds to

$$\partial D_\Omega^{i_0} \cap \left[\partial(\Omega \cup D_\Omega^{i_0}) \cap V_x \right] = \partial D_\Omega^{i_0} \cap \left[\partial(\Omega \cup (E \setminus \overline{D_\Omega^{i_0}})) \cap V_x \right],$$

since there they coincide with some local chart $g_x \in \mathcal{F}_\Omega^{i_0}$ of $\partial D_\Omega^{i_0}$.

Finally, we assume that there exists some constant $c > 0$ (which is independent of Ω) such that we have the implication

$$z \in \partial\Omega \setminus \cup_{x\in\partial\Omega\cap\partial D_\Omega} \left[\left\{(s, \tilde{g}_x(s)) : s \in \overline{B(0, r_\Omega^{i_0})}\right\} \cup \left\{(s, \hat{g}_x(s)) : s \in \overline{B(0, r_\Omega^{i_0})}\right\}\right]$$
$$\Longrightarrow \quad d\left(z, D_\Omega^{i_0}\right) \geq c > 0, \quad \text{for } 1 \leq i_0 \leq i_\Omega. \tag{2.3.51}$$

The family $\tilde{\mathcal{O}}$ now consists of all the domains Ω that satisfy the above conditions. Note that the inside and outside segment properties are also valid in (2.3.49), (2.3.50) with segments at least of length $a_\Omega > 0$. Notice also that the conditions (2.3.49)–(2.3.51) imply, in particular, that $\Omega \cup D_\Omega$ and $\Omega \cup (E \setminus \overline{D}_\Omega)$ are domains of class C.

Remark. The hypotheses (2.3.49), (2.3.50) should be understood in the context that the auxiliary open sets $D_\Omega = \cup_{i=1}^{i_\Omega} D_\Omega^i$ may be chosen just in order to specify that part of $\partial\Omega$ where the Dirichlet condition for (2.3.42) is valid. This gives considerable flexibility in verifying (2.3.49), (2.3.50). Simple examples in $\mathbf{R}^2$ show that if the intersection of Ω with small balls centered at $x \in \partial\Omega \cap \partial D_\Omega$ is convex, then (2.3.49) and (2.3.50) are fulfilled.

Next, we examine the convergence properties with respect to the Hausdorff–Pompeiu distance.

Theorem 2.3.16 *If $\{\Omega_n\} \subset \tilde{\mathcal{O}}$ satisfies $\lim_{n\to\infty} \tilde{d}_H(\Omega_n, \Omega) = 0$ for some open set $\Omega \subset E$, then $\Omega \in \tilde{\mathcal{O}}$, and the corresponding characteristic functions satisfy $\chi_{\Omega_n} \to \chi_\Omega$ a.e. in E, on a subsequence again indexed by n.*

Proof. Obviously, $\{\Omega_n\} \subset \tilde{\mathcal{O}} \subset \mathcal{O}$. Hence, thanks to the compactness properties of $\mathcal{O}$, we have $\Omega \in \mathcal{O}$, and owing to Theorem A3.9 in Appendix 3, $\chi_{\Omega_n} \to \chi_\Omega$ a.e. in E for a suitable subsequence indexed by n. Moreover, by virtue of Proposition A3.2(i), there is a subsequence, again indexed by n, such that
$$\lim_{n\to\infty} \tilde{d}_H(D_{\Omega_n}, D_\Omega) = 0.$$
Invoking Theorem A3.9 again, we find that all the connected components D_Ω^i of D_Ω are domains of class C that satisfy (2.3.47). Notice that the total number i_Ω of connected components of D_Ω is finite; indeed, the uniform interior segment property implies a lower bound for the measure of each component. Therefore, we may select yet another subsequence indexed by n such that
$$\chi_{D_{\Omega_n}^i} \to \chi_{D_\Omega^i} \quad \text{a.e. in } E, \quad 1 \leq i \leq i_\Omega.$$

Thus, it remains to show the persistence of the compatibility conditions (2.3.49)–(2.3.51) under the passage to the limit. To this end, let $x \in \partial\Omega \cap \partial D_\Omega^{i_0}$, $1 \leq i_0 \leq i_\Omega$, be arbitrary. By the definition of the Hausdorff–Pompeiu convergence, there exist $x_n \in \partial\Omega_n$, $n \in \mathbf{N}$, such that $x_n \to x$.

Assume now that there is a subsequence, again indexed by n, such that x_n cannot be represented in terms of the "restricted local charts" (i.e., by those defined on balls of radius $r > 0$, compare the comments following Definition A3.1 in Appendix 3), for all $n \in \mathbf{N}$. Then, in view of (2.3.47) and (2.3.51), it follows that $d(x_n, D_{\Omega_n}) \geq c$ for all $n \in \mathbf{N}$. But this is impossible, since $x \in \partial D_\Omega$ can also be approximated by points in ∂D_{Ω_n}. In conclusion, there must be some $n(x) \in \mathbf{N}$ such that for $n \geq n(x)$ we have

$$x_n = (s_n, g_n(s_n)), \quad s_n \in \overline{B(0,r)}, \tag{2.3.52}$$

with $g_n : B(0,k) \to \mathbf{R}$ being some appropriate local compatibility chart around $\partial\Omega_n \cap \partial D_{\Omega_n}$, as defined in (2.3.49), (2.3.50). We consider both types of local charts arising in (2.3.49) and (2.3.50), denoting them by $\tilde{g}_n$ and $\hat{g}_n$, respectively. They are both defined on $B(0,k)$, and (2.3.52) may be valid for both $g_n = \tilde{g}_n$ and $g_n = \hat{g}_n$, or only for one of them.

By the equicontinuity and equiboundedness assumptions on the family $\mathcal{F}$ from (2.3.48), we may by the Arzelà–Ascoli theorem assume that $\tilde{g}_n \to \tilde{g}$ and $\hat{g}_n \to \hat{g}$ uniformly in $B(0,k)$, and thus also $\hat{g}(s) \leq \tilde{g}(s)$ for all $s \in B(0,k)$.

Now recall that under our general assumptions the sets $\Omega_n \cup D_{\Omega_n}$ are domains of class C, with uniformly continuous local charts and uniform constants given by inequality (2.3.47). Thanks to Theorem A3.9 in Appendix 3, the same is true for $\Omega \cup D_\Omega$. Thus, the above argument, and passage to the limit as $n \to \infty$ in (2.3.49), show that (2.3.49) is also valid for Ω.

Next, we consider (2.3.50). Applying Proposition A3.10 in Appendix 3, we see that the behavior of the complementary sets $E \setminus \overline{D}_\Omega$ is similar to that of D_Ω, so that we may argue as for (2.3.49).

Finally, the stability of (2.3.51) follows by a direct passage to the limit as $n \to \infty$. This ends the proof of the assertion. □

Remark. Suppose that the assumptions of Theorem 2.3.16 are satisfied, and that

$$\overline{E} \setminus \Omega_n \to \overline{E} \setminus \Omega, \quad \overline{E} \setminus D_{\Omega_n} \to \overline{E} \setminus D_\Omega,$$

in the Hausdorff–Pompeiu sense. If then $K \subset \Omega \cup D_\Omega$ is compact, we can conclude that there is some $n(K) \in \mathbf{N}$ such that $K \subset (\Omega_n \cup D_{\Omega_n})$ for $n \geq n(K)$. Indeed, this form of the Γ property follows from noticing that

$$\overline{E} \setminus (\Omega_n \cup D_{\Omega_n}) \to \overline{E} \setminus (\Omega \cup D_\Omega)$$

by the previous result, and then applying Proposition A3.8 in Appendix 3.

The next result provides an alternative characterization of the space $V(\Omega)$. It is an extension of Theorem 2.3.9, with which it should be compared. In the following proof we make use of techniques developed in the proof of Theorem 2.3.9, in particular of translations of the system of local axes along appropriate vectors and of the interior and exterior segment properties. Since we have elaborated on this there, we can afford only to sketch the argument here.

Theorem 2.3.17 *Let $\Omega \in \widetilde{\mathcal{O}}$ be given, and let $\tilde{v}$ denote the extension by zero of a function $v \in V(\Omega)$ to the set $(\Omega \setminus \overline{C}) \cup D_\Omega$. Then $\tilde{v} \in H^1((\Omega \setminus \overline{C}) \cup D_\Omega)$. Conversely, if $w \in H^1((\Omega \setminus \overline{C}) \cup D_\Omega)$ is such that $w = 0$ a.e. in $D_\Omega \setminus \Omega$, then $w|_{\Omega \setminus \overline{C}} \in V(\Omega)$.*

Proof. Let $v \in V(\Omega)$ be given. Then, by definition, there are $v_n \in C_0^\infty(\mathbf{R}^d)$, open sets Q_{v_n}, and compact sets K_{v_n} such that $\overline{D}_\Omega \subset Q_{v_n}$, $K_{w_n} \subset (\Omega \setminus \overline{C})$, $v_n = 0$ on $Q_{v_n} \setminus K_{v_n}$, for all $n \in \mathbf{N}$, and $v_n|_{\Omega \setminus \overline{C}} \to v$ in $H^1(\Omega \setminus \overline{C})$.

In particular, $v_n = 0$ in $D_\Omega \setminus K_{v_n}$ and in $D_\Omega \setminus \Omega$. Thus, $\left\{v_n|_{D_\Omega \cup (\Omega \setminus \overline{C})}\right\}$ is a Cauchy sequence in $H^1((\Omega \setminus \overline{C}) \cup D_\Omega)$. Let $\hat{v} \in H^1((\Omega \setminus \overline{C}) \cup D_\Omega)$ be its limit. Then $v = \hat{v}|_{\Omega \setminus \overline{C}}$ and $\hat{v}|_{D_\Omega \setminus \Omega} = 0$; hence $\hat{v} = \tilde{v}$.

Conversely, let $w \in H^1((\Omega \setminus \overline{C}) \cup D_\Omega)$ satisfy $w = 0$ a.e. in $D_\Omega \setminus \Omega$. We consider an open cover of $\partial\Omega$ given by a finite number of neighborhoods V_j, $1 \leq j \leq l$, that contains a covering of $\partial\Omega \cap \partial D_\Omega$ as indicated in (2.3.49), (2.3.50), by a finite number of open sets V_x. We choose another open set V_0 such that $\overline{V}_0 \subset \Omega$ and

$$\Omega \subset \bigcup_{j=0}^{l} V_j. \tag{2.3.53}$$

To this covering, we associate a partition of unity $\{\psi_j\}_{j=1}^{l} \subset C_0^\infty(\mathbf{R}^d)$ such that $0 \leq \psi_j \leq 1$ and $\operatorname{supp} \psi_j \subset V_j$, $1 \leq j \leq l$, and

$$\sum_{j=0}^{l} \psi_j(x) = 1 \quad \forall x \in \Omega. \tag{2.3.54}$$

Let

$$w_j(x) = \psi_j(x) w(x), \quad x \in \Omega, \quad 1 \leq j \leq l.$$

Obviously, $w_0 \in V(\Omega)$. We aim to show that also $w_j \in V(\Omega)$, $1 \leq j \leq l$. Once this is proved, the assertion follows from the observation that $w(x) = \sum_{j=0}^{l} w_j(x)$ for all $x \in \Omega$.

We have to distinguish among three situations.

Case 1: Assume that $d(\overline{V}_j, \overline{D}_\Omega) > 0$. Let $\tilde{w}_j$ denote the extension of w_j by zero to $\mathbf{R}^d$. Then $\tilde{w}_j \in H^1(\mathbf{R}^d \setminus (V_j \cap \partial\Omega))$. Similarly as in the proof of Theorem 2.3.9, we shift the system of local axes corresponding to the local chart V_j in the direction of the outside segment to $\partial\Omega$. This is possible since Ω is of class C.

Let $\tilde{w}_j^t \in H^1(\mathbf{R}^d \setminus (V_j \cap \Omega)_t)$ denote the shifted mapping, where $(V_j \cap \partial\Omega)_t$ is a translation of $V_j \cap \partial\Omega$ in the same direction and $t > 0$ denotes the length of the translation vector. For sufficiently small values $t > 0$ we then have $d((V_j \cap \partial\Omega)_t, \overline{D}_\Omega) > 0$ and $d((V_j \cap \partial\Omega)_t, \overline{\Omega}) > 0$.

Then $\tilde{w}_j^t|_\Omega \in H^1(\Omega)$, and $\lim_{t \to 0} \tilde{w}_j^t|_\Omega = \tilde{w}_j|_\Omega = w_j$ in $H^1(\Omega \setminus \overline{C})$ by the continuity in the mean of the integral.

Now let $\tilde{w}_j^{t,\varepsilon}$ denote the convolution of $\tilde{w}_j^t$ with ρ_ε, where ρ_ε is the standard Friedrichs mollifier of order $\varepsilon > 0$. Then $\tilde{w}_j^{t,\varepsilon} \in C_0^\infty(\mathbf{R}^d)$, and owing to the above-mentioned positivity of distances, the above convergences are preserved since $\tilde{w}_j^{t,\varepsilon} \to \tilde{w}_j^t$ strongly in $H^1(\Omega \setminus \overline{C})$ for $\varepsilon \searrow 0$. Moreover, $\tilde{w}_j^{t,\varepsilon} \in V(\Omega)$, for sufficiently small $\varepsilon > 0$, by construction. This proves that $w_j \in V(\Omega)$ in this case.

Case 2: Suppose that V_j is of type V_x, $x \in \partial\Omega \cap \partial D_\Omega$. This time, we perform a translation in the "vertical" segment direction of the local chart provided by (2.3.49), (2.3.50). The obtained shifted mapping is again denoted by $\tilde{w}_j^t \in H^1(\mathbf{R}^d \setminus (\partial\Omega \cap V_j)_t)$. It follows that its support has a positive distance from $\overline{D}_\Omega \setminus \Omega$ for sufficiently small $t > 0$. Then $\tilde{w}_j^t|_\Omega \in H^1(\Omega)$, and $\tilde{w}_j^{t,\varepsilon}$, its convolution with ρ_ε, will inherit these properties for sufficiently small $\varepsilon > 0$. The argument then follows as above.

Case 3: In the final case, we may assume that $\overline{V}_j \subset D_\Omega$ and, by the assumption of the theorem, that $\operatorname{supp} w_j \subset \overline{D_\Omega \cap \Omega}$. The extension by zero of w_j to $\mathbf{R}^d$, again denoted by $\tilde{w}_j$, satisfies $\tilde{w}_j \in H^1(\mathbf{R}^d)$. We perform a translation in the direction of the interior segment to $\Omega \cup (E \setminus \overline{D}_j)$, provided by (2.3.50). Then a smoothing using a Friedrichs mollifier can be performed such that the support of the resulting mapping $\tilde{w}_j^{t,\varepsilon} \in C_0^\infty(\mathbf{R}^d) \cap V(\Omega)$ remains at a positive distance from $\overline{D}_\Omega \setminus \Omega$. Again, passage to the limit as $\varepsilon \searrow 0$ yields that $w_j \in V(\Omega)$. This concludes the proof of the assertion. □

Theorem 2.3.18 *Let $\{\Omega_n\} \subset \tilde{\mathcal{O}}$ satisfy $\lim_{n\to\infty} \tilde{d}_H(\Omega_n, \Omega) = 0$ and $\chi_{\Omega_n} \to \chi_\Omega$ a.e. Let $y_n \in \Lambda(\Omega_n)$, $n \in \mathbf{N}$, and $y \in \Lambda(\Omega)$ denote the solutions of (2.3.42) corresponding to Ω_n, $n \in \mathbf{N}$, and Ω, respectively. Then for any open set K with $\overline{K} \subset \Omega \setminus \overline{C}$ there is some $n(K) \in \mathbf{N}$ such that $\overline{K} \subset (\Omega_n \setminus \overline{C})$ for $n \geq n(K)$. Moreover, $y_n|_K \to y|_K$ weakly in $H^1(K)$.*

Proof. Let K be a fixed open set satisfying $\overline{K} \subset (\Omega \setminus \overline{C})$. Then the first assertion is a consequence of the Γ property; see Proposition A3.8 in Appendix 3. From now on, let $n \geq n(K)$. We rewrite (2.3.42) in the form

$$\int_{K\setminus C} [\nabla y_n \cdot (\nabla y_n - \nabla v) + y_n (y_n - v)]\, dx - \int_{K\setminus C} (y_n - v)\, f\, dx$$
$$\leq - \int_{(\Omega_n\setminus C)\setminus K} [\nabla y_n \cdot (\nabla y_n - \nabla v) + y_n(y_n - v)]\, dx + \int_{(\Omega_n\setminus C)\setminus K} (y_n - v) f\, dx,$$
$$\text{for all } v \in \Lambda(\Omega_n). \tag{2.3.55}$$

In the following, we will work with test functions belonging to $C^1(\overline{E}) \cap \Lambda(\Omega)$ that vanish in a small neighborhood of $\overline{D}_\Omega \setminus \Omega$. Their restrictions to $\Omega_n \setminus \overline{C}$ belong to $\Lambda(\Omega_n)$ for sufficiently large n; indeed, we can infer from the convergence properties of Ω_n and D_{Ω_n} that they vanish in $D_{\Omega_n} \setminus \Omega_n$ for sufficiently large n, and we can apply the previous theorem.

Now notice that the sequence $\{|y_n|_{H^1(\Omega_n \setminus \overline{C})}\}$ is bounded, thanks to the coercivity of the elliptic operator. Hence, for any open set G with $\overline{G} \subset (\Omega \setminus \overline{C})$ there is a subseqence of $\{y_n\}$ depending on G that converges weakly in $H^1(G)$.

Now choose an increasing sequence $\{G_k\}_{k\in\mathbf{N}}$ of open and bounded sets such that $\overline{G}_k \subset (\Omega \setminus \overline{C})$, $k \in \mathbf{N}$, and $\cup_{k=1}^{\infty} \overline{G}_k = \Omega \setminus \overline{C}$. Using the same procedure as in the proof of Theorem 2.3.12, we can construct a mapping $\tilde{y} \in H^1(\Omega \setminus \overline{C})$ having the property that for any $k \in \mathbf{N}$ there is a subsequence $\{y_{k_n}\}$ of $\{y_n\}$ such that

$$y_{k_n}|_{G_k} \to \tilde{y}|_{G_k} \quad \text{as } n \to \infty, \quad \text{weakly in } H^1(G_k). \tag{2.3.56}$$

We aim to show that $\tilde{y} = y$. To this end, choose $k \in \mathbf{N}$ so large that $\overline{K} \subset G_k$. By (2.3.56), there is a subsequence, again indexed by n, such that $y_n|_K \to \tilde{y}|_K$ weakly in $H^1(K)$. Employing the weak lower semicontinuity of quadratic forms and (2.3.56), we infer that

$$\begin{aligned} \liminf_{n\to\infty} \int_{K\setminus C} &[\nabla y_n \cdot (\nabla y_n - \nabla v) + y_n\,(y_n - v)]\,dx \\ &\geq \int_{K\setminus C} [\nabla \tilde{y} \cdot (\nabla \tilde{y} - \nabla v) + \tilde{y}\,(\tilde{y} - v)]\,dx, \end{aligned} \tag{2.3.57}$$

$$\lim_{n\to\infty} \int_{K\setminus C} (y_n - v) f\,dx = \int_{K\setminus C} (\tilde{y} - v) f\,dx. \tag{2.3.58}$$

For $v \in C^1(E) \cap \Lambda(\Omega)$ the right-hand side of (2.3.55) can be estimated as follows:

$$\begin{aligned} -\int_{(\Omega_n\setminus C)\setminus K} &[\nabla y_n \cdot (\nabla y_n - \nabla v) + y_n\,(y_n - v)]\,dx + \int_{(\Omega_n\setminus C)\setminus K} (y_n - v) f\,dx \\ &\leq \int_{(\Omega_n\setminus C)\setminus K} (\nabla y_n \cdot \nabla v + y_n\, v)\,dx + \int_{(\Omega_n\setminus C)\setminus K} (y_n - v) f\,dx, \end{aligned}$$

since the quadratic terms are nonnegative. Hölder's inequality yields

$$\begin{aligned} \int_{(\Omega_n\setminus C)\setminus K} &(\nabla y_n \cdot \nabla v + y_n\, v + y_n\, f - v\, f)\,dx \\ &\leq |v|_{C^1(\overline{E})} \int_{(\Omega_n\setminus C)\setminus K} (|\nabla y_n| + |y_n| + |f|)\,dx \\ &\leq M\,|v|_{C^1(\overline{E})}\,(\text{meas}[(\Omega_n \setminus C) \setminus K])^{\frac{1}{2}}, \end{aligned} \tag{2.3.59}$$

where $M > 0$ is independent of both K and n. Thanks to the pointwise a.e. convergence of the characteristic functions, we also have

$$\lim_{n\to\infty} \text{meas}[(\Omega_n \setminus C) \setminus K] = \text{meas}[(\Omega \setminus C) \setminus K]. \tag{2.3.60}$$

Thus, combining (2.3.57)–(2.3.60), we can pass to the limit as $n \to \infty$ in (2.3.55) to arrive at the inequality

$$\int_{K\setminus C} [\nabla \tilde{y} \cdot (\nabla \tilde{y} - \nabla v) + \tilde{y}\,(\tilde{y} - v)]\, dx - \int_{K\setminus C} (\tilde{y} - v) f\, dx$$
$$\leq M\, |v|_{C^1(\overline{E})}\, (\mathrm{meas}[(\Omega \setminus C) \setminus K])^{\frac{1}{2}}, \qquad (2.3.61)$$

which is valid for any open set K satisfying $\overline{K} \subset (\Omega \setminus \overline{C})$. Choosing $K = G_k$, $k \in \mathbf{N}$, and letting $k \to \infty$, we conclude that for any $v \in C^1(\overline{E}) \cap \Lambda(\Omega)$ vanishing in a small neighborhood of $\partial\Omega \cap D_\Omega$,

$$\int_{\Omega\setminus C} [\nabla \tilde{y} \cdot (\nabla \tilde{y} - \nabla v) + \tilde{y}\,(\tilde{y} - v)]\, dx \leq \int_{\Omega\setminus C} f\,(\tilde{y} - v)\, dx. \qquad (2.3.62)$$

Since C is a fixed Lipschitz domain, the trace theorem and a passage to the limit yield that $\tilde{y}|_{\partial C} = 1$. Since y_n is nonnegative for every $n \in \mathbf{N}$, we may conclude that $\tilde{y} \in H^1(\Omega \setminus C)_+$.

Finally, according to Theorem 2.3.17, the functions $y_n \in V(\Omega_n)$ may be extended by zero to mappings $\hat{y}_n \in (\Omega_n \setminus \overline{C}) \cup D_{\Omega_n}$. Moreover, since

$$\tilde{d}_H((\Omega_n \setminus \overline{C}) \cup D_\Omega, (\Omega \setminus \overline{C}) \cup D_\Omega) \to 0 \quad \text{as } n \to \infty,$$

we may again construct a mapping $\hat{y} \in H^1((\Omega \setminus \overline{C}) \cup D_\Omega)$ such that for any open set K with $\overline{K} \subset (\Omega \setminus \overline{C}) \cup D_\Omega$, we have $\hat{y}_n|_K \to \hat{y}|_K$ on a subsequence depending on K. Obviously, $\hat{y}|_{D_\Omega \setminus \Omega} = 0$ and $\hat{y}|_{\Omega \setminus \overline{C}} = \tilde{y}$. But then the second assertion in Theorem 2.3.17 yields that also $\tilde{y} \in V(\Omega)$, i.e., $\tilde{y} \in \Lambda(\Omega)$.

Now observe that by the definition of $V(\Omega)$, the set $C^1(\overline{E}) \cap \Lambda(\Omega)$ is dense in $\Lambda(\Omega)$: indeed, taking the positive part of a function is a continuous operation in H^1, and smoothing preserves nonnegativity. Hence, we may pass to the limit with respect to v in (2.3.62) to conclude that $\tilde{y}$ solves (2.3.42) and therefore coincides with y.

To conclude the proof, let K be any open set such that $\overline{K} \subset (\Omega \setminus \overline{C})$. Then, as shown above, there is a subsequence depending on K such that $y_n|_K \to y|_K$ weakly in $H^1(K)$. But since the limit y is uniquely determined, the convergence is valid for the entire sequence, which finishes the proof of the assertion. □

Remark. By virtue of Theorem 2.3.18, variational inequalities of obstacle type (and their associated free boundary problems; see Examples 1.2.5 and 1.2.6 in Chapter 1) are well-posed with respect to domain perturbations, even in the case of mixed boundary conditions.

Corollary 2.3.19 *The shape optimization problem (2.3.45), (2.3.46), (2.3.42) has at least one minimizer* $\Omega^* \in \widetilde{\mathcal{O}}$.

Remark. It ought to be clear that the above result remains valid for more general cost functionals.

2.3.4 Partial Extensions

In this paragraph, we demonstrate that very general families of unknown sets and of cost functionals may be considered in shape optimization problems governed by certain differential systems.

We first discuss the elastoplastic torsion problem considered in Example 1.2.5 in Chapter 1. To this end, let $E \subset \mathbf{R}^d$ be a fixed open and bounded set, and let $f \in L^2(E)$ be given. We put $\mathcal{O} = \{\Omega \subset E : \Omega \text{ is a domain}\}$. For any $\Omega \in \mathcal{O}$ we consider the variational inequality

$$\int_\Omega \nabla y(x) \cdot (\nabla y(x) - \nabla v(x))\, dx \le \int_\Omega f(x)(y(x) - v(x))\, dx \quad \forall\, v \in C_\Omega, \tag{2.3.63}$$

$$C_\Omega = \Big\{ y \in H_0^1(\Omega) : |\nabla y(x)| \le 1 \text{ a.e. in } \Omega \Big\}. \tag{2.3.64}$$

By the Lions–Stampacchia theorem (see Theorem A2.3 in Appendix 2), the problem (2.3.63), (2.3.64) has for every admissible domain $\Omega \in \mathcal{O}$ a unique solution $y = y_\Omega$. Thanks to (2.3.64), $y_\Omega \in C_\Omega$ can be extended by zero to a function $\tilde{y}_\Omega \in W_0^{1,\infty}(E)$.

Now assume that $J : W_0^{1,\infty}(E) \to \mathbf{R}$ is a weakly lower semicontinuous functional. We then consider the optimal design problem

$$\operatorname*{Min}_{\Omega \in \mathcal{O}} J(\tilde{y}_\Omega). \tag{2.3.65}$$

Theorem 2.3.20 *The optimal design problem (2.3.63)–(2.3.65) has at least one optimal domain $\Omega^* \in \mathcal{O}$.*

Proof. Let $\{\Omega_n\} \subset \mathcal{O}$ be a minimizing sequence. In view of Proposition A3.2 in Appendix 3, we may assume that there is some $\Omega^* \in \mathcal{O}$ with

$$\lim_{n\to\infty} \tilde{d}_H(\Omega_n, \Omega^*) = 0.$$

Using the abbreviations $y_n = y_{\Omega_n}$, $\tilde{y}_n = \tilde{y}_{\Omega_n}$, inserting $v \equiv 0$ in (2.3.63), and recalling (2.3.64), we can infer that $\{\tilde{y}_n\}$ is bounded in $W_0^{1,\infty}(E)$. Hence, there is some $\tilde{y} \in W_0^{1,\infty}(E)$ such that $\tilde{y}_n \to \tilde{y}$ weakly* in $W_0^{1,\infty}(E)$, on a subsequence again indexed by n. Now observe that

$$C_E = \Big\{ y \in H_0^1(E) : |\nabla y(x)| \le 1 \text{ a.e. in } E \Big\}$$

is convex and, by Theorem A1.6 in Appendix 1, weakly closed in $W^{1,\infty}(E)$. Consequently, $\tilde{y} \in C_E$. Moreover, the Arzelà–Ascoli theorem implies that $\tilde{y}_n \to \tilde{y}$ uniformly in $\overline{E}$.

Now let $x \in \overline{E} \setminus \Omega^*$. Since $\tilde{d}_H(\Omega_n, \Omega^*) \to 0$, there exist $x_n \in \overline{E} \setminus \Omega_n$, $n \in \mathbf{N}$, such that $x_n \to x$. We have

$$\lim_{n\to\infty} |\tilde{y}(x) - \tilde{y}_n(x_n)| \le \lim_{n\to\infty} \big(|\tilde{y}(x) - \tilde{y}(x_n)| + |\tilde{y}(x_n) - \tilde{y}_n(x_n)|\big) = 0. \tag{2.3.66}$$

Consequently, since $\tilde{y}_n|_{\overline{E}\setminus\Omega_n} \equiv 0$, we have $\tilde{y}|_{\overline{E}\setminus\Omega^*} \equiv 0$.

Next, we define the sets

$$\Omega^k = \Omega^* \setminus \left\{ x \in \Omega^* : d(x, \partial\Omega^*) \le \frac{1}{k} \right\}, \quad k \in \mathbf{N},$$

which are nonempty and open (but not necessarily connected) provided that $k \in \mathbf{N}$ is sufficiently large (which we assume). Let

$$\tilde{y}_+^k = \left(\tilde{y}_+ - \frac{1}{k} \right)_+ ,$$

with $(\cdot)_+$ denoting the positive part. Recalling that $\tilde{y}$ is by construction a Lipschitz continuous function of rank 1, we can easily show that $\operatorname{supp} \tilde{y}_+^k \subset \overline{\Omega}^k$, so that $\tilde{y}_+^k \in W_0^{1,\infty}(\Omega^*)$. We also have $|\nabla \tilde{y}_+^k(x)| \le |\nabla \tilde{y}(x)|$ a.e. in Ω^*, by the rules for computing the gradient of the maximum of two functions. The sequence $\{\tilde{y}_+^k\}$ is thus bounded in $W_0^{1,\infty}(\Omega^*)$. With obvious notation, we can derive the same conclusions for $\tilde{y}_-^k = (\tilde{y}_- - \frac{1}{k})_+$ and for $\tilde{y}^k = \tilde{y}_+^k - \tilde{y}_-^k$, where the inequality

$$|\nabla \tilde{y}^k(x)| \le |\nabla \tilde{y}(x)| \quad \text{a.e. in } \Omega^* \tag{2.3.67}$$

follows from the fact that the supports of $\tilde{y}_+^k$ and $\tilde{y}_-^k$ are disjoint. It is obvious that

$$\tilde{y}^k \to \tilde{y}, \quad \text{pointwise in } \Omega^*, \text{ and weakly* in } W_0^{1,\infty}(\Omega^*).$$

This proves that $\tilde{y} \in W_0^{1,\infty}(\Omega^*)$, and, a fortiori, that $\tilde{y} \in C_{\Omega^*}$.

Now let $v \in C_{\Omega^*}$ be arbitrarily fixed. We construct the functions v_+^k, v_-^k, and $v^k = v_+^k - v_-^k$ as above, obtaining the same boundedness and convergence properties. Since $\operatorname{supp} v^k \subset \overline{\Omega}^k \subset \Omega^*$, the Γ property (cf. Proposition A3.8 in Appendix 3) shows that there is some $n(k) \in \mathbf{N}$ such that $\operatorname{supp} v^k \subset \Omega_n$ for $n \ge n(k)$. Thus, invoking (2.3.67), $v^k \in C_{\Omega_n}$ for $n \ge n(k)$, and we can use v^k as test functions in (2.3.63). It follows that

$$\int_{\Omega_n} \nabla y_n \cdot (\nabla y_n - \nabla v^k)\, dx \le \int_{\Omega_n} f\,(y_n - v^k)\, dx, \quad n \ge n(k), \tag{2.3.68}$$

or, rewriting (2.3.68) in terms of the associated extensions by zero to E,

$$\int_E \nabla \tilde{y}_n \cdot (\nabla \tilde{y}_n - \nabla \tilde{v}^k)\, dx \le \int_E f\,(\tilde{y}_n - \tilde{v}^k)\, dx, \quad n \ge n(k).$$

Passage to the limit as $n \to \infty$, using the weak lower semicontinuity of quadratic forms, then yields that

$$\int_E \nabla \tilde{y} \cdot (\nabla \tilde{y} - \nabla \tilde{v}^k)\, dx \le \int_E f\,(\tilde{y} - \tilde{v}^k)\, dx. \tag{2.3.69}$$

Now let $k \to \infty$ in (2.3.69). Together with the previously established properties, we conclude that $\tilde{y}|_{\Omega^*} = y_{\Omega^*}$. Moreover, since J is weakly lower semicontinuous, we see that

$$J(\tilde{y}_{\Omega^*}) \le \liminf_{n\to\infty} J(\tilde{y}_n) = \operatorname*{Min}_{\Omega\in\mathcal{O}} J(\tilde{y}_\Omega).$$

That is, $\Omega^* \in \mathcal{O}$ is an optimal domain for the problem (2.3.63)–(2.3.65). □

Remark. The continuity property with respect to the domain on which the previous proof relies is an exemplification of Theorem A3.14 in Appendix 3, although in a slightly different setting.

The next result refers to the classical transmission conditions problem; compare Example 1.2.7 in Chapter 1. The physical background of this example renders it interesting to admit just measurable sets as admissible candidates for the regions giving the optimal shape in the material distribution problem.

Let $E \subset \mathbf{R}^d$ describe a given body made up by two different materials occupying the measurable sets $\Omega \subset E$ and $E \setminus \Omega$. The physical properties of the two regions are assumed different, which is reflected by taking different coefficients in the governing elliptic equations in these two regions. If Ω is Lipschitz, we may formally cast this into the form

$$-a_1 \Delta y_1 + b_1 y_1 = f \quad \text{in } \Omega, \tag{2.3.70}$$

$$-a_2 \Delta y_2 + b_2 y_2 = f \quad \text{in } E \setminus \Omega, \tag{2.3.71}$$

$$a_1 \frac{\partial y_1}{\partial n} = a_2 \frac{\partial y_2}{\partial n}, \quad y_1 = y_2, \quad \text{in } \partial\Omega \setminus (\partial\Omega \cap \partial E), \tag{2.3.72}$$

$$a_i \frac{\partial y_i}{\partial n} = 0 \quad \text{in } \Gamma_1, \quad y_i = 0 \quad \text{in } \Gamma_2, \quad i = 1, 2, \tag{2.3.73}$$

where Γ_i, $i = 1, 2$, with $\Gamma_1 \cup \Gamma_2 = \partial E$ and $\Gamma_1 \cap \Gamma_2 = \emptyset$, are assumed to be nonempty Lipschitz boundary pieces, a_i, b_i, $i = 1, 2$, are positive constants, and $f \in L^2(E)$ is given. Then, with

$$y_\Omega(x) = \begin{cases} y_1(x) & \text{in } \Omega, \\ y_2(x) & \text{in } E \setminus \Omega, \end{cases} \tag{2.3.74}$$

the weak formulation of (2.3.70)–(2.3.73) takes the form

$$\int_E \{ [a_1 \chi_\Omega + a_2(1 - \chi_\Omega)] \nabla y_\Omega \cdot \nabla v + [b_1 \chi_\Omega + b_2(1 - \chi_\Omega)] y_\Omega \, w \} (x) \, dx$$

$$= \int_E f(x) \, v(x) \, dx \quad \forall v \in V, \tag{2.3.75}$$

$$V = \left\{ y \in H^1(E) : y = 0 \text{ in } \Gamma_2 \right\}. \tag{2.3.76}$$

Here, χ_Ω is the characteristic function of Ω in E. For any measurable set $\Omega \subset E$ the bilinear form governing (2.3.75) is bounded and coercive in V, so that (2.3.75), (2.3.76) admits a unique weak solution $y_\Omega \in V$ satisfying (2.3.70)–(2.3.73) formally.

In the case of material distribution (layout) problems (cf. Example 1.2.7 in Chapter 1), it makes sense and is of interest to consider the sets Ω and $E\setminus\Omega$ occupied by the respective material as merely measurable. In this respect, the present discussion continues the investigation of the problem (R) in §2.3.1 above.

We denote by $\mathcal{O}$ the family of all the admissible choices of sets $\Omega \subset E$ (to be specified below), and assume that a fixed measurable set $D \subset E$ and a desired target function $z_d \in L^2(D)$ are given. We then consider the following model of structural optimization problems:

$$\operatorname*{Min}_{\Omega\in\mathcal{O}} \left\{ \int_D |y_\Omega(x) - z_d(x)|^2\, dx \right\}, \tag{2.3.77}$$

with y_Ω defined by (2.3.75), (2.3.76).

In Pironneau [1984], the problem (2.3.77) is studied by interpreting the characteristic functions χ_Ω directly as a control into coefficients parameter, i.e., by imposing the (nonconvex) constraint that the control should attain only the values 0 and 1 in the whole domain E.

Next, we specify the family $\mathcal{O}$ of admissible sets $\Omega \subset E$. To this end, let

$$H(x) = \begin{cases} \{0\}, & \text{for } x < 0, \\ [0,1], & \text{for } x = 0, \\ \{1\}, & \text{for } x > 0, \end{cases}$$

denote the (maximal monotone) extension of the Heaviside graph. Justified by (2.3.81) below, we then postulate that the characteristic mapping χ_Ω is of the form

$$\chi_\Omega = H(p_\Omega) \tag{2.3.78}$$

for $p_\Omega \in U_{ad} \subset H^1_{\mathrm{loc}}(E)$, where the admissible controls $p \in U_{ad}$ are defined by the conditions

$$|p(x)| + |\nabla p(x)|_{\mathbf{R}^d} \geq \nu > 0, \quad \text{a.e. in } E, \quad \text{for some } \nu > 0, \tag{2.3.79}$$

$$|p|_{H^{1+\theta_K}(K)} \leq M_K\,, \quad \text{for all compact sets } K \subset E, \quad \text{with constants } M_K > 0 \text{ and } \theta_K > 0 \text{ depending on } K. \tag{2.3.80}$$

If $\operatorname{meas}(\partial\Omega) = 0$, the signed distance functions d_Ω introduced in Appendix 3 satisfy (2.3.79), according to Clarke [1983, p. 66]. The condition (2.3.80) is a weak local regularity assumption on $\partial\Omega$, which will play the role of a compactness condition, replacing the uniform continuity hypotheses from the previous paragraphs. Under stronger regularity assumptions, inequality (2.3.79) ensures the applicability of the implicit function theorem, providing a local description of the "boundary set" $\{x \in E : p(x) = 0\}$. However, the requirements (2.3.79), (2.3.80) are still very weak; for instance, the assumptions of the Clarke [1983, p. 255] implicit function theorem for Lipschitz mappings are not fulfilled.

In arbitrary dimension, the sets $\Omega = \Omega_p \subset E$ resulting from (2.3.78) for $p \in U_{ad}$ are just measurable and defined only up to a set of measure zero. According to Brézis [1983, p. 195], by (2.3.79), and since $p \in H^1_{\text{loc}}(E)$, we have

$$\text{meas}(\{x \in E : p(x) = 0\}) = 0. \tag{2.3.81}$$

This shows that $H(p)$ is indeed a characteristic function and justifies (2.3.78). We thus may rewrite (2.3.75)–(2.3.77) in the form

$$\operatorname*{Min}_{p \in U_{ad}} \left\{ \int_D |y_p(x) - z_d(x)|^2 \, dx \right\}, \tag{2.3.82}$$

$$\begin{aligned} &\int_E \{[a_1 H(p) + a_2(1 - H(p))] \, \nabla y_p \cdot \nabla v \, + \, [b_1 H(p) + b_2(1 - H(p))] \, y_p \, v\} \, dx \\ &\quad = \int_E f(x) \, v(x) \, dx, \quad \forall \, v \in V. \end{aligned} \tag{2.3.83}$$

Theorem 2.3.21 *Suppose that the functions $p \in U_{ad}$ satisfy the conditions (2.3.79) and (2.3.80). Then the problem (2.3.82), (2.3.83) has at least one optimal pair $[y^*, p^*] \in V \times U_{ad}$.*

Proof. Obviously, $U_{ad} \neq \emptyset$, since the constant functions belong to U_{ad}. Let $[y_n, p_n] \in V \times U_{ad}$ be a minimizing sequence, with y_n being the solution to (2.3.83) corresponding to p_n. We take an increasing sequence $\{G_k\}_{k \in \mathbf{N}}$ of open sets satisfying $\overline{G}_k \subset E$, $k \in \mathbf{N}$, and $\cup_{k=1}^{\infty} \overline{G}_k = E$. In view of (2.3.80), the sequence $\{p_n|_{G_k}\}_{n \in \mathbf{N}}$ forms a compact subset of $H^1(G_k)$, for any $k \in \mathbf{N}$.

Therefore, selecting subsequences iteratively, we can conclude that there is some $p \in H^1_{\text{loc}}(E)$ such that for any $k \in \mathbf{N}$ there is a subsequence $\{p_{k_n}\}_{n \in \mathbf{N}}$ satisfying $p_{k_n} \to p$ strongly in $H^1(G_k)$. In particular, we may assume that

$$p_{k_n} \to p \quad \text{and } \nabla p_{k_n} \to \nabla p, \qquad \text{a.e. in } G_k.$$

Consequently, p satisfies (2.3.79) and (2.3.80) with the same constants; that is, we have $p \in U_{ad}$. Now observe that

$$a_1 H(p_n) + a_2(1 - H(p_n)) \geq \min(a_1, a_2) > 0,$$

$$b_1 H(p_n) + b_2(1 - H(p_n)) \geq \min(b_1, b_2) > 0,$$

almost everywhere in E. Thus, inserting $v = y_n$ in (2.3.83) for any $n \in \mathbf{N}$, we readily see that $\{y_n\}$ is bounded in $H^1(E)$, and, extracting a suitable subsequence indexed by n, we may assume that $y_n \to y$ weakly in $H^1(E)$.

Moreover, since $\{H(p_n)\}$ is bounded in $L^\infty(E)$, there is some $w \in L^\infty(E)$ such that $H(p_n) \to w$ weakly* in $L^\infty(E)$, for another subsequence indexed by n. Invoking the demiclosedness of maximal monotone operators (see Proposition A1.4 in Appendix 1), we can infer that $w \in H(p)$. Since $p \in U_{ad}$, and owing to (2.3.81), $H(p)$ is a characteristic function in E, and we may write $w = H(p)$.

We also have that $H(p_n) \to H(p)$ a.e. in E, since we know by (2.3.79) that $p_n(x) \to p(x) \neq 0$ a.e. in E. If $p(x) > 0$, then $p_n(x) > 0$ for $n \geq n(x)$, and thus $H(p_n(x)) = H(p(x)) = 1$. By the same token, if $p(x) < 0$, then $H(p_n(x)) = H(p(x)) = 0$ for $n \geq n(x)$. Therefore, invoking Lebesgue's theorem of dominated convergence, we can conclude that $H(p_n) \to H(p)$ strongly in $L^s(E)$ for all $s \geq 1$.

Since $[y_n, p_n]$ satisfies (2.3.83) for $n \in \mathbf{N}$, the above convergence properties allow one to pass to the limit as $n \to \infty$ to arrive at the conclusion that $[y_p, p]$, with $y_p = y$, also satisfies (2.3.83); that is, $[y, p] \in V \times U_{ad}$ is an admissible pair for the problem (2.3.82), (2.3.83). Finally, observe that the weak convergence of $\{y_n\}$ to y in $H^1(E)$ implies that

$$\lim_{n\to\infty} \int_D |y_n(x) - z_d(x)|^2\, dx = \int_D |y(x) - z_d(x)|^2\, dx,$$

which shows that $[y, p]$ is an optimal pair. □

Remark. The above argument applies to any weakly lower semicontinuous functional on $H^1(E)$. For instance, functionals defined on manifolds of codimension one are allowed in this setting. One example of this type is given by

$$\int_\Gamma |y(\sigma) - z_0(\sigma)|^2\, d\sigma,$$

with some (relatively) open part Γ of ∂E and some $z_0 \in L^2(\Gamma)$. Other boundary conditions may also be imposed on ∂E. It is also possible to postulate (2.3.80) only for an open subset K of $E \setminus C$, where C is a closed set of zero measure. This permits cracks in the corresponding boundaries $\partial\Omega$ (cf. Bucur and Zolésio [1994]).

Remark. Considering the measurable sets $\Omega_n = \{x \in E : p_n(x) > 0\}$, the proof of Theorem 2.3.21 uses a variant of parametric convergence (compare Appendix 3) that is based on the a.e. in E convergence of p_n and of ∇p_n.

Remark. In Pironneau [1984, p. 134] it is mentioned that by taking $a_2 \searrow 0$ and $b_2 \searrow 0$ in (2.3.70)–(2.3.73) the Neumann boundary value problem is approximated. It is as yet unknown under what minimal hypotheses such passages to the limit are successful. In Chapter 5, we will also discuss relaxation and approximation procedures that further develop the techniques from this paragraph.

We close this chapter with some comments on the mathematical literature devoted to the existence question in shape optimization problems.

The counterexample of Murat [1971] shows that for the problem of control into coefficients for second-order elliptic operators certain boundedness conditions on the gradient of the leading coefficient are necessary in order to guarantee a continuous dependence coefficient $\mapsto$ solution in appropriate topologies. A survey bringing various compactness conditions and corresponding continuity results is due to Bendsøe [1984]. In Section 2.1, we have included several abstract theorems of this type, related to optimal control problems governed by strongly nonlinear equations.

For coefficients that occur only in the lower-order terms of the differential operator, or for special cost functionals, weaker hypotheses suffice for the proof of existence, as follows froms the works of Tahraoui [1992], Céa and Malanowski [1979], and Tiihonen and Gonzalez de Paz [1994]. This has been discussed in §2.2.1. Sprekels and Tiba [1998/1999] have shown that the situation is completely different in the setting of fourth-order elliptic operators. Namely, boundedness in L^∞ suffices for existence in problems governed by fourth-order equations with control coefficients appearing in the leading terms (recall §2.2.2).

For variable domain problems a turning point was marked by the example given by Cioranescu and Murat [1982], who showed that high oscillations in the boundaries of the unknown domains may lead to a very singular behavior of the sequence of associated solutions. It should be noted that such examples were initially motivated by questions from homogenization theory; see Cioranescu and Donato [1999], Zhikov, Kozlov, and Oleinik [1994]. Classical existence results for variable domain optimization problems are due to Murat and Simon [1976], using the mapping method that reduces the problem to a control into coefficients problem (under appropriate regularity assumptions; see Section 5.3 in Chapter 5 below). The work of Chenais [1975] established the first direct approach to the existence question, by using the uniform extension property in Lipschitz domains, i.e., in domains with corners. An important survey, including supplementary results and examples, is provided by the textbook of Pironneau [1984]. Delfour and Zolésio [2001] have demonstrated that similar arguments generalize to what they call the "uniform cusp condition."

Recent results of Tiba [1999], [2003], Liu, Neittaanmäki, and Tiba [2000], [2003], Neittaanmäki and Tiba [2000], which are included in paragraphs 2.3.2 and 2.3.3 of Section 2.3, replace the uniform Lipschitz assumption by uniform continuity conditions. From a geometric point of view, this is related to the segment property and allows for cusps on the boundary of the unknown domains. There are no requirements concerning the dimension of the underlying Euclidean space. In special boundary value problems or for special cost functionals, even families of just measurable sets may be taken into account; see Gonzalez de Paz [1982], [1994], Mäkinen, Neittaanmäki, and Tiba [1992]; see paragraphs 2.3.1 and 2.3.4.

By using potential theory arguments, Sverak [1993], Chambolle and Doveri [1997], Zhong [1997], and Bucur and Varchon [2000], proved continuity and existence for arbitrary open sets, under the compactness assumption that the number of the connected components of the complementary set is limited from above. While this condition is very general and easy to check, the dimension of the space is connected with the type of the elliptic operator (for the Laplace operator, the dimension should be two).

In the monograph by Sokołowski and Zolésio [1992], an existence theory based on the notion of *generalized perimeter of open sets* is presented. Further results using potential theory and conditions are due to Bucur and Zolésio [1994a], [1994b], [1996]. In the work of Henrot [1994], a survey of continuity results of solutions to Laplace's equation with respect to their domain of definition has been presented. The textbook of Delfour and Zolésio [2001] gives a unifying presentation in this respect and includes further interesting results.

Chapter 3

Optimality Conditions

A central problem in the theory of optimal control is that of deriving first-order necessary conditions. It plays an outstanding role in many respects: regularity and/or bang-bang properties of optimal pairs, gradient methods, feedback controllers. This chapter is mainly devoted to the investigation of several methods to recover the optimality conditions in various types of control problems. Relevant examples and applications will be included in each paragraph.

3.1 Abstract Approaches

3.1.1 The Convex Case. Subdifferential Calculus

We begin our exposition by examining the linear case in connection with the problem (P) defined by (1.1.1)–(1.1.4) in the first section of Chapter 1. To this end, let U, V, Z denote reflexive Banach spaces that together with their respective duals U^*, V^*, Z^* are strictly convex, and H a Hilbert space that is identified with its dual and satisfies $V \subset H \subset V^*$ with continuous and dense embedding. In addition, let $f \in Z$ and operators $A \in L(V, Z)$ and $B \in L(U, Z)$ be given.

Let $C \subset H$, $U_{ad} \subset U$ be given nonempty, convex, and closed (with respect to the corresponding topologies) sets. We include the constraints (1.1.3), (1.1.4) in the cost criterion by using the cost functional (compare (1.1.5))

$$L : V \times U \to]-\infty, +\infty], \quad L(y,u) = \tilde{L}(y,u) + I_{C\times U_{ad}}(y,u),$$

where $\tilde{L} : V \times U \to]-\infty, +\infty]$ is a proper, convex, and lower semicontinuous functional. We then consider the control problem (P):

$$\inf \left\{ L(y,u) : (y,u) \in V \times U \right\} \quad \text{subject to } Ay = Bu + f.$$

Without loss of generality, we may assume that $f = 0$. Indeed, if $[y_0, u_0]$ is an admissible pair, that is, if $L(y_0, u_0) < +\infty$ and $Ay_0 = Bu_0 + f$, then it is easily verified that a pair $[y^*, u^*] \in V \times U$ is a solution to (P) if and only if the shifted pair $[\hat{y}, \hat{u}] = [y^* - y_0, u^* - u_0]$ solves the shifted problem

$$\inf \left\{ \tilde{L}(\hat{y} + y_0, \hat{u} + u_0) + I_{(C-y_0)\times(U_{ad}-u_0)}(\hat{y}, \hat{u}) : (\hat{y}, \hat{u}) \in V \times U \right\}$$
$$\text{subject to } A\hat{y} = B\hat{u}.$$

In the following, we will consider problem (P) with $f = 0$.

We now introduce the notion of *extremal pairs*: we say that the pair $[y^*, u^*]$ is extremal for (P) if there is some $p^* \in Z^*$ such that

$$Ay^* = Bu^*, \tag{3.1.1}$$
$$A^*p^* \in -(\partial L)_1(y^*, u^*), \tag{3.1.2}$$
$$B^*p^* \in (\partial L)_2(y^*, u^*). \tag{3.1.3}$$

Here, $A^* \in L(Z^*, V^*)$ and $B^* \in L(Z^*, U^*)$ denote the dual operators of A and B, respectively, and $(\partial L)_i$ denotes the ith component of the subdifferential ∂L, $i = 1, 2$, and *not* the partial subdifferential with respect to the ith argument of L. Hence, we have $\partial L = [(\partial L)_1, (\partial L)_2]$.

Now let $G(\Lambda) \subset U \times V$ denote the graph of some operator $\Lambda \in L(U, V)$ satisfying

$$G^{-1}(\Lambda) \cap \mathrm{dom}(L) \neq \emptyset.$$

Then the function $\Gamma : U \to]-\infty, +\infty]$,

$$\Gamma(u) = L(\Lambda u, u) \tag{3.1.4}$$

is proper, convex, and lower semicontinuous. We determine the subdifferential $\partial\Gamma$ of Γ, using Theorem A1.12 in Appendix 3. To this end, we put, in the terminology of Theorem A1.12,

$$X = V \times U, \quad X_1 = V \times \{0\}, \quad X_2 = \{0\} \times U,$$

and we introduce the linear and bounded operator $\mathcal{A} : V \times U \to V \times U$,

$$\mathcal{A}(v, u) = [\Lambda u, u], \quad \text{for } [v, u] \in V \times U.$$

Observe that for all $[v, u] \in V \times U$ and $[z, w] \in V^* \times U^*$,

$$\begin{aligned} ([z, w], \mathcal{A}(v, u))_{(V^*\times U^*)\times(V\times U))} &= (z, \Lambda u)_{V^*\times V} + (w, u)_{U^*\times U} \\ &= (\Lambda^* z + w, u)_{U^*\times U}, \end{aligned}$$

so that the dual operator $\mathcal{A}^* \in L(V^* \times U^*, V^* \times U^*)$ is given by

$$\mathcal{A}^*(z, w) = [0, \Lambda^* z + w], \quad \text{for } [z, w] \in V^* \times U^*.$$

Moreover, since L plays the role of the mapping φ in Theorem A1.12, we have

$$R(\mathcal{A}) \cap \text{int}_1 [\text{dom}(\varphi) \cap X_1] \neq \emptyset$$

if we assume that

$$G^{-1}(\Lambda) \cap \left[\text{int}\left(\text{dom}(L(\cdot\,, 0))\right) \times \{0\}\right] \neq \emptyset, \tag{3.1.5}$$

where int_1 denotes the interior with respect to X_1.

Next, we need to check that the operator $\mathcal{A}^*|_{X_1^0} : X_1^0 \to X^*$, where

$$X_1^0 = \{x^* \in X^* : (x^*, x)_{X^* \times X} = 0 \quad \forall x \in X\}$$

denotes the polar subspace of X_1, has a bounded inverse. Obviously, $X_1^0 = \{0\} \times U^*$, and thus $\mathcal{A}^*(z, w) = [0, w]$ for all $[z, w] = [0, w] \in X_1^0$. Hence, $\mathcal{A}^*|_{X_1^0}$ coincides with the restriction of the identity mapping to X_1^0 and therefore has a bounded inverse.

Under the above assumptions, we thus may apply Theorem A1.12 to arrive at the conclusion that

$$\partial\Gamma(u) = \mathcal{A}^* \left((\partial L)(\Lambda u, u)\right) = \Lambda^*((\partial L)_1(\Lambda u, u)) + (\partial L)_2(\Lambda u, u). \tag{3.1.6}$$

Remark. The presence of $\{0\}$ in the assumption (3.1.5) is not essential. Indeed, if $\tilde{u} \in U$ is admissible and if

$$G^{-1}(\Lambda) \cap \left[\text{int}\left(\text{dom}(L(\cdot\,, \tilde{u}))\right) \times \{\tilde{u}\}\right] \neq \emptyset, \tag{3.1.5$'$}$$

then we may replace L by the function

$$L_1 : V \times U \to]-\infty, +\infty], \quad L_1(v, u) = L(v + \Lambda\tilde{u}, u + \tilde{u}),$$

which satisfies (3.1.5), so that (3.1.6) follows by the same argument as above.

We have the following result.

Theorem 3.1.1 *Let the above hypothesis be fulfilled, and suppose, in addition, that A has a bounded inverse $A^{-1} : Z \to V$. Then a pair $[y^*, u^*]$ is optimal for (P) if and only if it is extremal.*

Proof. Let $\Lambda : U \to V$ be given by $\Lambda u = A^{-1}Bu$. A control u^* is optimal if and only if

$$0 \in \partial\Gamma(u^*), \tag{3.1.7}$$

that is, owing to (3.1.6), if and only if

$$0 \in B^*(A^*)^{-1}\left[(\partial L)_1(A^{-1}Bu^*, u^*)\right] + (\partial L)_2(A^{-1}Bu^*, u^*). \tag{3.1.8}$$

Suppose now that $[y^*, u^*]$ is extremal. Then, by (3.1.1), $A^{-1}Bu^* = y^*$, and thus, using (3.1.2), (3.1.3), and (3.1.6),

$$\partial\Gamma(u^*) = B^*(A^*)^{-1}[(\partial L)_1(y^*, u^*)] + (\partial L)_2(y^*, u^*) \ni -B^*(A^*)^{-1}A^*p^* + B^*p^*,$$

that is, $0 \in \partial\Gamma(u^*)$, and $[y^*, u^*]$ is optimal.

Conversely, if $[y^*, u^*]$ is optimal, then, trivially, (3.1.1) holds. Moreover, (3.1.8) implies that there is some

$$z \in [-B^*(A^*)^{-1}((\partial L)_1(y^*, u^*))] \cap [(\partial L)_2(y^*, u^*)].$$

In particular, there is some $q \in -(\partial L)_1(y^*, u^*)$ such that $z = B^*(A^*)^{-1}q$. With the choice $p^* = (A^*)^{-1}q$, then (3.1.2) and (3.1.3) are fulfilled; that is, $[y^*, u^*]$ is extremal. □

Remark. In the terminology of control theory, the optimality condition (3.1.2) is usually referred to as the *adjoint state equation* corresponding to the state equation (3.1.1), while relation (3.1.3) is called the *(Pontryagin) maximum principle*.

Example 3.1.2 Let $\Omega \subset \mathbf{R}^d$ be a bounded domain, and suppose that

$$g : \Omega \times \mathbf{R} \times \mathbf{R} \to]-\infty, +\infty]$$

is a function such that $g(x, \cdot, \cdot)$ is convex, lower semicontinuous, and proper, for a.e. $x \in \Omega$. In addition, let g be measurable with respect to the σ-field of $\Omega \times \mathbf{R} \times \mathbf{R}$ generated by the product of the Lebesgue σ-field in Ω and the Borel σ-field in $\mathbf{R} \times \mathbf{R}$ (see the definition of a normal convex integrand in Appendix 1).

We suppose that there is some $u_0 \in \mathbf{R}$ such that

$$|g(x, y, u_0)| \le \alpha(x)\, |y|^p + \beta(x), \quad \text{for a.e. } x \in \Omega \quad \text{and for all } y \in \mathbf{R}, \tag{3.1.9}$$

with suitable functions $\alpha \in L^\infty(\Omega)$ and $\beta \in L^p(\Omega)$, where $1 < p < +\infty$. We consider the optimal control problem

$$\inf \left\{ \int_\Omega g(x, y(x), u(x))\, dx \right\}, \tag{3.1.10}$$

subject to $y \in L^p(\Omega)$, $u \in L^p(\Omega)$, and

$$\Delta y = Bu \quad \text{a.e. in } \Omega, \tag{3.1.11}$$

$$y = 0 \quad \text{on } \partial\Omega, \tag{3.1.12}$$

where $B : L^p(\Omega) \to L^p(\Omega)$ is a linear bounded operator. We specify the involved spaces and operators by putting

$$U = L^p(\Omega) = Z = W, \quad V = W^{2,p}(\Omega) \cap W_0^{1,p}(\Omega), \quad A : V \to Z, \quad Ay = \Delta y.$$

The functional $L : W \times U \to]-\infty, +\infty]$ is given by

$$L(y, u) = \int_\Omega g(x, y(x), u(x))\, dx. \tag{3.1.13}$$

The problem (P) is defined by (3.1.11)–(3.1.13). We assume that the natural embedding $i : V \to W$ is the corresponding observation operator as in §1.2.1. From (3.1.9) we infer that $\operatorname{dom}(L(\cdot\,, u_0)) = L^p(\Omega) = W$, and with $\Lambda = A^{-1}B$ the hypothesis (3.1.5)′ is fulfilled. Moreover, it follows from Proposition A1.1 in Appendix 1 that

$$\begin{aligned} \partial L(y,u) \;=\; & \Big\{[h, v] \in L^q(\Omega) \times L^q(\Omega) : 1/p + 1/q = 1, \\ & [h(x), v(x)] \in \partial g(x, y(x), u(x)) \text{ for a.e. } x \in \Omega\Big\}. \end{aligned} \tag{3.1.14}$$

By virtue of Theorem 3.1.1, we have the following result.

Corollary 3.1.3 *Suppose (3.1.9) holds. Then a pair $[y^*, u^*] \in V \times U$ is optimal for (3.1.11)–(3.1.13) if and only if there is some $p^* \in L^q(\Omega)$, where $\frac{1}{p} + \frac{1}{q} = 1$, such that with $\partial g = [(\partial g)_1, (\partial g)_2]$,*

$$\begin{aligned} \Delta y^* &= Bu^*, \quad \text{in } \Omega, \\ -\Delta p^* &\in i^*[(\partial g)_1(x, y^*(x), u^*(x)], \quad \text{in } \Omega, \\ B^* p^* &\in (\partial g)_2(x, y^*(x), u^*(x)), \quad \text{in } \Omega, \\ y^* &= 0, \quad p^* = 0, \quad \text{on } \partial\Omega. \end{aligned}$$

Here, the differential equation and the boundary condition for p^* have to be understood in the transposition sense (cf. Example A2.7 in Appendix 2, where the special case $p = 2$ is considered).

Remark. If also constraints $u \in U_{ad} \subset U$, $y \in C \subset V$, have to be respected in this example, then (3.1.5)′ or (3.1.9) shows that the condition $\operatorname{int}(C) \neq \emptyset$

in W has to be imposed for the above argument to hold (compare with (1.1.5) as well). This condition is very restrictive, since simple constraint sets such as

$$C = \{y \in L^p(\Omega) : y(x) \geq c \text{ a.e. in } \Omega\}$$

have empty interior in $L^p(\Omega)$. More about such *interiority conditions* will be said in the next sections. They are mainly related to the set C of state constraints, while the set U_{ad} of control constraints may be a general convex and closed subset of U.

3.1.2 The Convex Case. Mathematical Programming

We consider again the abstract control problem (1.1.1), (1.1.2) of Chapter 1, but this time with general mixed state-control constraints given in inequality form,

$$h(y, u) \leq 0, \tag{3.1.15}$$

where $h : V \times U \to \mathbf{R}$ is a convex and continuous function. Regarding the constraints, this formulation is more general and more explicit than the previous one.

We assume that V and U are Hilbert spaces, $Z = V^*$, $A \in L(V, V^*)$ is a bijective operator with $A^{-1} \in L(V^*, V)$, $B \in L(U, V^*)$, and L is a convex and continuous function on $V \times U$. By shifting the domains of L and h as in §3.1.1 above, if necessary, and redenoting the resulting mappings again by L and h, respectively, we may assume that $f = 0$. We denote by

$$K = \{(y, u) \in V \times U : \; Ay = Bu\}$$

the closed subspace of $V \times U$ defined by the state equation. We include the state equation in the cost functional by replacing L by $L + I_K$, where I_K is the indicator function of the subspace K in $V \times U$ (this is similar to the approach used in Section 1.1 of Chapter 1, where the constraints have been implicitly included in the cost).

We have thus reformulated the control problem as an infinite-dimensional convex programming problem,

$$\text{Min} \left\{ L(y, u) + I_K(y, u) \right\} \tag{3.1.16}$$

subject to (3.1.15).

Let $[y^*, u^*] \in V \times U$ denote an optimal pair for the problem (3.1.15), (3.1.16). Such a pair exists under appropriate admissibility and coercivity and/or boundedness conditions resembling those given in Theorems 2.1.2 and 2.1.3 in Chapter 2.

We also impose the *interiority condition:* there is some $[\overline{y}, \overline{u}]$ that is admissible for (3.1.15), (3.1.16) and satisfies

$$[\overline{y}, \overline{u}] \in \text{int}(D), \quad \text{where } D = \{[y,u] \in V \times U : \ h(y,u) \le 0\}.$$

A possible relaxation of this hypothesis, which is also connected to a penalization of the state equation, will be discussed in §3.2.2 below.

In order to obtain the first-order optimality conditions (the so-called *Karush–Kuhn–Tucker theorem*) in the setting of mathematical programming, one needs a so-called *constraint qualification* condition. It has been proved in Tiba [1995a], [1996] that a very weak form of such a condition is the following one:

$$\begin{aligned}&\text{For every bounded set } M \subset V \times U \text{ with } M \setminus D \neq \emptyset \text{ there is some}\\ &c_M > 0 \text{ such that } h(y,u) \ge c_M\, d([y,u], D) \quad \forall\, [y,u] \in M \setminus D.\end{aligned} \tag{3.1.17}$$

In finite-dimensional spaces, the (*metric regularity*) condition (3.1.17) is even necessary for the Karush–Kuhn–Tucker theorem to hold; see Tiba [1996]. A comprehensive generalization of these abstract results can be found in Tiba and Zălinescu [2004]. In order to keep the exposition at a reasonable length, we only indicate a direct argument that is adapted to the present situation.

Theorem 3.1.4 *Under the above assumptions there exists some $\lambda \ge 0$ such that*

$$[0,0] \in \partial L(y^*, u^*) + \partial I_K(y^*, u^*) + \lambda\, \partial h(y^*, u^*), \tag{3.1.18}$$

$$\lambda\, h(y^*, u^*) = 0. \tag{3.1.19}$$

For the proof of Theorem 3.1.4, we need the following auxiliary result.

Lemma 3.1.5 *Under the above assumptions, the multivalued operator $N \subset (V \times U) \times (V \times U)^*$ given by*

$$N(y,u) = \begin{cases} \{[0,0]\}, & \text{if } h(y,u) < 0,\\ \{\lambda w : \ \lambda \ge 0, \ w \in \partial h(y,u)\}, & \text{if } h(y,u) = 0,\\ \emptyset, & \text{if } h(y,u) > 0, \end{cases} \tag{3.1.20}$$

is maximal monotone, and $N = \partial I_D$.

Proof. First, we show that $N \subset \partial I_D$ if h is continuous and convex. Clearly, $N(y,u) = \partial I_D(y,u) = \emptyset$ whenever $h(y,u) > 0$. If $h(y,u) < 0$, then $(y,u) \in$

$\mathrm{int}(D)$ by the continuity of h, and we again have $N(y,u) = \partial I_D(y,u)$. Suppose now that $h(y,u) = 0$. Then, by the definition of the subdifferential (cf. Appendix 1),

$$\begin{aligned}\partial h(y,u) &= \{w = [w_1, w_2] \in V^* \times U^* : -h(y_0,u_0) \le (w_1, y - y_0)_{V^*\times V} \\ &\qquad + (w_2, u - u_0)_{U^*\times U}, \quad \forall\, (y_0,u_0) \in V \times U\}.\end{aligned}$$

Now let $\lambda \ge 0$ and $w = [w_1, w_2] \in \partial h(y,u)$ be arbitrary. If $(y_0,u_0) \in D$, then $h(y_0,u_0) \le 0$, and thus

$$(\lambda w_1, y - y_0)_{V^*\times V} + (\lambda w_2, u - u_0)_{U^*\times U} \;\ge\; -\lambda\, h(y_0,u_0) \;\ge\; 0,$$

that is, $\lambda w \in \partial I_D(y,u)$. In conclusion, we have $N \subset \partial I_D$, and $N \subset (V \times U) \times (V \times U)^*$ is a monotone operator. We apply the Minty characterization (Theorem A1.3 in Appendix 1) to prove that N is even maximal monotone, whence $N = \partial I_D$ would follow.

To this end, we show that $R(N + F) = V^* \times U^*$, where $F : V \times U \to V^* \times U^*$ denotes the duality mapping (the Riesz isomorphism in the Hilbert space $V \times U$). Now let $[y,u] \in V \times U$ be arbitrary. We have to find some $[\tilde{y}, \tilde{u}] \in \mathrm{dom}\,(N) = D$ such that

$$F([\tilde{y},\tilde{u}]) + N(\tilde{y},\tilde{u}) \ni F([y,u]). \tag{3.1.21}$$

By Theorem A1.2 in Appendix 1, the operator $\partial h \subset (V \times U) \times (V^* \times U^*)$ is maximal monotone, and for every $\lambda > 0$ there is a unique solution $[y_\lambda, u_\lambda] \in V \times U$ to the problem (cf. the definitions of J_λ and A_λ in Appendix 1)

$$F([y_\lambda,u_\lambda]) + \lambda\,[z_\lambda, w_\lambda] = F([y,u]), \quad [z_\lambda, w_\lambda] = \partial h_\lambda(y,u) \in \partial h(y_\lambda,u_\lambda), \tag{3.1.22}$$

where h_λ is the Yosida–Moreau regularization of h, and where $\partial h_\lambda(y,u)$ is the Fréchet derivative of h_λ at $[y,u]$. Multiplying both sides of the first equality in (3.1.22) by $[y_\lambda,u_\lambda]$, and invoking the monotonicity of ∂h, we readily find that

$$|[y_\lambda,u_\lambda]|_{V\times U} \;\le\; |[y,u]|_{V\times U}\,;$$

that is, the set $M = \{[y_\lambda,u_\lambda] : \lambda > 0\}$ is bounded in $V \times U$. Here, possibly shifting the domain of ∂h appropriately, one can also assume and use that $[0,0] \in \partial h(0,0)$.

We have, thanks to Theorem A1.5(ii) in Appendix 1,

$$\begin{aligned}h_\lambda(y,u) &= h(y_\lambda,u_\lambda) + \frac{1}{2\lambda}\left|\,[y,u] - [y_\lambda,u_\lambda]\right|^2_{V\times U} \\ &= h(y_\lambda,u_\lambda) - \frac{\lambda}{2}\left|\,[z_\lambda,w_\lambda]\right|^2_{V^*\times U^*}.\end{aligned}$$

Hence,

$$\left| [z_\lambda, w_\lambda] \right|^2_{(V\times U)^*} = \frac{2}{\lambda}\,(h_\lambda(y,u) - h(y_\lambda,u_\lambda)) \le \frac{2}{\lambda}\,(h(y,u) - h(y_\lambda,u_\lambda)),$$

where the latter inequality follows from Theorem A1.5(i).

Assume now that $[y_\lambda, u_\lambda] \notin D$. Then $h(y_\lambda, u_\lambda) > 0$, and thus

$$|\,[z_\lambda, w_\lambda]|^2_{V^*\times U^*} \le \frac{2}{\lambda}\,h(y,u).$$

Now observe that

$$[z_\lambda, w_\lambda] \;=\; \partial h_\lambda(y,u) \in \partial h(y_\lambda, u_\lambda).$$

Hence, taking hypothesis (3.1.17) into account, we get, denoting by P_D the projection onto D in $V \times U$,

$$\begin{aligned} c_M\,|\,[y_\lambda,u_\lambda] - P_D[y_\lambda,u_\lambda]|_{V\times U} \;&\le\; h(y_\lambda,u_\lambda) - h(P_D[y_\lambda,u_\lambda]) \\ &\le |\,[z_\lambda,w_\lambda]|_{V^*\times U^*}|\,[y_\lambda,u_\lambda] - P_D[y_\lambda,u_\lambda]|_{V\times U}, \end{aligned} \tag{3.1.23}$$

where $c_M > 0$ is the constant in (3.1.17) corresponding to the bounded set M of all pairs $[y_\lambda, u_\lambda]$ obtained via (3.1.22), which is independent of $\lambda > 0$. Combining (3.1.22), (3.1.23), we see that

$$c_M^2 \;\le\; \frac{2}{\lambda}\,h(y,u),$$

which is a contradiction for sufficiently large $\lambda > 0$, since h is continuous and everywhere finite. This shows that $[y_\lambda, u_\lambda] \in D$ for sufficiently large $\lambda > 0$. Then, since N given by (3.1.20) is a cone, equation (3.1.22) can be rewritten as

$$F([y_\lambda,u_\lambda] - [y,u]) + N(y_\lambda,u_\lambda) \ni [0,0]. \tag{3.1.24}$$

Hence, we have (3.1.21) with $[\tilde{y},\tilde{u}] = [y_\lambda, u_\lambda]$, and the assertion of Lemma 3.1.5 is completely proved. □

Proof of Theorem 3.1.4. The problem (3.1.15), (3.1.16) is equivalent to the minimization of $L + I_K + I_D$; that is,

$$[0,0] \in \partial(L + I_K + I_D)(y^*, u^*). \tag{3.1.25}$$

By the interior admissibility of $[\overline{y}, \overline{u}]$, we have

$$\operatorname{int}(D) \cap \operatorname{dom}(L + I_K) \ne \emptyset.$$

Thus, by the additivity rule for the subdifferential (compare the remark following Theorem A1.11 in Appendix 1), and owing to Lemma 3.1.5,

$$[0,0] \in \partial(L+I_K)(y^*,u^*) + \partial I_D(y^*,u^*) \;=\; \partial(L+I_K)(y^*,u^*) + N(y^*,u^*). \tag{3.1.25$'$}$$

Since L is continuous on $V \times U$, we can use the additivity of the subdifferential in the first term, so that

$$[0,0] \in \partial L(y^*, u^*) + \partial I_K(y^*, u^*) + N(y^*, u^*).$$

Finally, (3.1.20) implies that there is some $\lambda \geq 0$ satisfying (3.1.18) and (3.1.19). □

Corollary 3.1.6 *There is some* $p^* \in V$ *that together with* $[y^*, u^*]$ *satisfies the optimality conditions*

$$-A^* p^* \in (\partial L)_1(y^*, u^*) + \lambda\, (\partial h)_1(y^*, u^*),$$

$$B^* p^* \in (\partial L)_2(y^*, u^*) + \lambda\, (\partial h)_2(y^*, u^*),$$

$$\lambda\, h(y^*, u^*) = 0, \quad \lambda \geq 0.$$

Proof. Since K is a subspace, we have $\partial I_K(y^*, u^*) = K^\perp$. We show that $K^\perp = \mathcal{M}$, where $\mathcal{M} = \{[A^* p, -B^* p] : \ p \in V\}$. To this end, let $p \in V$ be given, and $[y, u] \in K$ be arbitrary. Then

$$\begin{aligned}([y,u], [A^* p, -B^* p])_{(V \times U) \times (V^* \times U^*)} &= (y, A^* p)_{V \times V^*} - (u, B^* p)_{U \times U^*} \\ &= (Ay, p)_{V^* \times V} - (Bu, p)_{V^* \times V} = 0.\end{aligned}$$

Hence, $[A^* p, -B^* p] \in K^\perp$, and thus $\mathcal{M} \subset K^\perp$.

Conversely, if $[z, w] \in K^\perp$, then we have for any $[y, u] \in K$ that $(z, y)_{V^* \times V} + (w, u)_{U^* \times U} = 0$. Now recall that $A^{-1} \in L(V, V^*)$. Thus, $(A^*)^{-1}$ exists and is continuous. Putting $p = (A^*)^{-1} z$, we have $z = A^* p$, and we easily verify that $w = -B^* p$. Hence, $[z, w] \in \mathcal{M}$, and thus $K^\perp \subset \mathcal{M}$.

Now, having shown that $\mathcal{M} = K^\perp$, we obtain the assertion by a direct application of Theorem 3.1.4. □

Remark. From the works of Tiba [1996] and Tiba and Zălinescu [2004] it is known that the hypothesis (3.1.17) is fulfilled if the classical *Slater condition* is satisfied, i.e., there is $[\hat{y}, \hat{u}] \in D$ such that $h(\hat{y}, \hat{u}) < 0$. Since h is continuous, this also ensures that $\text{int}\,(D) \neq \emptyset$. Notice that Corollary 3.1.6 also includes the (last) *complementarity condition*, which is typical for inequality constraints.

3.1.3 The Differentiable Case

In this paragraph, we renounce the convexity hypothesis on L and the linearity of A. Instead, we require that $L : V \times U \to \mathbf{R}$ be Fréchet differentiable. Let $\nabla L = [\nabla_1 L, \nabla_2 L]$ denote its Fréchet derivative. The operator A is allowed to depend directly on $u \in U$, i.e., we consider an operator of the form $A : V \times U \to Y$, where Y is some Banach space. We postulate that A is Fréchet differentiable on $V \times U$ with Fréchet derivative $\nabla A = [\nabla_1 A, \nabla_2 A]$. Moreover, let a nonempty, closed, and convex set $U_{ad} \subset U$ be given. The control problem under investigation then reads as follows:

$$\inf\left\{L(y,u)\right\} \tag{3.1.26}$$

subject to $u \in U_{ad}$ and to

$$A(y,u) = 0. \tag{3.1.27}$$

Notice that in this situation the existence question may be very difficult. In general, only *local* optimal pairs can be found.

The problem (3.1.26), (3.1.27) may be interpreted as a constrained differentiable programming problem in infinite-dimensional spaces. We adopt the control approach here; namely, we assume that (3.1.27) defines uniquely a Gâteaux differentiable mapping $\theta : U_{ad} \to V$, that is, $A(\theta(u),u) = 0$ for any $u \in U_{ad}$.

Theorem 3.1.7 *Let $[y^*, u^*]$ be a local optimal pair for (3.1.26), (3.1.27), and suppose that the bounded linear operator $\nabla_1 A(y^*,u^*) : V \to Y$ is surjective. Then there is some $p^* \in Y^*$ that, together with y^* and u^*, satisfies the optimality conditions*

$$\nabla_1 L(y^*,u^*) + (\nabla_1 A(y^*,u^*))^* p^* \;=\; 0, \tag{3.1.28}$$

$$(\nabla_2 L(y^*,u^*) + (\nabla_2 A(y^*,u^*))^* p^*, v - u^*)_{U^* \times U} \;\geq\; 0 \quad \forall\, v \in U_{ad}. \tag{3.1.29}$$

Proof. We choose a variation of u^* of the form

$$u = u^* + \lambda(v - u^*), \quad \text{where } \lambda \in [0,1], \quad v \in U_{ad}. \tag{3.1.30}$$

Since U_{ad} is convex, we have $u \in U_{ad}$. Moreover, since u^* is a local minimum, there is some $0 < \lambda_0 \leq 1$ such that for $0 < \lambda \leq \lambda_0$,

$$L(y^*,u^*) = L(\theta(u^*),u^*) \leq L(\theta(u),u) = L(\theta(u^* + \lambda(v-u^*)), u^* + \lambda(v-u^*)). \tag{3.1.31}$$

We pass all the terms in (3.1.31) to the right-hand side, divide by $\lambda > 0$, and take the limit as $\lambda \searrow 0$. Owing to the differentiability assumptions and to the chain rule, we then find that

$$\nabla_2 L(y^*,u^*)(v-u^*) + \nabla_1 L(y^*,u^*)(z) \geq 0 \quad \forall\, v \in U_{ad}, \tag{3.1.32}$$

with $z = \lim_{\lambda \searrow 0} \frac{1}{\lambda}[\theta(u^* + \lambda(v-u^*)) - \theta(u^*)]$ in V.

Since $\nabla_1 A(y^*,u^*)$ is surjective, the dual operator $(\nabla_1 A(y^*,u^*))^* : Y^* \to V^*$ has a bounded inverse (cf. Yosida [1980, Chapter VII]), and thus equation (3.1.28) admits a unique solution, denoted by $p^* \in Y^*$.

We differentiate (3.1.27) with respect to u at the point u^* and in the direction $v - u^*$. The chain rule yields that

$$\nabla_1 A(y^*, u^*)(z) + \nabla_2 A(y^*, u^*)(v - u^*) = 0. \tag{3.1.33}$$

In view of (3.1.32) and (3.1.33), and by the definition of the dual operator, we can infer that

$$\begin{aligned}\nabla_1 L(y^*, u^*)(z) &= (\nabla_1 L(y^*, u^*), z)_{V^* \times V} = (-(\nabla_1 A(y^*, u^*))^*(p^*), z)_{V^* \times V} \\ &= -(p^*, \nabla_1 A(y^*, u^*)(z))_{Y^* \times Y} = (p^*, \nabla_2 A(y^*, u^*)(v - u^*))_{Y^* \times Y} \\ &= (\nabla_2 A(y^*, u^*)^*(p^*), v - u^*)_{U^* \times U}.\end{aligned}$$

Combining this with (3.1.32), we get (3.1.29), and the proof is finished. □

Remark. Under appropriate assumptions (and in the absence of state constraints) the convex problem (P) investigated in Chapter 1 may be put in the form (3.1.26), (3.1.27), so that Theorem 3.1.7 can be applied.

Example 3.1.8 As an application to nonlinear control problems, we examine a simple elliptic control problem arising in identification theory (see Falk [1990]). To this end, let $\Omega \subset \mathbf{R}^d$ be a (sufficiently smooth) bounded domain, and let $f \in L^2(\Omega)$ and $g \in L^2(\partial\Omega)$ be given. Moreover, we consider the set of admissible controls

$$U_{ad} = \Big\{ u \in L^\infty(\Omega) : \ a \le u(x) \le b, \ \text{a.e. in } \ \Omega \Big\},$$

where $0 < a < b$ are given. Clearly, U_{ad} is nonempty, convex, bounded, and compact with respect to the weak* topology of $L^\infty(\Omega)$. We then study the problem

$$\operatorname{Min} \left\{ \frac{1}{2} \int_{\partial\Omega} (y(\sigma) - g(\sigma))^2 \, d\sigma \right\} \tag{3.1.34}$$

subject to

$$-\Delta y + u\, y = f \quad \text{in } \Omega, \tag{3.1.35}$$

$$\frac{\partial y}{\partial n} = 0 \quad \text{on } \partial\Omega, \tag{3.1.36}$$

$$u \in U_{ad}. \tag{3.1.37}$$

It is well known (cf. Appendix 2) that (3.1.35), (3.1.36) has for every $u \in U_{ad}$ a unique weak solution $y = y(u) \in H^1(\Omega)$. Since u belongs to $L^\infty(\Omega)$, a simple direct argument yields that $\Delta y \in L^2(\Omega)$, and thus $y \in H^2(\Omega)$ since Ω is smooth.

A "practical" significance of the problem (3.1.34)–(3.1.37) is the following: we want to find a coefficient function u such that the trace of y on $\partial\Omega$ coincides with the given observed function $g \in L^2(\partial\Omega)$ (see (1.2.7)″, (1.2.8)″ in Example 1.2.6 in Chapter 1). This is an identification problem and also belongs to the category of control by coefficients problems (compare Ahmed and Teo [1981], Casas [1992], Lurie [1990]).

To fit the problem (3.1.34)–(3.1.37) into our abstract framework, we put

$$V = \left\{ y \in H^2(\Omega) : \frac{\partial y}{\partial n} = 0 \text{ on } \partial\Omega \right\}, \quad U = L^\infty(\Omega), \quad Y = L^2(\Omega),$$

$$A(y,u) = -\Delta y + u\,y - f, \quad L(y,u) = \frac{1}{2}\int_{\partial\Omega} (y(\sigma) - g(\sigma))^2\, d\sigma.$$

First, we show by a direct argument that (3.1.34)–(3.1.37) has at least one optimal pair $[y^*, u^*] \in V \times U_{ad}$. To this end, let $\{[y_n, u_n]\} \subset V \times U_{ad}$ be a minimizing sequence. Multiplying (3.1.35) for $n \in \mathbf{N}$ by y_n, and integrating over Ω and by parts, we find that

$$\int_\Omega \left(|\nabla y_n(x)|^2 + u_n(x)\,|y_n(x)|^2 \right) dx = \int_\Omega f(x)\, y_n(x)\, dx.$$

Thus, since $u_n \in U_{ad}$, we can infer that $\{y_n\}$ is bounded in $H^1(\Omega)$. From (3.1.35) it then follows that $\{\Delta y_n\}$ is bounded in $L^2(\Omega)$, and standard elliptic theory shows that $\{y_n\}$ is bounded in $H^2(\Omega)$. Therefore, there is some subsequence, again indexed by n, such that

$$y_n \to y \quad \text{weakly in } H^2(\Omega), \qquad u_n \to u \quad \text{weakly* in } L^\infty(\Omega).$$

Clearly, $u \in U_{ad}$. Moreover, by compact embedding, we may without loss of generality assume that

$$\frac{\partial y_n}{\partial n} \to \frac{\partial y}{\partial n} \quad \text{strongly in } L^2(\partial\Omega), \qquad y_n \to y \quad \text{strongly in } L^2(\partial\Omega).$$

Consequently, $y \in V$, and $L(y,u)$ is minimal, which shows that $[y^*, u^*] = [y, u]$ is an optimal pair.

Next, we apply Theorem 3.1.7 to derive the first-order necessary conditions of optimality. To this end, let $[y^*, u^*] \in V \times U_{ad}$ be an optimal pair. Again, we can assume (possibly making appropriate shifts) that $f = 0$.

First, observe that $A(\cdot, u^*) : V \to L^2(\Omega)$ is a linear and bounded operator. Obviously, A and L are Fréchet differentiable mappings, where

$$\nabla_1 A(y,u) : V \to Y; \quad \nabla_1 A(y,u)(z) = -\Delta z + u\,z \quad \forall\, z \in V,$$

$$\nabla_2 A(y,u) : U \to Y; \quad \nabla_2 A(y,u)(v) = v\,y \quad \forall\, v \in U,$$

$$\nabla_1 L(y,u) : V \to \mathbf{R}; \quad \nabla_1 L(y,u)(z) = \int_{\partial\Omega} (y - g)\, z\, d\sigma \quad \forall\, z \in V,$$

$$\nabla_2 L(y,u) = 0.$$

Obviously, $\nabla_1 A(y^*,u^*)$ is a surjective mapping from V onto Y. Now suppose that $p^* \in Y^* = L^2(\Omega)$ satisfies (3.1.28). Then we have

$$\begin{aligned} 0 &= (\nabla_1 L(y^*,u^*) + (\nabla_1 A(y^*,u^*))^*(p^*), z)_{V^*\times V} \\ &= \int_{\partial\Omega} (y^* - g)\, z\, d\sigma + (p^*, \nabla_1 A(y^*,u^*)(z))_{Y^*\times Y} \\ &= \int_{\partial\Omega} (y^* - g)\, z\, d\sigma + \int_\Omega p^*\,(-\Delta z + u\, z)\, dx \quad \forall\, z \in V. \end{aligned}$$

Applying the Riesz representation theorem to Y, and recalling the fact that V is a dense subspace of Y, we conclude that p^* is in fact the unique transposition solution (compare Appendix 2) to the adjoint system (3.1.28),

$$-\Delta p^* + u^*\, p^* = 0 \quad \text{in } \Omega, \qquad \frac{\partial p^*}{\partial n} = g - y^* \quad \text{on } \partial\Omega.$$

Clearly, we have $\Delta p^* = u^*\, p^* \in L^2(\Omega)$ in the sense of distributions. Standard elliptic regularity theory, using the fact that $g - y^* \in L^2(\partial\Omega)$, then shows that even $p^* \in H^1(\Omega)$. Moreover, it is easily seen that the dual operator $(\nabla_2 A(y^*,u^*))^* : Y^* = L^2(\Omega) \to U^* = L^\infty(\Omega)^*$ is given by

$$(\nabla_2 A(y^*,u^*))^*(p) = y^*\, p \quad \forall\, p \in L^2(\Omega).$$

Thus, the maximum principle (3.1.29) takes in this case the form

$$\int_\Omega y^*(x) p^*(x) (v(x) - u^*(x))\, dx \geq 0 \quad \forall\, v \in U_{ad}.$$

We finally remark that other examples along these lines may be found in Tröltzsch [1984].

3.1.3.1 Singular Control Problems

This class of control problems is characterized by not well-posed state systems and has been studied by Lions [1983] and Bonnans [1982]. Here, we give an example enjoying good differentiability properties; extensions to nondifferentiable cases are possible (cf. Komornik and Tiba [1985], Tiba [1990, Chapter II]).

Let $\Omega \subset \mathbf{R}^3$ be a smooth domain, and let $y_d \in L^6(\Omega)$ be given. We consider the problem

$$\operatorname{Min} \left\{ \frac{1}{6} \int_\Omega (y(x) - y_d(x))^6\, dx + \frac{1}{2} \int_\Omega u^2(x)\, dx \right\}, \tag{3.1.38}$$

$$-\Delta y - y^3 = u \quad \text{in } \Omega, \tag{3.1.39}$$

$$y = 0 \quad \text{on } \partial\Omega. \tag{3.1.40}$$

No explicit constraints are imposed, but control constraints will appear in an implicit manner. A direct application of Theorem 3.1.7 to the problem (3.1.38)–(3.1.40) is impossible, since the mapping θ is not well-defined. Indeed, the state system (3.1.39), (3.1.40) may have no solution or multiple solutions, owing to the presence of the nonmonotone term $-y^3$. Therefore, we will employ another approach using admissible pairs and the so-called *adapted penalization method*. Recall also that in §2.2.1 in Chapter 2 a related situation has been examined from the viewpoint of the existence of optimal pairs.

The first proposition, which follows the lines of Lions [1983, Chapter 3], shows that (3.1.39), (3.1.40) admits at least three solutions in certain situations.

Theorem 3.1.9 *The elliptic problem (3.1.39), (3.1.40) has at least three solutions for* $u = 0$.

Proof. Clearly, $y = 0$ is a solution, and if y is a nonzero solution, so is $-y$. We want to prove the existence of a positive solution. To this end, define the optimization problem

$$\inf \left\{ \int_\Omega |\nabla y(x)|^2 \, dx \right\} \tag{3.1.41}$$

subject to

$$y \in C = \left\{ y \in H_0^1(\Omega) : \int_\Omega |y(x)|^4 \, dx = 1 \right\}.$$

The constraint is meaningful since $H_0^1(\Omega)$ is continuously embedded in $L^6(\Omega)$ if $\Omega \subset \mathbf{R}^3$ (cf. Theorem A2.2 in Appendix 2).

Let $\{y_n\} \subset C$ be a minimizing sequence. Then it is bounded in $H_0^1(\Omega)$, and compact in $L^4(\Omega)$. Hence, we may assume that there is some $y_0 \in H_0^1(\Omega)$ such that $y_n \to y$ weakly in $H_0^1(\Omega)$ and strongly in $L^4(\Omega)$. Clearly, $y_0 \in C$, and y_0 minimizes (3.1.41). Then, by the chain rule, $|y_0| \in H_0^1(\Omega)$ and is again a solution of (3.1.41), and we may thus assume that $y_0 \geq 0$. Owing to the constraint, $y_0 \not\equiv 0$, and it is easily seen that y_0 minimizes the homogeneous functional

$$M(y) = \frac{1}{|y|^2_{L^4(\Omega)}} \int_\Omega |\nabla y(x)|^2 \, dx, \quad y \in H_0^1(\Omega).$$

We have

$$\frac{d}{d\lambda} M(y_0 + \lambda y)|_{\lambda=0} = 0 \quad \forall\, y \in H_0^1(\Omega),$$

and an elementary calculation shows that $y_0 \in H_0^1(\Omega)$ is a weak solution to the problem

$$-\Delta y_0 - |\nabla y_0|^2_{L^2(\Omega)}\, y_0^3 = 0 \quad \text{in } \Omega, \quad y_0 = 0 \quad \text{on } \partial\Omega. \tag{3.1.42}$$

Multiplying (3.1.42) by a constant $K^{-1} > 0$ such that $K^2|\nabla y_0|^2_{L^2(\Omega)} = 1$, we see that $\tilde{y}_0 = K^{-1} y_0$ is the desired solution to (3.1.39), (3.1.40). □

Remark. Bahri [1980], [1981] showed that the problem (3.1.39), (3.1.40) admits infinitely many solutions for every u in a dense subset of $L^2(\Omega)$.

We call $u \in L^2(\Omega)$ *admissible* if there is some $y \in H^2(\Omega)$ satisfying (3.1.39), (3.1.40). For such controls u the cost functional (3.1.38) is, by the Sobolev embedding theorem, well-defined, and the control problem (3.1.38)–(3.1.40) is nontrivial by the above results. We shall understand it as a minimization problem over the set of admissible pairs $[y, u]$ thus defined.

Although the state system is not well-posed, the optimal control problem still is well-posed in the following sense.

Theorem 3.1.10 *The problem (3.1.38)–(3.1.40) admits an optimal pair* $[y^*, u^*]$.

Proof. Let $\{[y_n, u_n]\}$ be a minimizing sequence of admissible pairs. Then, owing to (3.1.38), $\{y_n\}$ is bounded in $L^6(\Omega)$, and $\{u_n\}$ is bounded in $L^2(\Omega)$. Moreover, (3.1.39) shows that $\{y_n\}$ is bounded in $H^2(\Omega) \cap H_0^1(\Omega)$. Hence, we may without loss of generality assume that

$$y_n \to \overline{y} \quad \text{weakly in } H^2(\Omega) \cap H_0^1(\Omega) \cap L^6(\Omega), \qquad u_n \to \overline{u} \text{ weakly in } L^2(\Omega).$$

Moreover, thanks to the compactness of the embedding $H^2(\Omega) \subset L^2(\Omega)$, we may assume that $y_n \to \overline{y}$ pointwise a.e. in Ω. The boundedness of $\{y_n\}$ in $L^6(\Omega)$ and Egorov's theorem then imply that $y_n^3 \to \overline{y}^3$ weakly in $L^2(\Omega)$.

Passing to the limit as $n \to \infty$ in (3.1.39), we obtain that the pair $[\overline{y}, \overline{u}]$ is admissible for the problem (3.1.38)–(3.1.40), and the weak lower semicontinuity of the norm yields that $[y^*, u^*] = [\overline{y}, \overline{u}]$ is optimal. □

Remark. Since the problem (3.1.38)–(3.1.40) is not convex, the optimal pair is not necessarily unique.

Remark. The natural condition that u be admissible may be interpreted as an implicit control constraint. Its nonconvex character shows the difficulty of the problem. It is possible to consider explicit control constraints as well. In this case strong assumptions have to be imposed in order that the admissible set remain nonempty (cf. Lions [1983, Chapter III]).

Theorem 3.1.11 *If $[y^*, u^*] \in H^2(\Omega) \times L^2(\Omega)$ is an optimal pair for (3.1.38)–(3.1.40), then there is some $p^* \in L^2(\Omega)$ such that*

$$\begin{aligned} -\Delta y^* - y^{*3} &= u^* \quad \textit{in } \Omega, \\ -\Delta p^* - 3y^{*2}p^* &= (y^* - y_d)^5 \quad \textit{in } \Omega, \\ y^* = p^* &= 0 \quad \textit{on } \partial\Omega, \\ p^* + u^* &= 0 \quad \textit{in } \Omega. \end{aligned}$$

Remark. According to the argument below, the equation and the boundary conditions for p^* should be understood in the following (transposition) sense:

$$\begin{aligned} &\left(p^*, \Delta\xi + 3y^{*2}\xi\right)_{L^2(\Omega)} + \left((y^* - y_d)^5, \xi\right)_{L^{6/5}(\Omega)\times L^6(\Omega)} = 0, \\ &\forall\, \xi \in H^2(\Omega) \cap H_0^1(\Omega) \cap L^6(\Omega). \end{aligned}$$

Proof of Theorem 3.1.11. In view of the fact that the state system (3.1.39), (3.1.40) is not well-posed, we cannot apply Theorem 3.1.7. We therefore use a special *adapted penalization method*. To this end, let $\varepsilon > 0$ be arbitrary. We define the penalized functional

$$\begin{aligned} J_\varepsilon(z, v) \;=\; & \frac{1}{6}\int_\Omega (z - y_d)^6(x)\, dx + \frac{1}{2}\int_\Omega v^2(x)\, dx + \frac{1}{2\varepsilon}\left|-\Delta z - z^3 - v\right|^2_{L^2(\Omega)} \\ & + \frac{1}{2}\left|z - y^*\right|^2_{L^2(\Omega)} + \frac{1}{2}\left|v - u^*\right|^2_{L^2(\Omega)}, \end{aligned} \tag{3.1.43}$$

which is finite for every $[z, v] \in X = (H^2(\Omega) \cap H_0^1(\Omega) \cap L^6(\Omega)) \times L^2(\Omega)$. The last two terms in (3.1.43) reflect the "adapted" character of J_ε, i.e., we penalize around the optimal pair $[y^*, u^*]$.

First, let us prove that J_ε has for every $\varepsilon > 0$ a minimizer in X. For this purpose, choose a minimizing sequence $\{[z_n, v_n]\} \subset X$ for J_ε. Apparently, this sequence is bounded in X, and we may without loss of generality assume that $[z_n, v_n] \to [z_\varepsilon, v_\varepsilon]$ weakly in X. By the weak lower semicontinuity of norms, $[z_\varepsilon, v_\varepsilon]$ minimizes J_ε.

Now observe that for any $\varepsilon > 0$,

$$J_\varepsilon[z_\varepsilon, v_\varepsilon] \le J_\varepsilon[y^*, u^*] = \frac{1}{6}\left|y^* - y_d\right|^6_{L^6(\Omega)} + \frac{1}{2}\left|u^*\right|^2_{L^2(\Omega)},$$

which implies that $\{[z_\varepsilon, v_\varepsilon]\}_{\varepsilon>0}$ is a bounded subset of X.

Next, we notice that J_ε is Fréchet differentiable on the entire space X. Hence, writing

$$p_\varepsilon = \frac{1}{\varepsilon}(\Delta z_\varepsilon + z_\varepsilon^3 + v_\varepsilon),$$

we have for every $[\xi, \mu] \in X$ that

$$0 = (\nabla J_\varepsilon(z_\varepsilon, v_\varepsilon), [\xi, \mu])_{X^* \times X} = \int_\Omega (z_\varepsilon - y_d)^5 \, \xi \, dx + \int_\Omega v_\varepsilon \, \mu \, dx$$
$$+ \int_\Omega p_\varepsilon \left(\Delta\xi + 3z_\varepsilon^2 \, \xi + \mu\right) \, dx + \int_\Omega (z_\varepsilon - y^*) \, \xi \, dx + \int_\Omega (v_\varepsilon - u^*) \, \mu \, dx \,,$$

whence it follows that

$$\left(p_\varepsilon, \Delta\xi + 3z_\varepsilon^2\xi\right)_{L^2(\Omega)} + \left((z_\varepsilon - y_d)^5, \xi\right)_{L^{6/5}(\Omega)\times L^6(\Omega)}$$
$$+ (z_\varepsilon - y^*, \xi)_{L^2(\Omega)} = 0, \tag{3.1.44}$$
$$(p_\varepsilon + 2v_\varepsilon - u^*, \mu)_{L^2(\Omega)} = 0. \tag{3.1.45}$$

In particular, $p_\varepsilon = u^* - 2v_\varepsilon$, which implies that $\{p_\varepsilon\}_{\varepsilon>0}$ is bounded in $L^2(\Omega)$. It is then a standard argument (compare Lemma 3.2.2 in the next section) to show that on a subsequence, $v_\varepsilon \to u^*$ strongly in $L^2(\Omega)$, and $z_\varepsilon \to y^*$ strongly in $H^2(\Omega) \cap H_0^1(\Omega) \cap L^6(\Omega)$. Denoting by p^* the strong limit of $\{p_\varepsilon\}$, and passing to the limit as $\varepsilon \searrow 0$ in (3.1.44), (3.1.45), we obtain the asserted result. □

Example 3.1.12 We reconsider the plasma problem introduced in Example 1.2.6 in Chapter 1. In the a priori unknown plasma region $D \subset \Omega$ (Ω is the void chamber of the tokamak; see Figure 2.1 in Chapter 1), a semilinear nonsingular (since $x > 0$) elliptic equation is satisfied by the poloidal flux ψ (Blum [1989]),

$$-\frac{\partial}{\partial x}\Big(\frac{1}{x}\frac{\partial\psi}{\partial x}\Big) - \frac{\partial}{\partial y}\Big(\frac{1}{x}\frac{\partial\psi}{\partial y}\Big) = h(\psi) \quad \text{in } D, \tag{3.1.46}$$

where $h : \mathbf{R} \to \mathbf{R}$ is a continuous positive mapping that has also to be determined.

Assuming that the problem of determining the free boundary ∂D has already been solved using the least squares approach described by the relations (1.2.44), (1.2.41)′, (1.2.42)′, (1.2.45), in Chapter 1, it is a natural and important question to identify the function h in $C(\mathbf{R})_+$ from the boundary data on ∂D (which are now available since we assume that the problem is solved in $\Omega \setminus \overline{D}$). It thus makes sense to consider the following global reformulation in Ω:

Given the measurements $\psi = f$ and $\frac{1}{x}\frac{\partial\psi}{\partial n} = g$ on the outer boundary $\partial\Omega$ of the void chamber in a tokamak, find $h \in C(\mathbf{R})_+$ such that ψ solves the Cauchy problem

$$-\frac{\partial}{\partial x}\Big(\frac{1}{x}\frac{\partial\psi}{\partial x}\Big)-\frac{\partial}{\partial y}\Big(\frac{1}{x}\frac{\partial\psi}{\partial y}\Big) = h(\psi) \text{ in } \Omega, \tag{3.1.46$'$}$$

$$\psi = f \quad \text{on } \partial\Omega, \tag{3.1.47}$$

$$\frac{1}{x}\frac{\partial\psi}{\partial n} = g \quad \text{on } \partial\Omega. \tag{3.1.48}$$

If the solution of this identification problem is correctly found, then we will have $h(\psi) \equiv 0$ in some subregion of Ω, while $h(\psi) > 0$ in the remaining part of Ω. The free boundary ∂D is then given by the curve separating the two regions.

Let us apply the least squares method. We then obtain a singular control problem of the form

$$\text{Min}\left\{\frac{1}{2}\int_{\partial\Omega}\left|\frac{1}{x}\frac{\partial\psi}{\partial n}-g\right|^2 d\sigma\right\}, \tag{3.1.49}$$

subject to $h \in C(\mathbf{R})_+$, (3.1.46)$'$, and (3.1.47). The minimization parameter is no longer some datum entering the right-hand side of the equation, or the boundary conditions, or the coefficients. In problem (3.1.49), the control is just the unknown mapping $h \in C(\mathbf{R})_+$, which has to be identified. In view of the nonconvex character of the problem (3.1.49), it is difficult to find its global minimum and, a fortiori, to prove its uniqueness. If this can be done, we say that the *identifiability property* is valid for the problem (3.1.46)$'$, (3.1.47), (3.1.48). However, identifiability cannot be expected, in general, as the following simple counterexample shows.

We take $\Omega_1 = \Big\{(x_1, x_2) \in \mathbf{R}^2 : \ x_1^2 + x_2^2 \le 2\Big\}$ and consider the simplified equation

$$-\Delta\psi = h(\psi) \ge 0 \quad \text{in } \Omega_1, \tag{3.1.46$''$}$$

together with the Cauchy boundary conditions (3.1.47), (3.1.48). Moreover, we assume that another set of measurements is known on some curve Γ (see Figure 2.1 in Chapter 1) contained in Ω_1, for instance on

$$\Gamma = \Big\{(x_1, x_2) \in \Omega_1 : \ x_1^2 + x_2^2 = 1\Big\}.$$

In the applications, such a supplementary set of measurements is usually not available, but even if it is, h cannot be uniquely determined in general. To see this, we define $\Omega_0 = \Big\{(x_1, x_2) \in \Omega_1 : \ 1 \le x_1^2 + x_2^2 \le 2\Big\}$, and the functions

$$y(x_1, x_2) = -\Big(x_1^2 + x_2^2 - 1\Big)\Big(x_1^2 + x_2^2 - 2\Big) \quad \text{in } \Omega_0, \tag{3.1.50}$$

$$y_d = y + d\,y^2, \quad \text{for } d \in \mathbf{R}. \tag{3.1.51}$$

Obviously,

$$y_d = 0 \quad \text{on } \partial\Omega_0, \qquad \frac{\partial y_d}{\partial n} = \frac{\partial y}{\partial n} \quad \text{on } \partial\Omega_0, \qquad \forall\, d \in \mathbf{R}.$$

A simple calculation yields that

$$-\Delta y_d = 16\,w - 12 - 8\,d\left(8\,w^3 - 39\,w^2 + 26\,w - 6\right),$$

where we have set $w = x_1^2 + x_2^2$, which belongs to $[1, 2]$ for each $(x_1, x_2) \in \Omega_0$. Moreover, it is easy to see that

$$-\Delta y_d \geq 88\,d + 4 \geq 0 \quad \text{ if } d > 0.$$

From (3.1.50), (3.1.51) we infer that $0 \leq y(x_1, x_2) \leq \frac{1}{4}$ in Ω_0, and

$$w = \begin{cases} \dfrac{3 - \sqrt{1 - 4\,y}}{2} & \text{in } \left\{(x_1, x_2) \in \Omega_0;\ x_1^2 + x_2^2 \leq \frac{3}{2}\right\} = \Omega_0^1, \\ \dfrac{3 + \sqrt{1 - 4\,y}}{2} & \text{in } \left\{(x_1, x_2) \in \Omega_0;\ x_1^2 + x_2^2 \geq \frac{3}{2}\right\} = \Omega_0^2, \end{cases}$$

$$y = \frac{-1 + \sqrt{1 + 4\,d\,y_d}}{2d} \quad \text{in } \Omega_0.$$

This shows that there exist functions h_1^d and h_2^d such that for every $d > 0$, the corresponding functions y_d satisfy the same conditions on $\partial\Omega_0$, the positivity condition in Ω_0, and the same equation

$$-\Delta y_d = h_i^d(y_d), \quad \text{in } \Omega_0^i, \quad i = 1, 2.$$

This means that the functions h_i^d cannot be identified from the boundary measurements on $\partial\Omega_1$. Observe that the boundary $\partial\Omega_0^1 \cap \partial\Omega_0^2$ in the counterexample plays the role of the free boundary in the plasma example.

In the works of Beretta and Vogelius [1991], [1992], and Vogelius [1994], it is shown that the identifiability property for h is valid for domains Ω with corners. The identifiability question remains open for smooth domains Ω that are not radially symmetric (which was the case in the above counterexample).

Remark. In Neittaanmäki and Tiba [1994, Chapter IV.2], a parabolic singular control problem has been discussed. Although the situation examined there is essentially different, Theorem 3.1.7 may be applied.

3.2 Penalization

Penalization is an important method in optimization problems and has many variants: internal or external, of the constraints or of the state system, exact or adapted, and so on. In this section, we shall examine penalization techniques for the problem (P) defined by (1.1.1)–(1.1.4) in the first section of Chapter 1. Therefore, we again assume that U, V, Z are reflexive Banach spaces that together with their respective duals U^*, V^*, Z^* are strictly convex, and that H is a Hilbert space identified with its dual. In addition, let $f \in Z$ and operators $A \in L(V, Z)$ and $B \in L(U, Z)$ be given, and assume that A possesses an inverse $A^{-1} \in L(Z, V)$. We also assume that $C \subset H$, $U_{ad} \subset U$ are given nonempty, convex, and closed (with respect to the corresponding topologies) sets.

In the first paragraph of this section, we will discuss a standard approach to optimality conditions based on the (exterior) penalization of the constraints and on the use of the so-called *Slater interiority conditions*. In the second paragraph, we will use a penalization of the state equation and a new estimate on the adjoint state, which allows for a relaxation of the Slater assumption. Let us also recall that in Section 2.1 of Chapter 2 the "variational inequality" technique, which may be viewed as an internal penalization method, has already been briefly examined.

3.2.1 The Standard Approach

Let us consider the problem (P) as specified above. We assume the cost functional in the standard form (compare relation (1.2.1))

$$L(y, u) = \theta(y) + \psi(u),$$

where $\theta : V \to \mathbf{R}$ is convex, continuous and $\psi : U \to]-\infty, +\infty]$ denotes some proper, convex, and lower semicontinuous functional.

Let $[y_0, u_0]$ be an admissible pair for (P), which is generally assumed to exist. Defining $z = y - y_0$, $v = u - u_0$, we obtain the "shifted" problem

$$\text{Min}\left\{\theta(z + y_0) + \psi(v + u_0)\right\}, \tag{3.2.1}$$

subject to

$$Az = Bv, \tag{3.2.2}$$

$$z \in C - y_0, \tag{3.2.3}$$

$$v \in U_{ad} - u_0. \tag{3.2.4}$$

Redefining θ, ψ, C, U_{ad} according to the above transformation, we may again suppose that $f = 0$ and $[y_0, u_0] = [0, 0]$ in the problem (P), which simplifies the exposition. Next, observe that we can employ the same argument as in the proof of Theorem 2.1.2 in Chapter 2 to show the existence of at least one optimal pair $[y^*, u^*] \in C \times U_{ad}$ provided that ψ is such that L satisfies the coercivity condition (2.1.6).

We impose the classical *Slater condition*, which, upon taking the transformation (3.2.1)–(3.2.4) into account, has the following form:

$$0 \in \operatorname{int}(C \cap V) \quad (\text{in the topology of } V). \tag{3.2.5}$$

We have the following result.

Theorem 3.2.1 *Let the above general assumptions be fulfilled, and suppose that ψ is continuous at some point in U_{ad}, and that (3.2.5) is satisfied. Then $[y^*, u^*]$ is an optimal pair for (P) if and only if there is some $p^* \in Z^*$ such that*

$$A^*p^* \in \partial\theta(y^*) + \partial I_{C\cap V}(y^*), \tag{3.2.6}$$

$$-B^*p^* \in \partial I_{U_{ad}}(u^*) + \partial\psi(u^*). \tag{3.2.7}$$

Remark. We note that (3.2.6), (3.2.7) are special cases (formally) of (3.1.2), (3.1.3), when the constraints are included in the cost functional. The main difference is given by the formulation of the interiority assumptions (3.2.5), respectively (3.1.5), and by the respective proofs. Notice, however, that (3.2.6), (3.2.7) yield a more precise formulation of the optimality conditions under weaker assumptions.

Proof of Theorem 3.2.1. We first show the necessity of the optimality conditions. The basic idea consists in penalizing the state constraints. To this end, define the regularized problem (P_λ):

$$\operatorname{Min}\left\{\theta_\lambda(y) + \psi_\lambda(u) + (I_C)_\lambda(y) + \frac{1}{2}\,|u - u^*|_U^2\right\}, \tag{3.2.8}$$

$$Ay = Bu, \tag{3.2.9}$$

$$u \in U_{ad}, \tag{3.2.10}$$

where $(I_C)_\lambda$, θ_λ, ψ_λ are the Yosida–Moreau regularizations (cf. Appendix 1) of the indicator function I_C of C in H, and of θ and ψ in V, respectively.

Let us prove that for every $\lambda > 0$, the problem (P_λ) has a uniquely determined optimal pair $[y_\lambda^*, u_\lambda^*]$. To this end, let $\lambda > 0$ be fixed, and let $\{[y_n, u_n]\} \subset V \times U_{ad}$ be a minimizing sequence for (3.2.8)–(3.2.10).

We first show that $\{[y_n, u_n]\}$ is bounded in $V \times U$. For this purpose, we denote in the following by C_i, $i \in \mathbf{N}$, positive constants that do not depend on $n \in \mathbf{N}$ (but possibly on $\lambda > 0$). First observe that, owing to Theorem A1.5 in Appendix 1, the mapping $(I_C)_\lambda$ (respectively, θ_λ and ψ_λ) is Gâteaux differentiable on H (respectively, on V), and we have

$$\psi_\lambda(u_n) \geq \psi_\lambda(0) + (\nabla\psi_\lambda(0), u_n)_{U^*\times U} \geq -C_1\,(1 + |u_n|_U)\,.$$

Likewise, using the fact that $Ay_n = Bu_n$, that is, $|y_n|_V \leq C_2\,|u_n|_U$, by the assumptions on A and B,

$$\theta_\lambda(y_n) \geq -C_3\,(1 + |y_n|_V) \geq -C_4\,(1 + |u_n|_U)\,.$$

Finally, $(I_C)_\lambda$ is obviously nonnegative on H. Thus, invoking the boundedness of the cost functional (3.2.8) on the minimizing sequence, and the quadratic *adapted penalization* term $\frac{1}{2}|u - u^*|_U^2$ in (3.2.8), we conclude that $\{u_n\}$ is bounded in U. It then follows from (3.2.9), in view of the boundedness of B and of A^{-1}, that $\{y_n\}$ is bounded in V. Hence, a suitable subsequence of $\{[y_n, u_n]\}$ converges weakly in $V \times U$ to some $[y_\lambda^*, u_\lambda^*] \in V \times U_{ad}$ that satisfies $Ay_\lambda^* = Bu_\lambda^*$, and, owing to the weak lower semicontinuity, is a minimizer of the penalized cost functional.

To prove uniqueness, assume that two optimal pairs $[y_i, u_i]$, $i = 1, 2$, are given. Since U is strictly convex, it follows (again owing to the presence of the adapted penalization term) that the penalized cost functional is strictly convex with respect to u, which implies that $u_1 = u_2$. From (3.2.9), and from the invertibility of A, it then follows that also $y_1 = y_2$, as claimed.

Now let $v \in U_{ad} - u_\lambda^*$ be arbitrary. Then $u_\mu = u_\lambda^* + \mu v \in U_{ad}$ for all $\mu \in [0,1]$, by the convexity of U_{ad}. We put

$$y_\mu = A^{-1}Bu_\mu, \qquad r = \frac{1}{\mu}\,(y_\mu - y_\lambda^*) \in V\,.$$

Then $Ar = Bv$, and we obtain from the optimality of $[y_\lambda^*, u_\lambda^*]$, using the Gâteaux differentiability of the Yosida–Moreau regularizations, as well as the fact that

$$(\nabla(I_C)_\lambda(y_\lambda^*), r)_H = (\nabla(I_C)_\lambda(y_\lambda^*), r)_{V^*\times V}\,,$$

that

$$\begin{aligned}
0 \leq \lim_{\mu\searrow 0}\frac{1}{\mu}\Big[&\theta_\lambda(y_\mu) + \psi_\lambda(u_\mu) + (I_C)_\lambda(y_\lambda) + \frac{1}{2}|u_\mu - u^*|_U^2 \\
&- \theta_\lambda(y_\lambda^*) - \psi_\lambda(u_\lambda^*) - (I_C)_\lambda(y_\lambda^*) - \frac{1}{2}|u_\lambda^* - u^*|_U^2\Big] \\
= (\nabla\theta_\lambda(y_\lambda^*), r)_{V^*\times V} &+ (\nabla\psi_\lambda(u_\lambda^*), v)_{U^*\times U} + (\nabla(I_C)_\lambda(y_\lambda^*), r)_{V^*\times V} \\
&+ (F_U(u_\lambda^* - u^*), v)_{U^*\times U},
\end{aligned}$$

where $F_U : U \to U^*$ denotes the duality mapping (see Appendix 1).

We define $p_\lambda \in Z^*$ as the unique solution of the adjoint state equation

$$A^* p_\lambda = \nabla \theta_\lambda(y_\lambda^*) + \nabla (I_C)_\lambda(y_\lambda^*) \in V^*, \tag{3.2.11}$$

which exists since the dual operator A^* has a bounded inverse. Then it follows from (3.2.11) that

$$0 \le (A^* p_\lambda, r)_{V^* \times V} + (\nabla \psi_\lambda(u_\lambda^*), v)_{U^* \times U} + (F_U(u_\lambda^* - u^*), v)_{U^* \times U}.$$

Now observe that

$$(A^* p_\lambda, r)_{V^* \times V} = (p_\lambda, Ar)_{Z^* \times Z} = (p_\lambda, Bv)_{Z^* \times Z} = (B^* p_\lambda, v)_{U^* \times U}.$$

Thus, since $v \in U_{ad} - u_\lambda^*$ is arbitrary, we can infer that

$$B^* p_\lambda + \nabla \psi_\lambda(u_\lambda^*) + F_U(u_\lambda^* - u^*) \in -\partial I_{U_{ad}}(u_\lambda^*). \tag{3.2.12}$$

At this point, we stop briefly to establish the convergence of u_λ^*, y_λ^*, in order to pass to the limit as $\lambda \searrow 0$ in (3.2.12), (3.2.11). We have the following result.

Lemma 3.2.2 *As* $\lambda \searrow 0$,

$$u_\lambda^* \to u^* \quad \textit{strongly in } U, \tag{3.2.13}$$

$$y_\lambda^* \to y^* \quad \textit{strongly in } V. \tag{3.2.14}$$

Proof. In the following, we denote by C_i, $i \in \mathbf{N}$, generic constants that do not depend on $\lambda > 0$. By the minimum property, and since $[0,0]$ is admissible by assumption, we have

$$\theta_\lambda(y_\lambda^*) + \psi_\lambda(u_\lambda^*) + (I_C)_\lambda(y_\lambda^*) + \frac{1}{2}|u_\lambda^* - u^*|_U^2$$

$$\le \theta_\lambda(0) + \psi_\lambda(0) + \frac{1}{2}|u^*|_U^2 \le C_1. \tag{3.2.15}$$

Moreover, (3.2.2) implies that

$$|y_\lambda^*|_V \le C_2 \, |u_\lambda^*|_U.$$

Since $(I_C)_\lambda(y_\lambda^*) \ge 0$, and since θ_λ, ψ_λ are bounded from below (uniformly with respect to λ) by suitable affine functions, (3.2.15) shows that $\{u_\lambda^*\}_{\lambda>0}$ is bounded in U, and consequently, $\{y_\lambda^*\}_{\lambda>0}$ is bounded in V.

Hence, there is a subsequence $\lambda_n \searrow 0$ such that $u_{\lambda_n}^* \to \overline{u}$ weakly in U, and $y_{\lambda_n}^* \to \overline{y}$ weakly in V, where $\overline{u} \in U_{ad}$ and $A\overline{y} = B\overline{u}$.

Now observe that (3.2.15) also implies that

$$(I_C)_\lambda(y_\lambda^*) = \inf\left\{\frac{1}{2\lambda}|y_\lambda^* - y|_H^2 + I_C(y) : y \in H\right\}$$
$$= \frac{1}{2\lambda} \inf\left\{|y_\lambda^* - y|_H^2 : y \in C\right\} \leq C_3,$$

and consequently,

$$\lim_{\lambda \searrow 0} \inf\left\{|y_\lambda^* - y|_H^2 : y \in C\right\} = 0.$$

Now recall that V is continuously embedded in H. Therefore, $y_{\lambda_n} \to \overline{y}$ weakly in H, and the weak lower semicontinuity of $|\cdot|_H$ yields that also

$$\inf\left\{|\overline{y} - y|_H^2 : y \in C\right\} = 0,$$

or, since C is closed in H, that $\overline{y} \in C$. Consequently, the pair $[\overline{y}, \overline{u}]$ is admissible for (P).

Next, let J_λ^θ denote the *resolvent* operator associated with $\partial\theta : V \to V^*$ (cf. Appendix 1). By Theorem A1.5(ii), we have

$$\theta_\lambda(y_\lambda^*) = \frac{1}{2\lambda}\left|y_\lambda^* - J_\lambda^\theta(y_\lambda^*)\right|_V^2 + \theta\big(J_\lambda^\theta(y_\lambda^*)\big), \tag{3.2.16}$$

which, thanks to (3.2.15), is uniformly bounded in $\lambda > 0$. Since, owing to Theorem A1.5(i),

$$\theta\big(J_\lambda^\theta(y_\lambda^*)\big) \leq \theta_\lambda(y_\lambda^*),$$

which is again uniformly bounded by (3.2.15), we conclude that

$$J_{\lambda_n}^\theta(y_{\lambda_n}^*) \to \overline{y} \quad \text{weakly in } V.$$

By the same token,

$$J_{\lambda_n}^\psi(u_{\lambda_n}^*) \to \overline{u} \quad \text{weakly in } U,$$

with obvious notation.

Now observe that by assumption, $[y^*, u^*]$ is optimal for (P) and thus, in particular, admissible for (P_λ). Therefore, invoking the minimality property of $[y_\lambda^*, u_\lambda^*]$ for (P_λ) and Theorem A1.5(i), we have the estimate

$$\theta\big(J_\lambda^\theta(y_\lambda^*)\big) + \psi\big(J_\lambda^\psi(u_\lambda^*)\big) + \frac{1}{2}|u_\lambda^* - u^*|_U^2$$
$$\leq \theta_\lambda(y_\lambda^*) + \psi_\lambda(y_\lambda^*) + \frac{1}{2}|u_\lambda^* - u^*|^2 \leq \theta_\lambda(y^*) + \psi_\lambda(u^*).$$

Taking the above subsequence $\{\lambda_n\}$, and letting $n \to \infty$ in this inequality, we infer from the weak lower semicontinuity that

$$\theta(\overline{y}) + \psi(\overline{u}) + \frac{1}{2}|\overline{u} - u^*|_U^2 \le \theta(y^*) + \psi(u^*).$$

Thus, since $[\overline{y}, \overline{u}]$ is admissible for (P), we must have $\overline{u} = u^*$ and $\overline{y} = y^*$.

Now notice that the above estimate also implies that the adapted penalization term converges to zero. Hence, $u^*_{\lambda_n} \to u^*$ strongly in U, which entails that also $y^*_{\lambda_n} \to y^*$ strongly in V. Since the limit point is uniquely determined, these convergences hold for the whole families $\{y^*_\lambda\}_{\lambda>0}$ and $\{u^*_\lambda\}_{\lambda>0}$, respectively. This completes the proof of the lemma. □

Proof of Theorem 3.2.1. (continued)

We estimate p_λ, where we denote by C_i, $i \in \mathbf{N}$, positive constants that do not depend on $\lambda > 0$. By virtue of the Slater condition (3.2.5), there is some $\rho > 0$ such that $\rho v \in C$ for all $v \in V$ satisfying $|v|_V = 1$. Multiplying (3.2.11) by $y^*_\lambda - \rho v$, where $|v|_V = 1$, we obtain that

$$\begin{aligned}(A^* p_\lambda, y^*_\lambda - \rho v)_{V^*\times V} &= (\nabla\theta_\lambda(y^*_\lambda) + \nabla(I_C)_\lambda(y^*_\lambda), y^*_\lambda - \rho v)_{V^*\times V}\\ &\ge \theta_\lambda(y^*_\lambda) - \theta_\lambda(\rho v) + (I_C)_\lambda(y^*_\lambda) - (I_C)_\lambda(\rho v).\end{aligned}$$

Now observe that by definition, $(I_C)_\lambda(y^*_\lambda) \ge 0$ and $(I_C)_\lambda(\rho v) = 0$. Also, $\theta(\rho v)$ is bounded, since θ is continuous. Hence, using Theorem A1.5(i), we have

$$-\theta_\lambda(\rho v) \ge -\theta(\rho v) \ge -C_1 .$$

Also, using Theorem A1.5(i) again, and recalling that by Proposition A1.4(ii) the operator J^θ_λ is bounded, we have that

$$\theta_\lambda(y^*_\lambda) \ge \theta\big(J^\theta_\lambda(y^*_\lambda)\big) \ge -C_2\left(1 + \left|J^\theta_\lambda(y^*_\lambda)\right|_V\right) \ge -C_3 ,$$

since $\{y^*_\lambda\}_{\lambda>0}$ is bounded in V. Summarizing, we have

$$\begin{aligned}(A^* p_\lambda, \rho v)_{V^*\times V} &\le C_4 + (p_\lambda, A y^*_\lambda)_{Z^*\times Z} = C_4 + (p_\lambda, B u^*_\lambda)_{Z^*\times Z}\\ &= C_4 + (B^* p_\lambda, u^*_\lambda)_{U^*\times U} .\end{aligned}$$

Now notice that (3.2.12) implies the existence of some $w_\lambda \in \partial I_{U_{ad}}(u^*_\lambda)$ such that

$$B^* p_\lambda = -\nabla\psi_\lambda(u^*_\lambda) - F_U(u^*_\lambda - u^*) - w_\lambda .$$

Using the boundedness of $\{u^*_\lambda\}_{\lambda>0}$, the fact that $[0, 0]$ is admissible, and the boundedness of the operator J^ψ_λ, and invoking Theorem A1.5(i) once again, we find that

$$(B^* p_\lambda, u^*_\lambda)_{U^*\times U} \le C_5 - (w_\lambda + \nabla\psi_\lambda(u^*_\lambda), u^*_\lambda)_{U^*\times U}$$

$$\leq C_5 - I_{U_{ad}}(u_\lambda^*) + I_{U_{ad}}(0) - \psi_\lambda(u_\lambda^*) + \psi_\lambda(0)$$
$$\leq C_6 - \psi\Big(J_\lambda^\psi(u_\lambda^*)\Big) + \psi(0) \leq C_7.$$

Since $v \in V$ with $|v|_V = 1$ is arbitrary, the set $\{A^*p_\lambda\}_{\lambda>0}$ is bounded in V^* by the uniform boundedness principle; see Yosida [1980]. Since A^* has a continuous inverse, it follows that $\{p_\lambda\}_{\lambda>0}$ is bounded in Z^*, and there is a subsequence $\lambda_n \searrow 0$ such that $p_{\lambda_n} \to p^*$ weakly in Z^*, and thus $A^*p_{\lambda_n} \to A^*p^*$ weakly in V^*, for some $p^* \in Z^*$.

Now notice that by the definition of the subdifferential, (3.2.11) may be rewritten as

$$(A^*p_\lambda, y_{\lambda_n}^* - w)_{V^*\times V} \geq \theta_\lambda(y_{\lambda_n}^*) - \theta_{\lambda_n}(w) + (I_C)_{\lambda_n}(y_{\lambda_n}^*) \quad \forall\, w \in C \cap V.$$

Passing to the limit as $n \to \infty$, we obtain

$$(A^*p^*, y^* - w)_{V^*\times V} \geq \theta(y^*) - \theta(w) \quad \forall\, w \in C \cap V,$$

that is,

$$A^*p^* \in \partial(\theta + I_{C\cap V})(y^*),$$

whence (3.2.6) follows, since the continuity of θ implies the additivity of the subdifferential.

A similar argument, using (3.2.12), may be employed to obtain that

$$-B^*p^* \in \partial(I_{U_{ad}} + \psi)(u^*).$$

Since, by assumption, ψ is continuous at some point in U_{ad}, we again have the additivity of the subdifferential, and (3.2.8) is also proved. This concludes the proof of necessity.

Next, we turn to the proof of the sufficiency of the optimality conditions. Let $w \in C \cap V$ and $\ell \in U_{ad}$ be arbitrary. Then, by the definition of the subdifferential, we conclude from (3.2.6) and (3.2.7) that

$$(A^*p^*, y^* - w)_{V^*\times V} - (B^*p^*, u^* - \ell)_{U^*\times U} \geq \theta(y^*) - \theta(w) + \psi(u^*) - \psi(\ell).$$

Now suppose that $[w, \ell]$ is admissible for (P). Then $Aw = B\ell$, so that the left-hand side vanishes, which means that $[y^*, u^*]$ is optimal for (P). With this, the assertion is completely proved. □

Remark. We notice that the Slater assumption (3.2.5) was used only in the proof of necessity.

Remark. The above method is called *adapted penalization* and was used in

Barbu and Precupanu [1986], and Barbu [1984] for optimal control problems governed by abstract evolution equations. Although the convergence properties stated in Lemma 3.2.2 are strong, this approach is not constructive, since the unknown u^* occurs in the definition of (P_λ).

Example 3.2.3 In this example, we apply Theorem 3.2.1 to the problem of the *boundary control via Dirichlet conditions* discussed in Example 1.2.2 in Chapter 1 under very weak assumptions for the controls. To this end, let $\Omega \subset \mathbf{R}^d$ be a smooth domain, and assume that $a_0 \in L^\infty(\Omega)$ is nonnegative, and that $a_{ij} \in W^{1,\infty}(\Omega)$ for $1 \le i,j \le d$, where possibly $a_{ij} \neq a_{ji}$. We then consider the elliptic operator A defined in (1.2.6) and examine the control problem

$$\operatorname*{Min}_{u \in H^{-1/2}(\partial\Omega)} \left\{ \frac{1}{2} \int_\Omega (y(x) - y_d(x))^2 \, dx + \frac{1}{2} |u|^2_{H^{-1/2}(\partial\Omega)} \right\}$$

subject to the state system (compare (1.2.20), (1.2.24))

$$Ay = f \quad \text{in } \Omega, \qquad y = u \quad \text{on } \partial\Omega,$$

where $y_d \in L^2(\Omega)$ is a desired target. Since only $u \in U = H^{-1/2}(\partial\Omega)$, we have $y \in L^2(\Omega)$, and we have to consider A as an operator acting from $L^2(\Omega)$ into the space X^* where $X = H^2(\Omega) \cap H_0^1(\Omega)$. Accordingly, the state equation has to be interpreted in the transposition sense (cf. Example A2.7 in Appendix 2).

To derive the necessary conditions of optimality, we fit the problem into the setting of Theorem 3.2.1 by employing the transformation introduced in Example 1.2.2 in Chapter 1, where we may (possibly after making appropriate shifts) assume that $f \equiv 0$. We put

$$V = Z = H = V^* = Z^* = L^2(\Omega),$$

and let

$$\widetilde{A} : L^2(\Omega) \to L^2(\Omega), \quad \widetilde{A}y = y,$$

be the identity operator. The operator $\widetilde{A}$, which is obviously continuously invertible, will play the role of the operator A in the application of Theorem 3.2.1. Moreover, we define the operator $B \in L(H^{-1/2}(\partial\Omega), L^2(\Omega))$ by $Bu = y_u$, where $y_u \in L^2(\Omega)$ is the unique solution to

$$Ay_u = 0 \quad \text{in } \Omega, \tag{3.2.17}$$

$$y_u = u \quad \text{on } \partial\Omega, \tag{3.2.18}$$

in the transposition sense. This means that

$$\int_\Omega y_u \Big[\sum_{i,j=1}^d \frac{\partial}{\partial x_j} \left(a_{ij} \frac{\partial \varphi}{\partial x_i} \right) - a_0 \, \varphi \Big] \, dx = \int_{\partial\Omega} u \, \frac{\partial \varphi}{\partial n_{A^*}} \, d\sigma \quad \forall \varphi \in X, \tag{3.2.19}$$

where $\partial\varphi/\partial n_{A^*}$ denotes the outer conormal derivative (compare (1.2.12) in Chapter 1) associated with the dual operator $A^* : X \to L^2(\Omega)$,

$$A^*\varphi(x) = -\sum_{i,j=1}^{d} \frac{\partial}{\partial x_j}\left(a_{ij}(x)\frac{\partial\varphi}{\partial x_i}\right) + a_0(x)\,\varphi(x), \quad \varphi \in X.$$

With this notation, the governing state equation can be rewritten as $\widetilde{A}y = Bu$, which will play the role of the state equation in the application of Theorem 3.2.1.

Next, we calculate the dual operator $B^* : L^2(\Omega) \to H^{1/2}(\partial\Omega)$ of B. To this end, let $v \in L^2(\Omega)$ be arbitrary, and let $p \in X$ satisfy

$$A^*p = v \quad \text{in } \Omega, \qquad p = 0 \quad \text{on } \partial\Omega. \tag{3.2.20}$$

Then,

$$\begin{aligned}(u, B^*v)_{H^{-1/2}(\partial\Omega)\times H^{1/2}(\partial\Omega)} &= (Bu, v)_{L^2(\Omega)} = (y_u, v)_{L^2(\Omega)} \\ &= (y_u, A^*p)_{L^2(\Omega)} = -\int_{\partial\Omega} u\,\frac{\partial p}{\partial n_{A^*}}\,d\sigma;\end{aligned}$$

that is, we have

$$B^*v = -\frac{\partial p}{\partial n_{A^*}}\,.$$

Suppose now that $[y^*, u^*]$ is an optimal pair. By Theorem 3.2.1, there is some $v^* \in L^2(\Omega)$ such that (for $p^* \in X$ given by the solution to (3.2.20) for $v = v^*$)

$$\widetilde{A}^*v^*\,(= v^*) = y^* - y_d \quad \text{in } \Omega, \tag{3.2.21}$$

$$-\,B^*v^*\left(= \frac{\partial p^*}{\partial n_{A^*}}\right) = u^* \quad \text{on } \partial\Omega. \tag{3.2.22}$$

Eliminating the "auxiliary" adjoint state v^* from (3.2.20)–(3.2.22), we finally arrive at the standard optimality system with the "true" adjoint state $p^* \in H^2(\Omega) \cap H_0^1(\Omega)$,

$$A^*p^* = y^* - y_d \quad \text{in } \Omega, \qquad \frac{\partial p^*}{\partial n_{A^*}} = u^* \quad \text{on } \partial\Omega.$$

Remark. In the textbook by Barbu [1984, Chapter 3.8], Dirichlet boundary control problems are treated using the theory of variational inequalities. Our treatment above is a purely linear one. In Lions [1968, Chapter II.5] a direct argument is presented for such examples.

Example 3.2.4 Another variant of interest is related to *boundary observation problems*. As in Example 3.2.3 above, and again with $f = 0$, the control system is defined by (1.2.20), (1.2.24), but this time the cost functional is chosen as

$$L(y,u) = \frac{1}{2}\left|\frac{\partial y}{\partial n_A} - y_d\right|^2_{H^{1/2}(\partial\Omega)} + \frac{1}{2}\,|u|^2_{H^{3/2}(\partial\Omega)},$$

where $y_d \in H^{1/2}(\partial\Omega)$ is the desired target. In this case, we have maximal regularity; indeed, since $u \in U = H^{3/2}(\partial\Omega)$, it follows from standard elliptic regularity theory that $y \in H^2(\Omega)$.

We define the operator B as in Example 3.2.3, where this time we have $B \in L(H^{3/2}(\partial\Omega), H^2(\Omega))$. Moreover, we put $V = Z = H^2(\Omega)$, and $\widetilde{A} : V \to Z$ is the identity operator. Again, $\widetilde{A}^* : H^2(\Omega)^* \to H^2(\Omega)^*$ is the identity operator. Finally, we define the observation operator

$$D : V \to W = H^{1/2}(\partial\Omega)\,, \quad Dy = \frac{\partial y}{\partial n_A}\,.$$

Suppose now that $[y^*, u^*] \in V \times U$ is an optimal pair. By Theorem 3.2.1, there is some adjoint state $v^* \in H^2(\Omega)^*$ satisfying the adjoint state equation (3.2.6), which, with the canonical isomorphism $J : H^{1/2}(\partial\Omega) \to H^{-1/2}(\partial\Omega)$, takes the form

$$\widetilde{A}^* v^* = v^* = D^* J(Dy^* - y_d). \tag{3.2.23}$$

Since $H^2(\Omega)^*$ is not a space of distributions, $D^* \in L(H^{-1/2}(\partial\Omega), H^2(\Omega)^*)$ is difficult to describe explicitly. We remedy this by applying B^* to (3.2.23), which yields

$$B^* v^* = (DB)^* J(Dy^* - y_d). \tag{3.2.24}$$

The operator $DB \in L(H^{3/2}(\partial\Omega), H^{1/2}(\partial\Omega))$, $D\,B\,u = \frac{\partial y}{\partial n_A}$, is the *Dirichlet-to-Neumann mapping*, and its adjoint $(DB)^* \in L(H^{-1/2}(\partial\Omega), H^{-3/2}(\partial\Omega))$ is again the Dirichlet-to-Neumann mapping. Now, if $p^* \in L^2(\Omega)$ is the unique transposition solution to

$$A^* p^* = 0 \quad \text{in } \Omega, \tag{3.2.25}$$

$$p^* = J\left(\frac{\partial y^*}{\partial n_A} - y_d\right) \quad \text{on } \partial\Omega, \tag{3.2.26}$$

then, by formally multiplying (3.2.25) by $y = B\,u$ and integrating by parts, we have

$$\int_{\partial\Omega} \frac{\partial y}{\partial n_A}\, J\left(\frac{\partial y^*}{\partial n_A} - y_d\right) d\sigma = \int_{\partial\Omega} u\, \frac{\partial p^*}{\partial n_{A^*}}\, d\sigma.$$

This can be justified rigorously by using a smoothing of $J(\frac{\partial y^*}{\partial n_A} - y_d)$ in (3.2.26) and then passing to the limit in the resulting relations. We then have, owing to the definition of the dual operator,

$$(DB)^* J(Dy^* - y_d) = \frac{\partial p^*}{\partial n_{A^*}} \quad \text{on } \partial\Omega.$$

Since $p^* \in L^2(\Omega)$ and $A^* p^* = 0 \in L^2(\Omega)$, it follows from a well-known regularity result due to Lions and Magenes [1968, Chapter II] that $\partial p^* / \partial n_{A^*} \in H^{-1}(\partial\Omega)$.

Thanks to (3.2.24), we can determine $B^* v^*$ from (3.2.25), (3.2.26), and the maximum principle (3.2.7) becomes, with the canonical isomorphism $J_1 : H^{\frac{3}{2}}(\partial\Omega) \to H^{-\frac{3}{2}}(\partial\Omega)$,

$$-\frac{\partial p^*}{\partial n_{A^*}} = J_1 u^* \quad \text{on } \partial\Omega. \tag{3.2.27}$$

Again, the variable v^* obtained from Theorem 3.2.1 plays only an auxiliary role, and the optimality conditions with the "true" adjoint state p^* are given by (3.2.25)–(3.2.27).

Remark. In §5.1.4 of Chapter 5 (see (5.1.78)–(5.1.80)) the case of mixed boundary conditions will be briefly discussed in a different context.

Remark. In the above examples, we have intentionally omitted possible further constraints that would enter into the optimality system via suitable subdifferential operators (see Section 3.1). In the next paragraph, we will pay special attention to constraints by discussing the relaxation of the Slater condition.

3.2.2 Penalization of the State Equation

Again we study, with a different approach and under weaker assumptions, the general control problem (1.1.1), (1.1.2) from Chapter 1, where the spaces U, V, Z, and the cost functional L have the properties listed there, and where $f \in Z$ is given. Let $B \in L(U, Z)$, and assume that $A \in L(V, Z)$ has a bounded inverse. We consider the mixed (state-control) constraints

$$[y, u] \in D, \tag{3.2.28}$$

where $D \subset V \times U$ is nonempty, closed, and convex.

As always, we assume admissibility, that is, that there is some $[y, u] \in D$ satisfying $Ay = Bu + f$ and $L(y, u) < +\infty$. If L is coercive with respect to u uniformly in y (cf. (2.1.6)), then analogous reasoning as in the proof of Theorem 2.1.2 in Chapter 2 yields the existence of at least one optimal pair $[y^*, u^*] \in D$.

The Slater (interiority) condition (3.2.5) (see also condition (3.1.17) and the last remark in §3.1.2 in the previous section) can in this setting be reformulated in the form

$$\text{There is some feasible pair } [\overline{y},\overline{u}] \text{ such that } \overline{y}\in \operatorname{int}\Big(\{y\in V:[y,\overline{u}]\in D\}\Big) \text{ with respect to the topology of } V. \tag{3.2.29}$$

The main aim of this section is to weaken this classical constraint qualification. More precisely, we assume that instead of (3.2.29),

$$\text{There is some bounded set } M\subset D\subset V\times U \text{ such that } 0\in \operatorname{int}(T(M)) \text{ with respect to the topology of } Z, \tag{3.2.30}$$

where $T: V\times U\to Z$ is defined by

$$T(y,u) \;=\; Ay - Bu - f. \tag{3.2.31}$$

First, let us verify that condition (3.2.30) is indeed weaker than (3.2.29).

Proposition 3.2.5 *If (3.2.29) holds, then (3.2.30) is fulfilled.*

Proof. Condition (3.2.29) implies, in particular, that

$$A\overline{y} \;=\; B\overline{u} + f. \tag{3.2.32}$$

Let $\rho>0$, and let $\xi\in Z$ with $|\xi|_Z\le\rho$ be arbitrary. We denote by y_ξ the (unique) solution to

$$Ay_\xi \;=\; B\overline{u} + f + \xi. \tag{3.2.33}$$

Taking the difference between (3.2.32), (3.2.33), and invoking the boundedness of A^{-1}, we find that

$$|\overline{y}-y_\xi|_V \;\le\; \left|A^{-1}\right|_{L(Z,V)}\,\rho.$$

Hence, for sufficiently small $\rho>0$, it follows from (3.2.29) that $[y_\xi,\overline{u}]\in D$ for all $\xi\in Z$ with $|\xi|_Z\le\rho$, and it is then easily seen that (3.2.30) holds true with the choice

$$M=\{[y_\xi,\overline{u}]:\xi\in Z,\ |\xi|_Z\le\rho\}.$$

The assertion is thus proved. □

Remark. Assume that $U\subset Z$ with continuous embedding, and that $B:U\to Z$ is the canonical injection. Then one may postulate the following interiority condition for the control:

There is some feasible pair $[\tilde{y}, \tilde{u}]$ such that $\operatorname{int}\Big(\{u \in U : [\tilde{y}, u] \in D\}\Big) \neq \emptyset$ with respect to the topology of Z. (3.2.34)

Using a similar argument as above, one readily verifies that also (3.2.34) implies (3.2.30). However, in this case the Slater condition (3.2.29) need not be satisfied; that is, (3.2.30) is strictly weaker than (3.2.29).

Assume now that $[y^*, u^*]$ is an optimal pair. In order to derive the first-order necessary optimality conditions, we define the penalized problem

$$(\mathrm{P}_\lambda) \qquad \operatorname*{Min}_{[y,u]\in D} \Big\{ L_\lambda(y,u) + \frac{1}{2}\,|u - u^*|_U^2 + \frac{1}{2\lambda}\,|Ay - Bu - f|_Z^2 \Big\},$$

where L_λ denotes the Yosida–Moreau regularization (cf. Appendix 1) of the convex function L for $\lambda > 0$. We note that the use of this cost functional constitutes a combination of a penalization of the state equation with the adapted penalization method, and that the state equation disappears in (P_λ).

First, let us show that (P_λ) admits a unique optimal pair $[y_\lambda, u_\lambda] \in D$. To this end, let $\{[y_n, u_n]\} \subset D$ be a minimizing sequence. We show that $\{[y_n, u_n]\}$ is bounded in $V \times U$.

We denote by C_i, $i \in \mathbf{N}$, positive constants that do not depend on n. To begin with, notice that for $z_n = Ay_n - Bu_n - f$ we have the estimate

$$|y_n|_V = \left|A^{-1}(Bu_n + f + z_n)\right|_V \leq C_1\left(|u_n|_U + |z_n|_Z + 1\right) \quad \forall n \in \mathbf{N}. \tag{3.2.35}$$

Now observe that L_λ is bounded from below by a mapping that is affine in both y and u, uniformly in $\lambda > 0$. Hence, by (3.2.35) and since $\{[y_n, u_n]\}$ is a minimizing sequence, it follows that for every $\delta > 0$,

$$\begin{aligned}
C_2 &\geq L_\lambda(y_n, u_n) + \frac{1}{2}\,|u_n - u^*|_U^2 + \frac{1}{2\lambda}\,|z_n|_Z^2 \\
&\geq -C_3\left(|y_n|_V + |u_n|_U + 1\right) + \frac{1}{2}\,|u_n - u^*|_U^2 + \frac{1}{2\lambda}|z_n|_Z^2 \\
&\geq -C_4\left(|z_n|_Z + |u_n|_U + 1\right) + \frac{1}{2}\,|u_n - u^*|_U^2 + \frac{1}{2\lambda}\,|z_n|_Z^2 \\
&\geq -C_4\left(\frac{\delta}{4}\,|z_n|_Z^2 + \frac{\delta}{4}\,|u_n|_U^2 + \frac{2}{\delta} + 1\right) + \frac{1}{2}\,|u_n - u^*|_U^2 + \frac{1}{2\lambda}\,|z_n|_Z^2\,.
\end{aligned}$$

Choosing $\delta > 0$ sufficiently small, we find that $\{|u_n|_U\}$ and $\{|z_n|_Z\}$ are bounded, and (3.2.35) yields the boundedness of $\{|y_n|_V\}$.

Hence, there is some subsequence, again indexed by n, such that $[y_n, u_n] \to [y_\lambda, u_\lambda]$ weakly in $V \times U$. Thanks to the convexity of D, $[y_\lambda, u_\lambda] \in D$. The weak lower semicontinuity of the penalized cost functional then shows that $[y_\lambda, u_\lambda]$ is optimal. The uniqueness of the optimal pair follows from the fact that the penalized cost functional is strictly convex.

Proposition 3.2.6 *For* $\lambda \searrow 0$,

$$y_\lambda \to y^* \quad \textit{strongly in } V, \qquad u_\lambda \to u^* \quad \textit{strongly in } U, \tag{3.2.36}$$

$$\{\lambda^{1/2} r_\lambda\} \quad \textit{is bounded in } Z, \qquad \textit{where } r_\lambda = \frac{1}{\lambda}\,(Ay_\lambda - Bu_\lambda - f). \tag{3.2.37}$$

Proof. The optimality of $[y_\lambda, u_\lambda]$ and the properties of the Yosida–Moreau regularization (cf. Theorem A1.5(i)) yield that

$$\begin{aligned} L_\lambda(y_\lambda, u_\lambda) &+ \frac{1}{2}\,|u_\lambda - u^*|_U^2 + \frac{1}{2\lambda}\,|Ay_\lambda - Bu_\lambda - f|_Z^2 \\ &\le L_\lambda(y^*, u^*) \le L(y^*, u^*). \end{aligned} \tag{3.2.38}$$

Note that we have the relation

$$Ay_\lambda = Bu_\lambda + f + \lambda\, r_\lambda. \tag{3.2.39}$$

Thus, similarly as in (3.2.35), $|y_\lambda|_V$ is linearly bounded in terms of $|u_\lambda|_U$ and $|\lambda\, r_\lambda|_Z$. Since L_λ is bounded from below by an affine mapping (uniformly with respect to $\lambda > 0$), and thanks to (3.2.38) and (3.2.39), we find that $\{|u_\lambda|_U\}_{\lambda>0}$ and $\{|\lambda^{1/2}\, r_\lambda|_Z\}_{\lambda>0}$ are bounded, whence also the boundedness of $\{|y_\lambda|_V\}_{\lambda>0}$ follows.

Consequently, there is some subsequence $\lambda_n \searrow 0$ such that $[y_{\lambda_n}, u_{\lambda_n}] \to [\hat{y}, \hat{u}]$ weakly in $V \times U$ for some $[\hat{y}, \hat{u}] \in D$. Passing to the limit as $\lambda_n \searrow 0$ in (3.2.39), we see that $A\hat{y} = B\hat{u} + f$, that is, $[\hat{y}, \hat{u}]$ is feasible for (P). Moreover, using (3.2.38) and Theorem A1.5(ii) in Appendix 1, we conclude that

$$L(y^*, u^*) \ge L_\lambda(y_\lambda, u_\lambda) = L(J_\lambda(y_\lambda, u_\lambda)) + \frac{1}{2\lambda}|[y_\lambda, u_\lambda] - J_\lambda(y_\lambda, u_\lambda)|_{V\times U}^2,$$

where J_λ is the monotone resolvent associated with the maximal monotone mapping ∂L. But then we must have

$$[y_\lambda, u_\lambda] - J_\lambda(y_\lambda, u_\lambda) \to 0 \text{ strongly in } V \times U,$$

and thus $J_{\lambda_n}(y_{\lambda_n}, u_{\lambda_n}) \to [\hat{y}, \hat{u}]$ weakly in $V \times U$. Invoking the weak lower semicontinuity of L, we can pass to the limit as $\lambda_n \searrow 0$ in (3.2.38) and infer that

$$L(\hat{y}, \hat{u}) + \frac{1}{2}\,|\hat{u} - u^*|_U^2 \le L(y^*, u^*). \tag{3.2.40}$$

Therefore, in view of the optimality of $[y^*, u^*]$, we must have $[\hat{y}, \hat{u}] = [y^*, u^*]$, as well as $u_{\lambda_n} \to u^*$ strongly in U. But then also

$$\lim_{n\to\infty} |y_{\lambda_n} - y^*|_V \le \left|A^{-1}\right|_{L(Z,V)} \lim_{n\to\infty} |Bu_{\lambda_n} - Bu^* + \lambda_n r_{\lambda_n}|_Z = 0.$$

Finally, observe that the limit is uniquely determined, so that we have (3.2.36) not only on the subsequence $\{\lambda_n\}$ but generally for $\lambda \searrow 0$. This concludes the proof of the assertion. □

Next, we derive first-order optimality conditions for (P_λ). We have the following result.

Proposition 3.2.7 *For any* $[y, u] \in D$,

$$\begin{aligned}&(\nabla L_\lambda(y_\lambda, u_\lambda), [y_\lambda, u_\lambda] - [y, u])_{(V^*\times U^*)\times(V\times U)}\\&\quad + (F_U(u_\lambda - u^*), u_\lambda - u)_{U^*\times U} - (F_Z(r_\lambda), Ay - Bu - f)_{Z^*\times Z} \le 0,\end{aligned} \tag{3.2.41}$$

where $F_U : U \to U^*$, $F_Z : Z \to Z^*$ *are the duality mappings.*

Proof. Let $[y, u] \in D$ and $s \in [0, 1]$ be arbitrary. Then $[z, w] \in D$ for $z = y_\lambda + s\,(y - y_\lambda)$, $w = u_\lambda + s\,(u - u_\lambda)$. We thus have

$$\begin{aligned}L_\lambda(y_\lambda, u_\lambda) &+ \frac{1}{2}\,|u_\lambda - u^*|_U^2 + \frac{1}{2\lambda}\,|Ay_\lambda - Bu_\lambda - f|_Z^2\\&\le L_\lambda(z, w) + \frac{1}{2}\,|w - u^*|_U^2 + \frac{1}{2\lambda}\,|Az - Bw - f|_Z^2 .\end{aligned}$$

Passing all the terms to the left-hand side, dividing by $s > 0$, and letting $s \searrow 0$, we obtain after a little calculation that

$$\begin{aligned}&(\nabla L_\lambda(y_\lambda, u_\lambda), [y_\lambda, u_\lambda] - [y, u])_{(V^*\times U^*)\times(V\times U)} + (F_U(u_\lambda - u^*), u_\lambda - u)_{U^*\times U}\\&\quad - (F_Z(r_\lambda), Ay - Bu - f - \lambda\, r_\lambda)_{Z^*\times Z} \le 0.\end{aligned} \tag{3.2.42}$$

Since

$$\lambda\,(F_Z(r_\lambda), r_\lambda)_{Z^*\times Z} = \lambda\,|r_\lambda|_Z^2 \ge 0,$$

(3.2.41) follows. □

Remark. The condition (3.2.42) is also sufficient for the optimality of $[y_\lambda, u_\lambda]$, since the problem (P_λ) is obviously convex.

We now aim to derive a more familiar form of the optimality system for (P_λ). To this end, we consider a simplified (auxiliary) adjoint system for (P_λ) that does not take the constraints $[y, u] \in D$ into account, namely

$$A^* p_\lambda = \nabla_1 L_\lambda(y_\lambda, u_\lambda). \tag{3.2.43}$$

Here, $A^* \in L(Z^*, V^*)$ is the adjoint of A, and $\nabla_1 L_\lambda(y_\lambda, u_\lambda)$ and $\nabla_2 L_\lambda(y_\lambda, u_\lambda)$ are the components of $\nabla L_\lambda(y_\lambda, u_\lambda)$.

Taking the dual pairing with $y_\lambda - y$ in (3.2.43), we get the identity

$$(\nabla_1 L_\lambda(y_\lambda, u_\lambda), y_\lambda - y)_{V^* \times V} = (A^* p_\lambda, y_\lambda - y)_{V^* \times V} = (p_\lambda, Ay_\lambda - Ay)_{Z^* \times Z}, \tag{3.2.44}$$

whence, substituting in (3.2.42), we can infer that for any $[y, u] \in D$,

$$\begin{aligned} &(\nabla_2 L_\lambda(y_\lambda, u_\lambda), u_\lambda - u)_{U^* \times U} + (p_\lambda, Ay_\lambda - Ay)_{Z^* \times Z} \\ &\quad + (F_U(u_\lambda - u^*), u_\lambda - u)_{U^* \times U} - (F_Z(r_\lambda), Ay - Bu - f)_{Z^* \times Z} \\ &\quad \leq -\lambda\, |r_\lambda|_Z^2 \leq 0. \end{aligned} \tag{3.2.45}$$

Taking first $u = u_\lambda$, and then $y = y_\lambda$, we find that for $[y, u] \in D$,

$$(p_\lambda + F_Z(r_\lambda), Ay_\lambda - Ay)_{Z^* \times Z} \leq 0, \quad \text{whenever } [y, u_\lambda] \in D, \tag{3.2.46}$$

$$\begin{aligned} &(\nabla_2 L_\lambda(y_\lambda, u_\lambda), u_\lambda - u)_{U^* \times U} + (F_U(u_\lambda - u^*), u_\lambda - u)_{U^* \times U} \\ &\quad - (B^* F_Z(r_\lambda), u_\lambda - u)_{U^* \times U} \leq 0, \quad \text{whenever } [y_\lambda, u] \in D. \end{aligned} \tag{3.2.47}$$

Remark. The relations (3.2.46), (3.2.47) constitute a more familiar version of the optimality conditions for (P_λ). In particular, if $D = C \times U_{ad}$, where $C \subset V$ and $U_{ad} \subset U$ are nonempty, convex, and closed in V, respectively in U, and if $N(u_\lambda) = \partial I_{U_{ad}}(u_\lambda)$ denotes the normal cone to U_{ad} at $u_\lambda \in U_{ad}$, then (3.2.47) attains the form

$$\nabla_2 L_\lambda(y_\lambda, u_\lambda) + F_U(u_\lambda - u^*) + N(u_\lambda) \ni B^* F_Z(r_\lambda).$$

This is a standard form of the Pontryagin maximum principle; see Barbu and Precupanu [1986, Chapter IV], and Tiba [1990, Chapter II]. The term $F_Z(r_\lambda)$ is the "true" adjoint state variable for (P_λ).

Proposition 3.2.8 *Assume that L is continuous on $V \times U$. Then there are a subsequence $\lambda_n \searrow 0$, some $[w_1, w_2] \in \partial L(y^*, u^*)$, and some $p^* \in Z^*$ such that*

$$\nabla L_\lambda(y_{\lambda_n}, u_{\lambda_n}) \to [w_1, w_2] \text{ weakly in } V^* \times U^*, \tag{3.2.48}$$

$$p_{\lambda_n} \to p^* \text{ weakly in } Z^*, \tag{3.2.49}$$

$$A^* p^* = w_1. \tag{3.2.50}$$

Proof. Invoking Proposition A1.4(ii) in Appendix 1, we can infer that

$$\nabla L_\lambda(y_\lambda, u_\lambda) \in \partial L(J_\lambda(y_\lambda, u_\lambda)).$$

Using the same argument as in the proof of Proposition 3.2.6, we find that

$$J_\lambda(y_\lambda, u_\lambda) \to [y^*, u^*] \text{ strongly in } V \times U.$$

Since L is continuous, ∂L is defined everywhere in $V^* \times U^*$ and (cf. Appendix 1) locally bounded. Consequently, $\{\nabla L_\lambda(y_\lambda, u_\lambda)\}_{\lambda>0}$ is bounded in $V^* \times U^*$, and there is a subsequence $\lambda_n \searrow 0$ such that

$$\nabla L_\lambda(y_\lambda, u_\lambda) \to [w_1, w_2] \text{ weakly in } V^* \times U^*.$$

The demiclosedness of the maximal monotone operator ∂L shows (compare Proposition A1.4(iv) in Appendix 1) that $[w_1, w_2] \in \partial L(y^*, u^*)$.

Next, notice that A^* has a bounded inverse. Hence, in view of (3.2.43), $\{p_\lambda\}_{\lambda>0}$ is bounded in Z^*, and we may without loss of generality assume that $p_{\lambda_n} \to p^*$ weakly in Z^*. Then (3.2.50) is a direct consequence of (3.2.43), (3.2.48), and (3.2.49). □

Theorem 3.2.9 *Let the hypothesis (3.2.30) be fulfilled, let $[y^*, u^*]$ be an optimal pair for (P), and let $[w_1, w_2] \in \partial L(y^*, u^*)$ and $p^* \in Z^*$ be defined as in Proposition 3.2.8. Then there exists some $q^* \in Z^*$ such that*

$$(p^* + q^*, Ay^* - Ay)_{Z^* \times Z} \le 0, \quad \text{if } [y, u^*] \in D, \tag{3.2.51}$$

$$(w_2, u^* - u)_{U^* \times U} - (B^* q^*, u^* - u)_{U^* \times U} \le 0, \quad \text{if } [y^*, u] \in D. \tag{3.2.52}$$

Moreover, the inequality resulting from adding (3.2.51) and (3.2.52) is valid for any $[y, u] \in D$, and it is sufficient for the optimality of $[y^*, u^*]$.

Proof. We first show that the set $\{q_\lambda = F_Z(r_\lambda)\}_{\lambda>0}$ is bounded in Z^*. To this end, observe that (3.2.30) implies the existence of some $\rho > 0$ such that

$$\rho\xi \in T(M) \quad \text{for all } \xi \in Z \quad \text{with } |\xi|_Z = 1.$$

That is, there exist $[z_\xi, w_\xi] \in M \subset D$ satisfying $T(z_\xi, w_\xi) = \rho\xi$, and the assumed boundedness of M implies the boundedness of

$$\{[z_\xi, w_\xi] : \xi \in Z, \ |\xi|_Z = 1\}$$

in $V \times U$. Taking $[y, u] = [z_\xi, w_\xi]$ in (3.2.41), and invoking the boundedness of the other terms (proved in the above propositions), we conclude that there is some $C_1 > 0$, independent of λ, such that

$$\rho\,(q_\lambda, \xi)_{Z^* \times Z} \le C_1 \quad \text{for all } \xi \in Z \quad \text{with } |\xi|_Z = 1. \tag{3.2.53}$$

Hence, $\{q_\lambda\}_{\lambda>0}$ is bounded in Z^*. Therefore, we may find a subsequence of the sequence $\lambda_n \searrow 0$ constructed in Proposition 3.2.8, without loss of generality

$\{\lambda_n\}$ itself, such that $q_{\lambda_n} \to q^*$ weakly in Z^*. Passing to the limit as $\lambda_n \searrow 0$ in (3.2.46) and (3.2.47), respectively, we obtain (3.2.51) and (3.2.52). Similarly, we can pass to the limit in (3.2.41) to obtain the remaining asserted inequality, namely (3.2.54) below.

To prove the sufficiency, let $[y,u]$ be any feasible pair for (P). Then we have, adding (3.2.51) and (3.2.52),

$$(p^*+q^*, Ay^*-Ay)_{Z^*\times Z} + (w_2, u^*-u)_{U^*\times U} - (B^*q^*, u^*-u)_{U^*\times U} \le 0, \quad (3.2.54)$$

which, by the feasibility of $[y,u]$, simplifies to

$$(p^*, Ay^* - Ay)_{Z^*\times Z} + (w_2, u^* - u)_{U^*\times U} \le 0.$$

Then (3.2.50) yields that

$$(w_1, y^* - y)_{V^*\times V} + (w_2, u^* - u)_{U^*\times U} \le 0,$$

that is, owing to the definition of the subdifferential, $L(y^*,u^*) \le L(y,u)$. This ends the proof of the assertion. □

In order to prepare subsequent applications, we now derive a variant of Theorem 3.2.9. For this purpose, let Y be a (not necessarily reflexive) Banach space satisfying $Y \subset Z$ with continuous and dense embedding. We replace (3.2.30) by the (weaker) condition

$$\begin{array}{l}\text{There is some bounded set } M \subset D \subset V \times U \text{ such that}\\ 0 \in \operatorname{int}(T(M)) \text{ in the topology of } Y.\end{array} \quad (3.2.55)$$

We also assume the compatibility condition (which is automatically fulfilled in many examples)

$$(v,w)_{Y\times Y^*} = (v,w)_{Z\times Z^*},$$

whenever both terms are meaningful.

Then the same argument as in the derivation of (3.2.53) shows that under condition (3.2.55),

$$\rho\,(q_\lambda, \xi)_{Y^*\times Y} \le 0 \quad \text{for all } \xi \in Y \quad \text{with } |\xi|_Y = 1. \quad (3.2.56)$$

That is, $\{q_\lambda\}_{\lambda>0}$ is this time bounded in Y^* (instead of in Z^*, as above). Now let $q^* \in Y^*$ be any cluster point of $\{q_\lambda\}_{\lambda>0}$ with respect to the weak* topology of Y^*. We then have the following result.

Theorem 3.2.10 *A pair* $[y^*, u^*]$ *is optimal for (P) if and only if*

$$(w_2, u^* - u)_{U^* \times U} + (p^*, Ay^* - Ay)_{Z^* \times Z} - (q^*, Ay - Bu - f)_{Y^* \times Y} \leq 0$$
$$\text{for any } [y, u] \in D \text{ such that } T(y, u) \in Y. \tag{3.2.57}$$

Proof. The necessity is a direct consequence of (3.2.45) and (3.2.56), since one may pass to the limit in all the terms if $T(y, u) \in Y$. The sufficiency follows as in the proof of Theorem 3.2.9, since $T(y, u) = 0 \in Y$ for any $[y, u]$ that is feasible for (P). □

Remark. In the work of Bergounioux and Tiba [1996], the above technique is discussed in the context of abstract evolution equations.

Example 3.2.11 We analyze a special case of Example 1.2.1 in Chapter 1 in greater detail. To this end, we fix a bounded smooth domain $\Omega \subset \mathbf{R}^d$, a target function $z_d \in L^2(\Omega)$, and some $\beta \geq 0$, and consider the constrained distributed control problem

$$\text{Min} \left\{ \frac{1}{2} \int_\Omega (y(x) - z_d(x))^2 \, dx + \frac{\beta}{2} \int_\Omega u^2(x) \, dx \right\} \tag{3.2.58}$$

subject to

$$-\Delta y = f + u \quad \text{in } \Omega, \tag{3.2.59}$$

$$y = 0 \quad \text{on } \partial\Omega, \tag{3.2.60}$$

$$e(x) \leq y(x) \leq g(x) \quad \text{a.e. in } \Omega, \tag{3.2.61}$$

$$a(x) \leq u(x) \leq b(x) \quad \text{a.e. in } \Omega. \tag{3.2.62}$$

Here, $f, a, b \in L^\infty(\Omega)$, and $e, g \in C(\overline{\Omega})$ with $e(x) \leq 0 \leq g(x)$ on $\partial\Omega$ (compatibility) are given. Observe that according to Appendix 2, the state system (3.2.59), (3.2.60) has for every $u \in L^\infty(\Omega)$ a unique strong solution $y = y(u)$ that belongs to $W^{2,p}(\Omega)$ for any $p > 1$. By Theorem A2.2(iv) in Appendix 2, we then have $y(u) \in C(\overline{\Omega})$.

Notice also that if, for instance, $e(x) = 0 = g(x)$ on some part of $\partial\Omega$, then the closed and convex set

$$C = \left\{ y \in H_0^1(\Omega) : e \leq y \leq g \quad \text{a.e. in } \Omega \right\} \tag{3.2.63}$$

has an empty interior even in the topology of $L^\infty(\Omega)$. Likewise, the closed and convex set

$$U_{ad} = \left\{ u \in L^2(\Omega) : a \leq u \leq b \quad \text{a.e. in } \Omega \right\} \tag{3.2.64}$$

has an empty interior if a coincides with b on some subset of Ω having positive measure.

In these cases interiority assumptions like (3.1.5) or (3.1.5)′, the Slater condition (3.2.5) or its variant (3.2.29), cannot be applied. Instead, we postulate:

$$\begin{aligned}&\text{There exist some } \alpha > 0 \text{ and some } \tilde{u} \in U_{ad} \text{ such that for all } x \in \overline{\Omega},\\ &e(x) \le y(\tilde{u} - \alpha)(x) \le y(\tilde{u} + \alpha)(x) \le g(x).\end{aligned} \tag{3.2.65}$$

Here, $y(\tilde{u} - \alpha)$ and $y(\tilde{u} + \alpha)$ denote the solutions of (3.2.59), (3.2.60) corresponding to $\tilde{u} - \alpha$ and $\tilde{u} + \alpha$, respectively.

Notice that $\tilde{u} - \alpha$ and $\tilde{u} + \alpha$ need not belong to U_{ad}, i.e., they may be not admissible. On the other hand, the comparison theorem for elliptic equations (cf. Theorem A2.8 in Appendix 2) immediately shows that the pair $[\tilde{y}, \tilde{u}]$, where $\tilde{y} = y(\tilde{u})$, satisfies

$$y(\tilde{u} - \alpha) \le \tilde{y} \le y(\tilde{u} + \alpha) \text{ in } \overline{\Omega},$$

and is thus feasible for the problem (3.2.58)–(3.2.62). Consequently, condition (3.2.65) is stronger than just admissibility.

Notice also that as explained above in connection with (3.2.63), (3.2.64), the condition (3.2.65) is not an interiority condition; however, it ensures that the admissible pairs form a "rich" set and that the control problem (3.2.58)–(3.2.62) is nontrivial.

Remark. A stronger variant of (3.2.65) is to postulate the existence of two controls $\tilde{u}$, $\hat{u}$, that are feasible for (3.2.58)–(3.2.62) and can be separated by a positive constant.

Let us show that the condition (3.2.65) is stronger than (3.2.55). To this end, we fix the functional-analytic setting. We put

$$\begin{aligned}&D = C \times U_{ad}, \quad V = H_0^1(\Omega) \cap H^2(\Omega), \quad U = Z = L^2(\Omega), \quad Y = L^\infty(\Omega),\\ &A : V \to Z, \quad Ay = -\Delta y, \qquad B : U \to Z, \quad Bu = u.\end{aligned}$$

We now choose

$$M = \Big\{ [y(\tilde{u} + \alpha\,\xi), \tilde{u}] : \xi \in Y, \ |\xi|_Y \le 1 \Big\}.$$

Then we have for any $[y(\tilde{u} + \alpha\,\xi, \tilde{u}] \in M$ that

$$-\Delta y(\tilde{u} - \alpha) \le -\Delta y(\tilde{u} + \alpha\,\xi) \le -\Delta y(\tilde{u} + \alpha) \quad \text{a.e. in } \Omega,$$

and the maximum principle for elliptic equations yields that

$$y(\tilde{u} - \alpha) \le y(\tilde{u} + \alpha\,\xi) \le y(\tilde{u} + \alpha) \quad \text{in } \overline{\Omega}.$$

Hence, $M \subset D$. Moreover, we have $T(M) = \{\alpha\xi : \xi \in Y,\ |\xi|_Y \le 1\}$, so that $0 \in \text{int}\,(T(M))$, which shows that (3.2.55) is satisfied.

We therefore can apply Theorem 3.2.10 to the present situation, assuming that an optimal pair $[y^*, u^*] \in D$ exists. The auxiliary mapping p^* is by (3.2.43) defined as the unique solution to the problem

$$-\Delta p^* = y^* - z_d \quad \text{in } \Omega, \qquad p^* = 0 \quad \text{on } \partial\Omega,$$

and the optimality condition (3.2.57) takes the following form: there exists some $q^* \in Y^*$ such that for all $[y, u] \in D$ satisfying $\Delta y \in L^\infty(\Omega)$,

$$\int_\Omega \beta\, u^*(u^* - u)\, dx + \int_\Omega p^*(\Delta y - \Delta y^*)\, dx + (q^*, \Delta y + u + f)_{Y^* \times Y} \le 0. \quad (3.2.66)$$

Since the constraints are separated, we may decouple (3.2.66) by first choosing $[y^*, u]$, and then $[y, u^*]$. It then follows that

$$(\beta\, u^* - q^*, u^* - u)_{Y^* \times Y} \le 0 \quad \forall\, u \in U_{ad}, \quad (3.2.67)$$

$$(p^* + q^*, \Delta y - \Delta y^*)_{Y^* \times Y} \le 0 \quad \forall\, y \in C \text{ with } \Delta y \in Y. \quad (3.2.68)$$

Notice that (3.2.67) is the standard form of the maximum principle, while (3.2.68), together with the equation satisfied by p^*, gives a decoupled form of the usual adjoint equation for the adjoint state q^*, where the influence of the state constraints is expressed by the inequality (3.2.68). See (3.2.69) below.

We now demonstrate that the necessary and sufficient conditions (3.2.66) (or (3.2.67), (3.2.68)) are a powerful tool for gaining further insight into the control problem. For this purpose, we define the sets

$$\Omega_e = \{x \in \overline{\Omega} : y^*(x) = e(x)\}, \quad \Omega_g = \{x \in \overline{\Omega} : y^*(x) = g(x)\},$$
$$\Omega^0 = \Omega \setminus (\Omega_e \cup \Omega_g).$$

Since $y^*, e, g \in C(\overline{\Omega})$, the sets Ω_e and Ω_g are closed, while Ω^0 is open.

Let $\varphi \in \mathcal{D}(\Omega)$ be a test function with support $\text{supp}\,\varphi \subset \Omega^0$. Since $y^*, e, g \in C(\overline{\Omega})$, and by the compactness of $\text{supp}\,\varphi$, there is some $\rho > 0$ such that $y^* \pm \rho\varphi \in C$. Taking $y^* \pm \rho\varphi$ as test functions in (3.2.68), we conclude that

$$(p^* + q^*, \Delta\varphi)_{Y^* \times Y} = 0 \quad \forall\, \varphi \in \mathcal{D}(\Omega) \text{ with } \text{supp}\,\varphi \subset \Omega^0.$$

Taking the equation satisfied by p^* into account, we see that the Lagrange multiplier $q^* \in Y^*$ satisfies

$$-\Delta q^* + j = y^* - z_d \quad \text{in } \mathcal{D}'(\Omega), \quad (3.2.69)$$

where $j \in \mathcal{D}'(\Omega)$ is a distribution that is supported in $\Omega \setminus \Omega^0$, the *active constraints set*.

The relation (3.2.69) is a familiar form of the adjoint equation for state constrained control problems. In particular, it shows that $q^* \in W^{2,p}_{\text{loc}}(\Omega^0)$ if $z_d \in L^p(\Omega)$ for some $p > 1$. By the Sobolev embedding theorem (Theorem A2.2(iv) in Appendix 2), we even have $q^* \in C(\Omega^0)$ if $p > \frac{d}{2}$.

We are now prepared to prove a result on the structure of the optimal pair $[y^*, u^*]$ that is known in the literature as a *generalized bang-bang* result (see Tröltzsch [1984]).

Corollary 3.2.12 *Suppose that $\beta = 0$, and assume that $z_d \in L^p(\Omega)$ for some $p > \frac{d}{2}$. Then there is some set $N \subset \Omega^0$ having zero measure such that*

$$\Omega^0 \setminus N \subset \Big(\{x \in \Omega : y^*(x) = z_d(x)\} \cup \{x \in \Omega : u^*(x) = a(x)\} \\ \cup \{x \in \Omega : u^*(x) = b(x)\}\Big).$$

Proof. Since $p > \frac{d}{2}$, we have $q^* \in C(\Omega^0)$, and relation (3.2.67) shows that

$$\int_{\Omega^0} q^*(x)(u^*(x) - u(x))\, dx \ge 0, \tag{3.2.70}$$

for any $u \in U_{ad}$ with $u = u^*$ a.e. in $\Omega \setminus \Omega^0$. We have

$$\Omega^0 = \Big\{x \in \Omega^0 : q^*(x) > 0\Big\} \cup \Big\{x \in \Omega^0 : q^*(x) < 0\Big\} \cup \Big\{x \in \Omega^0 : q^*(x) = 0\Big\},$$

where the first two subsets are open. If they have positive measure, then (3.2.70) implies that $u^* = b$ a.e. on the first subset, and $u^* = a$ a.e. on the second one. If the last subset has positive measure, then (3.2.69) and the maximal regularity of q^* on Ω^0 imply that $y^* = z_d$ a.e. in this last subset. This concludes the proof. □

Remark. By virtue of the definition of Ω^0, it follows that a.e. in Ω at least one of the functions y^*, u^* attains one of the "extreme" values e, g, a, b or the target z_d.

Remark. A related example involving just one integral state constraint will be studied below in Section 4.2 in Chapter 4. In Proposition 4.2.6 a regularity property of the optimal control will be proved.

Example 3.2.13 We take the same cost functional and the same state equations as in Example 3.2.11, but this time the constraints are

$$|y| \le u \quad \text{a.e. in } \Omega. \tag{3.2.71}$$

Relation (3.2.71) is an example of mixed (state–control) constraints. It is related to the so-called *bottleneck problems* considered in Bellman [1957].

For simplicity, we assume that $\overline{\Omega}$ is a subset of the unit ball $B(0,1) \subset \mathbf{R}^d$, where $d \geq 2$. To fix the functional-analytic setting, we put

$$V = H^2(\Omega) \cap H_0^1(\Omega), \quad U = Z = L^2(\Omega), \quad Y = L^p(\Omega), \quad p > \frac{d}{2}.$$

The operators A and B are as in Example 3.2.11. We want to apply Theorem 3.2.10 to the present situation and thus need to verify that the hypothesis (3.2.55) is fulfilled.

To begin with, observe that since $f \in L^\infty(\Omega)$, we may choose some constant $\hat{c} > 0$ such that $f + \hat{c} \geq 0$ a.e. in Ω. Then, thanks to the maximum principle, the solution $\tilde{y}$ of the state system (3.2.59), (3.2.60) associated with the function $\tilde{u}$, where $\tilde{u} = \hat{c}$ a.e. in Ω, satisfies $\tilde{y} \geq 0$ a.e. in Ω.

Denote by y_f the solution of the state system corresponding to f when $u = 0$. Then there is some constant $m > 0$ such that $-m \leq y_f \leq m$ a.e. in Ω, again by the maximum principle. We consider the auxiliary function $w : \mathbf{R}^d \to \mathbf{R}$,

$$w(x) = \frac{\hat{c}}{2}\left(1 - |x|^2_{\mathbf{R}^d}\right).$$

Obviously, w vanishes on $\partial B(0,1)$, and $-\Delta w = d\,\hat{c} \geq 2\,\hat{c}$ in $B(0,1)$. Then we have

$$-\Delta(y_f + w - \tilde{y}) \geq \hat{c} > 0 \quad \text{a.e. in } \Omega, \quad (y_f + w - \tilde{y})|_{\partial\Omega} = w|_{\partial\Omega} \geq 0.$$

Hence, taking $\hat{c} \geq 4\,m$, recalling that $\tilde{u} = \hat{c}$ a.e. in Ω, and invoking the definition of w and the maximum principle once more, we have the chain of inequalities

$$\begin{aligned} 0 \leq \tilde{y} \leq y_f + w \leq m + w &= m + \tilde{u} - \frac{\hat{c}}{2} - \frac{\hat{c}}{2}\,|x|^2_{\mathbf{R}^d} \\ &\leq m + \tilde{u} - \frac{\hat{c}}{2} \leq \tilde{u} - m \end{aligned} \tag{3.2.72}$$

a.e. in Ω. The inequalities (3.2.72) show that the pair $[\tilde{y}, \tilde{u}]$ satisfies (3.2.71) in a stronger form and is admissible for (P). Now put

$$M = \{\, [y(f+v), \tilde{u}] : v \in B_p(\tilde{u}, \lambda)\},$$

where, for $\lambda > 0$,

$$B_p(\tilde{u}, \lambda) = \left\{ v \in L^p(\Omega) \,:\, |v - \tilde{u}|_{L^p(\Omega)} \leq \lambda \right\}.$$

By the continuity with respect to the right-hand side in elliptic equations, and since $p > \frac{d}{2}$ so that the Sobolev embedding theorem can be applied, we can choose some sufficiently small $\lambda > 0$ such that

$$|\tilde{y} - y(f+v)|_{C(\overline{\Omega})} \leq m \quad \forall\, v \in B_p(\tilde{u}, \lambda).$$

Then, $M \subset D$ by (3.2.72), and $T(M) = B_p(0, \lambda)$, so that $0 \in \operatorname{int}(T(M))$, and thus (3.2.55) holds. Notice that in the case $d \le 3$ we can take $p = 2$, so that then even the stronger postulate (3.2.30) is fulfilled.

We are now in a position to apply Theorem 3.2.10: there exists some $q^* \in L^{p'}(\Omega)$, where $\frac{1}{p} + \frac{1}{p'} = 1$, such that, with p^* given as in Example 3.2.11,

$$\beta\,(u^*, u^* - u)_{L^2(\Omega)} + (p^*, \Delta y - \Delta y^*)_{L^2(\Omega)} + (q^*, \Delta y + u + f)_{L^{p'}(\Omega)\times L^p(\Omega)}$$
$$\le 0 \quad \text{for all } [y,u] \in D \text{ satisfying } T(y,u) \in L^p(\Omega). \tag{3.2.73}$$

Also in this example a generalized bang-bang property for the optimal pair can be proved, as the following result shows.

Corollary 3.2.14 *Let $\beta = 0$, and assume that $y^*, u^* \in C(\overline{\Omega})$. Then there is some set $N \subset \Omega$ of zero measure such that*

$$\Omega \setminus N = \Big(\{x \in \Omega : \, |y^*(x)| = u^*(x)\} \cup \{x \in \Omega : \, y^*(x) = z_d(x)\}\Big). \tag{3.2.74}$$

Proof. Suppose that $\Omega^* = \{x \in \Omega : |y^*(x)| < u^*(x)\} \neq \emptyset$. By the continuity of y^* and u^*, Ω^* is open and has nonzero measure.

Now let $\varphi \in \mathcal{D}(\Omega)$ be arbitrary with $\operatorname{supp}\varphi \subset \Omega^*$. Since $y^*, u^* \in C(\overline{\Omega})$, and owing to the compactness of $\operatorname{supp}\varphi$, there is some sufficiently small $\lambda > 0$ such that

$$|y^*(x) \pm \lambda\,\varphi(x)| \le |y^*(x)| + \lambda\,|\varphi(x)| \le u^*(x) \quad \text{a.e. in } \Omega.$$

Inserting $[y,u] = [y^* \pm \lambda\,\varphi, u^*] \in D$ in (3.2.73), we obtain after a short calculation that

$$\int_\Omega (p^*(x) + q^*(x))\,\Delta\varphi(x)\,dx = 0.$$

Consequently, there is a distribution $j \in \mathcal{D}'(\Omega)$, which is supported on the active constraints set $\Omega \setminus \Omega^*$, such that

$$-\Delta q^* + j = y^* - z_d \quad \text{in } \mathcal{D}'(\Omega). \tag{3.2.75}$$

Since $z_d \in L^2(\Omega)$, this implies a regularity property for the Lagrange multiplier, namely that $q^* \in H^2_{\text{loc}}(\Omega^*)$.

Next, we insert the pairs $[y,u] = [y^*, u^* \pm \lambda\,\varphi] \in D$ in (3.2.73), where φ and $\lambda > 0$ are as above. We obtain

$$\int_\Omega q^*(x)\,\varphi(x)\,dx = 0. \tag{3.2.76}$$

Using (3.2.75), we infer that

$$\int_\Omega (y^*(x) - z_d(x))\,\varphi(x)\,dx \;=\; -\,(\Delta q^*, \varphi)_{\mathcal{D}'(\Omega)\times\mathcal{D}(\Omega)} \;=\; -\int_\Omega q^*(x)\Delta\varphi(x)\,dx \;=\; 0,$$

since $\Delta\varphi \in \mathcal{D}(\Omega)$ and $\operatorname{supp}\Delta\varphi \subset \Omega^*$, so that (3.2.76) applies. Hence, $y^* = z_d$ a.e. in Ω^*, and (3.2.74) is proved. □

Remark. The two sets occurring on the right-hand side of (3.2.74) need not be disjoint. The first one obviously corresponds to the situation that the constraint is active.

3.2.3 Semilinear Equations and Exact Penalization

In this paragraph, we examine the optimality conditions for a constrained distributed-boundary control problem governed by a general semilinear boundary value problem of elliptic type. We study only the case of integral state constraints, which are easier to handle. A very general situation including pointwise constraints and variational inequalities will be discussed below in §3.3.2.

To fix things, let $\Omega \subset \mathbf{R}^d$ denote a bounded domain having a smooth boundary Γ, and let $K_d \subset \mathbf{R}$ and $K_b \subset \mathbf{R}$ be bounded. For given $m \in \mathbf{R}$, we consider the problem (P_m) defined by

$$\text{Min}\left\{J(y,u) = \int_\Omega L(x, y(x), u_d(x))\,dx + \int_\Gamma \ell(\sigma, y(\sigma), u_b(\sigma))\,d\sigma\right\}, \tag{3.2.77}$$

subject to

$$Ay(x) \;=\; F(x, y(x), u_d(x)) \quad \text{in } \Omega, \tag{3.2.78}$$

$$\alpha\, y(\sigma) + \frac{\partial y}{\partial n_A}(\sigma) \;=\; f(\sigma, y(\sigma), u_b(\sigma)) \quad \text{on } \Gamma, \tag{3.2.79}$$

$$\int_\Omega h(x, y(x))\,dx \;\le\; m, \tag{3.2.80}$$

$$u \;=\; [u_d, u_b] \in U_{ad} \;=\; \Big\{u \in L^\infty(\Omega) \times L^\infty(\Gamma) \,:\, u_d(x) \in K_d \text{ a.e. in } \Omega,\; u_b(\sigma) \in K_b \text{ a.e. in } \Gamma\Big\}. \tag{3.2.81}$$

Here, $\alpha > 0$ is a given constant, and we assume that

$$Ay(x) \;=\; -\sum_{i,j=1}^{d} \frac{\partial}{\partial x_i}\Big(a_{ij}(x)\,\frac{\partial y}{\partial x_j}(x)\Big), \tag{3.2.82}$$

where $a_{ij} \in L^\infty(\Omega)$ and $a_{ij} = a_{ji}$ for all $1 \le i,j \le d$, and, with some fixed constant $\omega > 0$,

$$\sum_{i,j=1}^{d} a_{ij}(x)\,\xi_i\,\xi_j \geq \omega \sum_{i=1}^{d} \xi_i^2 \quad \forall \xi \in \mathbf{R}^d \quad \text{and a.e. } x \in \Omega. \tag{3.2.83}$$

As in the previous sections of this book,

$$\frac{\partial y}{\partial n_A} = \sum_{i,j=1}^{d} a_{ij}\,\frac{\partial y}{\partial x_i}\cos(n, x_j) \tag{3.2.84}$$

denotes the outer conormal derivative to Γ associated with A.

The mappings $F : \Omega \times \mathbf{R}^2 \to \mathbf{R}$, $L : \Omega \times \mathbf{R}^2 \to \mathbf{R}$, $h : \Omega \times \mathbf{R} \to \mathbf{R}$, $\ell : \Gamma \times \mathbf{R}^2 \to \mathbf{R}$, and $f : \Gamma \times \mathbf{R}^2 \to \mathbf{R}$, are assumed to be continuous functions that possess a continuous partial derivative with respect to the second variable on their respective domains of definition, denoted by F_y, L_y, h_y, ℓ_y, and f_y, respectively. Further hypotheses will be added below when the necessity arises.

Next, we introduce the functions

$$G(x,y,u) = -\int_0^y F(x,\xi,u)\,d\xi, \quad g(\sigma,y,u) = -\int_0^y f(\sigma,\xi,u)\,d\xi,$$

and define for fixed $u = [u_d, u_b] \in U_{ad}$ the functionals $\phi(u;\cdot)$, $\varphi(u;\cdot) : H^1(\Omega) \to]-\infty,+\infty]$,

$$\phi(u;y) = \begin{cases} \dfrac{1}{2}\displaystyle\int_\Omega \sum_{i,j=1}^{d} a_{ij}(x)\,\frac{\partial y}{\partial x_i}\,\frac{\partial y}{\partial x_j}\,dx + \int_\Omega G(x,y(x),u_d(x))\,dx, & \\ \qquad\qquad\qquad\qquad \text{if } G(\cdot\,,y(\cdot),u_d(\cdot)) \in L^1(\Omega), & \\ +\infty, \qquad\qquad\qquad \text{otherwise}, & \end{cases} \tag{3.2.85}$$

$$\varphi(u;y) = \begin{cases} \dfrac{\alpha}{2}\displaystyle\int_\Gamma y^2\,d\sigma + \int_\Gamma g(\sigma,y(\sigma),u_b(\sigma))\,d\sigma, & \text{if } g(\cdot,y(\cdot),u_b(\cdot)) \in L^1(\Gamma), \\ +\infty, & \text{otherwise}. \end{cases} \tag{3.2.86}$$

We generally assume that for all $(x,y,u) \in \Omega \times \mathbf{R}^2$,

$$|F(x,0,u)| \leq M_1(x) + \hat{C}\,|u|, \tag{3.2.87}$$

$$0 \leq -F_y(x,y,u) \leq \Big(M_2(x) + \hat{C}\,|u|\Big)\,\eta(|y|), \tag{3.2.88}$$

$$0 \leq -f_y(\sigma,y,u) \leq \Big(R_2(\sigma) + \hat{C}\,|u|\Big)\,\eta(|y|), \tag{3.2.89}$$

$$|f(\sigma,0,u)| \leq R_1(\sigma) + \hat{C}\,|u|. \tag{3.2.90}$$

Here, $\eta : \mathbf{R}_+ \to \mathbf{R}_+$ is a nondecreasing function, and $M_1 \in L^2(\Omega)$, $R_1 \in L^2(\Gamma)$, $M_2 \in L^s(\Omega)$, where $s > \frac{d}{2}$ and $R_2 \in L^\infty(\Gamma)$ are given functions. Moreover, above and in what follows, $\hat{C} > 0$ always denotes a given constant.

Observe that under these assumptions on F and f the functions G and g are convex and continuously differentiable with respect to y. Moreover, $F(\cdot, 0, u) \in L^2(\Omega)$ and $f(\cdot, 0, u) \in L^2(\Gamma)$ for any fixed $u \in \mathbf{R}$, and we have for every $(x, y, u) \in \Omega \times \mathbf{R}^2$ the inequality

$$\begin{aligned} G(x,y,u) &= -\int_0^y F(x,\xi,u)\,d\xi = -\int_0^y \Big(F(x,0,u) + F_y(x,\rho(\xi),u)\,\xi\Big)\,d\xi \\ &\geq -F(x,0,u)\,y, \end{aligned}$$

where $\rho(\xi) \in [0, \xi]$ is an intermediate point given by the mean value theorem. An analogous inequality holds for g. Moreover, $G(x, 0, u) \equiv 0$ and $g(\sigma, 0, u) \equiv 0$. Then, G and g are convex integrands with respect to y on $L^2(\Omega)$ and $L^2(\Gamma)$, respectively. Moreover, according to Proposition A1.1 in Appendix 1, these integrands possess subdifferentials given by $-F(x, y(x), u)$, respectively by $-f(\sigma, y(\sigma), u)$, whenever they belong to $L^2(\Omega)$, respectively to $L^2(\Gamma)$.

Proposition 3.2.15 *Let $u = [u_d, u_b] \in U_{ad}$ be given. If $y \in W^{2,s}(\Omega)$ is a strong solution to the boundary value problem (3.2.78), (3.2.79), then y minimizes the functional $\phi(u; \cdot) + \varphi(u; \cdot)$ on $H^1(\Omega)$.*

Proof. Since $s > \frac{d}{2}$, it follows from Theorem A2.2(iv) in Appendix 2 that $y \in C(\overline{\Omega})$. Moreover, $u_d(x) \in K_d$ a.e. in Ω, so that $u_d \in L^\infty(\Omega)$. Then (3.2.87) and (3.2.88) imply that $F(\cdot, y(\cdot), u_d(\cdot)) \in L^2(\Omega)$. Similarly, using (3.2.89) and (3.2.90), we infer that $f(\cdot, y(\cdot), u_b(\cdot)) \in L^2(\Gamma)$.

The previous discussion concerning the properties of G and g shows that the integrands have subdifferentials at y. Then also $\phi(u; \cdot)$ and $\psi(u; \cdot)$ are subdifferentiable at y, since the other terms occurring in their definitions (3.2.85), (3.2.86) are convex and continuous.

Now let $v \in H^1(\Omega)$ be arbitrary. By (3.2.78), (3.2.79), (3.2.82) and using integration by parts, we obtain that

$$\begin{aligned} &\int_\Omega F(x, y(x), u_d(x))(y(x) - v(x))\,dx \\ &\quad = \int_\Omega \sum_{i,j=1}^d a_{ij}(x)\,\frac{\partial}{\partial x_j} y(x)\,\frac{\partial}{\partial x_i}(y(x) - v(x))\,dx - \int_\Gamma \frac{\partial y}{\partial n_A}\,(y - v)\,d\sigma. \end{aligned}$$

Invoking Proposition A1.1 in Appendix 1, we can conclude that there is some $w \in \partial\phi(u; y)$ such that

$$-\,(w, y - v)_{H^1(\Omega)^* \times H^1(\Omega)} = \int_\Omega F(x, y(x), u_d(x))(y(x) - v(x))\,dx$$

$$- \int_{\Omega} \sum_{i,j=1}^{d} a_{ij}(x) \frac{\partial}{\partial x_j} y(x) \frac{\partial}{\partial x_i} (y(x) - v(x))\, dx$$

$$= \int_{\Gamma} (\alpha\, y(\sigma) - f(\sigma, y(\sigma), u_b(\sigma)))\, (y(\sigma) - v(\sigma))\, d\sigma$$

$$\geq \varphi(u; y) \; - \; \varphi(u; v).$$

Therefore, by the definition of the subdifferential, $-w \in \partial\varphi(u;y)$, and thus

$$0 \in \partial\phi(u;y) + \partial\varphi(u;y) \subset \partial(\phi(u;\cdot) + \varphi(u;\cdot))(y).$$

The assertion now follows from applying Theorem A1.6 in Appendix 1 to the present situation. □

Remark. In the above inclusion we have equality only under additional interiority hypotheses like those given in Theorems A1.10 and A1.11 in Appendix 1. Fortunately, such very restrictive conditions need not be imposed here.

Proposition 3.2.16 *For every* $u = [u_d, u_a] \in U_{ad}$ *the associated functional* $\phi(u;\cdot) + \varphi(u;\cdot)$ *has a unique minimizer on* $H^1(\Omega)$.

Proof. Under the above assumptions, the mapping $\phi(u;\cdot) + \varphi(u;\cdot)$ is strictly convex on $H^1(\Omega)$, since the mapping

$$y \;\mapsto\; \frac{1}{2} \int_{\Omega} \sum_{i,j=1}^{d} a_{ij}(x) \frac{\partial y}{\partial x_i}(x) \frac{\partial y}{\partial x_j}(x)\, dx + \frac{\alpha}{2} \int_{\Gamma} y^2(\sigma)\, d\sigma$$

has this property. The mapping $\phi(u;\cdot) + \varphi(u;\cdot)$ is also coercive on $H^1(\Omega)$, which follows from the trace theorem, Friedrichs' inequality, and since condition (iii) in the definition of convex integrands in Appendix 1 is fulfilled. The assertion now follows from Theorem A1.6 in Appendix 1 and the remark following it. □

Definition 3.2.17 For any $u = [u_d, u_b] \in U_{ad}$ the unique minimizer y of the associated functional $\phi(u;\cdot) + \varphi(u;\cdot)$ on $H^1(\Omega)$ is called the *weak* (or *variational*) solution of (3.2.78), (3.2.79), and we denote it by $y = y_u$.

Example 3.2.18 (Bonnans and Casas [1991])

Consider equation (3.2.78) with the Dirichlet boundary condition

$$y \;=\; 0 \;\text{ on }\; \Gamma. \tag{3.2.79$'$}$$

Then it turns out that under the conditions (3.2.87), (3.2.88) the state system (3.2.78), (3.2.79)′ has for any $u_d \in L^\infty(\Omega)$ a unique weak solution $y \in H_0^1(\Omega) \cap$

$C^\gamma(\overline{\Omega})$ for some $\gamma \in]0,1[$, that is, y is Hölder continuous of order γ. Moreover, there is some constant $\hat{C} > 0$ such that

$$|y|_{H_0^1(\Omega)} + |y|_{C(\overline{\Omega})} \leq \hat{C}\left(1 + |u|_{L^\infty(\Omega)}\right).$$

Recall that the continuity property of y played an essential role in the proof of Proposition 3.2.15.

Example 3.2.19 (Barbu [1993, Chapter 3.1])
We consider the boundary value problem

$$-\Delta y + cy = F \quad \text{in } \Omega, \qquad \frac{\partial y}{\partial n} + \beta(y) \ni 0 \quad \text{on } \Gamma,$$

where $\beta \subset \mathbf{R} \times \mathbf{R}$ is a maximal monotone graph, $c > 0$, and $F \in L^2(\Omega)$. Then one obtains that $y \in H^2(\Omega)$, and there is a constant $\hat{C} > 0$, independent of F, such that

$$|y|_{H^2(\Omega)} \leq \hat{C}\left(1 + |F|_{L^2(\Omega)}\right) \quad \forall F \in L^2(\Omega).$$

Note that for $d \leq 3$ we have the continuous embedding $H^2(\Omega) \subset C^\gamma(\overline{\Omega})$ for some $\gamma \in]0,1[$, so that in this case we may replace $|y|_{H^2(\Omega)}$ in the above inequality by $|y|_{C(\overline{\Omega})}$.

Remark. Other regularity results for semilinear and quasilinear second-order elliptic equations can be found in the works by Stampacchia [1965], Gilbarg and Trudinger [1983], and Ladyzenskaya and Uraltseva [1968].

In the following, we will always assume for the treatment of the optimization problem (3.2.77)–(3.2.81) that for any fixed $u \in U_{ad}$ the unique variational solution $y = y_u$ to the state system (3.2.78), (3.2.79) belongs to $H^1(\Omega) \cap C(\overline{\Omega})$ and satisfies, with a constant $\hat{c} > 0$ that is independent of $u = [u_d, u_b] \in U_{ad}$,

$$|y|_{H^1(\Omega)} + |y|_{C(\overline{\Omega})} \leq \hat{c}\left(1 + |u_d|_{L^\infty(\Omega)} + |u_b|_{L^\infty(\Gamma)}\right). \tag{3.2.91}$$

In order to deal with the state constraints, we again have to impose a constraint qualification. This time, we postulate a condition that in the terminology of Clarke [1983, Chapter VI] is called *calmness*. In order to formulate this condition, we have to introduce the notion of the *value function*, denoted by V, of the optimization problem (3.2.77)–(3.2.81), where we will only put into evidence its dependence on m. We define, for any $m > 0$,

$$V(m) = \inf\left\{J(y_u, u) \,:\, u \in U_{ad} \text{ and } y_u \text{ satisfies } (3.2.78) - (3.2.80)\right\},$$

where we put $V(m) = +\infty$ if (3.2.80) is violated by y_u for every $u \in U_{ad}$. Observe that, owing to the boundedness of K_d and K_b, and by virtue of (3.2.91) and the continuity of h, the state constraint (3.2.80) is satisfied automatically by y_u for any $u \in U_{ad}$ provided that $m \in \mathbf{R}$ is sufficiently large. We also note that the mapping $m \mapsto V(m)$ is decreasing on $\mathbf{R}$, so that

$$V(m') < +\infty \quad \text{for all } m' \geq m \text{ if } V(m) < +\infty.$$

We state the *calmness property*:

There exist constants $r > 0$ and $\varepsilon > 0$ such that

$$V(m') \geq V(m) - r\,|m' - m| \quad \forall\, m' \in [m - \varepsilon, m + \varepsilon]. \tag{3.2.92}$$

Since $m \mapsto V(m)$ is decreasing on $\mathbf{R}$, it follows that (3.2.92) is automatically satisfied if $V(m+\varepsilon) = +\infty$. On the other hand, if $V(m-\varepsilon) < +\infty$, then V is finite on the whole interval $[m-\varepsilon, m+\varepsilon]$ and, by the monotonicity property, a.e. differentiable there. It follows that the calmness property (3.2.92) has a generic character.

Next, we specify the assumptions on the functions L, ℓ, and h. We postulate, with the same notation as in (3.2.87)–(3.2.90),

$$|L(x,0,u)| \leq M_1(x) + \hat{C}\,|u|, \tag{3.2.93}$$

$$|L_y(x,y,u)| \leq \Big(M_1(x) + \hat{C}\,|u|\Big)\,\eta(|y|), \tag{3.2.94}$$

$$|\ell(\sigma,0,u)| \leq R_1(\sigma) + \hat{C}\,|u|, \tag{3.2.95}$$

$$|\ell_y(\sigma,y,u)| \leq \Big(R_1(\sigma) + \hat{C}\,|u|\Big)\,\eta(|y|), \tag{3.2.96}$$

$$|h_y(x,y)| \leq M_1(x)\,\eta(|y|). \tag{3.2.97}$$

Now let us introduce the *exact penalization* of (P_m), which we call the problem $(\mathrm{P}_{r,\varepsilon})$, corresponding to the constants r and ε appearing in (3.2.92):

$$\inf\left\{ J_r(u) = J(y_u,u) + r\Big(\int_\Omega h(x,y_u(x))\,dx - m\Big)_+ \;:\; u \in U_{ad}\right\}, \tag{3.2.98}$$

where $z_+ = \max\{z,0\}$ denotes the positive part of $z \in \mathbf{R}$ and y_u is the variational solution to (3.2.78), (3.2.79), and subject to the state constraint

$$\int_\Omega h(x,y_u(x))\,dx \leq m + \varepsilon. \tag{3.2.99}$$

Observe that in view of (3.2.91), (3.2.97), and the continuity of h, the state constraints (3.2.80) and (3.2.99) are meaningful.

Proposition 3.2.20 *Let $\delta > 0$, and suppose that $u_\delta \in U_{ad}$ is a δ-solution for (P_m); that is, y_{u_δ} satisfies (3.2.80) and $J(y_{u_\delta}, u_\delta) \le V(m) + \delta$. Then u_δ is a δ-solution for $(P_{r,\varepsilon})$; that is, y_{u_δ} satisfies (3.2.99) and $J_r(u_\delta) \le V(r,\varepsilon) + \delta$, where*

$$V(r,\varepsilon) = \inf\Big\{J_r(u) \,:\, u \in U_{ad}\,,\quad \textit{(3.2.99) holds}\Big\}$$

denotes the value function of $(P_{r,\varepsilon})$.

Proof. By virtue of (3.2.77) and (3.2.92), we obviously have

$$\begin{aligned} V(m) &= \inf\Big\{V(m') + r\,|m' - m| \,:\, m' \in [m-\varepsilon, m+\varepsilon]\Big\} \\ &= \inf\Big\{J(y_u, u) + r\,|m' - m| \,:\, \int_\Omega h(x, y_u(x))\,dx \le m', \\ &\qquad\qquad u \in U_{ad}\,,\ m' \in [m-\varepsilon, m+\varepsilon]\Big\} \\ &= \inf\Big\{J_r(u) \,:\, u \in U_{ad}\,,\ \text{(3.2.99) holds}\Big\} \;=\; V(r,\varepsilon). \end{aligned}$$

The last equality follows from the definition of the infinimum. Since u_δ is admissible for (P_m), it is admissible for $(\mathrm{P}_{r,\varepsilon})$, and $J_r(u_\delta) = J(y_{u_\delta}, u_\delta)$. Moreover, the above identity shows that

$$J_r(u_\delta) \;=\; J(y_{u_\delta}, u_\delta) \;\le\; V(m) + \delta \;=\; V(r,\varepsilon) + \delta,$$

and the proof is finished. □

Remark. Proposition 3.2.20 justifies the use of the term *exact penalization.*

We define the *pseudo-Hamiltonian* mappings associated with the distributed control and the boundary control, respectively:

$$H(x, y, u_d, p) \;=\; L(x, y, u_d) \;+\; p\,F(x, y, u_d), \tag{3.2.100}$$

$$j(\sigma, y, u_b, p) \;=\; \ell(\sigma, y, u_b) \;+\; p\,f(\sigma, y, u_b). \tag{3.2.101}$$

We also define an "intermediate" adjoint state $p_{u,v}$ corresponding to the controls $u = [u_d, u_b]$, $v = [v_d, v_b] \in U_{ad}$ in the following way:

Let $y_u, y_v \in H^1(\Omega) \cap C(\overline{\Omega})$ be the (unique) variational solutions to (3.2.78), (3.2.79) corresponding to u and v, respectively. Then the mean value theorem yields the existence of intermediate points $\xi_1(x)$, $\xi_2(x)$, $\xi_3(x)$ between $y_u(x)$ and $y_v(x)$, for a.e. $x \in \Omega$, and of intermediate points $\xi_4(\sigma)$, $\xi_5(\sigma)$ between $y_u(\sigma)$ and $y_v(\sigma)$, for a.e. $\sigma \in \Gamma$, respectively, such that we have the following set of relations (which we denote by $(*)$):

$$
\begin{aligned}
F(x,y_v(x),v_d(x)) - F(x,y_u(x),v_d(x)) &= F_y(x,\xi_1(x),v_d(x))\,(y_v(x)-y_u(x)),\\
L(x,y_v(x),v_d(x)) - L(x,y_u(x),v_d(x)) &= L_y(x,\xi_2(x),v_d(x))\,(y_v(x)-y_u(x)),\\
h(x,y_v(x)) - h(x,y_u(x)) &= h_y(x,\xi_3(x))\,(y_v(x)-y_u(x)),\\
f(\sigma,y_v(\sigma),v_b(\sigma)) - f(\sigma,y_u(\sigma),v_b(\sigma)) &= f_y(\sigma,\xi_4(\sigma),v_b(\sigma))\,(y_v(\sigma)-y_u(\sigma)),\\
\ell(\sigma,y_v(\sigma),v_b(\sigma)) - \ell(\sigma,y_u(\sigma),v_b(\sigma)) &= \ell_y(\sigma,\xi_5(\sigma),v_b(\sigma))\,(y_v(\sigma)-y_u(\sigma)).
\end{aligned}
$$

Next, let us fix any

$$
w \in \partial_+\left(\int_\Omega h(x,y_v(x))\,dx \;-\; m\right),
$$

where $\partial_+(\cdot)$ is the subdifferential of the convex mapping $(\,\cdot\,)_+ : \mathbf{R} \to \mathbf{R}$, $z \mapsto z_+$. We then consider the linear elliptic boundary value problem

$$
\begin{aligned}
Ap &= F_y(x,\xi_1(x),v_d(x))\,p + L_y(x,\xi_2(x),v_d(x))\\
&\qquad + r\,w\;h_y(x,\xi_3(x)) \quad \text{in } \Omega, \qquad (3.2.102)\\
\alpha\,p + \frac{\partial p}{\partial n_A} &= f_y(\sigma,\xi_4(\sigma),v_b(\sigma))\,p + \ell_y(\sigma,\xi_5(\sigma),v_b(\sigma)) \quad \text{on } \partial\Omega. \qquad (3.2.103)
\end{aligned}
$$

We have the following result.

Proposition 3.2.21 *Under the above assumptions, the system (3.2.102), (3.2.103) has a unique variational solution $p \in H^1(\Omega)$, denoted by $p = p_{u,v}$.*

Proof. The variational form of (3.2.102), (3.2.103) is

$$
\begin{aligned}
&\int_\Omega \sum_{i,j=1}^d a_{ij}(x)\,\frac{\partial p}{\partial x_i}\,\frac{\partial z}{\partial x_j}\,dx - \int_\Omega F_y\,p\,z\,dx - \int_\Omega L_y\,z\,dx - r\,w\int_\Omega h_y\,z\,dx\\
&\qquad + \alpha\int_\Gamma p\,z\,d\sigma - \int_\Gamma f_y\,p\,z\,d\sigma - \int_\Gamma \ell_y\,z\,d\sigma = 0 \quad \forall\, z \in H^1(\Omega), \qquad (3.2.104)
\end{aligned}
$$

where F_y, f_y, L_y, h_y, ℓ_y have to be evaluated at the same arguments as in (3.2.102), (3.2.103). The associated "energy" functional is given by

$$
\begin{aligned}
\psi_{u,v}(p) &= \frac{1}{2}\int_\Omega \sum_{i,j=1}^d a_{ij}(x)\,\frac{\partial p}{\partial x_i}\,\frac{\partial p}{\partial x_j}\,dx + \frac{\alpha}{2}\int_\Gamma p^2\,d\sigma - \frac{1}{2}\int_\Omega F_y\,p^2\,dx\\
&\qquad - \frac{1}{2}\int_\Gamma f_y\,p^2\,d\sigma - \int_\Omega L_y\,p\,dx - r\,w\int_\Omega h_y\,p\,dx - \int_\Gamma \ell_y p\,d\sigma, \qquad (3.2.105)
\end{aligned}
$$

where F_y, f_y, L_y, h_y, ℓ_y again have to be evaluated at the same arguments as in (3.2.102), (3.2.103).

Using the boundedness of K_d and K_b, and invoking (3.2.91) and the growth and sign conditions (3.2.87)–(3.2.90) and (3.2.93)–(3.2.97), we can easily check that the bilinear form associated with (3.2.104) is continuous and coercive on $H^1(\Omega)$. By the same token, and thanks to the trace theorem, the linear terms in (3.2.104) are bounded on $H^1(\Omega)$. By the Lax–Milgram lemma, (3.2.104) has a unique solution $p \in H^1(\Omega)$, which then is the unique minimizer of the functional (3.2.105). Since the formulation (3.2.105) corresponds to Definition 3.2.17, the proof is complete. □

Proposition 3.2.22 *Under the above assumptions,*

$$\begin{aligned} J_r(v) - J_r(u) &\le \int_\Omega H(x, y_u(x), v_d(x), p_{u,v}(x))\,dx - \int_\Omega H(x, y_u(x), u_d(x), p_{u,v}(x))\,dx \\ &\quad + \int_\Gamma j(\sigma, y_u(\sigma), v_b(\sigma), p_{u,v}(\sigma))\,d\sigma - \int_\Gamma j(\sigma, y_u(\sigma), u_b(\sigma), p_{u,v}(\sigma))\,d\sigma. \quad (3.2.106) \end{aligned}$$

Proof. In what follows, we omit the arguments of the involved functions. Using the identities $(*)$ introducing the functions ξ_i, $i = \overline{1,5}$, and recalling the definition of the subdifferential, we conclude that for any $w \in \partial_+ \left(\int_\Omega h(x, y_v)\,dx - m \right)$,

$$\begin{aligned} J_r(v) - J_r(u) &= \int_\Omega [L(x, y_v, v_d) - L(x, y_u, v_d) + L(x, y_u, v_d) - L(x, y_u, u_d)]\,dx \\ &\quad + \int_\Gamma [\ell(\sigma, y_v, v_b) - \ell(\sigma, y_u, v_b) + \ell(\sigma, y_u, v_b) - \ell(\sigma, y_u, u_b)]\,d\sigma \\ &\quad + r\left(\int_\Omega h(x, y_v)\,dx - m \right)_+ - r\left(\int_\Omega h(x, y_u)\,dx - m \right)_+ \\ &\le \int_\Omega L_y(x, \xi_2(x), v_d)(y_v - y_u)\,dx + \int_\Omega [L(x, y_u, v_d) - L(x, y_u, u_d)]\,dx \\ &\quad + \int_\Gamma \ell_y(\sigma, \xi_5(\sigma), v_b)(y_v - y_u)\,d\sigma + \int_\Gamma [\ell(\sigma, y_u, v_b) - \ell(\sigma, y_u, u_b)]\,d\sigma \\ &\quad + r\,w \int_\Omega h_y(x, \xi_3(x))\,(y_v - y_u)\,dx\,. \end{aligned}$$

Invoking (3.2.104), we thus can conclude that

$$\begin{aligned} J_r(v) - J_r(u) &\le \int_\Omega [L(x, y_u, v_d) - L(x, y_u, u_d)]\,dx + \int_\Gamma [\ell(\sigma, y_u, v_b) - \ell(\sigma, y_u, u_b)]\,d\sigma \end{aligned}$$

$$+\int_\Omega \sum_{i,j=1}^{d} a_{ij}(x)\,\frac{\partial p_{u,v}}{\partial x_i}\,\frac{\partial}{\partial x_j}(y_v - y_u)\,dx$$

$$-\int_\Omega F_y(x,\xi_1(x),v_d)\,(y_v - y_u)\,p_{u,v}\,dx + \alpha\int_\Gamma p_{u,v}\,(y_v - y_u)\,d\sigma$$

$$-\int_\Gamma f_y(\sigma,\xi_4(\sigma),v_b)\,p_{u,v}\,(y_v - y_u)\,dx\,. \qquad (3.2.107)$$

Recall that y_v and y_u, respectively, are minimizers of the corresponding functionals $\phi(v;\cdot) + \varphi(v;\cdot)$ and $\phi(u;\cdot) + \varphi(u;\cdot)$, respectively, and therefore satisfy the associated Euler equations. We thus have

$$0 = \int_\Omega \sum_{i,j=1}^{d} a_{ij}(x)\,\frac{\partial y_u}{\partial x_i}\,\frac{\partial z}{\partial x_i}\,dx - \int_\Omega F(x,y_u,u_d)\,z\,dx$$

$$+\,\alpha\int_\Gamma y_u\,z\,d\sigma - \int_\Gamma f(\sigma,y_u,u_b)\,z\,d\sigma \qquad \forall\, z\in H^1(\Omega), \quad (3.2.108)$$

and a similar identity holds for y_v. Therefore, invoking the identities $(*)$ once more, we obtain that

$$J_r(v) - J_r(u)$$

$$\le \int_\Omega [L(x,y_u,v_d) - L(x,y_u,u_d)]\,dx + \int_\Gamma [\ell(\sigma,y_u,v_b) - \ell(\sigma,y_u,u_b)]\,d\sigma$$

$$-\int_\Omega F_y(x,\xi_1(x),v_d)\,(y_v - y_u)\,p_{u,v}\,dx - \int_\Gamma f_y(\sigma,\xi_4(\sigma),v_b)\,(y_v - y_u)\,p_{u,v}\,d\sigma$$

$$+\int_\Omega F(x,y_v,v_d)\,p_{u,v}\,dx - \int_\Omega F(x,y_u,u_d)\,p_{u,v}\,dx$$

$$+\int_\Gamma f(\sigma,y_v,v_b)\,p_{u,v}\,d\sigma - \int_\Gamma f(\sigma,y_u,u_b)\,p_{u,v}\,d\sigma$$

$$= \int_\Omega L(x,y_u,v_b)\,dx - \int_\Omega L(x,y_u,u_b)\,dx + \int_\Gamma \ell(\sigma,y_u,v_b)\,d\sigma$$

$$-\int_\Gamma \ell(\sigma,y_u,u_b)\,d\sigma - \int_\Omega [F(x,y_v,v_d) - F(x,y_u,v_d)]\,p_{u,v}\,dx$$

$$-\int_\Gamma [f(\sigma,y_v,v_b) - f(\sigma,y_u,v_b)]\,p_{u,v}\,d\sigma$$

$$+\int_\Omega [F(x,y_v,v_d) - F(x,y_u,u_d)]\,p_{u,v}\,dx$$

$$+\int_\Gamma [f(\sigma,y_v,v_b) - f(\sigma,y_u,u_b)]\,p_{u,v}\,d\sigma\,. \qquad (3.2.109)$$

Thus, recalling (3.2.100), (3.2.101), we have proved the assertion. □

Let us now define the so-called *spike variations* of a control $u \in U_{ad}$. To this end, let $x_0 \in \Omega$, $\omega_0 \in \Gamma$, and, for k and $\bar{k}$ positive,

$$S_k(x_0) = \{x \in \Omega : |x - x_0| \le 1/k\}, \quad S_{\bar{k}}(\omega_0) = \{\sigma \in \Gamma : |\sigma - \omega_0| \le 1/\bar{k}\}.$$

Then the spike variations of u_d and u_b in $x_0 \in \Omega$ and $\omega_0 \in \Gamma$, respectively, are defined as all functions of the form

$$u_d^k(x) = \begin{cases} \overline{v}, & \text{if } x \in S_k(x_0), \\ u_d(x), & \text{otherwise,} \end{cases} \tag{3.2.110}$$

$$u_b^{\bar{k}}(\sigma) = \begin{cases} \tau, & \text{if } \sigma \in S_{\bar{k}}(\omega_0), \\ u_b(\sigma), & \text{otherwise,} \end{cases} \tag{3.2.111}$$

with some $\overline{v} \in K_d$ and $\tau \in K_b$, respectively.

In the following, we will always assume that the index $\bar{k}$ depends on k, $\bar{k} = \bar{k}(k)$, in such a way that

$$\bar{k} \to \infty \quad \text{if } k \to \infty .$$

We will specify the form of this dependence below after (3.2.122).

Proposition 3.2.23 *If u is admissible for (P_m) and $k \in \mathbf{N}$ is sufficiently large, then $u^k = [u_d^k, u_b^{\bar{k}}]$ is admissible for $(P_{r,\varepsilon})$.*

Proof. Obviously, by (3.2.110), (3.2.111), $u^k \in U_{ad}$ for all $k \in \mathbf{N}$. Let $y^k \in H^1(\Omega) \cap C(\overline{\Omega})$, $k \in \mathbf{N}$, denote the associated weak solutions of the state system (3.2.79), (3.2.78). We show that y^k satisfies (3.2.99) for sufficiently large k (depending on $\varepsilon > 0$).

Hypothesis (3.2.91) shows that $\{y^k\}_{k \in \mathbf{N}}$ is bounded in $H^1(\Omega) \cap C(\overline{\Omega})$, since $\{u^k\}_{k \in \mathbf{N}}$ is bounded in $L^\infty(\Omega) \times L^\infty(\Gamma)$, by (3.2.110), (3.2.111). Hence, for a subsequence again indexed by k, we have $y^k \to y$ weakly in $H^1(\Omega)$, and by compact embedding, strongly in $L^2(\Omega)$. Without loss of generality, we may thus assume that $y^k \to y$ pointwise a.e. in Ω. In view of the growth condition (3.2.97), we can apply Lebesgue's theorem of dominated convergence to obtain that

$$\lim_{k \to \infty} \int_\Omega h(x, y^k(x))\, dx = \int_\Omega h(x, y(x))\, dx. \tag{3.2.112}$$

Next, we show that $y = y_u$, that is, y is the (unique) variational solution of (3.2.78), (3.2.79) corresponding to $u = [u_d, u_b]$. This will imply that the above convergences, in particular (3.2.112), hold for the entire sequence $\{y^k\}$ and not only for the subsequence constructed above.

Let $\phi_k = \phi(u^k; \cdot)$ and $\varphi_k = \varphi(u^k; \cdot)$ denote the functionals (3.2.85), (3.2.86), associated with u^k. We then have

$$\phi_k(y^k) + \varphi_k(y^k) \le \phi_k(z) + \varphi_k(z) \quad \forall\, z \in H^1(\Omega). \tag{3.2.113}$$

Now notice that $y^k \to y$ a.e. in Ω, and $y^k \to y$ a.e. on Γ (by the trace theorem), as well as, by definition, $u_d^k \to u_d$ a.e. in Ω, and $u_b^{\bar{k}} \to u_b$ a.e. on Γ. Thus, by the continuity of G and g, and since $\bar{k} \to \infty$ as $k \to \infty$,

$$\begin{aligned}\lim_{k\to\infty} G(x, y^k(x), u_d^k(x)) &= G(x, y(x), u_d(x)) \quad \text{for a.e. } x \in \Omega,\\ \lim_{k\to\infty} g(\sigma, y^k(\sigma), u_b^{\bar{k}}(\sigma)) &= g(\sigma, y(\sigma), u_b(\sigma)) \quad \text{for a.e. } \sigma \in \Gamma.\end{aligned}$$

Using the boundedness of $\{y^k\}$ in $C(\overline{\Omega})$, of $\{u_d^k\}$ in $L^\infty(\Omega)$, and of $\{u_b^{\bar{k}}\}$ in $L^\infty(\Gamma)$, and invoking the growth conditions (3.2.87)–(3.2.90), we readily find that Lebesgue's theorem can again be applied to yield that for $k \to \infty$,

$$\int_\Omega G(x, y_k, u_d^k)\, dx + \int_\Gamma g(\sigma, y_k, u_b^{\bar{k}})\, d\sigma \to \int_\Omega G(x, y, y_d)\, dx + \int_\Gamma g(\sigma, y, u_b)\, d\sigma\,. \tag{3.2.114}$$

Moreover, since the functional

$$y \mapsto \frac{1}{2}\int_\Omega \sum_{i,j=1}^d a_{ij}(x)\,\frac{\partial y}{\partial x_i}(x)\,\frac{\partial y}{\partial x_j}(x)\, dx + \frac{\alpha}{2}\int_\Gamma y^2(\sigma)\, d\sigma$$

is weakly lower semicontinuous on $H^1(\Omega)$, it follows from (3.2.114) that

$$\liminf_{k\to\infty} \{\phi_k(y_k) + \varphi_k(y_k)\} \ge \phi(u; y) + \varphi(u; y). \tag{3.2.115}$$

Now observe that if $z \in L^\infty(\Omega) \cap H^1(\Omega)$ and $z|_\Gamma \in L^\infty(\Gamma)$, then one can also pass to the limit as $k \to \infty$ in the right-hand side of (3.2.113):

$$\begin{aligned}\lim_{k\to\infty} \{\phi_k(z) + \varphi_k(z)\} &= \lim_{k\to\infty} \Big\{\frac{1}{2}\int_\Omega \sum_{i,j=1}^d a_{ij}(x)\,\frac{\partial z}{\partial x_i}\frac{\partial z}{\partial x_j}\, dx + \frac{\alpha}{2}\int_\Gamma z^2\, d\sigma\\ &\quad + \int_\Omega G(x, z, u_d)\, dx + \int_\Gamma g(\sigma, z, u_b)\, d\sigma\\ &\quad + \int_{S_k(x_0)} [G(x, z, \bar{v}) - G(x, z, u_d)]\, dx\\ &\quad + \int_{S_{\bar{k}}(\omega_0)} [g(\sigma, z, \omega) - g(\sigma, z, u_b)]\, d\sigma\Big\}\\ &= \phi(u; z) + \varphi(u; z).\end{aligned}$$

Here, we have used the growth conditions (3.2.87)–(3.2.90) to conclude the necessary integrability properties of G and g from the boundedness of z.

Observe now that in light of assumption (3.2.91), it suffices to minimize $\phi(u; \cdot) + \varphi(u; \cdot)$ over $H^1(\Omega) \cap C(\overline{\Omega})$ in the definition of the variational solution.

Consequently, combining the above estimates, we have shown that $y = y_u$. Since $[y, u]$ is admissible for (P_m), the assertion follows from (3.2.112). □

Now let $\overline{u} = [\overline{u}_d, \overline{u}_b] \in U_{ad}$ be given, and let $u^k = [\overline{u}_d^k, \overline{u}_b^k]$, $k \in \mathbf{N}$, denote associated spike variations according to (3.2.110), (3.2.111). Moreover, denote by $\overline{y} = y_{\overline{u}}$ the variational solution of (3.2.78), (3.2.79) corresponding to $\overline{u}$, and by $p_k = p_{\overline{u},u^k} \in H^1(\Omega)$ the variational solution of (3.2.102), (3.2.103) associated with $u = \overline{u}$ and $v = u^k$, $k \in \mathbf{N}$.

Here, we have to notice that the intermediate points ξ_i, $1 \le i \le 5$, and the point w appearing in (3.2.102), (3.2.103) may vary with $k \in \mathbf{N}$. We indicate this dependence by a superimposed index k. We have the following result.

Proposition 3.2.24 *Under the above assumptions, $\{p_k\}$ is bounded in $H^1(\Omega)$, and there is some subsequence, again indexed by k, such that $p_k \to \overline{p}$ weakly in $H^1(\Omega)$, and $\overline{p} \in H^1(\Omega)$ is the variational solution of*

$$A\overline{p} = F_y(x,\overline{y},\overline{u}_d)\,\overline{p} + L_y(x,\overline{y},\overline{u}_d) + r\,w\,h_y(x,\overline{y}) \quad \text{in } \Omega, \tag{3.2.116}$$

$$\alpha\,\overline{p} + \frac{\partial \overline{p}}{\partial n_A} = f_y(\sigma,\overline{y},\overline{u}_b)\,\overline{p} + \ell_y(\sigma,\overline{y},\overline{u}_b) \quad \text{on } \Gamma, \tag{3.2.117}$$

with some $w \in \partial_+\Big(\int_\Omega h(x,\overline{y}(x))\,dx - m\Big)$.

Proof. Recall that $\{u^k\}$ and $\{y^k\}$ are bounded in $L^\infty(\Omega)\times L^\infty(\Gamma)$ and $H^1(\Omega)\cap C(\overline{\Omega})$, respectively. Moreover, we have $u^k \to \overline{u}$ pointwise a.e. in $\Omega\times\Gamma$, as well as $y^k \to \overline{y}$ pointwise a.e. in Ω, while p_k minimizes on $H^1(\Omega)$ the quadratic functional $\psi_k = \psi_{\overline{u},u^k}$ given by (3.2.105) for $u = \overline{u}$ and $v = u^k$. Then

$$\psi_k(p_k) \le \psi_k(z), \quad \forall z \in H^1(\Omega). \tag{3.2.118}$$

In particular, for $z = 0$,

$$\psi_k(p_k) \le 0 \quad \forall\, k \in \mathbf{N}. \tag{3.2.119}$$

Since $\{y^k\}$ and $\{u^k\}$ are uniformly bounded, and since the graph $\partial_+(\cdot)$ is bounded in $\mathbf{R}$, it follows from the growth conditions (3.2.93)–(3.2.97) and (3.2.87)–(3.2.90), together with the relations $(*)$, that the coefficient functions of ψ_k satisfy the following estimates:

$$\begin{aligned} z_1^k(x) &= \Big|F_y\big(x,\xi_1^k(x),\overline{u}_d^k(x)\big)\Big| \le \Big[M_2(x) + \hat{C}\,\big|\overline{u}_d^k(x)\big|\Big]\,\eta\Big(\big|\xi_1^k(x)\big|\Big) \\ &\le C_1\,(1 + M_2(x)) \quad \text{for a.e. } x \in \Omega, \\ z_2^k(x) &= \Big|L_y\big(x,\xi_2^k(x),\overline{u}_d^k(x)\big)\Big| \le \Big[M_1(x) + \hat{C}\,\big|\overline{u}_d^k(x)\big|\Big]\,\eta\Big(\big|\xi_2^k(x)\big|\Big) \end{aligned}$$

$$
\begin{aligned}
&\le C_2\,(1+M_1(x)) \quad \text{for a.e. } x\in\Omega\,,\\
z_3^k(x) \;&=\; \Big|w^k\,h_y\Big(x,\xi_3^k(x)\Big)\Big| \;\le\; |w^k|\,M_1(x)\,\eta\Big(\big|\xi_3^k(x)\big|\Big)\\
&\le C_3\,M_1(x) \quad \text{for a.e. } x\in\Omega\,,\\
z_4^k(\sigma) \;&=\; \Big|f_y\Big(\sigma,\xi_4^k(\sigma),\overline{u}_b^{\bar k}(\sigma)\Big)\Big| \;\le\; \Big[\,R_2(\sigma)+\hat C\,\big|\overline{u}_b^{\bar k}(\sigma)\big|\,\Big]\;\eta\Big(\big|\xi_4^k(\sigma)\big|\Big)\\
&\le C_4\,(1+R_2(\sigma)) \quad \text{for a.e. } \sigma\in\Gamma\,,\\
z_5^k(\sigma) \;&=\; \Big|\ell_y\Big(\sigma,\xi_5^k(\sigma),\overline{u}_b^{\bar k}(\sigma)\Big)\Big| \;\le\; \Big[\,R_1(\sigma)+\hat C\,\big|\overline{u}_b^{\bar k}(\sigma)\big|\,\Big]\;\eta\Big(\big|\xi_5^k(\sigma)\big|\Big)\\
&\le C_5\,(1+R_1(\sigma)) \quad \text{for a.e. } \sigma\in\Gamma\,,
\end{aligned}
$$

with constants $C_i>0$, $1\le i\le 5$, which do not depend on $k\in\mathbf{N}$. Consequently, there is some constant $C_6>0$ such that

$$
|z_1^k|_{L^s(\Omega)} \,+\, |z_2^k|_{L^2(\Omega)} \,+\, |z_3^k|_{L^2(\Omega)} \,+\, |z_4^k|_{L^\infty(\Gamma)} \,+\, |z_5^k|_{L^2(\Gamma)} \,\le\, C_6 \quad \forall\, k\in\mathbf{N}.
$$

Thus, invoking (3.2.83), (3.2.105), and the Schwarz and Friedrichs inequalities, and using the fact that $-F_y\ge 0$ and $-f_y\ge 0$, we can deduce that for all $k\in\mathbf{N}$ and all $p\in H^1(\Omega)$,

$$
\begin{aligned}
\psi_k(p) \;&\ge\; \frac{\omega}{2}\int_\Omega |\nabla(x)|^2\,dx \,+\, \frac{\alpha}{2}\int_\Gamma p^2(\sigma)\,d\sigma \,-\, C_6\,\Big(2\,|p|_{L^2(\Omega)}+|p|_{L^2(\Gamma)}\Big)\\
&\ge\; \delta\,|p|^2_{H^1(\Omega)} \,-\, C_7\,,
\end{aligned}
$$

with constants $\delta>0$ and $C_7>0$ that do not depend on k. Then (3.2.119) yields that $\{p_k\}$ is bounded in $H^1(\Omega)$, and there is a subsequence, again indexed by k, such that $p_k\to\overline{p}$ weakly in $H^1(\Omega)$, and, by compact embedding (cf. Theorem A2.2(ii) in Appendix 2), strongly in $L^q(\Omega)$, where $q\ge 1$ is arbitrary for $d=1$ and $1\le q<\frac{2d}{d-2}$ for $d\ge 2$.

Moreover, owing to Theorem A2.1(i) in Appendix 2, the trace operator is compact from $H^1(\Omega)$ into $L^2(\Gamma)$. Therefore, we have $p_k|_\Gamma\to\overline{p}|_\Gamma$ strongly in $L^2(\Gamma)$. Finally, we may assume that $w^k\to w$, for some $w\in[0,1]$, in view of the boundedness of $\partial_+(\cdot)$.

Invoking the a.e. convergences $y^k\to\overline{y}$ and $u^k\to\overline{u}$, and using Lebesgue's theorem, it is readily seen that for $k\to\infty$,

$$
\begin{aligned}
F_y\Big(\cdot,\xi_1^k(\cdot),\overline{u}_d^k(\cdot)\Big) \;&\to\; F_y(\cdot,\overline{y}(\cdot),\overline{u}_d(\cdot)) \quad \text{strongly in } L^s(\Omega),\\
L_y\Big(\cdot,\xi_2^k(\cdot),\overline{u}_d^k(\cdot)\Big) \;&\to\; L_y(\cdot,\overline{y}(\cdot),\overline{u}_d(\cdot)) \quad \text{strongly in } L^2(\Omega),\\
w^k\,h_y\Big(\cdot,\xi_3^k(\cdot)\Big) \;&\to\; w\,h_y(\cdot,\overline{y}(\cdot)) \quad \text{strongly in } L^2(\Omega),\\
f_y\Big(\cdot,\xi_4^k(\cdot),\overline{u}_b^{\bar k}(\cdot)\Big) \;&\to\; f_y(\cdot,\overline{y}(\cdot),\overline{u}_b(\cdot)) \quad \text{strongly in } L^t(\Gamma) \text{ for all } t>1,
\end{aligned}
$$

$$\ell_y\left(\cdot, \xi_5^k(\cdot), \overline{u}_b^k(\cdot)\right) \to \ell_y(\cdot, \overline{y}(\cdot), \overline{u}_b(\cdot)) \quad \text{strongly in } L^2(\Gamma).$$

In addition, the relation (3.2.112) is apparently fulfilled. Thus, using the demiclosedness of the maximal monotone operator $\partial_+(\cdot)$ on $\mathbf{R}$ (cf. Proposition A1.4(iv) in Appendix 1), we can conclude that $w \in \partial_+(\int_\Omega h(x, \overline{y}(x))\, dx - m)$.

Summarizing the above facts, and referring to (3.2.118) and to Definition 3.2.17, we infer that $\overline{p} \in H^1(\Omega)$ is the variational solution to (3.2.104), with the coefficients there replaced by the above limits. The assertion is thus proved.

□

In order to derive the first-order necessary conditions of optimality, we have to make a further assumption, namely that (3.2.91) is also valid for the adjoint equation (3.2.116), (3.2.117) in a slightly stronger form; this additional assumption will guarantee that the approximate adjoint states converge *uniformly.* More precisely, we postulate:

For every $\overline{u} = [\overline{u}_d, \overline{u}_b] \in U_{ad}$ the solution to (3.2.116), (3.2.117) belongs to $C^\gamma(\overline{\Omega})$ with some fixed $\gamma \in]0,1[$, and it is bounded with respect to the coefficient functions.

Remark. The equations considered in Example 3.2.18 and in Example 3.2.19 satisfy this stronger assumption. Furthermore, if in the growth conditions (3.2.87)–(3.2.90) we postulate that $M_1, M_2 \in L^\infty(\Omega)$ and $R_1 \in L^\infty(\Gamma)$, then the weak solution $\overline{p}$ of (3.2.116), (3.2.117) belongs to $W^{1,\mu}(\Omega)$ with $\mu > 1$ sufficiently large as to guarantee that $\overline{p} \in C^\gamma(\overline{\Omega})$ for some $\gamma > 0$. Further regularity results of this type have been proved in Gilbarg and Trudinger [1983, Chapter 6.7].

Theorem 3.2.25 *Let the above hypotheses be fulfilled, and suppose that* $[\overline{y}, \overline{u}] \in (H^1(\Omega) \cap C(\overline{\Omega})) \times U_{ad}$ *is an optimal pair for* (P_m), *and* $\overline{p} \in H^1(\Omega) \cap C^\gamma(\overline{\Omega})$ *the solution of (3.2.116), (3.2.117). Then*

$$H(x, \overline{y}(x), \overline{u}_d(x), \overline{p}(x)) \le H(x, \overline{y}(x), w, \overline{p}(x)) \quad \forall\, w \in K_d, \text{ for a.e. } x \in \Omega, \tag{3.2.120}$$

$$j(\sigma, \overline{y}(\sigma), \overline{u}_b(\sigma), \overline{p}(\sigma)) \le j(\sigma, \overline{y}(\sigma), \tilde{w}, \overline{p}(\sigma)) \quad \forall\, \tilde{w} \in K_b, \text{ for a.e. } \sigma \in \Gamma. \tag{3.2.121}$$

Proof. Let $x_0 \in \Omega$, $\omega_0 \in \Gamma$, and $k \in \mathbf{N}$ be given. We consider for arbitrary fixed $w \in K_d$ and $\tilde{w} \in K_b$ the spike variations $u^k = [\overline{u}_d^k, \overline{u}_b^k]$ given by (compare

(3.2.110), (3.2.111))

$$\overline{u}_d^k(x) = \begin{cases} w, & \text{if } x \in S_k(x_0), \\ \overline{u}_d(x), & \text{otherwise}, \end{cases}$$

$$\overline{u}_b^{\bar{k}}(\sigma) = \begin{cases} \widetilde{w}, & \text{if } \sigma \in S_{\bar{k}}(\omega_0), \\ \overline{u}_b(\sigma), & \text{otherwise}. \end{cases}$$

Now let $\varepsilon > 0$ be arbitrary. Then, thanks to Proposition 3.2.23, there is some $k_0(\varepsilon) \in \mathbf{N}$ such that u^k is admissible for $(\mathrm{P}_{r,\varepsilon})$ provided that $k \geq k_0(\varepsilon)$ (which will be assumed henceforth). Then it follows from Proposition 3.2.20 that $J_r(\overline{u}) \leq J_r(u^k)$, and putting $p_k = p_{\overline{u},u^k}$, we conclude from Proposition 3.2.22 that

$$\begin{aligned} &\int_\Omega H(x, \overline{y}(x), \overline{u}_d(x), p_k(x))\, dx + \int_\Gamma j(\sigma, \overline{y}(\sigma), \overline{u}_b(\sigma), p_k(\sigma))\, d\sigma \\ &\quad \leq \int_\Omega H\big(x, \overline{y}(x), \overline{u}_d^k(x), p_k(x)\big)\, dx + \int_\Gamma j\big(\sigma, \overline{y}(\sigma), \overline{u}_b^{\bar{k}}(\sigma), p_k(\sigma)\big)\, d\sigma. \end{aligned} \tag{3.2.122}$$

Notice that $\{p_k\}$ is bounded in $H^1(\Omega) \cap C^\gamma(\overline{\Omega})$, under the above regularity hypothesis. Hence, as in the proof of Proposition 3.2.24, we may select a subsequence, again indexed by k, such that $p_k \to \overline{p}$ weakly in $H^1(\Omega)$ and, owing to the Arzelà–Ascoli theorem, strongly in $C(\overline{\Omega})$.

Next, we specify how $\bar{k}$ depends on k; namely, we choose for given $k \in \mathbf{N}$ the corresponding $\bar{k} = \bar{k}(k) > 0$ in such a way that $\operatorname{meas}\big(S_k(x_0)\big) = \operatorname{meas}\big(S_{\bar{k}}(\omega_0)\big)$ (which is denoted by ν_k), as computed in $\mathbf{R}^d$, respectively on Γ. Clearly, $\nu_k \to 0$ as $k \to \infty$, and thus $\bar{k} \to \infty$ as $k \to \infty$, as required.

Dividing by $\nu_k > 0$, we may rewrite (3.2.122) in the form (where arguments are again omitted)

$$\begin{aligned} &\frac{1}{\nu_k} \int_{S_k(x_0)} H(x, \overline{y}, \overline{u}_d, \overline{p})\, dx + \frac{1}{\nu_k} \int_{S_{\bar{k}}(\omega_0)} j(\sigma, \overline{y}, \overline{u}_b, \overline{p})\, d\sigma \\ &\quad + \frac{1}{\nu_k} \int_{S_k(x_0)} (p_k - \overline{p})\, F(x, \overline{y}, \overline{u}_d)\, dx + \frac{1}{\nu_k} \int_{S_{\bar{k}}(\omega_0)} (p_k - \overline{p})\, f(\sigma, \overline{y}, \overline{u}_b)\, d\sigma \\ &\leq \frac{1}{\nu_k} \int_{S_k(x_0)} H(x, \overline{y}, w, \overline{p})\, dx + \frac{1}{\nu_k} \int_{S_{\bar{k}}(\omega_0)} j(\sigma, \overline{y}, \widetilde{w}, \overline{p})\, d\sigma \\ &\quad + \frac{1}{\nu_k} \int_{S_k(x_0)} (p_k - \overline{p})\, F(x, \overline{y}, w)\, dx + \frac{1}{\nu_k} \int_{S_{\bar{k}}(\omega_0)} (p_k - \overline{p}) f(\sigma, \overline{y}, \widetilde{w})\, d\sigma. \end{aligned} \tag{3.2.123}$$

The terms containing $p_k - \overline{p}$ converge to zero for $k \to \infty$. Therefore, we can pass to the limit as $k \to \infty$ in (3.2.123) assuming that x_0, ω_0 are Lebesgue points for the given mappings. Then (3.2.120), (3.2.121) follow, and the proof is finished.

□

Remark. In this section, we have pointed out the role of the $C(\overline{\Omega})$-regularity (respectively, $C^\gamma(\overline{\Omega})$-regularity) for the state and adjoint equations, and where it is needed. While $H^1(\Omega)$-regularity is easily shown, to prove the (Hölder) continuity of variational solutions is quite a difficult task. We have provided pertinent references to the literature.

Remark. Quasilinear elliptic optimization problems have been studied by Casas and Fernandez [1991]. Parabolic and hyperbolic quasilinear control problems were investigated in Tiba [1990, Chapter II] using different methods. The unstable (not well-posed) case can be found in the works of Bonnans and Casas [1989], Wang, G. [2002], and Wang, G. and Wang, L. [2003]. Pointwise state constraints were discussed in Bonnans and Casas [1989], [1991], and Casas and Fernandez [1993]. A survey of various stability concepts related to the calmness condition and their applications to optimal control can also be found in Bonnans and Casas [1992]. A second-order analysis for control problems associated with semilinear elliptic equations has been performed in the works of Bonnans [1998], Casas and Tröltzsch [1999], and Casas and Mateos [2002a].

3.3 Control of Variational Inequalities

This section continues the analysis of the first-order optimality conditions for the case that the state systems are governed by variational inequalities. In this situation, convexity and/or differentiability properties of the optimization problem are missing, and special smoothing and approximating techniques have to be employed.

We have seen above in §3.1.2 that the optimality conditions involve the calculation of the gradients of the nonlinear terms appearing in the state system. In the case of variational inequalities, the nonlinearity is given by the generalized gradient (the subdifferential) of a convex mapping. Therefore, a generalized second-order derivative of convex mappings ought to appear in a natural way in the derivation of first-order necessary conditions. We will address this problem below in §3.3.3.

3.3.1 An Abstract Result

In this paragraph, we will interpret the variational inequality as an (infinite) system of inequalities, aiming to derive necessary conditions resembling the classical so-called *Fritz–John conditions* known from differentiable nonlinear optimization theory. These are weaker than the Karush–Kuhn–Tucker optimality conditions discussed above in §3.1.2 in a different setting, since the Lagrange multiplier associated with the objective function L may vanish (compare with (3.1.18)).

To begin with, let V, H, U denote (not necessarily reflexive) Banach spaces such that the embeddings $H \subset H^*$ and $V \subset H \subset V^*$ exist and are continuous, and such that (compatibility of the duality pairing)

$$(v,h)_{H\times H^*} = (v,h)_{V\times V^*} \quad \text{whenever } v \in V, \quad h \in H.$$

We postulate that $A : V \to H$, $\theta : V \to \mathbf{R}$, and $\psi : U \to \mathbf{R}$ possess Fréchet derivatives ∇A, $\nabla\theta$, and $\nabla\psi$, respectively, on their domains of definition, and that $B \in L(U,H)$. The cost functional is given by

$$L(y,u) = \theta(y) + \psi(u), \tag{3.3.1}$$

and the state equation by the variational inequality

$$Ay + \partial\phi(y) \ni Bu. \tag{3.3.2}$$

Here, $\phi = I_K$ with some nonempty, closed, and convex cone $K \subset V$. We assume that the state equation (3.3.2) admits a unique solution $y \in V$ for every $u \in U$.

In addition, we impose the control constraints $u \in U_{ad} \subset U$ and the state constraints $y \in C \subset H$, where as usual, U_{ad} and C are nonempty, convex, and closed in U and H, respectively. Hence, we study a very general optimal control problem, but we will still be able to recover "nonqualified" first-order necessary conditions of Fritz–John type.

Since K is a convex cone, problem (3.3.1), (3.3.2) may be equivalently rewritten as

$$\operatorname{Min}\left\{\theta(y) + \psi(u)\right\}, \tag{3.3.3}$$

$$(y,u) \in \tilde{C} \times U_{ad}, \quad \text{where } \tilde{C} = K \cap C, \tag{3.3.4}$$

$$(Ay - Bu, y)_{V\times V^*} \le 0, \tag{3.3.5}$$

$$(Ay - Bu, z)_{V\times V^*} \ge 0, \quad \forall\, z \in K. \tag{3.3.6}$$

Indeed, (3.3.2) is by definition equivalent to the variational inequality

$$(Ay - Bu, y - v)_{V^*\times V} \le 0 \quad \forall\, v \in K,$$

which obviously follows from (3.3.5), (3.3.6). Conversely, we get (3.3.5) for $v = 0 \in K$, and (3.3.6) for $v = z + y \in K$ (notice that $w = \frac{1}{2}(z+y) \in K$ and thus $v = 2\,w \in K$).

Since $z \in K$ can be arbitrarily chosen, (3.3.6) constitutes an infinite system of independent inequalities (recall that $K \subset V$ and V is an infinite-dimensional space). In addition, if

$$K^0 = \{z \in H^* : (z, y)_{H^* \times H} \le 0 \quad \forall y \in K\}$$

denotes the polar cone of K in H and $\hat{K} = K^0 \cap H$, then (3.3.6) may be rewritten as

$$Ay - Bu \in -\hat{K}. \tag{3.3.7}$$

We thus obtain an infinite-dimensional differentiable programming problem in the independent variables y and u. In view of (3.3.5), it is essentially nonconvex. Let $[y^*, u^*]$ be an optimal pair for the problem (3.3.3)–(3.3.7), which we assume to exist. Then we have the following result.

Theorem 3.3.1 *Assume that* $\partial I_{\widetilde{C}} = \partial I_C + \partial I_K$ *and that* $\operatorname{int}(\hat{K}) \ne \emptyset$ *in* H. *Then there are* $\alpha \ge 0$, $p \in H^*$, *and* $\mu \ge 0$ *such that*

$$\alpha + |p|_{H^*} + \mu > 0, \tag{3.3.8}$$

$$(p, z)_{H^* \times H} \ge 0 \quad \forall z \in \hat{K}, \tag{3.3.9}$$

$$(p, Ay^* - Bu^*)_{H^* \times H} = 0, \tag{3.3.10}$$

$$\begin{aligned} 0 \in\ & \alpha \nabla\theta(y^*) + (\nabla A(y^*))^* p + \mu (\nabla A(y^*))^* y^* \\ & + \mu [Ay^* - Bu^* + \partial I_K(y^*)] + \partial I_C(y^*), \end{aligned} \tag{3.3.11}$$

$$0 \in \alpha \nabla\psi(u^*) - B^* p - \mu B^* y^* + \partial I_{U_{ad}}(u^*). \tag{3.3.12}$$

Proof. We use standard techniques from differentiable programming theory, where we refer the reader to Barbu and Precupanu [1986, Chapter 3.1]. Let $\mathcal{A}$ be the set of admissible pairs, and let

$$\begin{aligned} TC(\mathcal{A}; [y^*, u^*]) = \Big\{ & [y, u] \in V \times U : \exists [y_n, u_n] \in \mathcal{A} \text{ with } [y_n, u_n] \to [y^*, u^*] \\ & \text{in } V \times U, \text{ and } \lambda_n \nearrow \infty \text{ such that} \\ & \lambda_n([y_n, u_n] - [y^*, u^*]) \to [y, u] \Big\} \cup \Big\{ [0, 0] \Big\} \end{aligned} \tag{3.3.13}$$

be the so-called *tangent cone* at $\mathcal{A}$ in $[y^*, u^*]$. We note that under certain assumptions it coincides with the polar cone of the normal cone to $\mathcal{A}$ in $[y^*, u^*]$ (cf. Appendix 1 and Lemaréchal and Hiriart-Urruty [1993, Chapter III.5]). Thanks to the optimality of $[y^*, u^*]$, we have, with the above notations,

$$\begin{aligned} & \lim_{n \to \infty} \lambda_n \Big[\theta(y_n) + \psi(u_n) - \theta(y^*) - \psi(u^*) \Big] \\ & = (\nabla\theta(y^*), y)_{V^* \times V} + (\nabla\psi(u^*), u)_{U^* \times U} \\ & \ge 0 \quad \forall [y, u] \in \overline{\operatorname{conv}}\, TC(\mathcal{A}; [y^*, u^*]). \end{aligned} \tag{3.3.14}$$

We now use the assumption that $\operatorname{int}(\hat{K}) \ne \emptyset$ in order to show, as in Barbu and Precupanu [1986, p. 189], that

$$\Big[TC((C \cap K) \times U_{ad}; [y^*, u^*]) \cap S \Big] \subset TC(\mathcal{A}; [y^*, u^*]), \tag{3.3.15}$$

where $S \subset V \times U$ is the counterimage of the set $[-\text{int}(\mathbf{R}_+ \times \hat{K}) - R(y^*, u^*)]$ under the application of $\nabla R(y^*, u^*)$, where $R : V \times U \to \mathbf{R} \times H$ is the Fréchet differentiable mapping

$$R(y, u) = [(Ay - Bu, y)_{V \times V^*}, Ay - Bu].$$

To prove the inclusion (3.3.15), we may confine ourselves to consider some $[y, u] \in TC((C \cap K) \times U_{ad}; [y^*, u^*]) \cap S$ such that $[y, u] \neq [0, 0]$. Then there are $[y_n, u_n] \in (C \cap K) \times U_{ad}$ and $\lambda_n > 0$, $n \in \mathbf{N}$, having the corresponding properties stated in the definition (3.3.13). By the differentiability of R, we have

$$R(y_n, u_n) - R(y^*, u^*) = \nabla R(y^*, u^*)\,(y_n - y^*, u_n - u^*) + \mu\left(|(y_n - y^*, u_n - u^*)_{V \times U}|\right),$$

with a function μ satisfying $\lim_{h \searrow 0} \frac{\mu(h)}{h} = 0$. Multiplying by λ_n, we thus get

$$\lim_{n \to \infty} \lambda_n\,(R(y_n, u_n) - R(y^*, u^*)) = \nabla R(y^*, u^*)\,(y, u).$$

Now recall that $[y, u] \in S$, and thus, by the definition of S,

$$\nabla R(y^*, u^*)(y, u) \in -\text{int}\left(\mathbf{R}_+ \times \hat{K}\right) - R(y^*, u^*).$$

Since the latter set is open, we must have, for a sufficiently large $N_0 \in \mathbf{N}$,

$$\lambda_n\,(R(y_n, u_n) - R(y^*, u^*)) \in -\text{int}\,(\mathbf{R}_+ \times \hat{K}) - R(y^*, u^*) \quad \text{for } n \geq N_0,$$

whence

$$\lambda_n\, R(y_n, u_n) + (1 - \lambda_n)\, R(y^*, u^*) \in -\text{int}\,(\mathbf{R}_+ \times \hat{K}) \quad \text{for } n \geq N_0.$$

By (3.3.13), $\lambda_n \nearrow \infty$, so that we can assume that $\lambda_n > 1$. Thus, noticing that $R(y^*, u^*) \in -(\mathbf{R}_+ \times \hat{K})$ by the optimality of $[y^*, u^*]$, and recalling that $\mathbf{R}_+ \times \hat{K}$ is a cone, we can infer that $[y_n, u_n] \in \mathcal{A}$ for $n \geq N_0$, which finishes the proof of (3.3.15).

Combining (3.3.14) and (3.3.15), we realize that $[0, 0]$ is an optimal pair for the convex programming problem

$$\text{Min}\left\{(\nabla\theta(y^*), y)_{V^* \times V} + (\nabla\psi(u^*), u)_{U^* \times U}\right\} \tag{3.3.16}$$

subject to $[y, u] \in TC(\widetilde{C} \times U_{ad}; [y^*, u^*])$ and to

$$\nabla R(y^*, u^*)[y, u] + R(y^*, u^*) \in -(\mathbf{R}_+ \times \hat{K}). \tag{3.3.17}$$

Note that $TC(\widetilde{C} \times U_{ad}; [y^*, u^*])$ is a convex cone, since $\widetilde{C} \times U_{ad}$ is convex. We refer to (3.3.18) below for an equivalent definition of the tangent cone in the convex case.

Invoking the familiar Fritz–John multiplier rule from convex differentiable programming (see, for instance, Barbu and Precupanu [1986, Chapter 3.1]), we find that there exist $\alpha \geq 0$, $\mu \geq 0$, and $p \in H^*$ satisfying (3.3.8)–(3.3.10) and the variational inequality

$$\alpha\,(\nabla\theta(y^*), y)_{V^*\times V} + \alpha\,(\nabla\psi(u^*), u)_{U^*\times U} + \mu\,(\nabla A(y^*)y - Bu, y^*)_{V\times V^*}$$
$$+ \mu\,(Ay^* - Bu^*, y)_{V\times V^*} + (p, \nabla A(y^*)y - Bu)_{H^*\times H} \geq 0\,,$$
$$\forall\,[y,u] \in TC(\widetilde{C}\times U_{ad}; [y^*,u^*]) = \overline{\bigcup_{\lambda>0} \lambda(\widetilde{C}\times U_{ad} - [y^*,u^*])}\,. \qquad (3.3.18)$$

Since $\widetilde{C}\times U_{ad}$ is convex, and since $\partial I_C(y^*)$ is a closed convex cone, we easily recover (3.3.11) and (3.3.12) from (3.3.18) and from the definition of the subdifferential. This concludes the proof of the assertion. □

Remark. The term $Ay^*-Bu^*+\partial I_C(y^*)$ in (3.3.11) does not necessarily vanish, which is a typical situation for multiequations. The fact that $\alpha = 0$ is possible in Theorem 3.3.1 reflects the "nonqualified" character of the optimality system (3.3.8)–(3.3.12); that is, the cost functional may not appear at all in (3.3.11) and (3.3.12).

Example 3.3.2 We now discuss the case of an obstacle problem as state equation. To this end, let $\Omega \subset \mathbf{R}^d$ be a bounded domain with sufficiently smooth boundary, and let $\Lambda \in W^{2,\infty}(\Omega)$ with $\Lambda \leq 0$ on $\partial\Omega$ be given. Moreover, we make the choices $V = W^{2,\infty}(\Omega)\cap W_0^{1,\infty}(\Omega)$, $H = L^\infty(\Omega)$, and

$$K = \left\{y \in W^{2,\infty}(\Omega)\cap W_0^{1,\infty}(\Omega) : y \geq \Lambda \text{ in } \Omega\right\}, \qquad (3.3.19)$$

leaving U, U_{ad}, C, θ, ψ unspecified. Apparently, K is not a cone, but $\tilde{K} = K - \Lambda$ is one. We then study the problem

$$\text{Min}\left\{\theta(y) + \psi(u)\right\} \qquad (3.3.20)$$

subject to $y \in \tilde{K}\cap C$, $u \in U_{ad}$, and

$$-\Delta y + \beta(y-\Lambda) \ni Bu \quad \text{a.e. in } \Omega, \qquad (3.3.21)$$

where $\beta \subset \mathbf{R}\times\mathbf{R}$ is the maximal monotone graph

$$\beta(r) = \begin{cases} \{0\}, & r > 0,\\ ]-\infty, 0], & r = 0,\\ \emptyset, & \text{otherwise,} \end{cases}$$

which corresponds to $\partial I_{\tilde{K}}$.

Now put $\hat{K} = \tilde{K}^0 \cap H = \{v \in H : v \leq 0 \text{ a.e. in } \Omega\}$. Then the inclusion (3.3.21) may be equivalently rewritten as

$$\int_\Omega (y(x) - \Lambda(x))\,(-\Delta y(x) - Bu(x))\,dx \leq 0, \tag{3.3.22}$$

$$-\Delta y - Bu \in -\hat{K}. \tag{3.3.23}$$

With $A : V \to H$, $Ay = -\Delta y$, the problem (3.3.20), (3.3.21) then attains the form (3.3.3)–(3.3.7). We remark that by the definition of β, the sets of admissible pairs for the problems (3.3.20), (3.3.21) and (3.3.22), (3.3.23), (3.3.20) coincide, and the two problems are equivalent. Since $\operatorname{int}(\hat{K}) \neq \emptyset$, we may apply Theorem 3.1.4, and there are $\alpha \geq 0$, $\mu \geq 0$, $p \in H^*$, not all equal to zero, such that (3.3.9), (3.3.10) are satisfied along with

$$\begin{aligned} 0 &\in \alpha\nabla\theta(y^*) + A^*p + \mu A^*(y^* - \Lambda) + \mu\,[-\Delta y^* - Bu^* + \beta(y^* - \Lambda)] \\ &\quad + \partial I_C(y^*), \\ 0 &\in \alpha\nabla\psi(u^*) - B^*p - \mu B^*(y^* - \Lambda) + \partial I_{U_{ad}}(u^*). \end{aligned} \tag{3.3.24}$$

Notice that the term $-\Delta y^* - Bu^* + \beta(y^* - \Lambda)$ in (3.3.24) is not necessarily equal to zero. An important question left open is under what conditions the Lagrange multiplier α in (3.3.24) or in Theorem 3.3.1 is nonzero. Notice that in this example the classical Slater condition cannot be fulfilled, since the admissible pairs satisfy (3.3.22) with equality.

Remark. This section is based on the work of Barbu and Tiba [1990]. Recently, Voisei [2004] has developed a new and unifying approach to unconstrained control problems governed by general nonlinear equations, based on the notion of support functions for multivalued operators.

3.3.2 Semilinear Variational Inequalities

In this section, we study a general control problem governed by elliptic variational inequalities in a functional analytic setting. The used method applies the Ekeland [1979] variational principle (see Theorem A1.15 in Appendix 1) and has a general character.

Suppose that $\Omega \subset \mathbf{R}^d$ is a bounded domain having a smooth boundary $\partial\Omega$. We then consider the problem

$$\text{Min}\left\{J(y,u) = \int_\Omega L(x,y(x),u(x))\,dx\right\} \tag{3.3.25}$$

subject to

$$Ay(x) + \varphi(x,y(x),u(x)) + \beta(y(x)) \ni 0 \quad \text{in } \Omega, \tag{3.3.26}$$

$$y = 0 \quad \text{on } \partial\Omega, \tag{3.3.27}$$

$$u \in U_{ad}, \quad \text{with } U_{ad} = \{u \in L^\infty(\Omega) : u(x) \in K \quad \text{a.e. in } \Omega\}, \tag{3.3.28}$$

where $K \subset \mathbf{R}$ is some nonempty set, and where

$$Ay(x) = -\sum_{i,j=1}^{d} \frac{\partial}{\partial x_i}\left(a_{ij}(x)\,\frac{\partial y}{\partial x_j}(x)\right).$$

We denote by

$$\widetilde{A}y(x) = -\sum_{i,j=1}^{d} \frac{\partial}{\partial x_j}\left(a_{ij}(x)\,\frac{\partial y}{\partial x_i}(x)\right)$$

the formal adjoint of A. The coefficient functions $a_{ij} \in C^1(\overline{\Omega})$ of A are not necessarily symmetric and are assumed to satisfy the ellipticity condition

$$\sum_{i,j=1}^{d} a_{ij}(x)\,\xi_i\,\xi_j \geq \omega\sum_{i=1}^{d}\xi_i^2, \quad \forall\,\xi \in \mathbf{R}^d, \quad \forall\,x \in \Omega, \quad \text{with some } \omega > 0. \tag{3.3.29}$$

The operator $\beta \subset \mathbf{R} \times \mathbf{R}$ is assumed to be maximal monotone, and the mappings $\varphi : \Omega \times \mathbf{R} \times K \to \mathbf{R}$, $L : \Omega \times \mathbf{R} \times K \to \mathbf{R}$ are continuous and have continuous derivatives φ_y and L_y with respect to y, such that (compare the conditions (3.2.87), (3.2.88), (3.2.93), (3.2.94) in the previous paragraph)

$$|\varphi(x,0,u)| \leq M(x) + \hat{C}\,|u|, \tag{3.3.30}$$

$$0 \leq \varphi_y(x,y,u) \leq \left(M(x) + \hat{C}\,|u|\right)\eta(|y|), \tag{3.3.31}$$

$$|L(x,0,u)| \leq M(x) + \hat{C}\,|u|, \tag{3.3.32}$$

$$|L_y(x,y,u)| \leq \left(M(x) + \hat{C}\,|u|\right)\eta(|y|). \tag{3.3.33}$$

Here, $\eta : \mathbf{R}_+ \to \mathbf{R}_+$ is a nondecreasing function, $\hat{C} > 0$ is a fixed constant, and $M \in L^s(\Omega)$ with $s \geq \max\{2, \frac{d}{2}\}$ is a given function.

Now let $\varepsilon > 0$ and denote by β_ε an ε-uniform approximation of β (which differs from the Yosida–Moreau approximation of β) as defined in Example A1.13 in Appendix 1. We recall that β_ε has the following properties:

$$\beta(y+\varepsilon) \geq \beta_\varepsilon(y) \geq \beta(y-\varepsilon) \quad \forall\, y \in \mathbf{R}; \quad \text{that is,}$$

$$\xi \geq \eta \geq \nu, \quad \forall \xi \in \beta(y+\varepsilon),\ \forall \eta \in \beta_\varepsilon(y),\ \forall \nu \in \beta(y-\varepsilon)$$

(where we view β, β_ε as multivalued operators extended to $\mathbf{R}$ by $-\infty$ on the left of their domains and by $+\infty$ on the right of their domains), (3.3.34)

$$\operatorname{dom}(\beta) \subset \operatorname{dom}(\beta_\varepsilon). \tag{3.3.35}$$

According to Appendix 1, it is possible to construct a continuously differentiable ε-uniform approximation β_ε of β. We associate with it the approximating state system

$$A y^\varepsilon(x) + \varphi(x, y^\varepsilon(x), u(x)) + \beta_\varepsilon(y^\varepsilon(x)) = 0 \quad \text{in } \Omega, \tag{3.3.36}$$

$$y^\varepsilon = 0 \quad \text{on } \partial\Omega. \tag{3.3.37}$$

Invoking Theorem A2.10 in Appendix 2, we can infer that for any feasible $u \in U_{ad}$ the boundary value problems (3.3.26), (3.3.27), and (3.3.36), (3.3.37), have unique solutions $y = y_u$, and $y^\varepsilon = y_u^\varepsilon$, respectively, that belong to $W^{2,s}(\Omega) \cap H_0^1(\Omega)$ and satisfy

$$|y^\varepsilon - y|_{L^\infty(\Omega)} \leq \varepsilon. \tag{3.3.38}$$

Proposition 3.3.3 *For any $u \in U_{ad}$ there is some constant $\widetilde{C}(u) > 0$ such that*

$$|J(y_u^\varepsilon, u) - J(y_u, u)| \leq \widetilde{C}(u)\, \varepsilon. \tag{3.3.39}$$

If K is bounded, then the constant, redenoted by $\widetilde{C}$, does not depend on $u \in U_{ad}$.

Proof. In view of (3.3.33) and (3.3.38), the mean value theorem yields that

$$|J(y_u^\varepsilon, u) - J(y_u, u)| \leq \varepsilon\, \eta\big(|y_u|_{L^\infty(\Omega)} + 1\big) \int_\Omega \big(M(x) + \hat{C}\,|u(x)|\big)\, dx,$$

whence the assertion immediately follows. □

Remark. If K is bounded, then the inequality (3.3.39) is also valid for the optimal values of the problems (3.3.25)–(3.3.28), and (3.3.25), (3.3.28), (3.3.36), (3.3.37).

We define the *pseudo-Hamiltonian* of the problem (3.3.25)–(3.3.28) by (compare (3.2.100))

$$H(x,y,u,p) \;=\; L(x,y,u) \;-\; p\,\varphi(x,y,u). \tag{3.3.40}$$

We use the complete metric space (E,d) given by the Ekeland metric:

$$E \;=\; \Big\{u \in L^\infty(\Omega) : \, u \in U_{ad}\Big\}, \tag{3.3.41}$$

$$d(u,v) \;=\; \text{meas}\Big(\{x \in \Omega : u(x) \neq v(x)\}\Big). \tag{3.3.42}$$

Theorem 3.3.4 *Suppose that K is bounded. Let $\delta > 0$, and suppose that u is a δ-solution of the problem (3.3.25)–(3.3.28) (cf. Section 2.1 in Chapter 2). Put $\delta_\varepsilon \;=\; \delta + 2\,\widetilde{C}\,\varepsilon$, where $\widetilde{C} > 0$ is the constant from Proposition 3.3.3. Then there exists some $u_\varepsilon \in U_{ad}$ with $d(u,u_\varepsilon) \le \sqrt{\delta_\varepsilon}$ such that*

(i) *u_ε is a δ_ε-solution of the problem (3.3.25), (3.3.28), (3.3.36), (3.3.37).*

(ii) *Let $y^\varepsilon \;=\; y^\varepsilon_{u_\varepsilon}$ be obtained by (3.3.36), (3.3.37) with $u \;=\; u_\varepsilon$. If p_ε denotes the adjoint state corresponding to u_ε, defined as the unique solution to*

$$\widetilde{A}p_\varepsilon(x) \;+\; \varphi_y(x,y^\varepsilon(x),u_\varepsilon(x))\,p_\varepsilon \;+\; \beta'_\varepsilon(y^\varepsilon(x))\,p_\varepsilon$$
$$=\; L_y(x,y^\varepsilon(x),u_\varepsilon(x)) \quad \text{in } \Omega, \tag{3.3.43}$$

$$p_\varepsilon \;=\; 0 \quad \text{on } \partial\Omega, \tag{3.3.44}$$

then $p_\varepsilon \in W^{2,s}(\Omega) \cap H^1_0(\Omega)$, and for any $v \in K$,

$$H(x,y^\varepsilon(x),u_\varepsilon(x),p_\varepsilon(x)) \;\le\; H(x,y^\varepsilon(x),v,p_\varepsilon(x)) \;+\; \sqrt{\delta_\varepsilon} \quad \text{for a.e. } x \in \Omega. \tag{3.3.45}$$

Remark. Thanks to Theorem A2.2(iv), we also have $p_\varepsilon \in C(\overline{\Omega})$.

Proof of Theorem 3.3.4. We proceed in several steps.

Step 1: We at first prove that (3.3.43), (3.3.44) admits a unique solution $p_\varepsilon \in W^{2,s}(\Omega) \cap H^1_0(\Omega)$. To this end, notice that in view of (3.3.30)–(3.3.33) we have

$$\widetilde{\varphi}_y \;=\; \varphi_y(\cdot\,,y^\varepsilon(\cdot),u_\varepsilon(\cdot)) \in L^s(\Omega), \quad \widetilde{L}_y \;=\; L_y(\cdot\,,y^\varepsilon(\cdot),u_\varepsilon(\cdot)) \in L^s(\Omega),$$
$$\widetilde{\beta}'_\varepsilon \;=\; \beta'_\varepsilon(y^\varepsilon(\cdot)) \in L^\infty(\Omega),$$

since $u_\varepsilon(x) \in K$ a.e. in Ω. Moreover, owing to Theorem A2.2(iv) in Appendix 2, y^ε is continuous on $\overline{\Omega}$ and thus attains its extremal values.

Now recall that the embedding result of Theorem A2.2(i) in Appendix 2 implies that $H_0^1(\Omega)$ is continuously embedded in $L^q(\Omega)$, where $1 \le q < \infty$ for $d \le 2$ and $1 \le q \le 2d/(d-2)$ for $d \ge 3$. Hence, there is some constant $M > 0$ such that for all $u, v \in H_0^1(\Omega)$ we have, since $s > \max\{2, \frac{d}{2}\}$,

$$\left| \int_\Omega \widetilde{\varphi}_y(x)\, u(x)\, v(x)\, dx \right| \le |\widetilde{\varphi}_y|_{L^s(\Omega)}\, |u\, v|_{L^{s/(s-1)}(\Omega)} \le M\, |u|_{H_0^1(\Omega)}\, |v|_{H_0^1(\Omega)}.$$

Consequently, the bilinear form corresponding to the left-hand side of (3.3.43) is continuous on $H_0^1(\Omega)$. Moreover, since $\widetilde{\varphi}_y$ and $\widetilde{\beta}'_\varepsilon$ are nonnegative functions, it follows from (3.3.29) and (3.3.31), using Poincaré's inequality, that it is also coercive on $H_0^1(\Omega)$. In addition, the right-hand side belongs to $L^s(\Omega)$ and defines a linear and continuous functional on $H_0^1(\Omega)$. Hence, we can infer from the Lax–Milgram lemma that (3.3.43), (3.3.44) has a unique weak solution $p_\varepsilon \in H_0^1(\Omega)$.

Next, we show that $p_\varepsilon \in L^\infty(\Omega)$. To this end, let for $h > 0$,

$$\widetilde{\varphi}_y^h(x) = \int_{\mathbf{R}^d} \rho_h(x - \xi)\, \widetilde{\varphi}_y(\xi)\, d\xi,$$

where $\rho_h \in C_0^\infty(\mathbf{R}^d)$ is a Friedrichs mollifier for $h > 0$. Clearly, $\widetilde{\varphi}_y^h \in L^\infty(\Omega)$, and $\widetilde{\varphi}_y^h$ is nonnegative. It thus follows from Example A2.6 in Appendix 2 that the unique solution p_ε^h to the linear elliptic problem

$$\widetilde{A} p_\varepsilon^h + \widetilde{\varphi}_y^h(x)\, p_\varepsilon^h + \widetilde{\beta}_\varepsilon(x)\, p_\varepsilon^h = \widetilde{L}_y(x) \quad \text{in } \Omega, \qquad p_\varepsilon^h = 0 \quad \text{on } \partial\Omega,$$

belongs to $W^{2,s}(\Omega) \cap H_0^1(\Omega)$. Likewise, we have $z \in W^{2,s}(\Omega) \cap H_0^1(\Omega)$ for the unique solution to the linear problem

$$\widetilde{A} z = \widetilde{L}_y(x) \quad \text{in } \Omega, \qquad z = 0 \quad \text{on } \partial\Omega.$$

If $\widetilde{L}_y \ge 0$, then the maximum principle (cf. Theorem A2.8 in Appendix 2) implies that $p_\varepsilon^h(x) \ge 0$ in Ω. Subtracting the two equations from each other, we see that

$$\widetilde{A}(z - p_\varepsilon^h)(x) = \widetilde{\varphi}_y^h(x)\, p_\varepsilon^h(x) + \beta'_\varepsilon(y^\varepsilon(x)) \ge 0 \quad \text{for a.e. } x \in \overline{\Omega}.$$

Invoking the maximum principle once more, we find that

$$0 \le p_\varepsilon^h(x) \le z(x) \le |z|_{L^\infty(\Omega)},$$

which is bounded by the embedding result of Theorem A2.2(iv).

In the general case, we apply the above argumentation first to the positive and then to the negative part of $\widetilde{L}_y$. The linearity of the equation then yields that $|p_\varepsilon^h(x)| \le \hat{C}_1$ with some constant $\hat{C}_1 > 0$ that does not depend on $h > 0$.

Finally, testing the differential equation for p_ε^h by p_ε^h, and invoking (3.3.29), we readily see that $\{p_\varepsilon^h\}_{h>0}$ is bounded in $H_0^1(\Omega)$. Hence there are some sequence $h_n \searrow 0$ and some $\hat{p} \in H_0^1(\Omega)$ such that $p_\varepsilon^{h_n} \to \hat{p}$ weakly in $H_0^1(\Omega)$, strongly in $L^2(\Omega)$, and pointwise a.e. in Ω. Passing to the limit as $h_n \searrow 0$ in the equation for $p_\varepsilon^{h_n}$, and using the approximation properties of the Friedrichs mollifier, we find that $\hat{p} = p_\varepsilon$, and the above inequality implies that

$$|p_\varepsilon(x)| \leq \hat{C}_1, \quad \text{a.e. in } \Omega,$$

that is, we have shown that $p_\varepsilon \in L^\infty(\Omega)$. Now that this is verified, we are in a position to conclude (cf. Example A2.6 in Appendix 2) that $p_\varepsilon \in W^{2,s}(\Omega)$.

Step 2: We notice that by Proposition 3.3.3,

$$\begin{aligned} J(y_u^\varepsilon) &\leq J(y_u, u) + \widetilde{C}\,\varepsilon \leq \inf\{J(y_v, v) : v \in U_{ad}\} + \delta + \widetilde{C}\,\varepsilon \\ &\leq \inf\{J(y_v^\varepsilon, v) : v \in U_{ad}\} + \delta + 2\widetilde{C}\,\varepsilon; \end{aligned}$$

that is, u is a δ_ε-solution to the approximate optimal control problem (3.3.25), (3.3.28), (3.3.36), (3.3.37).

Next, we show that the mapping $u \mapsto y_u^\varepsilon$ given by (3.3.36), (3.3.37) is continuous from the metric space (E, d) to $W^{2,s}(\Omega) \cap H_0^1(\Omega)$ endowed with the weak topology. Let $u_k \to u$ in (E, d), and let y_k denote the solution of (3.3.36), (3.3.37) associated with u_k, $k \in \mathbf{N}$. Since K is bounded, it follows that $\{u_k\}$ is bounded in $L^\infty(\Omega)$, and thus, owing to Theorem A2.9 in Appendix 2, $\{y_k\}$ is bounded in $W^{2,s}(\Omega) \cap H_0^1(\Omega)$.

Consequently, there is a subsequence, again indexed by k, such that $y_k \to y$ weakly in $W^{2,s}(\Omega) \cap H_0^1(\Omega)$, and, by compact embedding, strongly in $C(\overline{\Omega})$. Then, thanks to Lebesgue's theorem and to (3.3.30), (3.3.31), $\varphi(\cdot\,, y_k(\cdot), u_k(\cdot)) \to \varphi(\cdot\, y(\cdot), u(\cdot))$ strongly in $L^s(\Omega)$, since we have convergence a.e. in Ω. In addition, $\beta_\varepsilon(y_k) \to \beta_\varepsilon(y)$ strongly in $L^\infty(\Omega)$, since β_ε is locally Lipschitz and $\{y_k\}$ is bounded in $L^\infty(\Omega)$. Passing to the limit as $k \to \infty$ in (3.3.36), (3.3.37), we find that $y = y_u^\varepsilon$. By the uniqueness of the limit, we have $y_k \to y_u^\varepsilon$ weakly in $W^{2,s}(\Omega) \cap H_0^1(\Omega)$ for the entire sequence, which finishes the proof of the claim.

Then it follows from the conditions (3.3.32), (3.3.33), and, once more, Lebesgue's theorem, that $J(y_k, u_k) \to J(y_u, u)$, whence we conclude that the mapping $u \mapsto J(y_u^\varepsilon, u)$ is continuous from (E, d) to $\mathbf{R}$. We also notice that in view of the discussions above, this mapping is bounded from below on (E, d).

Step 3: Let $x_0 \in \Omega$ and $v \in K$ be arbitrary. As above in §3.2.4 (compare (3.2.110)), we introduce the *spike variations* of a control $u \in U_{ad}$ at x_0 associated with v by putting

$$v_k(x) = \begin{cases} v, & \text{if } x \in S_k(x_0), \\ u(x), & \text{otherwise,} \end{cases}$$

where $S_k(x_0) = \left\{x \in \Omega : |x - x_0| \leq 1/k\right\}$. We set

$$m_k = \left[\operatorname{meas}(S_k(x_0))\right]^{-1}.$$

We have the following result.

Lemma 3.3.5 *Let $u \in U_{ad}$ and $v \in K$ be given. Then for a.e. $x_0 \in \Omega$,*

$$\lim_{k\to\infty} m_k \Big[J\left(y_k^\varepsilon, v_k\right) - J\left(y_u^\varepsilon, u\right) \Big]$$
$$= H\left(x_0, y_u^\varepsilon(x_0), v, p_u^\varepsilon(x_0)\right) - H\left(x_0, y_u^\varepsilon(x_0), u(x_0), p_u^\varepsilon(x_0)\right), \quad (3.3.46)$$

where v_k, $k \in \mathbf{N}$, are the spike variations at x_0 associated with v, where $y_k^\varepsilon = y_{v_k}^\varepsilon$, $k \in \mathbf{N}$, and where p_u^ε is the solution of (3.3.43), (3.3.44), associated with u and y_u^ε.

Proof. Let $x_0 \in \Omega$ be given. Then for every $k \in \mathbf{N}$,

$$J(y_k^\varepsilon, v_k) - J(y_u^\varepsilon, u) = \int_\Omega \Big[L(x, y_u^\varepsilon(x), v_k(x)) - L(x, y_u^\varepsilon(x), u(x))\Big]\, dx$$
$$+ \int_\Omega \Big[L(x, y_k^\varepsilon(x), v_k(x)) - L(x, y_u^\varepsilon(x), v_k(x))\Big]\, dx.$$

By the mean value theorem, there exist intermediate points $\xi_k^i(x)$, $i = 1, 2, 3$, $x \in \Omega$, such that

$$\varphi(x, y_k^\varepsilon(x), v_k(x)) - \varphi(x, y_u^\varepsilon(x), v_k(x)) = \varphi_y\Big(x, \xi_k^1(x), v_k(x)\Big)\left(y_k^\varepsilon(x) - y_u^\varepsilon(x)\right),$$
$$\beta_\varepsilon(y_k^\varepsilon(x)) - \beta_\varepsilon(y_u^\varepsilon(x)) = \beta_\varepsilon'\Big(\xi_k^2(x)\Big)\left(y_k^\varepsilon(x) - y_u^\varepsilon(x)\right),$$
$$L(x, y_k^\varepsilon(x), v_k(x)) - L(x, y_u^\varepsilon(x), v_k(x)) = L_y\Big(x, \xi_k^3(x), v_k(x)\Big)\left(y_k^\varepsilon(x) - y_u^\varepsilon(x)\right).$$

Now let $\tilde{p}_k \in W^{2,s}(\Omega) \cap H_0^1(\Omega)$ denote the unique solution to the problem (which is similar to (3.3.43), (3.3.44))

$$\tilde{A}\tilde{p}_k + \varphi_y\Big(x, \xi_k^1(x), v_k(x)\Big)\tilde{p}_k + \beta_\varepsilon'\Big(\xi_k^2(x)\Big)\tilde{p}_k = L_y\Big(x, \xi_k^3(x), v_k(x)\Big) \quad \text{in } \Omega,$$
$$\tilde{p}_k = 0 \quad \text{on } \partial\Omega.$$

Then, integrating by parts and invoking (3.3.36), the last integral above can be rewritten as

$$\int_\Omega \Big[L(x, y_k^\varepsilon(x), v_k(x)) - L(x, y_u^\varepsilon(x), v_k(x))\Big]\, dx$$
$$= \int_\Omega L_y\Big(x, \xi_k^3(x), v_k(x)\Big)\left(y_k^\varepsilon(x) - y_u^\varepsilon(x)\right) dx$$
$$= \int_\Omega \Big[\tilde{A}\tilde{p}_k(x) + \varphi_y\Big(x, \xi_k^1(x), v_k(x)\Big)\tilde{p}_k(x) + \beta_\varepsilon'\Big(\xi_k^2(x)\Big)\tilde{p}_k(x)\Big]$$

$$
\begin{aligned}
&\cdot (y_k^\varepsilon(x) - y_u^\varepsilon(x))\, dx \\
&= \int_\Omega \tilde{p}_k(x)\, A(y_k^\varepsilon - y_u^\varepsilon)(x)\, dx + \int_\Omega \beta_\varepsilon'\big(\xi_k^2(x)\big)\, \tilde{p}_k(x)\, (y_k^\varepsilon(x) - y_u^\varepsilon(x))\, dx \\
&\quad + \int_\Omega \varphi_y\big(x, \xi_k^1(x), v_k(x)\big)\, (y_k^\varepsilon(x) - y_u^\varepsilon(x))\, \tilde{p}_k(x)\, dx \\
&= \int_\Omega \big[\varphi(x, y_u^\varepsilon(x), u(x)) \;-\; \varphi(x, y_k^\varepsilon(x), v_k(x))\big]\, \tilde{p}_k(x)\, dx \\
&\quad + \int_\Omega \big[\beta_\varepsilon(y_u^\varepsilon(x)) \;-\; \beta_\varepsilon(y_k^\varepsilon(x))\big]\, \tilde{p}_k(x)\, dx \\
&\quad + \int_\Omega \big[\varphi(x, y_k^\varepsilon(x), v_k(x)) \;-\; \varphi(x, y_u^\varepsilon(x), v_k(x))\big]\, \tilde{p}_k(x)\, dx \\
&\quad + \int_\Omega \big[\beta_\varepsilon(y_k^\varepsilon(x)) - \beta_\varepsilon(y_u^\varepsilon(x))\big]\, \tilde{p}_k(x)\, dx \\
&= \int_\Omega \big[\varphi(x, y_u^\varepsilon(x), u(x)) - \varphi(x, y_u^\varepsilon(x), v_k(x))\big]\, \tilde{p}_k(x)\, dx.
\end{aligned}
$$

We thus obtain that

$$
\begin{aligned}
&J(y_k^\varepsilon, v_k) \;-\; J(y_u^\varepsilon, u) \\
&\quad = \int_\Omega \big[L(x, y_u^\varepsilon(x), v_k(x)) \;-\; \varphi(x, y_u^\varepsilon(x), v_k(x))\, \tilde{p}_k(x) \;-\; L(x, y_u^\varepsilon(x), u(x)) \\
&\qquad\qquad + \varphi(x, y_u^\varepsilon(x), u(x))\, \tilde{p}_k(x)\big]\, dx \\
&\quad = \int_{S_k(x_0)} \big[H(x, y_u^\varepsilon(x), v, \tilde{p}_k(x)) \;-\; H(x, y_u^\varepsilon(x), u(x), \tilde{p}_k(x))\big]\, dx. \qquad (3.3.47)
\end{aligned}
$$

We notice that $\{v_k\}$ is bounded in $L^\infty(\Omega)$, and that $v_k \to u$ a.e. in Ω. Thus, standard estimates for variational inequalities (see Appendix 2) yield that $\{y_k^\varepsilon\}$ is bounded in $W^{2,s}(\Omega) \cap H_0^1(\Omega)$.

Hence, there is some $\hat{y} \in W^{2,s}(\Omega) \cap H_0^1(\Omega)$ such that for a suitable subsequence, which is again indexed by k, $y_k^\varepsilon \to \hat{y}$ weakly in $W^{2,s}(\Omega) \cap H_0^1(\Omega)$ and, by compact embedding, strongly in $C(\overline{\Omega})$. Passing to the limit as $k \to \infty$, we realize that $\hat{y} = y_u^\varepsilon$, and the uniqueness of the limit shows that the above convergences hold for the entire sequence.

Consequently, we can conclude that $\xi_k^i \to y_u^\varepsilon$, $i = 1, 2, 3$, uniformly in Ω, and, in view of the assumptions (3.3.30)–(3.3.33),

$$
\begin{aligned}
&\varphi_y\big(\cdot, \xi_k^1(\cdot), v_k(\cdot)\big) \;\to\; \varphi_y(\cdot, y_u^\varepsilon(\cdot), u(\cdot)), \quad \beta_\varepsilon'\big(\xi_k^2\big) \;\to\; \beta_\varepsilon'(y_u^\varepsilon), \\
&L_y\big(\cdot, \xi_k^3(\cdot), v_k(\cdot)\big) \;\to\; L_y(\cdot, y_u^\varepsilon(\cdot), u(\cdot)), \quad \text{all strongly in } L^s(\Omega).
\end{aligned}
$$

Clearly, the sequence $\{\tilde{p}_k\}$ is also bounded in $W^{2,s}(\Omega) \cap H_0^1(\Omega)$. Using the above convergences, and invoking again the embedding result of Theorem

A2.2(iv) in Appendix 2, we can easily verify that

$$\tilde{p}_k \to p_u^\varepsilon \quad \text{strongly in } C(\overline{\Omega}),$$

where p_u^ε denotes the solution of (3.3.43), (3.3.44) associated with u and y_u^ε.

Next, (3.3.47) yields that

$$\begin{aligned}
J(y_k^\varepsilon, v_k) &- J(y_u^\varepsilon, u) \\
&= \int_{S_k(x_0)} \left[H(x, y_u^\varepsilon(x), v, \tilde{p}_k(x)) - H(x, y_u^\varepsilon(x), u(x), \tilde{p}_k(x))\right] dx \\
&= \int_{S_k(x_0)} \left[H(x, y_u^\varepsilon(x), v, p_u^\varepsilon(x)) - H(x, y_u^\varepsilon(x), u(x), p_u^\varepsilon(x))\right] dx \\
&+ \int_{S_k(x_0)} \varphi(x, y_u^\varepsilon(x), v)\,(p_u^\varepsilon - \tilde{p}_k)(x)\, dx \\
&+ \int_{S_k(x_0)} \varphi(x, y_u^\varepsilon(x), u(x))\,(\tilde{p}_k - p_u^\varepsilon)(x)\, dx. \qquad (3.3.48)
\end{aligned}$$

Multiplying by m_k, and taking into account that almost all points $x_0 \in \Omega$ are Lebesgue points for an integrable mapping, we get (3.3.46), since $\tilde{p}_k \to p_u^\varepsilon$ uniformly in $\overline{\Omega}$. This concludes the proof of the lemma. □

Continuation of the proof of Theorem 3.3.4.

Step 4: We apply the Ekeland variational principle (cf. Theorem A1.15 in Appendix 1) to the present situation, namely to u and the problem (3.3.25), (3.3.28), (3.3.36), (3.3.37). It follows that there exists a δ_ε-solution u_ε that satisfies $d(u, u_\varepsilon) \le \sqrt{\delta_\varepsilon}$ and

$$J(y^\varepsilon, u_\varepsilon) \le J(y_z^\varepsilon, z) + \sqrt{\delta_\varepsilon}\, d(u_\varepsilon, z) \quad \forall\, z \in (E, d), \qquad (3.3.49)$$

where y_z^ε is the solution of (3.3.36), (3.3.37) corresponding to z, and y^ε is defined in Theorem 3.3.4.

Now let $\{v_k\}$ be a sequence of spike variations of u_ε at $x_0 \in \Omega$ corresponding to $v \in K$. We insert $z = v_k$ in (3.3.49) and divide by $d(u_\varepsilon, v_k) = \text{meas}(S_k(x_0)) = m_k^{-1}$, $k \in \mathbf{N}$. Taking the limit as $k \to \infty$, and invoking Lemma 3.3.5, we obtain (3.3.45). This finishes the proof of the assertion. □

We are now in a position to prove the first-order necessary optimality conditions.

Theorem 3.3.6 *Suppose that K is bounded, and let u^* be a solution of the problem (3.3.25)–(3.3.28) and $y^* = y_{u^*}$ the corresponding state. Let, for $\varepsilon > 0$,*

$u_\varepsilon \in U_{ad}$ denote the control given by Theorem 3.3.4 that corresponds to $u = u^*$. Then for $\varepsilon \searrow 0$,

$$d(u_\varepsilon, u^*) \to 0, \tag{3.3.50}$$

$$y^\varepsilon \to y^* \quad \text{weakly in } W^{2,s}(\Omega) \cap H_0^1(\Omega), \quad \text{where } y^\varepsilon = y^\varepsilon_{u_\varepsilon}. \tag{3.3.51}$$

In addition, there exist $p^* \in H_0^1(\Omega) \cap L^\infty(\Omega)$, $\rho \in H^{-1}(\Omega)$, and a sequence $\varepsilon_n \searrow 0$ such that

$$p_{\varepsilon_n} \to p^*, \quad \text{weakly in } H_0^1(\Omega) \text{ and weakly* in } L^\infty(\Omega),$$

$$\beta'_{\varepsilon_n}(y^{\varepsilon_n})\, p_{\varepsilon_n} \to \rho, \quad \text{weakly in } H^{-1}(\Omega),$$

and such that

$$\widetilde{A}p^* + \varphi_y(\cdot, y^*(\cdot), u^*(\cdot))\, p^* + \rho = L_y(\cdot, y^*(\cdot), u^*(\cdot)) \quad \text{in } \Omega, \tag{3.3.52}$$

$$p^* = 0 \quad \text{on } \partial\Omega, \tag{3.3.53}$$

and, for all $v \in K$,

$$H(x, y^*(x), u^*(x), p^*(x)) \le H(x, y^*(x), v, p^*(x)) \quad \text{for a.e. } x \in \Omega. \tag{3.3.54}$$

Proof. The relation (3.3.50) is obvious. Moreover, the set $\{y^\varepsilon\}_{\varepsilon>0}$ is bounded in $W^{2,s}(\Omega) \cap H_0^1(\Omega)$. Hence, for a suitable sequence $\varepsilon_n \searrow 0$, $y^{\varepsilon_n} \to \hat{y}$ weakly in $W^{2,s}(\Omega) \cap H_0^1(\Omega)$ and, by compact embedding, strongly in $C(\overline{\Omega})$. From Theorem A2.10 in Appendix 2 we then infer that $\hat{y} = y^*$, and the convergence holds for the entire sequence.

In order to verify (3.3.52)–(3.3.54), notice that we can employ the same comparison technique as in Step 1 of the proof of Theorem 3.3.4 to see that $\{p_\varepsilon\}_{\varepsilon>0}$ is a bounded set in $H_0^1(\Omega) \cap L^\infty(\Omega)$. Notice, however, that one cannot expect to find a uniform bound for $\{p_\varepsilon\}_{\varepsilon>0}$ in $W^{2,s}(\Omega)$, since $\{\beta'_\varepsilon(y^\varepsilon)\}_{\varepsilon>0}$ is not bounded.

From (3.3.43) we then conclude that $\{\beta'_\varepsilon(y^\varepsilon)\, p_\varepsilon\}_{\varepsilon>0}$ is bounded in $H^{-1}(\Omega)$. Thus, there is a sequence $\varepsilon_n \searrow 0$ such that

$$p_{\varepsilon_n} \to \hat{p}, \quad \text{weakly in } H_0^1(\Omega) \text{ and weakly* in } L^\infty(\Omega),$$

$$\beta'_{\varepsilon_n}(y^{\varepsilon_n})\, p_{\varepsilon_n} \to \rho, \quad \text{weakly in } H^{-1}(\Omega).$$

Passing to the limit as $\varepsilon_n \searrow 0$ in (3.3.43), we find that $\hat{p} = p^*$ satisfies (3.3.52) and (3.3.53). Moreover, since the embedding $H_0^1(\Omega) \subset L^2(\Omega)$ is compact, we have, for a subsequence again indexed by ε_n, $p_{\varepsilon_n} \to p^*$ a.e. in Ω, and thus, for a.e. $x \in \Omega$,

$$H(x, y^{\varepsilon_n}(x), u_{\varepsilon_n}(x), p_{\varepsilon_n}(x)) \to H(x, y^*(x), u^*(x), p^*(x)),$$

$$H(x, y^{\varepsilon_n}(x), v, p_{\varepsilon_n}(x)) \to H(x, y^*(x), v, p^*(x)).$$

Since for $u = u^*$ in Theorem 3.3.4, $\delta_{\varepsilon_n} = [2\widetilde{C}\,\varepsilon_n]^{1/2} \to 0$ as $n \to \infty$, the relation (3.3.54) is a consequence of (3.3.45). This concludes the proof of the assertion. $\square$

Remark. The definition of the distribution ρ underlines one of the main difficulties occurring in control problems governed by variational inequalities: it seems to be impossible to introduce a procedure for the "differentiation" of maximal monotone operators. In the next paragraph, a partial result in this direction will be derived.

It is also possible to recover further information on the distribution ρ by an argument similar to that employed in the derivation of the generalized bang-bang results for the state-constrained problems considered above in §3.2.2. To this end, we assume that β is the maximal monotone extension of a monotone step function defined on some real interval. That is, the graph of β is composed of segments parallel to the axes. We set

$$D = \{r \in \mathbf{R} : r \text{ is a point of discontinuity of } \beta\}, \quad \Omega_0 = \{x \in \Omega : y^*(x) \notin D\}.$$

Notice that Ω_0 is an open (possibly empty) subset of Ω, since $y^* \in C(\overline{\Omega})$.

Corollary 3.3.7 *Under the above assumptions,* $\operatorname{supp} \rho \subset (\Omega \setminus \Omega_0)$.

Proof. Suppose that $\Omega_0 \neq \emptyset$ (otherwise the result is trivial), and let $\psi \in \mathcal{D}(\Omega)$ with $\operatorname{supp} \psi \subset \Omega_0$ be given. We choose some $c > 0$ such that

$$\inf \{\, d(y^*(x), D) \;:\; x \in \operatorname{supp} \psi \,\} \;\geq\; 2\,c,$$

which is possible since y^* is continuous and $\operatorname{supp} \psi$ is compact. Since $y^\varepsilon \to y^*$ strongly in $C(\overline{\Omega})$, there is some $\varepsilon_0 > 0$ such that for any $0 < \varepsilon \leq \varepsilon_0$,

$$\inf \{\, d(y^\varepsilon(x), D) \;:\; x \in \operatorname{supp} \psi \,\} \;\geq\; c > 0.$$

By definition (compare Appendix 1), $\beta_\varepsilon(r)$ is constant whenever $d(r, D) \geq c$ and $\varepsilon > 0$ is sufficiently small. Consequently, $\beta_\varepsilon'(y^\varepsilon(x)) = 0$ if $x \in \operatorname{supp} \psi$ and $\varepsilon > 0$ is sufficiently small. Therefore,

$$\rho(\psi) = \lim_{\varepsilon \searrow 0} \int_\Omega \beta'(y^\varepsilon(x))\, p_\varepsilon(x) \psi(x)\, dx = 0,$$

which concludes the proof of the assertion. $\square$

Remark. For an original investigation along these lines, we quote the works of Bonnans and Tiba [1991] and of Tiba [1990, Chapter III.5].

3.3.3 State Constraints and Penalization of the Equation

The treatment of state constraints in control problems involving variational inequalities is still not completely solved. Besides the results on Fritz–John-type necessary conditions derived in §3.3.1, we refer in this connection to the works of He [1987], and of Neittaanmäki and Tiba [1994, Chapter VI.1], where a parabolic case is studied using a different approach. In this section, we will prove a partial result of Kuhn–Tucker–Karush type employing an adapted penalization of the state system as in §3.2.2 and an ad hoc notion of a second-order derivative of indicator functions. We consider a typical example for an obstacle problem:

$$\text{Min}\left\{\frac{1}{2}\,|y-y_d|^2_{L^2(\Omega)}+\frac{1}{2}\,|u|^2_{H^{-1}(\Omega)}\right\} \tag{3.3.55}$$

subject to

$$-\Delta y+\partial I_K(y)\ni u \quad \text{in } \Omega, \tag{3.3.56}$$

$$y=0 \quad \text{on } \partial\Omega, \tag{3.3.57}$$

$$u\in U_{ad}\subset H^{-1}(\Omega), \tag{3.3.58}$$

$$y\in C\subset H^1_0(\Omega). \tag{3.3.59}$$

Here, U_{ad} denotes a nonempty and convex set that is closed in $H^{-1}(\Omega)$,

$$K=\left\{y\in H^1_0(\Omega):\ y\geq 0 \text{ a.e. in } \Omega\right\}$$

is the positive cone in $H^1_0(\Omega)$, and C is assumed to be a nonempty, convex, and compact subset of $H^1_0(\Omega)$. The target function $y_d\in L^2(\Omega)$ is given. In addition, $\Omega\subset\mathbf{R}^d$ is a bounded domain in $\mathbf{R}^d$ with smooth boundary $\partial\Omega$, and

$$\partial I_K\subset H^1_0(\Omega)\times H^{-1}(\Omega)$$

is the (maximal monotone) subdifferential of the indicator function of K.

Notice that if $u\in L^2(\Omega)$ in (3.3.56), then it follows from Theorem A2.9 in Appendix 2 that $y\in H^2(\Omega)\cap H^1_0(\Omega)$, and we have

$$\Delta y(x)+u(x)\in\beta(y(x)) \quad \text{for a.e. } x\in\Omega,$$

where $\beta\subset\mathbf{R}\times\mathbf{R}$ is the maximal monotone graph

$$\beta(y)=\begin{cases}\{0\}, & \text{if } y>0,\\ ]-\infty,0], & \text{if } y=0,\\ \emptyset, & \text{if } y<0.\end{cases} \tag{3.3.60}$$

As demonstrated above in §3.2.1, we may without loss of generality assume that $[0,0] \in C \times U_{ad}$, and it follows from (3.3.60) that $[0,0]$ is admissible for the problem (3.3.55)–(3.3.59). Further assumptions will be added later as the necessity arises.

For any $u \in H^{-1}(\Omega)$, the state system (3.3.56), (3.3.57) has a unique solution $y = y_u \in H_0^1(\Omega)$. The existence of at least one optimal pair $[y^*, u^*] \in C \times U_{ad}$ follows as in the proof of Theorem 2.1.3(b) in Chapter 2, where we notice that the special form of the cost functional in (3.3.55) implies that for any minimizing sequence $\{[y_n, u_n]\}$ the sequence $\{u_n\}$ is automatically bounded in $H^{-1}(\Omega)$.

In a first step toward the investigation of the optimality conditions, we use an approximation procedure. As motivation, we observe that

$$
\begin{aligned}
w \in \partial I_K(y) &\iff 0 \le (w, y - v)_{H^{-1}(\Omega) \times H_0^1(\Omega)} \quad \forall\, v \in K \\
&\iff y \in K, \;\; w \in K^0, \;\; (w, y)_{H^{-1}(\Omega) \times H_0^1(\Omega)} = 0, \qquad (3.3.61)
\end{aligned}
$$

where

$$K^0 = \left\{ w \in H^{-1}(\Omega) \,:\, (w, y)_{H^{-1}(\Omega) \times H_0^1(\Omega)} \le 0 \text{ for all } y \in K \right\}$$

is the polar cone of K. Since K is the positive cone in $H_0^1(\Omega)$, K^0 coincides with the negative cone K_- in $H^{-1}(\Omega)$. We adopt the usual convention to write $y \ge 0$ for $y \in K$, and $w \le 0$ for $w \in K_-$.

We now introduce for $\varepsilon > 0$ the penalized problem

$$
\begin{aligned}
\text{Min} \Big\{ &\frac{1}{2} \, |y - y_d|^2_{L^2(\Omega)} + \frac{1}{2} \, |u|^2_{H^{-1}(\Omega)} - \frac{1}{\varepsilon} \, (w, y)_{H^{-1}(\Omega) \times H_0^1(\Omega)} \\
&+ \frac{1}{2} \, |u - u^*|^2_{H^{-1}(\Omega)} + \frac{1}{2} \, |w - w^*|^2_{H^{-1}(\Omega)} \Big\}, \qquad (3.3.62)
\end{aligned}
$$

subject to

$$- \Delta y + w = u, \quad \text{in } \Omega, \qquad (3.3.63)$$

$$y \in C, \quad y \ge 0, \qquad (3.3.64)$$

$$w \in H^{-1}(\Omega), \quad w \le 0, \qquad (3.3.65)$$

$$u \in U_{ad}. \qquad (3.3.58)$$

Remark. The idea to introduce an additional control w is basically due to Saguez and Bermudez [1985] and to Mignot and Puel [1984].

Notice that $[0,0,0]$ is an admissible triple for (3.3.58), (3.3.62)–(3.3.65). We show the existence of at least one optimal triple $[y_\varepsilon, u_\varepsilon, w_\varepsilon] \in C \times U_{ad} \times H^{-1}(\Omega)$.

To this end, let $\{[y_n, u_n, w_n]\} \subset C \times U_{ad} \times H^{-1}(\Omega)$ be a minimizing sequence; that is, in particular, we have

$$w_n = \Delta y_n + u_n, \quad y_n \geq 0 \quad w_n \leq 0, \quad \text{for all } n \in \mathbf{N}.$$

Since

$$(w_n, y_n)_{H^{-1}(\Omega)\times H_0^1(\Omega)} \leq 0 \quad \forall n \in \mathbf{N},$$

it follows immediately that $\left\{|u_n|_{H^{-1}(\Omega)} + |w_n|_{H^{-1}(\Omega)}\right\}_{n\in\mathbf{N}}$ is bounded. Moreover, we have

$$\begin{aligned} -(w, y)_{H^{-1}(\Omega)\times H_0^1(\Omega)} &= -(\Delta y + u, y)_{H^{-1}(\Omega)\times H_0^1(\Omega)} \\ &= |y|^2_{H_0^1(\Omega)} - (u, y)_{H^{-1}(\Omega)\times H_0^1(\Omega)}, \end{aligned} \tag{3.3.66}$$

and the compactness of C implies that $\{y_n\}$ is convergent in $H_0^1(\Omega)$. Hence, we may without loss of generality assume that

$$\begin{aligned} &y_n \to \overline{y} \quad \text{strongly in } H_0^1(\Omega), \text{ and pointwise a.e. in } \Omega, \\ &u_n \to \overline{u} \quad \text{weakly in } H^{-1}(\Omega), \quad w_n \to \overline{w} \quad \text{weakly in } H^{-1}(\Omega), \end{aligned}$$

where obviously, $[\overline{y}, \overline{u}, \overline{w}]$ is admissible for the problem (3.3.58), (3.3.62)–(3.3.65). The lower semicontinuity of the cost functional (3.3.62) then shows that the triple $[y_\varepsilon, u_\varepsilon, w_\varepsilon] = [\overline{y}, \overline{u}, \overline{w}]$ is optimal. Notice that the optimal triple might not be unique, since the problem is nonconvex.

Lemma 3.3.8 *For* $\varepsilon \searrow 0$,

$$u_\varepsilon \to u^* \quad \textit{strongly in } H^{-1}(\Omega), \tag{3.3.67}$$

$$y_\varepsilon \to y^* \quad \textit{strongly in } H_0^1(\Omega), \tag{3.3.68}$$

$$w_\varepsilon \to w^* \quad \textit{strongly in } H^{-1}(\Omega). \tag{3.3.69}$$

Proof. Owing to (3.3.61) and to the optimality of $[y_\varepsilon, u_\varepsilon, w_\varepsilon]$, we have

$$\begin{aligned} &\frac{1}{2}|y_\varepsilon - y_d|^2_{L^2(\Omega)} + \frac{1}{2}|u_\varepsilon|^2_{H^{-1}(\Omega)} - \frac{1}{\varepsilon}(w_\varepsilon, y_\varepsilon)_{H^{-1}(\Omega)\times H_0^1(\Omega)} + \frac{1}{2}|u_\varepsilon - u^*|^2_{H^{-1}(\Omega)} \\ &+ \frac{1}{2}|w_\varepsilon - w^*|^2_{H^{-1}(\Omega)} \leq \frac{1}{2}|y^* - y_d|^2_{L^2(\Omega)} + \frac{1}{2}|u^*|^2_{H^{-1}(\Omega)}, \quad \forall \varepsilon > 0. \end{aligned} \tag{3.3.70}$$

Thus, $\{y_\varepsilon\}$, $\{u_\varepsilon\}$, $\{w_\varepsilon\}$ are bounded respectively in $L^2(\Omega)$, $H^{-1}(\Omega)$, $H^{-1}(\Omega)$, thanks to (3.3.64) and (3.3.65). Moreover, (3.3.63) implies that $\{y_\varepsilon\}$ is compact in $H_0^1(\Omega)$. Hence there is a sequence $\varepsilon_n \searrow 0$ such that the convergences (3.3.67)–(3.3.69) are valid as $\varepsilon_n \searrow 0$ with respect to the weak topologies of the indicated spaces, with some limit points $[\tilde{y}, \tilde{u}, \tilde{w}] \in H_0^1(\Omega) \times H^{-1}(\Omega) \times H^{-1}(\Omega)$.

By compactness, we may also assume that $y_{\varepsilon_n} \to \tilde{y}$ strongly in $H_0^1(\Omega)$ and pointwise a.e. in Ω.

Clearly, $[\tilde{y}, \tilde{u}, \tilde{w}]$ is an admissible triple, i.e., it satisfies (3.3.63)–(3.3.65) and (3.3.58). From (3.3.70), (3.3.64), (3.3.65), we also obtain that

$$(w_{\varepsilon_n}, y_{\varepsilon_n})_{H^{-1}(\Omega)\times H_0^1(\Omega)} \to 0 \quad \text{as } n \to \infty. \tag{3.3.71}$$

In particular,

$$(\tilde{w}, \tilde{y})_{H^{-1}(\Omega)\times H_0^1(\Omega)} = 0$$

must hold, and (3.3.61) shows that $\tilde{w} \in \partial I_K(\tilde{y})$; that is, the pair $[\tilde{y}, \tilde{u}]$ is admissible for (3.3.55)–(3.3.59).

Next, observe that letting $\varepsilon_n \searrow 0$, we obtain from (3.3.70) and from the weak lower semicontinuity of the involved norms that

$$\begin{aligned} &\frac{1}{2}\,|\tilde{y} - y_d|^2_{L^2(\Omega)} + \frac{1}{2}\,|\tilde{u}|^2_{H^{-1}(\Omega)} + \frac{1}{2}\,|\tilde{u} - u^*|^2_{H^{-1}(\Omega)} + \frac{1}{2}\,|\tilde{w} - w^*|^2_{H^{-1}(\Omega)} \\ &\quad \le \frac{1}{2}\,|y^* - y_d|^2_{L^2(\Omega)} + \frac{1}{2}\,|u^*|^2_{H^{-1}(\Omega)}. \end{aligned}$$

Thus, by the optimality of $[y^*, u^*]$, we must have $[\tilde{y}, \tilde{u}] = [y^*, u^*]$, and by the uniqueness of the limit, all the above convergences hold for the entire sequences as $\varepsilon \searrow 0$. In particular, (3.3.70) then implies that also (3.3.67) and (3.3.69) must be valid, which finishes the proof of the lemma. □

In order to derive the necessary conditions of optimality, we employ a variant of the technique used above in §3.2.2. To this end, we approximate the problem (3.3.58), (3.3.62)–(3.3.65) by adding a penalization of the state equation. More precisely, we consider for $\lambda > 0$ the problem

$$\begin{aligned} \text{Min}\Big\{ &\frac{1}{2}\,|y - y_d|^2_{L^2(\Omega)} + \frac{1}{2}\,|u|^2_{H^{-1}(\Omega)} - \frac{1}{\varepsilon}\,(w, y)_{H^{-1}(\Omega)\times H_0^1(\Omega)} \\ &+ \frac{1}{2}|u - u^*|^2_{H^{-1}(\Omega)} + \frac{1}{2}\,|w - w^*|^2_{H^{-1}(\Omega)} + \frac{1}{2\lambda}\,|\Delta y + u - w|^2_{H^{-1}(\Omega)} \\ &+ \frac{1}{2}\,|u - u_\varepsilon|^2_{H^{-1}(\Omega)} + \frac{1}{2}\,|y - y_\varepsilon|^2_{H_0^1(\Omega)} \Big\}, \end{aligned} \tag{3.3.72}$$

subject to (3.3.58), (3.3.64), and (3.3.65).

The existence of at least one optimal triple $[y^\lambda, u^\lambda, w^\lambda]$ for fixed $\varepsilon > 0$ follows easily from the compactness of C. In the following, we keep $\varepsilon > 0$ fixed, first deriving estimates with respect to $\lambda > 0$, and then passing to the limit as $\lambda \searrow 0$. We begin with the following result.

Proposition 3.3.9 *As* $\lambda \searrow 0$,

$$y^\lambda \to y_\varepsilon \quad \textit{strongly in } H_0^1(\Omega), \tag{3.3.73}$$

$$u^\lambda \to u_\varepsilon \quad \textit{strongly in } H^{-1}(\Omega), \tag{3.3.74}$$

$$w^\lambda \to w_\varepsilon \quad \textit{strongly in } H^{-1}(\Omega). \tag{3.3.75}$$

Moreover, if $q^\lambda = \frac{1}{\lambda}(\Delta y^\lambda + u^\lambda - w^\lambda)$, *then* $\left\{\lambda^{1/2} q^\lambda\right\}_{\lambda>0}$ *is bounded in* $H^{-1}(\Omega)$, *and thus*

$$\lambda\, q^\lambda = \Delta y^\lambda + u^\lambda - w^\lambda \to 0 \quad \textit{strongly in } H^{-1}(\Omega) \textit{ as } \lambda \searrow 0. \tag{3.3.76}$$

Proof. By the optimality of $[y^\lambda, u^\lambda, w^\lambda]$, we have

$$\begin{aligned}
&\frac{1}{2}\left|y^\lambda - y_d\right|^2_{L^2(\Omega)} + \frac{1}{2}\left|u^\lambda\right|^2_{H^{-1}(\Omega)} - \frac{1}{\varepsilon}\left(w^\lambda, y^\lambda\right)_{H^{-1}(\Omega)\times H_0^1(\Omega)} \\
&\quad + \frac{1}{2}\left|u^\lambda - u^*\right|^2_{H^{-1}(\Omega)} + \frac{1}{2}\left|w^\lambda - w^*\right|^2_{H^{-1}(\Omega)} + \frac{1}{2\lambda}\left|\Delta y^\lambda + u^\lambda - w^\lambda\right|^2_{H^{-1}(\Omega)} \\
&\quad + \frac{1}{2}\left|u^\lambda - u_\varepsilon\right|^2_{H^{-1}(\Omega)} + \frac{1}{2}\left|y^\lambda - y_\varepsilon\right|^2_{H_0^1(\Omega)} \\
&\le \frac{1}{2}|y - y_d|^2_{L^2(\Omega)} + \frac{1}{2}|u|^2_{H^{-1}(\Omega)} - \frac{1}{\varepsilon}(w, y)_{H^{-1}(\Omega)\times H_0^1(\Omega)} \\
&\quad + \frac{1}{2}|u - u^*|^2_{H^{-1}(\Omega)} + \frac{1}{2}|w - w^*|^2_{H^{-1}(\Omega)} + \frac{1}{2}|u - u_\varepsilon|^2_{H^{-1}(\Omega)} \\
&\quad + \frac{1}{2}|y - y_\varepsilon|^2_{H_0^1(\Omega)},
\end{aligned} \tag{3.3.77}$$

for any $[y, u, w]$ satisfying (3.3.58), (3.3.63)–(3.3.65). For instance, the triple $[0, 0, 0]$ satisfies all the conditions. Thus, (3.3.77) shows that $\{[y^\lambda, u^\lambda, w^\lambda]\}_{\lambda>0}$ is a bounded subset of $H_0^1(\Omega) \times H^{-1}(\Omega) \times H^{-1}(\Omega)$, and $\{\lambda^{1/2} q^\lambda\}_{\lambda>0}$ is bounded in $H^{-1}(\Omega)$, so that (3.3.76) is satisfied. Hence, there exists some sequence $\lambda_n \searrow 0$ such that

$$\left[y^{\lambda_n}, u^{\lambda_n}, w^{\lambda_n}\right] \to [\tilde{y}_\varepsilon, \tilde{u}_\varepsilon, \tilde{w}_\varepsilon] \quad \text{weakly in } H_0^1(\Omega) \times H^{-1}(\Omega) \times H^{-1}(\Omega),$$

and in view of the semicontinuity of the involved norms and of the compactness of C, we may pass to the limit as $\lambda_n \searrow 0$ and infer that

$$\begin{aligned}
&\frac{1}{2}|\tilde{y}_\varepsilon - y_d|^2_{L^2(\Omega)} + \frac{1}{2}|\tilde{u}_\varepsilon|^2_{H^{-1}(\Omega)} - \frac{1}{\varepsilon}(\tilde{w}_\varepsilon, \tilde{y}_\varepsilon)_{H^{-1}(\Omega)\times H_0^1(\Omega)} + \frac{1}{2}|\tilde{u}_\varepsilon - u^*|^2_{H^{-1}(\Omega)} \\
&\quad + \frac{1}{2}|\tilde{w}_\varepsilon - w^*|^2_{H^{-1}(\Omega)} + \frac{1}{2}|\tilde{u}_\varepsilon - u_\varepsilon|^2_{H^{-1}(\Omega)} + \frac{1}{2}|\tilde{y}_\varepsilon - y_\varepsilon|^2_{H_0^1(\Omega)} \\
&\le \frac{1}{2}|y - y_d|^2_{L^2(\Omega)} + \frac{1}{2}|u|^2_{H^{-1}(\Omega)} - \frac{1}{\varepsilon}(w, y)_{H^{-1}(\Omega)\times H_0^1(\Omega)} + \frac{1}{2}|u - u^*|^2_{H^{-1}(\Omega)} \\
&\quad + \frac{1}{2}|w - w^*|^2_{H^{-1}(\Omega)} + \frac{1}{2}|u - u_\varepsilon|^2_{H^{-1}(\Omega)} + \frac{1}{2}|y - y_\varepsilon|^2_{H_0^1(\Omega)},
\end{aligned} \tag{3.3.78}$$

for any $[y, u, w]$ satisfying (3.3.58), (3.3.63)–(3.3.65). In conclusion, we may use the triple $[y_\varepsilon, u_\varepsilon, w_\varepsilon]$ (which is optimal for the problem (3.3.58), (3.3.62)–(3.3.65)).

It then follows that $[\tilde{y}_\varepsilon, \tilde{u}_\varepsilon, \tilde{w}_\varepsilon] = [y_\varepsilon, u_\varepsilon, w_\varepsilon]$, as well as (3.3.73) and (3.3.74). But then $\Delta y^{\lambda_n} \to \Delta y_\varepsilon$ strongly in $H^{-1}(\Omega)$, so that in view of (3.3.76), $w^{\lambda_n} \to w_\varepsilon$ strongly in $H^{-1}(\Omega)$.

The uniqueness of the limit point yields that the convergences hold for the entire sequences, which finishes the proof of the assertion. □

Since the problem (3.3.72) is an optimization problem in the independent variables y, u, w that vary in convex sets, we may take "partial" variations in each variable around $[y^\lambda, u^\lambda, w^\lambda]$. It is easily verified that this leads to the following first-order necessary conditions (which, in view of the nonconvexity of the problem (3.3.72), may not be sufficient):

$$\begin{aligned}
&\left(y^\lambda - y_d, y^\lambda - z\right)_{L^2(\Omega)} - \frac{1}{\varepsilon}\left(w^\lambda, y^\lambda - z\right)_{H^{-1}(\Omega)\times H_0^1(\Omega)} \\
&\quad + \left(y^\lambda - y_\varepsilon, y^\lambda - z\right)_{H_0^1(\Omega)} + \left(q^\lambda, \Delta(y^\lambda - z)\right)_{H^{-1}(\Omega)} \le 0 \\
&\quad \text{for all } z \in C \subset H_0^1(\Omega) \text{ with } z \ge 0;
\end{aligned} \tag{3.3.79}$$

$$\begin{aligned}
&\left(u^\lambda, u^\lambda - v\right)_{H^{-1}(\Omega)} + \left(u^\lambda - u^*, u^\lambda - v\right)_{H^{-1}(\Omega)} + \left(q^\lambda, u^\lambda - v\right)_{H^{-1}(\Omega)} \\
&\quad + \left(u^\lambda - u_\varepsilon, u^\lambda - v\right)_{H^{-1}(\Omega)} \le 0 \quad \text{for all } v \in U_{ad};
\end{aligned} \tag{3.3.80}$$

$$\begin{aligned}
&-\frac{1}{\varepsilon}\left(w^\lambda - r, y^\lambda\right)_{H^{-1}(\Omega)\times H_0^1(\Omega)} + \left(w^\lambda - w^*, w^\lambda - r\right)_{H^{-1}(\Omega)} \\
&\quad - \left(q^\lambda, w^\lambda - r\right)_{H^{-1}(\Omega)} \le 0 \quad \text{for all } r \in H^{-1}(\Omega) \text{ with } r \le 0.
\end{aligned} \tag{3.3.81}$$

In particular, (3.3.81) implies that

$$s^\lambda = y^\lambda + \varepsilon\, J^{-1} q^\lambda - \varepsilon\, J^{-1}(w^\lambda - w^*) \ge 0 \quad \text{in } \Omega,$$

where $J : H_0^1(\Omega) \to H^{-1}(\Omega)$ is the canonical isomorphism (i.e., $J = -\Delta$). However, $s^\lambda \notin C$ in general, and in order to derive an estimate for $\left\{q^\lambda\right\}_{\lambda>0}$, we have to impose an interiority assumption:

$$\exists \rho > 0 \text{ such that } V_\rho = \left\{v \in H^{-1}(\Omega) : |v|_{H^{-1}(\Omega)} \le \rho\right\} \subset U_{ad}.$$

Under this assumption, we are able to prove the boundedness of $\{q^\lambda\}$ in $H^{-1}(\Omega)$: indeed, addition of the relations (3.3.79)–(3.3.81) leads to the inequality

$$
\begin{aligned}
0 \geq\ & \left(y^\lambda - y_d, y^\lambda - z\right)_{L^2(\Omega)} - \frac{1}{\varepsilon}\left(w^\lambda, y^\lambda - z\right)_{H^{-1}(\Omega)\times H_0^1(\Omega)} \\
&+ \left(y^\lambda - y_\varepsilon, y^\lambda - z\right)_{H_0^1(\Omega)} + \left(u^\lambda, u^\lambda - v\right)_{H^{-1}(\Omega)} \\
&+ \left(u^\lambda - u^*, u^\lambda - v\right)_{H^{-1}(\Omega)} + \left(u^\lambda - u_\varepsilon, u^\lambda - v\right)_{H^{-1}(\Omega)} \\
&- \frac{1}{\varepsilon}\left(w^\lambda - r, y^\lambda\right)_{H^{-1}(\Omega)\times H_0^1(\Omega)} + \left(w^\lambda - w^*, w^\lambda - r\right)_{H^{-1}(\Omega)} \\
&+ \left(q^\lambda, \lambda q^\lambda - \Delta z - v + r\right)_{H^{-1}(\Omega)}. \qquad (3.3.82)
\end{aligned}
$$

The sum of all the terms except the last one is obviously bounded by a constant that is independent of $\lambda > 0$ whenever $[z, v, r]$ varies over a bounded subset of $H_0^1(\Omega) \times H^{-1}(\Omega) \times H^{-1}(\Omega)$. Then there is some $m_\varepsilon > 0$ such that

$$m_\varepsilon \geq (q^\lambda, -\Delta z - v + r)_{H^{-1}(\Omega)}.$$

Choosing $z = 0$, $r = 0$, and $v \in V_\rho$, we get the boundedness of $\{|q^\lambda|_{H^{-1}(\Omega)}\}_{\lambda>0}$. Indeed, we have

$$
\begin{aligned}
\left|q^\lambda\right|_{H^{-1}(\Omega)} &= \sup\left\{\left(q^\lambda, v\right)_{H^{-1}(\Omega)} : v \in H^{-1}(\Omega),\ |v|_{H^{-1}(\Omega)} = 1\right\} \\
&= \rho^{-1} \sup\left\{\left(q^\lambda, v\right)_{H^{-1}(\Omega)} : v \in H^{-1}(\Omega),\ |v|_{H^{-1}(\Omega)} = \rho\right\} \\
&\leq \rho^{-1} m_\varepsilon.
\end{aligned}
$$

Consequently, there exist some $q_\varepsilon \in H^{-1}(\Omega)$ and a sequence $\lambda_n \searrow 0$ such that $q^{\lambda_n} \to q_\varepsilon$ weakly in $H^{-1}(\Omega)$, and passage to the limit as $\lambda_n \searrow 0$ in (3.3.82), using Proposition 3.3.9, yields that

$$
\begin{aligned}
0 \geq\ & (y_\varepsilon - y_d, y_\varepsilon - z)_{L^2(\Omega)} - \frac{1}{\varepsilon}(w_\varepsilon, y_\varepsilon - z)_{H^{-1}(\Omega)\times H_0^1(\Omega)} \\
&+ (u_\varepsilon, u_\varepsilon - v)_{H^{-1}(\Omega)} + (u_\varepsilon - u^*, u_\varepsilon - v)_{H^{-1}(\Omega)} \\
&- \frac{1}{\varepsilon}(w_\varepsilon - r, y_\varepsilon)_{H^{-1}(\Omega)\times H_0^1(\Omega)} + (w_\varepsilon - w^*, w_\varepsilon - r)_{H^{-1}(\Omega)} \\
&+ (q_\varepsilon, -\Delta z - v + r)_{H^{-1}(\Omega)}.
\end{aligned}
$$

We notice that $(w_\varepsilon, y_\varepsilon)_{H^{-1}(\Omega)\times H_0^1(\Omega)} \leq 0$ and can be neglected in the above inequality. Hence, there is a constant m, independent of ε, such that for all $\varepsilon > 0$,

$$m - \frac{1}{\varepsilon}(w_\varepsilon, z)_{H^{-1}(\Omega)\times H_0^1(\Omega)} - \frac{1}{\varepsilon}(r, y_\varepsilon)_{H^{-1}(\Omega)\times H_0^1(\Omega)} \geq (q_\varepsilon, -\Delta z - v + r)_{H^{-1}(\Omega)}.$$

Choosing $z = 0$, $r = 0$, $v \in V_\rho$ once more, we conclude that $\{q_\varepsilon\}_{\varepsilon>0}$ is bounded in $H^{-1}(\Omega)$.

It is now an easy task to pass to the limit as $\lambda_n \searrow 0$ in (3.3.79)–(3.3.81) for fixed $\varepsilon > 0$ and to infer that

$$(y_\varepsilon - y_d, y_\varepsilon - z)_{L^2(\Omega)} - \frac{1}{\varepsilon}(w_\varepsilon, y_\varepsilon - z)_{H^{-1}(\Omega)\times H_0^1(\Omega)} + (q_\varepsilon, \Delta(y_\varepsilon - z))_{H^{-1}(\Omega)} \le 0 \quad \forall z \in C \text{ with } z \ge 0. \tag{3.3.83}$$

$$(u_\varepsilon, u_\varepsilon - v)_{H^{-1}(\Omega)} + (u_\varepsilon - u^*, u_\varepsilon - v)_{H^{-1}(\Omega)} + (q_\varepsilon, u_\varepsilon - v)_{H^{-1}(\Omega)} \le 0 \quad \forall v \in U_{ad}. \tag{3.3.84}$$

$$-\frac{1}{\varepsilon}(w_\varepsilon - r, y_\varepsilon)_{H^{-1}(\Omega)\times H_0^1(\Omega)} + (w_\varepsilon - w^*, w_\varepsilon - r)_{H^{-1}(\Omega)} - (q_\varepsilon, w_\varepsilon - r)_{H^{-1}(\Omega)} \le 0 \quad \forall r \in H^{-1}(\Omega) \text{ with } r \le 0. \tag{3.3.85}$$

The convergence properties of $\{[y_\varepsilon, u_\varepsilon, w_\varepsilon]\}_{\varepsilon>0}$ are given in (3.3.67)–(3.3.69). Since $\{|q_\varepsilon|_{H^{-1}(\Omega)}\}_{\varepsilon>0}$ is bounded, there is some constant $M > 0$ such that

$$|J^{-1}q_\varepsilon - J^{-1}(w_\varepsilon - w^*)|_{H_0^1(\Omega)} \le M \quad \forall \varepsilon > 0.$$

Therefore, $s_\varepsilon = y_\varepsilon + \varepsilon J^{-1}q_\varepsilon - \varepsilon J^{-1}(w_\varepsilon - w^*) \in C_\varepsilon$, where

$$C_\varepsilon = \Big\{s \in H_0^1(\Omega) : \inf\{|s - y|_{H_0^1(\Omega)} : y \in C\} \le \varepsilon M\Big\}.$$

We now introduce an "ad hoc definition" for a generalized second-order derivative of the indicator function $I_{C\cap K}$ (where $w \in \partial I_{C\cap K}(y)$ and $p \in H_0^1(\Omega)$) by setting

$$\begin{aligned}
&\widetilde{D}\,\partial I_{C\cap K}(y, w)\,p \\
&= \Big\{\ell \in H^{-1}(\Omega) : \text{ there exist sequences } \varepsilon_n \searrow 0,\ \{r_{\varepsilon_n}\} \subset H_0^1(\Omega), \\
&\quad \{y_{\varepsilon_n}\} \subset C \cap K,\ \{w_{\varepsilon_n}\} \subset \partial I_{C\cap K}(y_{\varepsilon_n}), \text{ and} \\
&\quad \{\tilde{w}_{\varepsilon_n}\} \subset \partial I_{C_{\varepsilon_n}\cap K}(y_{\varepsilon_n} + \varepsilon_n r_{\varepsilon_n}), \text{ such that} \\
&\quad r_{\varepsilon_n} \to p \text{ weakly in } H_0^1(\Omega),\ w_{\varepsilon_n} \to w,\ \tilde{w}_{\varepsilon_n} \to w \text{ strongly in } H^{-1}(\Omega), \\
&\quad y_{\varepsilon_n} \to y \text{ strongly in } H_0^1(\Omega),\ \frac{1}{\varepsilon_n}(\tilde{w}_{\varepsilon_n} - w_{\varepsilon_n}) \to \ell \text{ weakly in } H^{-1}(\Omega)\Big\}.
\end{aligned} \tag{3.3.86}$$

With this definition, we have the following result.

Proposition 3.3.10 *Suppose that the above hypotheses, in particular the interiority condition, are satisfied. If* $[y^*, u^*] \in C \times U_{ad}$ *is an optimal pair for the optimal control problem (3.3.55)–(3.3.59), then there exists some* $p^* \in H_0^1(\Omega)$ *such that*

$$-\Delta p^* \in y^* - y_d - \widetilde{D}\,\partial I_{C\cap K}(y^*, w^*)\,p^*, \tag{3.3.87}$$

$$-p^* \in u^* + \partial I_{U_{ad}}(u^*). \tag{3.3.88}$$

Proof. We notice that in view of the above estimates,

$$y_\varepsilon + \varepsilon J^{-1} q_\varepsilon - \varepsilon J^{-1}(w_\varepsilon - w^*) \in C_\varepsilon \cap K.$$

Moreover, the relation (3.3.85) may be rewritten as

$$y_\varepsilon + \varepsilon J^{-1} q_\varepsilon - \varepsilon J^{-1}(w_\varepsilon - w^*) \in \partial I_{K_-}(w_\varepsilon),$$

or, equivalently,

$$w_\varepsilon \in \partial I_K\left(y_\varepsilon + \varepsilon J^{-1} q_\varepsilon - \varepsilon J^{-1}(w_\varepsilon - w^*)\right).$$

Now put $p_\varepsilon = J^{-1} q_\varepsilon$ for $\varepsilon > 0$. Since $\{q_\varepsilon\}_{\varepsilon>0}$ is bounded in $H^{-1}(\Omega)$, there are a sequence $\varepsilon_n \searrow 0$ and some $q^* \in H^{-1}(\Omega)$ such that $q_{\varepsilon_n} \to q^*$ weakly in $H^{-1}(\Omega)$. By the continuity of J^{-1}, it then follows that $p_{\varepsilon_n} \to p^* = J^{-1} q^*$ weakly in $H_0^1(\Omega)$.

By the above remark, we have, for any $\varepsilon > 0$,

$$w_\varepsilon \in \partial I_K(y_\varepsilon + \varepsilon p_\varepsilon - \varepsilon J^{-1}(w_\varepsilon - w^*)) + \partial I_{C_\varepsilon}(y_\varepsilon + \varepsilon p_\varepsilon - \varepsilon J^{-1}(w_\varepsilon - w^*)), \tag{3.3.89}$$

where we have used the summation rule for the subdifferential (compare Appendix 1). Similarly, (3.3.83) implies that

$$-\Delta p_\varepsilon + \frac{1}{\varepsilon} w_\varepsilon \in y_\varepsilon - y_d + \frac{1}{\varepsilon}\partial I_{C\cap K}(y_\varepsilon). \tag{3.3.90}$$

Now observe that (3.3.89), (3.3.90) entail that

$$-\Delta p_\varepsilon \in y_\varepsilon - y_d + \frac{1}{\varepsilon}\left(\partial I_{C\cap K}(y_\varepsilon) - \partial I_{C_\varepsilon \cap K}(y_\varepsilon + \varepsilon p_\varepsilon - \varepsilon J^{-1}(w_\varepsilon - w^*))\right),$$

and the adjoint equation (3.3.87) follows from the definition (3.3.86), since $p_{\varepsilon_n} \to p^*$ weakly in $H_0^1(\Omega)$. Finally, the maximum principle (3.3.88) is a direct consequence of (3.3.84). This concludes the proof of the assertion. □

Remark. In Barbu [1984] a systematic study of control problems governed by variational inequalities, including the case of nonlinear evolution equations, has been carried out. This section presented several supplementary techniques and recent results that are mainly due to the authors.

3.4 Thickness Optimization for Plates

In Example 1.1.7 in Chapter 1, some minimization problem associated with certain (simplified) plate models have been introduced (see (1.2.54), (1.2.57)–(1.2.59), for instance). Related existence questions were discussed in detail in §2.2.2 of Chapter 2. This type of problem belongs to the class of control by coefficients problems, governed by fourth-order elliptic equations.

In this section, we will examine the differentiability properties of the (nonlinear) mapping coefficient $\mapsto$ solution, and we will present several applications concerning the optimality conditions and bang-bang properties of the minimizers. We remark that the abstract framework described above in §3.1.3 may be viewed as a "genus proximum" for the results derived in this section. Moreover, in §3.4.2 we will also show that control-theoretical methods may be employed to develop more sophisticated variational arguments that apply to the type of equations considered here.

3.4.1 Simply Supported Plates

As shown in §2.2.2 of Chapter 2, we can reformulate certain thickness optimization problems, using an easy transformation (see Example 3.4.8 below), as convex distributed control problems for second-order elliptic equations. To fix things, let $\Omega \subset \mathbf{R}^d$ denote an open, bounded domain having a sufficiently smooth boundary $\partial\Omega$, and let $\tau > 0$ be fixed. We then consider the problem

$$\operatorname{Min}\left\{ \int_\Omega \left[\varphi(x, l(x)) + \psi(x, y(x))\right] dx \right\}, \tag{3.4.1}$$

subject to

$$\Delta y = z\, l \quad \text{in } \Omega, \tag{3.4.2}$$

$$y = 0 \quad \text{on } \partial\Omega, \tag{3.4.3}$$

$$l \in U_{ad} = \left\{ k \in L^\infty(\Omega) : 0 < M^{-3} \le k(x) \le m^{-3} \quad \text{a.e. in } \Omega \right\}, \tag{3.4.4}$$

$$y \in C = \left\{ \mu \in L^2(\Omega) \, : \, y(x) \ge -\tau \quad \text{a.e. in } \Omega \right\}. \tag{3.4.5}$$

Notice that C and U_{ad} are nonempty, convex, and closed subsets of $L^2(\Omega)$, where, a fortiori, $U_{ad} \subset L^\infty(\Omega)$.

With minor changes of notation, the relations (3.4.1)–(3.4.5) coincide with the relations (2.2.16)–(2.2.22) in Chapter 2, where $z \in W^{2,p}(\Omega) \cap H_0^1(\Omega)$ is given as the solution to the Poisson equation with zero Dirichlet boundary conditions, corresponding to a given right-hand side $f \in L^p(\Omega)$ (compare (2.2.19),(2.2.20)). For simplicity, we assume that $p > \frac{d}{2}$, so that $z \in C(\overline{\Omega})$, by the Sobolev embedding theorem (cf. Theorem A2.2(iv) in Appendix 2).We also postulate that φ and ψ are convex integrands as defined in Appendix 1.

To analyze the problem (3.4.1)–(3.4.5), we employ the technique developed above in §2.2.2. Since the setting is slightly different, we briefly indicate the main steps in the argument.

The constraint (3.4.4) ensures the existence of at least one optimal pair $[y^*, l^*]$ if the admissibility condition is satisfied (cf. Theorem 2.2.5 in Chapter 2 and the remarks following it). We notice that the pair $[0, 0]$, while satisfying (3.4.2), (3.4.3), and (3.4.5), is not admissible, since (3.4.4) is violated. However, if M is sufficiently large, then the pair $[y_M, M^{-3}]$, where y_M denotes the solution to (3.4.2), (3.4.3) associated with $l \equiv M^{-3}$, is admissible for the problem (3.4.1)–(3.4.5): indeed, since $p > \frac{d}{2}$, there are constants $C_1 > 0$, $C_2 > 0$, independent of z and $l \in U_{ad}$, such that

$$|y_M|_{C(\overline{\Omega})} \leq C_1 \, |y_M|_{W^{2,p}(\Omega)} \leq C_2 \, M^{-3} \, |z|_{L^p(\Omega)} \to 0 \quad \text{as } M \to \infty,$$

so that y_M satisfies (3.4.5) for sufficiently large M, which shows the admissibility. We will henceforth assume that M is sufficiently large in this sense.

Now let $[y^*, l^*]$ be any optimal pair for (3.4.1)–(3.4.5). In order to derive the necessary conditions of optimality, we then introduce the penalized optimization problem

$$\text{Min} \left\{ \phi(l) + \Psi(y) + \frac{1}{2} \, |l - l^*|^2_{L^2(\Omega)} + \frac{1}{2\,\varepsilon} \, |\Delta y - zl|^2_{L^2(\Omega)} \right\}, \tag{3.4.6}$$

subject to (3.4.4), (3.4.5), and $y \in H^2(\Omega) \cap H^1_0(\Omega)$.

Here, we have abbreviated by ϕ and Ψ the first and, respectively, second integral functional appearing in (3.4.1). Having the subsequent applications in mind, we assume that ϕ, Ψ are convex, continuous, bounded from below, and Gâteaux differentiable on $L^2(\Omega)$.

Let us at first demonstrate that the penalized problem has a unique optimal pair $[y_\varepsilon, l_\varepsilon] \in C \times U_{ad}$. Indeed, since we have admissibility, we may choose a minimizing sequence $\{[y_n, l_n]\} \subset C \times U_{ad}$, which, since ϕ and Ψ are bounded from below, is easily seen to be a bounded subset of $(H^2(\Omega) \cap H^1_0(\Omega)) \times L^\infty(\Omega)$. Hence, we may without loss of generality assume that

$$y_n \to \hat{y} \quad \text{weakly in } H^2(\Omega) \cap H^1_0(\Omega), \qquad l_n \to \hat{l} \quad \text{weakly* in } L^\infty(\Omega),$$

where clearly, $[\hat{y}, \hat{l}] \in C \times U_{ad}$, which is weakly closed. The weak lower semicontinuity of the cost functional (3.4.6) then shows that $[y_\varepsilon, l_\varepsilon] = [\hat{y}, \hat{l}]$ is optimal, and the uniqueness is a consequence of the strict convexity of the cost functional. We have the following result.

Proposition 3.4.1 *Let us set* $r_\varepsilon = \varepsilon^{-1}(\Delta y_\varepsilon - z\, l_\varepsilon) \in L^2(\Omega)$, *for* $\varepsilon > 0$. *Then*

$$y_\varepsilon \to y^* \quad \text{strongly in } H^2(\Omega) \cap H_0^1(\Omega), \tag{3.4.7}$$

$$l_\varepsilon \to l^* \quad \text{strongly in } L^p(\Omega), \text{ for all } p \geq 1, \tag{3.4.8}$$

$$\left\{\varepsilon^{\frac{1}{2}}\, r_\varepsilon\right\}_{\varepsilon>0} \quad \text{is bounded in } L^2(\Omega). \tag{3.4.9}$$

Proof. Since $[y^*, l^*]$ is obviously admissible for the penalized problems for any $\varepsilon > 0$, it follows that

$$\phi(l_\varepsilon) + \Psi(y_\varepsilon) + \frac{1}{2}\,|l_\varepsilon - l^*|_{L^2(\Omega)} + \frac{1}{2\,\varepsilon}\,|\Delta y_\varepsilon - z\, l_\varepsilon|^2_{L^2(\Omega)} \leq \phi(l^*) + \Psi(y^*). \tag{3.4.10}$$

Now recall that ϕ and Ψ are bounded from below, while $\{l_\varepsilon\}_{\varepsilon>0}$ is bounded in $L^\infty(\Omega)$. Hence, (3.4.10) implies that $\{\varepsilon^{\frac{1}{2}}\, r_\varepsilon\}_{\varepsilon>0}$ is bounded in $L^2(\Omega)$. By the definition of r_ε, then also $\{\Delta y_\varepsilon\}_{\varepsilon>0}$ is bounded in $L^2(\Omega)$, which implies that $\{y_\varepsilon\}_{\varepsilon>0}$ is bounded in $H^2(\Omega) \cap H_0^1(\Omega)$.

Hence, we may select a sequence $\varepsilon_n \searrow 0$ such that $l_{\varepsilon_n} \to \hat{l}$ weakly* in $L^\infty(\Omega)$ and $y_{\varepsilon_n} \to \hat{y}$ weakly in $H^2(\Omega) \cap H_0^1(\Omega)$. Since $\Delta y_{\varepsilon_n} - z\, l_{\varepsilon_n} = \varepsilon_n\, r_{\varepsilon_n} \to 0$ strongly in $L^2(\Omega)$, we see that $\Delta \hat{y} = z\, \hat{l}$. In addition, $[\hat{y}, \hat{l}] \in C \times U_{ad}$, in view of the weak closedness of C and U_{ad} in $L^2(\Omega)$.

In summary, the pair $[\hat{y}, \hat{l}]$ is admissible for the problem (3.4.1)–(3.4.5), and the weak lower semicontinuity of the cost functional and (3.4.10) yield that

$$\phi(\hat{l}) + \Psi(\hat{y}) + \frac{1}{2}\left|\hat{l} - l^*\right|^2_{L^2(\Omega)} \leq \phi(l^*) + \Psi(y^*).$$

Thus, $[\hat{y}, \hat{l}]$ is optimal for (3.4.1)–(3.4.5), and $\hat{l} = l^*$, $\hat{y} = y^*$. Moreover, it follows that $l_{\varepsilon_n} \to l^*$ strongly in $L^2(\Omega)$. The uniqueness of the limit entails that all the above convergences hold generally for $\varepsilon \searrow 0$ and not only for the sequence $\{\varepsilon_n\}$. Also, since $\{l_\varepsilon\}_{\varepsilon>0}$ is bounded in $L^\infty(\Omega)$, the relation (3.4.8) follows from Lebesgue's theorem. Finally, we have, as $\varepsilon \searrow 0$,

$$\Delta y_\varepsilon = z\, l_\varepsilon + \varepsilon\, r_\varepsilon \to z\, \hat{l} = \Delta \hat{y}, \quad \text{strongly in } L^2(\Omega),$$

and (3.4.7) follows from standard elliptic estimates. This finishes the proof of the assertion. □

Proposition 3.4.2 *The optimal pair* $[y_\varepsilon, l_\varepsilon]$ *of the penalized problem satisfies the following necessary and sufficient optimality condition:*

$$(\nabla\phi(l_\varepsilon), k - l_\varepsilon)_{L^2(\Omega)} + (\nabla\Psi(y_\varepsilon), \mu - y_\varepsilon)_{L^2(\Omega)} + (l_\varepsilon - l^*, k - l_\varepsilon)_{L^2(\Omega)}$$
$$+ (r_\varepsilon, \Delta\mu - z\, k - \varepsilon\, r_\varepsilon)_{L^2(\Omega)} \geq 0, \quad \forall\, k \in U_{ad},\ \forall\, \mu \in C \cap H^2(\Omega) \cap H_0^1(\Omega). \tag{3.4.11}$$

Proof. For any $\lambda \in]0,1]$, $\widetilde{k} \in U_{ad} - l_\varepsilon$, $\widetilde{\mu} \in (C - y_\varepsilon) \cap H^2(\Omega) \cap H_0^1(\Omega)$, it holds that $\left[y_\varepsilon + \lambda\widetilde{\mu}, l_\varepsilon + \lambda\widetilde{k}\right]$ is admissible, and thus

$$\begin{aligned}\phi(l_\varepsilon) + \Psi(y_\varepsilon) + \frac{1}{2}|l_\varepsilon - l^*|^2_{L^2(\Omega)} + \frac{1}{2\,\varepsilon}|\Delta y_\varepsilon - z\,l_\varepsilon|^2_{L^2(\Omega)} \\ \le \phi\left(l_\varepsilon + \lambda\,\widetilde{k}\right) + \Psi(y_\varepsilon + \lambda\,\widetilde{\mu}) + \frac{1}{2}\left|l_\varepsilon + \lambda\,\widetilde{k} - l^*\right|^2_{L^2(\Omega)} \\ + \frac{1}{2\,\varepsilon}\left|\Delta y_\varepsilon + \lambda\,\Delta\widetilde{\mu} - z\,l_\varepsilon - \lambda\,z\,\widetilde{k}\right|^2_{L^2(\Omega)}.\end{aligned}$$

Dividing by $\lambda > 0$, and passing to the limit as $\lambda \searrow 0$, we conclude from the Gâteaux differentiability that

$$\begin{aligned}\left(\nabla\phi(l_\varepsilon), \widetilde{k}\right)_{L^2(\Omega)} + (\nabla\Psi(y_\varepsilon), \widetilde{\mu})_{L^2(\Omega)} + \int_\Omega (l_\varepsilon - l^*)\,\widetilde{k}\,dx \\ + \frac{1}{\varepsilon}\int_\Omega (\Delta y_\varepsilon - z\,l_\varepsilon)\,(\Delta\mu - z\,\widetilde{k})\,dx \ge 0.\end{aligned}$$

Putting $k = \widetilde{k} + l_\varepsilon$, $\mu = \widetilde{\mu} + y_\varepsilon$, we obtain (3.4.11). The sufficiency of (3.4.11) follows as in the subsequent proof of Theorem 3.4.3. □

Remark. Obviously, $[y_\varepsilon, l_\varepsilon]$ also satisfies the following necessary condition:

$$\begin{aligned}(\nabla\phi(l_\varepsilon), k - l_\varepsilon)_{L^2(\Omega)} + (\nabla\Psi(y_\varepsilon), \mu - y_\varepsilon)_{L^2(\Omega)} + (l_\varepsilon - l^*, k - l_\varepsilon)_{L^2(\Omega)} \\ + (r_\varepsilon, \Delta\mu - z\,k)_{L^2(\Omega)} \ge 0 \quad \forall k \in U_{ad},\ \forall\mu \in C \cap H^2(\Omega) \cap H_0^1(\Omega). \qquad (3.4.11)'\end{aligned}$$

Theorem 3.4.3 *If M is sufficiently large, then there is a sequence $\varepsilon_n \searrow 0$ such that*

$$\nabla\phi(l_{\varepsilon_n}) \to \nabla\phi(l^*) \quad \text{weakly in } L^2(\Omega), \qquad (3.4.12)$$

$$\nabla\Psi(y_{\varepsilon_n}) \to \nabla\Psi(y^*) \quad \text{weakly in } L^2(\Omega), \qquad (3.4.13)$$

$$r_{\varepsilon_n} \to r^* \quad \text{weakly in } L^s(\Omega),$$

$$\text{where } 1 \le s < \infty, \text{ for } d \le 2, \text{ and } 1 \le s < \frac{2d}{d-2}, \text{ for } d \ge 3. \qquad (3.4.14)$$

The optimal pair $[y^*, l^*]$ satisfies the necessary and sufficient optimality condition

$$\begin{aligned}(\nabla\phi(l^*), k - l^*)_{L^2(\Omega)} + (\nabla\Psi(y^*), \mu - y^*)_{L^2(\Omega)} \\ + (r^*, \Delta\mu - z\,k)_{L^s(\Omega)\times L^p(\Omega)} \ge 0, \quad \frac{1}{s} + \frac{1}{p} = 1, \\ \forall\,k \in U_{ad},\ \forall\,\mu \in C \cap W^{2,p}(\Omega) \cap H_0^1(\Omega). \qquad (3.4.15)\end{aligned}$$

Proof. The Gâteaux derivatives $\nabla\phi$ and $\nabla\Psi$ define maximal monotone operators on $L^2(\Omega)$ and are therefore locally bounded on the interior of their respective domains (cf. Appendix 1). Hence, by (3.4.7), (3.4.8) there is some sequence $\varepsilon_n \searrow 0$ such that

$$\nabla\phi(l_{\varepsilon_n}) \to q_1 \text{ weakly in } L^2(\Omega), \quad \nabla\Psi(y_{\varepsilon_n}) \to q_2 \text{ weakly in } L^2(\Omega).$$

Using Proposition 3.4.1, and recalling the fact that maximal monotone operators are demiclosed (cf. Proposition A1.4(iv) in Appendix 1), we infer that $q_1 = \nabla\phi(l^*)$ and $q_2 = \nabla\Psi(y^*)$, which establishes (3.4.12) and (3.4.13). Moreover, since U_{ad} is bounded in $L^\infty(\Omega)$ and all the other terms are convergent in some appropriate topology, it follows from (3.4.11)′ that there is some constant $C_1 > 0$, independent of $n \in \mathbf{N}$, such that

$$(r_{\varepsilon_n}, \Delta\mu - z\,k)_{L^s(\Omega)\times L^p(\Omega)} \geq -C_1, \tag{3.4.16}$$

provided that μ varies in a bounded subset of $W^{2,p}(\Omega)$.

Let us first take $k = 0$ in (3.4.16) (although $0 \notin U_{ad}$). We choose μ as the solution to the boundary value problem

$$\Delta\mu = \rho \text{ in } \Omega, \qquad \mu = 0 \text{ in } \partial\Omega.$$

It follows from the Sobolev embedding theorem (recall that $p > \frac{d}{2}$) that $\mu(x) \geq -\frac{\tau}{2}$ in $\overline{\Omega}$, provided that

$$\rho \in B_\delta(0) = \left\{\xi \in L^p(\Omega) \,:\, |\xi|_{L^p(\Omega)} \leq \delta\right\}$$

for a sufficiently small $\delta > 0$. Therefore, such a μ is admissible in (3.4.16), and it follows that

$$(r_{\varepsilon_n}, \rho)_{L^s(\Omega)\times L^p(\Omega)} \geq -C_2 \quad \text{for all } \rho \in B_\delta(0), \tag{3.4.17}$$

with a constant $C_2 > 0$ that is independent of $n \in \mathbf{N}$. Therefore, $\{r_{\varepsilon_n}\}$ is bounded in $L^s(\Omega)$, and we may without loss of generality assume that $r_{\varepsilon_n} \to r^*$ weakly in $L^s(\Omega)$.

This argument carries over to the situation that M is sufficiently large and $k \equiv M^{-3}$, by perturbing the ball $B_\delta(0)$, which in view of the definition of C is possible since $\mu(x) \geq -\frac{\tau}{2}$ in $\overline{\Omega}$ for $\rho \in B_\delta(0)$.

Summarizing the above arguments, we can claim that (3.4.12)–(3.4.14) hold for the sequence $\{\varepsilon_n\}$ constructed above. Invoking Proposition 3.4.1, we then get the validity of (3.4.15) by passing to the limit as $\varepsilon_n \searrow 0$ in (3.4.11)′, which proves the necessity of (3.4.15).

The sufficiency of (3.4.15) can be checked directly. Indeed, any admissible pair $[y, l]$ can serve as a test element in (3.4.15), and such a pair satisfies $\Delta y = z\,l$. Therefore,

$$(\nabla\phi(l^*), l - l^*)_{L^2(\Omega)} + (\nabla\Psi(y^*), y - y^*)_{L^2(\Omega)} \geq 0,$$

that is,

$$\phi(l^*) + \Psi(y^*) \le \phi(l) + \Psi(y),$$

by the definition of the subdifferential. This ends the proof. □

Remark. The above argument is a variant of the proof of Theorem 3.2.9 and is based on the duality-type estimate (3.4.17). The assumption "M sufficiently large" is typical for problems related to the optimization of plates. Notice that in this setting it is not necessary to introduce an auxiliary function p^* as in (3.2.50).

Remark. If $\Omega \subset \mathbf{R}^3$, then we may take $s = p = 2$.

Remark. By choosing in turn $k = l^*$ and $\mu = y^*$, we see that (3.4.15) is equivalent to the system

$$(\nabla\Psi(y^*), \mu - y^*)_{L^2(\Omega)} + (r^*, \Delta\mu - \Delta y^*)_{L^s(\Omega)\times L^p(\Omega)} \ge 0,$$

$$\forall \mu \in C \cap W^{2,p}(\Omega) \cap H_0^1(\Omega), \tag{3.4.18}$$

$$(\nabla\phi(l^*), k - l^*)_{L^2(\Omega)} + (zr^*, l^* - k)_{L^s(\Omega)\times L^p(\Omega)} \ge 0 \quad \forall k \in U_{ad}. \tag{3.4.19}$$

Example 3.4.4 We choose

$$\phi(l) \equiv 0, \quad \Psi(y) = \frac{1}{2}\int_\Omega (y(x) - y_d(x))^2\, dx,$$

where $y_d \in L^p(\Omega)$ is a "desired" or observed deflection of the plate. The problem (3.4.1)–(3.4.5) corresponds to identification-type problems such as (1.2.60) in Chapter 1. Since $y^* \in C(\overline{\Omega})$, the set

$$\Omega_1 = \{x \in \Omega \, : \, y^*(x) = -\tau\}$$

is closed, while $\Omega \setminus \Omega_1$ is open (and nonempty). Now take any $d \in \mathcal{D}(\Omega \setminus \Omega_1)$. Then, $\operatorname{supp} d \subset \Omega \setminus \Omega_1$ is compact, and $y^*(x) \ge -\tau + \lambda$ on $\operatorname{supp} d$ for some $\lambda > 0$, by the definition of Ω_1. Hence, for sufficiently small $\sigma > 0$ the functions $\mu = y^* \pm \sigma d$ are admissible in (3.4.18), whence we obtain that

$$\pm(y^* - y_d, \sigma\, d)_{L^2(\Omega)} \pm (r^*, \sigma\, \Delta d)_{L^2(\Omega)} \ge 0,$$

that is,

$$-\Delta r^* = y^* - y_d \quad \text{in } \mathcal{D}'(\Omega \setminus \Omega_1). \tag{3.4.20}$$

The interior regularity properties for (3.4.20) (cf. Appendix 2) show that $r^* \in W^{2,p}_{\mathrm{loc}}(\Omega \setminus \Omega_1)$. Consequently, $r^* \in C(\Omega \setminus \Omega_1)$, and the sets

$$\Omega_2 = \{x \in \Omega \setminus \Omega_1 \, : \, z(x) r^*(x) > 0\}, \quad \Omega_3 = \{x \in \Omega \setminus \Omega_1 \, : \, z(x) r^*(x) < 0\},$$

are open in Ω, with the closed complement $\Omega_4 = (\Omega \setminus \Omega_1) \setminus (\Omega_2 \cup \Omega_3)$.

If we choose $k = l^*$ on $\Omega \setminus \Omega_2$, then

$$(z\,r^*, l^* - k)_{L^s(\Omega_2)\times L^p(\Omega_2)} \geq 0,$$

that is, $l^* = m^{-3}$ a.e. in Ω_2. Analogously, $l^* = M^{-3}$ a.e. in Ω_3.

Now assume that $z(x) \neq 0$ in $\Omega \setminus \Omega_1$. Then, $r^*(x) = 0$ a.e. in Ω_4 if $\text{meas}(\Omega_4) > 0$. The maximal local regularity of r^* in $\Omega \setminus \Omega_1$ shows that its derivatives also vanish a.e. in Ω_4 (cf. Brézis [1983, p. 195]). Relation (3.4.20) then yields that $y^*(x) = y_d(x)$ a.e. in Ω_4.

We thus have proved the following result:

Corollary 3.4.5 *Suppose that the above assumptions are fulfilled. Then for any optimal pair $[y^*, u^*]$ of the problem (1.2.60) in Chapter 1 there exists a set $N \subset \Omega$ having zero measure such that for any $x \in \Omega \setminus N$ at least one of the four possibilities $y^*(x) = -\tau$, $y^*(x) = y_d(x)$, $u^*(x) = m$, $u^*(x) = M$, is valid.*

Remark. If $z(x) \neq 0$ in Ω, that is, owing to its continuity, if z has constant sign, then (3.4.2) implies that $\text{meas}(\Omega_1) = 0$, since Δy^* vanishes a.e. on each subset of Ω_1 having positive measure (compare Brézis [1983, p. 195]). This situation arises whenever f has constant sign in Ω, so that the strong maximum principle (cf. Theorem A2.8 in Appendix 2) can be applied to z.

Example 3.4.6 We now study a thickness optimization problem. In the notation introduced in Example 1.2.7 of Chapter 1, the functional to be minimized has the general form (recall that $u = l^{-\frac{1}{3}}$)

$$\text{Min}\left\{\int_\Omega \theta(u(x))\,dx\right\}, \tag{3.4.21}$$

subject to (3.4.2)–(3.4.5), where $\theta :]0, +\infty[\to \mathbf{R}$ denotes an increasing real mapping. In the present example, we fix $\theta(u) = -u^{-3}$, which satisfies the above monotonicity condition. The advantage over the standard functional considered in (1.2.57) in Chapter 1 is that by means of the transformation $l = u^{-3}$ in (3.4.1) we can choose $\phi(l) = -\displaystyle\int_\Omega l(x)\,dx$ and $\Psi(y) = 0$ on $L^2(\Omega)$. Using the same argument as in Example 3.4.4, we find that

$$\Delta r^* = 0 \quad \text{in } \mathcal{D}'(\Omega \setminus \Omega_1), \tag{3.4.22}$$

and therefore, $r^* \in C^\infty(\Omega \setminus \Omega_1)$. Then (3.4.19) becomes

$$(z\,r^* + 1, l^* - k)_{L^s(\Omega)\times L^p(\Omega)} \geq 0 \quad \forall k \in U_{ad}. \tag{3.4.23}$$

As in Example 3.4.4, we conclude that $l^*(x) = m^{-3}$ and $l^*(x) = M^{-3}$ a.e. in the open sets $\{x \in \Omega : z(x)\, r^*(x) + 1 > 0\}$ and $\{x \in \Omega : z(x)\, r^*(x) + 1 < 0\}$, respectively.

Now assume that the closed set $\Omega_5 = \{x \in \Omega : z(x)\, r^*(x) + 1 = 0\}$ has positive measure. Then we have in Ω_5

$$\Delta r^* = \frac{z\, \Delta z - 2\, |\nabla z|^2}{z^3(x)}, \tag{3.4.24}$$

where we notice that obviously $z(x) \neq 0$ in Ω_5. Therefore, $z\, \Delta z \geq 0$ a.e. in $\Omega_5 \setminus \Omega_1$.

Assume now that f has a constant sign in Ω. The previous remark then shows that $y^*(x) = -\tau$ can occur only on Ω_1, which has zero measure. Moreover, again by virtue of the maximum principle, $z\, f < 0$ a.e. in Ω. Consequently, $z\, \Delta z < 0$ a.e. in Ω, which contradicts (3.4.24). Hence, also meas(Ω_5) $= 0$. We thus have proved the following statement.

Corollary 3.4.7 *Suppose that the above assumptions are fulfilled and that f has constant sign in Ω. Then one of the two possibilities $u^*(x) = m$, $u^*(x) = M$ is valid for a.e. $x \in \Omega$.*

Remark. Corollary 3.4.7 is a classical bang-bang result due to Sprekels and Tiba [1998], while Corollary 3.4.5 gives a generalized bang-bang property, as defined in Tröltzsch [1984]. In the optimization of plates, the occurrence of a bang-bang solution was for the first time noticed in the numerical experiments reported by Arnăutu et al. [2000].

3.4.2 Clamped Plates and Control Variational Methods

We begin with a simple example that offers a different viewpoint in connection with the problems discussed in the previous paragraph.

Example 3.4.8 Using the same notation as in (3.4.2), we introduce, for fixed $l \in U_{ad}$ and $z \in W^{2,p}(\Omega)$, a distributed control problem with controls $h \in L^2(\Omega)$ and corresponding states $y = y_h \in H^2(\Omega) \cap H_0^1(\Omega)$, namely,

$$\text{Min} \left\{ \frac{1}{2} \int_\Omega l(x)\, h^2(x)\, dx \right\}, \tag{3.4.25}$$

subject to

$$\Delta y = l\, z + l\, h \quad \text{in } \Omega, \tag{3.4.26}$$

$$y = 0 \quad \text{on } \partial\Omega. \tag{3.4.27}$$

It is clear that (3.4.25)–(3.4.27) has the trivial solution $h^* \equiv 0$, where the corresponding state y_{h^*} solves (3.4.2), (3.4.3). Since $l \in U_{ad}$, h^* is unique. In fact, (3.4.25)–(3.4.27) is just a reformulation of the standard Dirichlet principle (energy minimization) for the simply supported plate. Indeed, integration by parts and the definition of z yield

$$\begin{aligned} &\operatorname*{Min}_{y \in H^2(\Omega) \cap H_0^1(\Omega)} \left\{ \frac{1}{2} \int_\Omega \frac{1}{l(x)} \left(\Delta y(x) - l(x)\, z(x) \right)^2 dx \right\} \\ &= \operatorname*{Min}_{y \in H^2(\Omega) \cap H_0^1(\Omega)} \left\{ \frac{1}{2} \int_\Omega u^3(x) (\Delta y(x))^2 \, dx - \int_\Omega y(x)\, f(x)\, dx \right\} \\ &\quad + \frac{1}{2} \int_\Omega l(x)\, z^2(x)\, dx, \end{aligned} \tag{3.4.28}$$

and the last integral in (3.4.28) does not depend on y. The right-hand side in (3.4.28) is the usual energy functional used in the variational formulation of the simply supported plate.

While this example just demonstrates how variational principles may be interpreted via control theory, there are other cases to be considered here, and in Section 6.1 of Chapter 6, that will benefit in an essential manner from this type of control approach.

We now analyze the fourth-order inhomogeneous boundary value problem

$$\Delta \left(u^3 \, \Delta y \right) = f \quad \text{in } \Omega, \tag{3.4.29}$$

$$y = \zeta \quad \text{on } \partial\Omega, \tag{3.4.30}$$

$$\frac{\partial y}{\partial n} = \frac{\partial \zeta}{\partial n} \quad \text{on } \partial\Omega, \tag{3.4.31}$$

where $f \in L^2(\Omega)$, $u \in U_{ad}$, and $\zeta \in H^2(\Omega)$ are given. Notice that for $\zeta = 0$ we obtain the usual boundary conditions for a clamped plate. To give (3.4.31) a proper meaning, we will henceforth assume that Ω is at least of class $C^{1,1}$, so that the trace theorem (cf. Theorem A2.1(ii) in Appendix 2) can be applied.

As above in Example 3.4.8, we now introduce some optimal control problems that will turn out to be equivalent to (3.4.29)–(3.4.31). These alternative formulations will enable us to gain new insights in corresponding thickness optimization problems.

The following constrained control problem is associated with (3.4.29)–(3.4.31):

$$\operatorname*{Min}_{h \in L^2(\Omega)} \left\{ \frac{1}{2} \int_\Omega l(x)\, h^2(x)\, dx \right\}, \tag{3.4.32}$$

subject to the state system

$$\Delta y = l\, h + l\, z \quad \text{in } \Omega, \tag{3.4.33}$$

$$y = \zeta \quad \text{on } \partial\Omega, \tag{3.4.34}$$

and to the state constraint

$$\frac{\partial y}{\partial n} = \frac{\partial \zeta}{\partial n} \quad \text{on } \partial\Omega. \tag{3.4.35}$$

Apparently, the pair $[y, h] = [\zeta, u^3 (\Delta\zeta - l\,z)]$ is admissible for the problem (3.4.32)–(3.4.35). Moreover, (3.4.32)–(3.4.35) has a unique optimal pair $[y^*, h^*] \in L^2(\Omega) \times H^2(\Omega)$. Indeed, if $\{[y_n, h_n]\} \subset H^2(\Omega) \times L^2(\Omega)$ is a minimizing sequence, then $l \in U_{ad}$ implies that $\{|h_n|_{L^2(\Omega)}\}$ is bounded, whence, in view of (3.4.33) and (3.4.34), the boundedness of $\{|y_n|_{H^2(\Omega)}\}$ follows. Thus, using the trace theorem, we may without loss of generality assume that

$$y_n \to \hat{y} \text{ weakly in } H^2(\Omega), \quad h_n \to \hat{h} \text{ weakly in } L^2(\Omega),$$

where $\hat{y}$ satisfies (3.4.33)–(3.4.35). Hence, $[\hat{y}, \hat{h}]$ is admissible, and the weak lower semicontinuity of the cost functional shows that $[y^*, h^*] = [\hat{y}, \hat{h}]$ is optimal. The uniqueness follows from the strict convexity of the cost functional.

It is obvious that $y - \zeta \in H_0^2(\Omega)$ for any admissible state y, and thus

$$\left\{h : h \in l^{-1}\,\Delta\left(H_0^2(\Omega)\right) + l^{-1}\,\Delta\zeta - z\right\} \subset L^2(\Omega) \tag{3.4.36}$$

provides a complete description of the set of admissible controls. Thanks to (3.4.36), the control problem (3.4.32)–(3.4.35) can be reformulated as a mathematical programming problem, namely as

$$\operatorname*{Min}_{s \in l^{-1}\Delta\left(H_0^2(\Omega)\right)} \left\{\frac{1}{2}\int_\Omega l(x)\,\left(s + l^{-1}\,\Delta\zeta - z\right)^2(x)\,dx\right\}. \tag{3.4.37}$$

We introduce the new unknown

$$\xi = l^{\frac{1}{2}}\,s \in l^{-\frac{1}{2}}\,\Delta\left(H_0^2(\Omega)\right).$$

Then a simple transformation yields the solution of (3.4.37), namely

$$s^* = l^{-\frac{1}{2}}\,\xi^*, \quad \text{with } \xi^* = P_Z\left(l^{\frac{1}{2}}\,z - l^{-\frac{1}{2}}\,\Delta\zeta\right), \tag{3.4.38}$$

where P_Z denotes the orthogonal projection operator onto the closed linear subspace $Z = l^{-\frac{1}{2}}\Delta\left(H_0^2(\Omega)\right)$ of $L^2(\Omega)$ with respect to the inner product $(\cdot,\cdot)_{L^2(\Omega)}$. From (3.4.38) it follows that

$$h^* = s^* + l^{-1}\Delta\zeta - z = l^{-\frac{1}{2}}\,P_{Z^\perp}\left(l^{-\frac{1}{2}}\Delta\zeta - l^{\frac{1}{2}}z\right), \tag{3.4.39}$$

where $P_{Z^\perp}$ is the orthogonal projection operator onto the orthogonal complement $Z^\perp$ of Z with respect to the inner product in $L^2(\Omega)$. We have the following characterization for the optimal control h^*:

Theorem 3.4.9 *h^* is the optimal state of the unconstrained boundary control problem*

$$\operatorname*{Min}_{\tau \in H^{-\frac{1}{2}}(\partial\Omega)} \left\{ \frac{1}{2} \int_\Omega l(x) \left(h + z - l^{-1}\Delta\zeta\right)^2 (x)\, dx \right\}, \tag{3.4.40}$$

$$\Delta h = 0 \quad in\ \Omega, \tag{3.4.41}$$

$$h = \tau \quad on\ \partial\Omega, \tag{3.4.42}$$

where the solution of (3.4.41), (3.4.42) has to be understood in the transposition sense (cf. Appendix 2).

Proof. It is a simple exercise to verify that $Z^\perp = l^{\frac{1}{2}} \{\Delta (H_0^2(\Omega))\}^\perp$ and

$$\left\{\Delta \left(H_0^2(\Omega)\right)\right\}^\perp = \left\{w \in L^2(\Omega) : \Delta w = 0 \text{ in } \mathcal{D}'(\Omega)\right\}. \tag{3.4.43}$$

Moreover, owing to (3.4.39), h^* solves

$$\operatorname*{Min}_{h \in \{\Delta(H_0^2(\Omega))\}^\perp} \left\{ \frac{1}{2} \int_\Omega l(x) \left(h + z - l^{-1}\Delta\zeta\right)^2 (x)\, dx \right\}. \tag{3.4.44}$$

Since the relations (3.4.43)–(3.4.44) may be reformulated as (3.4.40)–(3.4.42), the assertion is proved. □

Remark. The problem (3.4.40)–(3.4.42) does not have constraints. The relations (3.4.33) and (3.4.39) provide an explicit reduction of the fourth-order problem (3.4.29)–(3.4.31) to second-order elliptic equations. One may also check that (3.4.32) is a reformulation of the usual energy minimization approach; compare Example 3.4.8.

Theorem 3.4.10 *There is a one-to-one correspondence between the solutions to the three problems (3.4.40)–(3.4.42), (3.4.32)–(3.4.35), and (3.4.29)–(3.4.31).*

Proof. By Theorem 3.4.9, the problems (3.4.40)–(3.4.42) and (3.4.32)–(3.4.35) are equivalent. It thus suffices to show that the solution y of (3.4.29)–(3.4.31) can be recovered from the solution h^* to the problem (3.4.40)–(3.4.42) (or, equivalently, to the problem (3.4.44)).

To this end, take any $k \in \{\Delta (H_0^2(\Omega))\}^\perp$, and consider variations of the form $h^* + \lambda k$, $\lambda \in \mathbf{R}$. Letting $\lambda \to 0$, and recalling that h^* solves (3.4.44), we readily see that

$$\int_\Omega l(x) \left(h^* + z - l^{-1}\Delta\zeta\right)(x)\, k(x)\, dx = 0 \quad \forall\, k \in \left\{\Delta \left(H_0^2(\Omega)\right)\right\}^\perp. \tag{3.4.45}$$

We define the adjoint system to the problem (3.4.40)–(3.4.42) by

$$\Delta p^* = l\left(h^* + z - l^{-1}\Delta\zeta\right) \quad \text{in } \Omega, \tag{3.4.46}$$

$$p^* = 0 \quad \text{on } \partial\Omega. \tag{3.4.47}$$

Clearly, $p^* \in H^2(\Omega) \cap H_0^1(\Omega)$, and we have, by the definition of the transposition solution,

$$0 = \int_\Omega k(x)\,\Delta p^*(x)\,dx = \int_{\partial\Omega} k\,\frac{\partial p^*}{\partial n}\,d\sigma = \left(v, \frac{\partial p^*}{\partial n}\right)_{H^{-\frac{1}{2}}(\partial\Omega)\times H^{\frac{1}{2}}\partial\Omega)}, \tag{3.4.48}$$

where $v \in H^{-\frac{1}{2}}(\partial\Omega)$ denotes the "trace" of k on $\partial\Omega$ in the sense of Lions [1968, §4.2]. Since $v \in H^{-\frac{1}{2}}(\partial\Omega)$ is arbitrary, it follows from (3.4.48) that

$$\frac{\partial p^*}{\partial n} = 0 \quad \text{on } \partial\Omega. \tag{3.4.49}$$

A straightforward calculation, based on the definition of l, z, ζ, then shows that $p^* + \zeta$ coincides with the (unique) solution to (3.4.29)–(3.4.31), which concludes the proof of the assertion. □

Remark. Related arguments, using a penalization of (3.4.35) (or, equivalently, of (3.4.31)) have been employed by Sprekels and Tiba [1999a], [1998/1999], and by Arnăutu, Langmach, Sprekels, and Tiba [2000]. The constraint (3.4.35) in the problem (3.4.32)–(3.4.34) is affine, but not finite-dimensional. Therefore, the problem (3.4.40)–(3.4.42) remains infinite-dimensional and thus cannot be solved explicitly. The relations (3.4.33)–(3.4.35), and (3.4.41), (3.4.42), may be interpreted as the optimality system for either the problem (3.4.32) or (3.4.40), and they are, in turn, equivalent to (3.4.29)–(3.4.31).

Remark. The transformations used here and in §2.2.2 in Chapter 2 have the property that the solution y remains unchanged. Some examples of this type appear in the book of Haftka, Kamat, and Gürdal [1990] and are called there *resizing rules*. For the case of beams, we refer the reader in this respect to Sprekels and Tiba [1999b].

We shall now discuss the optimization problems associated with (3.4.29)–(3.4.31) under the assumption (see (3.4.4)) that

$$0 \le m \le u(x) \le M \quad \text{a.e. in } \Omega.$$

As a first step, we analyze the differentiability properties of the mapping $l \mapsto y$ defined by (3.4.32)–(3.4.35). Notice that although the relation $l = u^{-3}$ between

u and l is very simple, no continuity properties are valid in the weak* topology of $L^\infty(\Omega)$, for instance. The reformulations (3.4.32)–(3.4.35) and (3.4.40)–(3.4.42) have the advantage that they introduce l as the main unknown to remove this inconvenience. For shape optimization problems it suffices to analyze the behavior with respect to $l \in L^\infty(\Omega)$ and then just to transform the result into the language of $u \in L^\infty(\Omega)$, the thickness of the plate. Indeed, as explained above, one immediately recovers (3.4.29)–(3.4.31) from (3.4.33)–(3.4.35) and (3.4.41), and vice versa.

Theorem 3.4.11 *The mappings $l \mapsto y$ and $l \mapsto h$ are Fréchet differentiable for $l \in U_{ad} \subset L^\infty(\Omega)$ into $H^2(\Omega)$ and $L^2(\Omega)$, respectively, and the corresponding pair of directional derivatives at l in the direction v, denoted by $[\bar{y}, \bar{h}] \in H_0^2(\Omega) \times L^2(\Omega)$, is given as the unique solution pair to the system*

$$\Delta \bar{y} = l\,\bar{h} + v\,(h+z) \quad in\ \Omega, \tag{3.4.50}$$

$$\bar{y} = \frac{\partial \bar{y}}{\partial n} = 0 \quad on\ \partial\Omega, \tag{3.4.51}$$

$$\Delta \bar{h} = 0 \quad in\ \mathcal{D}'(\Omega). \tag{3.4.52}$$

Proof. Let $v \in L^\infty(\Omega)$ be arbitrary, and let $l \in L^\infty(\Omega)$ satisfy (3.4.4). Then there is some sufficiently small $\lambda_0 > 0$ such that

$$l + \lambda\, v \ge \frac{1}{2} M^{-3} > 0 \quad \text{a.e. in } \Omega, \quad \text{whenever } 0 < |\lambda| \le \lambda_0.$$

We denote by y_λ, h_λ the mappings associated with $l + \lambda\, v$ through the system (3.4.33)–(3.4.35), (3.4.41). We then have

$$\begin{aligned}
\Delta\left(\frac{y_\lambda - y}{\lambda}\right) &= (l + \lambda\, v)\,\frac{h_\lambda - h}{\lambda} + v\,(h+z) \quad \text{a.e. in } \Omega,\\
\frac{y_\lambda - y}{\lambda} &= 0, \quad \frac{\partial}{\partial n}\left(\frac{y_\lambda - y}{\lambda}\right) = 0 \quad \text{on } \partial\Omega,\\
\Delta\left(\frac{h_\lambda - h}{\lambda}\right) &= 0 \quad \text{in } \mathcal{D}'(\Omega).
\end{aligned}$$

Relation (3.4.43) shows that $\lambda^{-1}(h_\lambda - h) \in \{\Delta\,(H_0^2(\Omega))\}^\perp$, while from above, we get $\lambda^{-1}(y_\lambda - y) \in H_0^2(\Omega)$. Consequently,

$$\int_\Omega \Delta\left(\frac{y_\lambda - y}{\lambda}\right)\frac{h_\lambda - h}{\lambda}\, dx = 0.$$

Multiplying the equation for $\frac{y_\lambda - y}{\lambda}$ by $\frac{h_\lambda - h}{\lambda}$, we can infer that $\left\{\frac{h_\lambda - h}{\lambda}\right\}$ is bounded in $L^2(\Omega)$. In fact, we have

$$\left|\frac{h_\lambda - h}{\lambda}\right|_{L^2(\Omega)} \le 2\,M^3\,|v|_{L^\infty(\Omega)}\,|h+z|_{L^2(\Omega)}\,.$$

Then it follows immediately from the equation for $\frac{y_\lambda - y}{\lambda}$ that $\left\{\frac{y_\lambda - y}{\lambda}\right\}$ is bounded in $H_0^2(\Omega)$. Hence, there is some sequence $\lambda_n \to 0$ such that

$$\frac{y_{\lambda_n} - y}{\lambda_n} \to \hat{y} \text{ weakly in } H_0^2(\Omega), \qquad \frac{h_{\lambda_n} - h}{\lambda_n} \to \hat{h} \text{ weakly in } L^2(\Omega).$$

Obviously, $[\hat{y}, \hat{h}]$ solves (3.4.50)–(3.4.52). Now suppose that $[\tilde{y}, \tilde{h}]$ is a second solution. Then $\xi = \hat{y} - \tilde{y} \in H_0^2(\Omega)$ satisfies $\Delta\xi = l\,\eta$ in Ω, and $\eta \in \{\Delta\,(H_0^2(\Omega))\}^\perp$. Hence, dividing by l and testing by $\Delta\xi$, we find that

$$\int_\Omega l(x)^{-1}\,|\Delta\xi(x)|^2\,dx = \int_\Omega \eta(x)\,\Delta\xi(x)\,dx = 0,$$

whence $\xi = 0$ a.e. in Ω follows. This entails $\eta = 0$ a.e. in Ω, since $l \in U_{ad}$. This shows the uniqueness of the solution, and the above convergences are generally true for $\lambda \to 0$ and not only for the selected sequence.

Now that the directional derivatives have been identified, it remains to show the Fréchet differentiability at $l \in U_{ad} \subset L^\infty(\Omega)$. To this end, let $v \in L^\infty(\Omega)$ satisfy

$$|v|_{L^\infty(\Omega)} \le \frac{1}{2}\,M^{-3}, \quad \text{so that } v + l \ge \frac{1}{2}\,M^{-3} \text{ a.e. in } \Omega.$$

We denote by $[y_{l+v}, h_{l+v}]$ and $[y_l, h_l]$, respectively, the (unique) solutions of (3.4.33)–(3.4.35), (3.4.41) associated with $l+v$ and l. Subtracting the equations for y_l and for $\bar{y}$ from that for y_{l+v}, we obtain that $y_{l+v} - y_l - \bar{y} \in H_0^2(\Omega)$ and $h_{l+v} - h_l - \bar{h} \in \{\Delta\,(H_0^2(\Omega))\}^\perp$ satisfy the identity

$$\Delta(y_{l+v} - y_l - \bar{y}) = (l+v)\,h_{l+v} - l\,h_l - l\,\bar{h} - v\,h_l \quad \text{in } \Omega.$$

Testing by $h_{l+v} - h_l - \bar{h}$, we find that

$$0 = \int_\Omega \left[(l+v)\,h_{l+v} - l\,h_l - l\,\bar{h} - v\,h_l\right]\left[h_{l+v} - h_l - \bar{h}\right] dx,$$

whence

$$\begin{aligned}
M^{-3}\int_\Omega \left|h_{l+v} - h_l - \bar{h}\right|^2 dx &\le \int_\Omega l\,\left(h_{l+v} - h_l - \bar{h}\right)^2 dx \\
&= -\int_\Omega v\,(h_{l+v} - h_l)\left(h_{l+v} - h_l - \bar{h}\right) dx \\
&\le |v|_{L^\infty(\Omega)}\,|h_{l+v} - h_l|_{L^2(\Omega)}\left|h_{l+v} - h_l - \bar{h}\right|_{L^2(\Omega)}.
\end{aligned}$$

Next, we show that

$$|h_{l+v} - h_l|_{L^2(\Omega)} \leq \widehat{C}\, |v|_{L^\infty(\Omega)},$$

with a constant $\widehat{C} > 0$ that does not depend on v. Once this has been shown, the Fréchet differentiability follows immediately from the previous estimate.

To this end, we subtract one equation (3.4.33) for y_{l+v} and another for y_l from each other, multiply by $h_{l+v} - h_l$, and use the orthogonality once more to see that

$$0 = \int_\Omega (l+v)\, |h_{l+v} - h_l|^2\, dx + \int_\Omega v\,(h_l + z)\,(h_{l+v} - h_l)\, dx,$$

whence, using the Schwarz inequality and the fact that $l + v \geq \frac{1}{2} M^{-3}$ a.e. in Ω,

$$|h_{l+v} - h_l|^2_{L^2(\Omega)} \leq 2\, M^3\, |v|_{L^\infty(\Omega)}\, |h_l + z|_{L^2(\Omega)}\, |h_{l+v} - h_l|_{L^2(\Omega)}.$$

This concludes the proof of the assertion. □

Remark. Equations (3.4.50), (3.4.52) may be formally rewritten as the fourth-order equation

$$\Delta(u^3 \Delta \bar{y}) = \Delta(u^3 v(h + z)) \quad \text{in } \Omega,$$

together with the boundary conditions (3.4.51). It should be noted that the right-hand side is nonsmooth, since $u \in L^\infty(\Omega)$, $v \in L^\infty(\Omega)$.

Example 3.4.12 We study some optimal shape design problems in which the minimization parameter is the thickness $u \in L^\infty(\Omega)$ or, equivalently, $l \in L^\infty(\Omega)$:

$$\underset{l \in U_{ad}}{\text{Min}} \left\{ \frac{1}{2} \int_\Omega y^2(x) dx \right\}, \tag{3.4.53}$$

subject to (3.4.33)–(3.4.35), (3.4.41), where U_{ad} is given by (3.4.4). The existence of at least one optimal pair $[y^*, l^*]$ for this problem is a direct consequence of Theorem 2.2.7 in Chapter 2. We have the following result.

Proposition 3.4.13 *If $f \neq 0$ a.e. in Ω, then $u^*(x) \in \{m, M\}$ a.e. in Ω, where $u^*(x) = l^*(x)^{-\frac{1}{3}}$ is the optimal thickness for the problem (3.4.53).*

Proof. We introduce the adjoint system, namely

$$\Delta p = y^* \quad \text{in } \Omega, \qquad \Delta q = l^*\, p \quad \text{in } \Omega, \qquad q = \frac{\partial q}{\partial n} = 0 \quad \text{on } \partial\Omega.$$

Notice that the adjoint system has a unique solution $[p, q] \in L^2(\Omega) \times H_0^2(\Omega)$; indeed, the clamped plate problem

$$\Delta\left(\frac{1}{l^*}\Delta q\right) = y^* \text{ in } \Omega, \qquad q = \frac{\partial q}{\partial n} = 0 \text{ on } \partial\Omega,$$

has a unique weak solution $q \in H_0^2(\Omega)$, and the claim follows by putting $p = l^{*-1}\Delta q \in L^2(\Omega)$.

To continue, observe that for any $\lambda \in [0,1]$ and $w \in U_{ad}$,

$$l_\lambda = l^* + \lambda v \in U_{ad}, \text{ where } v = w - l^*.$$

Now denote by $[y_\lambda, h_\lambda]$ the solution pair of (3.4.33)–(3.4.35), (3.4.41) associated with l_λ. Since $l_\lambda \to l^*$ strongly in $L^\infty(\Omega)$, we have (compare the proof of Theorem 3.4.11) that $y_\lambda \to y^*$ strongly in $L^2(\Omega)$ as $\lambda \searrow 0$. Therefore, multiplying by $\bar{y} \in H_0^2(\Omega)$ in the first equation, and integrating by parts (which may be justified by regularizing p), we obtain

$$\begin{aligned}
0 &\leq \lim_{\lambda \searrow 0} \frac{1}{2\lambda}\int_\Omega \left(y_\lambda^2 - y^{*2}\right) dx = \int_\Omega y^*\,\bar{y}\,dx = \int_\Omega \Delta p\,\bar{y}\,dx = \int_\Omega p\,\Delta\bar{y}\,dx \\
&= \int_\Omega p\,l^*\,\bar{h}\,dx + \int_\Omega p\,v\,(h^* + z)\,dx = \int_\Omega \Delta q\,\bar{h}\,dx + \int_\Omega p\,v\,(h^* + z)\,dx \\
&= \int_\Omega p\,v\,(h^* + z)\,dx,
\end{aligned}$$

since $\bar{h}$ is orthogonal to $\Delta\left(H_0^2(\Omega)\right)$ in $L^2(\Omega)$. The Pontryagin maximum principle for the problem (3.4.53) therefore reads

$$0 \leq \int_\Omega p(x)(w(x) - l^*(x))\,(h^*(x) + z(x))\,dx \quad \forall\, w \in U_{ad},$$

or, equivalently,

$$-p\,(h^* + z) \in \partial I_{U_{ad}}(l^*),$$

where $\partial I_{U_{ad}}$ is the subdifferential of the indicator function $I_{U_{ad}}$ of U_{ad} in $L^\infty(\Omega)$.

Since $f \neq 0$ a.e. in Ω and $\Delta(h^* + z) = f$, we have $h^* + z \neq 0$ a.e. in Ω, by the interior regularity properties of h^* and the maximal regularity of z (see Brézis [1983, p. 195]). Indeed, otherwise we would have $\Delta(h^* + z) = 0$ on a subset of positive measure, which contradicts the assumption on f.

Then, owing to (3.4.33), $\Delta y^* \neq 0$ a.e. in Ω, and hence $y^* \neq 0$ a.e. in Ω, whence also $\Delta p \neq 0$ a.e. in Ω, and thus $p \neq 0$ a.e. in Ω. In consequence,

$p(h^* + z) \neq 0$ a.e. in Ω. The Pontryagin maximum principle and (3.4.53) then yield that

$$l^*(x) \in \left\{M^{-3}, m^{-3}\right\} \text{ a.e. in } \Omega,$$

since $\partial I_{U_{ad}}$ differs from $\{0\}$ only in the endpoints of the constraints interval. With this, the proof is finished. □

Remark. Proposition 3.4.13 is a bang-bang result for the problem (3.4.53). A more realistic example involving pointwise state constraints has been investigated in Sprekels and Tiba [2003].

3.4.2.1 Penalization and Variational Inequalities

We indicate now a variant of the control problem (3.4.32)–(3.4.35) in which the state constraint (3.4.35) is penalized. We assume that $l \in U_{ad}$, $\zeta = 0$, and consider for $\varepsilon > 0$ the problem

$$\text{Min}\left\{\frac{1}{2\varepsilon}\int_{\partial\Omega}\left(\frac{\partial y}{\partial n}(\sigma)\right)^2 d\sigma + \frac{1}{2}\int_\Omega l(x)\,h^2(x)\,dx\right\}, \tag{3.4.54}$$

$$\Delta y = l\,(h+z) \quad \text{in } \Omega, \tag{3.4.55}$$

$$y = 0 \quad \text{on } \partial\Omega. \tag{3.4.56}$$

It is well known in the penalization approach (see also §3.2.2 above) that the unique optimal pair $[y_\varepsilon, h_\varepsilon] \in H^2(\Omega) \times L^2(\Omega)$ of the problem (3.4.54)–(3.4.56) will approximate the solution of the constrained problem (3.4.32)–(3.4.35) for $\varepsilon \searrow 0$ and, implicitly, the deflection of the clamped plate. Here, we will investigate a more complex situation that corresponds to the case of a partially clamped plate respecting some unilateral conditions of obstacle type on the "free" part of the boundary.

That is, we shall indicate a control approach as in Example 3.4.8, applicable to a variational inequality associated with a clamped plate. As in the classical variational principle, the minimization of energy (of the cost functional) has to be performed on a convex subset, and not over the entire space. Correspondingly, we will add to (3.4.54)–(3.4.56) appropriate constraints that are not penalized in the cost functional.

We assume that the (smooth) boundary satisfies $\partial\Omega = \Gamma_1 \cup \Gamma_2$, where $\overline{\Gamma}_1 \cap \overline{\Gamma}_2 = \emptyset$ and $\text{meas}(\Gamma_i) > 0$, $i = 1, 2$. A typical example for a domain Ω satisfying these conditions is an annulus in $\mathbf{R}^2$.

We introduce a supplementary control parameter $v \in L^2(\Gamma_2)$ by replacing the boundary condition (3.4.56) by

$$y = 0 \quad \text{on } \Gamma_1, \tag{3.4.57}$$

$$y = v \quad \text{on } \Gamma_2, \tag{3.4.58}$$

with the constraint

$$v \in V_{ad} = \Big\{ w \in L^\infty(\Gamma_2) \,:\, -r \le w \le r \quad \text{a.e. on } \Gamma_2 \Big\}, \quad \text{for some } r > 0. \tag{3.4.59}$$

Relation (3.4.59) may either be interpreted as a control constraint or, in view of (3.4.58), as a state constraint.

The solution of the boundary value problem (3.4.55), (3.4.57), (3.4.58) has to be understood in the transposition sense (see Example A2.7 in Appendix 2); that is,

$$\int_\Omega y(x)\,\Delta\varphi(x)\,dx = \int_\Omega l(x)\,(z(x)+h(x))\,\varphi(x)\,dx + \int_{\Gamma_2} v(\sigma)\,\frac{\partial\varphi}{\partial n}(\sigma)\,d\sigma, \tag{$*$}$$

for all $\varphi \in H^2(\Omega) \cap H_0^1(\Omega)$. We recall from Appendix 2 that the mapping $T : \psi \mapsto \varphi = T\psi$, where φ is the unique solution to

$$-\Delta\varphi = \psi \text{ in } \Omega, \quad \varphi = 0 \text{ on } \partial\Omega,$$

is an isomorphism between the spaces $L^2(\Omega)$ and $H^2(\Omega) \cap H_0^1(\Omega)$; moreover, we only have $y \in L^2(\Omega)$ (actually, it is even true that $y \in H^{1/2}(\Omega)$, but this additional regularity will not be needed in the following). However, we have $y \in H^2(\mathcal{V})$ for some open neighborhood $\mathcal{V}$ of Γ_1 (cf. Necas [1967]). Hence the following cost functional is well-defined (in fact, the trace theorem implies that $\frac{\partial y}{\partial n} \in H^{\frac{1}{2}}(\Gamma_1)$):

$$\operatorname{Min}\left\{ \frac{1}{2\,\varepsilon} \int_{\Gamma_1} \left(\frac{\partial y}{\partial n}(\sigma) \right)^2 d\sigma + \frac{1}{2} \int_\Omega l(x)\,h^2(x)dx \right\}, \quad \text{for } \varepsilon > 0. \tag{3.4.60}$$

Let us first show that the control problem (3.4.55), (3.4.57)–(3.4.60) admits at least one optimal triple $[y_\varepsilon, h_\varepsilon, v_\varepsilon]$ in $L^2(\Omega) \times L^2(\Omega) \times L^2(\Gamma_2)$. To this end, notice first that the triple $[y, h, 0]$, where $[y, h]$ is the unique solution to (3.4.55), (3.4.56), is admissible.

Next, let $\{[y_n, h_n, v_n]\} \subset L^2(\Omega) \times L^2(\Omega) \times L^2(\Gamma_2)$ be a minimizing sequence. Then, obviously, $\{h_n\}$ and $\{\partial y_n/\partial n\}$ are bounded in $L^2(\Omega)$ and $L^2(\Gamma_1)$, respectively, and $\{v_n\}$ is bounded in $L^\infty(\Gamma_2)$. Moreover, it follows from $(*)$ above, applied to $[y_n, h_n, v_n]$, $n \in \mathbf{N}$, that

$$\int_\Omega y_n\,\Delta\varphi\,dx = \int_\Omega l\,(z+h_n))\,\varphi\,dx + \int_{\Gamma_2} v_n\,\frac{\partial\varphi}{\partial n}\,d\sigma \quad \forall\,\varphi \in H^2(\Omega) \cap H_0^1(\Omega).$$

Hence, using the isomorphism T, there are constants $C_1 > 0$, $C_2 > 0$, independent of ψ, such that

$$\left|\int_\Omega y_n\,\psi\,dx\right| = \left|\int_\Omega y_n\,\Delta\varphi\,dx\right| \le C_1\,|\varphi|_{H^2(\Omega)} \le C_2\,|\psi|_{L^2(\Omega)} \quad \forall\,\psi\in L^2(\Omega),$$

which proves that $\{y_n\}$ is bounded in $L^2(\Omega)$. Also, it obviously follows that $\Delta y_n = l\,(z+h_n)$ in the sense of distributions, so that $\{\Delta y_n\}$ is also bounded in $L^2(\Omega)$.

From the above consideration it follows that we may without loss of generality assume that

$$h_n \to \hat{h} \text{ weakly in } L^2(\Omega), \qquad \frac{\partial y_n}{\partial n} \to \hat{\zeta} \text{ weakly in } L^2(\Gamma_1),$$

$$v_n \to \hat{v} \text{ weakly* in } L^\infty(\Gamma_2), \quad y_n \to \hat{y} \text{ weakly in } L^2(\Omega).$$

Obviously, $\hat{v}$ obeys (3.4.59). Also, passing to the limit as $n\to\infty$ in the above relation, we find that

$$\int_\Omega \hat{y}\,\Delta\varphi\,dx = \int_\Omega l\,(z+\hat{h})\,\varphi\,dx + \int_{\Gamma_2} \hat{v}\,\frac{\partial\varphi}{\partial n}\,d\sigma \quad \forall\,\varphi\in H^2(\Omega)\cap H_0^1(\Omega),$$

that is, $\hat{y}$ is the transposition solution of

$$\Delta\hat{y} = l\,(z+\hat{h}) \text{ in } \Omega, \qquad \hat{y} = 0 \text{ on } \Gamma_1, \qquad \hat{y} = \hat{v} \text{ on } \Gamma_2.$$

In particular, we have $\Delta\hat{y} = l\,(z+\hat{h})$ in the sense of distributions, that is, $\Delta\hat{y}\in L^2(\Omega)$. Summarizing the convergence properties of $\{y_n\}$, we can apply Theorem 6.5 in Chapter II in Lions and Magenes [1968] to conclude that

$$\frac{\partial y_n}{\partial n} \to \frac{\partial\hat{y}}{\partial n} \quad \text{weakly in } H^{-\frac{3}{2}}(\partial\Omega).$$

But then we must have $\hat{\zeta} = \frac{\partial\hat{y}}{\partial n}$, by the uniqueness of the limit, and thus

$$\frac{\partial y_n}{\partial n} \to \frac{\partial\hat{y}}{\partial n} \quad \text{weakly in } L^2(\partial\Omega).$$

Using the lower semicontinuity of the cost functional, we then conclude that the triple $[y_\varepsilon, h_\varepsilon, v_\varepsilon] = [\hat{y}, \hat{h}, \hat{v}]$ is optimal.

Proposition 3.4.14 *The optimal triple* $[y_\varepsilon, h_\varepsilon, v_\varepsilon]\in L^2(\Omega)\times L^2(\Omega)\times L^\infty(\Gamma_2)$ *is unique.*

Proof. Suppose that $[\hat{y}_\varepsilon, \hat{h}_\varepsilon, \hat{v}_\varepsilon]$ is another optimal triple for the problem (3.4.55), (3.4.57)–(3.4.60). Then, thanks to the strict convexity of the norm, we get

$$\frac{\partial y_\varepsilon}{\partial n} = \frac{\partial \hat{y}_\varepsilon}{\partial n} \quad \text{in } \Gamma_1, \tag{3.4.61}$$

$$h_\varepsilon = \hat{h}_\varepsilon \quad \text{in } \Omega. \tag{3.4.62}$$

Then, (3.4.55), (3.4.57), (3.4.61), (3.4.62) show that y_ε and $\hat{y}_\varepsilon$ are solutions of the Cauchy problem for elliptic equations, and by the unique solvability of the Cauchy problem, it follows that $y_\varepsilon = \hat{y}_\varepsilon$ a.e. in Ω. Then also $v_\varepsilon = \hat{v}_\varepsilon$ in $L^2(\Gamma_2)$, which follows from the definition of the transposition solution, and from the fact that (Lions [1968]; see Theorem 5.2.12)

$$\left\{ \frac{\partial \varphi}{\partial n}|_{\Gamma_2} : \varphi \in H^2(\Omega) \cap H_0^1(\Omega) \right\}$$

is dense in $L^2(\Gamma_2)$. □

Proposition 3.4.15 *The set $\{[y_\varepsilon, h_\varepsilon, v_\varepsilon]\}_{\varepsilon>0}$ is bounded in $L^2(\Omega) \times L^2(\Omega) \times L^\infty(\Gamma_2)$. Moreover,*

$$\frac{\partial y_\varepsilon}{\partial n} \to 0, \text{ strongly in } L^2(\Gamma_1) \text{ as } \varepsilon \searrow 0.$$

Proof. The triple $[0, -z, 0]$ is admissible for any $\varepsilon > 0$, and thus

$$\frac{1}{2\varepsilon} \int_{\Gamma_1} \left(\frac{\partial y_\varepsilon}{\partial n} \right)^2 d\sigma + \frac{1}{2} \int_\Omega l\, h_\varepsilon^2 \, dx \le \frac{1}{2} \int_\Omega l\, z^2 \, dx \quad \forall \varepsilon > 0.$$

Hence, since $l \in U_{ad}$, the set $\{h_\varepsilon\}_{\varepsilon>0}$ is bounded in $L^2(\Omega)$, and

$$\frac{\partial y_\varepsilon}{\partial n} \to 0 \text{ strongly in } L^2(\Gamma_1).$$

By (3.4.59), $\{v_\varepsilon\}_{\varepsilon>0}$ is bounded in $L^\infty(\Gamma_2)$. Therefore, we may argue as above in the existence proof, invoking the relation (*) applied to $[y_\varepsilon, h_\varepsilon, v_\varepsilon]$, to conclude that $\{y_\varepsilon\}_{\varepsilon>0}$ is bounded in $L^2(\Omega)$. □

Proposition 3.4.16 *The optimal triple $[y_\varepsilon, h_\varepsilon, v_\varepsilon]$ satisfies the first-order optimality conditions given by (3.4.55), (3.4.57), (3.4.58), and*

$$h_\varepsilon = -p_\varepsilon \quad \text{a.e. in } \Omega,$$

$$\int_{\Gamma_2} \frac{\partial p_\varepsilon}{\partial n} (v - v_\varepsilon) \, d\sigma \ge 0 \quad \forall v \in U_{ad},$$

where $p_\varepsilon \in H^1(\Omega)$ is the unique solution to the adjoint problem

$$\Delta p_\varepsilon = 0 \quad \text{in } \Omega, \qquad p_\varepsilon = \frac{1}{\varepsilon} \frac{\partial y_\varepsilon}{\partial n} \quad \text{on } \Gamma_1, \qquad p_\varepsilon = 0 \quad \text{on } \Gamma_2.$$

Remark. According to (3.4.60) and its preceeding remark, we have $\frac{\partial y_\varepsilon}{\partial n} \in H^{\frac{1}{2}}(\Gamma_1)$, which justifies the regularity $p_\varepsilon \in H^1(\Omega)$. Notice also that the inequality expressing Pontryagin's maximum principle is meaningful: indeed, as mentioned above, the transposition solution p_ε satisfies $p_\varepsilon \in H^2(\mathcal{W})$ for some open neighborhood $\mathcal{W}$ of Γ_2, and thus $\frac{\partial p_\varepsilon}{\partial n}$ belongs to $L^2(\Gamma_2)$.

Proof of Proposition 3.4.16. Let $\xi \in L^2(\Omega)$ and $w \in V_{ad} - v_\varepsilon$ be given, and denote by $z_1 \in L^2(\Omega)$ the transposition solution to

$$\Delta z_1 = l\,\xi \quad \text{in } \Omega, \tag{3.4.63}$$

$$z_1 = 0 \quad \text{on } \Gamma_1, \tag{3.4.64}$$

$$z_1 = w \quad \text{on } \Gamma_2. \tag{3.4.65}$$

Then the triple $[y_\varepsilon + \lambda\, z_1, h_\varepsilon + \lambda\,\xi, v_\varepsilon + \lambda\, w]$ is admissible for any $\lambda \in [0,1]$, and thus

$$\frac{1}{2\,\varepsilon}\int_{\Gamma_1}\left(\frac{\partial y_\varepsilon}{\partial n}\right)^2 d\sigma + \frac{1}{2}\int_\Omega l\,h_\varepsilon^2\,dx$$
$$\leq \frac{1}{2\,\varepsilon}\int_{\Gamma_1}\left(\frac{\partial y_\varepsilon}{\partial n} + \lambda\frac{\partial z_1}{\partial n}\right)^2 d\sigma + \frac{1}{2}\int_\Omega l(h_\varepsilon + \lambda\xi)^2\,dx.$$

It follows that

$$\frac{1}{\varepsilon}\int_{\Gamma_1}\frac{\partial y_\varepsilon}{\partial n}\frac{\partial z_1}{\partial n}\,d\sigma + \int_\Omega l\,h_\varepsilon\,\xi\,dx \geq 0, \quad \forall\,\xi \in L^2(\Omega), \quad \forall\, w \in V_{ad} - v_\varepsilon. \tag{3.4.66}$$

Notice that according to Necas [1967], we have $z_1 \in H^2(\mathcal{V})$ with some neighborhood $\mathcal{V} \subset \Omega$ of Γ_1, as before. We take smooth approximations $z_1^\delta \in H^2(\Omega)$ and $p_\varepsilon^\delta \in H^2(\Omega)$ of z_1 and p_ε, respectively, for any $\delta > 0$. By virtue of general properties of the approximation, and owing to the trace theorem, we may assume that as $\delta \searrow 0$,

$$p_\varepsilon^\delta \to p_\varepsilon \text{ in } H^1(\Omega), \qquad z_1^\delta \to z_1 \text{ in } L^2(\Omega), \qquad \Delta p_\varepsilon^\delta = 0 \text{ in } \Omega,$$
$$\Delta z_1^\delta \to l\,\xi = \Delta z_1 \text{ in } L^2(\Omega), \qquad z_1^\delta \to z_1 \text{ in } H^2(\mathcal{V}),$$
$$p_\varepsilon^\delta \to p_\varepsilon \text{ in } H^2(\mathcal{W}), \qquad \frac{\partial z_1^\delta}{\partial n} \to \frac{\partial z_1}{\partial n} \text{ in } H^{\frac{1}{2}}(\Gamma_1),$$
$$p_\varepsilon^\delta \to \frac{1}{\varepsilon}\frac{\partial y_\varepsilon}{\partial n} \text{ in } H^{\frac{1}{2}}(\Gamma_1), \qquad \frac{\partial p_\varepsilon^\delta}{\partial n} \to \frac{\partial p_\varepsilon}{\partial n} \text{ in } H^{\frac{1}{2}}(\Gamma_2).$$

Now observe that by multiplying the equation of p_ε^δ by z_1^δ, and by using integration by parts,

$$\int_{\Gamma_1} p_\varepsilon^\delta \frac{\partial z_1^\delta}{\partial n}\, d\sigma = \int_\Omega \Delta z_1^\delta\, p_\varepsilon^\delta\, dx + \int_{\Gamma_2} z_1^\delta \frac{\partial p_\varepsilon^\delta}{\partial n}\, d\sigma.$$

Moreover, in view of the above convergences, and thanks to Lions and Magenes [1968, Chapter II, Theorem 6.5], we also have $z_1^\delta|_{\Gamma_2} \to w$ strongly in $H^{-\frac{1}{2}}(\Gamma_2)$. Therefore, we can pass to the limit as $\delta \searrow 0$ in the above relation, and using (3.4.66), we arrive at the inequality

$$\int_{\Gamma_2} \frac{\partial p_\varepsilon}{\partial n}\, w\, d\sigma + \int_\Omega l\,(p_\varepsilon + h_\varepsilon)\,\xi\, dx \ge 0, \quad \forall\, \xi \in L^2(\Omega),\ \forall\, w \in V_{ad} - v_\varepsilon,$$

whence the assertion follows. □

We now recover from the optimality system given in Proposition 3.4.16 an additional property of the optimal triple. To this end, recall that $l = u^{-3}$ (u is the thickness of the plate), and put

$$V = \left\{ \tilde{w} \in H^2(\Omega) \,:\, \tilde{w} = \frac{\partial \tilde{w}}{\partial n} = 0 \text{ on } \Gamma_1 \right\},$$

$$K = \{ \tilde{w} \in V \,:\, \tilde{w}|_{\Gamma_2} \in V_{ad} \}.$$

Next, we consider approximations $y_\varepsilon^\delta \in H^2(\Omega)$ of y_ε, the optimal state for (3.4.55), (3.4.56), (3.4.60) for $\delta > 0$, having the property that for $\delta \searrow 0$,

$$y_\varepsilon^\delta \to y_\varepsilon \text{ in } L^2(\Omega), \qquad \Delta y_\varepsilon^\delta \to l\,(z + h_\varepsilon) \text{ in } L^2(\Omega),$$

$$y_\varepsilon^\delta \to y_\varepsilon \text{ in } H^2(\mathcal{V}), \qquad \frac{\partial y_\varepsilon^\delta}{\partial n} \to \frac{\partial y_\varepsilon}{\partial n} \text{ in } H^{\frac{1}{2}}(\partial\Gamma_1),$$

$$y_\varepsilon^\delta \to v_\varepsilon \text{ in } H^{-\frac{1}{2}}(\Gamma_2).$$

This is similar to the previous argument for z_1^δ and again makes use of Theorem 6.5 in Lions and Magenes [1968, Chapter II].

Rewriting (3.4.55) in the form $u^3 \Delta y_\varepsilon = z + h_\varepsilon$, and testing by $\Delta y_\varepsilon^\delta - \Delta\tilde{w}$, where $\tilde{w} \in K$ is arbitrary, we infer from Proposition 3.4.16 that

$$\begin{aligned}
\int_\Omega u^3\, \Delta y_\varepsilon\, (\Delta y_\varepsilon^\delta - \Delta\tilde{w})\, dx &= \int_\Omega (z - p_\varepsilon)\,(\Delta y_\varepsilon^\delta - \Delta\tilde{w})\, dx \\
&= -\int_\Omega \nabla(z - p_\varepsilon)\cdot\nabla(y_\varepsilon^\delta - \tilde{w})\, dx + \int_{\partial\Omega} (z - p_\varepsilon)\,\frac{\partial}{\partial n}(y_\varepsilon^\delta - \tilde{w})\, d\sigma \\
&= \int_\Omega \Delta(z - p_\varepsilon)\,(y_\varepsilon^\delta - \tilde{w})\, dx + \int_{\partial\Omega} (z - p_\varepsilon)\,\frac{\partial}{\partial n}(y_\varepsilon^\delta - \tilde{w})\, d\sigma \\
&\quad - \int_{\partial\Omega} \frac{\partial}{\partial n}(z - p_\varepsilon)\,(y_\varepsilon^\delta - \tilde{w})\, d\sigma.
\end{aligned}$$

Now assume that $z \in H^2(\Omega)$ satisfies $z = \frac{\partial z}{\partial n} = 0$ in Γ_2. Then it follows that

$$\begin{aligned}\int_\Omega u^3\, \Delta y_\varepsilon\, (\Delta y_\varepsilon^\delta - \Delta \tilde{w})\, dx &= \int_\Omega \Delta z\, (y_\varepsilon^\delta - \tilde{w})\, dx + \int_{\Gamma_1} (z - p_\varepsilon)\, \frac{\partial y_\varepsilon^\delta}{\partial n}\, d\sigma \\ &\quad + \int_{\Gamma_2} (y_\varepsilon^\delta - \tilde{w})\, \frac{\partial p_\varepsilon}{\partial n}\, d\sigma - \int_{\Gamma_1} \frac{\partial}{\partial n}(z - p_\varepsilon)\, y_\varepsilon^\delta\, d\sigma.\end{aligned} \tag{3.4.67}$$

Hence, passing to the limit in (3.4.67), we conclude that

$$\begin{aligned}\int_\Omega u^3\, \Delta y_\varepsilon\, (\Delta y_\varepsilon - \Delta \tilde{w})\, dx &= \int_\Omega \Delta z\, (y_\varepsilon - \tilde{w})\, dx + \int_{\Gamma_1} (z - p_\varepsilon)\, \frac{\partial y_\varepsilon}{\partial n}\, d\sigma \\ &\quad + \int_{\Gamma_2} (v_\varepsilon - \tilde{w})\, \frac{\partial p_\varepsilon}{\partial n}\, d\sigma.\end{aligned}$$

From this, and from the maximum principle in Proposition 3.4.16, we infer the inequality

$$\begin{aligned}\int_\Omega u^3\, \Delta y_\varepsilon\, (\Delta y_\varepsilon - \Delta \tilde{w})\, dx &\le \int_\Omega \Delta z\, (y_\varepsilon - \tilde{w})\, dx + \int_{\Gamma_1} (z - p_\varepsilon)\, \frac{\partial y_\varepsilon}{\partial n}\, d\sigma \\ &= \int_\Omega \Delta z\, (y_\varepsilon - \tilde{w})\, dx + \int_{\Gamma_1} z\, \frac{\partial y_\varepsilon}{\partial n}\, d\sigma - \frac{1}{\varepsilon} \int_{\Gamma_1} \left(\frac{\partial y_\varepsilon}{\partial n}\right)^2 d\sigma \\ &\le \int_\Omega \Delta z\, (y_\varepsilon - \tilde{w})\, dx + \int_{\Gamma_1} z\, \frac{\partial y_\varepsilon}{\partial n}\, d\sigma.\end{aligned} \tag{3.4.68}$$

We obtain the following result.

Theorem 3.4.17 *Assume that the load $f \in L^2(\Omega)$ has the form $f = \Delta z$ with some $z \in H^2(\Omega)$ satisfying $z = \frac{\partial z}{\partial n} = 0$ on Γ_2. Then the variational inequality*

$$\int_\Omega u^3(x)\, \Delta y(x)\, (\Delta y - \Delta \mu)(x)\, dx \le \int_\Omega f(x)\, (y - \mu)(x)\, dx \quad \forall\, \mu \in K_1,$$

$$K_1 = \left\{ \mu \in L^2(\Omega) \,:\, \Delta \mu \in L^2(\Omega) \text{ with } \mu = \frac{\partial \mu}{\partial n} = 0 \text{ on } \Gamma_1,\ \mu|_{\Gamma_2} \in V_{ad} \right\}, \tag{3.4.69}$$

has a unique solution $\hat{y} \in K_1$, and $y_\varepsilon \to \hat{y}$ weakly in $L^2(\Omega)$.

Proof. By Proposition 3.4.15, there is some sequence $\varepsilon_n \searrow 0$ such that

$$\begin{aligned} &y_{\varepsilon_n} \to \hat{y} \text{ weakly in } L^2(\Omega), && \Delta y_{\varepsilon_n} \to \Delta \hat{y} \text{ weakly in } L^2(\Omega), \\ &h_{\varepsilon_n} \to \hat{h} \text{ weakly in } L^2(\Omega), && v_{\varepsilon_n} \to \hat{v} \text{ weakly* in } L^\infty(\Omega), \\ &\frac{\partial y_{\varepsilon_n}}{\partial n} \to \frac{\partial \hat{y}}{\partial n} = 0 \text{ strongly in } L^2(\Gamma_1). \end{aligned}$$

Clearly, $\hat{v} \in V_{ad}$, and it is readily verified that $\hat{y}$ is the transposition solution to the problem

$$\Delta\hat{y} = l(z + \hat{h}) \text{ in } \Omega, \qquad \hat{y} = 0 \text{ on } \Gamma_1, \qquad \hat{y} = \hat{v} \text{ on } \Gamma_2.$$

Passing to the limit as $\varepsilon_n \searrow 0$ in (3.4.68), recalling Proposition 3.4.15, we obtain (3.4.69) for any $\tilde{w} \in K$. Since K is dense in K_1 with respect to the $H^2(\Omega)$ norm, the inequality remains valid for all $\mu \in K_1$, which proves the existence.

Suppose now that $y_1, y_2 \in K_1$ are two solutions of (3.4.69). By choosing in turn $y = y_1$, $\mu = y_2$, and $y = y_2$, $\mu = y_1$, we obtain that

$$\int_\Omega u^3 \left(\Delta y_1 - \Delta y_2\right)^2 dx = 0,$$

i.e., $\Delta(y_1 - y_2) = 0$ a.e. in Ω. Since $y_1 - y_2$ satisfies Cauchy-type boundary conditions on Γ_1 by the definition of K_1, we get $y_1 = y_2$ a.e. in Ω, and the uniqueness follows. Notice that the uniqueness entails that the above convergences hold generally for $\varepsilon \searrow 0$, and not only for the sequences selected above. This concludes the proof of the assertion. □

Remark. Using regularity properties of transposition solutions (cf. Lions and Magenes [1968, Chapter II]), one can prove that $\{y_\varepsilon\}_{\varepsilon>0}$ is bounded in $H^{\frac{1}{2}}(\Omega)$, and thus, by the compactness of the embedding $H^{\frac{1}{2}}(\Omega) \subset L^2(\Omega)$, that $y_\varepsilon \to y$ strongly in $L^2(\Omega)$.

Remark. Relation (3.4.69) models a plate that is partially clamped on Γ_1 and subjected to condition (3.4.59) on Γ_2 (the deflection y should remain between the obstacles $-r$ and r). The approach via control problems governed by second-order elliptic equations presented in this section is constructive and can be employed for the numerical approximation of the solution to (3.4.69).

Remark. The intrinsic difficulty in the variational inequality (3.4.69) comes from the fact that the corresponding bilinear form

$$a(y, w) = \int_\Omega u^3(x)\, \Delta y(x)\, \Delta w(x)\, dx$$

is not coercive on the space V. Indeed, otherwise the second-order elliptic equation with Cauchy conditions on Γ_1 would be well-posed, which is known to be false in general. This fact demonstrates the flexibility and broad range of applicability of the control approach proposed here.

Chapter 4

Discretization

4.1 Finite Element Approximation of Elliptic Equations

The Finite Element Method (FEM) is a general discretization technique for the numerical solution of partial differential equations. In this chapter, we show how this techique can be employed to solve optimal control problems for elliptic equations. We underline that the scientific literature devoted to FEM applications in optimization problems involving partial differential equations is still under active development. Many questions dealing with the efficient implementation of the FEM, like a posteriori estimates or adaptivity, which are well understood for partial differential equations, are the subject of very recent research papers on the numerical approximation of control problems. We will limit our present exposition to the *a priori estimates* technique which we consider to be already well established in this field of the mathematical literature. We also notice that the first-order optimality conditions examined in the previous chapter are the basic tool in our approach.

Some basic features of the FEM are:

- The physical region of the problem is subdivided into nonoverlapping subregions, the *finite elements*.
- The solution of the governing equations is over each individual element approximated by a polynomial function (constant, linear, quadratic, etc.); the elementwise defined local polynomial approximations are then "glued together" to form a global piecewise polynomial approximation on the whole domain. The coefficients of this piecewise polynomial function then become the unknowns of the discretized problem.
- Substitution of the approximation into the governing equations yields a system of algebraic equations in the unknown parameters whose matrix,

by construction, is "sparse." Its solution yields the coefficients of the piecewise polynomial approximate solution to the original problem.

4.1.1 The Finite Element Method

When applying the FEM to a problem defined on a general domain $\Omega \subset \mathbf{R}^2$, we first approximate the shape of the domain by polygons and then divide the obtained polygonal domains Ω_h into small triangles, as shown in Figure 1.1. The vertices of the boundary $\partial\Omega_h$ of Ω_h are chosen on $\partial\Omega$ in such a way that each of them coincides with the vertices of some of the small triangles. The individual triangles are called the triangular *elements* of the FEM.

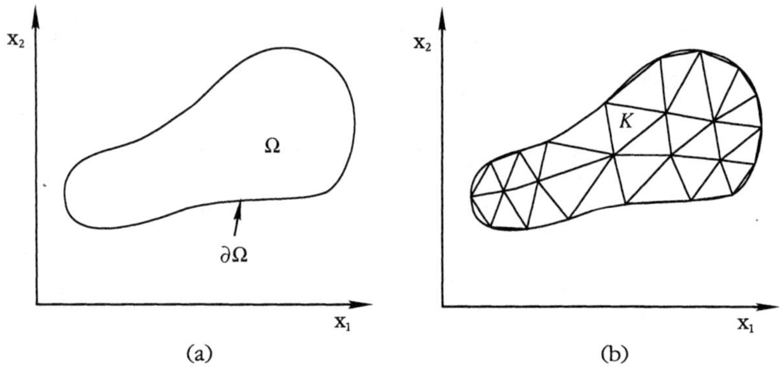

Figure 1.1. (a) A two-dimensional domain Ω, and (b) Ω_h and its division into triangular elements K.

Of course, triangulation is not the only way to subdivide a domain. For example, if the given domain is a rectangle, then it is possible to divide it into small subrectangles and to construct the approximation on these subrectangles. Also, combinations of different element types are possible. In the literature several general schemes concerning the generation of FE-meshes have been proposed; in this connection, we refer the reader to the works of George [1991], Kardestuncer and Norrie [1987], Soni, Thompson, and Weatherill [1999]. We notice that while there is a technical difference between subdivision into rectangles and triangulation, there is no essential difference from the viewpoint of principle. Hence, we will generally confine ourselves to triangulations; in higher dimensions simplex elements are used.

Throughout this chapter, if not specified otherwise, we assume that Ω is a polyhedral domain in $\mathbf{R}^d$, $d \in \mathbf{N}$. We now establish more rigorously a *triangulation* $\mathcal{T}_h$ of Ω: we assume that the set $\overline{\Omega}$ is subdivided into a finite number of subsets K (called *elements*) such that the following properties hold:

(1) $\overline{\Omega} = \bigcup_{K \in \mathcal{T}_h} K$.

(2) Every element $K \in \mathcal{T}_h$ is a polyhedron in $\mathbf{R}^d$.

(3) For each distinct $K_1, K_2 \in \mathcal{T}_h$, $\operatorname{int} K_1 \cap \operatorname{int} K_2 = \emptyset$.

(4) Every face of any polyhedron $K_1 \in \mathcal{T}_h$ is either a face of another polyhedron $K_2 \in \mathcal{T}_h$ or a part of the boundary $\partial\Omega$.

The *discretization (triangulation) parameter* $h > 0$ is the maximal diameter of all $K \in \mathcal{T}_h$. Note that there is a certain ambiguity in the meaning of the symbol $\mathcal{T}_h$; indeed, for given sufficiently small $h > 0$ it is obviously possible to construct many different triangulations $\mathcal{T}_h$ of Ω. Nevertheless, we will henceforth keep this commonly used notation. Sometimes $\mathcal{T}_h$ is called a *partition* or a *decomposition* of $\overline{\Omega}$ into elements (especially when the elements $K \in \mathcal{T}_h$ are not triangles). Also, the terms *division*, *grid,* and *mesh* are commonly used.

A set $\mathcal{T} = \{\mathcal{T}_h\}$ of triangulations of $\overline{\Omega}$ is said to be *a family of triangulations* if for any $\varepsilon > 0$ there exists some $\mathcal{T}_h \in \mathcal{T}$ with $h < \varepsilon$.

In order to analyze FEM approximations, one may postulate some additional properties. We call a family $\mathcal{T} = \{\mathcal{T}_h\}_{h>0}$ of triangulations *regular* if there exists a constant $\kappa > 0$ such that for any triangulation $\mathcal{T}_h \in \mathcal{T}$ and for any element $K \in \mathcal{T}_h$ there exists a ball $\mathcal{B}_K$ of radius $\varrho_K > 0$ such that

$$\mathcal{B}_K \subset K \text{ and } \kappa\, h_K \leq \varrho_K \text{ where } h_K = \operatorname{diam}(K).$$

Roughly speaking, the regularity of a family $\mathcal{T}$ means that the elements of $\mathcal{T}_h \in \mathcal{T}$ cannot "degenerate" for $h \searrow 0$. If $\kappa\, h \leq \varrho_K$ for any $K \in \mathcal{T}_h$, then the family is said to be *strongly regular*. However, in general this does not have to be true for regular families.

Having found such a triangulation $\mathcal{T}_h$ of the domain Ω, it is possible to construct finite-dimensional function spaces that approximate the function spaces (for example Sobolev spaces) governing the problem. We now give two simple examples.

Example 4.1.1 In this example, we describe the simplest *finite element space* in arbitrary space dimensions, namely the space of piecewise linear mappings. Linear elements are a special case of the so-called *Lagrange elements* and are extensively used in this monograph and, generally, in the scientific literature.

To this end, let x_i, $i = 1, \ldots, N$, be the (finite) set of all vertices of elements (assumed to be simplices) of $\mathcal{T}_h$, which are numbered in a given order. A *basis* of the corresponding finite element space V_h is then given by the set $\{\varphi_i\}_{i=1}^N$, where φ_i, $1 \le i \le N$, satisfies

$$\varphi_i(x_j) = \delta_{ij} \quad \text{for } i, j = 1, \ldots, N. \tag{4.1.1}$$

Notice that each of the φ_i has a "small" support in Ω; indeed, φ_i vanishes identically in every element of the grid not having a vertex at x_i.

Obviously, $V_h \subset C(\overline{\Omega})$, and V_h is N-dimensional. For any given pointwise mapping $y : \Omega \to \mathbf{R}$, one defines the (uniquely determined) interpolant $y_I \in V_h$ by requiring that

$$y_I(x_i) = y(x_i), \quad i = 1, \ldots, N. \tag{4.1.2}$$

It is easily seen from (4.1.1) and (4.1.2) that

$$y_I(x) = \sum_{i=1}^{N} y_i \, \varphi_i(x), \quad \text{where } y_i = y(x_i), \quad i = 1, \ldots, N. \tag{4.1.3}$$

If y is an integrable mapping that is not specified pointwise, then $y(x_i)$ may be replaced, for instance, by the average of y over all elements having a vertex at x_i. It is customary to identify y_I with the vector $(y_i)_{i=1}^N \in \mathbf{R}^N$.

If the Sobolev space to be discretized is of the type $H_0^1(\Omega)$, then the requirement that the traces of φ_i, $i = 1, \ldots, N$, be zero on the boundary $\partial\Omega$ should be included in the definition of V_h. Then (4.1.1) holds only for vertices in the interior of Ω.

Example 4.1.2 Let us consider the one-dimensional case in which $\Omega = (a, b)$ for $-\infty < a < b < +\infty$. We introduce a special type of finite element that belongs to the class of the so-called *Hermite splines*. For fixed $N \in \mathbf{N}$, we put $h = (b - a)/N$ and define the grid points $x_i = a + ih$, $i = 0, \ldots, N$. Let $m \in \mathbf{N} \cup \{0\}$, and define the finite element spaces

$$V_h^m = \left\{ v_h \in C^m[a, b] \, : \, v_h|_{[x_i, x_{i+1}]} \in P_{2m+1}[x_i, x_{i+1}], \; i = 0, \ldots, N-1 \right\}, \tag{4.1.4}$$

where $P_{2m+1}[x_i, x_{i+1}]$ is the space of all polynomials of maximal degree $2m+1$ over $[x_i, x_{i+1}]$. It is well known that a polynomial of maximal degree $2m + 1$ is uniquely determined by its values and by the values of its derivatives up to mth order in two distinct points. If the space to be discretized is of the type $H_0^1(a, b)$, the boundary conditions $v_h(a) = v_h(b) = 0$ have to be added to the definition of V_h.

Now consider the interpolation operator $\Pi_h : C^m[a, b] \to V_h^m$ defined by

$$\Pi_h y \in V_h^m, \quad (\Pi_h y)^{(j)}(x_i) = y^{(j)}(x_i), \quad j = 0, \ldots, m, \; i = 0, \ldots, N. \tag{4.1.5}$$

Then the following simple error estimate holds (see Ciarlet [1990]):

$$|y - \Pi_h y|_{H^1(a,b)} \le C \, h^{2m+1} \, |y^{(2m+2)}|_{C[a,b]} \quad \forall \, y \in C^{2m+2}[a, b],$$

where the constant $C > 0$ depends only on m.

For arbitrary $d \in \mathbf{N}$ and any triangulation $\mathcal{T}_h$ of a bounded set $\Omega \subset \mathbf{R}^d$, the *piecewise constant* interpolation operator $\widetilde{\Pi}_h : L^2(\Omega) \to L^\infty(\Omega)$,

$$(\widetilde{\Pi}_h y)(x) = \frac{1}{\text{meas}(K)} \int_K y(x)\, dx \quad \forall\, x \in K,$$

has the orthogonality property

$$\left(v_h, \widetilde{\Pi}_h y - y\right)_{L^2(\Omega)} = 0, \quad \forall\, y \in L^2(\Omega),\ \forall\, v_h \in \widetilde{\Pi}_h(L^2(\Omega)). \tag{4.1.6}$$

One reason for using complex higher-order finite element spaces is given by their better approximation properties. In this connection we cite without proof a classical interpolation result from Oden and Reddy [1976], which is due to Ciarlet and Raviart [1972].

Theorem 4.1.3 *Let $\Omega \subset \mathbf{R}^2$ be a polygonal domain, and let $k, m \in \mathbf{N}$ be fixed with $0 \le m \le k+1$. Suppose that the operator $\Pi \in L(H^{k+1}(\Omega), H^m(\Omega))$ has the property that its restriction to the subspace $\mathcal{P}_k(\Omega)$ of polynomials of maximal degree k over Ω is the identity operator. Then there exist constants $C > 0$ and $h_0 > 0$ such that for every triangulation $\mathcal{T}_h$ with $0 < h \le h_0$,*

$$|v - \Pi v|_{H^m(\Omega)} \le C\, \frac{h^{k+1}}{d^m}\, |v|_{H^{k+1}(\Omega)} \quad \forall\, v \in H^{k+1}(\Omega). \tag{4.1.7}$$

Here $d = \text{Min}_{K \in \mathcal{T}_h} d_K > 0$, where for every $K \in \mathcal{T}_h$ the quantity $d_K > 0$ denotes the maximal diameter of all open balls that are contained in K. If, in particular, the triangulation $\mathcal{T}_h$ is strongly regular, then (4.1.7) becomes

$$|v - \Pi v|_{H^m(\Omega)} \le C\, h^{k-m+1}\, |v|_{H^{k+1}(\Omega)} \quad \forall\, v \in H^{k+1}(\Omega). \tag{4.1.8}$$

Finally, let us remark that finite element approximations may be *internal* (or *conforming*) in the sense that $V_h \subset V$ for every $h > 0$. This is in contrast to *finite difference methods*, for which the associated discrete mappings do not necessarily belong to V. Such approximations are called *external* (or *nonconforming*). If $V_h \subset V$, then V_h may be endowed with the same ("trace") norm and inner product as V. Since V_h is finite-dimensional, it can also be equipped with any other finite-dimensional norm, for instance the Euclidean norm. Any of these norms is equivalent to the trace norm on V_h, but the corresponding constants may be unbounded as $h \searrow 0$.

It should also be noticed that there is an intrinsic relationship between Π and $\mathcal{T}_h$. Indeed, the maximal degree k of polynomials that are left invariant under the operator Π depends on the type of elements (triangles, rectangles, etc.) constituting $\mathcal{T}_h$.

4.1.2 Error Estimates for the FE Equations

In this paragraph, we study the variational equation

$$\text{(E)} \qquad \text{Find } y \in V \text{ such that } a(y,v) = F(v) \quad \forall\, v \in V.$$

Here, V is a real Hilbert space, $F \in V^*$, and $a : V \times V \to \mathbf{R}$ is a continuous bilinear form satisfying

$$a(y,y) \ge \gamma\, |y|_V^2 \quad \forall\, y \in V, \qquad \text{with some fixed constant } \gamma > 0.$$

Then, by the Lax–Milgram lemma, (E) has a unique solution $y \in V$.

Now let V_h denote an arbitrary finite-dimensional subspace of V endowed with the same norm, where $h > 0$ stands for a discretization parameter. The *Galerkin method* for approximating the solution of the elliptic variational problem (E) is introduced by

$$(\mathrm{E}_h) \qquad \text{Find } y_h \in V_h \text{ such that } a(y_h, v_h) = F(v_h) \quad \forall\, v_h \in V_h.$$

By the Lax–Milgram lemma, (E_h) also has a unique solution $y_h \in V_h$, called the *discrete solution*. If $a(\cdot\,,\cdot)$ is symmetric, then y_h and y, respectively, are also characterized by the properties

$$J(y_h) = \inf_{v_h \in V_h} \big\{ J(v_h) \big\}, \quad J(y) = \inf_{v \in V} \big\{ J(v) \big\},$$

where J is the associated "energy functional" given by

$$J(v) = \frac{1}{2}\, a(v,v) - F(v).$$

This alternative definition of the solution is known as the *Ritz method*. Note that

$$J(y) \le J(y_h),$$

since $V_h \subset V$. Moreover, from (E) and (E_h) we obtain the orthogonality relation

$$a(y - y_h, v_h) = 0 \quad \forall\, v_h \in V_h, \tag{4.1.9}$$

which states that the error $y - y_h$ is orthogonal to V_h with respect to the product $a(\cdot\,,\cdot)$, even if $a(\cdot,\cdot)$ is not symmetric. If $a(\cdot,\cdot)$ is symmetric, putting $v_h = y_h$, we obtain

$$\begin{aligned} 0 \;\le\; a(y - y_h, y - y_h) \;&=\; a(y,y) - 2\,a(y,y_h) + 2\,a(y_h,y_h) - a(y_h,y_h) \\ &=\; a(y,y) - a(y_h,y_h), \end{aligned}$$

that is,

$$a(y,y) \geq a(y_h,y_h) \quad \text{and} \quad F(y) \geq F(y_h).$$

Obviously, the orthogonality relation (4.1.9) (compare with (4.1.6)) implies that y_h is the projection of y onto V_h with respect to the scalar product $a(\cdot\,,\cdot)$, that is,

$$a(y-y_h, y-y_h) = \inf_{v_h \in V_h} a(y-v_h, y-v_h).$$

Hence the Ritz method yields the best approximation with respect to the *energy norm*, given by $a(\cdot,\cdot)$.

Now let $\{\varphi_i\}_{i=1}^N$ be a basis of the finite-dimensional space V_h. We look for the discrete solution y_h in the form of a linear combination of the basis elements,

$$y_h = \sum_{j=1}^N Y_j\, \varphi_j. \tag{4.1.10}$$

Then it follows from (E_h) that

$$a\Big(\sum_{j=1}^N Y_j\, \varphi_j, \varphi_i\Big) = F(\varphi_i), \quad i = 1, \ldots, N,$$

whence we arrive at the following system of algebraic equations for the unknowns $Y_1, \ldots, Y_N$:

$$\sum_{j=1}^N a\,(\varphi_j, \varphi_i)\, Y_j = F(\varphi_i), \quad i = 1, \ldots, N. \tag{4.1.11}$$

The matrix $A = (a(\varphi_j, \varphi_i))_{i,j=1}^N$ and the vector $(F(\varphi_i))_{i=1}^N$ are often called (by reference to problems in elasticity) the *stiffness matrix* and the *load vector*, respectively. The matrix A is positive definite and thus nonsingular; indeed, we have for any $v = \sum_i \xi_i\, \varphi_i \in V_h \setminus \{0\}$ that

$$\begin{aligned}(A\xi,\xi)_{\mathbf{R}^N} &= \sum_{i,j=1}^N a(\varphi_j,\varphi_i)\,\xi_i\,\xi_j = a\Big(\sum_{j=1}^N \xi_j\,\varphi_j, \sum_{i=1}^N \xi_i\,\varphi_i\Big) = a(v,v)\\ &\geq \gamma\,|v|_V^2 > 0 \quad \forall\, \xi = (\xi_1,\ldots,\xi_N)^T \in \mathbf{R}^N,\ \xi \neq 0,\end{aligned}$$

since $\{\varphi_i\}_{i=1}^N$ is a basis.

In practice, the space V_h and its basis are generated by the finite element method as described previously.

Example 4.1.4 Consider the case $m = 0$ in Example 4.1.2, that is,

$$V_h^0 = \Big\{v_h \in C[a,b] : \ v_h\big|_{[x_i,x_{i+1}]} \in P_1[x_i,x_{i+1}], \ i = 0,\ldots,N-1,$$
$$v_h(a) = v_h(b) = 0\Big\},$$

for $V = H_0^1(a,b)$ and

$$a(y,v) = \int_a^b \big[y'(x)\,v'(x) + \alpha\, y(x)\,v(x)\big]\,dx \quad \forall\, y,v \in V, \tag{4.1.12}$$

where $\alpha > 0$ is a constant, and

$$F(v) = \int_a^b f(x)\,v(x)\,dx \quad \forall\, v \in V,$$

for some given $f \in L^2(a,b)$. The basis of V_h^0 has the form

$$\varphi_i(x) = \begin{cases} \dfrac{x - x_{i-1}}{h}, & \text{for } x \in [x_{i-1}, x_i], \\ \dfrac{x_{i+1} - x}{h}, & \text{for } x \in [x_i, x_{i+1}], \\ 0, & \text{otherwise}, \end{cases} \tag{4.1.13}$$

for $i = 1, \ldots, N-1$. Hence, $\dim(V_h^0) = N-1$, and for y_h given by (4.1.10) the linear system (4.1.11) becomes

$$\sum_{j=1}^{N-1} a_{ij}\, Y_j = b_i, \quad i = 1, \ldots, N-1, \tag{4.1.14}$$

with

$$a_{ij} = a(\varphi_i, \varphi_j) = \int_a^b \big[\varphi_i'(x)\,\varphi_j'(x) + \alpha\, \varphi_i(x)\,\varphi_j(x)\big]\,dx, \quad i,j = 1, \ldots, N-1, \tag{4.1.15}$$

$$b_i = \int_a^b f(x)\,\varphi_i(x)\,dx, \quad i = 1, \ldots, N-1. \tag{4.1.16}$$

The matrix of the system (4.1.14) is symmetric, tridiagonal, and positive definite, since (see above) the bilinear form (4.1.12) satisfies

$$a(y,y) \geq \min\{1, \alpha\}\, |y|_V^2 \quad \forall\, y \in V.$$

Using (4.1.13), (4.1.15), one easily verifies that

$$\begin{aligned} a_{ii} &= \frac{2}{h} + \frac{2\alpha}{3} h, & i &= 1, \ldots, N-1, \\ a_{i,i+1} &= -\frac{1}{h} + \frac{\alpha}{6} h, & i &= 1, \ldots, N-2, \\ a_{i,i-1} &= -\frac{1}{h} + \frac{\alpha}{6} h, & i &= 2, \ldots, N-1, \\ a_{ij} &= 0, & |i-j| &> 1. \end{aligned}$$

Moreover, this matrix is ***strictly diagonally dominant***, that is,

$$|a_{ii}| > \sum_{\substack{j=1 \\ j \neq i}}^{N-1} |a_{ij}|, \quad i = 1, \ldots, N-1,$$

which ensures that (4.1.14) can be effectively solved by either elimination techniques or iterative methods (see Varga [1962], Duff et al. [1987]).

Finally, observe that the evaluation of (4.1.16) usually requires that one employ a numerical integration formula in order to approximate b_i for general functions f.

To continue the discussion of the general situation, let us now suppose that there exists a family of subspaces $\{V_h\}_{h>0} \subset V$ such that

$$\lim_{h \to 0} |y - y_h|_V = 0, \tag{4.1.17}$$

where y and y_h denote the solutions of (E) and (E_h), respectively. Then we say that the associated family $\{(\mathrm{E}_h)\}_{h>0}$ of discrete problems is *convergent*.

The convergence question is of paramount interest also in the general case in which the coefficients of a and of F are arbitrary functions and numerical integration has to be used for practical evaluation. In this case, we have to deal with an approximating bilinear form a_h and an approximating functional F_h rather than with the original a and F, which complicates the convergence analysis.

We suppose that the approximating bilinear form $a_h : V_h \times V_h \to \mathbf{R}$ is continuous (but not necessarily symmetric) and that there is some $\tilde{\gamma} > 0$, independent of $h > 0$, such that

$$a_h(v_h, v_h) \geq \tilde{\gamma}\, |v_h|_V^2 \quad \forall\, v_h \in V_h.$$

Usually, the latter property of the family $\{a_h\}_{h>0}$ can be deduced from the corresponding property of the bilinear form a provided that a sufficiently accurate numerical integration method is used (see Engels [1980]). We also assume that the approximation F_h is linear and continuous from V_h into $\mathbf{R}$, and consider the problem

$$(\tilde{E}_h) \qquad \text{Find } \tilde{y}_h \in V_h \text{ such that } a_h(\tilde{y}_h, v_h) = F_h(v_h) \quad \forall\, v_h \in V_h.$$

This constitutes an abstract model for the case that the bilinear form a and/or the linear form F cannot be exactly evaluated since numerical integration has to be used.

The problem $(\widetilde{E}_h)$ of course has a unique solution $\tilde{y}_h \in V_h$ for any $h > 0$, and we call the family $\{(\widetilde{E}_h)\}_{h>0}$ *convergent* if (4.1.17) holds with y_h replaced by $\tilde{y}_h$. We have the following error estimate (cf. Dautray and Lions [1988, Chapter XII, §1.5]):

Theorem 4.1.5 *Let y be the solution of (E) and $\tilde{y}_h$ the solution of $(\widetilde{E}_h)$. Then there exists a constant $C > 0$, independent of $h > 0$, such that*

$$|y - \tilde{y}_h|_V \le C \Bigg[\inf_{v_h \in V_h} \Big\{ |y - v_h|_V + \sup_{w_h \in V_h} \frac{|a(v_h, w_h) - a_h(v_h, w_h)|}{|w_h|_V} \Big\} + \sup_{w_h \in V_h} \Big\{ \frac{|F(w_h) - F_h(w_h)|}{|w_h|_V} \Big\} \Bigg]. \tag{4.1.18}$$

Proof. For any $v_h \in V_h$,

$$\begin{aligned}
\tilde{\gamma}\, |\tilde{y}_h - v_h|_V^2 &\le a_h(\tilde{y}_h - v_h, \tilde{y}_h - v_h) = a_h(\tilde{y}_h, \tilde{y}_h - v_h) - a_h(v_h, \tilde{y}_h - v_h) \\
&= F_h(\tilde{y}_h - v_h) - a_h(v_h, \tilde{y}_h - v_h) + a(y - v_h, \tilde{y}_h - v_h) \\
&\quad + a(v_h, \tilde{y}_h - v_h) - F(\tilde{y}_h - v_h) \\
&= a(y - v_h, \tilde{y}_h - v_h) + [a(v_h, \tilde{y}_h - v_h) - a_h(v_h, \tilde{y}_h - v_h)] \\
&\quad + [F_h(\tilde{y}_h - v_h) - F(\tilde{y}_h - v_h)].
\end{aligned}$$

We now use the continuity of a, where $M > 0$ is the corresponding constant, and we divide by $|\tilde{y}_h - v_h|_V$ (which is without loss of generality assumed to be nonzero) to obtain that

$$\tilde{\gamma}\, |\tilde{y}_h - v_h|_V \le M\, |y - v_h|_V + \frac{|a(v_h, \tilde{y}_h - v_h) - a_h(v_h, \tilde{y}_h - v_h)|}{|\tilde{y}_h - v_h|_V} + \frac{|F_h(\tilde{y}_h - v_h) - F(\tilde{y}_h - v_h)|}{|\tilde{y}_h - v_h|_V},$$

whence, taking into account that $\tilde{y}_h - V_h = V_h$,

$$\tilde{\gamma}\, |\tilde{y}_h - v_h|_V \le M |y - v_h|_V + \sup_{w_h \in V_h} \frac{|a(v_h, w_h) - a_h(v_h, w_h)|}{|w_h|_V} + \sup_{w_h \in V_h} \frac{|F_h(w_h) - F(w_h)|}{|w_h|_V}.$$

Since $|y - \tilde{y}_h|_V \le |y - v_h|_V + |\tilde{y}_h - v_h|_V$, the assertion follows from taking the infimum over $v_h \in V_h$. □

Remark. The first two terms in the right-hand side of (4.1.18) account for the distance of y from V_h and for the consistency of the approximation of the bilinear form a. They can be bounded by the expression

$$|y - \pi_h y|_V + \sup_{w_h \in V_h} \frac{|a(\pi_h y, w_h) - a_h(\pi_h y, w_h)|}{|w_h|_V},$$

where π_h is any interpolating operator from V into V_h. The last term in the right-hand side of (4.1.18) reflects the consistency of the approximation of F.

If no numerical integration scheme is needed, that is, if we can deal with (E_h) instead of with ($\widetilde{E}_h$), then (4.1.18) simplifies to the following statement:

Corollary 4.1.6 (Céa's lemma) *Suppose that y is the solution of (E) and y_h is the solution of (E_h). Then there exists a constant $C > 0$, which is independent of $h > 0$, such that*

$$|y - y_h|_V \le C \inf_{v_h \in V_h} |y - v_h|_V. \tag{4.1.19}$$

Consequently, a sufficient condition for convergence is that the family $\{V_h\}_{h>0}$ of subspaces of V satisfy

$$\lim_{h \searrow 0} \inf_{v_h \in V_h} |v - v_h|_V = 0,$$

i.e. $\bigcup_{h>0} V_h$ is dense in V with respect to the norm of V.

We continue with error estimates for the problems (E_h) and ($\widetilde{E}_h$), respectively, that are direct consequences of the above results.

Theorem 4.1.7 *Let $\mathcal{T} = \{\mathcal{T}_h\}$ be a regular family of triangulations of a polygon $\overline{\Omega} \subset \mathbf{R}^2$, let the solution $y \in V$ of (E) belong to $H^2(\Omega)$, and let $V \subset H^1(\Omega)$ be equipped with the norm of $H^1(\Omega)$. Suppose also that the solution of (E_h) satisfies $y_h \in V_h$, where*

$$V_h = \left\{ v_h \in V \cap C(\overline{\Omega}) : v_h|_K \in P_1(K) \quad \forall K \in \mathcal{T}_h \right\}.$$

Then there exist constants $h_0 > 0$ and $C > 0$ such that for any triangulation $\mathcal{T}_h \in \mathcal{T}$ with $0 < h \le h_0$ the following error estimate holds:

$$|y - y_h|_V \le C\, h\, |y|_{H^2(\Omega)}. \tag{4.1.20}$$

Proof. In the following, we denote by C_i, $i \in \mathbf{N}$, positive constants that do not depend on $h > 0$. To begin with, let us define the piecewise linear interpolation $\Pi_h y \in V_h$ of y over the whole domain Ω by

$$\Pi_h y|_K = \Pi_K y \quad \forall K \in \mathcal{T}_h,$$

where $\Pi_K y$ is the uniquely determined affine function defined on K that attains the same values in the three vertices of K as $y \in H^2(\Omega)$ (which is continuous by the Sobolev embedding theorem). Thanks to the interpolation inequality (4.1.7), we find that

$$|v - \Pi_K v|_{H^1(K)} \le C_1\, h_K\, |v|_{H^2(K)} \quad \forall\, v \in H^2(K). \tag{4.1.21}$$

Squaring and summing (4.1.21) for $v = y$, we obtain that

$$\begin{aligned} |y - \Pi_h y|^2_{H^1(\Omega)} &= \sum_{K\in\mathcal{T}_h} |y - \Pi_K y|^2_{H^1(K)} \le C_1^2 \sum_{K\in\mathcal{T}_h} h_K^2\, |y|^2_{H^2(K)} \\ &\le C_1^2\, h^2\, |y|^2_{H^2(\Omega)}. \end{aligned} \tag{4.1.22}$$

Hence, we can infer from Céa's lemma (cf. (4.1.19)) that

$$|y - y_h|_{H^1(\Omega)} \le C_2\, |y - \Pi_h y|_{H^1(\Omega)} \le C_3\, h\, |y|_{H^2(\Omega)}, \tag{4.1.23}$$

which concludes the proof of the assertion. □

Next, we will investigate the order of convergence of the linear triangular elements in the $L^2(\Omega)$ norm. Using the so-called *Aubin–Nitsche trick*, we shall prove that

$$|y - y_h|_{L^2(\Omega)} = \mathcal{O}(h^2)$$

under $H^2(\Omega)$ regularity.

Theorem 4.1.8 *Suppose that the assumptions of Theorem 4.1.7 are fulfilled and that $a(\cdot\,,\cdot)$ is symmetric. In addition, suppose that whenever $F(v) = (f, v)_{L^2(\Omega)}$ with some $f \in L^2(\Omega)$, the solution y to the problem (E) belongs to $H^2(\Omega)$ and satisfies, with some $\hat{c} > 0$ that is independent of f,*

$$|y|_{H^2(\Omega)} \le \hat{c}\, |f|_{L^2(\Omega)}. \tag{4.1.24}$$

Then there exist constants $h_0 > 0$ and $C > 0$ such that for any $\mathcal{T}_h \in \mathcal{T}$ with $0 < h \le h_0$,

$$|y - y_h|_{L^2(\Omega)} \le C\, h^2\, |y|_{H^2(\Omega)}.$$

Proof. Let $e = y - y_h \in V \subset L^2(\Omega)$ be the error function. We define $z \in V$ and $z_h \in V_h$ as the unique solutions to the problems

$$a(z, v) = (e, v)_{L^2(\Omega)} \;\; \forall\, v \in V, \quad a(z_h, v_h) = (e, v_h)_{L^2(\Omega)} \;\; \forall\, v_h \in V_h. \tag{4.1.25}$$

By assumption, $z \in H^2(\Omega)$, and we have, by Theorem 4.1.7 and (4.1.24),

$$|z - z_h|_{H^1(\Omega)} \le C_1\, h\, |z|_{H^2(\Omega)} \le \hat{c}\, C_1\, h |e|_{L^2(\Omega)}, \tag{4.1.26}$$

where we again denote by C_i, $i \in \mathbf{N}$, positive constants that do not depend on $h > 0$. Invoking to the symmetry of $a(\cdot\,,\cdot)$, and recalling the fact that $y - y_h$ is orthogonal to V_h, we obtain that

$$a(z_h, e) = a(y - y_h, z_h) = 0. \tag{4.1.27}$$

Thus, we can infer from (4.1.25), (4.1.26), (4.1.27), and (4.1.20), that

$$\begin{aligned} |y - y_h|^2_{L^2(\Omega)} &= |e|^2_{L^2(\Omega)} = (e, e)_{L^2(\Omega)} = a(z, e) = a(z - z_h, e) \\ &\le C_2\, |z - z_h|_{H^1(\Omega)}\, |e|_{H^1(\Omega)} \le C_3\, h^2\, |e|_{L^2(\Omega)}\, |y|_{H^2(\Omega)}, \end{aligned}$$

whence the assertion follows. □

Next, we consider the effect of numerical integration (in the coefficients of the bilinear form a and of F) on the FE solution. We solve a problem of the type $(\widetilde{E}_h)$, given by the modified system (let $\dim V_h = m$, with the basis $\{\varphi_i\}_{i=1}^m$)

$$\sum_{j=1}^m a_h(\varphi_j, \varphi_i)\, \widetilde{Y}_j = F_h(\varphi_i), \quad i = 1, \ldots, m, \quad \text{with } \tilde{y}_h = \sum_{j=1}^m \widetilde{Y}_j\, \varphi_j. \tag{4.1.28}$$

The application of numerical integration formulas leads to expressions of the form (see Křížek and Neittaanmäki [1990, Chapter 5])

$$\begin{aligned} a_h(\varphi_j, \varphi_i) &= \sum_{K \in \mathcal{T}_h} \sum_{q=1}^{Q} c_q^K \sum_{r,s=1}^{d} \Big(a_{rs}\, \frac{\partial \varphi_j}{\partial x_r}\, \frac{\partial \varphi_i}{\partial x_s} \Big)(x_q^K), \\ F_h(\varphi_i) &= \sum_{K \in \mathcal{T}_h} \sum_{q=1}^{Q} c_q^K\, f(x_q^K)\, \varphi_i(x_q^K), \end{aligned}$$

where c_q^K are the weights and x_q^K the nodes of the quadrature formula. In order that a_h and F_h be meaningful, we assume for simplicity that the coefficients a_{rs} of the bilinear form a and the right-hand side f of (E) are continuous in $\overline{\Omega}$. We have the general estimate

$$|y - \tilde{y}_h|_{H^1(\Omega)} = \mathcal{O}(h^k),$$

provided that the quadrature formula used is exact for all polynomials of degree up to $2k - 2$ and y, a_{rs}, f are sufficiently smooth. The proof of this result is based on the Bramble–Hilbert lemma and can be found in Ciarlet [1978, p. 199]. More precisely, if the family $\mathcal{T}_h$ is regular, then there exists a constant $C > 0$ that is independent of $h > 0$ such that for every solution $y \in C^{k+2}(\overline{\Omega})$ we have

$$|y - \tilde{y}_h|_{H^1(\Omega)} \le C\, h^k\, |y|_{C^{k+2}(\overline{\Omega})}.$$

Example 4.1.9 (Crouzeix and Rappaz [1990, Chapter 2]) The variational form for

$$-\Delta y = f \quad \text{on } \Omega \subset \mathbf{R}^2, \quad y = 0 \quad \text{on } \partial\Omega, \tag{4.1.29}$$

is (E) with $V = H_0^1(\Omega)$, and

$$a(y,v) = \int_\Omega \nabla y(x) \cdot \nabla v(x)\, dx \quad \forall\, y, v \in V, \quad F(v) = \int_\Omega f(x)\, v(x)\, dx \quad \forall\, v \in V.$$

Now let $\Omega \subset \mathbf{R}^2$ be a polygon, and let V_h be defined using polynomials in $P_1(K)$ for every $K \in \mathcal{T}_h$. The bilinear form a_h is just the restriction of a to $V_h \times V_h$, and

$$F_h(v_h) = \sum_{K \in \mathcal{T}_h} \left[\frac{1}{3} S_K \sum_{i=1}^{3} f(x_{iK})\, v_h(x_{iK}) \right],$$

where S_K is the area of K, and the $\{x_{iK}\}$ denote the vertices of K. Here, $f \in H^2(\Omega)$ and if $\tilde{y}_h$ denotes the solution to the corresponding problem $(\widetilde{E}_h)$, then there is some $C > 0$, which is independent of $h > 0$, such that

$$|y - \tilde{y}_h|_{L^2(\Omega)} \le C\, h^2\, |f|_{H^2(\Omega)}. \tag{4.1.30}$$

4.2 Error Estimates in the Finite Element Discretization of Control Problems

We consider the optimization problem (P) introduced in Chapter 1 (compare (1.1.1)–(1.1.4), (1.2.1)) in a Hilbert space setting:

$$\text{Min } \left\{ \theta(y) + \psi(u) \right\}, \tag{4.2.1}$$

subject to

$$Ay = Bu + f, \tag{4.2.2}$$

$$y \in C, \tag{4.2.3}$$

$$u \in U_{ad}. \tag{4.2.4}$$

Here, H, U, V are Hilbert spaces (not necessarily identified with their duals) such that the embeddings $V \subset H \subset V^*$ are continuous and compact, that is, in particular, there is a constant $\hat{c} > 0$ such that

$$|v|_H \le \hat{c}\, |v|_V \quad \forall\, v \in V.$$

Moreover, let $B \in L(U, V^*)$ and $f \in V^*$ be given, and suppose that $A \in L(V, V^*)$ has the property that the associated bilinear form

$$a(y,v) = (Ay, v)_{V^* \times V}$$

is coercive on V. Hence, there is some constant $\gamma > 0$ such that

$$a(y,y) \geq \gamma\,|y|_V^2 \quad \forall\, y \in V.$$

Notice that under these assumptions the operator $A^{-1}: V^* \to V$ exists and is continuous. Finally, we assume that $C \subset H$ and $U_{ad} \subset U$ are given nonempty and convex sets that are closed in H and U, respectively, and that the functionals $\theta : H \to \mathbf{R}_+$ and $\psi : U \to \mathbf{R}_+$ are convex and continuous. Notice that this setting allows for both distributed or boundary control problems, as in the Examples 1.2.1 and 1.2.2 in Chapter 1. Since the constraints (4.2.3), (4.2.4) are kept explicit, the assumed continuity of θ, ψ is not a strong limitation from the viewpoint of applications (compare with (1.1.5) in Chapter 1); on the other hand, it ensures that the subdifferentials $\partial\theta$, $\partial\psi$ exist at any point in H, U, respectively.

Let $[y_0, u_0]$ be an admissible pair for the problem (P), which is again assumed to exist. Using the shifts introduced in the relations (3.2.1)–(3.2.4) in Chapter 3, if necessary, we may henceforth without loss of generality assume that $f = 0$ and $[y_0, u_0] = [0,0]$ in the problem (4.2.1)–(4.2.4). This will simplify the exposition.

We impose further conditions:

$\partial\psi : U \to U^*$ is bounded, that is, maps bounded sets into bounded sets, and strongly monotone, that is, there is some $\alpha > 0$ such that

$$(w_1 - w_2, u_1 - u_2)_{U^* \times U} \geq \alpha\,|u_1 - u_2|_U^2 \quad \forall\, w_i \in \partial\psi(u_i),\ u_i \in U,\ i = 1,2. \tag{4.2.5}$$

$$0 \in \operatorname{int}(C) \text{ in the topology of } H \qquad \text{(Slater condition)}. \tag{4.2.6}$$

Observe that in view of (4.2.5), ψ is coercive in the sense that

$$\lim_{|u|_U \to \infty} \frac{\psi(u)}{|u|_U} = +\infty,$$

and Theorem 2.1.2 in Chapter 2 yields the existence of at least one optimal pair $[y^*, u^*]$. Since ψ is also strictly convex by (4.2.5), this optimal pair is unique. A typical example is $\psi(u) = \frac{1}{2}|u|_U^2$, in which case $\partial\psi = J : U \to U^*$ is the canonical (Riesz) isomorphism.

We consider some finite-dimensional subspaces $V_h \subset V$, $U_h \subset U$, where $h > 0$ is some discretization parameter, and where the norms and the scalar products in V_h and U_h coincide with those in H and U, respectively. We assume that

$$R_h^H : V \to V_h, \quad R_h^U : U \to U_h,$$

are given linear and bounded (so-called *restriction*) operators such that $R_h^H|_{V_h} : V_h \to V_h$ and $R_h^U|_{U_h} : U_h \to U_h$ are the identity operators. In practice, these operators may be some projection or interpolation operators (see (4.1.5), (4.1.6) in the previous section for particular cases). We define

$$\theta_h : V_h \to \mathbf{R}_+, \quad \theta_h = \theta|_{V_h}, \qquad \psi_h : U_h \to \mathbf{R}_+, \quad \psi_h = \psi|_{U_h}.$$

The finite-dimensional versions of the constraint sets are given by

$$U_{ad}^h = R_h^U(U_{ad}) \subset U_h, \quad C_h = R_h^H(C) \subset V_h.$$

Notice that the finite-dimensional constraint sets are again nonempty, convex, and closed in the respective spaces.

After these preparations, we can define the completely discretized control problem:

$$\text{Min}\left\{\theta_h(y_h) + \psi_h(u_h)\right\}, \tag{4.2.7}$$

$$u_h \in U_{ad}^h, \tag{4.2.8}$$

$$y_h \in C_h, \tag{4.2.9}$$

$$a_h(y_h, \varphi_h) \;=\; (B_h u_h, \varphi_h)_{V^* \times V} \quad \forall\, \varphi_h \in V_h, \tag{4.2.10}$$

with $a_h(\cdot\,,\cdot) = a(\cdot\,,\cdot)|_{V_h \times V_h}$ and $B_h = B|_{U_h}$. Note that also the finite-dimensional bilinear form a_h is coercive, since

$$a_h(v_h, v_h) \;\geq\; \gamma\,|v_h|_V^2 \;\geq\; \gamma\,\hat{c}^{-2}\,|v_h|_H^2 \;=\; \gamma\,\hat{c}^{-2}\,|v_h|_{V_h}^2 \quad \forall\, v_h \in V_h.$$

The Lax–Milgram lemma ensures the existence of a unique solution $y_h \in V_h$ of (4.2.10) for any $u_h \in U_h$. We also notice that the Slater condition (4.2.6) remains valid in the finite-dimensional setting, "uniformly" with respect to $h > 0$. Indeed, we have

$$C_h \;=\; R_h^H(C) \;\supset\; (C \cap V_h) \;\supset\; (S_H(0,\rho) \cap V_h) \;=\; S_{V_h}(0,\rho) \tag{4.2.11}$$

with some $\rho > 0$ that is independent of $h > 0$, where we set

$$S_H(0,\rho) = \{w \in H \,:\, |w|_H \leq \rho\}.$$

Here, we have used the assumption that $R_h^H|_{V_h}$ is the identity operator.

In particular, the pair $[0,0] \in C_h \times U_{ad}^h$ is admissible for (4.2.7)–(4.2.10) for all $h > 0$, and the existence of a unique discrete optimal pair $[y_h^*, u_h^*]$ follows as for the continuous problem (P), since ψ_h is strictly convex and coercive.

Proposition 4.2.1 *The set* $\{[y_h^*, u_h^*]\}_{h>0}$ *is bounded in* $V \times U$.

Proof. We have $\theta_h(y_h^*) + \psi_h(u_h^*) \le \theta_h(0) + \psi_h(0) = \theta(0) + \psi(0)$, whence, since θ is nonnegative,

$$\psi(u_h^*) \le \theta(0) + \psi(0),$$

and the coercivity of ψ on U shows that $\{u_h^*\}_{h>0}$ is bounded in U. Moreover, we can infer from (4.2.10) that

$$\alpha\, |y_h^*|_V^2 \;\le\; a_h(y_h^*, y_h^*) \;=\; (B_h u_h^*, y_h^*)_{V^*\times V},$$

whence the boundedness of $\{y_h^*\}_{h>0}$ in V follows. This finishes the proof of the assertion. □

A fundamental tool for the derivation of error estimates is given by the first-order optimality conditions. In the discretized setting (4.2.7)–(4.2.10), it is possible to employ arguments from mathematical programming. However, we will take a direct control approach instead, since this is more in the spirit of this text and has the advantage of being directly comparable with the continuous case.

For this purpose, we keep $h > 0$ fixed and define the penalized and regularized discrete control problem

$$\text{Min}\left\{\theta_h(y_h) + \psi_h(u_h) + (I_{C_h})_\lambda(y_h)\right\}, \tag{4.2.12}$$

subject to (4.2.8) and (4.2.10).

Here, $I_{C_h} : V_h \to]-\infty, \infty]$ is the indicator function of C_h in V_h, and $(I_{C_h})_\lambda$ denotes its Yosida–Moreau regularization for $\lambda > 0$ (compare Appendix 1).

First, observe that $[y_h^*, u_h^*] \in V_h \times U_{ad}^h$ is obviously admissible for the problem (4.2.8), (4.2.10), (4.2.12). It is then easily verified that (4.2.8), (4.2.10), (4.2.12) has a unique optimal pair $[y_{h\lambda}^*, u_{h\lambda}^*] \in V_h \times U_{ad}^h$. Moreover, Theorem A1.5(i) in Appendix 1 implies that

$$(I_{C_h})_\lambda(y_h^*) \;\le\; I_{C_h}(y_h^*) \;=\; 0,$$

and thus

$$\begin{aligned}\theta_h(y_{h\lambda}^*) + \psi_h(u_{h\lambda}^*) + (I_{C_h})_\lambda(y_{h\lambda}^*) \;&\le\; \theta_h(y_h^*) + \psi_h(u_h^*) + (I_{C_h})_\lambda(y_h^*)\\ &\le\; \theta_h(y_h^*) + \psi_h(u_h^*) + (I_{C_h})(y_h^*) \le\; \theta(0) + \psi(0).\end{aligned} \tag{4.2.13}$$

The same argument as in the proof of Proposition 4.2.1 then shows that the set $\{[y_{h\lambda}^*, u_{h\lambda}^*]\}_{\lambda>0}$ is bounded in $V \times U$.

Proposition 4.2.2 *There is some $p_{h\lambda} \in V_h$ such that the following optimality conditions are satisfied:*

$$a_h(y_{h\lambda}^*, \varphi_h) \;=\; (B_h u_{h\lambda}^*, \varphi_h)_{V^*\times V} \quad \forall\, \varphi_h \in V_h, \tag{4.2.14}$$

$$
\begin{aligned}
a_h(\varphi_h, p_{h\lambda}) &= (t_{h\lambda} + \nabla(I_{C_h})_\lambda(y^*_{h\lambda}), \varphi_h)_{V_h} \quad \forall \varphi_h \in V_h, \\
&\qquad \textit{for some } t_{h\lambda} \in \partial\theta_h(y^*_{h\lambda}), \qquad (4.2.15)\\
-B^*_h p_{h\lambda} &\in \partial I_{U^h_{ad}}(u^*_{h\lambda}) + \partial\psi_h(u^*_{h\lambda}). \qquad (4.2.16)
\end{aligned}
$$

Proof. We take variations $u^*_{h\lambda} + \mu\, w_h$, where $\mu \in [0,1]$ and $w_h = u_h - u^*_{h\lambda} \in U^h_{ad} - u^*_{h\lambda}$, and define associated $y_{\mu h}$ as the unique solutions to

$$
a_h(y_{\mu h}, \varphi_h) = (B_h u^*_{h\lambda} + \mu B_h w_h, \varphi_h)_{V^*\times V} \quad \forall \varphi_h \in V_h. \qquad (4.2.17)
$$

Observe that by construction,

$$
a_h(y^*_{h\lambda}, \varphi_h) = (B_h u^*_{h\lambda}, \varphi_h) \quad \forall \varphi_h \in V_h,
$$

and therefore,

$$
\gamma\, \hat{c}^{-2}\, |y_{\mu h} - y^*_{h\lambda}|^2_{V_h} \le a_h(y_{\mu h} - y^*_{h\lambda}, y_{\mu h} - y^*_{h\lambda}) \le \mu\, |B_h w_h|_{V_h}\, |y_{\mu h} - y^*_{h\lambda}|_{V_h}.
$$

Consequently, $y_{\mu h} \to y^*_{h\lambda}$ in V_h for $\mu \searrow 0$. We have

$$
\begin{aligned}
&\theta_h(y^*_{h\lambda}) + \psi_h(u^*_{h\lambda}) + (I_{C_h})_\lambda(y^*_{h\lambda}) \le \theta_h(y_{\mu h}) + \psi_h(u^*_{h\lambda} + \mu w_h) + (I_{C_h})_\lambda(y_{\mu h}),\\
&\quad \forall\, w_h \in U^h_{ad} - u^*_{h\lambda},\ \forall\, \mu \in [0,1].
\end{aligned}
$$

By the definition of the subdifferential, for any $t_{\mu h} \in \partial\theta_h(y_{\mu h})$ and any $s_{\mu h} \in \partial\psi_h(u^*_{h\lambda} + \mu\, w_h)$,

$$
\begin{aligned}
&(t_{\mu h}, y_{\mu h} - y^*_{h\lambda})_{V_h} + (s_{\mu h}, \mu\, w_h)_{U_h} + (\nabla(I_{C_h})_\lambda(y_{\mu h}), y_{\mu h} - y^*_{h\lambda})_{V_h} \ge 0,\\
&\quad \forall\, \mu \in [0,1],\ \forall\, w_h \in U^h_{ad} - u^*_{h\lambda}. \qquad (4.2.18)
\end{aligned}
$$

Now observe that $\partial\theta_h$ and $\partial\psi_h$ are bounded on bounded sets, and closed in finite-dimensional spaces. Hence, $\{t_{\mu h}\}_{\mu>0}$ and $\{s_{\mu h}\}_{\mu>0}$ are bounded in V_h and U_h, respectively, and we may select a sequence $\mu_n \searrow 0$ such that

$$
t_{\mu_n h} \to t_{h\lambda} \in \partial\theta_h(y^*_{h\lambda}) \quad \text{in } V_h, \qquad s_{\mu_n h} \to s_{h\lambda} \in \partial\psi_h(u^*_{h\lambda}) \quad \text{in } U_h.
$$

Hence, dividing by $\mu_n > 0$ and passing to the limit as $\mu_n \searrow 0$ in (4.2.18), we get

$$
0 \le (t_{h\lambda}, \ell_h)_{V_h} + (s_{h\lambda}, w_h)_{U_h} + (\nabla(I_{C_h})_\lambda(y^*_{h\lambda}), z_h)_{V_h}, \qquad (4.2.19)
$$

where $\ell_h \in V_h$ is the unique solution to

$$
a_h(\ell_h, \varphi_h) = (B_h w_h, \varphi_h)_{V^*\times V} \quad \forall \varphi_h \in V_h, \qquad (4.2.20)
$$

and where $w_h \in U^h_{ad} - u^*_{h\lambda}$ is arbitrary. By the definition of the adjoint system (4.2.15), and owing to (4.2.20), we may rewrite (4.2.19) as

$$(B_h w_h, p_{h\lambda})_{V^* \times V} + (s_{h\lambda}, w_h)_{U_h} \geq 0 \quad \forall\, w_h \in U^h_{ad} - u^*_{h\lambda}. \tag{4.2.21}$$

Since (4.2.16) is a direct consequence of (4.2.21), the proof is complete. □

Let us now examine the convergence properties for $\lambda \searrow 0$. In this setting the situation is simple, since we work in finite-dimensional spaces. As already noticed, $h > 0$ is fixed, and there is a sequence $\lambda_n \searrow 0$ such that

$$u^*_{h\lambda_n} \to \tilde{u}_h, \quad y^*_{h\lambda_n} \to \tilde{y}_h, \quad \text{as } n \to \infty.$$

Obviously, $\tilde{u}_h \in U^h_{ad}$, and $[\tilde{y}_h, \tilde{u}_h]$ satisfies (4.2.10).

Next, we derive a bound for $\{p_{h\lambda}\}_{\lambda>0}$. Recall that θ_h and $(I_{C_h})_\lambda$ are convex and nonnegative, and that θ is bounded on the closed ball of radius $\rho > 0$ in H, where $\rho > 0$ is given by the Slater assumption (4.2.11). It also follows from (4.2.16) that there are $s_{h\lambda} \in \partial\psi_h(u^*_{h\lambda})$ and $z_{h\lambda} \in \partial I_{U^h_{ad}}(u^*_{h\lambda})$ such that

$$B^*_h p_{h\lambda} = -s_{h\lambda} - z_{h\lambda}.$$

It follows that for any $\varphi_h \in V_h$ with $|\varphi_h|_{V_h} = 1$,

$$\begin{aligned}
&(t_{h\lambda} + \nabla(I_{C_h})_\lambda(y^*_{h\lambda}), y^*_{h\lambda} - \rho\,\varphi_h)_{V_h} \\
&\quad \geq \theta_h(y^*_{h\lambda}) - \theta_h(\rho\,\varphi_h) + (I_{C_h})_\lambda(y^*_{h\lambda}) - (I_{C_h})_\lambda(\rho\,\varphi_h) \\
&\quad \geq -\theta_h(\rho\,\varphi_h) + (I_{C_h})_\lambda(y^*_{h\lambda}) \geq -\theta(\rho\,\varphi_h) \geq -C_1,
\end{aligned} \tag{4.2.22}$$

with a constant $C_1 > 0$ that is independent of $\lambda > 0$, $h > 0$. Thus, inserting $y^*_{h\lambda} - \rho\,\varphi_h$ in (4.2.15), we find that

$$\begin{aligned}
\rho\, a_h(\varphi_h, p_{h\lambda}) &\leq C_1 + a_h(y^*_{h\lambda}, p_{h\lambda}) = C_1 + (B_h u^*_{h\lambda}, p_{h\lambda})_{V^* \times V} \\
&= C_1 + (u^*_{h\lambda}, B^*_h p_{h\lambda})_{U_h} = C_1 - (s_{h\lambda}, u^*_{h\lambda})_{U_h} - (z_{h\lambda}, u^*_{h\lambda})_{U_h} \\
&\leq C_1 - \psi_h(u^*_{h\lambda}) + \psi_h(0) - I_{U^h_{ad}}(u^*_{h\lambda}) + I_{U^h_{ad}}(0) \\
&= C_1 - \psi_h(u^*_{h\lambda}) + \psi_h(0) \leq C_2,
\end{aligned} \tag{4.2.23}$$

where $C_2 > 0$ is independent of $h > 0$, $\lambda > 0$. Consequently,

$$a_h(\varphi_h, p_{h\lambda}) \leq C_3, \quad \forall\, \varphi_h \in V_h, \ |\varphi_h|_{V_h} = 1,$$

with $C_3 > 0$ independent of $\lambda > 0$, $h > 0$. By fixing $\varphi_h = p_{h\lambda}/|p_{h\lambda}|_{V_h}$, this implies that

$$|p_{h\lambda}|_{V_h} \leq C_4 \quad \forall\, \lambda > 0, \quad \forall\, h > 0. \tag{4.2.24}$$

We also infer from (4.2.15) that

$$\begin{aligned} |\nabla(I_{C_h})_\lambda(y^*_{h\lambda})|^2_{V_h} &= a_h(\nabla(I_{C_h})_\lambda(y^*_{h\lambda}), p_{h\lambda}) - (t_{h\lambda}, \nabla(I_{C_h})_\lambda(y^*_{h\lambda}))_{V_h} \\ &\leq c_h \, |\nabla(I_{C_h})_\lambda(y^*_{h\lambda})|_{V_h}, \end{aligned} \tag{4.2.25}$$

with a constant $c_h > 0$ that does not depend on $\lambda > 0$. Consequently,

$$|\nabla(I_{C_h})_\lambda(y^*_{h\lambda})|_{V_h} \leq c_h,$$

and we may without loss of generality assume that for the sequence constructed above,

$$\nabla(I_{C_h})_{\lambda_n}(y^*_{h\lambda_n}) \to v^*_h \ \text{ in } \ V_h.$$

Recalling that $y^*_{h\lambda_n} \to \tilde{y}_h$, and invoking the maximal monotonicity of ∂I_{C_h}, we can conclude from Proposition A1.4(iv) in Appendix 1 that $v^*_h \in \partial I_{C_h}(\tilde{y}_h)$. Passing to the limit as $\lambda_n \searrow 0$ in (4.2.13), we find that

$$\theta_h(\tilde{y}_h) + \psi_h(\tilde{u}_h) + I_{C_h}(\tilde{y}_h) \leq \theta_h(y^*_h) + \psi_h(u^*_h).$$

But then we must have $\tilde{y}_h \in C_h$, and $[\tilde{y}_h, \tilde{u}_h]$ is admissible for the problem (4.2.7)–(4.2.10). By the uniqueness, $\tilde{y}_h = y^*_h$, $\tilde{u}_h = u^*_h$. Moreover, the above convergences hold generally for $\lambda \searrow 0$, and not only for the sequence selected above.

Now recall that $\{|p_{h\lambda}|_{V_h}\}_{\lambda>0}$ is bounded, so that there is some sequence $\lambda_n \searrow 0$ such that $p_{h\lambda_n} \to p_h$ in V_h. We also notice that by (4.2.16),

$$z_{h\lambda} \in -B^*_h p_{h\lambda} - \partial\psi_h(u^*_{h\lambda}), \ \text{ for some } z_{h\lambda} \in \partial I_{U^h_{ad}}(u^*_{h\lambda}).$$

Since $\partial\psi_h$ is bounded on bounded subsets of U_h, it follows that $\{|z_{h\lambda}|_{U_h}\}_{\lambda>0}$ is bounded, and we may without loss of generality assume that

$$z_{h\lambda_n} \to z_h \ \text{ in } U_h.$$

Thus, invoking the maximal monotonicity of $\partial I_{U^h_{ad}}$ and of $\partial\psi_h$, we conclude from Proposition A1.4(iv) in Appendix 1, letting $\lambda_n \searrow 0$, that

$$-B^*_h p_h \in \partial I_{U^h_{ad}}(u^*_h) + \partial\psi_h(u^*_h).$$

In summary, we have proved the following result:

Proposition 4.2.3 *There are some $p_h \in V_h$, and elements $t^*_h \in \partial\theta_h(y^*_h)$ and $v^*_h \in \partial I_{C_h}(y^*_h)$, such that the following optimality conditions are satisfied by $[y^*_h, u^*_h]$:*

$$\begin{aligned} a_h(y^*_h, \varphi_h) &= (B_h u^*_h, \varphi_h)_{V^* \times V}, \qquad \forall\, \varphi_h \in V_h, \\ a_h(\varphi_h, p_h) &= (t^*_h + v^*_h, \varphi_h)_{V_h}, \qquad \forall\, \varphi_h \in V_h, \\ -B^*_h p_h &\in \partial I_{U^h_{ad}}(u^*_h) + \partial\psi_h(u^*_h). \end{aligned}$$

Let us now recall the optimality conditions for the problem (4.2.1)–(4.2.4), as given in Theorem 3.2.1 in Chapter 3, where we put $Z = V^*$ and take $f = 0$ and $0 \in \operatorname{int}(C)$ in H and V as mentioned in (4.2.6). The pair $[y^*, u^*]$ is optimal if and only if there is some $p \in V$ such that

$$A^* p \in \partial\theta(y^*) + \partial I_C(y^*)\,, \tag{4.2.26}$$

$$-B^* p \in \partial I_{U_{ad}}(u^*) + \partial\psi(u^*)\,. \tag{4.2.27}$$

Let us suppose now that $U_h \subset (U^* \cap U)$ and that

$$(m^*, m)_{U^* \times U} = (m^*, m)_{U_h} \quad \text{if } m^*, m \in U_h.$$

By Proposition 4.2.3 and (4.2.27), there are $s \in \partial\psi(u^*)$ and $s_h \in \partial\psi_h(u_h^*)$ such that

$$z = -B^* p - s \in \partial I_{U_{ad}}(u^*) \subset U^*, \tag{4.2.28}$$

$$z_h = -B_h^* p_h - s_h \in \partial I_{U_{ad}^h}(u_h^*) \subset U_h. \tag{4.2.29}$$

In particular,

$$z + B^* p + s = z_h + B_h^* p_h + s_h = 0. \tag{4.2.30}$$

We now define some "interpolation" variables $[r_h, q_h]$ as any solution pair to the system

$$a_h(r_h, \varphi_h) = (Bu^*, \varphi_h)_{V^* \times V} \quad \forall\, \varphi_h \in V_h, \tag{4.2.31}$$

$$a_h(\varphi_h, q_h) = (t_h + v_h^*, \varphi_h)_{V_h} \quad \forall\, \varphi_h \in V_h, \tag{4.2.32}$$

where $t_h \in \partial\theta_h(r_h)$ is fixed arbitrarily and v_h^* is defined in Proposition 4.2.3.

Owing to the monotonicity of $\partial\theta$, and invoking the discrete state and adjoint equations, we can estimate:

$$\begin{aligned}
&(B^* p_h - B^* q_h, u_h^* - u^*)_{U^* \times U} \\
&\quad = (p_h - q_h, B_h u_h^* - Bu^*)_{V \times V^*} = a_h(y_h^* - r_h, p_h - q_h) \\
&\quad = (t_h^* + v_h^* - t_h - v_h^*, y_h^* - r_h)_{V_h} \geq 0.
\end{aligned} \tag{4.2.33}$$

We now postulate that the following additional compatibility condition is satisfied:

$$U_{ad}^h = R_h^U(U_{ad}) \subset U_{ad} \quad \forall\, h > 0.$$

Since $z \in \partial I_{U_{ad}}(u^*)$, it then follows from the definition of the subdifferential that

$$(z, u^* - u_h^*)_{U^* \times U} \geq I_{U_{ad}}(u^*) - I_{U_{ad}}(u_h^*) = 0,$$

and thus

$$(z_h - z, u_h^* - u^*)_{U^* \times U} = (z_h, u_h^* - u^*)_{U^* \times U} - (z, u_h^* - u^*)_{U^* \times U} \geq (z_h, u_h^* - u^*)_{U^* \times U}. \tag{4.2.34}$$

Moreover, in view of (4.2.5) and of Proposition 4.2.1, there is some $\widetilde{C} > 0$, independent of $h > 0$, such that for any $w_h \in \partial\psi(u_h^*)$, and for s, s_h as in (4.2.28), (4.2.29),

$$(s_h - s, u_h^* - u^*)_{U^* \times U} = (w_h - s, u_h^* - u^*)_{U^* \times U} + (s_h - w_h, u_h^* - u^*)_{U^* \times U} \geq \alpha\, |u_h^* - u^*|_U^2 - \widetilde{C}\, |s_h - w_h|_{U^*}. \tag{4.2.35}$$

We have the following result.

Theorem 4.2.4 *Suppose that (4.2.5) and (4.2.6) are fulfilled, and that $U_{ad}^h \subset U_{ad}$ and $U_h \subset (U^* \cap U)$. Then there is a constant $\hat{C} > 0$, depending on the optimal pair $[y^*, u^*]$ but not on $h > 0$, such that for any $w_h \in \partial\psi(u_h^*)$,*

$$\alpha\, |u_h^* - u^*|_U^2 \leq \hat{C}\, |s_h - w_h|_{U^*} + \hat{C}\, \left| R_h^U u^* - u^* \right|_U + |B^* p_h - B_h^* p_h + B^* p - B^* q_h|_{U^*}\, |u_h^* - u^*|_U. \tag{4.2.36}$$

Moreover, if $\psi(u) = \frac{1}{2}\, |u|_{L^2(\Omega)}^2$, and if R_h^U satisfies the orthogonality property (compare (4.1.6))

$$\left(v, R_h^U u - u\right)_{U^* \times U} = 0 \quad \forall v \in U_h, \quad \forall u \in U,$$

then

$$\alpha\, |u_h^* - u^*|_U \leq |B^* p_h - B_h^* p_h + B^* p - B^* q_h|_{U^*}\,. \tag{4.2.36$'$}$$

Proof. We denote by $C_i > 0$, $i \in \mathbf{N}$, constants that do not depend on $h > 0$. To begin with, recall that $\{|u_h^*|_U\}_{h>0}$ is by Proposition 4.2.1 bounded. Since ψ is continuous and $\partial\psi : U \to U^*$ is a bounded operator, and by virtue of the chain rule (compare Theorem A1.12 in Appendix 1)

$$s_h \in \partial\psi_h(u_h^*) = \partial(\psi \circ i_h)(u_h^*) = (i_h^* \circ \partial\psi \circ i_h)(u_h^*),$$

where $i_h : U_h \to U$ is the canonical embedding and $i_h^* : U^* \to U_h$ its dual (both having the operator norm unity), it then follows that $\{|s_h|_{U_h}\}_{h>0}$ is bounded. Hence, using (4.2.29) and recalling that $\{|p_h|_{V_h}\}_{h>0}$ is by (4.2.24) bounded, we conclude that $\{|z_h|_{U_h}\}_{h>0}$ is bounded.

Moreover, we have $R_h^U u^* \in U_{ad}^h$. Since $z_h \in \partial I_{U_{ad}^h}(u^*)$, it follows that

$$\left(z_h, u_h^* - R_h^U u^*\right)_{U_h} \geq 0.$$

Summarizing, we conclude that

$$\begin{aligned}(z_h, u_h^* - u^*)_{U^*\times U} &= \left(z_h, u_h^* - R_h^U u^*\right)_{U_h} + \left(z_h, R_h^U u^* - u^*\right)_{U^*\times U} \\ &\geq \left(z_h, R_h^U u^* - u^*\right)_{U^*\times U} \geq -C_1 \left|R_h^U u^* - u^*\right|_U . \qquad (4.2.37)\end{aligned}$$

We obtain, using (4.2.33)–(4.2.35) and (4.2.37),

$$\begin{aligned}&(B^*p_h - B^*q_h + s_h - s + z_h - z, u_h^* - u^*)_{U^*\times U} \\ &\geq \alpha \left|u_h^* - u^*\right|_U^2 - C_2 \left|s_h - s\right|_{U^*} - C_1 \left|R_h^U u^* - u^*\right|_U . \qquad (4.2.38)\end{aligned}$$

By (4.2.30), the left-hand side in (4.2.38) may be rewritten as

$$(B^*p_h - B_h^* p_h + B^*p - B^*q_h, u_h^* - u^*)_{U^*\times U} ,$$

whence (4.2.36) follows.

Concerning (4.2.36)′, we notice that the first term in the right-hand side of (4.2.36) vanishes for $\psi(u) = \frac{1}{2}\left|u\right|_U^2$. Moreover, owing to the orthogonality property, inequality (4.2.37) gives

$$(z_h, u_h^* - u^*)_U \geq 0. \qquad (4.2.37)'$$

Then, the last two terms in (4.2.38) disappear, and the proof is finished. □

Remark. The last term in (4.2.36) is an interpolation error associated with the adjoint equation. The quality of the bound for the term $\left|R_h^U u^* - u^*\right|_U$ depends on the regularity of u^*. Estimate (4.2.36)′ refers to distributed control problems.

Example 4.2.5 We now consider an example involving both state and control constraints, to which we apply the theory developed above:

$$\text{Min}\left\{\frac{1}{2}\int_\Omega (y(x) - y_d(x))^2\,dx + \frac{1}{2}\int_\Omega u^2(x)\,dx\right\}, \qquad (4.2.39)$$

subject to

$$-\Delta y = u \quad \text{in } \Omega, \qquad (4.2.40)$$

$$y = 0 \quad \text{on } \partial\Omega, \qquad (4.2.41)$$

$$u \in U_{ad} = \left\{w \in L^2(\Omega) \,:\, \alpha(x) \leq w(x) \leq \beta(x) \text{ for a.e. } x \in \Omega\right\}, \qquad (4.2.42)$$

$$y \in C = \left\{z \in L^2(\Omega) \,:\, \int_\Omega z(x)\,dx \leq 1\right\}. \qquad (4.2.43)$$

Here, $\Omega \subset \mathbf{R}^d$ is a bounded domain, α, β, $y_d \in L^2(\Omega)$ are given, and

$$\alpha \leq 0 \leq \beta \quad \text{a.e. in } \Omega. \tag{4.2.44}$$

To fit the problem (4.2.39)–(4.2.43) in our general setting, we put

$$V = H_0^1(\Omega), \quad U = U^* = H = H^* = L^2(\Omega), \quad V^* = H^{-1}(\Omega),$$
$$A = -\Delta, \quad B = \text{canonical embedding of } L^2(\Omega) \text{ in } H^{-1}(\Omega).$$

By (4.2.44), the pair $[0,0]$ is admissible (but, due to the presence of y_d, not necessarily optimal). It also ensures that the Slater condition (4.2.6) is fulfilled. Moreover, we have

$$a(u,v) = \int_\Omega \nabla u(x) \cdot \nabla v(x)\, dx \quad \forall\, u, v \in H_0^1(\Omega),$$
$$\theta(y) = \frac{1}{2} \int_\Omega (y(x) - y_d(x))^2\, dx, \quad \psi(u) = \frac{1}{2}\, |u|_U^2.$$

In particular, θ and ψ are nonnegative, convex, and continuous on $H_0^1(\Omega)$ and $L^2(\Omega)$, respectively, where $\partial\psi : L^2(\Omega) \to L^2(\Omega)$ is the identity that satisfies (4.2.5). Observe also that the state system (4.2.40), (4.2.41) has for any $u \in U_{ad}$ a unique weak solution $y = y_u$ that belongs to $H^2(\Omega) \cap H_0^1(\Omega)$ if Ω is smooth or convex.

Next, observe that the cost functional fulfills the coercivity condition (2.1.6). Hence, it follows from Theorem 2.1.2 in Chapter 2 that the problem (4.2.39)–(4.2.43) admits an optimal pair $[y^*, u^*]$, which, owing to the strict convexity of the cost functional, is unique.

In order to derive error estimates for the optimal controls, we first need to recover further regularity properties of u^*. Recalling Example 3.2.11 and Corollary 3.2.12 in Chapter 3, one may expect that under the constraint (4.2.42) generalized bang-bang properties for $[y^*, u^*]$ hold. We will therefore give a direct argument to analyze the regularity of u^*. We have the following result.

Proposition 4.2.6 *Suppose that $\alpha, \beta \in H^1(\Omega) \cap L^\infty(\Omega)$. Then $u^* \in H^1(\Omega)$.*

Proof. We use a specific penalization technique. We define for $\varepsilon > 0$ the penalized problem

$$\text{Min}\left\{\frac{1}{2}\int_\Omega (y(x) - y_d(x))^2\, dx + \frac{1}{2}\int_\Omega u^2(x)\, dx + \frac{1}{2\varepsilon}\Big(\int_\Omega y(x)\, dx - 1\Big)_+^2\right\}, \tag{4.2.45}$$

subject to (4.2.40)–(4.2.42).

It follows from Theorem 2.1.2 in Chapter 2 that the penalized problem (4.2.40)–(4.2.42), (4.2.45) has an optimal pair $[y_\varepsilon, u_\varepsilon] \in H_0^1(\Omega) \times U_{ad}$, which, by the strict convexity of the cost functional, is unique. In addition, we have

$$\frac{1}{2}\int_\Omega (y_\varepsilon(x) - y_d(x))^2\,dx + \frac{1}{2}\int_\Omega u_\varepsilon(x)^2\,dx + \frac{1}{2\varepsilon}\Big(\int_\Omega y_\varepsilon(x)\,dx - 1\Big)_+^2$$
$$\leq \frac{1}{2}\int_\Omega (y^*(x) - y_d(x))^2\,dx + \frac{1}{2}\int_\Omega u^*(x)^2\,dx,$$

from which, using the state system, we can conclude that the set $\{[y_\varepsilon, u_\varepsilon]\}_{\varepsilon>0}$ is bounded in $H_0^1(\Omega) \times L^2(\Omega)$. Moreover,

$$\lim_{\varepsilon \searrow 0}\Big(\int_\Omega y_\varepsilon(x)\,dx - 1\Big)_+ = 0.$$

Therefore, there is some sequence $\varepsilon_n \searrow 0$ such that $y_{\varepsilon_n} \to \overline{y}$ weakly in $H_0^1(\Omega)$ and strongly in $L^2(\Omega)$, and such that $u_{\varepsilon_n} \to \overline{u} \in U_{ad}$ weakly in $L^2(\Omega)$. But then, in view of the above convergence property, we must have

$$\int_\Omega \overline{y}(x)\,dx \leq 1,$$

that is, $[\overline{y}, \overline{u}]$ is admissible for (4.2.39)–(4.2.43). Passage to the limit as $\varepsilon \searrow 0$ in the above inequality, invoking the lower semicontinuity of the cost functional, then yields that $[\overline{y}, \overline{u}] = [y^*, u^*]$. This entails that the above convergences hold generally for $\varepsilon \searrow 0$, and not only for the selected sequence. Also, since obviously

$$|y_\varepsilon - y_d|^2_{L^2(\Omega)} + |u_\varepsilon|^2_{L^2(\Omega)} \to |y^* - y_d|^2_{L^2(\Omega)} + |u^*|^2_{L^2(\Omega)},$$

we even have strong convergence in $L^2(\Omega)$, and it follows that for $\varepsilon \searrow 0$,

$$y_\varepsilon \to y^* \text{ strongly in } H_0^1(\Omega), \quad u_\varepsilon \to u^* \text{ strongly in } L^2(\Omega).$$

Next, we exploit the first-order necessary optimality conditions for the penalized system, which by virtue of Theorem 3.1.7 in Chapter 3, takes the form

$$-\Delta p_\varepsilon = y_\varepsilon - y_d + \frac{1}{\varepsilon}\Big(\int_\Omega y_\varepsilon(x)\,dx - 1\Big)_+ \quad \text{in } \Omega, \tag{4.2.46}$$

$$p_\varepsilon(x) = 0 \quad \text{on } \partial\Omega, \tag{4.2.47}$$

$$0 \leq \int_\Omega (u_\varepsilon(x) + p_\varepsilon(x))(v(x) - u_\varepsilon(x))\,dx \quad \forall\, v \in U_{ad}. \tag{4.2.48}$$

The relations (4.2.46), (4.2.47) form the adjoint system which has a unique weak solution $p_\varepsilon \in H_0^1(\Omega)$. It is an easy exercise to verify that the maximum principle (4.2.48) implies that

$$u_\varepsilon(x) = \min\left\{\beta(x)\,,\, \max\{\alpha(x), -p_\varepsilon(x)\}\right\} \quad \text{for a.e. } x \in \Omega. \qquad (4.2.48)'$$

Then (4.2.48)′ and the lattice properties of $H^1(\Omega)$, where we invoke the regularity assumptions for α and β, yield that $u_\varepsilon \in H^1(\Omega)$.

Next, we derive bounds for $\{p_\varepsilon\}_{\varepsilon>0}$. Invoking the adjoint state equation, as well as the boundedness of $\left\{|y_\varepsilon|_{L^2(\Omega)}\right\}_{\varepsilon>0}$ and of $\left\{|u_\varepsilon|_{L^2(\Omega)}\right\}_{\varepsilon>0}$, we find that

$$\begin{aligned}
&\frac{1}{\varepsilon}\Big(\int_\Omega y_\varepsilon(x)\,dx - 1\Big)_+ \\
&\quad\le \frac{1}{\varepsilon}\Big(\int_\Omega y_\varepsilon(x)\,dx - 1\Big)_+\Big(\int_\Omega y_\varepsilon(x)\,dx - 1\Big) + \frac{1}{\varepsilon}\Big(\int_\Omega y_\varepsilon(x)\,dx - 1\Big)_+ \\
&\quad= \frac{1}{\varepsilon}\int_\Omega\Big(\int_\Omega y_\varepsilon(x)\,dx - 1\Big)_+ y_\varepsilon(x)\,dx \\
&\quad= \int_\Omega \nabla p_\varepsilon(x)\cdot\nabla y_\varepsilon(x)\,dx - \int_\Omega (y_\varepsilon(x) - y_d(x))\,y_\varepsilon(x)\,dx \\
&\quad\le -\int_\Omega u_\varepsilon^2(x)\,dx - \int_\Omega (y_\varepsilon(x) - y_d(x))\,y_\varepsilon(x)\,dx \,\le\, \hat{C},
\end{aligned}$$

with some $\hat{C} > 0$ that does not depend on $\varepsilon > 0$. The last inequality is a consequence of (4.2.40) and (4.2.48) with $v = 0$. Taking (4.2.46), (4.2.47) into account, we see that $\{p_\varepsilon\}_{\varepsilon>0}$ is bounded in $H_0^1(\Omega)$. Hence, there is some sequence $\varepsilon_n \searrow 0$ such that $p_{\varepsilon_n} \to p$ weakly in $H_0^1(\Omega)$ and pointwise a.e. in Ω. Since

$$y_\varepsilon \to y^*, \qquad u_\varepsilon \to u^*, \qquad \text{both strongly in } L^2(\Omega),$$

the passage to the limit as $\varepsilon \searrow 0$ in (4.2.48)′ yields that

$$u^* = \min\left\{\beta\,,\, \max\{\alpha\,,\, -p\}\right\} \quad \text{a.e. in } \Omega,$$

and the lattice properties of $H^1(\Omega)$ imply that $u^* \in H^1(\Omega)$, which concludes the proof. □

We now discretize the problem (4.2.39)–(4.2.43) using a triangulation $\mathcal{T}_h$, $h > 0$, of Ω, which is assumed to be a polygon in $\mathbf{R}^2$. We also assume that α and β are constant functions. We choose $V_h = R_h^H(V)$ as the set of piecewise linear finite elements associated with $\mathcal{T}_h$ (compare Example 4.1.1), and $U_h = R_h^U(U)$ as the set of piecewise constant functions associated with $\mathcal{T}_h$, where

$$R_h^U(u)|_K(x) = \frac{1}{\text{meas}(K)}\int_K u(x)\,dx, \quad \forall\, x \in K, \quad \forall\, u \in L^2(\Omega), \quad \forall\, K \in \mathcal{T}_h.$$

With these choices, we have

$$R_h^U(U_{ad}) \subset U_{ad}, \quad U_h \subset (U^* \cap U) = U = L^2(\Omega),$$

and in view of (4.1.6), the orthogonality condition of Theorem 4.2.4 is fulfilled. Now observe that in the notation of Theorem 4.2.4 we have

$$s_h = w_h = \partial\psi_h(u_h^*) = \partial\psi(u_h^*), \quad B^* p_h = B_h^* p_h. \tag{4.2.49}$$

Therefore, the error estimate (4.2.36)′ reduces to

$$|u_h^* - u^*|_{L^2(\Omega)} \leq \hat{C}_1 \, |B^*(p - q_h)|_{L^2(\Omega)}, \tag{4.2.50}$$

with some $\hat{C}_1 > 0$.

Another variant is to take piecewise linear finite elements also in the control space. Denoting the corresponding interpolation operator by $\widetilde{R}_h^U$, we have

$$\widetilde{R}_h^U(U_{ad} \cap H^1(\Omega)) \subset U_{ad}, \quad U_h = \widetilde{R}_h^U(H^1(\Omega)) \subset L^2(\Omega) = (U^* \cap U).$$

In this case the orthogonality condition of Theorem 4.2.4 is violated, and since (4.2.49) is still valid, we have the error estimate

$$|u_h^* - u^*|_{L^2(\Omega)} \leq \hat{C}_2 \left(|B^*(p - q_h)|_{L^2(\Omega)} + \left|\widetilde{R}_h^U u^* - u^*\right|_{L^2(\Omega)} \right). \tag{4.2.51}$$

Now recall that thanks to Proposition 4.2.6, $u^* \in H^1(\Omega)$. Hence, using the classical interpolation properties of finite elements (see (4.1.7) and Theorem 4.1.3), we have

$$\left|\widetilde{R}_h^U u^* - u^*\right|_{L^2(\Omega)} = \mathcal{O}(h). \tag{4.2.52}$$

It remains to estimate the term $|p - q_h|_{L^2(\Omega)}$. It follows from Theorem 4.1.7 that also this error is $\mathcal{O}(h)$ provided that $p \in H^2(\Omega)$. This is the case if Ω is a convex polygon, as one can see by passing to the limit in (4.2.46), (4.2.47), which is possible in view of the boundedness of the penalization term. Notice that in this example q_h is defined via the system

$$\int_\Omega \nabla r_h \cdot \nabla \varphi_h \, dx = \int_\Omega u^* \varphi_h \, dx \quad \forall \varphi_h \in V_h,$$

$$\int_\Omega \nabla q_h \cdot \nabla \varphi_h \, dx = \int_\Omega (r_h - y_d^h + m) \, \varphi_h \, dx \quad \forall \varphi_h \in V_h,$$

where y_d^h is an appropriate discretization of y_d (which has to belong to $H^1(\Omega)$), and $m \in \mathbf{R}$ is the cluster point of the set $\left\{\frac{1}{\varepsilon}(\int_\Omega y_\varepsilon(x)\, dx - 1)_+\right\}_{\varepsilon > 0}$, which is also used in the definition of p.

In summary, making the regularity hypothesis $y_d \in H^1(\Omega)$, and assuming that $\Omega \subset \mathbf{R}^2$ is a convex polygon, we arrive at the desired error estimate

$$|u_h^* - u^*|_{L^2(\Omega)} = \mathcal{O}(h), \tag{4.2.53}$$

if R_h^U is used, and

$$|u_h^* - u^*|_{L^2(\Omega)} = \mathcal{O}(h^{\frac{1}{2}}), \tag{4.2.54}$$

if $\tilde{R}_h^U$ is used.

The estimate (4.2.53) is optimal in the sense that it has the same order as the interpolation error estimate for u^*, according to Theorem 4.1.3 with $k = m = 0$.

Remark. A similar approach for parabolic control problems is due to Tiba and Tröltzsch [1996], Neittaanmäki and Tiba [1994, Chapter V]. An abstract variant was studied by Malanowski [1982]. The orthogonality argument from (4.2.36)′ is due to Casas and Tröltzsch [2003]. In Casas and Tröltzsch [2002], error estimates based on the theory of second-order optimality conditions were derived. Arnăutu and Neittaanmäki [1998] investigated higher-order discretization/interpolation techniques, including finite element and spectral methods. Adaptive finite element methods for control problems have been studied by Becker, Kapp, and Rannacher [2000], and by Li, Liu, Ma, and Tang [2002].

Remark. The discussion of Example 4.2.5 shows that our technique allows us to establish error estimates both for the adjoint state and for the optimal value. Taking into account the investigation of δ-solutions from Section 2.1 in Chapter 2, the last point is particularly significant for optimal control theory.

4.3 Semidiscretization

The semidiscretization approach (or *method of lines*) is well known for evolution equations. By discretizing only with respect to the space variables, the original parabolic or hyperbolic equation is approximated by a system of ordinary differential equations in the time variable. The advantage of this procedure is that it minimizes the discretization effort and allows us to use powerful numerical software for ordinary differential equations. This philosophy may also be applied to elliptic systems by discretizing only with respect to some of the independent variables; it has been developed for very general domains especially in the works of Xanthis and Schwab [1991], [1992] under the name of *method of arbitrary lines* (MAL). In this case the original elliptic equation is approximated by a boundary value problem for an ordinary differential system.

We will present here the basic ideas when applied to boundary control problems governed by elliptic systems. Proofs will not be provided or just sketched in this section since the arguments are very similar to those employed in the previous section. We encourage the reader to complete the proofs. For the sake of simplicity, we take $\Omega =]0,1[\times]-1,1[\subset \mathbf{R}^2$. We then study the optimization problem

$$\operatorname{Min}\left\{\frac{1}{2}\int_{\Omega}(y(x_1,x_2)-y_d(x_1,x_2))^2\,dx_1\,dx_2+\frac{1}{2}\int_{\Gamma_N}u^2(\sigma)\,d\sigma\right\}, \tag{4.3.1}$$

subject to

$$-\Delta y = f \quad \text{in } \Omega, \tag{4.3.2}$$

$$y = 0 \quad \text{on } \Gamma_D, \tag{4.3.3}$$

$$\frac{\partial y}{\partial n} = u \quad \text{on } \Gamma_N, \tag{4.3.4}$$

$$u \in U_{ad}. \tag{4.3.5}$$

Here, we have put $\Gamma_D = \Gamma_D^1 \cup \Gamma_D^2$, $\Gamma_N = \Gamma_N^1 \cup \Gamma_N^2$, where

$$\begin{aligned}
\Gamma_D^1 &= \left\{(x_1,x_2)\in\overline{\Omega}:\, x_2=1,\ x_1\in\,]0,1[\right\},\\
\Gamma_D^2 &= \left\{(x_1,x_2)\in\overline{\Omega}:\, x_1=0,\ x_2\in\,]-1,1[\right\},\\
\Gamma_N^1 &= \left\{(x_1,x_2)\in\overline{\Omega}:\, x_2=-1,\ x_1\in\,]0,1[\right\},\\
\Gamma_N^2 &= \left\{(x_1,x_2)\in\overline{\Omega}:\, x_1=1,\ x_2\in\,]-1,1[\right\},
\end{aligned}$$

and it is assumed that y_d, $f \in L^2(\Omega)$, while $U_{ad} \subset U = L^2(\Gamma_N)$ is a nonempty, closed, and convex set defining the control constraints. Notice that for any $u \in U_{ad}$ the state system (4.3.2)–(4.3.4) has a unique weak solution

$$y = Au \in V \quad \left(= H_D^1(\Omega) = \left\{w \in H^1(\Omega) : w|_{\Gamma_D} = 0\right\}\right)$$

satisfying

$$\int_\Omega \nabla y\cdot\nabla w\,dx_1\,dx_2 = \int_\Omega f\,w\,dx_1\,dx_2 + \int_{\Gamma_N} u\,w\,d\sigma \quad \forall\, w \in H_D^1(\Omega), \tag{4.3.6}$$

where $A : L^2(\Gamma_N) \to H_D^1(\Omega)$ is the operator $u \mapsto y$ given by (4.3.2)–(4.3.4). Clearly, A has a linear and bounded inverse A^{-1}. Next, we define the linear and bounded operator $B : L^2(\Gamma_N) \to H_D^1(\Omega)^*$ by

$$(Bu, w)_{H_D^1(\Omega)^*\times H_D^1(\Omega)} = \int_{\Gamma_N} u\,w\,d\sigma \quad \forall\, w \in H_D^1(\Omega). \tag{4.3.7}$$

Then the state system (4.3.2)–(4.3.4) (respectively, (4.3.6)) can be rewritten as $Ay = Bu + f$, and since we obviously have admissibility and a cost functional that is continuous, strictly convex, and coercive in the sense of (2.1.6), it follows from Theorem 2.1.2 in Chapter 2 that the control problem (4.3.1)–(4.3.5) has a unique optimal pair $[y^*, u^*] \in H^1_D(\Omega) \times U_{ad}$. Moreover, invoking Theorem 3.2.1 in Chapter 3, we have the optimality conditions

$$-\Delta p^* = y^* - y_d \quad \text{in } \Omega, \tag{4.3.8}$$

$$p^* = 0 \quad \text{on } \Gamma_D, \tag{4.3.9}$$

$$\frac{\partial p^*}{\partial n} = 0 \quad \text{on } \Gamma_N, \tag{4.3.10}$$

$$-B^* p^* \in u^* + \partial I_{U_{ad}}(u^*) \quad \text{on } \Gamma_N, \tag{4.3.11}$$

where $B^* : H^1_D(\Omega) \to L^2(\Gamma_N)$ is the dual operator of B.

For the following, we postulate that $u^* \in H^1(\Gamma_N)$. This regularity condition is fulfilled for a large class of constraint sets U_{ad}. A standard example is the obstacle case

$$U_{ad} = \{v \in L^2(\Gamma_N) : -1 \leq v \leq 1 \text{ a.e. in } \Gamma_N\}.$$

Indeed, in this case (4.3.11) can be rewritten as

$$\int_{\Gamma_N} (u^*(\sigma) + p^*(\sigma))\,(v(\sigma) - u^*(\sigma))\,d\sigma \geq 0 \quad \forall v \in U_{ad},$$

and the required regularity follows as in the previous Example 4.2.5 (recall the proof of Proposition 4.2.6).

To continue, observe that in the present setting the method of arbitrary lines reduces to the standard method of lines, and we can discretize with respect to x_2 while keeping x_1 as a continuous variable. To this end, fix some finite element basis $\{Q_\ell\}_{\ell=\overline{1,\nu}} \subset H^1(-1,1)$, $\nu \in \mathbf{N}$, and approximate the data and the unknown mappings in (4.3.1)–(4.3.5) by

$$y^\nu(x_1, x_2) = \sum_{\ell=1}^{\nu} y_\ell(x_1)\, Q_\ell(x_2)\,, \tag{4.3.12}$$

$$f^\nu(x_1, x_2) = \sum_{\ell=1}^{\nu} f_\ell(x_1)\, Q_\ell(x_2)\,, \tag{4.3.13}$$

$$y_{d\nu}(x_1, x_2) = \sum_{\ell=1}^{\nu} y_d^\ell(x_1)\, Q_\ell(x_2)\,, \tag{4.3.14}$$

$$u^\nu|_{\Gamma_N^2}(x_1, x_2) = \sum_{\ell=1}^{\nu} u_\ell\, Q_\ell(x_2)\,, \quad u_\ell \in \mathbf{R},\ \ell = \overline{1,\nu}, \tag{4.3.15}$$

$$u^\nu|_{\Gamma_N^1} = \overline{u}_1(x_1) \qquad \text{(notation)}. \tag{4.3.16}$$

We introduce the matrices

$$\widetilde{A} = (a_{\ell k})_{\ell,k=\overline{1,\nu}}, \quad a_{\ell k} = \int_{-1}^{1} Q_\ell(x_2)\, Q_k(x_2)\, dx_2, \tag{4.3.17}$$

$$B = (b_{\ell k})_{\ell,k=\overline{1,\nu}}, \quad b_{\ell k} = \int_{-1}^{1} Q'_\ell(x_2)\, Q'_k(x_2)\, dx_2. \tag{4.3.18}$$

Observe that (4.3.12)–(4.3.16) allow u to be discontinuous at the corner $\overline{\Gamma^1_N} \cap \overline{\Gamma^2_N}$. Also, the choice of the finite element basis should comply with the Dirichlet condition on Γ^1_D, that is,

$$Q_\ell(1) = 0, \quad \ell = \overline{1,\nu}, \tag{4.3.19}$$

must hold. Then also the compatibility with the Dirichlet boundary condition at the corner $\overline{\Gamma^1_D} \cap \overline{\Gamma^2_N}$ is guaranteed. We also require that

$$Q_1(-1) = 1, \quad Q_\ell(-1) = 0, \quad \ell = \overline{2,\nu}, \tag{4.3.20}$$

but this is not essential.

Substituting (4.3.12)–(4.3.20) into the weak formulation (4.3.6) of the state system, we finally arrive at the semidiscretized state system (δ_{1k} is the Kronecker symbol)

$$-\sum_{\ell=1}^{\nu} a_{\ell k}\, y''_\ell + \sum_{\ell=1}^{\nu} b_{\ell k}\, y_\ell = \sum_{\ell=1}^{\nu} a_{\ell k}\, f_\ell + \overline{u}_1\, \delta_{1k}, \quad k = \overline{1,\nu}, \tag{4.3.21}$$

$$y_k(0) = 0, \quad k = \overline{1,\nu}, \tag{4.3.22}$$

$$\sum_{\ell=1}^{\nu} a_{\ell k}\, y'_\ell(1) = \sum_{\ell=1}^{\nu} u_\ell\, a_{\ell k}, \quad k = \overline{1,\nu}. \tag{4.3.23}$$

We note that in this formulation the boundary control becomes a distributed control on Γ^1_n and a boundary control on the discretization nodes on Γ^2_N. It is here that we use (4.3.20); otherwise some coefficient would appear in the last term of (4.3.21).

Similarly, for the approximating cost functional to be minimized, we have

$$\operatorname{Min}\left\{\frac{1}{2}\int_0^1 \sum_{\ell,k=1}^{\nu} (y_\ell(x_1) - y_d^\ell(x_1))\, a_{\ell k}\, (y_k(x_1) - y_d^k(x_1))\, dx_1 \right.$$
$$\left. + \frac{1}{2}\int_0^1 \overline{u}_1^2(x_1)\, dx_1 + \frac{1}{2}\sum_{\ell,k=1}^{\nu} u_\ell\, a_{\ell k}\, u_k \right\}. \tag{4.3.24}$$

It is easily seen that the system (4.3.21)–(4.3.23) admits a unique solution $\tilde{y}^\nu$ in the space

$$V_\nu = \left\{ v = \sum_{\ell=1}^{\nu} v_\ell\, Q_\ell : \; v_\ell \in H^1(0,1), \; v_\ell(0) = 0, \; \ell = \overline{1,\nu} \right\}. \tag{4.3.25}$$

In the following, we will identify the function $\tilde{y}^\nu$ with the vector of functions $\{y_\ell(x_1)\}_{\ell=\overline{1,\nu}}$. The control mapping $\tilde{u}_\nu = \left\{\overline{u}_1(x_1), (u_\ell)_{\ell=\overline{1,\nu}}\right\}$ ranges in the space

$$U_\nu = \left\{u \in L^2(\Gamma_N) : u|_{\Gamma_N^1} \in L^2(\Gamma_N^1),\ u|_{\Gamma_N^2} = \sum_{\ell=1}^{\nu} u_\ell Q_\ell,\ u_\ell \in \mathbf{R},\ \ell = \overline{1,\nu}\right\}. \tag{4.3.26}$$

Obviously, $V_\nu \subset V = H_D^1(\Omega)$ and $U_\nu \subset U = L^2(\Gamma_N)$ are closed subspaces that inherit the scalar products and norms from $H_D^1(\Omega)$ and $L^2(\Gamma_N)$, respectively.

Let $R_\nu : U \to U_\nu$ denote a linear and bounded "restriction" operator. We define the semidiscrete control constraints

$$\tilde{u}_\nu \in U_{ad}^\nu = R_\nu(U_{ad}) \subset U_\nu, \tag{4.3.27}$$

and we assume that

$$U_{ad}^\nu \subset U_{ad} \tag{4.3.28}$$

is a closed and convex set.

Now observe that the matrix $\widetilde{A}$ is positive definite, since the set $\{Q_\ell\}_{\ell=\overline{1,\nu}}$ is linearly independent. It then follows that the discretized cost functional (4.3.24) is continuous, strictly convex, and coercive, and therefore the discretized control problem (4.3.21)–(4.3.24), (4.3.27) has a uniquely determined optimal pair $[y_\nu^*, u_\nu^*]$. Using the standard variation technique as demonstrated in Chapter 3, we can derive the characterization of $[y_\nu^*, u_\nu^*]$, given in form of the first-order optimality conditions:

Theorem 4.3.1 *Necessary and sufficient optimality conditions for* $[y_\nu^*, u_\nu^*]$ *are given by (4.3.21)–(4.3.23), by the adjoint system* ($\ell = \overline{1,\nu}$)

$$-\sum_{k=1}^{\nu} p_k''\, a_{\ell k} + \sum_{k=1}^{\nu} p_k\, b_{\ell k} = \sum_{k=1}^{\nu} a_{\ell k}\,(y_k^* - y_d^k), \tag{4.3.29}$$

$$-\sum_{k=1}^{\nu} p_k'(1)\, a_{\ell k} = 0, \tag{4.3.30}$$

$$p_\ell(0) = 0, \tag{4.3.31}$$

and by the maximum principle

$$-\left[p_1, \left(\sum_{k=1}^{\nu} p_k(1)\, a_{\ell k}\right)_{\ell=\overline{1,\nu}}\right] \in u_\nu^* + \partial I_{U_{ad}^\nu}(u_\nu^*). \tag{4.3.32}$$

The proof follows the lines of Section 4.2 above and is omitted. The subdifferential of $I_{U_{ad}^\nu}$ has to be computed in U_ν, and the left-hand side in (4.3.32) is in fact the semidiscretization of the dual operator B^* appearing in (4.3.11).

Let us briefly indicate how error estimates can be established in this problem. We give a sketch of the argument, which again closely parallels that from Section 4.2. We define

$$s = -B^*p - u^* \in \partial I_{U_{ad}}(u^*), \tag{4.3.33}$$

$$s_\nu = -\left[p_1, \left(\sum_{k=1}^{\nu} p_k(1)\, a_{\ell k}\right)_{\ell=\overline{1,\nu}}\right] - u_\nu^* \in \partial I_{U_{ad}^\nu}(u^*). \tag{4.3.34}$$

Then, clearly,

$$s + B^*p + u^* = s_\nu + \left[p_1, \left(\sum_{k=1}^{\nu} p_k(1)\, a_{\ell k}\right)_{\ell=\overline{1,\nu}}\right] + u_\nu^* = 0 \tag{4.3.35}$$

in $L^2(\Gamma_N)$.

We define the interpolating semidiscrete variables r_ℓ and q_k, which satisfy the system (4.3.21)–(4.3.23), respectively (4.3.29)–(4.3.31), but with different inputs:

$$-\sum_{\ell=1}^{\nu} a_{\ell k}\, r_\ell'' + \sum_{\ell=1}^{\nu} b_{\ell k}\, r_\ell = \sum_{\ell=1}^{\nu} a_{\ell k}\, f_\ell + u^*|_{\Gamma_N^1} \cdot \delta_{1k}, \tag{4.3.36}$$

$$r_k(0) = 0, \tag{4.3.37}$$

$$\sum_{\ell=1}^{\nu} a_{\ell k}\, r_\ell'(1) = \sum_{\ell=1}^{\nu} \tilde{u}_\ell^*\, a_{\ell k}, \tag{4.3.38}$$

for $k = \overline{1,\nu}$, and

$$-\sum_{k=1}^{\nu} q_k''\, a_{\ell k} + \sum_{k=1}^{\nu} q_k\, b_{\ell k} = \sum_{k=1}^{\nu} a_{\ell k}\,(r_k - y_d^k), \tag{4.3.39}$$

$$q_\ell(0) = 0, \tag{4.3.40}$$

$$\sum_{k=1}^{\nu} q_k'(1)\, a_{\ell k} = 0, \tag{4.3.41}$$

for $\ell = \overline{1,\nu}$. In (4.3.38), the scalars $\{\tilde{u}_\ell^*\}_{\ell=\overline{1,\nu}}$ are chosen to provide some interpolation of $u^*|_{\Gamma_N^2}$. Namely, if $\varepsilon > 0$ is given, we require that

$$\left| \sum_{\ell=1}^{\nu} \tilde{u}_\ell^*\, Q_\ell - u^*|_{\Gamma_N^2} \right|_{L^2(\Gamma_N^2)} < \varepsilon. \tag{4.3.42}$$

The validity of (4.3.42) for sufficiently large $\nu \in \mathbf{N}$ is ensured by the assumed regularity property $u^* \in H^1(\Gamma_N)$, provided that the basis $\{Q_\ell\}_{\ell=\overline{1,\nu}}$ is appropriately chosen.

The following inequalities may be obtained:

$$\Big(\Big[p_1, \Big(\sum_{k=1}^{\nu} p_k(1)\, a_{\ell k}\Big)_{\ell=\overline{1,\nu}}\Big] - \Big[q_1, \Big(\sum_{k=1}^{\nu} q_k(1)\, a_{\ell k}\Big)_{\ell=\overline{1,\nu}}\Big],$$

$$\Big[\overline{u}_1^*, (u_\ell^*)_{\ell=\overline{1,\nu}}\Big] - \Big[u^*|_{\Gamma_N^1}, (\tilde{u}_\ell^*)_{\ell=\overline{1,\nu}}\Big]\Big)_{U_\nu}$$

$$= \int_0^1 \sum_{\ell=1}^{\nu}(y_\ell^* - r_\ell) \sum_{k=1}^{\nu} a_{\ell k}\,(y_k^* - r_k) \ \geq\ 0, \tag{4.3.43}$$

by the properties of the matrix $\widetilde{A}$ in (4.3.17), as well as

$$(s_\nu - s, u_\nu^* - u^*)_{L^2(\Gamma_N)} \ \geq\ -\hat{C}\,\varepsilon, \tag{4.3.44}$$

for some constant $\hat{C} > 0$ depending neither on $\varepsilon > 0$ nor on $\nu \in \mathbf{N}$, and $\{s_\nu\}_{\nu\in\mathbf{N}}$ is bounded.

Combining (4.3.42)–(4.3.44) and (4.3.35)–(4.3.41), and arguing along the same lines as in the previous section, we can prove the desired error estimate (the details of the argument are omitted):

Theorem 4.3.2 *Let (4.3.42) be satisfied. Then there is some constant $\hat{C} > 0$, which is independent of both $\varepsilon > 0$ and $\nu \in \mathbf{N}$, such that*

$$|u_\nu^* - u^*|_{L^2(\Gamma_N)} \ \leq\ \hat{C}\,\varepsilon + C\,|p - q_\nu|_{H_D^1(\Omega)}.$$

Remark. In Tiba and Xanthis [1994], more general problems have been examined along these lines.

4.4 Optimal Control Problems Governed by Elliptic Variational Inequalities

This section deals with the convergence of approximations for control problems governed by elliptic variational inequalities. In this nondifferentiable and nonconvex setting, error estimates can be established only in some special cases, as we will see in the next section. Convergence results have been discussed by Haslinger and Neittaanmäki [1988, Chapter 10], [1996], for optimal design problems, by Arnăutu [1982] for parabolic variational inequalities, by Tiba [1990] for free boundary problems, by Neittaanmäki and Tiba [1994], and others. Since in this situation many local or global minimum points may occur, convergence of the discretized optimal solutions can be established only for suitable subsequences (see Theorem 4.4.2 below). Compared to the error estimates established in the previous sections, this result is weaker; however, its importance lies in the generality of the hypotheses under which it is proved.

To begin with, consider reflexive Banach spaces V and U; nonempty and convex sets $U_{ad} \subset U$ and $C \subset V$ that are closed in U and V, respectively, and let $f \in V^*$, $B \in L(U, V^*)$; and a linear operator $A : V \to V^*$ be given. In addition, let $\varphi : V \to]-\infty, +\infty]$ denote a proper, convex, and lower semicontinuous mapping such that $0 \in \operatorname{dom}(\varphi)$. The variational inequality under study is

$$Ay + \partial\varphi(y) \ni Bu + f, \tag{4.4.1}$$

where we suppose that the bilinear form $a : V \times V \to \mathbf{R}$ associated with A,

$$a(y, v) = (Ay, v)_{V^* \times V} \quad \forall\, y, v \in V, \tag{4.4.2}$$

is symmetric, continuous, and coercive; that is, there exist $M > 0$ and $\gamma > 0$ such that

$$|a(y, v)| \leq M\,|y|_V\,|v|_V \quad \forall\, y, v \in V, \qquad a(v, v) \geq \gamma\,|v|_V^2 \quad \forall\, v \in V.$$

The variational form of (4.4.1) is to find some $y \in V$ such that

$$a(y, v - y) + \varphi(v) - \varphi(y) \geq (Bu + f, v - y)_{V^* \times V} \quad \forall\, v \in V. \tag{4.4.3}$$

Under the above assumptions, A has a bounded inverse $A^{-1} : V^* \to V$, and the operator $A + \partial\varphi$ is by Theorem A1.10 in Appendix 1 maximal monotone in $V \times V^*$. Then Theorem A2.4 in Appendix 2 yields that (4.4.1) (or equivalently, (4.4.3)) has for every $u \in U$ a solution $y \in V$, which by the strict monotonicity of A is unique. We set $y = y(u)$ for $u \in U$.

Now let a proper, convex, and lower semicontinuous cost functional $J : V \times U \to \mathbf{R}$ be given. We then consider the optimal control problem

$$\text{(P)} \qquad \operatorname{Min}\,\big\{J(y, u)\big\} \quad \text{subject to (4.4.1) (or (4.4.3))}\,, \quad y \in C \text{ and } u \in U_{ad}.$$

By virtue of the general existence results established in Section 2.1 of Chapter 2, the existence of at least one optimal pair for (P) follows under the assumption that the admissible set

$$X = \{u \in U_{ad} \,:\, y(u) \in C\}$$

is nonempty, provided that J is coercive in the sense of (2.1.6) or that U_{ad} is bounded in U.

4.4.1 The Ritz–Galerkin Approximation

Let us for the following suppose that the constraints set U_{ad} is a compact subset of U, and let $h > 0$ be the discretization parameter (see Section 4.1). We assume that for any $h > 0$ there are given:

- a finite-dimensional subspace V_h of V with associated inner product $(\cdot,\cdot)_h$, and a finite-dimensional subspace U_h of U;

- nonempty and convex sets C_h and U_{ad}^h that are closed subsets of V_h and U_{ad}, respectively;

- a bilinear form $a_h : V_h \times V_h \to \mathbf{R}$;

- $f_h \in V_h$, $B_h \in L(U_h, V_h)$;

- proper, convex, and lower semicontinuous mappings $\varphi_h : V_h \to]-\infty, +\infty]$ with $0 \in \mathrm{dom}(\varphi_h)$, and $J_h : V_h \times U_h \to \mathbf{R}$.

We then can introduce the approximate variational inequality corresponding to (4.4.3), namely, to find $y_h \in V_h$ such that

$$a_h(y_h, v_h - y_h) + \varphi_h(v_h) - \varphi_h(y_h) \geq (B_h u_h + f_h, v_h - y_h)_h \quad \forall\, v_h \in V_h. \tag{4.4.4}$$

We also define the discrete set of admissible controls,

$$X_h = \left\{ u_h \in U_{ad}^h \,:\, \text{(4.4.4) has a solution } y_h \in C_h \right\},$$

and we suppose that $X_h \neq \emptyset$ for every $h > 0$. We state the approximate control problem

$$(\mathrm{P}_h) \quad \mathrm{Min} \left\{ J_h(y_h, u_h) \,:\, u_h \in X_h \right\} \quad \text{subject to (4.4.4)}, \;\; y_h \in C_h, \;\text{and}\; u_h \in U_{ad}^h.$$

Again, it follows from the general existence results established in Section 2.1 of Chapter 2 that (P_h) admits at least one optimal pair $[y_h^*, u_h^*]$ if J_h is coercive in the sense of (2.1.6) or U_{ad}^h is bounded in U_h.

We impose the following general hypotheses for our approximation scheme (where in (H3)–(H6) we always assume that $y_h, v_h \in V_h$, $h > 0$):

(H1) The family $\{a_h\}_{h>0}$ is *uniformly coercive* on $\{V_h\}_{h>0}$; that is, there is some $\alpha > 0$ such that (compare §4.1.2)

$$a_h(v_h, v_h) \geq \alpha\, |v_h|_V^2 \quad \forall\, v_h \in V_h, \; \forall\, h > 0.$$

(H2) The family $\{a_h\}_{h>0}$ is *uniformly bounded* on $\{V_h\}_{h>0}$, that is, there is some $\beta > 0$ such that

$$|a_h(y_h, v_h)| \le \beta\, |y_h|_V\, |v_h|_V \quad \forall\, y_h, v_h \in V_h,\ \forall\, h > 0.$$

(H3) If $y_h \to y$ weakly in V and $v_h \to v$ strongly in V, then $a_h(y_h, v_h) \to a(y, v)$.

(H4) If $y_h \to y$ weakly in V, then $\liminf_{h\to 0} a_h(y_h, y_h) \ge a(y, y)$.

(H5) If $v_h \to v$ strongly in V, then $\varphi_h(v_h) \to \varphi(v)$.

(H6) If $v_h \to v$ weakly in V, then $\liminf_{h\to 0} \varphi_h(v_h) \ge \varphi(v)$.

(H7) The family $\{\varphi_h\}_{h>0}$ is uniformly bounded from below by an affine function, that is, there exist $\lambda \in V^*$ and $\mu \in \mathbf{R}$ such that

$$\varphi_h(v_h) \ge (\lambda, v_h)_{V^*\times V} + \mu, \quad \forall\, v_h \in V_h,\ \forall\, h > 0.$$

(H8) There is some $\hat{C}_1 > 0$ with $|(f_h, v_h)_h| \le \hat{C}_1\, |v_h|_V \quad \forall\, v_h \in V_h,\ \forall\, h > 0.$

(H9) If $v_h \to v$ weakly in V, then $(f_h, v_h)_h \to (f, v)_{V^*\times V}$.

(H10) If $u_h \in U_{ad}^h$ for every $h > 0$ and $\sup_{h>0} |u_h|_U \le \rho$ for some $\rho > 0$, then there is some $\hat{C}_2 > 0$ such that

$$|(B_h u_h, v_h)_h| \le \hat{C}_2\, |v_h|_V \quad \forall\, v_h \in V_h,\ \forall\, h > 0.$$

(H11) If $(v_h, u_h) \in V_h \times U_h$, $h > 0$, and if $u_h \to u$ strongly in U and $v_h \to v$ weakly in V, then $(B_h u_h, v_h)_h \to (Bu, v)_{V^*\times V}$.

(H12) For any $v \in V$ there are $v_h \in V_h$, $h > 0$, with $v_h \to v$ strongly in V.

These hypotheses seem only at first glance to be restrictive and difficult to satisfy. We will see below in a typical example that this is in fact not the case. We also notice that under the above assumptions Theorem A2.4 in Appendix 2 guarantees that the discrete variational inequality (4.4.4) has for any $u_h \in U_h$ a unique solution $y_h = y_h(u_h) \in V_h$. We have the following convergence result for the state equation.

Theorem 4.4.1 *Suppose that the general assumptions of this paragraph and the hypotheses (H1)–(H12) are fulfilled. If $u_h \in U_h$ for any $h > 0$, and if $u_h \to u$ strongly in U, then $y_h(u_h) \to y(u)$ strongly in V.*

Proof. Since $\{u_h\}_{h>0}$ converges, there is some $\rho > 0$ such that $|u_h|_U \le \rho$ for all $h > 0$. Inserting $v_h = 0$ in (4.4.4), and invoking (H1), (H7), (H8), and

(H10), we conclude that

$$\begin{aligned}\alpha\,|y_h(u_h)|_V^2 \;&\le\; a_h(y_h(u_h), y_h(u_h))\\ &\le\; \varphi_h(0) - \varphi_h(y_h(u_h)) + (B_h u_h + f_h, y_h(u_h))_h\\ &\le\; \varphi_h(0) + \Big(|\lambda|_{V^*} + \hat{C}_2 + \hat{C}_1\Big)\; |y_h(u_h)|_V + |\mu|.\end{aligned}$$

Observe that $\varphi_h(0) < +\infty$ since $0 \in \mathrm{dom}(\varphi_h)$, and, similarly, $\varphi(0) < +\infty$. Hence, (H5) implies that $\{\varphi_h(0)\}_{h>0}$ is bounded, whence the boundedness of $\{|y_h(u_h)|_V\}_{h>0}$ follows. Consequently, there is a sequence $h_n \searrow 0$ such that $y_{h_n}(u_{h_n}) \to \overline{y}$ weakly in V.

Now let $v \in V$ be arbitrarily chosen. Thanks to (H12), there exist $v_{h_n} \in V_{h_n}$, $n \in \mathbf{N}$, such that $v_{h_n} \to v$ strongly in V. We insert this v_{h_n} in (4.4.4) for $h = h_n$, $n \in \mathbf{N}$. Passing to the limit as $n \to \infty$, and invoking the hypotheses (H3), (H5), (H6), (H9), and (H11), we can infer that

$$a(\overline{y}, v - \overline{y}) + \varphi(v) - \varphi(u) \ge (Bu + f, v - \overline{y})_{V^* \times V}.$$

Since $v \in V$ was arbitrary, we conclude that $\overline{y} = y(u)$, and the uniqueness of the limit entails that generally $y_h(u_h) \to y(u)$ weakly in V for $h \searrow 0$, and not only for the sequence selected above.

It remains to show that the convergence is in fact strong. To this end, we apply (H12) and choose $z_h \in V_h$, $h > 0$, such that $|z_h - y(u)|_V \to 0$. Then, obviously,

$$y_h(u_h) - z_h \to 0 \quad \text{weakly in } V.$$

Using (H1) and (4.4.4), we find that

$$\begin{aligned}\alpha\,|y_h(u_h) - z_h|_V^2 \;&\le\; a_h(y_h(u_h) - z_h, y_h(u_h) - z_h)\\ &=\; - a_h(y_h(u_h), z_h - y_h(u_h)) - a_h(z_h, y_h(u_h) - z_h)\\ &\le\; (B_h u_h + f_h, y_h(u_h) - z_h)_h + \varphi_h(z_h)\\ &\qquad - \varphi_h(y_h(u_h)) - a_h(z_h, y_h(u_h) - z_h).\end{aligned}$$

Passing to the limit as $h \searrow 0$, invoking the general hypotheses, we then obtain that

$$\lim_{h \to 0} |y_h(u_h) - z_h|_V = 0,$$

and the assertion follows from $|z_h - y(u)|_V \to 0$. □

To prove convergence of the approximating discrete optimal control problems, we need to impose further conditions that guarantee the consistency of the approximation of the cost functional and of the state and control constraints:

(U1) The family $\{U_{ad}^h\}_{h>0}$ is compact in the following sense: whenever

$u_h \in U_{ad}^h$, $h > 0$, are given, then there exists a sequence $h_n \searrow 0$ such that $u_{h_n} \to u \in U_{ad}$ strongly in U.

(U2) If $y_h \to y$ strongly in V and $y_h \in C_h$ for every $h > 0$, then $y \in C$.

(U3) For any $u \in X$ there are $u_h \in X_h$, $h > 0$, such that $u_h \to u$ strongly in U.

(U4) If $[y_h, u_h] \in V_h \times U_h$, $h > 0$, and $[y_h, u_h] \to [y, u]$ strongly in $V \times U$, then $J_h(y_h, u_h) \to J(y, u)$.

Theorem 4.4.2 *Suppose that the general assumptions of this section are satisfied, and suppose that (H1)–(H12) and (U1)–(U4) hold true. Let $[y_h^*, u_h^*]$ be an optimal pair for the problem (P_h) for any $h > 0$. Then there exist a sequence $h_n \searrow 0$ and some $[y^*, u^*] \in C \times U_{ad}$ such that $[y_{h_n}^*, u_{h_n}^*] \to [y^*, u^*]$ strongly in $V \times U$, and $[y^*, u^*]$ is an optimal pair for the problem (P).*

Proof. Thanks to (U1), there is a sequence $h_n \searrow 0$ such that $u_{h_n}^* \to u^* \in U_{ad}$ strongly in U. Then, by Theorem 4.4.1, $y_{h_n}^* \to y^* = y(u^*)$ strongly in V. By (U2), $y^* \in C$, and thus $u^* \in X$.

Now let $u \in X$ be arbitrary. Owing to (U3), there exist $u_h \in X_h$, $h > 0$, with $|u_h - u|_U \to 0$. Invoking Theorem 4.4.1 once more, we see that $y_h(u_h) \to y(u)$ strongly in V. The optimality of $[y_{h_n}^*, u_{h_n}^*]$ for (P_{h_n}) yields that

$$J_{h_n}(y_{h_n}^*, u_{h_n}^*) \le J_{h_n}(y_{h_n}(u_{h_n}), u_{h_n}),$$

whence, invoking (U4),

$$J(y^*, u^*) \le J(y(u), u).$$

Since $u \in X$ is arbitrary, the pair $[y^*, u^*]$ is optimal for (P). □

Remark. In the proofs of Theorems 4.4.1 and 4.4.2 the uniformity of the continuity of $\{a_h\}_{h>0}$ expressed by hypothesis (H2) has not been used. Indeed, we have just applied the continuity of each individual a_h in order to guarantee the solvability of (4.4.4). However, (H2) will be needed below.

In the sequel, we consider a somewhat different functional-analytic framework and a more explicit setting of the approximation. To this end, let H be a Hilbert space that is endowed with the inner product $(\cdot,\cdot)_H$ and identified with its dual, and let V be a reflexive Banach space satisfying $V \subset H \subset V^*$, where the embedding $V \subset H$ is dense, continuous, and compact. In particular, there is some constant $c_0 > 0$ such that

$$|v|_H \le c_0\, |v|_V \quad \forall\, v \in V. \tag{4.4.5}$$

We assume that U is a Hilbert space and that $B \in L(U, H)$. The other hypotheses concerning A and f are the same as above, and we assume that

$\varphi = I_C$, so that (4.4.1) takes the form

$$Ay + \partial I_C(y) \ni Bu + f. \tag{4.4.6}$$

Then (P) becomes a control problem without state constraints,

(P) $\qquad$ $\text{Min}\,\Big\{J(y,u)\Big\}$ subject to (4.4.6) and to $u \in U_{ad}$.

The existence of at least one optimal pair $[y^*, u^*] \in C \times U_{ad}$ for problem (P) is assumed and can be inferred by the techniques of Section 2.1 in Chapter 2.

The Ritz–Galerkin approximation is the same as above. Moreover, we assume that the inner product $(\cdot\,,\cdot)_h$ in V_h is just the inner product $(\cdot\,,\cdot)_H$ of H. We also introduce the operator $A_h : V_h \to V_h$ given by

$$(A_h y_h, v_h)_H = a_h(y_h, v_h) \quad \forall\, y_h, v_h \in V_h. \tag{4.4.7}$$

With these specifications, the approximate form of (4.4.6) reads

$$A_h y_h + \partial I_{C_h}(y_h) \ni B_h u_h + f_h, \tag{4.4.8}$$

and the problem (P_h) attains the form

(P_h) $\qquad$ $\text{Min}\,\Big\{J_h(y_h, u_h)\Big\}$ subject to (4.4.8) and to $u_h \in U_{ad}^h$.

The hypotheses (H5)–(H7) are removed, and we replace (H10) by the hypothesis

(H10)′ The family of operators $\{B_h\}_{h>0}$ is uniformly bounded with respect to $\{L(U_h, H)\}_{h>0}$; that is, there is some $\hat{C}_2 > 0$ such that

$$\sup_{h>0} |B_h|_{L(U_h,H)} \le \hat{C}_2.$$

Assume now that some linear and continuous restriction (interpolation or projection) operators $r_h : H \to V_h$ and $s_h : U \to U_h$ are given. Then (H12) may be naturally replaced by the hypothesis

(H12)′ $r_h v \to v$ strongly in V for every $v \in V$.

In addition, we assume

(S1) The family of operators $\{r_h\}_{h>0}$ is uniformly bounded in $L(H,V)$.

Remark. The natural way to define B_h is to put $B_h = r_h B$, that is, $B_h u_h = r_h(Bu_h)$, and (H10)′ is satisfied if (S1) holds.

It is natural to postulate similar assumptions for $\{s_h\}_{h>0}$:

(S2) $s_h u \to u$ strongly in U for every $u \in U$, and the family of operators $\{s_h\}_{h>0}$ is uniformly bounded in $L(U, U)$.

Next, we discuss the properties of C_h and U_{ad}^h. For this purpose, we recall the definition of the convergence of sets in the sense of Mosco (Mosco [1971, p. 595]). Consider the notation

$$\begin{aligned} \text{s-}\liminf C_h &= \{v \in V : \exists\, v_h \in C_h,\ \forall\, h > 0, \text{ such that } v_h \to v \\ &\qquad \text{strongly in } V\}, \\ \text{w-}\limsup C_h &= \{v \in V : \exists\, h_n \searrow 0 \text{ and } v_{h_n} \in C_{h_n},\ \forall\, n \in \mathbf{N}, \text{ such that} \\ &\qquad v_{h_n} \to v \text{ weakly in } V\}. \end{aligned}$$

Definition 4.4.3 We say that $\lim_{h\to 0} C_h = C$ *in the sense of Mosco* if

$$\text{w-}\limsup C_h \subset C \subset \text{s-}\liminf C_h.$$

We now define the family of sets

$$C_h = r_h(C), \quad U_{ad}^h = s_h(U_{ad}), \quad \text{for } h > 0. \tag{4.4.9}$$

Proposition 4.4.4 *The following statements are true:*

(i) *If (H12)′ and (S1) are fulfilled, and if* $C_h \subset C$ *for all* $h > 0$*, then* $\lim_{h\to 0} C_h = C$ *in the sense of Mosco.*

(ii) *If (S2) is true, and if* $U_{ad}^h \subset U_{ad}$ *for all* $h > 0$*, then* $\lim_{h\to 0} U_{ad}^h = U_{ad}$ *in the sense of Mosco.*

Proof. We show only (i); the proof of (ii) is similar. To begin with, let $v \in C$. We put $v_h = r_h v$ for any $h > 0$. By (4.4.9), then $v_h \in C_h$, $h > 0$, and (H12)′ implies that $v_h \to v$ strongly in V, that is, $v \in \text{s-}\liminf C_h$. Since $v \in C$ is arbitrary, we have shown that $C \subset \text{s-}\liminf C_h$.

To show the other inclusion in Definition 4.4.3, let $v \in \text{w-}\limsup C_h$ be arbitrary. Then there are $h_n \searrow 0$ and $v_{h_n} \in C_{h_n}$, $n \in \mathbf{N}$, such that $v_{h_n} \to v$ weakly in V. Since $C_{h_n} \subset C$ for any $n \in \mathbf{N}$, it follows that $\{v_{h_n}\}_{n\in\mathbf{N}} \subset C$, which is weakly closed in V. Hence, we can infer that $v \in C$, which concludes the proof of the assertion. □

In the following, we will derive another convergence result resembling Theorem 4.4.2 without using the hypotheses (U1)–(U3), which are replaced by the following assumptions:

(S3) There are constants $\delta_1 > 0$, $\delta_2 > 0$, and $\delta_3 \in \mathbf{R}$ such that

$$J(y,u) \ge \delta_1 |u|_U^2 - \delta_2 |y|_H + \delta_3 \quad \forall\, u \in U,\ \forall\, y \in H. \tag{4.4.10}$$

(S3h) There are constants $\tilde{\delta}_1 > 0$, $\tilde{\delta}_2 > 0$, and $\tilde{\delta}_3 \in \mathbf{R}$ such that

$$J_h(y_h,u_h) \ge \tilde{\delta}_1 |u_h|_U^2 - \tilde{\delta}_2 |y_h|_H + \tilde{\delta}_3 \quad \forall\, u_h \in U_h,\ \forall\, y_h \in V_h. \qquad (4.4.10)'$$

Remark. If (S3) is fulfilled, one can simply take $J_h = J|_{V_h \times U_h}$ in order to satisfy (S3h).

Another possibility is to assume:

(S3)′ The function $u \mapsto J(y,u)$ is coercive uniformly in $y \in V$.

(S3h)′ For every $h > 0$ the function $u_h \mapsto J_h(y_h,u_h)$ is coercive uniformly in $y_h \in V_h$.

The proof of the result similar to Theorem 4.4.2 is performed in several steps.

Step 1: We first examine the convergence properties of the sequences $\{u_h^*\}_{h>0}$ and $\{y_h^*\}_{h>0}$ if $[y_h^*, u_h^*]$ is an optimal pair for the problem (P_h), $h > 0$.

Let us denote by $C_i > 0$, $i \in \mathbf{N}$, constants not depending on $h > 0$. From (4.4.8) we infer, invoking (H1), (H2), (H8), and (H10)′, that for every $v_h \in C_h$,

$$\begin{aligned} \alpha\, |y_h^*|_V^2 &\le a_h(y_h^*, y_h^*) \le a_h(y_h^*, v_h) + (B_h u_h^* + f_h, y_h^* - v_h)_h \\ &\le \beta\, |y_h^*|_V\, |v_h|_V + C_1\, (1 + |u_h^*|_U)\, (|y_h^*|_V + |v_h|_V)\,. \end{aligned}$$

Now choose $v_h = r_h v_0$ with some fixed $v_0 \in V$. Then, using (S1), we obtain for every $\varepsilon > 0$ the estimate

$$\begin{aligned} |y_h^*|_V^2 &\le C_1\, |u_h^*|_U \cdot |y_h^*|_V + C_2\, (1 + |y_h^*|_V + |u_h^*|_U) \\ &\le \frac{C_1\,\varepsilon}{2} |y_h^*|_V^2 + \frac{C_1}{2\,\varepsilon} |u_h^*|_U^2 + C_2 \left(\frac{\varepsilon}{2} \left(|y_h^*|_V^2 + |u_h^*|_U^2 \right) + 1 + \frac{1}{\varepsilon} \right). \end{aligned}$$

Choosing $0 < \varepsilon < 2/(C_1 + C_2)$, we obtain that

$$|y_h^*|_V^2 \le C_3 + C_4\, |u_h^*|_U^2. \tag{4.4.11}$$

Since $[y_h^*, u_h^*]$ is an optimal pair for (P_h), we have

$$J_h(y_h^*, u_h^*) \le J_h(y_h(u_h), u_h) \quad \forall\, u_h \in U_{ad}^h. \tag{4.4.12}$$

Suppose now first that hypothesis (S3h) holds. Fixing $u_h = s_h u \in U_{ad}^h$ in (4.4.12), and invoking (U4) and Theorem 4.4.1, we get

$$J_h(y_h^*, u_h^*) \le C_5.$$

Next, we use (4.4.10)′ and (4.4.5) to obtain that

$$\begin{aligned} \tilde{\delta}_1\, |u_h^*|_U^2 \;&\le\; \tilde{\delta}_2\, |y_h^*|_H + |\tilde{\delta}_3| \,+\, J(y_h^*, u_h^*) \\ &\le\; c_0\, \tilde{\delta}_2\, |y_h^*|_V \,+\, |\tilde{\delta}_3| \,+\, C_5. \end{aligned} \tag{4.4.13}$$

From (4.4.11) and (4.4.13) we then can infer that $\{y_h^*\}_{h>0}$ is bounded in V and, in view of (4.4.5), also in H. Thus, by (4.4.13), $\{u_h^*\}_{h>0}$ is bounded in U.

Next, assume that hypothesis (S3h)′ is fulfilled instead of (S3h). Fixing u_h in (4.4.12), we can infer directly from (S3h)′ that $\{|u_h^*|_U\}_{h>0}$ is bounded. Therefore, thanks to (4.4.11), $\{y_h^*\}_{h>0}$ is bounded in V and, invoking (4.4.5), also in H.

In any of the cases considered above, there exists a sequence $h_n \searrow 0$ such that

$$u_{h_n}^* \;\to\; u^* \quad \text{weakly in } U, \tag{4.4.14}$$

$$y_{h_n}^* \;\to\; y^* \quad \text{weakly in } V \text{ and strongly in } H. \tag{4.4.15}$$

Step 2: We show that $[y^*, u^*]$ satisfies the variational inequality (4.4.6). To this end, recall that for every $v_{h_n} \in C_{h_n}$,

$$a_{h_n}(y_{h_n}^*, y_{h_n}^*) \le a_{h_n}(y_{h_n}^*, v_{h_n}) + (B_{h_n} u_{h_n}^* + f_{h_n}, y_{h_n}^* - v_{h_n})_{h_n}.$$

We take $v_{h_n} = r_{h_n} v$ with arbitrarily fixed $v \in C$ to obtain that

$$\begin{aligned} a_{h_n}(y_{h_n}^*, y_{h_n}^*) \;\le\; & a_{h_n}(y_{h_n}^*, r_{h_n} v) + (B_{h_n} u_{h_n}^*, y_{h_n}^*)_{h_n} - (B_{h_n} u_{h_n}^*, r_{h_n} v)_{h_n} \\ & + (f_{h_n}, y_{h_n}^*)_{h_n} - (f_{h_n}, r_{h_n} v)_{h_n}. \end{aligned} \tag{4.4.16}$$

To pass to the limit as $n \to \infty$ in this inequality, we invoke (H3) and (H4) to handle the terms involving a_{h_n}. For the terms involving f_{h_n} a weaker hypothesis than (H9) suffices, namely

(H9)′ If $v_h \to v$ strongly in V, then $(f_h, v_h)_h \to (f, v)_{V^* \times V}$.

On the other hand, we need to replace (H11) by a stronger hypothesis, namely

(H11)′ If $u_h \to u$ weakly in U and $v_h \to v$ strongly in V, then $(B_h u_h, v_h)_h \to (Bu, v)_{V^* \times V}$.

Remark. If $U = H$, and if B is a self-adjoint operator, then the passage to the limit can be achieved assuming (H11).

Now under the above assumptions, the passage to the limit as $n \to \infty$ yields that

$$a(y^*, v - y^*) \geq (Bu^* + f, v - y^*)_{V^* \times V} \quad \forall v \in C,$$

that is, $[y^*, u^*]$ satisfies (4.4.6).

Step 3: Next, we show that $[y^*, u^*]$ is optimal for (P). For this purpose, recall that $[y_h^*, u_h^*]$ is optimal for (P_h), so that

$$J_h(y_h^*, u_h^*) \leq J_h(y_h(s_h \overline{u}), s_h \overline{u}), \tag{4.4.17}$$

whenever $[\overline{y}, \overline{u}]$ is an optimal pair of (P). We now impose the additional hypothesis

(U4)′ If $u_h \to u$ weakly in U and $y_h \to y$ strongly in H, then $\liminf_{h \to 0} J_h(y_h, u_h) \geq J(y, u)$.

Invoking (S1), (S2), (U4), and (U4)′, and observing that the convergence of the right-hand side of (4.4.17) can be shown just as in the proof of Theorem 4.4.1, we may pass to the limit as $h_n \searrow 0$ in (4.4.17) to find that

$$J(y^*, u^*) \leq J(\overline{y}, \overline{u}).$$

Since $[\overline{y}, \overline{u}]$ is optimal for (P), and $[y^*, u^*]$ is admissible for (P) (cf. Step 2), we conclude that $[y^*, u^*]$ is an optimal pair for (P).

Step 4: We finally show that $y_{h_n}^* \to y^*$ strongly in V, where $\{h_n\}_{n \in \mathbf{N}}$ is the sequence satisfying (4.4.14), (4.4.15). Using (H1), we have, for any $v_h \in V_h$, $h > 0$,

$$\alpha\, |y_h^* - v_h|_V^2 \leq a_h(y_h^* - v_h, y_h^* - v_h) = -a_h(y_h^*, v_h - y_h^*) - a_h(v_h, y_h^* - v_h).$$

Now, owing to (4.4.8),

$$-a_h(y_h^*, v_h - y_h^*) \leq -(B_h u_h^* + f_h, v_h - y_h^*)_h,$$

whence, taking $v_h = r_h y^*$,

$$\alpha \, |y_h^* - r_h y^*|_V^2 \leq (B_h u_h^*, y_h^* - r_h y^*)_h + (f_h, y_h^* - r_h y^*)_h - a_h(r_h y^*, y_h^* - r_h y^*). \tag{4.4.18}$$

Recall the convergence relations (4.4.14) and(4.4.15), and that $r_h y^* \to y^*$ strongly in V (and H) by (H12)$'$. Invoking (H3), (H9)$'$, and (H11)$'$, we may pass to the limit as $h_n \searrow 0$ in (4.4.18) to find that

$$\lim_{n\to\infty} |y_{h_n}^* - r_{h_n} y^*|_V = 0.$$

But then

$$\lim_{n\to\infty} |y_{h_n}^* - y^*|_V \leq \lim_{n\to\infty} \Big(|y_{h_n}^* - r_{h_n} y^*|_V + |r_{h_n} y^* - y^*|_V \Big) = 0.$$

Summarizing, we have demonstrated the following convergence result:

Theorem 4.4.5 *In the new functional-analytic framework, let the following hypotheses be satisfied: $\varphi = I_C$, (4.4.9), (H1)–(H4), (H8), (H9)$'$, (H11)$'$, (H12)$'$, (S1)–(S3), (S3h) (or (S3h)$'$), (U4) and (U4)$'$. In addition, suppose that $[y_h^*, u_h^*]$ is an optimal pair for the problem (P_h), $h > 0$. Then there exist a sequence $h_n \searrow 0$ and an optimal pair $[y^*, u^*]$ for problem (P) such that $u_{h_n}^* \to u^*$ weakly in U and $y_{h_n}^* \to y^*$ strongly in V and in H.*

Let us now discuss the various hypotheses made in this section, which have been stated in an abstract framework, from the viewpoint of a typical example. For this purpose, we consider the case that $H = L^2(\Omega)$ and $V = H_0^1(\Omega)$. The problem given in Example 4.1.9, and also in Example 4.2.5 (cf. (4.2.40), (4.2.41)), leads to the bilinear form

$$a(y,v) = \int_\Omega \nabla y(x) \cdot \nabla v(x) \, dx \quad \forall \, y, v \in V.$$

In this case the hypotheses (H1)–(H4) are satisfied if we make the natural choice $a_h = a|_{V_h \times V_h}$. Likewise, (H5)–(H7) hold with the choice $\varphi_h = \varphi|_{V_h}$. The hypotheses (H8) and (H9)$'$ can be fulfilled by using an appropriate numerical integration formula if we set, for $f \in L^2(\Omega)$,

$$(f,v)_{V^* \times V} = \int_\Omega f(x) \, v(x) \, dx \quad \forall \, v \in V.$$

Moreover, (H10) follows from (S1). The assumptions (H11) or (H11)$'$ are more delicate, depending on the actual operator B (for instance, $B = id$ is a good choice). (H12)$'$ is natural for the finite element method. Finally, (U4) and (U4)$'$ can be satisfied by the choice $J_h = J|_{V_h \times U_h}$.

Remark. Additional approximation results for the control of (elliptic) variational inequalities can be found in the works of Davideanu [1989], and French and King [1991]. If convexity properties are valid, also error estimates may be obtained, as will be demonstrated in the next paragraph. In the works of Casas and Mateos [2002b], and of Casas [2002], error estimates for the discretization of constrained nonconvex elliptic optimal control problems have been derived. These results have to be understood in the sense that local optimal controls with a certain regularity have an attractor-type property: for any such minimum $\overline{u}$ there is some neighborhood within which the discretized optimization problems associated with the discretization parameters $h > 0$ have solutions $\overline{u}_h$ such that $\overline{u}_h \to \overline{u}$ as $h \searrow 0$. The error estimates provide the order of this convergence with respect to $h > 0$. The case under study corresponds to semilinear elliptic state equations, and the approach is essentially based on second-order sufficient optimality conditions. For a posteriori estimates in some control problems, we refer to Liu and Yan [2002], [2003].

4.5 Error Estimates in the Discretization of Control Problems with Nonlinear State Equation

In this section, we show that the results from Section 4.4 may be improved to obtain error estimates in some cases when the optimization problem is convex although the state system is nonlinear. The estimates are of the same type as those established above in Section 4.2.

In order to demonstrate that such convexity properties occur quite frequently in nonlinear problems, we first introduce several examples, including variational inequalities (with unilateral conditions in the domain or on the boundary), semilinear equations, control into coefficients problems, and higher-order elliptic systems.

Example 4.5.1 The classical formulation of the Signorini problem (cf. Example 1.2.5 in Chapter 1) reads

$$\sum_{i,j=1}^{d} \frac{\partial}{\partial x_i}\left(a_{ij}(x)\,\frac{\partial y}{\partial x_j}\right) + a_0(x)\,y = f \quad \text{in } \Omega, \tag{4.5.1}$$

$$\frac{\partial y}{\partial n_A} \ge 0, \quad y \ge 0, \quad \text{on } \partial\Omega, \quad \int_{\partial\Omega} y\,\frac{\partial y}{\partial n_A}\,d\sigma = 0. \tag{4.5.2}$$

Here, $\Omega \subset \mathbf{R}^d$ is a bounded and smooth domain, a_{ij}, $a_0 \in L^\infty(\Omega)$, and $f \in$

$L^2(\Omega)$ are given. With

$$C = \left\{ y \in H^1(\Omega) \,:\, y|_{\partial\Omega} \geq 0 \right\}$$

the problem (4.5.1), (4.5.2) admits a weak formulation as a variational inequality: Find $y \in C$ such that

$$\int_\Omega \sum_{i,j=1}^d a_{ij}(x) \frac{\partial y}{\partial x_j}(x) \left(\frac{\partial y}{\partial x_i} - \frac{\partial v}{\partial x_i} \right)(x)\, dx + \int_\Omega a_0(x)\, y(x)\, (y-v)(x)\, dx$$

$$\leq \int_\Omega f(x)(y-v)(x)\, dx \quad \forall\, v \in C. \tag{4.5.3}$$

The existence of a unique weak solution $y \in C$ of (4.5.3) follows from the Lions–Stampacchia theorem (cf. Theorem A2.3 in Appendix 2), provided that $a_0(x) \geq \alpha_1 > 0$ for a.e. $x \in \Omega$ and that there is some $\alpha_2 > 0$ such that

$$\sum_{i,j=1}^d a_{ij}(x)\, \xi_i\, \xi_j \,\geq\, \alpha_2\, |\xi|^2, \quad \forall\, \xi \in \mathbf{R}^d, \;\text{ for a.e. } x \in \Omega.$$

Now let $f_1, f_2 \in L^2(\Omega)$ be given, and $f_\lambda = \lambda f_1 + (1-\lambda) f_2$, for $\lambda \in [0,1]$. We denote by y_1, y_2, $y_\lambda \in C$ the corresponding solutions to (4.5.3). Since C is a convex cone, (4.5.3) is equivalent to

$$\int_\Omega \sum_{i,j=1}^d a_{ij}(x) \frac{\partial y}{\partial x_j} \frac{\partial y}{\partial x_i}\, dx + \int_\Omega a_0(x)\, y^2(x)\, dx \;=\; \int_\Omega f(x)\, y(x)\, dx, \quad y \in C, \tag{4.5.4}$$

$$\int_\Omega \sum_{i,j=1}^d a_{ij}(x) \frac{\partial y}{\partial x_j} \frac{\partial v}{\partial x_i}\, dx + \int_\Omega a_0(x) y(x) v(x)\, dx \;\geq\; \int_\Omega f(x)\, v(x)\, dx, \quad \forall\, v \in C. \tag{4.5.5}$$

Multiplying the relations (4.5.5) corresponding to f_1 and f_2 by λ and $(1-\lambda)$, respectively, and adding, we obtain that

$$\sum_{i,j=1}^d \int_\Omega a_{ij} \frac{\partial[\lambda y_1 + (1-\lambda) y_2]}{\partial x_j} \frac{\partial v}{\partial x_i}\, dx + \int_\Omega a_0\, [\lambda y_1 + (1-\lambda) y_2]\, v\, dx$$

$$\geq \int_\Omega [\lambda f_1 + (1-\lambda) f_2]\, v\, dx \quad \forall\, v \in C. \tag{4.5.6}$$

Now insert

$$v \,=\, y_\lambda - [y_\lambda - \lambda y_1 - (1-\lambda) y_2]_+ \,=\, \min\{y_\lambda, \lambda y_1 + (1-\lambda) y_2\} \in C$$

in the inequality (4.5.3) corresponding to the datum f_λ, and set

$$z \,=\, [y_\lambda - \lambda y_1 - (1-\lambda) y_2]_+ \in C.$$

We then have

$$\int_\Omega \sum_{i,j=1}^d a_{ij}(x)\,\frac{\partial y_\lambda}{\partial x_j}(x)\,\frac{\partial z}{\partial x_i}(x)\,dx + \int_\Omega a_0(x)\,y_\lambda(x)\,z(x)\,dx \le \int_\Omega f_\lambda(x)\,z(x)\,dx. \tag{4.5.7}$$

In fact, owing to (4.5.5) and since $z \in C$, even equality holds in (4.5.7). Therefore, inserting $v = z$ in (4.5.6), and subtracting (4.5.7) from the resulting inequality, we find that

$$\sum_{i,j=1}^d \int_\Omega a_{ij}\,\frac{\partial[\lambda y_1 + (1-\lambda)y_2 - y_\lambda]}{\partial x_j}\,\frac{\partial z}{\partial x_i}\,dx + \int_\Omega a_0\,[\lambda y_1 + (1-\lambda)y_2 - y_\lambda]\,z\,dx \ge 0, \tag{4.5.8}$$

whence, by the definition of z, it follows that $z = 0$ a.e. in Ω, and consequently,

$$y_\lambda(x) \le \lambda y_1(x) + (1-\lambda)y_2(x), \quad \text{for a.e. } x \in \Omega. \tag{4.5.9}$$

In conclusion, the nonlinear operator $T : L^2(\Omega) \to H^1(\Omega)$, $f \mapsto Tf = y$, is convex with respect to the a.e. ordering in Ω. By virtue of Theorem A2.4 in Appendix 2, T is also Lipschitz continuous.

It is possible to analyze the obstacle problem with unilateral conditions in the domain (compare (1.2.36) in Chapter 1) in a similar manner. In this case, we have to find some $y \in \widetilde{C} = \left\{z \in H_0^1(\Omega) \,:\, z \ge 0 \text{ a.e. in } \Omega\right\}$ such that

$$\sum_{i,j=1}^d \int_\Omega a_{ij}(x)\,\frac{\partial y}{\partial x_j}(x)\left(\frac{\partial y}{\partial x_i} - \frac{\partial v}{\partial x_i}\right)(x)\,dx \le \int_\Omega f(x)(y-v)(x)\,dx \quad \forall v \in \widetilde{C}.$$

Remark. Such convexity properties were first noticed by Lions [1972] and further studied by Lemaire [1985]. The optimization problems discussed in this section may be viewed as special cases of §3.3.1 in Chapter 3. However, the convexity properties may not be valid in that abstract case. For instance, in §3.3.1 the existence of an ordering relation is not assumed; moreover, the cost functional in Example 3.3.2 in Chapter 3 is not necessarily increasing (compare below the assumption on θ occurring in (4.5.21)).

Example 4.5.2 Consider now the semilinear elliptic boundary value problem

$$-\Delta y + j(y) = f \quad \text{in } \Omega, \tag{4.5.10}$$

$$y = 0 \quad \text{on } \partial\Omega, \tag{4.5.11}$$

where $\Omega \subset \mathbf{R}^d$ is a bounded domain. We assume that $j : \mathbf{R} \to \mathbf{R}$ is increasing and concave. Then (cf. Barbu and Precupanu [1986, Chapter II, §1]) j is continuous on $\mathbf{R}$ and (cf. Appendix 1) defines a maximal monotone graph in $\mathbf{R} \times \mathbf{R}$. Thus, by virtue of Theorem A2.4 in Appendix 2, the boundary value problem (4.5.10), (4.5.11) has for any given $f \in L^2(\Omega)$ a unique weak solution $y \in H_0^1(\Omega)$.

Using notation similar to that in Example 4.5.1, we can write

$$-\Delta y_\lambda + j(y_\lambda) = f_\lambda, \quad \text{in } \Omega, \tag{4.5.12}$$

$$-\Delta(\lambda y_1 + (1-\lambda)y_2) + \lambda j(y_1) + (1-\lambda)j(y_2) = f_\lambda, \quad \text{in } \Omega. \tag{4.5.13}$$

Then, using the concavity of j,

$$-\Delta(y_\lambda - \lambda y_1 - (1-\lambda)y_2) + j(y_\lambda) - j(\lambda y_1 + (1-\lambda)y_2) \le 0. \tag{4.5.14}$$

Multiplying (4.5.14) by $[y_\lambda - \lambda y_1 - (1-\lambda)y_2]_+$, and invoking the monotonicity of j, we can infer that

$$\int_\Omega \left|\nabla(y_\lambda - \lambda y_1 - (1-\lambda)y_2)_+\right|^2(x)\,dx = 0,$$

which implies that the inequality (4.5.9) holds also in this case.

As a related situation, let us now consider the fourth-order elliptic boundary value problem (compare §2.2.2 in Chapter 2)

$$\Delta(u^3\,\Delta y) = f, \quad \text{in } \Omega, \tag{4.5.15}$$

$$y = \Delta y = 0, \quad \text{on } \partial\Omega, \tag{4.5.16}$$

with a bounded and smooth domain $\Omega \subset \mathbf{R}^2$, which models the deflection of a simply supported plate located at Ω. If $h \in H^2(\Omega) \cap H_0^1(\Omega)$ denotes the solution to the elliptic boundary value problem

$$\Delta h = f \text{ in } \Omega, \qquad h = 0 \text{ on } \partial\Omega,$$

then the solution of (4.5.15), (4.5.16) satisfies

$$\Delta y = h\,u^{-3} \text{ in } \Omega, \qquad y = 0 \text{ on } \partial\Omega. \tag{4.5.17}$$

Assuming that $f \ge 0$ in Ω, we obtain from the maximum principle (cf. Theorem A2.8 in Appendix 2) that $h \le 0$ in Ω. Since u represents the thickness of the plate, it is natural to assume that u is positive. Then the right-hand side of (4.5.17) is concave with respect to u. Therefore, as in (4.5.12)–(4.5.14) (recall also §2.2.2 in Chapter 2), it follows that

$$y_\lambda(x) \le \lambda y_1(x) + (1-\lambda)y_2(x) \quad \text{for a.e. } x \in \Omega,$$

where we have used the same notation as in Example 4.5.1. That is, (4.5.9) is again satisfied.

This convexity property was observed by Kawohl and Lang [1997]. Results for equations with general semilinear terms, depending both on the solution and the independent variables, may be proved similarly.

One important such case is given by systems with control into coefficients. To this end, suppose that $\Omega \subset \mathbf{R}^d$ is a bounded and smooth domain. We consider the boundary value problem

$$-\sum_{i,j=1}^{d} \frac{\partial}{\partial x_i}\left(a_{ij}(x)\frac{\partial y}{\partial x_j}\right) = \varphi(x, y(x), u(x)) \quad \text{in } \Omega, \quad y = 0 \quad \text{on } \partial\Omega, \tag{4.5.18}$$

where the coefficients $a_{ij} \in C^1(\overline{\Omega})$ satisfy the ellipticity condition of Example 4.5.1 above. The function $\varphi : \Omega \times \mathbf{R} \times K \to \mathbf{R}$ (where $K \subset \mathbf{R}$ is some nonempty and closed interval) is, together with its partial derivative φ_y with respect to y, assumed to be continuous and to satisfy the following conditions:

$$|\varphi(x, 0, u)| \leq M(x) + \hat{C}\,|u| \quad \forall\,(x, u) \in \Omega \times K, \tag{4.5.19}$$

$$0 \geq \varphi_y(x, y, u) \geq -(M(x) + \hat{C}\,|u|)\,\eta(|y|) \quad \forall\,(x, y, u) \in \Omega \times \mathbf{R} \times K. \tag{4.5.20}$$

Here, $\hat{C} > 0$ is a fixed constant, $M \in L^s(\Omega)$, $s \geq \max\left\{2, \frac{d}{2}\right\}$, and $\eta : \mathbf{R}_+ \to \mathbf{R}_+$ is a nondecreasing function.

By virtue of Theorem A2.9 in Appendix 2, (4.5.18) has under the conditions (4.5.19), (4.5.20), for any $u \in L^\infty(\Omega)$ satisfying $u(x) \in K$ a.e. in Ω, a unique solution $y \in W^{2,s}(\Omega) \cap H_0^1(\Omega)$. By the embedding result of Theorem A2.2(iv) in Appendix 2, $y \in C(\overline{\Omega})$.

We assume that $\varphi(x, \cdot, \cdot)$ is a convex mapping. Then it follows, arguing as in (4.5.12)–(4.5.14), that (4.5.9) is again valid; in this situation, y_λ, y_1, y_2 are the solutions to (4.5.18) associated with $u_\lambda = \lambda u_1 + (1-\lambda)u_2$, respectively with u_1, u_2.

Finally, we also remark that Example 4.5.1 may be considered as a limiting case of this situation, since variational inequalities can be approximated by semilinear equations if the subdifferential is replaced by its Yosida–Moreau approximation. For instance, if $\beta \subset \mathbf{R} \times \mathbf{R}$ is given by

$$\beta(r) = \begin{cases} \emptyset, & \text{if } r < 0, \\ ]-\infty, 0], & \text{if } r = 0, \\ \{0\}, & \text{if } r > 0, \end{cases}$$

then the Yosida–Moreau regularization $\beta_\lambda(r) = -\lambda^{-1}(r)_-$ is increasing and concave as required in (4.5.10).

After these introductory examples, we will study in the remainder of this paragraph the following abstract control problem without state constraints:

$$\text{Min}\ \left\{\theta(y) + \psi(u)\right\}, \tag{4.5.21}$$

subject to

$$y = TBu, \tag{4.5.22}$$

$$u \in U_{ad}. \tag{4.5.23}$$

The main difference in comparison with the problem (4.2.1)–(4.2.4) studied above is that the operator T may be nonlinear. This renders the use of subdifferential calculus impossible and makes an alternative approach necessary.

The notation and the assumptions are as in Section 4.2: V is a Hilbert space, not necessarily identified with its dual, and U and H are Hilbert spaces identified with their duals. We assume that the embeddings $V \subset H \subset V^*$ are continuous and compact. In addition, $\psi : U \to \mathbf{R}$ and $\theta : H \to \mathbf{R}_+$ are convex and continuous mappings, $U_{ad} \subset U$ is nonempty, closed, and convex, and $B : U \to V^*$ is a linear and bounded operator. Finally, assume that $T : B(U_{ad}) \subset V^* \to V$ is a Lipschitz continuous mapping (in applications, T may be defined on a larger set).

A key assumption is the following:

(A1) The mapping $u \mapsto G(u) = \theta(TBu)$ is convex on U_{ad}.

This is clearly true for linear T; also, Examples 4.5.1, 4.5.2 demonstrate in function spaces (via (4.5.9), and assuming that θ is convex and increasing) that G may be convex in many other cases. We extend G by $+\infty$ outside of U_{ad}.

We again require the boundedness and the strong monotonicity of $\partial\psi$ (compare (4.2.5)), which here takes the following form (recall that U is identified with U^*):

There is some $\alpha > 0$ such that

$$(w_1 - w_2, u_1 - u_2)_U \geq \alpha\, |u_1 - u_2|_U^2 \quad \forall\, w_i \in \partial\psi(u_i),\ u_i \in U_i,\ i = 1, 2. \tag{4.5.24}$$

Hence, $\partial\psi$ is coercive on U, and it follows from Theorem A1.8 in Appendix 1 that ψ satisfies the coercivity condition

$$\lim_{|u|_V \to +\infty} \frac{\psi(u)}{|u|_U} = +\infty.$$

Now suppose that finite-dimensional subspaces $V_h \subset V$, $U_h \subset U$ are given, where $h > 0$ is some discretization parameter, and where the scalar products and norms in V_h and U_h coincide with those in H and U, respectively.

We put $B_h = B|_{U_h}$ and assume that a linear and bounded restriction operator $R_h^U : U \to U_h$ is given that satisfies the following conditions (with I_{U_h} denoting the identity operator on U_h):

(A2) $U_{ad}^h = R_h^U(U_{ad}) \subset U_h, \quad R_h^U|_{U_h} = I_{U_h}, \quad R_h^U u \to u$ strongly in U as $h \searrow 0, \quad \forall\, u \in U$.

Moreover, we assume that there is some $\rho > 0$ such that for

$$\mathcal{K}_h = U_{ad}^h + S_h(0,\rho) = \Big\{ w = u_h + z_h :\ u_h \in U_{ad}^h,\ z_h \in U_h,\ |z_h|_U \le \rho \Big\},$$

finite-dimensional projections $T_h : B_h(\mathcal{K}_h) \to V_h$ of T can be constructed for $h > 0$ such that $\{T_h B_h\}_{h>0}$ is uniformly Lipschitz continuous, that is, there is some $L > 0$ such that

$$\left| T_h B_h u_1^h - T_h B_h u_2^h \right|_V \le L \left| u_1^h - u_2^h \right|_{U_h} \quad \forall\, u_1^h, u_2^h \in \mathcal{K}_h, \quad \forall\, h > 0.$$

For the approximation properties of $\{T_h\}_{h>0}$ we assume that

(A3) $\qquad T_h B_h R_h^U u \to TBu \quad$ strongly in $V \quad \forall\, u \in U_{ad}$.

We also introduce the functions $G_h = \theta_h(T_h B_h)$ on $\mathcal{K}_h$, for $h > 0$, assumed to be convex, and we put $\theta_h = \theta|_{V_h}$ and $\psi_h = \psi|_{U_h}$.

We then study the following approximation of (4.5.21)–(4.5.23):

$$\operatorname{Min} \Big\{ \theta_h(y_h) + \psi_h(u_h) \Big\}, \tag{4.5.25}$$

$$y_h = T_h B_h u_h\,, \tag{4.5.26}$$

$$u_h \in U_{ad}^h\,. \tag{4.5.27}$$

Observe that ψ_h inherits the coercivity properties of ψ, uniformly in $h > 0$. Hence, since θ and θ_h are nonnegative, the cost functionals in (4.5.21) and (4.5.25) satisfy the coercivity conditions

$$\lim_{|u|_U \to +\infty} \frac{\theta(y) + \psi(u)}{|u|_U} = +\infty, \quad \text{uniformly in } y \in H,$$

$$\lim_{|u_h|_U \to +\infty} \frac{\theta_h(y_h) + \psi_h(u_h)}{|u_h|_U} = +\infty, \quad \text{whenever } [y_h, u_h] \in V_h \times U_h, \; h > 0.$$

The existence of (unique) optimal pairs $[y^*, u^*]$ and $[y_h^*, u_h^*]$ for the problems (4.5.21)–(4.5.23) and (4.5.25)–(4.5.27), respectively, follows from Theorem 2.1.2 in Chapter 2 and the strict convexity of the respective cost functionals. Obviously, we have for $h > 0$ the estimate

$$\theta_h(y_h^*) + \psi_h(u_h^*) \le \theta\big(T_h B_h R_h^U u^*\big) + \psi\big(R_h^U u^*\big).$$

In view of (A2) and (A3), the right-hand side is bounded. Hence, using the nonnegativity of θ_h and the coercivity of ψ_h, we can infer that $\{|u_h^*|_U\}_{h>0}$ is bounded. Then also $\{|y_h^*|_V\}_{h>0}$ is bounded; indeed, the uniform Lipschitz continuity of the family $\{T_h B_h\}_{h>0}$ yields, invoking (A2) and (A3) once more, that

$$\begin{aligned} |y_h^*|_V &= |T_h B_h u_h^*|_V \le \left|T_h B_h u_h^* - T_h B_h R_h^U u^*\right|_V + \left|T_h B_h R_h^U u^*\right|_V \\ &\le L \left|u_h^* - R_h^U u^*\right|_U + \left|T_h B_h R_h^U u^*\right|_V \le C_1, \end{aligned}$$

where $C_1 > 0$ does not depend on $h > 0$.

Lemma 4.5.3 *There is some $\tau_0 > 0$ such that for any $\tau \ge \tau_0$ the mapping G is Lipschitz continuous on $S(0,\tau) \cap U_{ad}$, and the family $\{G_h\}_{h>0}$ is uniformly Lipschitz continuous on $\{S(0,\tau) \cap \mathcal{K}_h\}_{h>0}$, where $S(0,\tau) = \{u \in U : |u|_U \le \tau\}$.*

Proof. We choose $\tau_0 > 0$ so large that $u_h^* \in S(0,\tau_0)$ for all $h > 0$. Suppose now that $\tau \ge \tau_0$. Similarly as above, we obtain for any $u_h \in S(0,\tau) \cap \mathcal{K}_h$ that

$$|T_h B_h u_h|_V \le L\,|u_h - R_h^U u^*|_U + |T_h B_h R_h^U u^*|_V,$$

where the right-hand side is bounded, owing to (A2) and (A3) and since $|u_h|_U \le \tau$. Set

$$\Lambda = \bigcup_{h>0} T_h B_h (S(0,\tau) \cap \mathcal{K}_h),$$

which is a bounded subset of V. Since the embedding $V \subset H$ is compact, the closure $\overline{\Lambda} \subset H$ is compact in H.

Now recall that θ is continuous on H and convex. Therefore, we can infer from Theorem A1.9 in Appendix 1 that $\theta|_{\overline{\Lambda}}$ is Lipschitz continuous with a Lipschitz constant $\widetilde{L}_\tau > 0$. Then clearly, θ_h is Lipschitz continuous on $T_h B_h(S(0,\tau) \cap \mathcal{K}_h)$ with the same Lipschitz constant, for any $h > 0$. The assertion now follows from the uniform Lipschitz continuity of $\{T_h B_h\}_{h>0}$ on $\{\mathcal{K}_h\}_{h>0}$. □

Proposition 4.5.4 *Let $I_{U^h_{ad}}$ denote the indicator function of U^h_{ad}. Then*

$$0 \in \partial\psi(u^*) + \partial G(u^*), \tag{4.5.28}$$

$$0 \in \partial\psi_h(u_h^*) + \partial G_h(u_h^*) + \partial I_{U^h_{ad}}(u^*). \tag{4.5.29}$$

Proof. Since θ and ψ are continuous and convex, the assertion follows directly from combining Theorem A1.6 in Appendix 1 with the additivity rule for subdifferentials (cf. the remark following Theorem A1.11). In (4.5.29), we have to add $\partial I_{U^h_{ad}}(u_h^*)$ since G_h is defined on $\mathcal{K}_h \supset U^h_{ad}$. □

Theorem 4.5.5 *There is some constant $\hat{C} > 0$, independent of $h > 0$, and for any $h > 0$ there is some $q_h \in \partial\psi_h(u_h^*)$, such that*

$$\alpha\,|u^* - u_h^*|_U^2 \le \hat{C}\,\Big[\big|R_h^U u^* - u^*\big|_U + |(T - T_h)Bu_h^*|_H + |w_h - q_h|_U$$
$$+ \big|(T - T_h)BR_h^U u^*\big|_H\Big], \quad \forall\, w_h \in \partial\psi(u_h^*). \tag{4.5.30}$$

Proof. We first show that $\bigcup_{h>0} \partial G_h(u_h^*)$ is bounded in U. To this end, choose $\tau_0 > 0$ as in Lemma 4.5.3, and let $\mu_h \in \partial G_h(u_h^*)$ for some $h > 0$. Putting $\tau = \tau_0 + \rho$, with the constant $\rho > 0$ introduced in the definition of $\mathcal{K}_h$, we then have

$$u_h^* + z \in S(0,\tau) \cap \mathcal{K}_h \quad \forall\, z \in U_h, \quad |z|_U = \rho.$$

By virtue of Lemma 4.5.3, the family $\{G_h\}_{h>0}$ is uniformly Lipschitz continuous on $\{S(0,\tau) \cap \mathcal{K}_h\}_{h>0}$ with some Lipschitz constant $L_\tau > 0$. Hence, invoking the definition of the subdifferential, we have, for all $z \in U_h$ with $|z|_U = \rho$,

$$L_\tau\,|z|_U \ge G_h(u_h^* + z) - G_h(u_h^*) \ge (-\mu_h, z)_U.$$

Hence,

$$|\mu_h|_{U_h^*} = |\mu_h|_{U_h} = |\mu_h|_U \le \frac{L_\tau}{\rho},$$

and the claim is proved.

Next, observe that by (4.5.29) there is some $z_h \in \partial I_{U^h_{ad}}(u^*)$ such that

$$z_h \in -\partial\psi_h(u_h^*) + \partial G_h(u_h^*).$$

Thanks to the chain rule for subdifferentials (compare Theorem A1.12 in Appendix 1), $\partial\psi_h(u_h^*) = \partial(\psi \circ i_h)(u_h^*)$, where $i_h : U_h \to U$ is the canonical injection. Therefore, using the boundedness of $\partial\psi$, we can infer that also $\bigcup_{h>0} \partial\psi_h(u_h^*)$ is bounded in U, and thus

$$|z_h|_U \le C_1 \quad \forall\, h > 0,$$

where C_i, $i \in \mathbf{N}$, will denote positive constants that do not depend on $h > 0$.

Now notice that Proposition 4.5.4 implies that there are

$$w \in \partial\psi(u^*), \qquad q_h \in \partial\psi_h(u_h^*), \qquad g \in \partial G(u^*), \qquad g_h \in \partial G_h(u_h^*),$$

such that

$$0 = w - q_h + g - g_h - z_h \quad \text{in } U. \tag{4.5.31}$$

Next, we have, using the definition of the subdifferential and the boundedness of $\{|z_h|_U\}_{h>0}$,

$$\begin{aligned}(-z_h, u^* - u_h^*)_U &= \left(z_h, u_h^* - R_h^U u^*\right)_U + \left(z_h, R_h^U u^* - u^*\right)_U \\ &\geq \left(z_h, R_h^U u^* - u^*\right)_U \geq -C_1 \left|R_h^U u^* - u^*\right|_U . \end{aligned} \tag{4.5.32}$$

Moreover, using (4.5.24) and the boundedness of $\{u_h^*\}_{h>0}$, we find that for any $w_h \in \partial\psi(u_h^*)$,

$$\begin{aligned}(w - q_h, u^* - u_h^*)_U &= (w - w_h, u^* - u_h^*)_U + (w_h - q_h, u^* - u_h^*)_U \\ &\geq \alpha\, |u^* - u_h^*|_U^2 - C_2\, |w_h - q_h|_U . \end{aligned} \tag{4.5.33}$$

In addition, owing to the definition of the subdifferential,

$$\begin{aligned}(g - g_h, u^* - u_h^*)_U &\geq G(u^*) - G(u_h^*) + (g_h, u_h^* - u^*)_U \\ &= G(u^*) - G(u_h^*) + \left(g_h, u_h^* - R_h^U u^*\right)_U + \left(g_h, R_h^U u^* - u^*\right)_U \\ &\geq G(u^*) - G(u_h^*) + G_h(u_h^*) - G_h\left(R_h^U u^*\right) - C_1 \left|R_h^U - u^*\right|_U . \end{aligned} \tag{4.5.34}$$

Summarizing, we have proved the estimate

$$\begin{aligned}\alpha\, |u^* - u_h^*|_U^2 &\leq 2C_1 \left|R_h^U u^* - u^*\right|_U + C_2\, |w_h - q_h|_U - G(u^*) + G(u_h^*) \\ &\quad - G_h(u_h^*) + G_h\left(R_h^U u^*\right) \quad \forall\, w_h \in \partial\psi(u_h^*) . \end{aligned} \tag{4.5.35}$$

Next, recall that $B_h = B|_{U_h}$, so that

$$T_h B_h u_h^* = T_h B u_h^*, \qquad T_h B R_h^U u^* = T_h B_h R_h^U u^*.$$

Moreover, $\{TBu_h^*\}_{h>0}$ and $\{T_h B R_h^U u^*\}_{h>0}$ are bounded in V and hence relatively compact in H. Since $\partial\theta$ is by Theorem A1.9 in Appendix 1 locally bounded on H, we can infer that for any $t \in \partial\theta(TBu_h^*)$ and $t_h \in \partial\theta(T_hBR_h^U u^*)$ we have the inequalities

$$\begin{aligned} G_h(u_h^*) - G(u_h^*) &= \theta(T_hB_hu_h^*) - \theta(TBu_h^*) \geq (t, T_hBu_h^* - TBu_h^*)_H \\ &\geq -C_3\,|(T-T_h)Bu_h^*|_H\,, \end{aligned} \tag{4.5.36}$$

$$\begin{aligned} G(u^*) - G_h\big(R_h^U u^*\big) &= \theta(TBu^*) - \theta\big(T_hB_hR_h^U u^*\big) \geq \big(t_h, TBu^* - T_hBR_h^U u^*\big)_H \\ &\geq -C_4\,\big|u^* - R_h^U u^*\big|_U - C_5\,\big|(T-T_h)BR_h^U u^*\big|_H\,. \end{aligned} \tag{4.5.37}$$

Inserting (4.5.36), (4.5.37) in (4.5.35), we finally arrive at (4.5.30), and the proof of the assertion is complete. □

Remark. The constant $\hat{C} > 0$ in (4.5.30) depends on u^*. The right-hand side in (4.5.30) collects the contributions of the interpolation errors and of the error originating from the discretization of the state equation.

Remark. Further error estimates for the discretization of control problems associated with variational inequalities may be found in the work of Liu and Tiba [2001], on which this section is based.

Chapter 5

Unknown Domains

In this chapter, we will analyze in greater detail methods for the solution of problems involving variable or undetermined sets. Examples for this class of problems are *shape optimization* and *free boundary problems*. We start with the latter subject, which has intimate connections to the theory of variational inequalities and allows a smooth transition from the discussion in Section 3.3 of Chapter 3 to the present direction of investigation. We will also demonstrate that there is a strong interrelation between the two types of problems studied in this chapter.

5.1 Free Boundary Problems

In Examples 1.2.5–1.2.7 in Chapter 1, we have already discussed free boundary value problems originating from applications in physics, namely, the problem of elastoplastic torsion, the electrochemical machining process, and the location of a plasma inside a tokamak machine. Some of these problems can be formulated as obstacle problems, that is, as variational inequalities with unilateral conditions either in the domain of definition or on its boundary. The corresponding coincidence set, where the solution "touches" the obstacle, has great physical relevance; it is one of the unknowns of the problem. If this is known, then the variational inequality reduces to a classical partial differential equation. In Section 2.3 of Chapter 2, optimization problems associated with such systems have been studied, mainly from the point of view of the existence of optimal configurations, while a detailed study of a constrained control problem for the obstacle variational inequality was carried out in §3.3.3 of Chapter 3.

In this section, we present further examples and general methods for their solution. Optimization questions will also be addressed, and the connection to optimal design problems will be underlined. This relationship has been put into evidence in the works of Alt and Caffarelli [1981], Zolésio [1990], [1994], Okhezin [1992], and Hoffmann and Tiba [1995]. Numerical approaches based on these ideas are due to Männikkö, Neittaanmäki, and Tiba [1994], Tiihonen

[1997], and Kärkkäinen and Tiihonen [1999].

5.1.1 The Dam Problem

We consider a rectangular dam consisting of an isotropic homogeneous (porous) material, denoting by $\Omega =]0,\alpha[\times]0,\beta[$ the interior of its cross section. The dam separates two water reservoirs of heights $\beta > h_1 > h_2 > 0$, and we denote by W its portion that is wet (saturated), due to the water flow in the porous medium (see Figure 1.1). Notice that on one face of the dam the height of W is assumed to be h_1, while the height h_3 on the other face may satisfy $h_3 > h_2$ (which in physical reality is usually the case). The boundary S separating the wet (saturated) region W from the dry (unsaturated) region $\Omega \setminus W$ is the free boundary, which has to be determined. We assume that S can be represented as the graph of a function $\xi :]0,\alpha[\to \mathbf{R}$ satisfying $0 < \xi(x) \le h_1$ for all $x \in]0,\alpha[$.

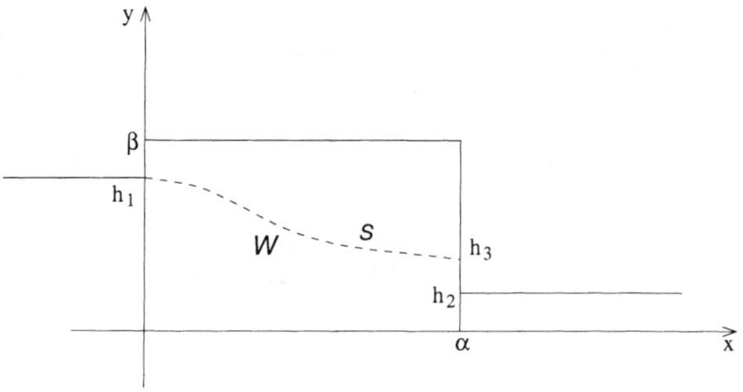

Figure 1.1. The rectangular dam Ω.

If $p(x,y)$ denotes the unknown hydraulic pressure at the point $(x,y) \in \Omega$, then Darcy's law (with normalized coefficients) yields that

$$\Delta p = 0 \quad \text{in } W, \tag{5.1.1}$$

and we have the boundary conditions

$$p(0,y) = h_1 - y, \quad \text{if } 0 \le y \le h_1, \tag{5.1.2}$$

$$p(\alpha,y) = h_2 - y, \quad \text{if } 0 \le y \le h_2, \tag{5.1.2$'$}$$

$$p(x,y) = 0, \quad \text{if } (x,y) \in S, \tag{5.1.3}$$

$$p(\alpha,y) = 0, \quad \text{if } h_2 \le y \le h_3, \tag{5.1.3$'$}$$

$$\frac{\partial p}{\partial y} = -1, \quad \text{if } y = 0, \tag{5.1.4}$$

$$\frac{\partial p}{\partial \nu} = -\nu_2, \quad \text{on } S, \tag{5.1.5}$$

where

$$\nu(x, \xi(x)) = (\nu_1(x, \xi(x)), \nu_2(x, \xi(x))) = (1 + \xi'^2(x))^{-1/2}(\xi'(x), -1)$$

is the outward unit normal to S at $(x, \xi(x)) \in S$. We extend p to the whole domain Ω by putting $p(x, y) = 0$ for $(x, y) \in \Omega \setminus W$, and we introduce the function

$$\tilde{p}(x, y) = p(x, y) + y \quad \forall\, (x, y) \in \Omega,$$

and the so-called *Baiocchi transformation of* p (see Baiocchi [1972])

$$z(x, y) = \int_y^{h_1} (\tilde{p}(x, s) - s)\, ds \quad \forall\, (x, y) \in \overline{\Omega}. \tag{5.1.6}$$

Since $\xi(x) \leq h_1$ in $]0, \alpha[$, we easily verify that

$$z(x, y) = \begin{cases} \displaystyle\int_y^{\xi(x)} p(x, s)\, ds, & \text{if } y < \xi(x), \\ 0, & \text{if } y \geq \xi(x). \end{cases}$$

We have the following result.

Proposition 5.1.1 *If* p *and* ξ *are sufficiently smooth, then*

$$\Delta z = \chi_W \quad \textit{in } \mathcal{D}'(\Omega), \tag{5.1.7}$$

where χ_W *is the characteristic function of* W *in* Ω.

Proof. We have for any test function $\varphi \in C_0^\infty(\Omega)$:

$$\begin{aligned}
\int_\Omega \Delta z(x, y)\, \varphi(x, y)\, dx\, dy &= -\int_\Omega \nabla z(x, y) \cdot \nabla \varphi(x, y)\, dx\, dy \\
&= -\int_W \left[\varphi_x(x, y) \int_y^{h_1} p_x(x, s)\, ds + \varphi_y(x, y)(y - \tilde{p}(x, y)) \right] dx\, dy \\
&= -\int_W \left\{ \left[\varphi(x, y) \int_y^{\xi(x)} \tilde{p}_x(x, s)\, ds \right]_x + [\varphi(x, y)(y - \tilde{p}(x, y))]_y \right\} dx\, dy \\
&\quad + \int_W \varphi(x, y) \left[\int_y^{\xi(x)} \tilde{p}_{xx}(x, s)\, ds + \xi'(x) \tilde{p}_x(x, \xi(x)) + 1 - \tilde{p}_y(x, y) \right] dx\, dy \\
&= \int_W \varphi(x, y)\, (\xi'(x) \tilde{p}_x(x, \xi(x)) - \tilde{p}_y(x, \xi(x)) + 1)\, dx\, dy \\
&= \int_0^\alpha (\xi'(x) \tilde{p}_x(x, \xi(x)) - \tilde{p}_y(x, \xi(x)) + 1) \int_0^{\xi(x)} \varphi(x, y)\, dy\, dx.
\end{aligned}$$

Here, we have used the divergence theorem to conclude that the second integral over W appearing in the above calculation vanishes; we have also used the fact that $\tilde{p}$ is harmonic in W, so that

$$\int_y^{\xi(x)} \tilde{p}_{xx}(x,s)\,ds = -\int_y^{\xi(x)} \tilde{p}_{yy}(x,s)\,ds = -\tilde{p}_y(x,\xi(x)) + \tilde{p}_y(x,y).$$

Now observe that in view of (5.1.5) we have for all $x \in]0,\alpha[$,

$$\begin{aligned}\xi'(x)\,\tilde{p}_x(x,\xi(x)) - \tilde{p}_y(x,\xi(x)) &= \xi'(x)\,p_x(x,\xi(x)) - (p_y(x,\xi(x)) + 1)\\ &= (1+\xi'^2(x))^{1/2}\left(\frac{\partial p}{\partial \nu}(x,\xi(x)) + \nu_2(x,\xi(x))\right)\\ &= 0.\end{aligned}$$

Consequently, we can infer that

$$\int_\Omega \Delta z(x,y)\,\varphi(x,y)\,dx\,dy = \int_\Omega \varphi(x,y)\,dx\,dy \quad \forall\,\varphi \in C_0^\infty(\Omega),$$

which concludes the proof of the assertion. □

Remark. It follows from the maximum principle that $p > 0$ in W. Notice that Theorem A2.8 in Appendix 2 yields $p \geq 0$ in W for the case that $p \geq 0$ on the whole boundary ∂W. Here, we only have $p \geq 0$ on $\partial W \setminus \{(x,0) : 0 < x < \alpha\}$; however, the maximum principle is still valid under the boundary condition (5.1.4) on $\{(x,0) : 0 < x < \alpha\}$, and it can even be shown that p is positive in W. Then also $z > 0$ in W, and since $z = 0$ in $\Omega \setminus W$, we may view z as the solution to the obstacle problem (in the classical formulation)

$$-\Delta z \geq -1, \quad z \geq 0, \quad \text{in } \Omega, \quad -\Delta z = -1, \quad \text{in } \{z > 0\}, \quad z = g, \quad \text{on } \partial\Omega, \tag{5.1.8}$$

where g can be explicitly constructed from the boundary conditions (5.1.2), (5.1.2)′, (5.1.3), (5.1.3)′, and from the above transformations. Notice that $z = g$ vanishes on the set

$$\Gamma = \{(0,y) : h_1 < y < \beta\} \cup \{(x,\beta) : 0 < x < \alpha\} \cup \{(\alpha,y) : h_3 < y < \beta\} \subset \partial\Omega.$$

Now let

$$V = \{v \in H^1(\Omega) : v = 0 \text{ on } \Gamma\}.$$

Then the variational inequality corresponding to (5.1.8) has the form

$$\int_\Omega \nabla z(x,y) \cdot \nabla(v-z)(x,y)\,dx\,dy \geq -\int_\Omega (v-z)(x,y)\,dx\,dy \quad \forall v \in K, \tag{5.1.9}$$

where

$$K = \{v \in V : v = g \text{ on } \partial\Omega, \ \ v \geq 0 \text{ in } \Omega\}.$$

Since the bilinear form associated with the left-hand side of (5.1.9) is coercive on V, it follows from the Lions–Stampacchia theorem (cf. Theorem A2.3 in Appendix 2) that (5.1.9) has a unique solution z in K.

Natori and Kawarada introduced in their classical paper [1981] an approximation/relaxation procedure for free boundary problems, called the *integrated penalty method*. To apply this method to (5.1.1)–(5.1.5), we assume for the moment that W and the characteristic function χ_W are known. We then define the "relaxed" boundary value problem

$$\begin{aligned} -\Delta(p_\varepsilon - y) + \frac{1}{\varepsilon}(1 - \chi_W)(p_\varepsilon - y) = 0 \quad \text{in } \Omega, \\ \frac{\partial(p_\varepsilon - y)}{\partial y} &= 0, \quad \text{if } y = \beta, \\ p_\varepsilon(0, y) - y &= 0, \quad \text{if } h_1 \leq y \leq \beta, \\ p_\varepsilon(\beta, y) - y &= 0, \quad \text{if } h_3 \leq y \leq \beta. \end{aligned} \tag{5.1.10}$$

We have denoted in (5.1.10) the unknown function by $p_\varepsilon - y$ in order to make the connection to the previous problem easier. On the remaining part of $\partial\Omega$, the function $p_\varepsilon - y$ satisfies (5.1.2), (5.1.2)′, (5.1.3)′, and (5.1.4) in the weak sense.

Proposition 5.1.2 *If W is a Lipschitz domain, then there is a sequence $\varepsilon_n \searrow 0$ such that $p_{\varepsilon_n} - y \to p$ weakly in $H^1(W)$ as $n \to \infty$.*

Proof. The function $p_\varepsilon - y - p$ satisfies (5.1.10) in the weak sense, with Δp on the right-hand side and with homogeneous mixed boundary conditions. Taking $p_\varepsilon - y - p$ as test function in the weak formulation, and using the fact that $(1 - \chi_W)p$ vanishes a.e. in Ω, we find that there is some $\hat{C} > 0$ that is independent of $\varepsilon > 0$ and satisfies

$$|p_\varepsilon - y - p|^2_{H^1(\Omega)} + \frac{1}{\varepsilon}\int_{\Omega \setminus W} (p_\varepsilon - y)^2 \, dx \, dy \, \leq \, \hat{C}. \tag{5.1.11}$$

Consequently, there are a sequence $\varepsilon_n \searrow 0$ and some $\hat{p} \in H^1(\Omega)$ such that $p_{\varepsilon_n} - y \to \hat{p}$ weakly in $H^1(\Omega)$ and, by compact embedding, strongly in $L^2(\Omega)$. Obviously, $\hat{p} = 0$ a.e. in $\Omega \setminus W$; the trace theorem then shows that $\hat{p}|_W$ satisfies (5.1.2), (5.1.2)′, (5.1.3), and (5.1.3)′.

Moreover, a simple distribution argument in (5.1.10) yields that $\Delta\hat{p} = 0$ in W, that is, $p = \hat{p}|_W$, and the proof is finished. □

Remark. The regularity of W is necessary in order to use the *Hedberg–Keldys stability property* (compare Theorem 2.3.9 in Chapter 2). In this respect, weaker assumptions may be imposed that are similar to those in Chapter 2. The requirement that W (or the free boundary) be known a priori means that the overdetermined condition on S given by (5.1.5) will also be satisfied by $\hat{p}$. In applications, S and W are unknown, and optimization algorithms may be constructed for their determination. In this way, shape optimization problems appear. Their connection to free boundary problems will be discussed in the next paragraph.

5.1.2 Free Boundary Problems and Optimal Design

We discuss some significant examples that demonstrate that there is a deep relationship (with many applications) between these two important classes of problems.

Example 5.1.3 First, we consider a free boundary problem arising in cryogenics or in plasma physics (compare Natori and Kawarada [1981] and Example 1.2.6 in Chapter 1; a related free boundary problem was discussed by Alt and Caffarelli [1981] via a shape variational principle).

To this end, let $\Omega \subset \mathbf{R}^2$ denote a (variable) domain in $\mathbf{R}^2$; we assume that Ω has the form of an annulus bounded by the fixed outer boundary Γ and the moving (unknown) inner boundary γ of length $l(\gamma)$. We then study the problem

$$\Delta u = 0 \quad \text{in } \Omega, \tag{5.1.12}$$

$$u = 0 \quad \text{on } \Gamma, \tag{5.1.13}$$

$$u = 1 \quad \text{on } \gamma, \tag{5.1.14}$$

$$|\nabla u|_{\mathbf{R}^2} = \hat{c} \quad \text{on } \gamma, \qquad \int_\gamma |\nabla u|_{\mathbf{R}^2}\, d\sigma = \hat{C}. \tag{5.1.15}$$

Here, $\hat{C} \in \mathbf{R}$ is a given constant, while the constant $\hat{c} \in \mathbf{R}$ is a priori unknown and has to be determined along with the solution u. If we have enough smoothness, and since by virtue of (5.1.14) the tangential component of ∇u vanishes on γ, we can reformulate (5.1.15) equivalently as

$$|\nabla u|_{\mathbf{R}^2} = \frac{\partial u}{\partial n} = \frac{\hat{C}}{l(\gamma)} \quad \text{on } \gamma. \tag{5.1.16}$$

We now transform the free boundary problem (5.1.12)–(5.1.15) into a shape optimization problem. For this purpose, we assume that the fixed outer boundary Γ can be represented as the graph of a (closed) curve parametrized in polar coordinates as $r = \tilde{\Gamma}(\theta)$ with a given positive function $\tilde{\Gamma}$ defined on $[0, 2\pi]$. We consider for a fixed "sufficiently small" $R > 0$ the set

$$U_{ad} = \{\eta \in C^1[0, 2\pi] : \eta(0) = \eta(2\pi), \;\; R \le \eta(\theta) \le \tilde{\Gamma}(\theta) \;\text{ for all } \theta \in [0, 2\pi]\},$$

and we associate with every $\eta \in U_{ad}$ the graph $\gamma(\eta)$ of the closed curve parametrized in polar coordinates as $r = \eta(\theta)$. Then we may replace the free boundary problem (5.1.12)–(5.1.15) by the following optimization problem

$$\underset{\eta \in U_{ad}}{\text{Min}} \left\{ \int_{\gamma(\eta)} \left| \frac{\partial y}{\partial n}(\sigma) - \frac{\hat{C}}{l(\gamma(\eta))} \right|^2 d\sigma \right\}, \tag{5.1.17}$$

subject to

$$\Delta y = 0 \quad \text{in } \Omega(\eta) \; = \; \left\{ (r, \theta) \in \mathbf{R}^2 : \; \eta(\theta) < r < \tilde{\Gamma}(\theta), \;\; 0 \le \theta \le 2\pi \right\}, \tag{5.1.18}$$

$$y \; = \; 0 \quad \text{on } \Gamma, \tag{5.1.19}$$

$$y \; = \; 1 \quad \text{on } \gamma(\eta). \tag{5.1.20}$$

Notice that in this formulation the free boundary has disappeared. Since we minimize directly over a prescribed class of domains, we have obtained a shape optimization problem; it is of the same type as problem (1.2.44), (1.2.41)′, (1.2.42)′, (1.2.45) discussed in Chapter 1.

Let us now discuss a free boundary problem arising in the study of junction semiconductor devices (see Bedivan [1997]). To this end, let $\Omega \subset \mathbf{R}^d$ be an (unknown) open set of the form

$$\Omega = \Omega(\alpha, \beta) = \left\{ [x, y] \in \mathbf{R}^d : \; y \in Q, \; \beta(y) < x < \alpha(y) \right\},$$

where $\overline{Q} \subset \mathbf{R}^{d-1}$ is a bounded Lipschitz domain and α, β are Lipschitz continuous functions in $\overline{Q}$, such that Ω is nonempty and connected. We then consider the free boundary problem

$$\begin{aligned} -\Delta z + z \; &= \; f \quad \text{in } \Omega, \\ z \; &= \; g_1 \quad \text{on } \Gamma_1, \\ z \; &= \; g_2 \quad \text{on } \Gamma_2, \\ \frac{\partial z}{\partial n} \; &= \; 0 \quad \text{on } \partial\Omega. \end{aligned}$$

Here, Γ_1 and Γ_2 are the (a priori unknown) parts of $\partial\Omega$ given by

$$\Gamma_1 = \left\{ [x, y] \in \mathbf{R}^d : \; x = \beta(y), \; y \in \overline{Q} \right\},$$

$$\Gamma_2 = \left\{ [x, y] \in \mathbf{R}^d : \; x = \alpha(y), \; y \in \overline{Q} \right\},$$

and $f \in L^2(\widehat{\Omega})$, $g_i \in H^1(\widehat{\Omega})$, $i = 1, 2$, are given functions on a prescribed bounded domain $\widehat{\Omega}$ that contains all the possible choices $\Omega = \Omega(\alpha, \beta)$. Note that on Γ_1, Γ_2 overdetermined boundary conditions are imposed that should determine the free boundary $\Gamma_1 \cup \Gamma_2$. We remark that in semiconductor theory u represents the electric potential in the pn junction Ω, while ∇u is the electric field in Ω.

Applying the same technique as in the previous example, we formulate the optimal design problem

$$\underset{\Omega}{\text{Min}} \left\{ \int_{\Gamma_1} (z - g_1)^2(\sigma)\, d\sigma + \int_{\Gamma_2} (z - g_2)^2(\sigma)\, d\sigma \right\},$$

subject to the above state equation with null Neumann conditions on $\partial\Omega$, and with $\Omega = \Omega(\alpha, \beta)$ playing the role of the minimizing parameter. A weighted penalization of this formulation (with weights $c_i > 0$, $i = \overline{1,3}$) is given by the quadratic cost functional

$$\begin{aligned} J(z, \phi) &= c_1 \left|\nabla z - \phi\right|^2_{L^2(\Omega)^d} + c_2 \left|\operatorname{div} \phi - z + f\right|^2_{L^2(\Omega)} \\ &\quad + c_3 \left|z - g_1\right|^2_{L^2(\Gamma_1)} + c_3 \left|z - g_2\right|^2_{L^2(\Gamma_2)}. \end{aligned}$$

Here, $z \in H^1(\Omega)$, and

$$\phi \in V = \left\{ \phi \in L^2(\Omega)^d : \operatorname{div} \phi \in L^2(\Omega),\ \ \phi \cdot n = 0 \ \text{on}\ \partial\Omega \right\},$$

where the outward unit normal n to $\partial\Omega$ exists in view of the Lipschitz assumption for Γ_i, $i = 1, 2$, and Q. If the minimum value of this functional is zero, then the corresponding function z solves the free boundary problem, and conversely. The optimization parameters are the mappings z, ϕ and the domain $\Omega = \Omega(\alpha, \beta) \subset \widehat{\Omega}$. In order to guarantee compactness, the admissible functions α, β are assumed to belong to a set of Lipschitz continuous functions on $\overline{Q}$ having a common Lipschitz constant.

We notice that the existence of minimizers of the above minimization problems can be obtained using the techniques presented in §2.3.1 of Chapter 2. It is clear that an interpretation of its solution in terms of the setting of the original problem is difficult if the optimal value is strictly positive.

We finally remark that this relaxation approach has been called in the work of Bedivan and Fix [1995] the *complete least squares formulation*. It is similar to the method of penalization of the state equation discussed in §3.2.2 of Chapter 3.

Example 5.1.4 We now discuss a similar approach proposed by Okhezin [1992] for the solution of one-dimensional two-phase Stefan problems. Although this example contains time-dependent operators, we include it in our exposition in order to demonstrate the generality of the ideas. In its simplest form, the classical formulation of the problem is to find (sufficiently smooth) functions y_1, y_2, and u such that

$$y_{1,t}(t,x) = \alpha_1 \, y_{1,xx}(t,x), \quad 0 < t < T, \; 0 < x < u(t), \tag{5.1.21}$$

$$y_{2,t}(t,x) = \alpha_2 \, y_{2,xx}(t,x), \quad 0 < t < T, \; u(t) < x < b, \tag{5.1.22}$$

$$y_1(0,x) = \varphi_1(x), \; x \in [0,u_0], \quad y_2(0,x) = \varphi_2(x), \, x \in [u_0,b], \tag{5.1.23}$$

$$y_1(t,0) = 0, \quad y_2(t,b) = 0, \tag{5.1.24}$$

$$y_1(t,u(t)) = y_2(t,u(t)) = 0, \quad 0 \le t \le T, \tag{5.1.25}$$

$$u_t(t) = \beta_2 \, y_{2,x}(t,u(t)) - \beta_1 \, y_{1,x}(t,u(t)), \quad 0 < t \le T, \tag{5.1.26}$$

$$u(0) = u_0 \in \,]0,b[\,. \tag{5.1.27}$$

Here, y_1, y_2 represent the "temperatures" in the two phases (solid and liquid, say), which coexist in the one-dimensional "container" $[0,b]$, separated by an interface located at time t at the space point $x = u(t)$, $0 \le t \le T$. The positive constants α_i, β_i, $i = 1,2$, are related to the physics of the problem (thermal conductivity, *Stefan condition* on the interface), and the initial temperature distributions φ_i, $i = 1,2$, in the phases and the initial location $u(0)$ of the interface are known.

Now let $v \in L^2(0,T)$ be a "control" function. We replace the Stefan condition (5.1.26) and (5.1.27) by

$$u'(t) \;=\; v(t) \quad \text{in } [0,T], \tag{5.1.28}$$

$$u(0) \;=\; u_0, \tag{5.1.29}$$

where v belongs to a prescribed set U_{ad} that usually is chosen compact in $C[0,T]$. In this way, a class of "admissible free boundaries" is defined. The original problem may then be reformulated as the optimal shape design problem

$$\operatorname*{Min}_{v \in U_{ad}} \left\{ \int_0^T \left[\beta_2 \, y_{2,x}(t,u(t)) - \beta_1 \, y_{1,x}(t,u(t)) - v(t)\right]^2 dt \right\}, \tag{5.1.30}$$

subject to the system (5.1.21)–(5.1.25) and (5.1.28), (5.1.29).

We remark that here the optimal shape design problem could be stated directly as an optimal control problem, since we considered the one-dimensional case. This is not possible in higher dimensions of space.

The above relationship between free boundary problems and shape optimization problems is reflected in the scientific literature by the use of similar methods. A survey along these lines may be found in the work of Hoffmann and Tiba [1995]. Several important techniques will be discussed below in this chapter.

5.1.3 Shape Optimization of Systems With Free Boundaries

As demonstrated in the first paragraph of this chapter, free boundary problems may be reformulated as variational inequalities using the Baiocchi transformation. The associated optimal control problems have been analyzed in detail in Section 3.3 of Chapter 3.

The study of domain optimization problems entails additional difficulties. An existence result under weak assumptions for the geometry has already been established for variational inequalities of obstacle type in Corollary 2.3.19 in Section 2.3. We now introduce a rather general procedure in this setting that allows us to transform such problems into distributed control problems. Applications to other types of equations or boundary conditions are possible (see §5.2.3.2 below).

To fix things, let $E \subset \overline{E} \subset \Omega \subset \mathbf{R}^2$ be given domains of class $C^{1,1}$, and let $\varphi \in C^2(\overline{\Omega})$ be a given function that satisfies

$$\varphi|_{\partial\Omega} < 0, \quad \varphi|_{\overline{\Omega}\setminus\overline{\Omega}_\varphi} < 0, \quad \varphi|_E > 0,$$

where

$$\Omega_\varphi = \{x \in \Omega : \varphi(x) > 0\} \quad (\supset E).$$

Notice that then $\partial\Omega \cap \partial\Omega_\varphi = \emptyset$. We denote by D a variable (unknown) domain of class $C^{1,1}$ such that $\Omega_\varphi \subset D \subset \Omega$. In D, we consider the obstacle problem

$$-\Delta y(x) + \beta(y(x) - \varphi(x)) \ni f(x) \quad \text{for a.e. } x \in D, \tag{5.1.31}$$

$$y = 0 \quad \text{on } \partial D, \tag{5.1.32}$$

where $f \in L^2(\Omega)$ is prescribed and $\beta \subset \mathbf{R} \times \mathbf{R}$ is the maximal monotone graph

$$\beta(r) = \begin{cases} \{0\}, & \text{if } r > 0, \\ ]-\infty, 0], & \text{if } r = 0, \\ \emptyset, & \text{if } r < 0. \end{cases} \tag{5.1.33}$$

Notice that the corresponding variational form of the obstacle problem (5.1.31)–(5.1.33) is given by

$$\int_D \nabla y(x) \cdot \nabla (v - y)(x)\, dx \ge \int_D f(x)\,(v(x) - y(x))\, dx \quad \forall\, v \in K_D,$$

where $K_D = \{v \in H_0^1(D) : v \ge \varphi \text{ a.e. in } D\}$. Thanks to the Lions–Stampacchia theorem (cf. Theorem A2.3 in Appendix 2), (5.1.31)–(5.1.33) admits a unique solution $y = y_D \in K_D$. Since D is of class $C^{1,1}$, we can infer from

Theorem A2.9 in Appendix 2 that $y_D \in H^2(D)$, and since we are in the two-dimensional case, it follows from the embedding theorem (see Theorem A2.2(iv) in Appendix 2) that y_D is continuous on $\overline{D}$.

We now introduce the so-called *optimal packaging problem*:

(PO) Find the domain $D \supset \Omega_\varphi$, of class $C^{1,1}$, having minimal area such that the *coincidence set* of (5.1.31), (5.1.32) satisfies

$$Z_D = \{x \in D : y_D(x) = \varphi(x)\} \supset E. \tag{5.1.34}$$

The form of the state constraint (5.1.34) is nonstandard and we assume admissibility, that is, that there is some set D of class $C^{1,1}$ such that $\Omega_\varphi \subset D \subset \Omega$ and $Z_D \supset E$. We also recall that (5.1.31)–(5.1.33) describe in the physical interpretation the vertical deflection y of an elastic membrane that is located over the domain D and clamped along the boundary ∂D, possibly in contact with the rigid obstacle given by the graph of φ. The membrane should cover ("pack") the obstacle over the region E.

We now aim to transform the shape optimization (PO) into a distributed control problem. For this purpose, we first extend y_D in a suitable way to $\overline{\Omega}$. To this end, assume that $\overline{D} \subset \Omega$. We then consider on $\Omega \setminus \overline{D}$ the controlled obstacle variational inequality

$$-\Delta z(x) + \beta(z(x) - \varphi(x)) \ni v(x), \quad \text{for a.e. } x \in \Omega \setminus \overline{D}, \tag{5.1.35}$$

$$z = 0, \quad \text{on } \partial D \cup \partial\Omega, \tag{5.1.36}$$

where $v \in L^2(\Omega \setminus \overline{D})$ is regarded as a control variable. Owing to the regularity assumptions made for the boundaries of D and Ω, it follows from Theorem A2.9 in Appendix 2 that (5.1.35), (5.1.36) admits a unique solution

$$z = z_{\Omega\setminus\overline{D},v} \in H^2(\Omega \setminus \overline{D}) \cap H_0^1(\Omega \setminus \overline{D}).$$

Again, Theorem A2.2(iv) yields that $z_{\Omega\setminus\overline{D},v} \in C(\overline{\Omega \setminus D})$. We have the following result.

Theorem 5.1.5 *Under the given assumptions for the domain D, there is some $v \in L^2(\Omega \setminus \overline{D})$ such that $z_{\Omega\setminus\overline{D},v}$, given by (5.1.35), (5.1.36), satisfies*

$$\frac{\partial z_{\Omega\setminus\overline{D},v}}{\partial n} = -\frac{\partial y_D}{\partial \nu} \quad \text{on } \partial D, \tag{5.1.37}$$

where n and ν are the normals to ∂D directed toward the interior and the exterior of D, respectively.

Proof. The trace theorem (cf. Theorem A2.1(ii) in Appendix 2) implies that the normal derivatives occurring in (5.1.37) belong to $H^{\frac{1}{2}}(\partial D)$, and there is some

$\hat{y} \in H^2(\Omega \setminus \overline{D}) \cap H_0^1(\Omega \setminus \overline{D})$, which is again by Theorem A2.2(iv) continuous on $\overline{\Omega \setminus D}$, such that

$$\frac{\partial \hat{y}}{\partial n} = -\frac{\partial y_D}{\partial \nu} \quad \text{on } \partial D.$$

Next, we define the function $\overline{y}$ by

$$\overline{y}(x) = \max\{\hat{y}(x), \varphi(x)\}, \quad x \in \Omega \setminus \overline{D}. \tag{5.1.38}$$

Then $\overline{y} \in H^1(\Omega \setminus \overline{D})$, by the lattice property of this space. Moreover, we have $\varphi(x) < 0$ in $\overline{\Omega} \setminus D$, and thus $\overline{y} = 0$ on $\partial(\Omega \setminus \overline{D})$. We also notice that $\hat{y}(x) > \varphi(x)$ for any $x \in \partial D$. By the continuity of $\hat{y} - \varphi$, then $\hat{y}(\xi) > \varphi(\xi)$ (and thus $\overline{y}(\xi) = \hat{y}(\xi)$) for all ξ in a set $\mathcal{N}(x)$ that contains x and is relatively open in $\Omega \setminus D$. This implies that also

$$\frac{\partial \overline{y}}{\partial n} = -\frac{\partial y_D}{\partial \nu} \quad \text{on } \partial D.$$

Now observe that also $\hat{y}(x) > \varphi(x)$ for all $x \in \partial\Omega$. Arguing by continuity again, we therefore can conclude that $\{\hat{y} \le \varphi\}$ is a compact subset of $\Omega \setminus \overline{D}$ that has a positive Hausdorff–Pompeiu distance from $\partial(\Omega \setminus \overline{D}) = \partial\Omega \cup \partial D$. Therefore, we can construct a nonnegative function $\eta \in C_0^\infty(\Omega \setminus \overline{D})$ such that

$$\operatorname{supp} \eta \supset \{\hat{y} \le \varphi\}, \quad \eta(x) \ge \varphi(x) - \hat{y}(x) \quad \text{in } \{\hat{y} \le \varphi\}.$$

Replacing $\hat{y}$ by $\hat{y} + \eta$ in (5.1.38), we then have $\overline{y} = \hat{y}$ and thus $\overline{y} \in H^2(\Omega \setminus \overline{D}) \cap H_0^1(\Omega \setminus \overline{D})$. Then $v = -\Delta \overline{y} \in L^2(\Omega \setminus \overline{D})$ is the desired control, since $\beta(\overline{y} - \varphi)$ vanishes in $\Omega \setminus \overline{D}$. □

Remark. Theorem 5.1.5 is an example of an *exact controllability* result. It allows one to extend the solution y_D of (5.1.31), (5.1.32) to a function $\tilde{y}_D \in H^2(\Omega) \cap H_0^1(\Omega) \cap C(\overline{\Omega})$ defined on the whole set Ω that still satisfies the same boundary value problem in Ω, with a right-hand side that is a modification of f. Indeed, we need only to put

$$\tilde{y}_D(x) = \begin{cases} y_D(x), & \text{if } x \in \overline{D}, \\ z_{\Omega \setminus \overline{D}, v}(x), & \text{if } x \in \Omega \setminus \overline{D}. \end{cases}$$

Such "controllability" results involving elliptic equations were first established by Beckert [1960] and Lions [1968] in an approximating sense. Sometimes one calls them *geometric controllability properties*, since certain boundaries or subdomains play an important role (and not the "final time," as is usual in evolution systems). For further results along these lines, we refer to §5.2.3.2 below.

We now consider for $u \in L^2(\Omega)$ the variational inequality

$$-\Delta y(x) + \beta(y(x) - \varphi(x)) \ni u(x) \quad \text{for a.e. } x \in \Omega, \tag{5.1.39}$$

$$y = 0 \quad \text{on } \partial\Omega. \tag{5.1.40}$$

Again, we can infer that (5.1.39), (5.1.40) has for any $u \in L^2(\Omega)$ a unique solution $y = y(u) \in H^2(\Omega) \cap H_0^1(\Omega) \cap C(\overline{\Omega})$. Clearly, $y(u) = y_\Omega$ if $u = f$. Also, Theorem 5.1.5 guarantees that for any domain $D \subset \overline{D} \subset \Omega$ that is admissible for (PO) there is at least one $\hat{u} \in L^2(\Omega)$ such that $\hat{u}|_D = f|_D$ and $y(\hat{u})|_D = y_D$.

We define for any $u \in L^2(\Omega)$ the set

$$\mathcal{M}(u) = \Big\{ E(u) \subset \mathbf{R}^2 : E(u) \text{ is a domain of class } C^{1,1} \text{ such that}$$
$$\Omega_\varphi \subset E(u) \subset \Omega \text{ and } y(u) = 0 \text{ on } \partial E(u) \Big\}.$$

Obviously, $\Omega \in \mathcal{M}(u)$, so that $\mathcal{M}(u) \neq \emptyset$. However, $\mathcal{M}(u)$ may also contain domains $E(u)$ that differ from Ω. Notice also that the (unknown) domain D that minimizes (PO) belongs to one of the sets $\mathcal{M}(u)$.

We now consider for every $n \in \mathbf{N}$ the following distributed control problem:

(P$_n$)
$$\operatorname*{Min}_{u \in L^2(\Omega)} \left\{ \inf_{E(u) \in \mathcal{M}(u)} \left[\int_{E(u)} (1 + n\,(u(x) - f(x))^2)\, dx \right] \right\}$$

subject to (5.1.39), (5.1.40), and to the constraint

$$\{x \in E(u) : y(u)(x) = \varphi(x)\} \supset E. \tag{5.1.41}$$

Remark. In the control problems (P$_n$) the "optimization parameter" D of problem (PO) has been replaced by the distributed controls u. Notice that

$$\int_{E(u)} (1 + n\,(u(x) - f(x))^2)\, dx = \operatorname{meas}(E(u)) + n\,|u - f|^2_{L^2(E(u))},$$

that is, the cost functional resembles the original objective that area be minimized, up to a penalty term that is intended to recover the correct right-hand side (namely, f) on $E(u)$.

The connection between the problems (PO) and (P$_n$) is given by the following result.

Theorem 5.1.6 *The following assertions hold:*

(i) *Let the domain* $D \subset \mathbf{R}^2$ *be admissible for (PO). Then there exist for every* $n \in \mathbf{N}$ *a function* $u_n \in L^2(\Omega)$ *and a corresponding domain* $E(u_n) \in \mathcal{M}(u_n)$ *that satisfy (5.1.39)–(5.1.41) and*

$$\mathrm{meas}(D) \geq \int_{E(u_n)} (1 + n\,(u_n(x) - f(x))^2)\,dx \;-\; \frac{1}{n}.$$

In particular, with obvious notation,

$$\inf(PO) \geq \inf(P_n) \quad \forall n \in \mathbf{N}. \tag{5.1.42}$$

(ii) *Let* $\delta > 0$ *and* $n \in \mathbf{N}$ *be given. If some* $u \in L^2(\Omega)$ *and a domain* $E(u) \in \mathcal{M}(u)$ *are admissible and* δ*-suboptimal for* (P_n)*, then* $E(u)$ *is suboptimal for (PO) in the following sense:*

$$\begin{aligned} \mathrm{meas}(E(u)) &\leq \inf(PO) + \delta, \\ |y_{E(u)} - \varphi|_{H^1(E)} &\leq \frac{(1+\hat{c})^{\frac{1}{2}}\,(\inf(PO)+\delta)^{\frac{1}{4}}}{n^{\frac{1}{4}}}, \end{aligned} \tag{5.1.43}$$

with a constant $\hat{c} > 0$ *that depends only on the domain* Ω*.*

Proof. (i): Let $D \subset \overline{D} \subset \Omega$ be admissible for (PO), and let $n \in \mathbf{N}$ be given. As already noticed, there is some $\hat{u} \in L^2(\Omega)$ such that $\hat{u}|_D = f|_D$ and $y(\hat{u})|_D = y_D$. Then (5.1.39)–(5.1.41) are obviously fulfilled with

$$u_n = \hat{u}, \quad y(u_n) = y(\hat{u}), \quad E(u_n) = D,$$

and we have

$$\mathrm{meas}(D) = \mathrm{meas}(E(u_n)) = \int_{E(u_n)} (1 + n\,(u_n(x) - f(x))^2)\,dx.$$

If D is just admissible for (PO), then an approximation argument (see Theorem 5.2.22 and its proof) gives that $\mathrm{meas}(D) \geq \inf(P_n)$ for any n. This proves (5.1.42).

(ii): Let $\delta > 0$, $n \in \mathbf{N}$, and $u \in L^2(\Omega)$ be given, $\hat{y} = y(u)$, and let the domain $E(u) \in \mathcal{M}(u)$ satisfy the constraint that $\{x \in E(u) : y(u)(x) = \varphi(x)\} \supset E$. By virtue of (5.1.42),

$$\int_{E(u)} (1 + n\,(u(x) - f(x))^2)\,dx \leq \inf(P_n) + \delta \leq \inf(PO) + \delta,$$

and therefore

$$\mathrm{meas}(E(u)) \leq \inf(PO) + \delta, \quad |u - f|^2_{L^2(E(u))} \leq \frac{\inf(PO)+\delta}{n}. \tag{5.1.44}$$

Now observe that by the definition of $E(u)$ the function $\hat{y} = y(u)$ satisfies the variational inequality

$$\int_{E(u)} \nabla \hat{y}(x) \cdot \nabla (v - \hat{y})(x)\, dx \ge \int_{E(u)} u(x)\, (v(x) - \hat{y}(x))\, dx \quad \forall\, v \in K_{E(u)}, \tag{5.1.45}$$

where $K_{E(u)} = \{v \in H_0^1(E(u)) : v \ge \varphi \text{ a.e. in } E(u)\}$. Recalling that by the definition of $y_{E(u)}$ in (5.1.31), (5.1.32),

$$\int_{E(u)} \nabla y_{E(u)}(x) \cdot \nabla (v - y_{E(u)})(x)\, dx \ge \int_{E(u)} f(x)\, (v(x) - y_{E(u)})\, dx \quad \forall\, v \in K_{E(u)},$$

we easily find that

$$\begin{aligned} \int_{E(u)} |\nabla (\hat{y} - y_{E(u)})(x)|^2\, dx &\le \int_{E(u)} (u(x) - f(x))\, (\hat{y}(x) - y_{E(u)}(x))\, dx \\ &\le n^{-1/2} (\inf(PO) + \delta)^{1/2}\, |\hat{y} - y_{E(u)}|_{L^2(E(u))}. \end{aligned} \tag{5.1.46}$$

Next, we use the fact (obtained by zero extension and Poincaré's inequality in Ω) that there is a constant $\hat{c} > 0$, which only depends on Ω, such that for all open subsets Ω_0 of Ω,

$$|z|^2_{L^2(\Omega_0)} \le \hat{c}\, |z|^2_{H_0^1(\Omega_0)} \quad \forall\, z \in H_0^1(\Omega_0). \tag{5.1.47}$$

Hence, invoking Young's inequality and (5.1.46), we find that

$$|\hat{y} - y_{E(u)}|^2_{H^1(E(u))} \le (1 + \hat{c})\, |\hat{y} - y_{E(u)}|^2_{H_0^1(E(u))} \le \frac{1 + \hat{c}}{n^{1/2}} (\inf(PO) + \delta)^{1/2}.$$

Since $\hat{y}$ coincides with φ on E by definition, and since $E(u) \supset E$, we have

$$|y_{E(u)} - \varphi|^2_{H^1(E)} \le |y_{E(u)} - \hat{y}|^2_{H^1(E(u))},$$

whence the assertion follows. □

Remark. Theorem 5.1.6(ii) states that domains $E(u)$ that are admissible and δ-suboptimal for (P_n), with "small" $\delta > 0$, can be regarded as good approximations for the domain D minimizing (PO): in fact, their area deviates from the optimal value by at most δ, while the state constraint is not "too strongly" violated. In the practical solution of (P_n), one also penalizes the state constraint (5.1.41), which gives a suboptimal solution with a weaker evaluation of the same type as in Theorem 5.1.6.

Remark. Other approaches to the optimal packaging problem have been derived by Tiba [1990, Chapter III.5], and by Hoffmann, Haslinger, and Kocvara

[1993]. The problem was originally introduced by Benedict, Sokołowski, and Zolésio [1984] and treated there via the penalty method.

A very similar approach was discussed in Tiba [1992] for the optimization of the electrochemical machining process. Using the notation introduced in Example 1.2.7 of Chapter 1, we consider nonempty domains $C \subset E \subset D \subset \Omega$ of class $C^{1,1}$, where $\overline{C} \subset E$ and D is variable. In $D \setminus \overline{C}$ (which gives the shape of the machine), we solve the obstacle problem

$$-\Delta y(x) + \beta(y(x)) \ni f(x) \quad \text{for a.e. } x \in D \setminus \overline{C}, \tag{5.1.48}$$

$$y = 0 \quad \text{on } \partial C, \quad y = 1 \quad \text{on } \partial D, \tag{5.1.49}$$

where $f \in L^2(\Omega)$ is given. The variational form of (5.1.48), (5.1.49) is (where in comparison with the previous example we have $\varphi \equiv 0$)

$$\int_{D\setminus\overline{C}} \nabla y(x) \cdot \nabla(v - y)(x)\, dx \geq \int_{D\setminus\overline{C}} f(x)\, (v(x) - y(x))\, dx \quad \forall\, v \in K_D,$$

where

$$K_D = \left\{ v \in H^1(D \setminus \overline{C}) : v|_{\partial C} = 0, \;\; v|_{\partial D} = 1, \;\; v \geq 0 \text{ a.e. in } D \setminus \overline{C} \right\}.$$

The Lions–Stampacchia theorem (Theorem A2.3 in Appendix 2) yields that (5.1.48), (5.1.49) admits a unique solution $y = y_D$ in K_D. Again, it follows from Theorems A2.9 and A2.2(iv) in Appendix 2 that $y_D \in H^2(D \setminus \overline{C}) \cap C(\overline{D \setminus C})$.

The boundaries ∂C and ∂D represent the electrodes, and the condition $y = 1$ on ∂D signifies that some constant voltage is applied (cf. Ockendon and Elliott [1982]). The coincidence set

$$Z_D = \{ x \in D \setminus \overline{C} : y_D(x) = 0 \}$$

then gives the final shape of the metal piece obtained around ∂C. According to Barbu and Friedman [1991], the following design problem is of interest: Find a domain D of class $C^{1,1}$ such that

$$E \subset D \subset \Omega, \quad Z_D \supset E \setminus C. \tag{5.1.50}$$

Since y_D ought to vanish on $E \setminus C$, the least squares approach leads to the minimization problem (compare with (1.2.51) in Chapter 1)

$$\text{Min} \left\{ \frac{1}{2} \int_{E\setminus C} y_D^2(x)\, dx \; : \; E \subset D \subset \Omega; \;\; D \text{ is of class } C^{1,1} \right\}. \tag{5.1.51}$$

Similarly as above, we may extend y_D to a function $\tilde{y}_D \in H^2(\Omega \setminus \overline{C}) \cap C(\overline{\Omega \setminus C})$, where this time we request the boundary condition $\tilde{y}_D = 1$ on $\partial D \cup \partial\Omega$. To this end, we consider for $v \in L^2(\Omega\setminus\overline{D})$ the unique solution $z = z_{\Omega\setminus\overline{D},v} \in H^2(\Omega\setminus\overline{D})$ to the obstacle problem (compare (5.1.35), (5.1.36))

$$-\Delta z(x) + \beta(z(x)) \ni v(x) \quad \text{for a.e. } x \in \Omega \setminus \overline{D}, \tag{5.1.35$'$}$$

$$z = 1 \quad \text{on } \partial\Omega \cup \partial D. \tag{5.1.36$'$}$$

As in Theorem 5.1.5, we can show the existence of some $v \in L^2(\Omega\setminus\overline{D})$ satisfying (5.1.37), and then define $\tilde{y}_D$ as in the remark prior to (5.1.39).

Next, we introduce for every "control" $u \in L^2(\Omega\setminus\overline{C})$ the variational inequality (compare (5.1.39), (5.1.40))

$$-\Delta y(x) + \beta(y(x)) \ni u(x) \quad \text{for a.e. } x \in \Omega \setminus \overline{C}, \tag{5.1.52}$$

$$y = 0 \quad \text{on } \partial C, \qquad y = 1 \quad \text{on } \partial\Omega, \tag{5.1.53}$$

and denote its unique solution by $y = y(u) \in H^2(\Omega \setminus \overline{C})$. Putting

$$\begin{aligned}\mathcal{M}(u) \;=\; \big\{&E(u) \subset \mathbf{R}^2 : \, E(u) \text{ is a domain of class } C^{1,1} \text{ such that} \\ &E \subset E(u) \subset \Omega \text{ and } y(u) = 1 \text{ on } \partial E(u)\big\},\end{aligned}$$

we can for any $n \in \mathbf{N}$ approximate the problem (5.1.48)–(5.1.51) by the following one (which resembles (P_n)):

$$\operatorname*{Min}_{u\in L^2(\Omega)} \left\{ \inf_{E(u)\in\mathcal{M}(u)} \left[\frac{1}{2}\int_{E\setminus C} y(u)^2(x)\,dx + \int_{E(u)} \left(1 + n\,(u(x) - f(x))^2\right) dx \right] \right\}, \tag{5.1.54}$$

subject to (5.1.52), (5.1.53), and to the state constraint

$$\{x \in E(u) : \, y(u)(x) = 0\} \supset E \setminus C.$$

It ought to be clear that a result of the same type as Theorem 5.1.6 can be proved for this situation.

Remark. As in the previous problem, the cost functional in (5.1.54) reflects the objective that the area of the domain (of the machine) be minimized, up to a penalty term intended to recover f as the right-hand side.

As a final contribution to this paragraph, we now indicate that in some cases in which the unknown domains D have a simple structure it is even possible to reduce the shape optimization problems to boundary control problems. For a typical example, let $0 < a < b$ and $c > 0$ be fixed,

$$\begin{aligned}\Omega \;=\; &(]0,b[\times[0,1]) \cup \left\{ (x_1,x_2) \in \mathbf{R}^2 : \, (x_2 - \tfrac{1}{2})^2 + x_1^2 \le \tfrac{1}{4} \right\} \\ &\cup \left\{ (x_1,x_2) \in \mathbf{R}^2 : \, (x_2 - \tfrac{1}{2})^2 + (x_1 - b)^2 \le \tfrac{1}{4} \right\},\end{aligned}$$

and let the obstacle be defined by a function $\varphi \in H^2(\Omega)$ satisfying $\varphi \le 0$ in $\{(x_1, x_2) \in \Omega : x_1 \ge a\}$. We consider the variable sets

$$D(\alpha) = \{x = (x_1, x_2) \in \Omega : x_1 < \alpha(x_2)\}, \tag{5.1.55}$$

where $E \subset D(\alpha) \subset \Omega$ with a given Lipschitz domain E, and where

$$\begin{aligned} \alpha \in U_{ad} = & \Big\{\alpha \in W^{1,\infty}(0,1) : a \le \alpha(x_2), \ (\alpha(x_2), x_2) \in \Omega, \ \forall x_2 \in [0,1], \\ & |\alpha'(x_2)| \le c \text{ for a.e. } x_2 \in (0,1)\Big\}. \end{aligned} \tag{5.1.56}$$

We then consider the obstacle problem (5.1.31)–(5.1.33) with $D = D(\alpha)$ and denote its unique solution by $y = y_{D(\alpha)}$. Notice that we only have $y_{D(\alpha)} \in K_{D(\alpha)} \subset H_0^1(D(\alpha))$ in this situation, since $D(\alpha)$ is only a Lipschitz domain by assumption, and not of class $C^{1,1}$. Now let

$$Z_{D(\alpha)} = \{x \in D(\alpha) : y_{D(\alpha)}(x) = \varphi(x)\}$$

denote the coincidence set of $D(\alpha)$. The state constraint associated with (5.1.34) then reads $Z_{D(\alpha)} \supset E$ and can be penalized by

$$\underset{\alpha \in U_{ad}}{\text{Min}} \left\{ \int_E (y_{D(\alpha)}(x) - \varphi(x))^2 \, dx \right\}. \tag{5.1.57}$$

We consider the optimal shape design problem (5.1.31)–(5.1.33), (5.1.57) with $D(\alpha)$ given by (5.1.55), (5.1.56). The boundedness of U_{ad} in $W^{1,\infty}(0,1)$ ensures the existence of at least one optimal domain $D(\alpha^*)$ (see Example 2.3.2 in Chapter 2). We associate with this shape optimization problem the following boundary control problem in the fixed domain Ω (where we assume $f = 0$, for simplicity):

$$\underset{w \in W}{\text{Min}} \left\{ \int_E (y(w)(x) - \varphi(x))^2 \, dx \right\}, \tag{5.1.58}$$

$$-\Delta y(w) + \beta(y(w) - \varphi) \ni 0 \quad \text{in } \Omega, \tag{5.1.59}$$

$$y(w) = w \quad \text{on } \partial\Omega, \tag{5.1.60}$$

$$y(w) \ge 0 \quad \text{in } \Omega_a = \{(x_1, x_2) \in \Omega : x_1 \le a\}, \tag{5.1.61}$$

$$W = \left\{ v \in H^{\frac{3}{2}}(\partial\Omega) : v = 0 \text{ on } \partial\Omega_a \cap \partial\Omega, \ \varphi \le v \le 0 \text{ on } \partial\Omega \setminus \partial\Omega_a \right\}. \tag{5.1.62}$$

Observe that $y_{D(\alpha)} \ge 0$ in $D(\alpha)$ if $f = 0$: indeed, this follows immediately from inserting $z = (y_{D(\alpha)})_+$ in the variational form of (5.1.31)–(5.1.33), which is given by

$$\int_{D(\alpha)} \nabla y_{D(\alpha)}(x) \cdot \nabla (z - y_{D(\alpha)})(x)\, dx \geq 0,$$

$$\forall\, z \in H_0^1(D(\alpha)), \ \ z \geq \varphi \text{ a.e. in } D(\alpha).$$

Since this property is not necessarily satisfied by $y(w) \in H^2(\Omega)$, we have imposed the constraint (5.1.61). Now, owing to (5.1.61), (5.1.62) and to the continuity of $y(w)$, the set

$$\Lambda = \{(x_1, x_2) \in \Omega \setminus \Omega_a \,:\, y(w)(x_1, x_2) = 0\}$$

is nonempty, and we may find some $\alpha \in U_{ad}$ whose graph is contained in Λ. However, the regularity of α still needs to be proved (or the assumptions on U_{ad} may be relaxed). If the problem (5.1.58)–(5.1.62) is approximated via standard finite element methods as in Chapter 4, then Lipschitz properties are automatically valid for the numerical approximation of α. We say that the boundary control problem (5.1.58)–(5.1.62) is "embedded" in the optimal design problem (5.1.31)–(5.1.33), (5.1.57).

Notice that the variational inequality structure of the state system (5.1.31)–(5.1.33) is helpful in the above argument. In more general problems like those to be discussed in the next sections, the situation is different, and comparison results for partial differential equations have to be employed.

Remark. The results of this paragraph are due to Tiba, Neittaanmäki, Mäkinen, and Tiihonen [1990], Tiba [1992], Tiba and Neittaanmäki [1995], and Hoffmann and Tiba [1995].

5.1.4 Controllability of the Coincidence Set

Motivated by the control approaches mentioned above, we indicate here controllability results for the free boundary in the obstacle problem that are mainly due to the work of Barbu and Tiba [1991].

To this end, let Ω and D denote bounded domains of class C^{m+2} in $\mathbf{R}^d$ satisfying $\overline{D} \subset \Omega$ and $\partial(\Omega \setminus \overline{D}) = \partial\Omega \cup \partial D$, where $\partial\Omega \cap \partial D = \emptyset$. We denote by ν the inward unit normal to ∂D and by n the outward unit normal to $\partial\Omega$. Notice that then ν and n, respectively, coincide with the outward unit normal to the domain $\Omega \setminus \overline{D}$ on ∂D and $\partial\Omega$, respectively.

We assume generally that $m > d/2$, where $d \geq 2$. Then, by virtue of Theorem A2.2(iv) in Appendix 2, the space $H^{m+k}(\Omega \setminus \overline{D})$ is for any $k \in \mathbf{N} \cup \{0\}$ continuously and compactly embedded in $C^k(\overline{\Omega \setminus D})$.

We define the obstacle by the function $\varphi \in H^{m+2}(\Omega)$, and we consider for $u \in H^{m-\frac{1}{2}}(\partial\Omega)$ the problem

$$\int_\Omega \nabla y(x) \cdot \nabla (v - y)(x)\, dx \geq 0, \quad \forall\, v \in K_u, \tag{5.1.63}$$

where the closed and convex set

$$K_u = \left\{ y \in H^1(\Omega) : \ y \geq \varphi \ \text{in} \ \Omega, \ \ y = u \ \text{on} \ \partial\Omega \right\} \tag{5.1.64}$$

is nonempty if $u \geq \varphi|_{\partial\Omega}$. We thus can conclude that (5.1.63), (5.1.64) has for any such $u \in H^{m-\frac{1}{2}}(\partial\Omega)$ a unique solution $y = y(u) \in K_u$. The controllability problem to be studied here is given by the following

$$\text{Find } u \in H^{m-\frac{1}{2}}(\partial\Omega) \text{ such that } D \subset E_u,$$

where $E_u = \{x \in \Omega : \ y(u)(x) = \varphi(x)\}$ denotes the coincidence set associated with (5.1.63), (5.1.64).

In the previous paragraph a least squares approach was introduced in connection with the above controllability problem. Here, we reformulate the problem as follows:

$$\text{Find } u \in H^{m-\frac{1}{2}}(\partial\Omega) \text{ such that } y^u = \varphi \text{ on } \partial D \text{ and } y^u \geq \varphi \text{ in } \Omega \setminus \overline{D}. \tag{5.1.65}$$

In this connection, $y^u \in H^m(\Omega \setminus \overline{D}) \cap C(\overline{\Omega \setminus D})$ denotes the unique solution to the linear elliptic boundary value problem

$$\Delta y^u = 0 \quad \text{in } \Omega \setminus \overline{D}, \tag{5.1.66}$$

$$\frac{\partial y^u}{\partial \nu} = \frac{\partial \varphi}{\partial \nu} \quad \text{on } \partial D, \qquad y^u = u \quad \text{on } \partial\Omega. \tag{5.1.67}$$

We notice that there exist constants $\hat{C}_i > 0$, $i = 1, 2, 3$, which do not depend on u, such that for all $u \in H^{m-\frac{1}{2}}(\partial\Omega)$,

$$\left| \int_{\partial D} y^u(\sigma)\, d\sigma \right| \leq \hat{C}_1 \left| y^u \right|_{C(\overline{\Omega \setminus D})} \leq \hat{C}_2 \left| y^u \right|_{H^m(\Omega \setminus \overline{D})} \leq \hat{C}_3 \left(1 + |u|_{H^{m-\frac{1}{2}}(\partial\Omega)} \right). \tag{5.1.68}$$

Proposition 5.1.7 *Suppose that $\Delta\varphi(x) \leq 0$ for all $x \in \overline{D}$. If the solution y^u of (5.1.66), (5.1.67) satisfies (5.1.65), then the function*

$$\tilde{y}(x) = \begin{cases} y^u(x), & \text{if } x \in \Omega \setminus \overline{D}, \\ \varphi(x), & \text{if } x \in \overline{D}, \end{cases} \tag{5.1.69}$$

is the solution to (5.1.63), (5.1.64).

Proof. By virtue of (5.1.69), and in view of the smoothness of ∂D, we have

$$\frac{\partial \tilde{y}}{\partial x_i} = \frac{\partial y^u}{\partial x_i} \quad \text{a.e. in } \Omega \setminus \overline{D}, \qquad \frac{\partial \tilde{y}}{\partial x_i} = \frac{\partial \varphi}{\partial x_i} \quad \text{a.e. in } D, \quad 1 \leq i \leq d.$$

Green's formula yields that for all $z \in K_u$,

$$\int_\Omega \nabla \tilde{y} \cdot \nabla(\tilde{y} - z)\, dx = \int_{\Omega\setminus\overline{D}} \nabla y^u \cdot \nabla(y^u - z)\, dx + \int_D \nabla\varphi \cdot \nabla(\varphi - z)\, dx$$

$$= -\int_{\Omega\setminus\overline{D}} \Delta y^u\,(y^u - z)\, dx - \int_{\partial D} \frac{\partial y^u}{\partial\nu}\,(y^u - z)\, d\sigma - \int_D \Delta\varphi\,(\varphi - z)\, dx$$

$$+ \int_{\partial D} \frac{\partial\varphi}{\partial\nu}\,(\varphi - z)\, d\sigma$$

$$= -\int_D \Delta\varphi\,(\varphi - z)\, dx \le 0,$$

which proves the assertion. □

Remark. Proposition 5.1.7 shows that the problem (5.1.65)–(5.1.67) and the original controllability problem are equivalent. Since the problem (5.1.65)–(5.1.67) is, owing to the linear structure of the state equations (5.1.66), (5.1.67), easier to handle, we will examine it in the following.

For what follows, we make the following assumption:

$$\Delta\varphi \le 0 \quad \text{in } D, \qquad \frac{\partial\varphi}{\partial\nu} \ge 0 \quad \text{on } \partial D. \tag{5.1.70}$$

We consider for every $n \in \mathbf{N}$ the optimal control problem

$$(\mathrm{O}_n) \qquad \operatorname*{Min}_{u\in H^{m-\frac{1}{2}}(\partial\Omega)} \left\{ \int_{\partial D} y^u(\sigma)\, d\sigma + \frac{1}{2n} |u|^2_{H^{m-\frac{1}{2}}(\partial\Omega)} \right\},$$

subject to (5.1.66), (5.1.67), and to the state constraint

$$y^u(x) \ge \varphi(x) \quad \text{in } \Omega \setminus \overline{D}. \tag{5.1.71}$$

We have the following result.

Lemma 5.1.8 *For any $n \in \mathbf{N}$, the problem (O_n) has a unique optimal control $u_n \in H^{m-\frac{1}{2}}(\partial\Omega)$ with the corresponding optimal state $y_n = y^{u_n} \in H^m(\Omega \setminus \overline{D})$ solving (5.1.66), (5.1.67). Moreover, there are a function $p_n \in W^{1,q}(\Omega \setminus \overline{D})$, where $1 \le q < d/(d-1)$, and a Borel measure $\mu_n \in C(\overline{\Omega \setminus D})^*$ that satisfy*

$$-\Delta p_n = \mu_n \quad \text{in } \Omega \setminus \overline{D}, \tag{5.1.72}$$

$$\frac{\partial p_n}{\partial\nu} = 1 \quad \text{on } \partial D, \qquad p_n = 0 \quad \text{on } \partial\Omega, \tag{5.1.73}$$

$$\frac{\partial p_n}{\partial n} = \frac{1}{n} F u_n \quad \text{on } \partial\Omega, \tag{5.1.74}$$

$$(\mu_n, y_n - z)_{C(\overline{\Omega\setminus D})^* \times C(\overline{\Omega\setminus D})} \geq 0 \quad \forall z \in C(\overline{\Omega \setminus D}) \text{ with } z \geq \varphi. \tag{5.1.75}$$

Here, $F : H^{m-\frac{1}{2}}(\partial\Omega) \to H^{-m+\frac{1}{2}}(\partial\Omega)$ *denotes the canonical isomorphism of the Hilbert space* $H^{m-\frac{1}{2}}(\partial\Omega)$ *onto its dual.*

Remark. The relations (5.1.72)–(5.1.74) have to be interpreted in the following weak sense:

$$(\mu_n, \psi)_{C(\overline{\Omega\setminus D})\times C(\overline{\Omega\setminus D})^*} - \int_{\Omega\setminus \overline{D}} \nabla p_n \cdot \nabla\psi \, dx + \int_{\partial D} \psi \, d\sigma$$

$$+ n^{-1} (F u_n, \psi)_{H^{-m+\frac{1}{2}}(\partial\Omega)\times H^{m-\frac{1}{2}}(\partial\Omega)} = 0 \quad \forall \psi \in H^m(\Omega \setminus \overline{D}), \tag{5.1.76}$$

$$p_n = 0 \quad \text{on } \partial\Omega. \tag{5.1.77}$$

Notice that by Theorem A2.1(i) in Appendix 2 the trace of p_n belongs to the space $W^{1-\frac{1}{q},q}(\partial\Omega)$, which implies that (5.1.73) and (5.1.77) are meaningful. Moreover, since $m > d/2$, the space $H^m(\Omega \setminus \overline{D})$ is continuously embedded in $C(\overline{\Omega \setminus D})$, so that also (5.1.76) makes sense.

Proof of Lemma 5.1.8. We first construct a pair that is admissible for (O_n) for every $n \in \mathbf{N}$. For this purpose, we choose some $\bar{u} \in H^{m+\frac{3}{2}}(\partial\Omega)$ such that

$$\min_{\sigma\in\partial\Omega} \bar{u}(\sigma) > \max_{x\in\overline{\Omega\setminus D}} \varphi(x).$$

Then it follows for $\bar{y} = y^{\bar{u}}$ that $\bar{y} \in H^{m+2}(\Omega \setminus \overline{D})$ and thus $\bar{y} \in C^2(\overline{\Omega \setminus D})$. The assumption that $\frac{\partial\varphi}{\partial\nu} \geq 0$ on ∂D (recall (5.1.70)) and Hopf's lemma (cf. Renardy and Rogers [1993, Lemma 4.7]) imply that the minimum of $\bar{y}$ can be attained only on $\partial\Omega$, so that

$$\min_{x\in\overline{\Omega\setminus D}} \bar{y}(x) \geq \min_{\sigma\in\partial\Omega} \bar{u}(\sigma) > \max_{x\in\overline{\Omega\setminus D}} \varphi(x);$$

that is, $[\bar{u}, \bar{y}]$ is an admissible pair for (O_n), for all $n \in \mathbf{N}$.

For what follows, let $n \in \mathbf{N}$ be fixed. We denote by C_i, $i \in \mathbf{N}$, positive constants that do not depend on n. We first show that (O_n) has a unique optimal pair. Indeed, if $\{[u_k, y^{u_k}]\}_{k\in\mathbf{N}}$ is a minimizing sequence for (O_n), we can assume, without loss of generality, that

$$\int_{\partial D} \varphi(\sigma)\, d\sigma + \frac{1}{2n} |u_k|^2_{H^{m-\frac{1}{2}}(\partial\Omega)} \leq \int_{\partial D} y^{u_k}(\sigma)\, d\sigma + \frac{1}{2n} |u_k|^2_{H^{m-\frac{1}{2}}(\partial\Omega)}$$

$$\leq \int_{\partial D} \bar{y}(\sigma)\, d\sigma + \frac{1}{2n} |\bar{u}|^2_{H^{m-\frac{1}{2}}(\partial\Omega)} \leq C_1 \quad \forall k \in \mathbf{N}.$$

Using (5.1.68), we thus can infer that

$$\frac{1}{2n}\left|u_k\right|^2_{H^{m-\frac{1}{2}}(\partial\Omega)} \le C_2, \quad \left|y^{u_k}\right|_{H^m(\Omega\setminus\overline{D})} \le C_2\,(1+\sqrt{n}), \quad \forall\, k \in \mathbf{N}.$$

Invoking the techniques developed in Chapter 2 (selecting weakly convergent subsequences and using the weak lower semicontinuity of the cost functional of (O_n)), we then can infer the existence of at least one optimal pair of (O_n). Since the cost functional is obviously strictly convex with respect to u, the optimal pair, which we denote by $[u_n, y_n]$, is unique. Moreover, we have the estimates

$$\frac{1}{2n}\left|u_n\right|^2_{H^{m-\frac{1}{2}}(\partial\Omega)} \le C_2, \quad \left|y_n\right|_{H^m(\Omega\setminus\overline{D})} \le C_2\,(1+\sqrt{n}), \quad \forall\, n \in \mathbf{N}. \tag{5.1.78}$$

Next, we introduce for every $\lambda > 0$ the penalized problem (Q_λ) given by

$$\underset{u\in H^{m-\frac{1}{2}}(\partial\Omega)}{\mathrm{Min}} \left\{ \int_{\partial D} y^u(\sigma)\,d\sigma + \frac{1}{2n}|u|^2_{H^{m-\frac{1}{2}}(\partial\Omega)} + \frac{1}{2\lambda}\int_{\Omega\setminus\overline{D}} |(y^u(x)-\varphi(x))_-|^2\,dx \right\}, \tag{5.1.79}$$

subject to (5.1.66), (5.1.67) (here, $z_- = -\,\mathrm{Min}\{z,0\} \ge 0$ is the negative part of z).

We notice first that for any $\lambda > 0$ the pair $[u_n, y_n]$ is admissible for (Q_λ). Arguing along the same lines as above for (O_n), we easily conclude the existence of a unique optimal pair $[u_\lambda, y_\lambda] \in H^{m-\frac{1}{2}}(\partial\Omega) \times H^m(\Omega\setminus\overline{D})$ that satisfies

$$\begin{aligned}&\int_{\partial D} y_\lambda(\sigma)\,d\sigma + \frac{1}{2n}\left|u_\lambda\right|^2_{H^{m-\frac{1}{2}}(\partial\Omega)} + \frac{1}{2\lambda}\int_{\Omega\setminus\overline{D}} |(y_\lambda(x)-\varphi(x))_-|^2\,dx \\ &\quad \le \int_{\partial D} y_n(\sigma)\,d\sigma + \frac{1}{2n}\left|u_n\right|^2_{H^{m-\frac{1}{2}}(\partial\Omega)} \le C_3 \quad \forall\,\lambda > 0.\end{aligned} \tag{5.1.80}$$

In particular, it follows from (5.1.68) that

$$\begin{aligned}&\frac{1}{2n}\left|u_\lambda\right|^2_{H^{m-\frac{1}{2}}(\partial\Omega)} - \hat{C}_3\left(1+\left|u_\lambda\right|_{H^{m-\frac{1}{2}}(\partial\Omega)}\right) + \frac{1}{2\lambda}\int_{\Omega\setminus\overline{D}} |(y_\lambda(x)-\varphi(x))_-|^2\,dx \\ &\le C_3,\end{aligned} \tag{5.1.81}$$

from which we can conclude that $\{u_\lambda\}_{\lambda>0}$ is bounded in $H^{m-\frac{1}{2}}(\partial\Omega)$, $\{y_\lambda\}_{\lambda>0}$ is bounded in $H^m(\Omega\setminus\overline{D})$, and $(y_\lambda-\varphi)_- \to 0$ strongly in $L^2(\Omega\setminus\overline{D})$ as $\lambda \searrow 0$. Moreover, there is a sequence $\lambda_n \searrow 0$ such that $u_{\lambda_n} \to \hat{u}$ weakly in $H^{m-\frac{1}{2}}(\partial\Omega)$ and $y_{\lambda_n} \to \hat{y}$ weakly in $H^m(\Omega\setminus\overline{D})$.

Clearly, $\hat{y} = y^{\hat{u}}$. Moreover, we have $\hat{y} \geq \varphi$ a.e. in $\Omega \setminus \overline{D}$; that is, $[\hat{u}, \hat{y}]$ is an admissible pair for (O_n). Passage to the limit as $\lambda_n \searrow 0$ in (5.1.80), using the weak lower semincontinuity of norms, then shows that $[\hat{u}, \hat{y}] = [u_n, y_n]$. The uniqueness of the limit point then entails that the weak convergences hold generally for $\lambda \searrow 0$ and not only for $\{\lambda_n\}$.

Next, we derive the maximum principle for u_λ. To this end, we perform variations around $[u_\lambda, y_\lambda]$ of the form $[u_\lambda + \delta v, y + \delta z]$, $\delta \in \mathbf{R}$, where $[v, z] \in H^{m-\frac{1}{2}}(\partial\Omega) \times H^m(\Omega \setminus \overline{D})$ satisfy the system

$$\Delta z = 0 \quad \text{in } \Omega \setminus \overline{D}, \tag{5.1.66$'$}$$

$$\frac{\partial z}{\partial \nu} = 0 \text{ on } \partial D, \qquad z = v \text{ on } \partial\Omega. \tag{5.1.67$'$}$$

We have, for any $\delta \in \mathbf{R}$,

$$\begin{aligned}\int_{\partial D} y_\lambda(\sigma)\, d\sigma &+ \frac{1}{2n}\, |u_\lambda|^2_{H^{m-\frac{1}{2}}(\partial\Omega)} + \frac{1}{2\lambda} \int_{\Omega\setminus\overline{D}} |(y_\lambda(x) - \varphi(x))_-|^2\, dx \\ &\leq \int_{\partial D} (y_\lambda(\sigma) + \delta\, z(\sigma))\, d\sigma + \frac{1}{2n}\, |u_\lambda + \delta\, v|^2_{H^{m-\frac{1}{2}}(\partial\Omega)} \cdot \\ &\quad + \frac{1}{2\lambda} \int_{\Omega\setminus\overline{D}} |(y_\lambda(x) + \delta\, z(x) - \varphi(x))_-|^2\, dx,\end{aligned}$$

and a straightforward calculation yields the Euler equation

$$0 = \int_{\partial D} z(\sigma)\, d\sigma + \frac{1}{n}\, (u_\lambda, v)_{H^{m-\frac{1}{2}}(\partial\Omega)} - \frac{1}{\lambda} \int_{\Omega\setminus\overline{D}} (y_\lambda(x) - \varphi(x))_-\, z(x)\, dx.$$

Now define the adjoint state $p_\lambda \in H^2(\Omega \setminus \overline{D})$ as the unique solution to the boundary value problem

$$-\Delta p_\lambda = -\lambda^{-1}(y_\lambda - \varphi)_- \quad \text{in } \Omega \setminus \overline{D}, \tag{5.1.82}$$

$$\frac{\partial p_\lambda}{\partial \nu} = 1 \quad \text{on } \partial D, \qquad p_\lambda = 0 \quad \text{on } \partial\Omega. \tag{5.1.83}$$

Multiplying (5.1.82) by z, integrating by parts, and invoking (5.1.66)$'$, (5.1.67)$'$, and (5.1.83), we find that

$$\begin{aligned}-\frac{1}{\lambda} \int_{\Omega\setminus\overline{D}} (y_\lambda(x) - \varphi(x))_-\, z(x)\, dx &= -\int_{\Omega\setminus\overline{D}} \Delta p_\lambda(x)\, z(x)\, dx \\ &= \int_{\Omega\setminus\overline{D}} \nabla p_\lambda(x) \cdot \nabla z(x)\, dx - \int_{\partial D} z(\sigma)\, d\sigma \\ &\quad - \int_{\partial\Omega} \frac{\partial p_\lambda}{\partial n}(\sigma)\, v(\sigma)\, d\sigma \\ &= -\int_{\partial D} z(\sigma)\, d\sigma - \int_{\partial\Omega} \frac{\partial p_\lambda}{\partial n}(\sigma)\, v(\sigma)\, d\sigma.\end{aligned}$$

Combining this identity with the Euler equation established above, and recalling that $v \in H^{m-\frac{1}{2}}(\partial\Omega)$ is arbitrary, we obtain the maximum principle (compare Example 3.2.3 and Example 3.2.4 in Chapter 3)

$$\frac{\partial p_\lambda}{\partial n} = \frac{1}{n} F u_\lambda \quad \text{on } \partial\Omega. \tag{5.1.84}$$

Moreover, we infer from (5.1.82), (5.1.83), invoking (5.1.81) and (5.1.84), that

$$\begin{aligned} |\Delta p_\lambda|_{L^1(\Omega\setminus\overline{D})} &= \left|\lambda^{-1}(y_\lambda - \varphi)_-\right|_{L^1(\Omega\setminus\overline{D})} = \int_{\Omega\setminus\overline{D}} \lambda^{-1}(y_\lambda(x) - \varphi(x))_-\,dx \\ &= \int_{\Omega\setminus\overline{D}} \Delta p_\lambda(x)\,dx = \int_{\partial D} 1\,d\sigma + \left(n^{-1} F u_\lambda, 1\right)_{H^{-m+\frac{1}{2}}(\partial\Omega)\times H^{m-\frac{1}{2}}(\partial\Omega)} \\ &\le C_5\left(1 + n^{-1}|u_\lambda|_{H^{m-\frac{1}{2}}(\partial\Omega)}\right) \le C_6. \end{aligned} \tag{5.1.85}$$

Invoking the results established in Brézis and Strauss [1973], we then can infer that $\{p_\lambda\}_{\lambda>0}$ is bounded in $W^{1,q}(\Omega\setminus\overline{D})$, whenever $1 < q < d/(d-1)$.

Hence, there is some sequence $\lambda_k \searrow 0$ such that $p_{\lambda_k} \to p_n$ weakly in $W^{1,q}(\Omega\setminus\overline{D})$ and such that $-\lambda_k^{-1}(y_{\lambda_k} - \varphi)_- \to \mu_n$ weakly star in $C(\overline{\Omega\setminus D})^*$. Passing to the limit as $\lambda_k \searrow 0$ in the sense of distributions in (5.1.82)–(5.1.84), we infer the validity of (5.1.72)–(5.1.74).

Finally, observe that for any $z \in C(\overline{\Omega\setminus D})$ with $z \ge \varphi$,

$$-\lambda^{-1}(y_\lambda - \varphi)_-\,(y_\lambda - z) = -\lambda^{-1}(y_\lambda - \varphi)_-\,(y_\lambda - \varphi) - \lambda^{-1}(y_\lambda - \varphi)_-\,(\varphi - z) \ge 0.$$

Thus, passage to the limit as $\lambda_k \searrow 0$ in the above inequality, in the dual pairing between $C(\overline{\Omega\setminus D})^*$ and $C(\overline{\Omega\setminus D})$, yields the validity of (5.1.75). This concludes the proof of the assertion. □

Lemma 5.1.9 *There is a constant* $\hat{C} > 0$ *such that*

$$n^{-\frac{1}{2}}|u_n|_{H^{m-\frac{1}{2}}(\partial\Omega)} + |\mu_n|_{C(\overline{\Omega\setminus D})^*} + |p_n|_{W^{1,q}(\Omega\setminus\overline{D})} \le \hat{C} \quad \forall n \in \mathbf{N}.$$

Proof. The bound for $|\mu_n|_{C(\overline{\Omega\setminus D})^*}$ follows immediately from passage to the limit as $\lambda_k \to 0$ in (5.1.85), and then the bound for $|p_n|_{W^{1,q}(\Omega\setminus\overline{D})}$ is a consequence of the results established in Brézis and Strauss [1973]. The boundedness of $\{n^{-1/2} u_n\}$ in $H^{m-\frac{1}{2}}(\partial\Omega)$ follows from (5.1.78). □

Theorem 5.1.10 *Let* $\Gamma \subset \partial D$ *be smooth, and suppose that* $\Pi \subset \Omega\setminus\overline{D}$ *is a domain such that* $\partial\Pi \cap \partial D = \Gamma$ *and such that* $\partial\Pi \cap \partial\Omega \ne \emptyset$ *is a smooth submanifold of* $\partial\Omega$. *Then there is no subsequence* $\{y_{n_\ell}\}$ *of* $\{y_n\}$ *such that*

$$y_{n_\ell}(x) > \varphi(x) \quad \forall x \in \Pi. \tag{5.1.86}$$

Proof. We argue by contradiction. Assume that there is some Π such that (5.1.86) is valid for some subsequence $\{y_{n_\ell}\}$. Thanks to (5.1.75), we have

$$\operatorname{supp} \mu_n \subset \left\{ x \in \overline{\Omega \setminus D} : \, y_n(x) = \varphi(x) \right\}.$$

Indeed, for any compact subset K of the open set $\{x \in \Omega \setminus D : \, y_n(x) > \varphi(x)\}$ and any $w \in C_0^\infty(\Omega \setminus \overline{D})$ with $\operatorname{supp} w \subset K$ we may choose $z(x) = y_n(x) \pm \rho\, w(x)$ in (5.1.75) provided that $\rho > 0$ is sufficiently small. Then it follows that

$$(\mu_n, w)_{C(\overline{\Omega \setminus D}) \times C(\overline{\Omega \setminus D})^*} = 0;$$

that is, μ_n vanishes on K, as claimed. Consequently, (5.1.72)–(5.1.74) show that

$$-\Delta p_{n_\ell} = 0 \quad \text{in } \Pi, \tag{5.1.87}$$

$$\frac{\partial p_{n_\ell}}{\partial \nu} = 1 \quad \text{on } \partial\Pi \cap \partial D, \qquad p_{n_\ell} = 0 \quad \text{on } \partial\Pi \cap \partial\Omega, \tag{5.1.88}$$

$$\frac{\partial p_{n_\ell}}{\partial n} = n_\ell^{-1} F u_{n_\ell} \quad \text{on } \partial\Pi \cap \partial\Omega. \tag{5.1.89}$$

By virtue of Lemma 5.1.9, we may without loss of generality assume that $p_{n_\ell} \to p$ weakly in $W^{1,q}(\Omega \setminus \overline{D})$. Moreover, Lemma 5.1.9 implies that $n_\ell^{-1} F u_{n_\ell} \to 0$ strongly in $H^{-m+\frac{1}{2}}(\partial\Omega)$. Passing to the limit as $n_\ell \to \infty$ in (5.1.87)–(5.1.89), we obtain that

$$\Delta p = 0 \quad \text{in } \Pi, \tag{5.1.90}$$

$$\frac{\partial p}{\partial \nu} = 1 \quad \text{on } \partial\Pi \cap \partial D, \qquad p = 0 \quad \text{on } \partial\Pi \cap \partial\Omega, \tag{5.1.91}$$

$$\frac{\partial p}{\partial n} = 0 \quad \text{on } \partial\Pi \cap \partial\Omega. \tag{5.1.92}$$

By the local and boundary regularity of p (cf. Lions and Magenes [1968]), and by its analyticity in Π, the Holmgren uniqueness theorem (cf. Renardy and Rogers [1993]) shows that $p \equiv 0$ in Π, since the Cauchy data on $\partial\Pi \cap \partial\Omega$ are zero. This contradicts the boundary condition on $\partial\Pi \cap \partial D$, which concludes the proof. □

Remark. With the notation $\Omega_n = \left\{ x \in \Omega \setminus \overline{D} : \, y_n(x) = \varphi(x) \right\}$, Theorem 5.1.10 shows that for every subsequence $n_\ell \to \infty$ the set $\bigcap_{n_\ell} \Omega_{n_\ell}^c$ does not contain domains of the type of Π. Roughly speaking, this means that the sets Ω_n "asymptotically" cover ∂D. Combining this with Proposition 5.1.7, it becomes clear that Theorem 5.1.10 gives in fact an approximate controllability result for the coincidence set defined in (5.1.64).

Another property of this kind is indicated in the next theorem, where the above assumptions on D, Ω, and φ, in particular (5.1.70), are fulfilled.

Theorem 5.1.11 *For every* $\varepsilon > 0$ *there are a connected open set* $Q_\varepsilon \subset \Omega\backslash\overline{D}$, *with* $(\Omega\backslash\overline{D})\backslash Q_\varepsilon$ *connected and satisfying* $\mathrm{meas}((\Omega\backslash\overline{D})\backslash Q_\varepsilon) < \varepsilon$, *and a sequence* $\{\tilde{u}_n\} \subset H^{m-\frac{1}{2}}(\partial\Omega)$ *such that the corresponding solutions* $\tilde{y}_n = y^{\tilde{u}_n} \in H^m(\Omega\backslash\overline{D})$ *of (5.1.66), (5.1.67) satisfy*

$$\tilde{y}_n(x) \geq \varphi(x) \quad \forall x \in Q_\varepsilon, \tag{5.1.93}$$

$$\tilde{y}_n \to \varphi \quad \text{strongly in } L^2(\partial D). \tag{5.1.94}$$

Proof. Since the line of argumentation closely resembles that employed above, we may only indicate the main ideas. First, choosing $\delta > 0$ sufficiently small, we can find a connected open set $Q_\varepsilon \subset \Omega \setminus \overline{D}$, with $(\Omega \setminus \overline{D}) \setminus Q_\varepsilon$ connected and satisfying $\mathrm{meas}((\Omega \setminus \overline{D}) \setminus Q_\varepsilon) < \varepsilon$, and

$$\Big\{x \in \Omega \setminus \overline{D} : \, d(x, \partial D) < \delta\Big\} \cap Q_\varepsilon = \emptyset, \tag{5.1.95}$$

$$\Big\{x \in \Omega \setminus \overline{D} : \, d(x, \partial\Omega) < \delta\Big\} \cap Q_\varepsilon = \emptyset. \tag{5.1.96}$$

We introduce for any $n \in \mathbf{N}$ the optimal control problem

$$\operatorname*{Min}_{u \in H^{m-\frac{1}{2}}(\partial\Omega)} \left\{ \frac{1}{2} \int_{\partial D} |y^u(\sigma) - \varphi(\sigma)|^2 \, d\sigma + \frac{1}{2n} |u|^2_{H^{m-1/2}(\partial\Omega)} \right\}, \tag{5.1.97}$$

subject to (5.1.66), (5.1.67), and to the state constraint

$$y^u \geq \varphi \quad \text{in } Q_\varepsilon. \tag{5.1.98}$$

We also consider its penalization for $\lambda > 0$:

$$\begin{aligned} \operatorname*{Min}_{u \in H^{m-\frac{1}{2}}(\partial\Omega)} & \left\{ \frac{1}{2} \int_{\partial D} |y^u(\sigma) - \varphi(\sigma)|^2 \, d\sigma + \frac{1}{2n} |u|^2_{H^{m-\frac{1}{2}}(\partial\Omega)} \right. \\ & \left. + \frac{1}{2\lambda} \int_{Q_\varepsilon} |(y^u(x) - \varphi(x))_-|^2 \, dx \right\}, \end{aligned} \tag{5.1.99}$$

subject to (5.1.66), (5.1.67).

Again, the pair $[\bar{u}, \bar{y}]$ constructed in the beginning of the proof of Lemma 5.1.8 is admissible for (5.1.97), (5.1.98) for all $n \in \mathbf{N}$. We may then argue as there to conclude that (5.1.97), (5.1.98) and (5.1.99) have unique optimal pairs

$[\tilde{u}_n, \tilde{y}_n]$ and $[\tilde{u}_\lambda, \tilde{y}_\lambda]$, respectively, such that $\tilde{u}_\lambda \to \tilde{u}_n$ weakly in $H^{m-\frac{1}{2}}(\partial\Omega)$ and $\tilde{y}_\lambda \to \tilde{y}_n$ weakly in $H^m(\Omega \setminus \overline{D})$, as $\lambda \searrow 0$.

If χ_ε is the characteristic function of Q_ε in $\Omega \setminus D$, then the adjoint state for (5.1.99) is given by

$$-\Delta \tilde{p}_\lambda = -\chi_\varepsilon \frac{1}{\lambda}(\tilde{y}_\lambda - \varphi)_- \quad \text{in } \Omega \setminus \overline{D}, \tag{5.1.100}$$

$$\frac{\partial \tilde{p}_\lambda}{\partial \nu} = \tilde{y}_\lambda - \varphi \quad \text{on } \partial D, \qquad p_\lambda = 0 \quad \text{on } \partial\Omega, \tag{5.1.101}$$

$$\frac{\partial \tilde{p}_\lambda}{\partial n} = n^{-1} F \tilde{u}_\lambda \quad \text{on } \partial\Omega. \tag{5.1.102}$$

Estimating as in (5.1.85), we find that $\left\{\chi_\varepsilon \lambda^{-1} (\tilde{y}_\lambda - \varphi)_-\right\}_{\lambda>0}$ is bounded in $L^1(\Omega \setminus \overline{D})$, which then also holds for $\{\Delta\tilde{p}_\lambda\}_{\lambda>0}$. Invoking the results of Brézis and Strauss [1973] once more, we conclude that $\{\tilde{p}_\lambda\}_{\lambda>0}$ is bounded in $W^{1,q}(\Omega \setminus \overline{D})$. Passing to the limit as $\lambda_n \searrow 0$ in (5.1.100)–(5.1.102) on a suitable subsequence, we can find $\tilde{p}_n \in W^{1,q}(\Omega \setminus \overline{D})$ and $\tilde{\mu}_n \in C(\overline{\Omega \setminus D})^*$ such that

$$-\Delta \tilde{p}_n = -\chi_\varepsilon \tilde{\mu}_n \quad \text{in } \Omega \setminus \overline{D}, \tag{5.1.103}$$

$$\frac{\partial \tilde{p}_n}{\partial \nu} = \tilde{y}_n - \varphi \quad \text{on } \partial D, \qquad \tilde{p}_n = 0 \quad \text{on } \partial\Omega, \tag{5.1.104}$$

$$\frac{\partial \tilde{p}_n}{\partial n} = n^{-1} F \tilde{u}_n \quad \text{on } \partial\Omega. \tag{5.1.105}$$

Since the statement of Lemma 5.1.9 holds true correspondingly, we may without loss of generality assume that $\tilde{p}_n \to \tilde{p}$ weakly in $W^{1,q}(\Omega \setminus \overline{D})$ and $\tilde{\mu}_n \to \tilde{\mu}$ in $C(\overline{\Omega \setminus D})^*$ as $n \to \infty$. Passage to the limit as $n \to \infty$ in (5.1.103)–(5.1.105), using the properties of $\tilde{u}_n$ and $\tilde{p}_n$, then leads to the conclusion that $\tilde{p}$ and $\tilde{\mu}$ satisfy the system

$$-\Delta \tilde{p} = \chi_\varepsilon \tilde{\mu} \quad \text{in } \Omega \setminus \overline{D}, \tag{5.1.106}$$

$$\frac{\partial \tilde{p}}{\partial \nu} = g \quad \text{on } \partial D, \qquad \tilde{p} = 0 \quad \text{on } \partial\Omega, \tag{5.1.107}$$

$$\frac{\partial \tilde{p}}{\partial n} = 0 \quad \text{on } \partial\Omega, \tag{5.1.108}$$

where g is the limit of $\{(\tilde{y}_n - \varphi)\}$ in the weak topology of $L^2(\partial D)$.

The relations (5.1.106)–(5.1.108) are to be understood in the sense that the trace of $\tilde{p}$ is null on $\partial\Omega$ and

$$(\tilde{\mu}\,\chi_\varepsilon, \psi)_{C(\overline{\Omega\setminus D})^* \times C(\overline{\Omega\setminus D})} = \int_{\Omega\setminus D} \nabla\tilde{p}(x) \cdot \nabla\psi(x)\, dx - \int_{\partial D} \psi(\sigma)\, g(\sigma)\, d\sigma$$

$$\forall \psi \in H^m(\Omega \setminus \overline{D}). \tag{5.1.109}$$

Clearly, we also have

$$\Delta\tilde{p} = 0 \quad \text{in } (\Omega \setminus \overline{D}) \setminus \overline{Q}_\varepsilon, \qquad \tilde{p} = 0, \quad \frac{\partial \tilde{p}}{\partial n} = 0 \quad \text{on } \partial\Omega,$$

and the Holmgren uniqueness theorem implies that $\tilde{p} = 0$ in $(\Omega \setminus \overline{D}) \setminus \overline{Q}_\varepsilon$.

By choosing in (5.1.109) an arbitrary ψ such that $\operatorname{supp}\psi$ is contained in a neighborhood of ∂D as indicated in (5.1.95), we infer from (5.1.109) that $g = 0$. By taking an appropriate convex combination of $\tilde{u}_n$ (and of $\tilde{y}_n$, defined by (5.1.66), (5.1.67)), we obtain that (5.1.93) is valid, and Mazur's theorem (see Yosida [1980]) ensures the strong convergence asserted in (5.1.94). In view of the definition of g, this finishes the proof of the assertion. □

Remark. Under appropriate hypotheses, namely if $-\Delta\varphi + \varphi \geq 0$ in D and $\frac{\partial\varphi}{\partial\nu} \geq 0$ on ∂D, the obstacle problem for the differential operator $-\Delta y + y$ with boundary control via Neumann conditions on $\partial\Omega$ has similar controllability properties for the coincidence set. We refer the reader to Barbu and Tiba [1991] for further details.

By the same technique as in Theorem 5.1.5 above, one may establish an exact controllability result for the coincidence set in the case of distributed controls. We have, if $\varphi \equiv 0$ for simplicity, the following result.

Theorem 5.1.12 *Let D and Ω be given as above, and let $\beta \subset \mathbf{R} \times \mathbf{R}$ be the maximal monotone graph defined in (5.1.33). Then there is some (not necessarily unique) $u \in L^2(\Omega)$ such that the variational inequality*

$$-\Delta y + \beta(y) \ni u \quad \text{in } \Omega, \tag{5.1.110}$$

$$y = 1 \quad \text{on } \partial\Omega, \tag{5.1.111}$$

has $\overline{D}$ as coincidence set.

Proof. We may define a mapping $\hat{y} \in L^\infty(\Omega)$ such that $\hat{y} \equiv 1$ in a smooth neighborhood of $\partial\Omega$ that has a positive distance from $\overline{D}_\varepsilon$, where $D_\varepsilon = \{x \in \Omega : d(x, \overline{D}) < \varepsilon\}$, and satisfies $\hat{y}(x) = (d(x, \overline{D}_\varepsilon))^2$ in the remainder of Ω. Clearly, $\hat{y}|_{\overline{D}_\varepsilon} \equiv 0$ and $\hat{y}(x) > 0$ in $\Omega \setminus \overline{D}_\varepsilon$. We consider the smoothing of $\hat{y}$ given by

$$y_\varepsilon(x) = \int_{B(0,1)} \hat{y}(x - \varepsilon r)\, \rho(r)\, dr,$$

where $\rho \in C_0^\infty(B(0,1))$ is a (nonnegative) Friedrichs mollifier. We assume that $\hat{y}$ is extended by 1 outside $\overline{\Omega}$. If $\varepsilon > 0$ is small enough, this construction yields that $y_\varepsilon(x) \equiv 1$ in a neighborhood of $\partial\Omega$, $y_\varepsilon(x) \equiv 0$ in $\overline{D}$, $y_\varepsilon(x) > 0$ in $\Omega \setminus \overline{D}$, and $y_\varepsilon \in C^\infty(\Omega)$. Then, we may choose $\beta(y_\varepsilon) \equiv 0$ in Ω and $u(x) = -\Delta y_\varepsilon$ in Ω, and (5.1.110), (5.1.111) are satisfied with the coincidence set given by $\overline{D}$. □

Remark. The above proof also ensures that $\beta(y)$ (the unknown reaction of the obstacle) may be taken as zero in Ω.

5.2 Direct Approaches

In this section, we review several ideas allowing simple direct approaches to shape optimization problems or their reduction to various types of control problems. While some of these procedures are rather general, others may be applied just to special classes of optimal design problems. The similarity with some of the methods considered in the previous section will be obvious. Since the practical solution of geometric optimization problems may prove to be a very difficult task, it is our aim to introduce the reader to a variety of solution techniques. In general, the successive application of several methods may be very helpful in obtaining efficient results in nonconvex minimization problems. As an important advantage, many of the methods introduced in this section are fixed domain methods that do not require remeshing in each step of the numerical solution when the finite element method is applied.

5.2.1 An Algorithm of Céa, Gioan, and Michel

The algorithm of Céa, Gioan, and Michel [1973] is one of the first contributions to the theory of optimal shape design. We recall it here (with some modifications), since it has a very abstract character and gives important insights into recent mathematical research devoted to the subject; cf. Masmoudi [1987], Guillaume and Idris [2002], and others (see also §5.3.2 below).

Let $\mathcal{A}$ be a collection of open subsets Ω of some given compact set with nonempty interior $D \subset \mathbf{R}^d$, and suppose that $J : \mathcal{A} \to \mathbf{R}$ is some cost functional. Usually, J is defined via some state function y_Ω obtained in each $\Omega \in \mathcal{A}$ as the solution of a partial differential equation.

We denote by $\Delta\Omega = \Delta\Omega^+ \cup \Delta\Omega^-$ some "variation" of Ω, distinguishing between the set $\Delta\Omega^+ \subset D \setminus \Omega$, which is "added" to Ω, and the set $\Delta\Omega^- \subset \Omega$, which is "subtracted" from Ω. We indicate this through the notation $\Omega + \Delta\Omega \in \mathcal{A}$, where

$$\Omega + \Delta\Omega = \Delta\Omega^+ \cup (\Omega \setminus \Delta\Omega^-). \tag{5.2.1}$$

We measure the order of magnitude of the variation by means of the quantity

$$\text{meas}(\Delta\Omega) = \text{meas}(\Delta\Omega^+) + \text{meas}(\Delta\Omega^-).$$

Notice that if $\Omega,\ \mathcal{O} \in \mathcal{A}$ are arbitrary, we have

$$\mathcal{O} = (\mathcal{O} \setminus \Omega) \cup (\mathcal{O} \cap \Omega) = (\mathcal{O} \setminus \Omega) \cup (\Omega \setminus (\Omega \setminus \mathcal{O})),$$

that is, $\mathcal{O} = \Omega + \Delta\Omega$ with $\Delta\Omega^+ = \mathcal{O} \setminus \Omega$ and $\Delta\Omega^- = \Omega \setminus \mathcal{O}$. Moreover, the variations defined in (5.2.1) may change the topological type of Ω, i.e., may introduce additional or "fill in" existing holes. In fact, relation (5.2.1) allows for all admissible variations of Ω, i.e., satisfying $\Omega + \Delta\Omega \in \mathcal{A}$.

The abstract optimal design problem to be studied here reads as follows:

$$\operatorname*{Min}_{\Omega\in\mathcal{A}} \left\{ J(\Omega) \right\}. \tag{5.2.2}$$

For the purposes of this paragraph, we will need only the following general hypothesis for the cost functional:

(H1) For every $\Omega \in \mathcal{A}$ there exist constants $\varepsilon_\Omega > 0$ ("small"), $C_\Omega > 0$, and a function $G_\Omega \in L^\infty(D)$ such that for any variation $\Delta\Omega$ of Ω with $\text{meas}(\Delta\Omega) \leq \varepsilon_\Omega$,

$$J(\Omega + \Delta\Omega) \;=\; J(\Omega) + T_1(\Omega; \Delta\Omega) + T_2(\Omega; \Delta\Omega), \tag{5.2.3}$$

where (notation)

$$T_1(\Omega; \Delta\Omega) = \int_{\Delta\Omega} G_\Omega(x)\, dx = \int_{\Delta\Omega^+} G_\Omega(x)\, dx - \int_{\Delta\Omega^-} G_\Omega(x)\, dx \tag{5.2.4}$$

and

$$|(T_2(\Omega; \Delta\Omega)| \leq C_\Omega \, (\text{meas}(\Delta\Omega))^2. \tag{5.2.5}$$

The asymptotic development (5.2.3)–(5.2.5) assumed in (H1) is inspired by the classical results of Hadamard [1968]. In Delfour and Zolésio [2001, Chapter 8], integral functionals satisfying (5.2.3) are studied in detail. Simple examples satisfying (5.2.3)–(5.2.5) are obtained when J depends directly on Ω, for instance, if $J(\Omega) = (\text{meas}(\Omega))^k$. More complex situations involving partial differential equations will be discussed below in §5.3.2.

Remark. Notice that the mapping $\Delta\Omega \mapsto T_1(\Omega; \Delta\Omega)$ is "additive" in the following sense: for any two disjoint variations $\Delta\Omega_1$ and $\Delta\Omega_2$ of Ω it is obvious that

$$T_1(\Omega; \Delta\Omega_1 \cup \Delta\Omega_2) = T_1(\Omega; \Delta\Omega_1) + T_2(\Omega; \Delta\Omega_2).$$

Moreover, $T_1(\Omega; \Delta\Omega)$ varies continuously with respect to $\Delta\Omega$ if the variation is measured in terms of $\text{meas}(\Delta\Omega)$. In this sense, the above mapping plays the role of the "differential" of J at the "point" Ω, while the mapping G_Ω represents the "gradient" of J at the "point" Ω.

Definition 5.2.1

(i) J has a *local minimum* at $\Omega \in \mathcal{A}$ if and only if there is some $r > 0$ such that for any variation $\Delta\Omega$ with $\Omega + \Delta\Omega \in \mathcal{A}$ and $\text{meas}(\Delta\Omega) \leq r$,

$$J(\Omega) \leq J(\Omega + \Delta\Omega). \tag{5.2.6}$$

(ii) Let J satisfy (5.2.3). $\Omega \in \mathcal{A}$ is called a *critical point* of J if there is some $r > 0$ such that for any variation $\Delta\Omega$ with $\Omega + \Delta\Omega \in \mathcal{A}$ and $\text{meas}(\Delta\Omega) \leq r$,

$$T_1(\Omega; \Delta\Omega) \geq 0. \tag{5.2.7}$$

Proposition 5.2.2 *Assume that $\mathcal{A}$ is the set of all open subsets of D, and assume that hypothesis (H1) holds. Then $\Omega \in \mathcal{A}$ is a critical point for J if and only if*

$$G_\Omega(x) \geq 0 \quad \textit{for a.e. } x \in D \setminus \Omega \quad \textit{and} \quad G_\Omega(x) \leq 0 \quad \textit{for a.e. } x \in \Omega. \tag{5.2.8}$$

Proof. According to (5.2.4), (5.2.7), we have for a critical point Ω that

$$\int_{\Delta\Omega^+} G_\Omega(x)\,dx - \int_{\Delta\Omega^-} G_\Omega(x)\,dx \geq 0,$$

for any choice $\Delta\Omega$ as given in Definition 5.2.1. Then (5.2.8) immediately follows from the Lebesgue points property. □

Proposition 5.2.3 *If J satisfies hypothesis (H1) and has a local minimum at $\Omega \in \mathcal{A}$, then Ω is a critical point of J.*

Proof. Relations (5.2.3) and (5.2.6) give that

$$\int_{\Delta\Omega^+} G_\Omega(x)\,dx - \int_{\Delta\Omega^-} G_\Omega(x)\,dx + T_2(\Omega; \Delta\Omega) \geq 0,$$

for any $\Delta\Omega^+ \subset D \setminus \Omega$, $\Delta\overline{\Omega} \subset \Omega$ with $\text{meas}(\Delta\Omega) \leq r$. Then, (5.2.5) gives

$$-\int_{\Delta\Omega^+} G_\Omega(x)\,dx + \int_{\Delta\Omega^-} G_\Omega(x)\,dx \leq C_\Omega \,\text{meas}(\Delta\Omega)^2 \tag{5.2.9}$$

for any $\Delta\Omega^+$ and $\Delta\Omega^-$ as above that satisfy $\Omega + \Delta\Omega \in \mathcal{A}$. Again, the Lebesgue point property and (5.2.9) yield that $G_\Omega \geq 0$ in $D \setminus \Omega$ and $G_\Omega \leq 0$ in Ω. Hence, by Proposition 5.2.2, Ω is a critical point for J. □

In the most interesting case, in which J depends on Ω via some state function y_Ω, the numerical approximation of y_Ω is usually performed using the finite element method as in Chapter 4. This requires, in particular, a division of Ω. Since Ω is an arbitrary open subset of D, we rather have to construct a partition $\mathcal{T} = \{\omega_i\}_{i\in I}$ of D into the elements ω_i, $i \in I$, where $I = \{1, \ldots, N\}$. The discretized unknown domains $\Omega \subset D$ are then obtained as all possible finite unions of elements of $\mathcal{T}$. This is just a numerical point of view, since the optimal domain need not be of this form in general; however, this will suffice or even be the case in many practical applications.

Then indicating some particular domain Ω is equivalent to prescribing a set $I_\Omega \subset I$ of indices,

$$\Omega = \operatorname{int}\left\{\bigcup_{i\in I_\Omega} \overline{\omega}_i\right\}. \tag{5.2.10}$$

We notice that all the discretized sets Ω are of Carathéodory type; that is, they satisfy $\partial\Omega = \partial(\overline{\Omega})$ if D is Carathéodory. This facilitates the treatment of edges, faces, etc., of the elements of $\mathcal{T}$. We also assume for the sake of simplicity that

$$\operatorname{meas}(\omega_i) = h > 0 \quad \forall i \in I.$$

We remark at this place that the above setting shows that the original shape optimization problem (5.2.2) may be "approximated" by an optimization problem over a finite set, namely over the collection of all subsets of the set of indices. However, this idea is usually unrealistic, since the cardinality of this finite set will be very large already for a coarse triangulation $\mathcal{T}$.

Based on the previous results, it is now possible to develop an algorithm for the approximate solution of (5.2.2) directly on the discrete level. To this end, the collection of all open sets $\mathcal{A}$ is replaced by the family $\mathcal{A}_d$ of all open sets of the form (5.2.10).

Definition 5.2.4 Let $\varepsilon > 0$ be given. $\Omega \in \mathcal{A}_d$ is called an *ε-critical point* of J if

$$\int_{\omega_i} G_\Omega(x)\, dx \geq -\varepsilon \operatorname{meas}(\omega_i) \quad \forall\, \omega_i \subset D \setminus \Omega, \tag{5.2.11}$$

$$\int_{\omega_i} G_\Omega(x)\, dx \leq \varepsilon \operatorname{meas}(\omega_i) \quad \forall\, \omega_i \subset \Omega. \tag{5.2.12}$$

This definition is based on the intuitive idea that the relations (5.2.11), (5.2.12) approximate (5.2.8) for $\varepsilon \searrow 0$.

Suppose now that some open subset Ω_m of D corresponding to the index set $I_m \subset I$ has already been constructed for some $m \in \mathbf{N}$. We select two new index sets $I_m^+ \subset I \setminus I_m$ and $I_m^- \subset I_m$ and define

$$I_{m+1} = I_m^+ \cup (I_m \setminus I_m^-),$$

as well as the associated open sets $\Delta\Omega_m^+$, $\Delta\Omega_m^-$, which in turn define

$$\Delta\Omega_m = \Delta\Omega_m^+ \cup \Delta\Omega_m^-, \qquad \Omega_{m+1} = \Delta\Omega_m^+ \cup (\Omega_m \setminus \Delta\Omega_m^-).$$

The selection of the sets I_m^+, I_m^- can be made in various ways; of course, the sets should be nonempty, if possible. In particular, assume that the constant appearing in (5.2.5) is independent of $\Omega \in \mathcal{A}_d$; that is, $C_\Omega = C > 0$. We then try to choose $\Delta\Omega_m$ such that

$$T_1(\Omega_m, \Delta\Omega_m) \le -(1+\theta)\, C \,\text{meas}(\Delta\Omega_m)^2, \tag{5.2.13}$$

where $\theta > 0$ is some fixed number. We then get the following result.

Theorem 5.2.5 *Suppose that $\mathcal{A}_d$ is the family of all finite unions of the elements ω_i of $\mathcal{T}$, where* $\text{meas}(\omega_i) = h$ *for all $i \in \mathbf{N}$, and suppose that (H1) holds with $C_\Omega = C > 0$ for all $\Omega \in \mathcal{A}_d$. Then there is some $k \in \mathbf{N}$, $k \ge 1$, such that the above construction of the set Ω_m satisfying (5.2.13) is possible for $0 \le m \le k-1$ and such that Ω_k is a $(1+\theta)\, C\, h$-critical point for J.*

Proof. From (5.2.3), (5.2.13) we infer that for any $m \in \mathbf{N}$,

$$J(\Omega_{m+1}) \le J(\Omega_m) - (1+\theta)\, C \,\text{meas}(\Delta\Omega_m)^2 + C\,\text{meas}(\Delta\Omega_m)^2, \tag{5.2.14}$$

that is, $J(\Omega_{m+1}) < J(\Omega_m)$ if $\text{meas}(\Delta\Omega_m) \ne 0$. Since I is a finite set, and since no cycles can appear in the choices satisfying (5.2.14), there is some $k \in \mathbf{N}$ such that $I_k^+ = \emptyset = I_k^-$, or in other words, for any choice $\Delta\Omega$ satisfying $\Omega_k + \Delta\Omega \in \mathcal{A}_d$,

$$T_1(\Omega_k, \Delta\Omega) > -(1+\theta)\, C\,\text{meas}(\Delta\Omega)^2.$$

In particular, for all $\omega_i \subset \Omega_k$ we must have

$$\int_{\omega_i} G_{\Omega_k}(x)\, dx < (1+\theta)\, C\, h^2,$$

while for all $\omega_i \subset D \setminus \Omega_k$, we must have

$$\int_{\omega_i} G_{\Omega_k}(x)\, dx > -(1+\theta)\, C\, h^2.$$

Comparison with (5.2.11), (5.2.12) finishes the proof of the assertion. □

Remark. The assumptions made for $\mathcal{A}$, $\mathcal{A}_d$ in Proposition 5.2.2 and Theorem 5.2.5, respectively, are not essential in the sense that similar (but weaker) results can be derived in more general situations (see Céa, Gioan, and Michel [1973]).

Remark. Since $\text{meas}(\omega_i) = h$ for all $i \in I$, relation (5.2.14) may be rewritten as

$$J(\Omega_{m+1}) + \theta\, C\, h^2\, |I_m^+ \cup I_m^-|^2 \leq J(\Omega_m), \tag{5.2.14'}$$

where $|I_m^+ \cup I_m^-|$ denotes the cardinal number of $I_m^+ \cup I_m^-$. Owing to (5.2.4) and (5.2.13), we also have

$$\int_{\omega_i} G_{\Omega_m}(x)\, dx \leq -(1+\theta)\, C\, h^2 \quad \text{if } i \notin I_m,$$

$$\int_{\omega_i} G_{\Omega_m}(x)\, dx \geq (1+\theta)\, C\, h^2 \qquad \text{if } i \in I_m.$$

In the general case, when (5.2.13) is not necessarily valid and Ω_m has already been constructed, we define

$$s_m^i := \frac{\delta_m^i}{(1+\theta)\, C\, h^2} \int_{\omega_i} G_{\Omega_m}(x)\, dx \quad \forall i \in I,$$

where

$$\delta_m^i = \begin{cases} +1, & \text{if } i \in I_m, \\ -1, & \text{if } i \notin I_m. \end{cases} \tag{5.2.15}$$

In each step of the algorithm, i.e., for any given Ω_m, one can compute the s_m^i, $i \in I$, according to the above definition and order them in decreasing order,

$$s_m^{j_1} \geq s_m^{j_2} \geq \cdots \geq s_m^{j_N},$$

where $(j_1, j_2, \ldots, j_N)$ is a permutation of $(1, 2, \ldots, N)$.

Since s_m^i may be computed for $i \in I$, it is enough to identify some $q \in \mathbf{N}$ and q indices $\{j_i\}_{i=\overline{1,q}} \subset I$ such that

$$\sum_{i=1}^{q} s_m^{j_i} \geq q^2. \tag{5.2.16}$$

The subset $\{j_i\}_{i=\overline{1,q}}$ satisfying (5.2.16) may be nonunique or empty (in which case the process terminates).

Once the choice (5.2.16) is fixed, we may take

$$I_m^+ = \{j_1, \ldots, j_q\} \setminus I_m, \quad I_m^- = \{j_1, \ldots, j_q\} \cap I_m.$$

Relation (5.2.14)′ suggests that the maximal estimated descent is achieved by choosing the largest possible q.

In view of (5.2.3), the algorithm presented above is a transposition of gradient-type optimization algorithms (compare Appendix 1). Constraints like

$$\tilde{\Omega} \subset \Omega \subset \hat{\Omega}$$

with given subsets $\widetilde{\Omega}$, $\widehat{\Omega}$ of D can also be incorporated. A variant of the above method is obtained as follows: define

$$t_m^i = \int_{\omega_i} G_{\Omega_m}(x)\, dx, \qquad i \in I,$$

and introduce the index set $\tilde{I}_{m+1}$ and the open set $\widetilde{\Omega}_{m+1} \subset D$ by putting

$$i \in \tilde{I}_{m+1} \text{ if and only if } t_m^i \leq 0, \qquad \widetilde{\Omega}_{m+1} = \mathrm{int}\Big(\bigcup_{i \in \tilde{I}_{m+1}} \omega_i \Big). \tag{5.2.17}$$

This is suggested by Proposition 5.2.2. If $J(\widetilde{\Omega}_{m+1}) < J(\Omega_m)$, then we take $I_{m+1} = \tilde{I}_{m+1}$ and $\Omega_{m+1} = \widetilde{\Omega}_{m+1}$. Otherwise, a step of type (5.2.16) has to be performed before the iteration step (5.2.17) is tried again. Notice that the cost functional has to be evaluated twice in order to compare the values $J(\widetilde{\Omega}_{m+1})$ and $J(\Omega_m)$. This is not necessary in the algorithm (5.2.16) since (5.2.14)′ automatically guarantees the descent property.

Remark. If $J(\Omega)$ is introduced via the solution y_Ω to some partial differential equation defined in Ω, then the assumption $G_\Omega \in L^\infty(D)$ in (5.2.4) makes necessary high regularity properties for y_Ω and, implicitly, for Ω (cf. Appendix 2). Other approaches are then needed in order to remove this drawback or to allow an analysis of the continuous case as well. We recall here the classical *mapping method* of Murat and Simon [1976] and the *speed method* of Zolésio [1979], [1981]. More about this will be said in Section 5.3 in connection with the sensitivity analysis of shape optimization problems.

5.2.2 Characteristic Functions

One simple and important idea to avoid the difficulties related to the direct treatment of complicated geometric situations is to use characteristic functions. They provide a complete description (up to sets of measure zero) of the configurations under study, which has the advantage that a geometric setting is replaced by an analytic one to which the powerful methods of analysis can be applied. However, the inherent difficulties of the original shape optimization problems show up again in various ways in the functional-analytic reformulation. For instance, it is known (cf. Pironneau [1984]) that the weak-star limit in L^∞ of a sequence of characteristic functions does not need to be a characteristic function again. In general, such a limit would be a function taking any value in the interval [0,1]. This situation forms the basis of a relaxation procedure used in optimal design theory for material distribution problems (cf. Bendsøe [1995]) that will be discussed below in the next paragraph. On the other hand, if additional uniform regularity assumptions are imposed on the boundaries of the unknown sets (see Chapter 2), then one can hope for convergence almost everywhere so that the limit is a characteristic function as well.

In this subsection, we will investigate several techniques for the approximation and/or relaxation of domain optimization problems that are based on the use of characteristic functions and that are due to the works of Bendsøe and Kikuchi [1988], Buttazzo and Dal Maso [1991], and Tiba et al. [1992]. This approach can be compared with the integrated penalty method of Kawarada and Natori [1981] described above in §5.1.1.

5.2.2.1 Structural Material Optimization

Many papers, both in the engineering and the mathematical literature, are devoted to problems entering the subject of this paragraph and to various methods for their solution. We will limit our exposition to only some of these approaches, beginning with the optimal layout of materials, which has already been addressed in Example 1.2.7 of Chapter 1 and in §2.3.4 of Chapter 2. We adopt the notation introduced there.

Let $E \subset \mathbf{R}^d$ describe a given solid body made up of two different materials occupying, respectively, the measurable sets $\Omega \subset E$ and $E\backslash\Omega$. We assume that $\partial E = \Gamma_1 \cup \Gamma_2$, where $\Gamma_1 \cap \Gamma_2 = \emptyset$ and where Γ_i, $i = 1, 2$, are Lipschitz boundary pieces. For given $f \in L^2(E)$ and given positive constants a_1, a_2, b_1, b_2 we then consider the transmission boundary value problem (2.3.70)–(2.3.73).

Now let (cf. (2.3.76))

$$V = \Big\{ w \in H^1(E) : \; w = 0 \; \text{ on } \; \Gamma_2 \Big\} .$$

We denote by H the maximal monotone extension of the Heaviside graph in $\mathbf{R} \times \mathbf{R}$ and introduce as admissible set U_{ad} the set of all functions $p \in H^1_{\text{loc}}(E)$ satisfying the conditions (2.3.79) and (2.3.80). We recall that according to (2.3.81),

$$\text{meas}(\{x \in E : \, p(x) = 0\}) = 0 \quad \forall \, p \in U_{ad}, \tag{5.2.18}$$

so that $H(p)$ is a characteristic function for any $p \in U_{ad}$. It is thus justified to introduce the family of measurable subsets Ω satisfying $\chi_\Omega = H(p_\Omega)$ for some $p_\Omega \in U_{ad}$. The optimal layout problem then has the form (compare (2.3.82), (2.3.83))

$$\underset{p \in U_{ad}}{\text{Min}} \left\{ \int_D |y_p(x) - z_d(x)|^2 \, dx \right\} , \tag{5.2.19}$$

$$\int_E \{ [a_1 H(p) + a_2(1 - H(p))] \, \nabla y_p \cdot \nabla w + [b_1 H(p) + b_2(1 - H(p))] \, y_p \, w \} \, dx$$
$$= \int_E f(x) \, w(x) \, dx \quad \forall \, w \in V, \tag{5.2.20}$$

where $D \subset E$ is some given measurable set, $z_d \in L^2(D)$ is the desired (observed) state, and y_p is the unique solution to the variational problem (5.2.20) corresponding to the choice $p \in U_{ad}$.

We now introduce a first relaxation/approximation scheme related to the use of characteristic functions. To this end, we approximate H by its Yosida–Moreau approximation (Appendix 1) for $\varepsilon > 0$,

$$H_\varepsilon(r) = \begin{cases} 1, & \text{if } r > \varepsilon, \\ r\,\varepsilon^{-1}, & \text{if } r \in [0,\varepsilon], \\ 0, & \text{if } r < 0. \end{cases} \tag{5.2.21}$$

We then replace (5.2.20) by

$$\int_E \Big\{ [a_1 H_\varepsilon(p) + a_2(1 - H_\varepsilon(p))]\, \nabla y_p^\varepsilon \cdot \nabla w \,+\, [b_1 H_\varepsilon(p) + b_2(1 - H_\varepsilon(p))]\, y_p^\varepsilon\, w \Big\}\, dx$$

$$= \int_E f(x)\, w(x)\, dx \quad \forall\, w \in V. \tag{5.2.22}$$

The problem (5.2.19), (5.2.22) is a relaxed version of the optimal layout problem (5.2.19), (5.2.20). By a variant of Theorem 2.3.21 in Chapter 2, we get the existence of at least one optimal pair $[y_\varepsilon, p_\varepsilon] \in H^1(E) \times U_{ad}$ for (5.2.19), (5.2.22), where $y_\varepsilon = y^\varepsilon_{p_\varepsilon}$, with y_p^ε denoting the solution to (5.2.22) corresponding to $p \in U_{ad}$.

Theorem 5.2.6 *For any open set K with $\overline{K} \subset E$ there is some sequence $\varepsilon_n \searrow 0$ such that*

$$y_{\varepsilon_n} \to \hat{y} \text{ weakly in } H^1(E), \qquad p_{\varepsilon_n} \to \hat{p} \text{ strongly in } H^1(K),$$

where $[\hat{y}, \hat{p}]$ is an optimal pair for the problem (5.2.19), (5.2.20).

Proof. Thanks to condition (2.3.80), there is some sequence $\varepsilon_n \searrow 0$ such that $p_{\varepsilon_n} \to \hat{p}$ strongly in $H^1(K)$. Without loss of generality, we may assume that $p_{\varepsilon_n} \to \hat{p}$ and $\nabla p_{\varepsilon_n} \to \nabla \hat{p}$ pointwise a.e. in K.

Now choose any sequence $\{K_j\}_{j \in \mathbf{N}}$ of open sets such that

$$\overline{K} \subset \overline{K}_j \subset \overline{K}_{j+1} \subset E \quad \forall\, j \in \mathbf{N}, \qquad \bigcup_{j=1}^{\infty} \overline{K}_j = E.$$

Applying the above argument successively with incresing j to the sets K_j, $j \in \mathbf{N}$, we deduce that the function $\hat{p}$ can be defined on the whole set E, where it turns out that $\hat{p} \in H^1_{\text{loc}}(E)$ and $p_{\varepsilon_n} \to \hat{p}$ and $\nabla p_{\varepsilon_n} \to \nabla \hat{p}$ pointwise a.e. in E. In particular, it then follows that $\hat{p}$ satisfies (2.3.79) and (2.3.80), hence belongs to U_{ad}. Notice that the function $\hat{p}$ is not necessarily uniquely determined, since it may depend on K and the sequence $\{K_j\}_{j \in \mathbf{N}}$.

Since $1 \geq H_\varepsilon(p) \geq 0$ a.e. in E, it follows from inserting $w = y_\varepsilon = y^\varepsilon_{p_\varepsilon}$ in (5.2.22) that $\{y_\varepsilon\}_{\varepsilon>0}$ is bounded in $H^1(E)$. We thus may assume without loss of generality that $y_{\varepsilon_n} \to \hat{y}$ weakly in $H^1(E)$. Moreover, $\{H_\varepsilon(p_\varepsilon)\}_{\varepsilon>0}$ is bounded in $L^2(E)$, so that we may assume that $H_{\varepsilon_n}(p_{\varepsilon_n}) \to z$ weakly in $L^2(E)$ and thus weakly in $L^2(K)$. On the other hand, $p_{\varepsilon_n} \to \hat{p}$ strongly in $L^2(K)$. Using the demiclosedness (cf. Proposition A1.4(iv) in Appendix 1) of the maximal monotone operator in $L^2(K)$ induced by the Heaviside graph, we infer that $z = H(\hat{p})$ a.e. in K. Since this argument can be repeated for any of the sets K_j, $j \in \mathbf{N}$, we even have $z = H(\hat{p})$ a.e. in E. Since $\hat{p} \in U_{ad}$, it follows from (5.2.18) that $H(\hat{p})$ is a characteristic function.

Next, we show that $H_{\varepsilon_n}(p_{\varepsilon_n}) \to H(\hat{p})$ pointwise a.e. in E. Indeed, if $\hat{p}(x) > 0$, then $p_{\varepsilon_n}(x) > \frac{1}{2}\hat{p}(x) > \varepsilon_n$ for $n \geq N_1(x)$, whence, by (5.2.21), $H_{\varepsilon_n}(p_{\varepsilon_n}(x)) = H(\hat{p}(x)) = 1$. Likewise, if $\hat{p}(x) < 0$, then $p_{\varepsilon_n}(x) < 0$ for $n \geq N_2(x)$, which implies that $H_{\varepsilon_n}(p_{\varepsilon_n}(x)) = H(\hat{p}(x)) = 0$. By (5.2.19), and since $\hat{p} \in U_{ad}$, one of these two situations occurs for a.e. $x \in E$.

Combining these pointwise a.e. convergence results with Lebesgue's theorem, we obtain that $H_{\varepsilon_n}(p_{\varepsilon_n}) \to H(\hat{p})$ strongly in $L^s(E)$ for all $s \geq 1$. Since all the convergences derived above are valid on a common subsequence, we may pass to the limit as $n \to \infty$ in (5.2.22) to find that $\hat{y} = y_{\hat{p}}$, that is, $[\hat{y}, \hat{p}]$ is an admissible pair for the problem (5.2.19), (5.2.20). To see that it is optimal, observe that

$$\int_D |y_{\varepsilon_n}(x) - z_d(x)|^2\,dx \leq \int_D |y_p^{\varepsilon_n}(x) - z_d(x)|^2\,dx \quad \forall\, p \in U_{ad}. \tag{5.2.23}$$

By an argument of the same type as above, one can prove that $y_p^{\varepsilon_n} \to y_p$ weakly in $H^1(E)$, on a subsequence (which may without loss of generality be assumed to coincide with the subsequence constructed above), where y_p solves (5.2.20) for $p \in U_{ad}$. Taking the limit as $n \to \infty$ in (5.2.23), we obtain that $[\hat{y}, \hat{p}]$ is indeed an optimal pair for the problem (5.2.19), (5.2.20). This concludes the proof of the assertion. □

Corollary 5.2.7 *There is a sequence* $\varepsilon_n \searrow 0$ *such that*

$$y_{\varepsilon_n} \to \hat{y} \ \textit{strongly in}\ H^1(E) \ \textit{as}\ n \to \infty. \tag{5.2.24}$$

Proof. There is some constant $c > 0$ such that for every $\varepsilon > 0$,

$$\begin{aligned} c\,|y_\varepsilon - \hat{y}|^2_{H^1(E)} &\leq \int_E \Big[a_1 H_\varepsilon(p_\varepsilon) + a_2(1 - H_\varepsilon(p_\varepsilon))\Big]\, |\nabla(y_\varepsilon - \hat{y})|^2_{\mathbf{R}^d}\,dx \\ &\quad + \int_E \Big[b_1 H_\varepsilon(p_\varepsilon) + b_2(1 - H_\varepsilon(p_\varepsilon))\Big]\,(y_\varepsilon - \hat{y})^2\,dx \\ &= \int_E \Big\{\Big[a_1 H_\varepsilon(p_\varepsilon) + a_2(1 - H_\varepsilon(p_\varepsilon))\Big]\,\nabla y_\varepsilon \cdot \nabla(y_\varepsilon - \hat{y}) \end{aligned}$$

$$
\begin{aligned}
&\qquad + \Big[b_1 H_\varepsilon(p_\varepsilon) + b_2(1 - H_\varepsilon(p_\varepsilon))\Big]\, y_\varepsilon\,(y_\varepsilon - \hat{y})\Big\}\, dx \\
&- \int_E \Big\{\Big[a_1 H_\varepsilon(p_\varepsilon) + a_2(1 - H_\varepsilon(p_\varepsilon))\Big]\, \nabla\hat{y}\cdot\nabla(y_\varepsilon - \hat{y}) \\
&\qquad + \Big[b_1 H_\varepsilon(p_\varepsilon) + b_2(1 - H_\varepsilon(p_\varepsilon))\Big]\, \hat{y}\,(y_\varepsilon - \hat{y})\Big\}\, dx \\
= &- \int_E \Big\{\Big[a_1 H_\varepsilon(p_\varepsilon) + a_2(1 - H_\varepsilon(p_\varepsilon))\Big]\, \nabla\hat{y}\cdot\nabla(y_\varepsilon - \hat{y}) \\
&\qquad + \Big[b_1 H_\varepsilon(p_\varepsilon) + b_2(1 - H_\varepsilon(p_\varepsilon))\Big]\, \hat{y}\,(y_\varepsilon - \hat{y})\Big\}\, dx \\
&+ \int_E f\,(y_\varepsilon - \hat{y})\, dx \;=\; I_1^\varepsilon + I_2^\varepsilon.
\end{aligned}
$$

By virtue of Theorem 5.2.6, there is some sequence $\varepsilon_n \searrow 0$ such that $I_2^{\varepsilon_n} \to 0$. Concerning $I_1^{\varepsilon_n}$, we first estimate the expression

$$
\int_E [a_1 H_{\varepsilon_n}(p_{\varepsilon_n}) + a_2(1 - H_{\varepsilon_n}(p_{\varepsilon_n}))]\, \nabla\hat{y}\cdot\nabla(y_{\varepsilon_n} - \hat{y})\, dx. \tag{5.2.25}
$$

From Theorem 5.2.6 we know that

$$
\nabla\hat{y}\cdot\nabla(y_{\varepsilon_n} - \hat{y}) \to 0 \quad \text{weakly in } L^1(E).
$$

Moreover, the sequence $\{a_1 H_{\varepsilon_n}(p_{\varepsilon_n}) + a_2(1 - H_{\varepsilon_n}(p_{\varepsilon_n}))\}$ is bounded in $L^\infty(E)$ and strongly convergent in $L^s(E)$ for all $s \geq 1$. Hence there is a subsequence of $\{\varepsilon_n\}$, which is again indexed by n, such that as $n \to \infty$,

$$
[a_1 H_{\varepsilon_n}(p_{\varepsilon_n}) + a_2(1 - H_{\varepsilon_n}(p_{\varepsilon_n}))]\, \nabla(y_{\varepsilon_n} - \hat{y}) \to u \quad \text{weakly in } L^2(E)^d.
$$

By Egorov's theorem, there exists for any $\delta > 0$ some measurable subset $E_\delta \subset E$ with $\text{meas}(E \setminus E_\delta) < \delta$ such that

$$
a_1 H_{\varepsilon_n}(p_{\varepsilon_n}) + a_2(1 - H_{\varepsilon_n}(p_{\varepsilon_n})) \to a_1 H(\hat{p}) + a_2(1 - H(\hat{p})) \quad \text{uniformly in } E_\delta.
$$

Combining this with the fact that $\nabla(y_{\varepsilon_n} - \hat{y}) \to 0$ weakly in $L^2(E)^d$, we see that $u = 0$ a.e. in E. Hence, the expression in (5.2.25) approaches zero as $n \to \infty$. Since the same argument applies to the remainder of the expression $I_1^{\varepsilon_n}$, we have $I_1^{\varepsilon_n} \to 0$ as $n \to \infty$. This concludes the proof of the assertion. □

Remark. It is possible to extend the above results to more general cost functionals than (5.2.20) as long as these are continuous with respect to the weak topology of $H^1(E)$. The results of Theorem 5.2.6 and of Corollary 5.2.7 are due to Liu, Neittaanmäki, and Tiba [2000], [2003].

Remark. By the above transformations, we have succeeded in replacing the unknown domains Ω by the unknown control mappings $p \in U_{ad}$, i.e., we have reduced the shape optimization problem to a control in the coefficients problem. The mapping H_ε is Lipschitzian, and in fact, additional regularity can be achieved for it by performing a smoothing via a Friedrichs mollifier. The mapping method (see Section 5.3 and Example 1.2.7 in Chapter 1) produces a similar reduction; however, (5.2.19), (5.2.20) obviously has several advantages: the control appears in an algebraic form (no derivatives of p enter the coefficients), the topological type of the optimal $\widehat{\Omega}$ corresponding to the optimal $\hat{p}$ is "free" (for instance, $\widehat{\Omega}$ does not need to be connected a priori), and the regularity assumptions on $p \in U_{ad}$, and implicitly on the unknown domains Ω, are rather mild. As a drawback, we mention that the applicability of this method is restricted. But this can be further extended, as will be demonstrated at the end of this paragraph.

Now let us assume that H_ε is a C^1-approximation of H. Let $\theta_\varepsilon : L^\infty(E) \to L^2(E)$ be the (state) mapping $p \mapsto y$ defined by (5.2.22). For what follows we assume that $U_{ad} \subset L^\infty(E)$, which is no restriction from the viewpoint of the original problem. The next result and the numerical example demonstrate the constructive character of the present approach. For simplicity, we assume that $b_1 = b_2 = a_0 > 0$ and we take zero Neumann boundary conditions on ∂E.

Theorem 5.2.8 *Suppose that $H_\varepsilon \in C^1(\mathbf{R})$ for some $\varepsilon > 0$. Then the mapping $\theta_\varepsilon : L^\infty(E) \to L^2(E)$ is Gâteaux differentiable, and for every $v \in L^\infty(E)$ the directional derivative $r = \nabla\theta_\varepsilon(p)v$ at $p \in L^\infty(E)$ in the direction $v \in L^\infty(E)$ belongs to $H^1(E)$ and satisfies*

$$\int_E \Big[(a_1 - a_2)H_\varepsilon'(p)\,\nabla y \cdot \nabla w + (a_1 H_\varepsilon(p) + a_2(1 - H_\varepsilon(p)))\,\nabla r \cdot \nabla w + a_0\, r\, w\Big]\, dx = 0 \quad \forall\, w \in H^1(E), \tag{5.2.26}$$

where $y = \theta_\varepsilon(p)$.

Remark. If p satisfies the conditions (2.3.79), (2.3.80) with strict inequalities, and if v is sufficiently smooth, then $p + \lambda v$ will still satisfy (2.3.79), (2.3.80) provided that $|\lambda| > 0$ is sufficiently small. This shows that it is possible to perform variations around certain points p and in certain directions v. In general, the question of directional differentiability can be studied independently of approximation or existence properties, and one can simply assume that U_{ad} is a convex and closed subset of $L^\infty(E)$.

Taking into account that by the above method $H(p)$ and $H(p + \lambda v)$ represent the characteristic functions of the domains associated with p and $p + \lambda v$, respectively, it is clear that the domain variations constructed in this way are very general (see Example 5.2.9 below, and compare with Section 5.3).

Proof of Theorem 5.2.8. We set $y_\lambda = \theta_\varepsilon(p+\lambda v)$ for $\lambda > 0$. Subtraction of the equations corresponding to y and y_λ yields, for any $w \in H^1(E)$,

$$0 = \int_E \Big\{ \big[a_1 H_\varepsilon(p+\lambda v) - a_1 H_\varepsilon(p) + a_2 H_\varepsilon(p) - a_2 H_\varepsilon(p+\lambda v)\big] \nabla y \cdot \nabla w$$

$$+ \big[a_1 H_\varepsilon(p+\lambda v) + a_2(1 - H_\varepsilon(p+\lambda v))\big] \nabla(y_\lambda - y)\cdot \nabla w + a_0\,(y_\lambda - y)\,w \Big\}\,dx.$$

Dividing by $\lambda > 0$, choosing $w = \lambda^{-1}(y_\lambda - y)$, and using the fact that H_ε obviously has the Lipschitz constant $1/\varepsilon$, we find that

$$a \left| \frac{y_\lambda - y}{\lambda} \right|^2_{H^1(E)} \le |a_1 - a_2|\,\frac{1}{\varepsilon}\,|v|_{L^\infty(E)} |\nabla y|_{L^2(E)} \left| \nabla\left(\frac{y_\lambda - y}{\lambda}\right)\right|_{L^2(E)},$$

with $a = \min\{a_0, a_1, a_2\} > 0$. Hence, $\{\lambda^{-1}(y_\lambda - y)\}_{\lambda>0}$ is bounded in $H^1(E)$, and there is a sequence $\lambda_n \searrow 0$ such that $\lambda_n^{-1}(y_{\lambda_n} - y) \to r$ weakly in $H^1(E)$ and, by compact embedding, strongly in $L^2(E)$. Dividing the above equation for $\lambda = \lambda_n$ by λ_n, and passing to the limit as $n \to \infty$, we easily verify that (5.2.26) holds, which concludes the proof. □

We now define the operator $T \in L(L^2(E), L^1(E))$ by putting $Tq = l$ with

$$l = (a_1 - a_2)\, H_\varepsilon'(p)\, \nabla z \cdot \nabla y, \tag{5.2.27}$$

where $z \in H^1(E)$ is the (unique) solution to the "adjoint" equation

$$\int_E \Big\{ \big[a_1 H_\varepsilon(p) + a_2(1 - H_\varepsilon(p))\big] \nabla z \cdot \nabla w + a_0\, z\, w + q\, w \Big\}\, dx = 0 \quad \forall\, w \in H^1(E). \tag{5.2.28}$$

It turns out that the linear and continuous operator $S : L^\infty(E) \to L^2(E)$, $Sv = \nabla\theta_\varepsilon(p)v = r$, is just the adjoint of T, which can easily be verified from the definition and by putting $w = r$ in (5.2.28) and $w = z$ in (5.2.26). It enables us to compute the gradient of the cost functional of the problem (5.2.19), (5.2.22) (which we denote by $\hat{J}(p)$, where $y_p = \theta_\varepsilon(p)$). We have

$$\begin{aligned} \lim_{\lambda\to 0} \frac{1}{\lambda}(\hat{J}(p+\lambda v) - \hat{J}(p)) &= 2\int_D r(x)(y(x) - z_d(x))\,dx \\ &= 2\int_E r(x)\,\chi_D(x)(y - z_d)(x)\,dx \\ &= 2\,(Sv, \chi_D\,(y - z_d))_{L^2(E)} \\ &= 2\,(v, T(\chi_D\,(y - z_d)))_{L^\infty(E)\times L^1(E)}. \end{aligned} \tag{5.2.29}$$

By choosing $q = \chi_D\,(y - z_d)$ in (5.2.28), (5.2.27), the computation of the directional derivative of $\hat{J}$ at p in the direction v is finished.

Example 5.2.9 We briefly discuss the numerical treatment of the problem (5.2.19), (5.2.22) with $a_1 = 10$, $a_2 = a_0 = 1$, $E = D =]0,1[\times]0,1[\subset \mathbf{R}^2$, and $f(x_1, x_2) = x_1^2 + x_2^2$.

Following Mäkinen, Neittaanmäki, und Tiba [1992], the Heaviside mapping is approximated by the mapping $H_\varepsilon \in C^1(\mathbf{R})$,

$$H_\varepsilon(p) = \begin{cases} 1 - \dfrac{1}{2}\exp(-p/\varepsilon), & \text{if } p \geq 0, \\ \dfrac{1}{2}\exp(p/\varepsilon), & \text{if } p < 0, \end{cases} \qquad \text{with } \varepsilon = \frac{1}{10}. \tag{5.2.30}$$

Such an approximation was later introduced in image reconstruction problems by Chan and Vese [1997], [2001] and is sometimes called the *Chan–Vese regularization*.

The regularized state problem (5.2.22) has been discretized by using four-node quadrilateral Lagrangian elements with a regular rectangular mesh. For the sake of simplicity, the control parameter p has been taken piecewise constant. Notice that then $p \notin H^1_{\text{loc}}(E)$, and consequently $p \notin U_{ad}$ in general. However, the relaxed problem (5.2.19), (5.2.22) also makes sense in this case, and meaningful approximations could be obtained with this simplified approach.

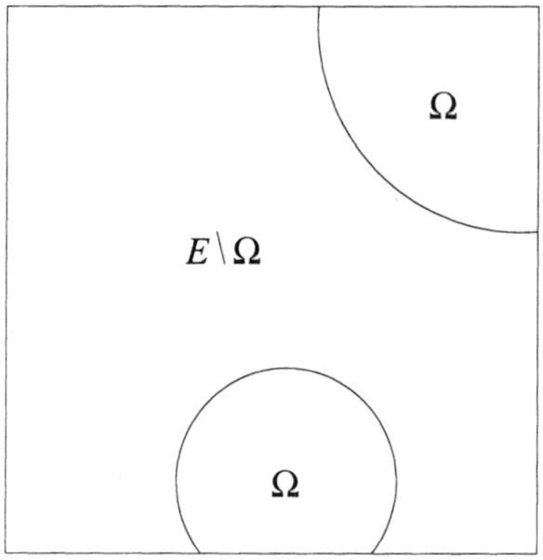

Figure 2.1. Domain defining y_d in Example 5.2.9.

The discrete analogue of (5.2.22) is the linear-algebraic system

$$K(p)q(p) = f, \tag{5.2.31}$$

where $K(p)$ is the "stiffness" matrix (which depends on the control variable p) and $q(p)$ is the vector of the nodal values of the solution, i.e., the discretization of the solution to (5.2.22). The approximating cost becomes

$$\min_p \{(q(p) - q_d)^T M(q(p) - q_d)\}, \tag{5.2.32}$$

where q_d is the vector corresponding to z_d and M is the "mass" matrix.

Problem (5.2.31), (5.2.32) constitutes the discrete analogue of (5.2.19), (5.2.22). It has been solved using the conjugate gradient algorithm E04DGF of the NAG subroutine library. The gradient is obtained by numerical approximation of (5.2.26), (5.2.28), (5.2.29).

In Figure 2.1, we have depicted a choice for Ω and for $E \setminus \Omega$. Let us denote the solution to the transmission conditions problem (5.2.18) corresponding to the geometry of Figure 2.1 by z_d. By the formulation of the shape optimization problem (5.2.19), (5.2.20), it is obvious that Figure 2.1 gives a global solution to the problem introduced in this example, with the above choice of z_d.

In Figures 2.2 and 2.3 the obtained numerical solution has been depicted for a discretization with 100 and 900 elements, respectively. As initial guess of the algorithm, we took $p \equiv 0$ in E. The numerical data are gathered in Table 2.1. The results are very close to the true global solution.

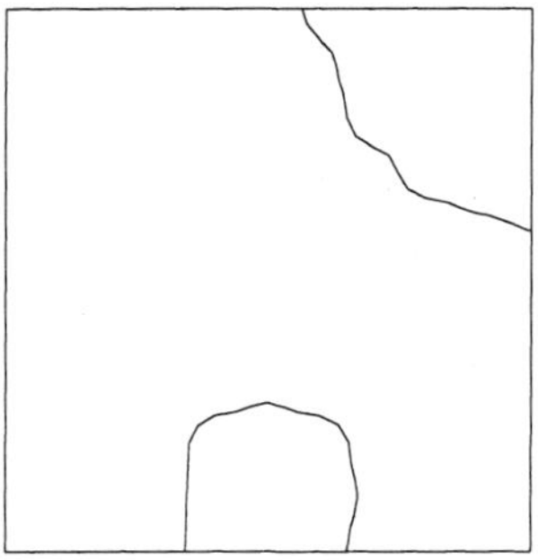

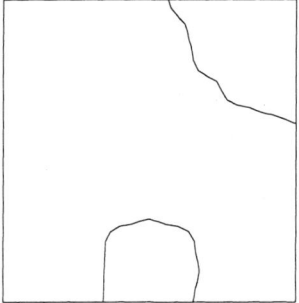

Figure 2.2. Boundary of Ω using 100 elements.

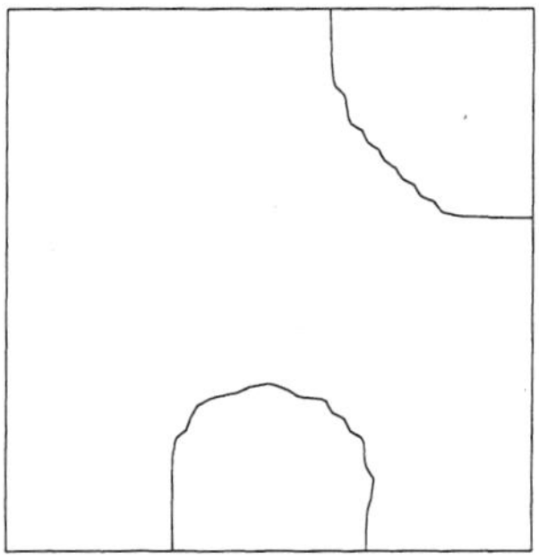

Figure 2.3. Boundary of Ω using 900 elements.

Number of elements	Initial cost	Final cost	Iterations
100	23.1	1.41×10^{-3}	15
900	24.9	1.11×10^{-3}	23

Table 2.1. Results of Example 5.2.9.

It is known that in nonconvex optimization problems the initial guess may have a big influence on the quality of the obtained result. If we choose in this example the initial p as

$$p(x_1, x_2) = \begin{cases} +0.2, & \text{if } (x_1, x_2) \in \Omega^0, \\ -0.2, & \text{if } (x_1, x_2) \in E \setminus \Omega^0, \end{cases}$$

with $\Omega^0 = \left\{]0,1[\times]0, \frac{1}{4}[\right\} \cup \left\{]0,1[\times]\frac{3}{4}, 1[\right\}$, then the obtained numerical solution is depicted in Figure 2.4 (400 elements). Only a local minimum was achieved in this case by the conjugate gradient method. The cost decreased from 11.2 to 0.0194 in 43 iterations. However, the topology changed during the algorithm.

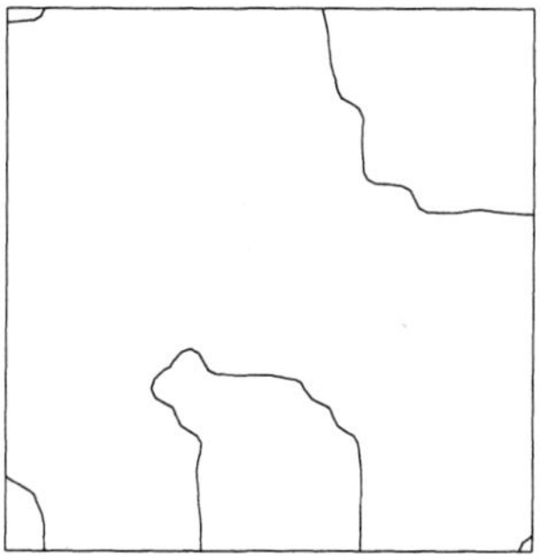

Figure 2.4. Boundary of Ω using the second initial guess for p.

Remark. In the works of Bendsøe and Kikuchi [1988] and Bendsøe [1995], the *material distribution* method or the *topology optimization* method was introduced. In the case of layout problems, one simply replaces the characteristic functions by functions $p \in L^\infty(E)$ that may attain values in the entire interval $[0, 1]$. For instance, the relaxed formulation of (5.2.18) reads, with the data from Example 5.2.9,

$$-\operatorname{div}(l\,\nabla y) + a_0\, y = f \quad \text{in } E, \qquad l\,\frac{\partial y}{\partial n} = 0 \quad \text{on } \partial E. \tag{5.2.33}$$

Here, $l \in L^\infty(E)$ may attain values in the interval $[a_2, a_1] = [1, 10]$. The problem (5.2.33), (5.2.19) is a standard control in the coefficients problem with constraints. It remains nonconvex, but it is simpler than both the optimal layout problem (5.2.18), (5.2.20) and its approximation (5.2.20), (5.2.22). Its difficulty lies in the fact that the relationship between the relaxed problem (5.2.33), (5.2.20) and the original one is just on an intuitive level: the result of the optimization procedure will yield a mapping $l^* \in L^\infty(E)$, and the set $\{x \in E : l^*(x) = 10\}$ will be contained in the searched domain, while the set $\{x \in E : l^*(x) = 1\}$ will be contained in its complement. However, the set where intermediate values $1 < l^*(x) < 10$ are attained has, in general, positive measure and is difficult to interpret. One idea to minimize this set is to add a penalty term of the form

$$\lambda \int_E (1 - l(x))\,(l(x) - 10)\,dx, \quad \text{with "big" } \lambda > 0,$$

to the cost functional, when (5.2.33) is taken into account. It is also possible to impose an additional constraint of the form

$$\int_E (1 - l(x))\,(l(x) - 10)\,dx \leq \varepsilon$$

with some "small" $\varepsilon > 0$ (cf. Borrvall and Petersson [2001]).

We underline that the two material constituents in the layout problem may be just substance and void, and consequently, general shape optimization problems may be discussed via variants of this approach. The possible appearance of composite materials with microscopic material constituents shows the connection with the homogenization theory as well. We refer the reader to the works of Cioranescu and Donato [1999], Zhikov, Kozlov, and Oleinik [1994], and Allaire [2001].

The above presentation shows that the topology optimization method provides only a rough form of the optimal structure. Further refinements via the *boundary variation* technique (see Section 5.3 below) are necessary. The combination of both approaches is sometimes called the *integrated topology* method and is frequently used in shape optimization (cf. Bendsøe [1995]).

5.2.2.2 Bang-Bang Controls and Characteristic Functions

In this paragraph, we examine a relaxation method of the above type for the capacity optimization problem (R) introduced in §2.3.1 of Chapter 2. In this case, the relaxation procedure even provides a solution to the original problem, which is not true in general situations.

We briefly recall the formulation of the problem (R). To this end, let $\Omega \subset \mathbf{R}^d$ denote a given bounded domain, and let $A \subset \Omega$ be a measurable subset with characteristic function χ_A. We consider the problem

$$\text{Min}\left\{\int_\Omega |\nabla u(x)|^2\,dx \,:\, u \in H_0^1(\Omega),\ \int_\Omega \chi_A(x)\,(u(x) - 1)^2\,dx = 0\right\}. \tag{5.2.34}$$

The problem (5.2.34) has a unique minimizer $u_A \in H_0^1(\Omega)$. We denote by

$$E_d(A) = \int_\Omega |\nabla u_A(x)|^2\,dx$$

the corresponding optimal value, called the *energy* or *capacity* of the set A with respect to the domain Ω. The shape optimization problem (R) is then

$$\operatorname{Min}\left\{E_d(A) : \operatorname{meas}(A) = v\right\}, \tag{5.2.35}$$

where v is a prescribed "volume" in the interval $[0, \operatorname{meas}(\Omega)]$.

The relation (5.2.34) defines the state system. If A is sufficiently regular, it follows that u_A satisfies Laplace's equation in $\Omega \setminus A$ (compare §2.3.1 in Chapter 2). A penalization of (5.2.34) is obtained by

$$\operatorname*{Min}_{u \in H_0^1(\Omega)} \left\{\int_\Omega |\nabla u(x)|^2 \, dx + \int_\Omega f(x)(u(x) - 1)^2 \, dx\right\}, \tag{5.2.36}$$

for any $f \in L^\infty(\Omega)$ satisfying $f \geq 0$ a.e. in Ω (and with A corresponding to the set of points where f is strictly positive). One may compare (5.2.36) with the *integrated penalty approach* of Natori and Kawarada [1981]; see §5.1.1.

We denote by $\phi(f)$ the functional assigning to every $f \in L^\infty(\Omega)_+$ the minimal value in (5.2.36). Notice that (compare the proof of Proposition 2.2.3 in §2.2.1 in Chapter 2) ϕ is a concave functional, as lower envelope of a family of affine mappings. The relaxed formulation of the problem (5.2.35) is then

$$\operatorname{Min}\left\{\phi(f) : 0 \leq f(x) \leq 1 \text{ a.e. in } \Omega, \ \int_\Omega f(x) \, dx = v\right\}. \tag{5.2.37}$$

We remark that in this formulation the characteristic functions have been replaced by mappings attaining arbitrary values in $[0, 1]$, as in the approach of Bendsøe mentioned in the previous paragraph. The volume constraint is maintained.

Theorem 5.2.10 *The problem (5.2.37) admits at least one solution f^*.*

Proof. The convex set

$$C = \left\{f \in L^\infty(\Omega) : 0 \leq f(x) \leq 1 \text{ a.e. in } \Omega, \ \int_\Omega f(x) \, dx = v\right\}$$

is weakly* compact in $L^\infty(\Omega)$. Denoting by u_f the solution of (5.2.36) associated with $f \in L^\infty(\Omega)_+$, we obtain that

$$\int_\Omega |\nabla u_f(x)|^2 \, dx \leq \phi(f) \leq \int_\Omega f(x) \, dx = v, \tag{5.2.38}$$

which corresponds to choosing $u \equiv 0$ in (5.2.36). It thus suffices to consider in (5.2.36) the minimization over the closed ball $\overline{B(0, v)}$ in $H_0^1(\Omega)$. The following continuity property holds for ϕ: take any functions $f, f_0 \in C$ such that $|f - f_0|_{L^\infty(\Omega)} \leq \delta$. Then we can infer, using the minimal property of u_f, that

$$\begin{aligned}
&\phi(f) - \phi(f_0) \\
&\quad = \int_\Omega |\nabla u_f|^2\,dx + \int_\Omega f(u_f - 1)^2\,dx - \int_\Omega |\nabla u_{f_0}|^2\,dx - \int_\Omega f_0(u_{f_0} - 1)^2\,dx \\
&\quad \le \int_\Omega |\nabla u_{f_0}|^2\,dx + \int_\Omega f(u_{f_0} - 1)^2\,dx - \int_\Omega |\nabla u_{f_0}|^2\,dx - \int_\Omega f_0(u_{f_0} - 1)^2\,dx \\
&\quad = \int_\Omega (u_{f_0} - 1)^2(f - f_0)\,dx \le c\,\delta, \qquad (5.2.39)
\end{aligned}$$

where $c > 0$ is given by (5.2.38) via the embedding property of Sobolev spaces. Since the same inequality holds if we reverse the order of f and f_0, we see that

$$|\phi(f) - \phi(f_0)| \le c\,\delta \quad \forall\, f, f_0 \in C, \quad |f - f_0|_{L^\infty(\Omega)} \le \delta. \qquad (5.2.40)$$

The last inequality in (5.2.39) also shows that (5.2.40) holds for any $f, f_0 \in C$ satisfying the condition

$$\int_\Omega w^2(f - f_0)\,dx \le c\,\delta, \qquad (5.2.41)$$

when

$$w \in H_0^1(\Omega), \quad |w|^2_{H_0^1(\Omega)} \le v. \qquad (5.2.42)$$

Now, by virtue of the compactness of the embedding of $H_0^1(\Omega)$ in $L^2(\Omega)$ (cf. Theorem A2.2(i) in Appendix 2), the set

$$\left\{ z \in L^1(\Omega) : z = w^2 \text{ for some } w \text{ satisfying (5.2.42)} \right\}$$

is compact in $L^1(\Omega)$. Then (5.2.41) defines a weak neighborhood of f_0 in $L^\infty(\Omega) \cap C$. Consequently, (5.2.40) proves the continuity of the functional ϕ in this topology on the weakly* compact set C. The assertion now follows from the Weierstrass theorem. □

Corollary 5.2.11 *Among the minimizers f^* of ϕ on C there is a characteristic function of a set having the measure v.*

Proof. Since ϕ is concave and C is convex, the set of minimizers of ϕ over C contains at least one extremal point of C. According to Castaing and Valadier [1977], the extremal points of C are characteristic functions of sets of measure v. □

Remark. Such an f^* is a bang-bang minimizer of ϕ and a solution of the original problem (R). The set $A = \operatorname{supp} f^*$ may be further studied by means of the first-order optimality conditions; see Gonzalez de Paz [1982], [1989], [1994].

5.2.3 Controllability and Fictitious Domains Approaches

We discuss in this paragraph a class of methods termed in the mathematical literature *embedding*, *fictitious domains*, or *controllability* approaches. There are many variants, as will become clear from our brief survey. One of the origins of this methodology is the numerical treatment of partial differential equations in complicated domains by using finite differences (see Astrakmantsev [1978]). There, the equation is "extended" by various procedures to a larger simple domain on which finite differences may easily be applied. The solution of the initial problem is obtained, for instance, as a restriction to the original domain. These ideas have successfully been applied to exterior domain problems (for example, for the Helmholtz equation, see Proskurowski and Windlund [1979] and Atamian [1991]), and to variable domain problems. One example of this type is the plasma problem discussed in §1.2.3 of Chapter 1 (cf. Chapter I.2 in Blum [1989]).

Let us notice that in the case of shape optimization problems the disadvantage of solving "bigger" problems (in larger domains) naturally disappears by working in the largest admissible domain. Moreover, as in the cases discussed in the previous paragraphs of this section, we shall remain within the setting of fixed domain methods, thus avoiding the extremely time-consuming operations of remeshing the domain and of recomputing the stiffness matrix in each iteration of the algorithm.

We also underline that the scientific literature related to the subject of this paragraph is very large and embraces a wide range of properties and applications. However, it is our feeling that at the core of these developments are various "geometric" controllability properties of the considered boundary value problems. In the following, we will stress this point of view. Let us mention that among the first authors who studied such approximation and controllability results for elliptic equations are H. Beckert [1960], J.-L. Lions [1968], and A. Göpfert [1971].

5.2.3.1 Boundary Controllability

In the book of Lions [1968], the following approximate controllability result has been established:

Theorem 5.2.12 *Let $\Omega \subset \mathbf{R}^d$ be a bounded domain with smooth boundary $\partial\Omega = \Gamma_1 \cup \Gamma_2$, where $\Gamma_1 \cap \Gamma_2 = \emptyset$. For any $u \in L^2(\Gamma_1)$, let the function $y = y_u \in L^2(\Omega)$ be the unique solution (in the transposition sense) to the elliptic problem*

$$-\Delta y = 0 \quad in\ \Omega, \tag{5.2.43}$$

$$y = u \quad on\ \Gamma_1, \qquad y = 0 \quad on\ \Gamma_2. \tag{5.2.44}$$

Then $\left\{\frac{\partial y_u}{\partial n} : u \in L^2(\Gamma_1)\right\}$ forms a dense subspace of $H^{-1}(\Gamma_2)$.

Remark. The transposition solution (for its definition, recall Example A2.7 in Appendix 2) actually belongs to the space $H^{\frac{1}{2}}(\Omega)$. Since $\Delta y_u = 0 \in L^2(\Omega)$, it is possible (cf. Lions [1968]) to define $\frac{\partial y_u}{\partial n} \in H^{-1}(\Gamma_2)$. We also notice that an analogous result holds true for general second-order elliptic operators. In Theorem 5.2.13 below, an alternative (constructive) approach is discussed.

In connection with the plasma identification problem introduced in §1.2.3 of Chapter 1, a similar controllability problem arises. We consider a simply connected bounded domain $\Omega \subset \mathbf{R}^2$ with smooth boundary $\partial\Omega$ and a closed smooth curve $\Gamma \subset \Omega$, and we denote by Ω_0 the domain bounded by Γ and $\partial\Omega$. Moreover, we assume that a function $f \in H^{\frac{3}{2}}(\partial\Omega)$ and a "target" function $g \in H^{\frac{1}{2}}(\partial\Omega)$ are measured, representing the trace and the normal derivative of the poloidal flux ψ.

The question is then to find a control $u \in H^{\frac{3}{2}}(\Gamma)$ such that the unique solution $\psi = \psi_u \in H^2(\Omega_0)$ to the boundary value problem (which is elliptic, since $x > 0$ is positively bounded away from zero; see Figure 2.1 in Chapter 1),

$$-\frac{\partial}{\partial x}\left(\frac{1}{x}\frac{\partial\psi}{\partial x}\right) - \frac{\partial}{\partial y}\left(\frac{1}{x}\frac{\partial\psi}{\partial y}\right) = 0 \quad \text{in } \Omega_0, \tag{5.2.45}$$

$$\psi = f \quad \text{on } \partial\Omega, \qquad \psi = u \quad \text{on } \Gamma, \tag{5.2.46}$$

satisfies

$$\frac{1}{x}\frac{\partial\psi_u}{\partial n} = g \quad \text{on } \partial\Omega. \tag{5.2.47}$$

In general, the problem (5.2.45)–(5.2.47) has no solution in Ω_0, since on $\partial\Omega$ Cauchy data are imposed on ψ.

The least squares approximation method is formulated as a control problem with a supplementary Tikhonov regularization term. We consider for $\varepsilon > 0$ the problem

$$\operatorname*{Min}_{u\in H^{\frac{3}{2}}(\Gamma)} \left\{\frac{1}{2}\left|\frac{1}{x}\frac{\partial\psi_u}{\partial n} - g\right|^2_{H^{\frac{1}{2}}(\partial\Omega)} + \frac{\varepsilon}{2}\,|u|^2_{H^{\frac{3}{2}}(\Gamma)}\right\}, \tag{5.2.48}$$

subject to (5.2.45), (5.2.46).

Notice that (up to obvious modifications) the control problem (5.2.45), (5.2.46), (5.2.48) is a special case of the problem considered previously in Example 3.2.4 of Chapter 3. In the setting used there, it immediately follows from the techniques introduced in Section 2.1 of Chapter 2 that for any $\varepsilon > 0$ it has a unique optimal pair $[u_\varepsilon, \psi_\varepsilon] \in H^{\frac{3}{2}}(\Gamma) \times H^2(\Omega_0)$, where $\psi_\varepsilon = \psi_{u_\varepsilon}$. The following approximate controllability result is similar to that of Theorem 5.2.12, but has a constructive character.

Theorem 5.2.13

$$\frac{1}{x}\frac{\partial \psi_\varepsilon}{\partial n} \to g \text{ strongly in } H^{\frac{1}{2}}(\partial\Omega) \text{ as } \varepsilon \searrow 0. \tag{5.2.49}$$

Proof. By taking $u = 0$ in (5.2.45), (5.2.46), and by denoting the solution thus obtained by ψ_0, (5.2.48) yields that for every $\varepsilon > 0$,

$$\frac{1}{2}\left|\frac{1}{x}\frac{\partial \psi_\varepsilon}{\partial n} - g\right|^2_{H^{\frac{1}{2}}(\partial\Omega)} + \frac{\varepsilon}{2}|u_\varepsilon|^2_{H^{\frac{3}{2}}(\Gamma)} \le \frac{1}{2}\left|\frac{1}{x}\frac{\partial \psi_0}{\partial n} - g\right|^2_{H^{\frac{1}{2}}(\partial\Omega)},$$

that is, $\left\{\frac{1}{x}\frac{\partial \psi_\varepsilon}{\partial n}\right\}_{\varepsilon>0}$ and $\left\{\varepsilon^{\frac{1}{2}} u_\varepsilon\right\}_{\varepsilon>0}$ are bounded in $H^{\frac{1}{2}}(\partial\Omega)$ and $H^{\frac{3}{2}}(\Gamma)$, respectively. Hence, there is a sequence $\varepsilon_n \searrow 0$ such that $\left(\frac{1}{x}\frac{\partial \psi_{\varepsilon_n}}{\partial n} - g\right) \to \ell$ weakly in $H^{\frac{1}{2}}(\partial\Omega)$.

We define the adjoint system (compare (3.2.25), (3.2.26)) by

$$Lp_\varepsilon = 0 \quad \text{in } \Omega_0, \tag{5.2.50}$$

$$p_\varepsilon = 0 \quad \text{on } \Gamma, \qquad p_\varepsilon = J\left(\frac{1}{x}\frac{\partial \psi_\varepsilon}{\partial n} - g\right) \quad \text{on } \partial\Omega, \tag{5.2.51}$$

where L is the operator given by the left-hand side of (5.2.45), and where J denotes the canonical isomorphism between $H^{\frac{1}{2}}(\partial\Omega)$ and its dual $H^{-\frac{1}{2}}(\partial\Omega)$. The solution p_ε has to be understood in the transposition sense (compare Example A2.7 in Appendix 2); it belongs to the space $L^2(\Omega_0)$. From (3.2.27) we infer that Pontryagin's maximum takes the form, with $J_1 : H^{\frac{3}{2}}(\Gamma) \to H^{-\frac{3}{2}}(\Gamma)$ being the canonical isomorphism,

$$-\frac{1}{x}\frac{\partial p_\varepsilon}{\partial n} = \varepsilon\, J_1 u_\varepsilon \quad \text{on } \Gamma. \tag{5.2.52}$$

Since the solution operator of the adjoint state system (5.2.50), (5.2.51) is linear and continuous from $H^{-\frac{1}{2}}(\partial\Omega)$ into $L^2(\Omega_0)$, we can conclude that $p_{\varepsilon_n} \to \overline{p}$ weakly in $L^2(\Omega_0)$, where

$$L\overline{p} = 0 \quad \text{in } \Omega_0, \tag{5.2.53}$$

$$\overline{p} = 0 \quad \text{on } \Gamma, \qquad \overline{p} = J(\ell) \quad \text{on } \partial\Omega. \tag{5.2.54}$$

Passage to the limit as $\varepsilon_n \searrow 0$ in (5.2.52), using the boundedness of $\left\{\varepsilon^{\frac{1}{2}} u_\varepsilon\right\}_{\varepsilon>0}$ in $H^{\frac{3}{2}}(\partial\Omega)$, also yields that

$$\frac{1}{x}\frac{\partial \overline{p}}{\partial n} = 0 \quad \text{on } \Gamma. \tag{5.2.55}$$

The Holmgren uniqueness theorem for the Cauchy problem for $\overline{p}$ around Γ gives $\overline{p} = 0$. Consequently, we must have $\frac{1}{x}\frac{\partial \psi_{\varepsilon_n}}{\partial n} \to g$ weakly in $H^{\frac{1}{2}}(\partial\Omega)$.

To prove strong convergence, we invoke Mazur's theorem (cf. Yosida [1980]), which yields that g is the strong limit of a sequence of suitable convex combinations of the functions $\frac{1}{x}\frac{\partial\psi_{\varepsilon_n}}{\partial n}$. Hence, we may assume that for any $n \in \mathbf{N}$ there exist some $m_n \in \mathbf{N}$, real numbers $\alpha_{n,j} > 0$, $1 \le j \le m_n$, such that $\sum_{j=1}^{m_n} \alpha_{n,j} = 1$, and a set $\{\varepsilon_{n_j} : 1 \le j \le m_n\} \subset \{\varepsilon_n\}_{n\in\mathbf{N}}$ such that

$$\sum_{j=1}^{m_n} \alpha_{n,j}\,\frac{1}{x}\,\frac{\partial\psi_{\varepsilon_{n_j}}}{\partial n} \to g \ \text{ strongly in } H^{\frac{1}{2}}(\partial\Omega) \ \text{ as } n \to \infty.$$

Now observe that

$$\sum_{j=1}^{m_n} \alpha_{n,j}\,\psi_{\varepsilon_{n_j}} = \psi_{\overline{u}_n} \ \text{ for } \overline{u}_n = \sum_{j=1}^{m_n} \alpha_{n,j}\,u_{\varepsilon_{n_j}}, \quad \forall\, n \in \mathbf{N}.$$

Hence, we may for any $k \in \mathbf{N}$ insert the admissible pair $[\overline{u}_n, \psi_{\overline{u}_n}]$ in the cost functional (5.2.48) to obtain that

$$\begin{aligned}
&\frac{1}{2}\left|\frac{1}{x}\,\frac{\partial\psi_{\varepsilon_k}}{\partial n} - g\right|^2_{H^{\frac{1}{2}}(\partial\Omega)} + \frac{\varepsilon_k}{2}\,\left|u_{\varepsilon_k}\right|^2_{H^{\frac{3}{2}}(\partial\Omega)} \\
&\qquad \le \frac{1}{2}\left|\frac{1}{x}\sum_{j=1}^{m_n} \alpha_{n,j}\,\frac{\partial\psi_{\varepsilon_{n_j}}}{\partial n} - g\right|^2_{H^{\frac{1}{2}}(\partial\Omega)} + \frac{\varepsilon_k}{2}\left|\sum_{j=1}^{m_n} \alpha_{n,j}\,u_{\varepsilon_{n_j}}\right|^2_{H^{\frac{3}{2}}(\partial\Omega)}.
\end{aligned}$$

Now let $\delta > 0$ be given. We then may fix some $n \in \mathbf{N}$ such that

$$\frac{1}{2}\left|\frac{1}{x}\sum_{j=1}^{m_n} \alpha_{n,j}\,\frac{\partial\psi_{\varepsilon_{n_j}}}{\partial n} - g\right|^2_{H^{\frac{1}{2}}(\partial\Omega)} < \frac{\delta^2}{4}\,.$$

Recalling that

$$M = \sup\left\{\varepsilon_j\,\left|u_{\varepsilon_j}\right|^2_{H^{\frac{3}{2}}(\partial\Omega)} \ : \ j \in \mathbf{N}\right\} < +\infty,$$

we find that

$$\begin{aligned}
\frac{\varepsilon_k}{2}\left|\sum_{j=1}^{m_n} \alpha_{n,j}\,u_{\varepsilon_{n_j}}\right|^2_{H^{\frac{3}{2}}(\partial\Omega)} &\le \sum_{j=1}^{m_n} \frac{\varepsilon_k}{2}\,\alpha_{n,j}\,\left|u_{\varepsilon_{n_j}}\right|^2_{H^{\frac{3}{2}}(\partial\Omega)} \\
&\le \frac{M}{2}\sum_{j=1}^{m_n} \alpha_{n,j}\,\frac{\varepsilon_k}{\varepsilon_{n_j}} < \frac{\delta^2}{4},
\end{aligned}$$

provided that $k \ge k(\delta)$, where $k(\delta) \in \mathbf{N}$ is chosen so large that

$$\varepsilon_{k(\delta)} < \frac{\delta^2}{2M}\,\mathrm{Min}\left\{\varepsilon_{n_1}, \ldots, \varepsilon_{n_{m_n}}\right\}.$$

Summarizing the above estimates, we have shown that for $k \ge k(\delta)$,

$$\left|\frac{1}{x}\frac{\partial\psi_{\varepsilon_k}}{\partial n} - g\right|_{H^{\frac{1}{2}}(\partial\Omega)} < \delta\,,$$

which proves the asserted strong convergence. Since the limit point is uniquely determined, this convergence holds generally for $\varepsilon \searrow 0$ and not only for $\{\varepsilon_n\}$. This concludes the proof of the assertion. □

Remark. In practical applications (see Blum [1989]) the cost functional in (5.2.48) may be replaced by the simpler least squares expression

$$J_\varepsilon(u) = \frac{1}{2}\left|\frac{1}{x}\frac{\partial\psi_u}{\partial n} - g\right|^2_{L^2(\partial\Omega)} + \frac{\varepsilon}{2}\,|u|^2_{L^2(\Gamma)}. \qquad (5.2.48)'$$

Instead of (5.2.49), we then obtain that $\frac{1}{x}\frac{\partial\psi_\varepsilon}{\partial n} \to g$ strongly in $L^2(\partial\Omega)$.

By a similar technique as above, namely applying the unique solvability of the Cauchy problem to only a "small" subset Γ_δ of Γ, approximate controllability results involving sign constraints on the control may also be established for general elliptic operators, different boundary conditions, etc. We have the following result (which can be extended to dimensions of space $d \geq 3$):

Proposition 5.2.14 *Let $\Omega \subset \mathbf{R}^2$ and $\Gamma \subset \Omega$ be given as above, and suppose that A is a general second-order elliptic operator of the type (1.2.6) on Ω. Moreover, let $h \in L^2(\Omega_0)$ be fixed, and let $\delta > 0$ be given. Then there exist a smooth set $\Gamma_\delta \subset \Gamma$ with* $\mathrm{meas}(\Gamma_\delta) < \delta$ *and a sequence $\{u_n\} \subset H^{\frac{3}{2}}(\Gamma)$ such that*

$$u_n \leq 0 \quad on\ \Gamma \setminus \Gamma_\delta, \qquad (5.2.56)$$

$$\psi_{u_n} \to 0 \quad strongly\ in\ L^2(\partial\Omega), \qquad (5.2.57)$$

where $\psi_{u_n} \in H^2(\Omega_0)$ is the unique solution to

$$A\psi_{u_n} = h \quad in\ \Omega_0, \qquad (5.2.58)$$

$$\psi_{u_n} = u_n \quad on\ \Gamma, \qquad \frac{\partial\psi_{u_n}}{\partial n} = 0 \quad on\ \partial\Omega. \qquad (5.2.59)$$

Remark. Notice that the inequality (5.2.56) is meaningful, since $u_n \in C(\Gamma)$ because we also have $\psi_{u_n} \in C(\overline{\Omega}_0)$. In the case that $\Omega \subset \mathbf{R}^d$ with $d \geq 3$, this is guaranteed if we take $\{u_n\} \subset H^{m-\frac{1}{2}}(\Gamma)$ with $m > \frac{d}{2}$. A special boundary controllability result for free boundary problems has been reported above in §5.1.4.

We now turn our attention to the application of controllability results in shape optimization problems. To this end, let E and Ω be given bounded smooth domains in $\mathbf{R}^d$ with $\overline{E} \subset \Omega$, and let D denote an unknown domain satisfying $E \subset D \subset \Omega$. Moreover, let $h \in H^{m-2}(\Omega)$ and $y_d \in L^2(\Omega)$ be given, where $m > \frac{d}{2}$. We then study the following model problem (cf. Pironneau [1984], Chapters III and IV):

$$\operatorname*{Min}_{E\subset D\subset\Omega} \left\{ J(D) = \int_E |y_D(x) - y_d(x)|^2 \, dx \right\}, \tag{5.2.60}$$

subject to

$$-\Delta y_D = h \quad \text{in } D, \tag{5.2.61}$$

$$y_D = 0 \quad \text{on } \partial D. \tag{5.2.62}$$

The (unique) solution $y_D \in H_0^1(D)$ to (5.2.61), (5.2.62) has to be understood in the weak sense (cf. Appendix 2), since no regularity assumptions are imposed on ∂D.

We associate with the optimal design problem (5.2.60)–(5.2.62) the following control problem with constraints:

$$\operatorname*{Min}_{u\in U_{ad}} \left\{ \tilde{J}(u) = \int_E |y_u(x) - y_d(x)|^2 \, dx \right\}, \tag{5.2.63}$$

subject to

$$-\Delta y_u = h \quad \text{in } \Omega, \tag{5.2.64}$$

$$y_u = u \quad \text{on } \partial\Omega, \tag{5.2.65}$$

$$U_{ad} = \left\{ u \in H^{m-\frac{1}{2}}(\partial\Omega) \ : \ u \le 0 \text{ on } \partial\Omega \right\}, \tag{5.2.66}$$

$$y_u \ge 0 \quad \text{in } \overline{E}. \tag{5.2.67}$$

Notice that $y_u \in H^m(\Omega)$, and since $m > \frac{d}{2}$, it follows from the embedding result of Theorem A2.2(iv) in Appendix 2 that $y_u \in C(\overline{\Omega})$, so that (5.2.66) and (5.2.67) are meaningful pointwise.

Theorem 5.2.15 *If $h \ge 0$, then the problem (5.2.63)–(5.2.67) is embedded in the problem (5.2.60)–(5.2.62).*

Proof. For every $u \in H^{m-\frac{1}{2}}(\partial\Omega)$, we define

$$\tilde{D}_u = \operatorname{int}\left(\{x \in \Omega : \ y_u(x) \ge 0\}\right),$$

which, thanks to (5.2.67), is an open set such that $E \subset \tilde{D}_u \subset \Omega$. We define $D_u \subset \Omega$ to be the connected component of $\tilde{D}_u$ containing E. Then $y_u|_{D_u}$ is the unique solution to the Dirichlet problem (5.2.61), (5.2.62) in D_u.

Through the correspondence $u \mapsto D_u$, we assign to each control $u \in H^{m-\frac{1}{2}}(\partial\Omega)$ satisfying (5.2.66), (5.2.67) a domain $D_u \subset \Omega$ such that $J(D_u) = \tilde{J}(u)$. Observe that no regularity can be guaranteed for D_u in general; however, the existence of a weak solution y_{D_u} for (5.2.61), (5.2.62) with $D = D_u$ is ensured. Since obviously $y_{D_u} = y_u|_{D_u}$, it is in fact a strong solution.

The correspondence $u \mapsto D_u$ is not empty, since the weak maximum principle implies that we may take $u \equiv 0$ and $D_u = \Omega$. It is also injective, since if u_1 and u_2 produce the same subdomain $\widetilde{D}$ by the above construction, then $y_1 - y_2 \equiv 0$ in $\widetilde{D}$, whence, in view of the analyticity of $y_1 - y_2$ in Ω, $y_1 - y_2 \equiv 0$ in Ω. Consequently, $u_1 = u_2$ on $\partial\Omega$. This ends the proof. □

For the converse statement we shall need the following approximate constrained controllability-type hypothesis (compare with Proposition 5.2.14):

(H) Let Ω be a smooth domain and $h \geq 0$ in Ω. For any subdomain $D \subset \Omega$ there is a sequence $\{u_n\} \subset H^{m-\frac{1}{2}}(\partial\Omega)$ with $u_n \leq 0$ on $\partial\Omega$ such that the corresponding solutions y_{u_n} to (5.2.64), (5.2.65) satisfy

$$y_{u_n}|_D \to y_D \quad \text{strongly in } L^2(D), \tag{5.2.68}$$

where y_D is the unique weak solution to (5.2.61), (5.2.62).

Example 5.2.16 To fix ideas, we first consider the one-dimensional case, taking $\Omega =]0,1[$. Let $a, b \in]0,1[$ with $a < b$ be arbitrary, and let h_1 be negative in $]0,1[$. Then the function

$$y(x) = \int_a^x (x-t)h_1(t)\,dt + \frac{x-a}{a-b}\int_a^b (b-t)h_1(t)\,dt, \quad x \in [0,1],$$

is the unique solution to the boundary value problem

$$y''(x) = h_1(x), \quad \text{in }]0,1[, \quad y(a) = y(b) = 0,$$

and we obviously have

$$\begin{aligned} y(0) &= \int_0^a t h_1(t)\,dt + \frac{a}{b-a}\int_a^b (b-t)h_1(t)\,dt \leq 0, \\ y(1) &= (1-b)\int_a^b h_1(t)\left(1 - \frac{b-t}{b-a}\right)dt + \int_b^1 (1-t)h_1(t)\,dt \leq 0. \end{aligned}$$

Thus, in the one-dimensional case (H) is satisfied as an exact controllability property, which is a consequence of the fact that $y''(x) = h_1(x) \leq 0$ on $]0,1[$, that is, of the concavity of y.

Next, we study a two-dimensional case, taking $\Omega = B\left(1, \frac{3}{2}\right)$, the disk of radius $r = \frac{3}{2}$ centered at $(1,0)$ (see Figure 2.5). For $(x_1, x_2) \in \Omega$, let

$$y(x_1,x_2) = \begin{cases} \frac{1}{4}x_1^2 - x_2^2, & \text{if } x_1 \le 1, \\ \frac{1}{4}x_1^2 - x_2^2 - (x_1-1)^4, & \text{if } x_1 \ge 1, \end{cases}$$

$$h(x_1,x_2) = \begin{cases} \frac{3}{2}, & \text{if } x_1 \le 1, \\ \frac{3}{2} + 12(x_1-1)^2, & \text{if } x_1 \ge 1. \end{cases}$$

We notice that $h \ge 0$ on Ω and that y solves (5.2.64). Moreover, if $D = \{(x_1,x_2) \in \Omega : x_1 > 0,\ y(x_1,x_2) > 0\}$, then $y = 0$ on ∂D. Since $u = y|_{\partial\Omega}$ satisfies $u > 0$ between the lines $\{x_2 = \pm\frac{1}{2}x\}$ for $x_1 < 0$, we conclude that (H) cannot be valid as an exact controllability property. This is due to the fact that in higher dimensions the positivity of Δy does not ensure the convexity of y. In fact, y is a saddle function for $x_1 \le 1$. However, there is certain numerical evidence (cf. Neittaanmäki and Tiba [1995], Example 5.1) that assumption (H) may yet be true.

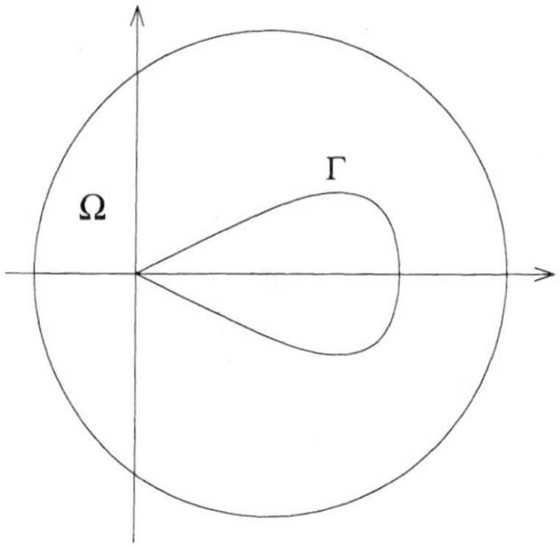

Figure 2.5.

Example 5.2.17 In the work of Göpfert [1971], a two-dimensional counterexample for approximate controllability properties for elliptic equations under sign constraints has been given. While this example neither contradicts hypothesis (H) nor Proposition 5.2.14 directly, it is worth mentioning in order to put into evidence the inherent difficulty of the problem.

In the unit disk in the plane, parametrized by the polar coordinates (r, φ), $r \in [0,1]$, $\varphi \in [0, 2\pi]$, we consider the unique solution y_u to Laplace's equation with boundary conditions for $r = 1$ given by

$$y_u(1,\varphi) = \begin{cases} u(\varphi) \geq 0, & \text{if } -\frac{\pi}{6} \leq \varphi \leq \frac{\pi}{6}, \\ 0, & \text{otherwise.} \end{cases}$$

The question then is whether arbitrary positive regular mappings defined on the curve $\left\{(r,\varphi) : r = \frac{1}{2},\ -\frac{\pi}{6} \leq \varphi \leq \frac{\pi}{6}\right\}$ can on $[-\frac{\pi}{6}, \frac{\pi}{6}]$ be approximated in the uniform topology by functions $y_u(\frac{1}{2}, \cdot)$ generated by appropriately chosen, positive, $u = u(\varphi)$.

By the well-known Poisson formula (cf. Smirnov [1955, p. 525], Protter and Weinberger [1967, p. 107]), we have

$$y_u(r,\varphi) = \frac{1}{2\pi} \int_{-\pi}^{\pi} u(t) \frac{1 - r^2}{1 - 2r\cos(t-\varphi) + r^2}\, dt,$$

and therefore

$$y_u(\tfrac{1}{2},\varphi) = \frac{1}{2\pi} \int_{-\frac{\pi}{6}}^{\frac{\pi}{6}} \frac{\frac{3}{4}u(t)}{\frac{5}{4} - \cos(t-\varphi)}\, dt.$$

Clearly, $(t-\varphi) \in [-\frac{\pi}{3}, \frac{\pi}{3}]$, and thus $0 < \frac{1}{4} \leq \frac{5}{4} - \cos(t-\varphi) \leq \frac{3}{4}$. Consequently, since $u \geq 0$, we can infer that

$$\frac{1}{2\pi} \int_{-\frac{\pi}{6}}^{\frac{\pi}{6}} u(t)\, dt \leq y(\tfrac{1}{2},\varphi) \leq \frac{3}{2\pi} \int_{-\frac{\pi}{6}}^{\frac{\pi}{6}} u(t)\, dt,$$

whence

$$0 \leq d = \sup_{\varphi \in [-\frac{\pi}{6},\frac{\pi}{6}]} y(\tfrac{1}{2},\varphi) - \inf_{\varphi \in [-\frac{\pi}{6},\frac{\pi}{6}]} y(\tfrac{1}{2},\varphi) \leq \frac{1}{\pi} \int_{-\frac{\pi}{6}}^{\frac{\pi}{6}} u(t)\, dt.$$

Consider now $a(\varphi) \geq 0$ satisfying

$$\sup_{\varphi \in [-\frac{\pi}{6},\frac{\pi}{6}]} a(\varphi) - \inf_{\varphi \in [-\frac{\pi}{6},\frac{\pi}{6}]} a(\varphi) = d' > 0, \qquad \inf_{\varphi \in [-\frac{\pi}{6},\frac{\pi}{6}]} a(\varphi) = \frac{1}{4} d'.$$

Assuming that the uniform approximation of $a(\cdot)$ via $y_u(\frac{1}{2}, \cdot)$ is possible, we may choose $u \geq 0$ such that

$$\left| a(\varphi) - y_u(\tfrac{1}{2},\varphi) \right| \leq \varepsilon < \frac{d'}{16} \quad \forall \varphi \in [-\tfrac{\pi}{6}, \tfrac{\pi}{6}]. \tag{5.2.69}$$

Then,

$$|d' - d| \leq \left| \sup_{\varphi \in [-\frac{\pi}{6},\frac{\pi}{6}]} a(\varphi) - \sup_{\varphi \in [-\frac{\pi}{6},\frac{\pi}{6}]} y_u(\tfrac{1}{2},\varphi) \right|$$
$$+ \left| \inf_{\varphi \in [-\frac{\pi}{6},\frac{\pi}{6}]} a(\varphi) - \inf_{\varphi \in [-\frac{\pi}{6},\frac{\pi}{6}]} y_u(\tfrac{1}{2},\varphi) \right| \leq 2\,\varepsilon,$$

according to (5.2.69). Using the definitions of d and d', and invoking the fact that $y_u(\frac{1}{2}, \varphi)$ should be positive by the approximation property (5.2.69), we obtain that

$$\frac{d'}{2} - \varepsilon \le \frac{d}{2} \le y_u(\tfrac{1}{2}, \varphi) \le d \le d' + 2\,\varepsilon.$$

We thus arrive at the following contradiction:

$$\inf_{\varphi \in [-\frac{\pi}{6}, \frac{\pi}{6}]} y_u(\tfrac{1}{2}, \varphi) - \inf_{\varphi \in [-\frac{\pi}{6}, \frac{\pi}{6}]} a(\varphi) \ge -\frac{d'}{4} + \frac{d'}{2} - \varepsilon = \frac{d'}{4} - \varepsilon > 3\,\varepsilon.$$

We establish now a partial converse to Theorem 5.2.15.

Theorem 5.2.18 *Suppose that (H) is fulfilled, and let D with $E \subset D \subset \Omega$ be any fixed domain. Then for every $\varepsilon > 0$ there exists a control $u_{\varepsilon,D} \in H^{m-\frac{1}{2}}(\partial\Omega)$ that is admissible for (5.2.63)–(5.2.67) and satisfies*

$$|\tilde{J}(u_{\varepsilon,D}) - J(D)| = |J(D_{u_{\varepsilon,D}}) - J(D)| < \varepsilon. \tag{5.2.70}$$

Proof. We first consider the case $\overline{E} \subset D$. Assume that h is so smooth that the weak solution to the Poisson equation has C^2 interior regularity, which can be achieved by a smoothing of h, if necessary. Similarly, by possibly adding some constant $\lambda > 0$, we may suppose that $h(x) > 0$ a.e. in Ω. Then it follows from the strong maximum principle that the solution $y_D \in C^2(D)$ to (5.2.61), (5.2.62) satisfies $y_D(x) > 0$ for any $x \in D$. Hence there is some constant $\hat{c} > 0$ such that $y_D(x) \ge \hat{c} > 0$ in $\overline{E}$.

By virtue of hypothesis (H), we may find a sequence $\{u_n\} \subset H^{m-\frac{1}{2}}(\partial\Omega)$ with $u_n \le 0$ on $\partial\Omega$ such that the corresponding solutions $y_n = y_{u_n}$ of (5.2.64), (5.2.65) satisfy (5.2.68), that is, we have $y_n - y_D \to 0$ strongly in $L^2(D)$. Since $y_n - y_D$ is harmonic in D for any $n \in \mathbf{N}$, we can conclude from the solid mean property of harmonic functions that $y_n - y_D \to 0$ uniformly in $\overline{E}$.

We then obtain that there is some $n_0 \in \mathbf{N}$ such that $y_n(x) \ge \frac{\hat{c}}{2} > 0$ for all $x \in \overline{E}$ whenever $n \ge n_0$. Therefore, the pair $[y_n, u_n]$ is admissible for the problem (5.2.63)–(5.2.67) if $n \ge n_0$. Moreover, we obviously have $\tilde{J}(u_n) \to J(D)$ and $\tilde{J}(u_n) = J(D_{u_n})$. Thus, choosing $n \ge n_0$ so large that $|\tilde{J}(u_n) - J(D)| < \varepsilon$, we obtain (5.2.70) with the choice $u_{\varepsilon,D} = u_n$. This ends the proof for the case $\overline{E} \subset D$.

Suppose now that $\overline{E} \not\subset D$. We construct in two steps a perturbation of D that satisfies this condition and yields a sufficiently good approximation of the solution to the Dirichlet problem (5.2.61), (5.2.62).

We begin our construction by choosing a sequence $\{D_k\}_{k\in\mathbf{N}}$ of open bounded sets satisfying $E \subset D_k \subset D_{k+1} \subset D$, $k \in \mathbf{N}$, as well as $D_k \to D$ with respect to the Hausdorff–Pompeiu complementary distance. That is (cf. Appendix 3), $d_H(\overline{\Omega} \setminus D_k, \overline{\Omega} \setminus D) \to 0$ as $k \to \infty$. Since E is assumed to be smooth, the sets D_k may be chosen smooth as well. We now prove an auxiliary result.

Lemma 5.2.19 *Suppose that the sequence $\{D_k\}_{k\in\mathbf{N}}$ satisfies the conditions stated above, and let $\tilde{y}_k$ denote the extension by zero to D of the corresponding solutions $y_k = y_{D_k}$, $k \in \mathbf{N}$, to the problem (5.2.61), (5.2.62). Then $\tilde{y}_k \to y_D$ weakly in $H_0^1(D)$.*

Proof. Obviously we have, for any $k \in \mathbf{N}$ and any $\delta > 0$,

$$\begin{aligned}\int_D |\nabla \tilde{y}_k|^2 \, dx &= \int_{D_k} |\nabla y_k|^2 \, dx = \int_{D_k} h \, y_k \, dx = \int_D h \, \tilde{y}_k \, dx \\ &\leq \frac{\delta}{2} \int_D |\tilde{y}_k|^2 \, dx + \frac{1}{2\delta} \int_D h^2 \, dx.\end{aligned}$$

Choosing $\delta > 0$ small enough, we infer from Poincaré's lemma that $\{\tilde{y}_k\}_{k\in\mathbf{N}}$ is bounded in $H_0^1(D)$. Hence, there is a subsequence, again indexed by k, such that $\tilde{y}_k \to \tilde{y}$ weakly in $H_0^1(D)$. Passage to the limit as $k \to \infty$ in the relation

$$\int_D \nabla \tilde{y}_k \cdot \nabla \varphi \, dx = \int_{D_k} \nabla y_k \cdot \nabla \varphi \, dx = \int_{D_k} h \, \varphi \, dx = \int_D h \, \varphi \, dx,$$

for all $\varphi \in C_0^\infty(D)$ with $\operatorname{supp} \varphi \subset D_K$, then shows that $\tilde{y} = y_D$. The uniqueness of the limit point entails that the convergence holds for the entire sequence and not only for the selected subsequence. □

Proof of Theorem 5.2.18 (continued). From Lemma 5.2.19, we have $J(D_k) \to J(D)$ as $k \to \infty$. Therefore, there is some $k_0 \in \mathbf{N}$ such that

$$|J(D_{k_0}) - J(D)| < \frac{\varepsilon}{2}.$$

We now choose a sequence $\{\widehat{D}_m\}_{m\in\mathbf{N}}$ with $\overline{E} \subset \widehat{D}_m \subset \Omega$, $m \in \mathbf{N}$, such that

$$d_H(\overline{\Omega} \setminus \widehat{D}_m, \overline{\Omega} \setminus D_{k_0}) \to 0 \quad \text{as } m \to \infty$$

and $\chi_{\widehat{D}_m} \to \chi_{D_{k_0}}$ pointwise a.e. in Ω for the associated characteristic functions. See (5.3.29) for such a construction. Since ∂D_{k_0} is smooth, we may argue as in the proof of Theorem 2.3.12 in Chapter 2 to arrive at the following conclusion: if we put $\widehat{y}_m = y_{\widehat{D}_m}$, and if we denote by $\widetilde{y}_m$ its extension by zero to Ω, then we have

$$\widetilde{y}_m \to y_{D_{k_0}} \text{ weakly in } H_0^1(\Omega), \qquad \widehat{y}_m|_E \to y_{D_{k_0}}|_E \text{ strongly in } L^2(E).$$

See Lemma 2.3.3 and its proof as well.

Consequently, $J(\widehat{D}_m) \to J(D_{k_0})$ as $m \to \infty$, and we can fix some $m_0 \in \mathbf{N}$ such that $|\,J(\widehat{D}_{m_0}) - J(D_{k_0})| < \frac{\varepsilon}{2}$. Then $|J(\widehat{D}_{m_0}) - J(D)| < \varepsilon$, and we may employ the argument from the first part of the proof for $\widehat{D}_{m_0}$. This finishes the proof of the assertion. □

Remark. The above method extends to general cost functionals that are continuous with respect to the weak topology of H^1. For instance, integrals defined on ∂E may be taken into consideration. If $\overline{E} \subset D$, cost functionals that are continuous with respect to the strong topology of H^1 may also be taken into account. This can be seen from the first part of the proof of Theorem 5.2.18 and the convergence properties of harmonic functions.

Remark. Results of the above type were discussed first in Tiba [1990b] and further developed in Neittaanmäki and Tiba [1995]. In the work of Haslinger, Hoffmann, and Kocvara [1993], related ideas have been used in a two-dimensional setting, preserving the explicit action of the geometric unknowns. A survey of results based on controllability and fictitious domain approaches is due to Hoffmann and Tiba [1995]. Recent progress in boundary control approaches to domain embedding techniques, including the possibility of using finite-dimensional controls and also some numerical applications, has been reported in the work of Badea and Daripa [2001].

Remark. The method presented in this paragraph has several advantages: it is a fixed domain approach, the geometric parameters are replaced by control functions, the differential operator is not modified, and the finite element mesh and the stiffness matrices remain unchanged during all iteration steps. In Theorem 5.2.18, we have also underlined the convex character of the boundary control problem (5.2.63)–(5.2.67). Moreover, these ideas may be applied in a large setting of problems, including variational inequalities (cf. §5.1.4).

Example 5.2.20 Let $\Omega = B(1, \frac{3}{2})$ be as in the second part of Example 5.2.16 and $E =]-\frac{3}{10}, \frac{3}{10}[\times]-\frac{7}{10}, \frac{7}{10}[$. We consider the boundary control problem

$$\operatorname*{Min}_{u \in U_{ad}} \left\{ J(u) = \frac{1}{2} |y - y_d|^2_{H^1(E)} + \frac{1}{2\lambda} \int_E y_-^2(x)\, dx \right\},$$

with some $\lambda > 0$, subject to

$$\begin{aligned} -\Delta y_u &= h \quad \text{in } \Omega, \qquad y_u = u \quad \text{on } \partial\Omega, \\ U_{ad} &= \left\{ u \in H^{\frac{3}{2}}(\partial\Omega) :\ u \le 0 \text{ on } \partial\Omega \right\}, \end{aligned}$$

which is a variant of the problem (5.2.63)–(5.2.67), including a penalization of the state constraint in the cost. We have chosen $\lambda = 10^{-3}$, $y_d = \frac{3}{2} - 4x_1^2 - x_2^2$, and $h \equiv 10$.

The finite element discretization was performed using piecewise linear elements with 661 interior nodes and 120 boundary nodes (for the control variables $u_i = u(x_i)$). The control problem has been formulated as a mathematical programming method and solved by the sequential quadratic programming method (subroutine E04VDF of the NAG library). See Example 5.2.9. The state problem has been solved via Cholesky decomposition. As usual, the adjoint state technique has been employed to compute the gradient of the cost function.

The initial guess of the control was $u_i^0 = -4.5$ for $i = \overline{1,120}$ (in all the nodes on the boundary). This choice yields the value 3.35 for the cost. After 12 iterations, the value of the cost was reduced to $2 \cdot 10^{-4}$. The obtained control and the corresponding state are depicted in Figures 2.6 and 2.7 below. The domain lying inside the curve B is the solution of the shape optimization problem.

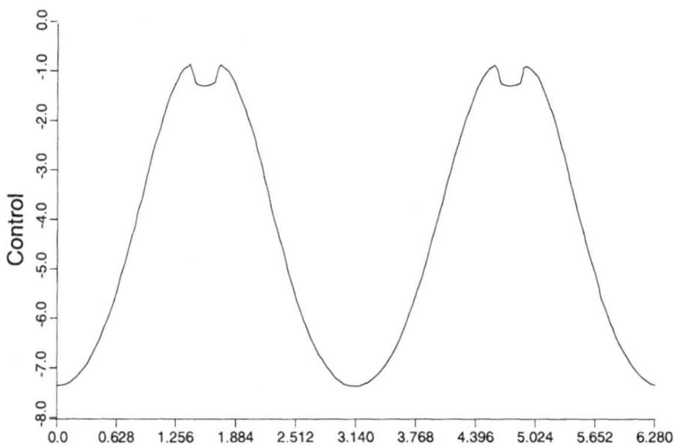

Figure 2.6. Obtained control in Example 5.2.20.

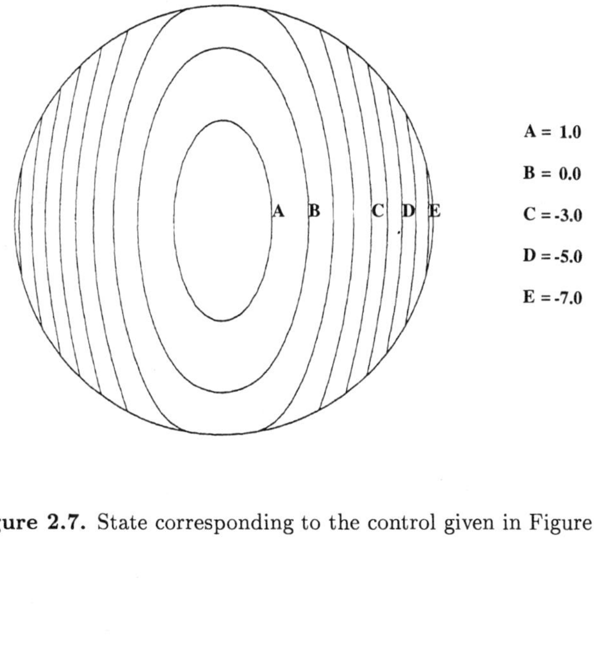

Figure 2.7. State corresponding to the control given in Figure 2.6.

The same problem has also been solved with the conventional *moving mesh technique*, where the radial coordinates of the boundary nodes were chosen as the control variables. The initial guess was a unit disk. The moving mesh approach leads to essentially the same optimal design, with fewer sequential quadratic programming iterations than the boundary control approach. But the total computational effort is larger, since the finite element mesh has to be updated at the beginning of each iteration step. Since the coefficients in the moving grid method depend on the control points on the boundary, one cannot use the same matrix (and its factorization) in all iteration steps as in the case of the boundary control approach.

5.2.3.2 Distributed Controls

In the case of distributed control problems, exact controllability properties can be derived. This has already been exploited in §5.1.3 for a nonlinear problem. In this paragraph, we will derive supplementary results for linear equations. We first consider the problem

$$-\Delta y = \chi_0 \, u \quad \text{in } \Omega, \qquad y = 0 \quad \text{on } \partial\Omega, \tag{5.2.71}$$

where Ω is a bounded smooth domain in $\mathbf{R}^d$ and χ_0 denotes the characteristic function of a smooth open set $\Omega_0 \subset \Omega$ satisfying $\partial\Omega \subset \overline{\Omega}_0$; that is, Ω_0 is a (relative) neighborhood of $\partial\Omega$.

Theorem 5.2.21 *For every $v \in H^{\frac{1}{2}}(\partial\Omega)$ there is some $u = u(v) \in L^2(\Omega)$ such that the corresponding (unique) solution $y = y_u$ to (5.2.71) belongs to $H^2(\Omega) \cap H_0^1(\Omega)$ and satisfies*

$$\frac{\partial y_u}{\partial n} = v \quad \textit{on } \partial\Omega. \tag{5.2.72}$$

Proof. We define the "adjoint" system

$$-\Delta z = 0 \quad \text{in } \Omega, \qquad z = \varphi \quad \text{on } \partial\Omega, \tag{5.2.73}$$

which for any $\varphi \in H^{-\frac{1}{2}}(\partial\Omega)$ has a unique transposition solution $z = z_\varphi \in L^2(\Omega)$; that is, we have (cf. Appendix 2)

$$\int_\Omega z_\varphi(x)\, \Delta w(x)\, dx = \left(\varphi, \frac{\partial w}{\partial n}\right)_{H^{-\frac{1}{2}}(\partial\Omega)\times H^{\frac{1}{2}}(\partial\Omega)} \qquad \forall\, w \in H^2(\Omega) \cap H_0^1(\Omega).$$

Taking $u = z_\varphi$ in (5.2.71), and putting $w = y_u = y_{z_\varphi}$, we find that

$$\int_{\Omega_0} z_\varphi^2(x)\,dx = -\int_\Omega z_\varphi(x)\,\Delta w(x)\,dx = -\left(z_\varphi, \frac{\partial w}{\partial n}\right)_{H^{-\frac{1}{2}}(\partial\Omega)\times H^{\frac{1}{2}}(\partial\Omega)}$$

$$= -\left(\varphi, \frac{\partial w}{\partial n}\right)_{H^{-\frac{1}{2}}(\partial\Omega)\times H^{\frac{1}{2}}(\partial\Omega)} \qquad \forall\,\varphi \in H^{-\frac{1}{2}}(\partial\Omega).$$

Let the linear and bounded operator $\Lambda : H^{-\frac{1}{2}}(\partial\Omega) \to H^{\frac{1}{2}}(\partial\Omega)$ be defined by $\Lambda\varphi = -\frac{\partial y_{z_\varphi}}{\partial n}$. It then follows that

$$(\varphi, \Lambda\varphi)_{H^{-\frac{1}{2}}(\partial\Omega)\times H^{\frac{1}{2}}(\partial\Omega)} = \int_{\Omega_0} z_\varphi^2(x)\,dx. \tag{5.2.74}$$

Now observe that we may consider in Ω_0 a similar boundary value problem as (5.2.73), which this time defines an isomorphism between $H^{-\frac{1}{2}}(\partial\Omega_0)$ and $L^2(\Omega_0)$. Since $z_\varphi|_{\Omega_0}$ may be considered as a solution for it, we have, in particular,

$$\hat{c}\,|z_\varphi|_{L^2(\Omega_0)} \geq |\varphi|_{H^{-\frac{1}{2}}(\partial\Omega)}, \tag{5.2.75}$$

with some $\hat{c} > 0$ that does not depend on φ. Then we can infer from (5.2.74) and (5.2.75) that Λ is a coercive and surjective operator. Consequently, for every $v \in H^{\frac{1}{2}}(\partial\Omega)$ there is some $\varphi \in H^{-\frac{1}{2}}(\partial\Omega)$ (and thus some $u = z = z_\varphi \in L^2(\Omega)$) such that

$$\frac{\partial y_u}{\partial n} = \frac{\partial y_{z_\varphi}}{\partial n} = v.$$

This ends the proof of the assertion. □

Remark. The method used in the above proof is called the *Hilbert uniqueness method (HUM)* and was introduced by Lions [1988]. It has been successfully applied in the study of exact controllability properties of linear and nonlinear evolution equations (mainly in the hyperbolic case). Notice that in the case $\Omega_0 = \Omega$ the trace theorem can directly be applied. The result of Theorem 5.2.21 completes that of Theorem 5.2.12, where boundary controls are acting.

We now turn our attention to another optimal shape design problem. To fix things, suppose that smooth bounded domains E, Ω with $\overline{E} \subset \Omega \subset \mathbf{R}^d$ and functions $y_d, h \in L^2(\Omega)$ are given. We aim to find a domain D of class $C^{1,1}$ with $\overline{E} \subset D \subset \Omega$ that solves the problem

$$\operatorname*{Min}_{\overline{E}\subset D\subset\Omega;\ D \text{ is of class } C^{1,1}} \left\{ J(D) = \int_E |y(x) - y_d(x)|^2\,dx \right\}, \tag{5.2.76}$$

subject to

$$-\sum_{i,j=1}^{d}\frac{\partial}{\partial x_i}\left(a_{ij}\frac{\partial y}{\partial x_j}\right)+a_0\,y=h \quad \text{in } D, \tag{5.2.77}$$

$$By=0 \quad \text{on } \partial D. \tag{5.2.78}$$

Here, $a_{ij}\in W^{1,\infty}(\Omega)$, $i,j=1,\ldots,d$, the coefficients matrix $A=(a_{ij})_{i,j=\overline{1,d}}$ satisfies the ellipticity condition (1.2.5), $a_0\in L^\infty(\Omega)$ is nonnegative, and B denotes a first-order linear boundary operator on the boundary. This formulation covers a variety of boundary conditions, namely Dirichlet ($B=id$), Neumann ($B=\frac{\partial}{\partial n_A}$), and Robin ($B=\frac{\partial}{\partial n_A}+\alpha\, id$, with some $\alpha>0$) conditions.

By the ellipticity of the matrix A, we may assume that the bilinear form defined by the left-hand side of (5.2.77), (5.2.78) is coercive on an appropriate subspace of $H^1(D)$ (depending on the boundary conditions) for any domain D of class $C^{1,1}$ satisfying $\overline{E}\subset D\subset\Omega$, so that (5.2.77), (5.2.78) has for any $h\in L^2(\Omega)$ a unique strong solution $y=y_h$ in an appropriate subspace of $H^2(D)$. As in §5.1.3, we associate with (5.2.76)–(5.2.78) a distributed control problem (with fixed $\eta>0$), namely

$$\operatorname*{Min}_{u\in L^2(\Omega)}\left\{J_\eta(u)=\int_E|y(x)-y_d(x)|^2\,dx+\eta\int_{E_y}|u(x)-h(x)|^2\,dx\right\}, \tag{5.2.79}$$

where $y=y_u$ is the unique strong solution to

$$-\sum_{i,j=1}^{d}\frac{\partial}{\partial x_i}\left(a_{ij}\frac{\partial y}{\partial x_j}\right)+a_0\,y=u \quad \text{in } \Omega, \tag{5.2.80}$$

$$By=0 \quad \text{on } \partial\Omega. \tag{5.2.81}$$

In the cost functional in (5.2.79) the set E_y is some subdomain of class $C^{1,1}$ (possibly of minimal area) such that $\overline{E}\subset E_y\subset\Omega$ and $By=0$ on ∂E_y. In light of (5.2.81), one such example is $E_y=\Omega$.

Unless additional compactness assumptions (see Chapter 2) are made, the existence of optimal solutions of either (5.2.76)–(5.2.78) or (5.2.79)–(5.2.81) cannot be guaranteed. However, we have the following general result that connects suboptimal solutions of the two problems.

Theorem 5.2.22 *For any domain D of class $C^{1,1}$ satisfying $\overline{E}\subset D\subset\overline{D}\subset\Omega$ and any $\eta\geq 0$ there is some $u_D\in L^2(\Omega)$ such that $J(D)=J_\eta(u_D)$. Moreover,*

$$\inf\left\{J_\eta(u):\,u\in L^2(\Omega)\right\}\leq\inf\left\{J(D):\,\overline{E}\subset D\subset\Omega;\ D \text{ is of class } C^{1,1}\right\}. \tag{5.2.82}$$

Conversely, whenever $\eta > 0$ *and* $\delta_\eta > 0$ *are given and* $[y_\eta, u_\eta]$ *(where* $y_\eta = y_{u_\eta}$ *) is a* δ_η*-optimal pair for (5.2.79), then there is some* $\varepsilon_\eta > 0$ *such that* E_{y_η} *is an* ε_η*-optimal domain for (5.2.76) and such that* $\varepsilon_\eta - \delta_\eta \to 0$ *as* $\eta \to \infty$.

Proof. Let $\eta \geq 0$, and let D be a domain of class $C^{1,1}$ with $\overline{E} \subset D \subset \Omega$. We first assume that $\overline{D} \subset \Omega$. We denote by y_D the corresponding unique strong solution to (5.2.77), (5.2.78), and for $v \in L^2(\Omega \setminus D)$ we define y_v as some strong solution to the problem

$$-\sum_{i,j=1}^{d} \frac{\partial}{\partial x_i}\left(a_{ij}\frac{\partial y_v}{\partial x_j}\right) + a_0\, y_v = v \quad \text{in } \Omega\setminus\overline{D}, \qquad B y_v = 0 \quad \text{on } \partial\Omega. \quad (5.2.83)$$

The trace theorem implies that there is some $\tilde{y} \in H^2(\Omega\setminus\overline{D})$ such that $\tilde{y} = y_D$ and $\frac{\partial \tilde{y}}{\partial \nu} = -\frac{\partial y_D}{\partial n}$ on ∂D. Here, ν and n denote the inward and outward, respectively, unit normal vectors to ∂D. Again by virtue of the trace theorem, we may assume that also $B\tilde{y} = 0$ on $\partial\Omega$.

Since $\tilde{y} \in H^2(\Omega\setminus\overline{D})$, we may choose $v \in L^2(\Omega\setminus D)$ using (5.2.83), by taking $y_v = \tilde{y}$. Furthermore, we define

$$u_D = \begin{cases} h & \text{in } \overline{D}, \\ v & \text{in } \Omega\setminus\overline{D}, \end{cases} \qquad \text{and} \qquad y_{u_D} = \begin{cases} y_D & \text{in } \overline{D}, \\ y_v & \text{in } \Omega\setminus\overline{D}. \end{cases}$$

Then y_{u_D} belongs to $H^2(\Omega)$ and solves (5.2.80), (5.2.81) with the right-hand side $u_D \in L^2(\Omega)$. Moreover, (5.2.78) yields that $By_{u_D} = 0$ on ∂D, so that we may choose $E_{y_{u_D}} \subset D$. Therefore, since $u_D = h$ in $E_{y_{u_D}}$,

$$\begin{aligned} J(D) = \int_E |y_D - y_d|^2\, dx &= \int_E |y_{u_D} - y_d|^2\, dx + \eta \int_{E_{y_{u_D}}} |u_D - h|^2\, dx = J_\eta(u_D) \\ &\geq \inf\left\{J_\eta(u) \,:\, u \in L^2(\Omega)\right\}. \end{aligned}$$

Next, we briefly indicate the argument for the general case $\overline{D} \not\subset \Omega$. We do not go into much detail since the different types of boundary operators B require a separate treatment, which would lead to a major detour without bringing new insights. The general idea is to choose a sequence $\{D_k\}_{k\in\mathbf{N}}$ of domains of class $C^{1,1}$ with $\overline{E} \subset D_k \subset \overline{D}_k \subset D$, $k \in \mathbf{N}$, that converge to D in the sense of the Hausdorff–Pompeiu complementary distance (cf. Appendix 3). See (5.3.29) for such a construction. Then it follows from the techniques introduced in Chapter 2 that the corresponding solutions y_{D_k} converge to y_D at least in a suitable local sense. This, in turn, suffices to guarantee that $J(D_k) \to J(D)$ as $k \to \infty$, and passage to the limit as $k \to \infty$ in the inequality

$$J(D_k) \geq \inf\left\{J_\eta(u) : \, u \in L^2(\Omega)\right\}$$

yields that (5.2.82) holds true.

Suppose now, conversely, that $\eta > 0$ and $\delta_\eta > 0$ are given and that $[y_\eta, u_\eta]$ is δ_η-optimal for (5.2.79). Then

$$J_\eta(u_\eta) \le \inf_{u\in L^2(\Omega)} J_\eta(u) + \delta_\eta \le \inf_D J(D) + \delta_\eta = \lambda_\eta .$$

Consequently,

$$|u_\eta - h|_{L^2(E_{u_\eta})} \le \lambda_\eta^{\frac{1}{2}}\, \eta^{-\frac{1}{2}} .$$

Now let $\tilde{y}_\eta$ denote the unique weak solution to

$$-\sum_{i,j=1}^{d} \frac{\partial}{\partial x_i}\left(a_{ij}\frac{\partial \tilde{y}_\eta}{\partial x_j}\right) + a_0\, \tilde{y}_\eta = h \quad \text{in } E_{y_\eta}, \qquad B\tilde{y}_\eta = 0 \quad \text{on } \partial E_{y_\eta} .$$

Then it is easily verified that there is some constant $\hat{c} > 0$, which is independent of η and δ_η, such that

$$|y_\eta - \tilde{y}_\eta|_{H^1(E_{y_\eta})} \le \hat{c}\, \lambda_\eta^{\frac{1}{2}}\, \eta^{-\frac{1}{2}} = \rho_\eta .$$

But then

$$\begin{aligned} J(E_{u_\eta}) &= |\tilde{y}_\eta - y_d|^2_{L^2(E)} \le \left(|\tilde{y}_\eta - y_\eta|_{L^2(E)} + |y_\eta - y_d|_{L^2(E)}\right)^2 \\ &\le \rho_\eta^2 + \lambda_\eta + 2\rho_\eta\sqrt{\lambda_\eta} = \inf_D J(D) + \varepsilon_\eta , \end{aligned}$$

where, obviously,

$$\varepsilon_\eta - \delta_\eta = \rho_\eta^2 + 2\rho_\eta\sqrt{\lambda_\eta} \to 0 \text{ as } \eta \to \infty .$$

This concludes the proof of the assertion. □

Remark. Since $\varepsilon_\eta - \delta_\eta \to 0$ as $\eta \to \infty$, a perturbed converse inequality to (5.2.82) is also valid. In applications, we fix $\eta > 0$ sufficiently large and determine some δ_η-solution $[y_\eta, u_\eta]$ to (5.2.79)–(5.2.81). Then the domain E_{y_η} will provide a suboptimal solution to (5.2.76)–(5.2.78).

Remark. The problem (5.2.79)–(5.2.81) is close to standard linear–quadratic optimal control problems, except for the presence of the unknown domain E_y. One difference with respect to the usual setting is that we need to determine E_y. If $d = 2$, this amounts to determining the curve defined by $By = 0$. In the case of the Dirichlet condition, this curve is just the level curve $\{y = 0\}$. In the case of the Neumann boundary condition $\frac{\partial y}{\partial n} = 0$ the curve is orthogonal to the level curves of y, while in the case of the Robin condition $\frac{\partial y}{\partial n} + \alpha\, y = 0$

the angle θ between the unknown curve and the level curves of y is given by the relation

$$\cos\theta(x) = \pm\frac{\alpha\, y(x)}{|\nabla y(x)|_{\mathbf{R}^2}} = \mp\frac{n(x)\cdot\nabla y(x)}{|\nabla y(x)|_{\mathbf{R}^2}}, \tag{5.2.84}$$

with the sign depending on the choice of the tangent direction to the curves.

In the following, we will discuss how the gradient of the mapping $y \mapsto E_y$ may be computed, which is necessary for gradient optimization algorithms. We first derive an abstract result along these lines and then give some theoretical examples and numerical applications.

To this end, let $PC(\overline{\Omega})$ denote the space of piecewise continuous and bounded functions on $\overline{\Omega}$ equipped with the $L^\infty(\Omega)$ norm, and suppose that $h, u \in PC(\overline{\Omega})$. We consider variations of the form $u + \lambda\mu$, where $\lambda \in \mathbf{R}$ and $\mu \in PC(\overline{\Omega})$.

Let $y = y_u$ and $y_\lambda = y_{u+\lambda\mu}$ be the corresponding solutions to (5.2.80), (5.2.81). Then $y_\lambda = y + \lambda\sigma$, where $\sigma = y_\mu$ is the unique solution to (5.2.80), (5.2.81) associated with μ. Furthermore, we assume that the corresponding domains E_y, E_{y_λ} of class $C^{1,1}$ are chosen in such a way that E_{y_λ} converges to E_y with respect to the complementary Hausdorff–Pompeiu distance, that is, we have (cf. Appendix 3)

$$\lim_{\lambda\to 0} d_H(\overline{\Omega}\setminus E_{y_\lambda}, \overline{\Omega}\setminus E_y) = 0.$$

This is a natural requirement without which differentiability cannot be expected.

We now choose functions $g_\lambda, g \in C^1(\overline{\Omega})$ such that $g_\lambda \to g$ strongly in $C^1(\overline{\Omega})$ and such that

$$g = 0 \quad \text{on } \partial E_y, \qquad g_\lambda = 0 \quad \text{on } \partial E_{y_\lambda}, \tag{5.2.85}$$

$$\frac{\partial g}{\partial n} \neq 0 \quad \text{on } \partial E_y. \tag{5.2.86}$$

As explained in the previous remark, the functions g, g_λ can be obtained directly from the (known) solutions y, y_λ. For instance, in arbitrary dimension, if $By = \frac{\partial y}{\partial n}$, then g may be constructed (up to a constant) as a solution of

$$\nabla g\cdot\nabla y = 0 \quad \text{in } \Omega.$$

Similar simple arguments can be employed for other types of boundary operators B. We make the following assumption:

(A) There is some $j \in C(\overline{\Omega})$ such that

$$\lim_{\lambda\to 0}\left|\frac{g_\lambda - g}{\lambda} - j\right|_{C(\overline{\Omega})} = 0. \tag{5.2.87}$$

Remark. The idea behind this formulation is that the boundaries ∂E_y and ∂E_{y_λ}, which are defined by the conditions $By = 0$ and $By_\lambda = 0$ (relating them to y and y_λ), are now determined as the level sets of g and g_λ, respectively. Notice that in structural optimization problems (compare §5.2.2.1 above) g (there denoted by p) appears explicitly as a control mapping and that the considered variations (in Theorem 5.2.8) satisfy (5.2.87) for $d = 2$.

Lemma 5.2.23 *Suppose that assumption (A) is fulfilled, and suppose that* $d = 2$. *Then*

$$\lim_{\lambda\to 0}\frac{1}{\lambda}\left[\int_{E_{y_\lambda}}|u(x)-h(x)|^2\,dx-\int_{E_y}|u(x)-h(x)|^2\,dx\right]$$
$$=-\int_{\partial E_y}\frac{|u(\sigma)-h(\sigma)|^2}{\frac{\partial g}{\partial n}(\sigma)}\,j(\sigma)\,d\sigma. \tag{5.2.88}$$

Proof. Since E_{y_λ} approximates E_y for $\lambda\to 0$ in the Hausdorff–Pompeiu sense, we may just determine the limit

$$I_1=\lim_{\lambda\to 0}\frac{1}{\lambda}\left[\int_{E_{y_\lambda}\cap\omega}|u(x)-h(x)|^2\,dx-\int_{E_y\cap\omega}|u(x)-h(x)|^2\,dx\right],$$

where ω is some neighborhood of an arbitrary point $M\in\partial E_y$. We define $\Gamma=\partial E_y\cap\omega$ and $\Gamma_\lambda=\partial E_{y_\lambda}\cap\omega$. If $|\lambda|$ is sufficiently small, we have $\Gamma_\lambda\neq\emptyset$.

Let us choose a new local system of coordinates with origin at M such that the x_1-axis is normal and the x_2-axis is tangent to ∂E_y. Thanks to (5.2.86), we have $\nabla g\neq 0$ on ∂E_y. The implicit function theorem shows the existence of a C^1 mapping α such that Γ is defined by the equation $x_1=\alpha(x_2)$. The curve Γ_λ can be expressed similarly: as $g_\lambda\to g$ strongly in $C^1(\overline{\Omega})$, we have $\nabla g_\lambda\neq 0$ in a neighborhood of M, for sufficiently small $|\lambda|\neq 0$. Thus, there is a C^1 mapping α_λ such that Γ_λ is given by the equation $x_1=\alpha_\lambda(x_2)$; in addition, since $E_{y_\lambda}\to E_y$ in the Hausdorff–Pompeiu sense, we can claim that $\alpha_\lambda\to\alpha$ uniformly.

Choosing ω smaller if necessary, we can assume that $\omega=]a,b[\times]c,d[$ with some constants $a<b$, $c<d$, and that α and α_λ are defined in $[c,d]$ for $|\lambda|\le\lambda_0$ with some $\lambda_0>0$. In the new coordinates, we then have

$$\begin{aligned}I_1&=\lim_{\lambda\to 0}\frac{1}{\lambda}\int_c^d\int_{\alpha(x_2)}^{\alpha_\lambda(x_2)}|u-h|^2(x_1,x_2)\,dx_1\,dx_2\\&=\int_c^d|u-h|^2(\alpha(x_2),x_2)\cdot\lim_{\lambda\to 0}\frac{\alpha_\lambda(x_2)-\alpha(x_2)}{\lambda}\,dx_2,\end{aligned}$$

using the absolute continuity of the inner integral and the chain rule for any fixed $x_2\in[c,d]$. The existence of the limit under the integral sign, uniformly in $[c,d]$, follows from (5.2.87) and from the relation

$$g_\lambda(\alpha_\lambda(x_2), x_2) - g(\alpha(x_2), x_2) = 0, \quad x_2 \in [c, d].$$

Indeed, adding and subtracting $g_\lambda(\alpha(x_2), x_2)$ and dividing the resulting identity by λ, we can infer from assumption (A) and a direct calculation that

$$0 = j(\alpha(x_2), x_2) + \frac{\partial g}{\partial x_1}(\alpha(x_2), x_2) \cdot \lim_{\lambda \to 0} \frac{\alpha_\lambda(x_2) - \alpha(x_2)}{\lambda}.$$

Here, the properties that $g_\lambda \to g$ strongly in $C^1(\overline{\Omega})$ and $\alpha_\lambda \to \alpha$ strongly in $C[c, d]$ have been essentially used. Therefore,

$$I_1 = -\int_c^d (u - h)^2(\alpha(x_2), x_2) \, \frac{j(\alpha(x_2), x_2)}{g_{x_1}(\alpha(x_2), x_2)} \, \frac{\sqrt{1 + \alpha'(x_2)^2}}{\sqrt{1 + \frac{g_{x_2}(\alpha(x_2), x_2)^2}{g_{x_1}(\alpha(x_2), x_2)^2}}} \, dx_2, \tag{5.2.89}$$

where we have used the fact that, thanks to the implicit function theorem,

$$\alpha'(x_2) = -\frac{g_{x_2}(\alpha(x_2), x_2)}{g_{x_1}(\alpha(x_2), x_2)},$$

where $g_{x_1} \neq 0$ (due to (5.2.86)) and the g_{x_2} denote the partial derivatives of g.

By virtue of the continuity of g, and owing to the choice of ω, we may assume that g has a constant sign in $E_y \cap \omega$, say that $g > 0$ in $E_y \cap \omega$. Then, again by (5.2.86), $g_{x_1}(\alpha(x_2), x_2) < 0$, and a detailed computation of the square root yields that

$$\begin{aligned} I_1 &= \int_c^d (u - h)^2(\alpha(x_2), x_2) \, \frac{j(\alpha(x_2), x_2)}{|\nabla g(\alpha(x_2), x_2)|_{\mathbf{R}^2}} \, \sqrt{1 + \alpha'(x_2)^2} \, dx_2 \\ &= \int_{(\partial E_y) \cap \omega} \frac{(u(\sigma) - h(\sigma))^2}{|\nabla g(\sigma)|} \, j(\sigma) \, d\sigma \; = \; -\int_{(\partial E_y) \cap \omega} \frac{(u(\sigma) - h(\sigma)^2}{\frac{\partial g}{\partial n}(\sigma)} \, j(\sigma) \, d\sigma. \end{aligned}$$

In the case $g < 0$ in $E_y \cap \omega$ the computation is similar, leading to the same result. This ends the proof of the assertion. □

Theorem 5.2.24 *Suppose that assumption (A) is fulfilled and that* $d = 2$. *Then* J_η *has a directional derivative at* $u \in PC(\overline{\Omega})$ *in the direction* $\mu \in PC(\overline{\Omega})$, *which is given by*

$$\begin{aligned} \nabla J_\eta(u)\mu &= 2 \int_E (y(x) - y_d(x)) \, y_\mu(x) \, dx + 2\,\eta \int_{E_y} (u(x) - h(x)) \, \mu(x) \, dx \\ &\quad -\eta \int_{\partial E_y} \frac{(u(\sigma) - h(\sigma))^2}{\frac{\partial g}{\partial n}(\sigma)} \, j(\sigma) \, d\sigma, \end{aligned} \tag{5.2.90}$$

where y_μ is the solution to (5.2.80), (5.2.81) associated with μ.

Proof. We have

$$\lim_{\lambda\to 0}\frac{1}{\lambda}\left(J_\eta(u+\lambda\mu)-J_\eta(u)\right)=\lim_{\lambda\to 0}\frac{1}{\lambda}\left[\int_E\left((y_\lambda-y_d)^2-(y-y_d)^2\right)dx\right.$$

$$\left.+\eta\int_{E_{y_\lambda}}(u+\lambda\mu-h)^2\,dx-\eta\int_{E_y}(u-h)^2\,dx\right]$$

$$=\lim_{\lambda\to 0}\left[2\int_E(y-y_d)\,y_\mu\,dx+\lambda\int_E y_\mu^2\,dx+2\,\eta\int_{E_{y_\lambda}}(u-h)\,\mu\,dx\right.$$

$$\left.+\lambda\eta\int_{E_{y_\lambda}}\mu^2\,dx+\frac{\eta}{\lambda}\int_{E_{y_\lambda}}(u-h)^2\,dx-\frac{\eta}{\lambda}\int_{E_y}(u-h)^2\,dx\right].$$

Now recall that E_{y_λ} converges to E_y with respect to the the Hausdorff–Pompeiu distance. Then it follows from Appendix 3 (cf. Proposition A3.6) for the associated characteristic functions that $\chi_{E_{y_\lambda}}\to\chi_{E_y}$ strongly in $L^1(\Omega)$. Hence,

$$\int_{E_{y_\lambda}}(u-h)\,\mu\,dx\to\int_{E_y}(u-h)\,\mu\,dx,$$

and the result follows from Lemma 5.2.23. □

Remark. It is possible to extend the above argument to higher dimensions.

Example 5.2.25 Let $\Omega\subset\mathbf{R}^2$ be a smooth domain, and assume that the state equation (5.2.80), (5.2.81) is of the form

$$-\Delta y+y=u\quad\text{in }\Omega,\qquad y=0\quad\text{on }\partial\Omega.\tag{5.2.91}$$

Then E_y is defined by the postulate $y=0$ on ∂E_y. We assume that $\nabla y\neq 0$ on ∂E_y. In view of Hopf's lemma, this condition is for instance fulfilled whenever u has a constant sign and differs from zero.

Next, we choose $g=y$, $g_\lambda=y_\lambda$, and $j=y_\mu$ (all defined previously). Since $u,\mu,h\in PC(\overline{\Omega})$, it follows from general elliptic regularity theory (cf. Appendix 2) that $y,y_\lambda\in W^{2,p}(\Omega)$ for any $p\geq 1$, and Theorem A2.2(iv) in Appendix 2 ensures that $y,y_\lambda\in C^1(\overline{\Omega})$, as required. Moreover, we obviously have $y_\lambda\to y$ strongly in $C^1(\overline{\Omega})$, and (5.2.85)–(5.2.87) are fulfilled.

In this example an explicit variant, in terms of an adjoint equation, of the formula (5.2.90) for the gradient of the cost functional can be derived.

Proposition 5.2.26 *For the state system (5.2.91), the gradient of the cost functional J_η at $u \in PC(\overline{\Omega})$ is given by*

$$\nabla J_\eta(u) = z + 2\,\eta(\,u - h)\,\chi_{E_y}, \tag{5.2.92}$$

where χ_{E_y} is the characteristic function of E_y and $z \in H_0^1(\Omega)$ is the solution to the adjoint system

$$\begin{aligned}\int_\Omega (\nabla z(x)\cdot\nabla w(x) + z(x)\,w(x))\,dx \;&=\; 2\int_E (y(x) - y_d(x))\,w(x)\,dx\\ -\,\eta\int_{\partial E_y} &\frac{(u(\sigma) - h(\sigma))^2}{\frac{\partial y}{\partial n}(\sigma)}\,w(\sigma)\,d\sigma \quad \forall\, w \in H_0^1(\Omega).\end{aligned} \tag{5.2.93}$$

Remark. Under regularity assumptions, the adjoint equation (5.2.93) may be formally interpreted as a transmission problem (compare with (2.3.70)–(2.3.73) in Chapter 2). Indeed, putting $z_1 = z|_{\Omega\setminus\overline{E}_y}$, $z_2 = z|_{E_y}$, we have

$$\begin{aligned}-\Delta z_1 + z_1 \;&=\; 0 \quad \text{in } \Omega\setminus\overline{E}_y,\\ -\Delta z_2 + z_2 \;&=\; 2\,\chi_E\,(y - y_d) \quad \text{in } E_y,\\ z_1 \;&=\; z_2 \quad \text{on } \partial E_y, \qquad z_2 = 0 \quad \text{on } \partial\Omega,\\ \frac{\partial z_1}{\partial n} + \frac{\partial z_2}{\partial \nu} \;&=\; -\frac{\eta\,(u-h)^2}{\frac{\partial y}{\partial n}} \quad \text{on } \partial E_y.\end{aligned}$$

Proof. Notice first that the Lax–Milgram lemma guarantees the unique existence of a solution $z \in H_0^1(\Omega)$ to (5.2.93). Now we obtain from (5.2.90), using (5.2.93), that

$$\begin{aligned}\nabla J_\eta(u)\mu \;&=\; 2\int_E (y - y_d)\,y_\mu\,dx + 2\,\eta\int_{E_y}(u-h)\,\mu\,dx - \eta\int_{\partial E_y}\frac{(u-h)^2}{\frac{\partial y}{\partial n}}\,y_\mu\,d\sigma\\ &=\; \int_\Omega (\nabla z\cdot\nabla y_\mu + z\,y_\mu)\,dx + 2\,\eta\int_{E_y}(u-h)\,\mu\,dx\\ &=\; \int_\Omega z\,\mu\,dx + 2\,\eta\int_\Omega \chi_{E_y}\,(u-h)\,\mu\,dx,\end{aligned}$$

since y_μ satisfies (5.2.91) with the right-hand side μ. □

Remark. The regularity of $u \in PC(\overline{\Omega})$ is preserved if a gradient-type optimization algorithm based on (5.2.92) is used. Here, we use the fact that z given by (5.2.93) satisfies $z \in H^{1+\varepsilon}(\Omega)$ for some $\varepsilon > 0$, according to Lions and Magenes [1968, Theorem 8.3, Chapter II], since the terms on the right-hand side of (5.2.93) define linear continuous functionals on $H_0^{1-\varepsilon}(\Omega)$. Since $d = 2$, this entails $z \in C(\overline{\Omega})$.

Remark. For ordinary differential equations, the assumption (A) can easily be checked for many types of boundary conditions. For example, let us consider the problem

$$-y'' + y = u \quad \text{in }]0,1[, \qquad y(0) = y'(1) + ay(1) = 0,$$

where a is some constant. If $E_y =]\alpha, \beta[$, and if we assume that $y'(\alpha) \neq 0$ and $y''(\beta) + ay'(\beta) \neq 0$, then we may choose $g = y$ around α and $g = y' + ay$ around β, and all regularity conditions are fulfilled. Also Proposition 5.2.26 applies in this setting.

Example 5.2.27 We give a numerical application of Proposition 5.2.26. To this end, fix $\Omega =]0,1[\times]0,1[$ and $E =]0.36, 0.64[\times]0.36, 0.64[$. The functions y_d and h are defined by

$$y_d(x_1, x_2) = \cos(c\,r), \qquad h(x_1, x_2) = (c^2 + 1)\cos(c\,r) + \frac{c}{r}\sin(c\,r),$$

where

$$c = \frac{25\,\pi}{12}, \qquad r(x_1, x_2) = \sqrt{(x_1 - 0.5)^2 + (x_2 - 0.5)^2}\,.$$

Then the domain enclosed by the boundary curve

$$\left\{(x_1, x_2) \in \mathbf{R}^2 : \ (x_1 - 0.5)^2 + (x_2 - 0.5)^2 = 0.24^2\right\}$$

is a solution to the original problem (5.2.76)–(5.2.78) with $y = y_d$. Furthermore, let $\eta = 10^4$.

The discretization was performed using piecewise linear finite elements. There were 289 nodes in the finite element mesh, which remained fixed in all iterations. The minimization of the cost function was done using the conjugate gradient method, where as initial guess for the control we made the very rough choice $u^{(0)} \equiv 1000$. The boundary curve defining E_y, given by the level line $\{y = 0\}$, and the domain E_y were approximated by the set that contains all discretization nodes inside of ∂E_y.

The discrete gradient has been computed by direct discretization in (5.2.92), (5.2.93). The numerical results are summarized in Table 2.2 and in Figures 2.8–2.12 below. Notice that in the first iteration $E_y = \Omega$, so that the boundary term vanishes. A good approximation of the desired domain was obtained already in iteration 3.

Remark. The use of distributed control problems in shape optimization has been discussed in Tiba [1992]. Further results and numerical examples can be found in Tiba and Männikkö [1995].

No. of iter.	$J_\eta(u)$	$\int_E (y - y_d)^2\,dx$	$\int_{E_y} (u - h)^2\,dx$
0	$0.86292 \cdot 10^{10}$	$0.43154 \cdot 10^{3}$	$0.86292 \cdot 10^{6}$
1	$0.23535 \cdot 10^{9}$	$0.18088 \cdot 10^{1}$	$0.23535 \cdot 10^{5}$
2	$0.88937 \cdot 10^{7}$	$0.18442 \cdot 10^{0}$	$0.88937 \cdot 10^{3}$
3	$0.14318 \cdot 10^{7}$	$0.40952 \cdot 10^{-2}$	$0.14318 \cdot 10^{3}$
4	$0.14246 \cdot 10^{6}$	$0.97820 \cdot 10^{-2}$	$0.14246 \cdot 10^{2}$
5	$0.12932 \cdot 10^{6}$	$0.11212 \cdot 10^{-1}$	$0.12932 \cdot 10^{2}$
6	$0.12394 \cdot 10^{6}$	$0.11773 \cdot 10^{-1}$	$0.12394 \cdot 10^{2}$
7	$0.12299 \cdot 10^{6}$	$0.11869 \cdot 10^{-1}$	$0.12299 \cdot 10^{2}$
8	$0.12267 \cdot 10^{6}$	$0.11900 \cdot 10^{-1}$	$0.12267 \cdot 10^{2}$
9	$0.12247 \cdot 10^{6}$	$0.11919 \cdot 10^{-1}$	$0.12247 \cdot 10^{2}$
10	$0.12244 \cdot 10^{6}$	$0.11922 \cdot 10^{-1}$	$0.12244 \cdot 10^{2}$
11	$0.12242 \cdot 10^{6}$	$0.11923 \cdot 10^{-1}$	$0.12242 \cdot 10^{2}$
12	$0.12242 \cdot 10^{6}$	$0.11924 \cdot 10^{-1}$	$0.12242 \cdot 10^{2}$
13	$0.12241 \cdot 10^{6}$	$0.11924 \cdot 10^{-1}$	$0.12241 \cdot 10^{2}$

Table 2.2. Convergence of the method.

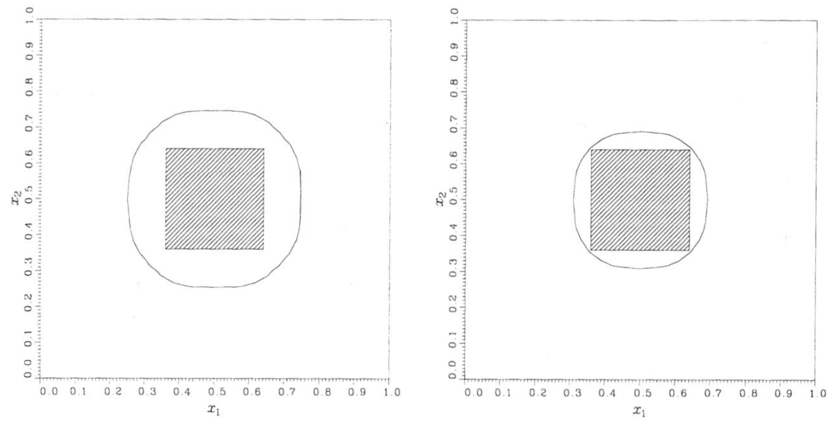

Figure 2.8. Domain E_y after 3 and 4 iterations.

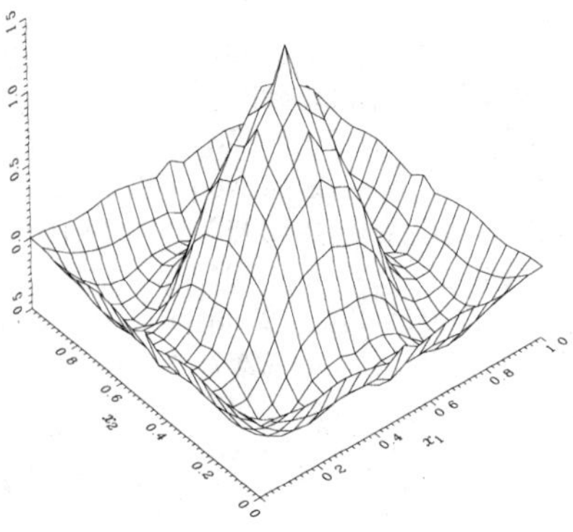

Figure 2.9. y after 3 iterations.

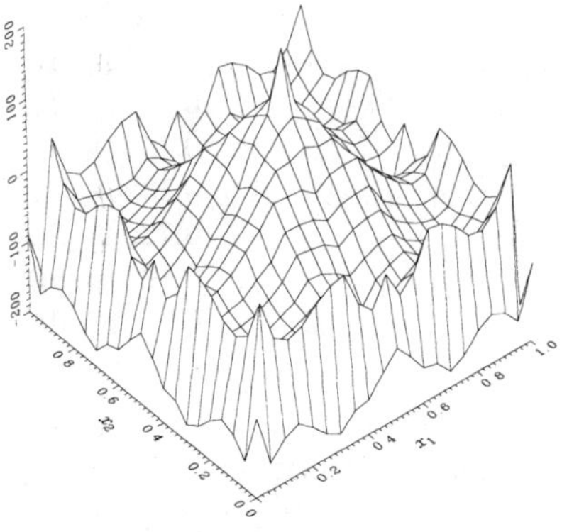

Figure 2.10. u after 3 iterations.

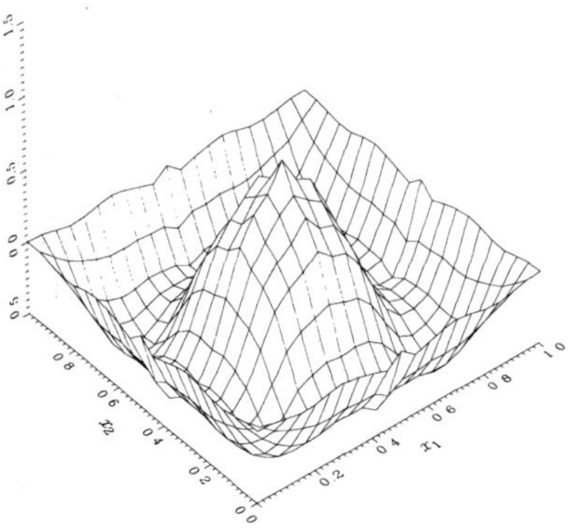

Figure 2.11. y after 13 iterations.

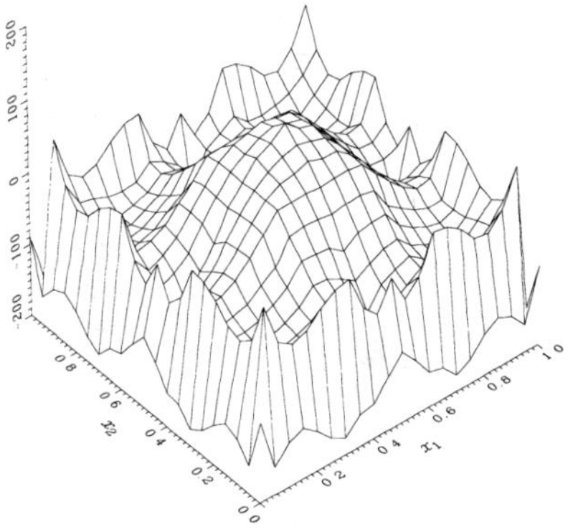

Figure 2.12. u after 13 iterations.

5.3 Domain Variations

This section is devoted to a short presentation of some methods that are frequently used in the study of domain optimization problems. In the first paragraph, we briefly review classical techniques known as *boundary variations*, *interior variations*, *speed method*, and *mapping method*. Since there exist already very good textbooks dealing with these approaches (cf. Pironneau [1984], Haslinger and Neittaanmäki [1988], [1996], Sokołowski and Zolésio [1992], and Delfour and Zolésio [2001]), we will confine ourselves to just pointing out several basic features in order to introduce the reader to this class of methods.

While these methods may be used both for numerical and theoretical studies (existence, optimality conditions, regularity properties), they also have several well-known drawbacks, such as the difficult and costly implementation on a computer and the strong regularity assumptions that have to be imposed on the data. Their possibly most severe limitation is that the topological type of the domain is preserved via all such variations, which means that the optimization is performed over quite a restricted class of domains. These remarks have motivated the development of many alternative ideas, some of which have been discussed in the previous section. In particular, the *integrated topology method* of Bendsøe [1995] specifies the combination of a relaxation procedure with the boundary variations technique, in order to achieve a good approximation of the optimal geometry.

The works of Schumacher [1995], Sverak [1993], Masmoudi [1987], [2001], Sokołowski and Zochowski [1999], [2003], Céa, Garreau, Guillaume, and Masmoudi [2000], Garreau, Guillaume, and Masmoudi [2001], Guillaume and Idris [2002], and Nazarow and Sokołowski [2003] take into account variations of the geometry that may change the topological type (i.e., the connectivity characteristics). These ideas will be discussed in the last paragraph of this section. We underline that one of their origins may be traced back to the pioneering paper of Céa, Gioan, and Michel [1973] discussed in §5.2.1 above.

Throughout this section, the main point of interest will be the sensitivity analysis of shape optimization problems (for existence questions, we refer to the detailed treatment in Chapter 2). Such differentiability properties with respect to domain variations were studied as early as in 1907 by Hadamard (see Hadamard [1968]). In Theorem 5.2.24, we have already established a result of this type in connection with a fixed domain method.

5.3.1 Classical Approaches

We start with the *mapping method* introduced by Murat and Simon [1976]. The basic assumption is that the unknown domain $\Omega \subset \mathbf{R}^d$ is obtained as the image of a reference domain $M \subset \mathbf{R}^d$; that is, $\Omega = T(M)$ with a bijection $T : \mathbf{R}^d \to \mathbf{R}^d$ satisfying $T, T^{-1} \in W^{k,\infty}(\mathbf{R}^d)^d$ for some appropriate $k \in \mathbf{N}$

and having the property that $|T - id|_{W^{k,\infty}(\mathbf{R}^d)}$ is "small," where $id : \mathbf{R}^d \to \mathbf{R}^d$ is the identity operator. Then, the true (unknown) input of the optimization problem defined on this set of admissible Ω will be the corresponding set of mappings $T : \mathbf{R}^d \to \mathbf{R}^d$ as specified above. Hence, a standard change of variables transports the problem to the fixed domain M.

Example 5.3.1 Let, with some fixed constants $\varepsilon > 0$, $a > 0$,

$$\Omega_u = \left\{(x_1, x_2, x_3) \in \mathbf{R}^3 : 0 < x_3 < 1,\ x_1^2 + x_2^2 < u(x_3)^2\right\},$$

$$u \in U_{ad} = \left\{u \in C^2[0,1] : u(x) \geq \varepsilon,\ u(0) = 1,\ |u|_{C^2[0,1]} \leq a\right\}.$$

We define

$$\Gamma_1 = \left\{(x_1, x_2, 0) \in \mathbf{R}^3 : x_1^2 + x_2^2 \leq 1\right\},$$

$$\Gamma_2 = \left\{(x_1, x_2, x_3) \in \mathbf{R}^3 : x_1^2 + x_2^2 = u(x_3)^2,\ x_3 \in [0,1]\right\},$$

$$\Gamma_3 = \left\{(x_1, x_2, 1) \in \mathbf{R}^3 : x_1^2 + x_2^2 \leq u(1)^2\right\}.$$

We consider the following shape design problem:

$$\text{Min}\left\{\int_0^1 u(x_3)\, dx_3\right\}, \tag{5.3.1}$$

subject to

$$\Delta y = 0 \quad \text{in } \Omega_u, \tag{5.3.2}$$

$$\frac{\partial y}{\partial n} = q_{\text{in}} \quad \text{on } \Gamma_1, \tag{5.3.3}$$

$$\frac{\partial y}{\partial n} = 0 \quad \text{on } \Gamma_2, \tag{5.3.4}$$

$$\frac{\partial y}{\partial n} + \alpha y = q_s \quad \text{on } \Gamma_3, \tag{5.3.5}$$

and to the constraints

$$u \in U_{ad}, \qquad y|_{\Gamma_1} \leq y_f. \tag{5.3.6}$$

The problem (5.3.1)–(5.3.6) is related to the design of a space radiator (cf. Delfour, Payne, and Zolésio [1983], Haslinger, Neittaanmäki, and Tiba [1987]). In this connection, $\alpha > 0$, and the positive constants q_{in}, q_s are related to the inward, respectively outward, thermal power flux at the source, respectively at the interface between the diffuser and the heatpipe saddle.

We scale the domain Ω_u such that it becomes the fixed cylinder

$$\Omega = \left\{(x_1, x_2, x_3) \in \mathbf{R}^3 : x_1^2 + x_2^2 < 1,\ x_3 \in [0,1]\right\}.$$

Then the Laplacian in (5.3.2) is transformed into the operator $A(u)$ generated by the following bilinear form on $H^1(\Omega) \times H^1(\Omega)$:

$$a_u(y, v) = \int_\Omega \left(\frac{\partial y}{\partial x_1}, \frac{\partial y}{\partial x_2}, u(x_3) \frac{\partial y}{\partial x_3} - \frac{\partial y}{\partial x_1} u'(x_3)\, x_1 - \frac{\partial y}{\partial x_2} u'(x_3)\, x_2 \right)$$

$$\cdot \left(\frac{\partial v}{\partial x_1}, \frac{\partial v}{\partial x_2}, u(x_3) \frac{\partial v}{\partial x_3} - \frac{\partial v}{\partial x_1} u'(x_3)\, x_1 - \frac{\partial v}{\partial x_2} u'(x_3)\, x_2 \right) dx_1\, dx_2\, dx_3$$

$$+ \alpha\, u^2(1) \int_{\{x_1^2 + x_2^2 \le 1\}} y\, v\, dx_1\, dx_2. \tag{5.3.7}$$

Above, "$\cdot$" is (as usual) the scalar product in $\mathbf{R}^3$. Relation (5.3.7) demonstrates that after the change of variables, the control u, together with its derivative, enters into the coefficients of the differential operator in the state equation. The constraints (5.3.6) and the cost functional (5.3.1) remain unchanged.

In the general case, when $\Omega \subset \mathbf{R}^d$ is some unknown domain, assume that a given cost functional

$$J(\Omega) = \int_\Omega j(y(\Omega)(x))\, dx \tag{5.3.8}$$

has to be minimized subject to the state system (in variational formulation)

$$\int_\Omega (\nabla y(\Omega)(x) \cdot \nabla w(x) - f(x)\, w(x))\, dx = 0 \quad \forall\, w \in H_0^1(\Omega). \tag{5.3.9}$$

If $\Omega = T(M)$ with some smooth transformation T as above, then, by a simple change of variables in the integral, (5.3.8), (5.3.9) may be rewritten as

$$J(\Omega) = J(T) = \int_M j((y(\Omega) \circ T)(x))\ |\det(\nabla T(x))|\, dx, \tag*{(5.3.8)$'$}$$

$$\int_M (\nabla (y(\Omega) \circ T) \cdot \nabla T^{-1} \nabla \hat{w} - (f \circ T) \hat{w})\ |\det(\nabla T)|\, dx = 0$$

$$\forall\, \hat{w} \in H_0^1(M). \tag*{(5.3.9)$'$}$$

Here, ∇T is the Jacobian of the transformation T.

As we have already seen from Example 5.3.1, the problem (5.3.8)$'$, (5.3.9)$'$ is a control into coefficients problem. One word of caution is in order: it may happen that many transformations T produce the same geometrical image while being different. For instance, a ball under any rotation about its center gives the ball. While this does not affect, in general, the search for an optimal domain, one has to pay attention to this aspect.

From the constraint (5.3.6) and the definition of U_{ad} one can infer that the coefficients in (5.3.7) are contained in a compact subset of $C[0, 1]$. Under corresponding boundedness assumptions, the same is true for the coefficients in (5.3.9)′ (given by T, its inverse, and their derivatives). Thus, Theorem 2.1.3(a) in Chapter 2 yields the existence of at least one optimal transformation T^* (or equivalently, domain $\Omega^* = T^*(M)$) for the shape design problem (5.3.8), (5.3.9). One should compare, however, the strong regularity assumptions imposed above with the general existence results proved in Chapter 2.

Such variations of M, given by $\Omega = T(M)$ for T not "too far" from the identity in $W^{k,\infty}(\mathbf{R}^d)^d$, are called *interior variations* in the literature and have been studied by Garabedian and Spencer [1952], and Daniljuk [1970], in a different context. With their help, first-order optimality conditions for the solutions of (5.3.8), (5.3.9) can be derived, simply by applying the chain rule of differentiation to the composed functional (5.3.8)′ in $W^{k,\infty}(M)^d$ (see Pironneau [1984, §8.2], and Céa [1986]). Even Fréchet differentiability may be proved for the functional (5.3.8) or for similar functionals defined on $\partial\Omega$; see Simon [1980].

We briefly describe now the *speed* (or *material derivative*) method introduced by Zolésio [1979], [1981]. Roughly speaking, this may be compared with the computation of a directional derivative in standard calculus. In the monograph of Delfour and Zolésio [2001] several variants of this approach, together with its relationship to mapping or level set methods, have been studied in detail.

We assume that a vector field (of "speeds" or "velocities") $V(x) \in \mathbf{R}^d$ is given for any $x \in M$. We choose as admissible variations of M the domains M_t obtained via the transformations

$$T_t(x) = x + t\,V(x), \quad x \in M, \ t > 0, \tag{5.3.10}$$

i.e., $M_t = T_t(M)$, $t > 0$. Correspondingly, using (5.3.10), (5.3.8)′, (5.3.9)′, one may introduce

$$y_t = y(M_t), \qquad J_t = J(T_t), \qquad t > 0.$$

It is important to compute $\frac{d}{dt}(J_t)(0)$, which corresponds to the variation around the original domain M in the direction V (to use the language of directional derivatives). In general, $\frac{d}{dt}(J_t)(0) = L(V)$ is a linear continuous functional of V if V belongs to some appropriate functional space X defined on M. Then, representations of the type

$$L(V) = (G, V)_{X^* \times X}\,, \tag{5.3.11}$$

with some $G \in X^*$ (the *gradient* of J at the "point" M), may be found. Under smoothness hypotheses, taking into account the development

$$J_t = J_0 + t\,L(V) + \cdots, \tag{5.3.12}$$

one can devise descent algorithms (compare Appendix 1) for the numerical solution of (5.3.8)–(5.3.9) or derive first-order optimality conditions,

$$L(V) \geq 0 \quad \forall V \in X_{ad}, \tag{5.3.13}$$

where $X_{ad} \subset X$ is some admissible set of variations.

More general transformations $T(x,t)$, $t \geq 0$, $x \in M$, with $T(x,0) = id$, may also be considered. Assuming differentiability in $t = 0$, we then can write

$$T(x,t) = T(x,0) + t\,T_t'(x,0) + \cdots,$$

for small $t > 0$. Choosing $V(x) = T_t'(x,0)$, we infer from the above relation that $T(x,t)$ produces the same domain transformation as (5.3.10), in a second-order approximation sense. The transformations $T(x,t)$ and (5.3.10) are also called *tangent* at the origin. Zolésio [1979] has shown that such transformations generate the same relations (5.3.11)–(5.3.13); that is, the speed field $T_t'(x,0) = V(x)$ at the origin carries all information of interest concerning $T(x,t)$.

After this rather formal presentation, let us concentrate on the rigorous sensitivity study of the problem (5.3.8), (5.3.9). One of the main theoretical difficulties, when taking variations of a domain M (even of the simplified form (5.3.10)), is that the corresponding solutions of (5.3.9), denoted by y_t, are defined on different sets, namely on $M_t = T_t(M)$, $t \geq 0$.

Now let K be some open set such that $\overline{K}$ is a compact subset of M. Since $M_t \to M$ as $t \searrow 0$ in the complementary Hausdorff–Pompeiu topology, it follows from Proposition A3.8 in Appendix 3 that there is some $t_0 > 0$ such that $\overline{K} \subset M_t$ for $0 \leq t \leq t_0$. Then y_t is for $0 \leq t \leq t_0$ defined in K, and it makes sense to ask for the differentiability of $y_t|_K$ with respect to t at $t = 0$. We will call this a *local differentiability* property, in the sense that it holds for any open set K such that $\overline{K}$ is a compact subset of M.

Clearly, the local derivative can be defined in the whole set M by taking an increasing sequence of sets K_m as above such that $\cup_{m\in\mathbf{N}}\overline{K}_m = M$. Alternative approaches use an extension of y_t to a larger fixed domain $\widetilde{M}$ that contains M_t for $0 \leq t \leq t_0$. For zero Dirichlet boundary conditions as in (5.3.9), one may use the zero extension. In general, sophisticated extension techniques like the Calderon extension (see Adams [1975]) have to be employed, and regularity of the boundary ∂M is required.

Notice that all the three points of view mentioned above appeared early in the scientific literature. Mignot, Murat, and Puel [1979] studied their interdependence for a specific problem and under appropriate smoothness assumptions.

We are now going to illustrate by means of an example using a differentiability property of the mapping $\Omega \mapsto y(\Omega)$ defined by (5.3.9). We assume that the admissible domains Ω are contained in a fixed bounded open set D and that $f \in L^2(D)$. In any point $x \in D$, a vector $V(x) \in \mathbf{R}^d$ is given such that $V \in W^{1,\infty}(D)^d$. Furthermore, we may assume that for sufficiently small $\lambda \in \mathbf{R}$, say for $|\lambda| < \lambda_0$, the set $\Omega_\lambda = (id + \lambda V(\cdot))^{-1}(\Omega)$ is defined and contained in D.

Let us denote by $y \in H_0^1(\Omega)$ and $y_\lambda \in H_0^1(\Omega_\lambda)$ the solutions to (5.3.9) associated with Ω and Ω_λ, respectively, and by $\tilde{y}$ and $\tilde{y}_\lambda$ their extensions by zero to the whole domain D. Under mild regularity assumptions (in fact, a uniform segment property of $\partial\Omega$, $\partial\Omega_\lambda$ suffices for this conclusion; see also Appendix 3) it follows from the chain of arguments developed in the proof of Theorem 2.3.12 in Chapter 2 that $\tilde{y}_\lambda \to \tilde{y}$ weakly in $H_0^1(D)$. Clearly, also $\Omega_\lambda \to \Omega$ with respect to the complementary Hausdorff–Pompeiu distance.

Finally, we postulate regularity for $\partial\Omega$ and $\partial\Omega_\lambda$, $|\lambda| < \lambda_0$, and the condition that there is some constant $\hat{C} > 0$ such that for all $|\lambda| < \lambda_0$,

$$\operatorname{meas}(\Omega \setminus \tilde{\Omega}_\lambda) \le \hat{C}\,\lambda, \quad \text{with } \tilde{\Omega}_\lambda = \{x \in \Omega : (id + \lambda V)^{-1}(x) \in \Omega\}. \tag{5.3.14}$$

One can easily construct examples where (5.3.14) is fulfilled.

Theorem 5.3.2 *Suppose that $f \in L^{r'}(D)$ for some $r' > d$, where $\frac{1}{r} + \frac{1}{r'} = 1$. Then there is some function $z \in L^r(\Omega)$ that is harmonic in Ω and satisfies*

$$\lim_{\lambda\to 0} \int_\Omega \frac{\nabla \tilde{y}_\lambda|_\Omega(x) - \nabla y(x)}{\lambda} \cdot \nabla\varphi(x)\,dx = -\int_\Omega z(x)\,\Delta\varphi(x)\,dx, \tag{5.3.15}$$

$$-\int_\Omega z(x)\,\Delta\varphi(x)\,dx + \int_\Omega \nabla y(x) \cdot \nabla(\nabla\varphi \cdot V)(x)\,dx$$
$$= \int_\Omega f(x)\left[\nabla\varphi(x) \cdot V(x)\right]dx \quad \forall\, \varphi \in W^{2,r'}(\Omega) \cap H_0^1(\Omega). \tag{5.3.16}$$

Proof. To keep the exposition at a reasonable length, we give only an "extended sketch" of the proof here. References to the relevant literature will also be provided. To begin with, let us denote for any $\varphi \in W^{2,r'}(\Omega) \cap H_0^1(\Omega)$ its extension by zero to D by $\tilde{\varphi}$. Obviously, $\tilde{\varphi} \in H_0^1(D)$. Likewise, we put

$$\varphi_\lambda(x) = \varphi((id + \lambda V)(x)) = \varphi(x + \lambda\, V(x)), \quad x \in \Omega_\lambda,$$

and denote its extension by zero to D by $\tilde{\varphi}_\lambda$. Then it follows from Litvinov [2000, p. 76] that $\varphi_\lambda \in H_0^1(\Omega_\lambda)$ and also $\tilde{\varphi}_\lambda \in H_0^1(D)$.

Writing (5.3.9) for $\tilde{y}$ and $\tilde{y}_\lambda$ with the test functions $\tilde{\varphi}$ and $\tilde{\varphi}_\lambda$, respectively, subtracting and dividing by λ, we find that

$$\int_D \frac{\nabla\tilde{y}_\lambda - \nabla\tilde{y}}{\lambda} \cdot \nabla\tilde{\varphi}\,dx + \int_D \nabla\tilde{y}_\lambda \cdot \frac{\nabla\tilde{\varphi}_\lambda - \nabla\tilde{\varphi}}{\lambda}\,dx = \int_D f\,\frac{\tilde{\varphi}_\lambda - \tilde{\varphi}}{\lambda}\,dx. \tag{5.3.17}$$

Now it is easily seen that under the regularity assumptions made,

$$\lambda^{-1}\,(\tilde{\varphi}_\lambda - \tilde{\varphi}) \to \nabla\tilde{\varphi} \cdot V \quad \text{strongly in } L^2(D).$$

Hence, the expression on the right side of (5.3.17) satisfies

$$\lim_{\lambda\to 0}\int_D f\,\frac{\tilde{\varphi}_\lambda-\tilde{\varphi}}{\lambda}\,dx = \int_D f\,(\nabla\tilde{\varphi}\cdot V)\,dx = \int_\Omega f\,(\nabla\varphi\cdot V)\,dx. \qquad (5.3.18)$$

In the second term on the left-hand side of (5.3.17), we perform the change of variables $t = x + \lambda V(x)$ to obtain that

$$\begin{aligned}
&\int_D \nabla\tilde{y}_\lambda\cdot\frac{\nabla\tilde{\varphi}_\lambda-\nabla\tilde{\varphi}}{\lambda}\,dx\\
&= \int_{(id+\lambda V)^{-1}(\Omega)} \nabla y_\lambda\cdot\frac{\nabla\varphi_\lambda-\nabla\tilde{\varphi}}{\lambda}\,dx\\
&= \int_\Omega \nabla y_\lambda((id+\lambda V)^{-1}(t))\cdot\frac{\nabla\varphi(t)-\nabla\tilde{\varphi}((id+\lambda V)^{-1}(t))}{\lambda}\\
&\qquad\qquad\cdot\left|\det\nabla(id+\lambda V)^{-1}(t)\right|\,dt\\
&\quad+\int_\Omega \nabla y_\lambda((id+\lambda V)^{-1}(t))\cdot[\nabla\varphi(t)\cdot\nabla V((id+\lambda V)^{-1}(t))]\\
&\qquad\qquad\cdot\left|\det\nabla(id+\lambda V)^{-1}(t)\right|\,dt,
\end{aligned} \qquad (5.3.19)$$

where ∇V and $\nabla(id+\lambda V)^{-1}$ denote the Jacobian matrices of the vectorial mappings V and $(id+\lambda V)^{-1}$, respectively, and where we have used the chain rule

$$\nabla\varphi_\lambda(x) = \nabla[\varphi((id+\lambda V)(x))] = \nabla\varphi((id+\lambda V)(x))\cdot[id+\lambda\nabla V(x)]. \quad (5.3.20)$$

The last product in (5.3.20) is the product of a row vector and a matrix.

The advantage of the formulation (5.3.19) is that now the integral and the functions are defined on the common fixed domain Ω.

Since we have $\tilde{y}_\lambda \to \tilde{y}$ weakly in $H_0^1(D)$ and $(id+\lambda V)^{-1} \to id$ strongly in $W^{1,\infty}(\Omega)$, one can show, using Friedrichs regularizations of $\tilde{y}_\lambda$ and $\tilde{y}$, that $\nabla y_\lambda((id+\lambda V)^{-1}) \to \nabla y$ weakly in $L^2(\Omega)^d$. Notice also that

$$\frac{\nabla\varphi(t)-\nabla\tilde{\varphi}((id+\lambda V)^{-1}(t))}{\lambda} \to \nabla^2\varphi(t)\cdot V(t) \quad \text{a.e. in } \Omega. \qquad (5.3.21)$$

This follows from the assumption that $\varphi\in W^{2,r'}(\Omega)$ and from the observation that $t\in\Omega$ implies $(id+\lambda V)^{-1}(t)\in\Omega$ for $|\lambda|$ "small," and consequently

$$\nabla\tilde{\varphi}((I+\lambda V)^{-1}(t)) = \nabla\varphi((I+\lambda V)^{-1}(t)).$$

For $t\in\Omega\setminus\widetilde{\Omega}_\lambda$, we have $\nabla\tilde{\varphi}((I+\lambda V)^{-1}(t)) = 0$. Moreover, since $r' > d$, it follows from Theorem A2.2(iv) in Appendix 2 that $\nabla\varphi\in L^\infty(\Omega)^d$. Thus, in view of (5.3.14), there is some $C>0$ such that

$$\frac{1}{\lambda}\int_{\Omega\setminus\widetilde{\Omega}_\lambda} |\nabla\varphi(t)|\,dt \le C.$$

The boundedness in $L^1(\Omega \cap \tilde{\Omega}_\lambda) = L^1(\tilde{\Omega}_\lambda)$ of the ratio defined in (5.3.21) is a consequence of the differentiability properties of φ. The argument uses a regularization of φ, the mean value theorem, and the continuity in the mean with respect to translations (see Hewitt and Stromberg [1965]), which may also be applied to the perturbations $id + \lambda V$ occurring here. Since this is rather technical and would lead to a major detour, we just quote the books and the articles of Sokołowski and Zolésio [1992, Ch. 2.14], Courant [1950, p. 292], Rousselet [1983], and Pascali and Sburlan [1978, p. 170] for relevant details.

Next observe that $f \in L^{r'}(D)$. Therefore, invoking the regularity assumptions for Ω, Ω_λ imposed via (5.3.14), we may conclude that $y_\lambda \in W^{2,r'}(\Omega_\lambda)$. From this, using Theorem A2.2(iv) in Appendix 2, we obtain that $\{\nabla y_\lambda((id + \lambda V)^{-1})\}$ is bounded in $L^\infty(\Omega)^d$. A fortiori, it converges uniformly to its limit. This can be proved by "transporting" the equation (5.3.9) from Ω_λ to Ω via the indicated change of variables.

By virtue of (5.3.21) and due to the above uniform convergence, it is possible to pass to the limit as $\lambda \to 0$ in (5.3.19) and to obtain that

$$\lim_{\lambda \to 0} \int_D \nabla \tilde{y}_\lambda \cdot \frac{\nabla \tilde{\varphi}_\lambda - \nabla \tilde{\varphi}}{\lambda}\, dx = \int_\Omega \nabla y \cdot [\nabla^2 \varphi \cdot V]\, dx + \int_\Omega \nabla y \cdot [\nabla \varphi \cdot \nabla V]\, dx, \quad (5.3.22)$$

since $|\det \nabla (id + \lambda V)^{-1}| \to 1$ in $L^\infty(\Omega)$. Relations (5.3.22) and (5.3.18) show that the first integral in (5.3.17) has a limit as well, for any $\varphi \in W^{2,r'}(\Omega) \cap H_0^1(\Omega)$.

Now recall (compare Appendix 2) that $W^{2,r'}(\Omega) \cap H_0^1(\Omega)$ is isomorphic to $L^{r'}(\Omega)$ via the operator $\varphi = T_1 p$, $p \in L^{r'}(\Omega)$, and φ given by

$$-\Delta \varphi = p \quad \text{in } \Omega, \qquad \varphi = 0 \quad \text{on } \partial\Omega. \quad (5.3.23)$$

We introduce the linear and continuous operator $T : L^{r'}(\Omega) \to H^1(\Omega)^d$, $Tp = \nabla \varphi$ with φ given by (5.3.23). Then we can write

$$\int_D \frac{\nabla \tilde{y}_\lambda - \nabla \tilde{y}}{\lambda} \cdot \nabla \tilde{\varphi}\, dx = \int_\Omega \frac{\nabla \tilde{y}_\lambda - \nabla y}{\lambda} \cdot \nabla \varphi\, dx = \int_\Omega \frac{\nabla \tilde{y}_\lambda - \nabla y}{\lambda} \cdot Tp\, dx. \quad (5.3.24)$$

Notice that the last integral in (5.3.24) defines a linear and bounded functional on $L^{r'}(\Omega)$ for any fixed λ. The Riesz theorem ensures the existence of a unique $z_\lambda \in L^r(\Omega)$ such that

$$\int_D \frac{\nabla \tilde{y}_\lambda - \nabla \tilde{y}}{\lambda} \cdot \nabla \tilde{\varphi}\, dx = \int_\Omega z_\lambda\, p\, dx. \quad (5.3.25)$$

The definition of weak convergence and the convergence of the first integral, for any $\varphi = T_1 p$, show that $z_\lambda \to z$ weakly in $L^r(\Omega)$. Therefore we get, by using (5.3.23),

$$\lim_{\lambda \to 0} \int_D \frac{\nabla \tilde{y}_\lambda - \nabla \tilde{y}}{\lambda} \cdot \nabla \tilde{\varphi}\, dx = \int_\Omega z\, p\, dx = -\int_\Omega z\, \Delta \varphi\, dx, \tag{5.3.26}$$

for any $\varphi \in W^{2,r'}(\Omega) \cap H_0^1(\Omega)$. This proves (5.3.15).

Next observe that any $\hat{\varphi} \in C_0^\infty(\Omega)$ has compact support in Ω and in Ω_λ, for sufficiently small $|\lambda|$. Hence, it is admissible as test function in (5.3.9) for both Ω and Ω_λ, and we can infer that

$$\begin{aligned} \int_D \frac{\nabla \tilde{y}_\lambda - \nabla \tilde{y}}{\lambda} \cdot \nabla \hat{\varphi}\, dx &= \frac{1}{\lambda} \left[\int_{\Omega_\lambda} \nabla y_\lambda \cdot \nabla \hat{\varphi}\, dx - \int_\Omega \nabla y \cdot \nabla \hat{\varphi}\, dx \right] \\ &= \frac{1}{\lambda} \left[\int_{\operatorname{supp} \hat{\varphi}} f \hat{\varphi}\, dx - \int_{\operatorname{supp} \hat{\varphi}} f \hat{\varphi}\, dx \right] = 0. \end{aligned} \tag{5.3.27}$$

From (5.3.26), (5.3.27) it follows that z is harmonic in Ω in the sense of distributions. Invoking (5.3.26), (5.3.22), and (5.3.18), we can pass to the limit in (5.3.17) to obtain (5.3.16). This ends the proof of the assertion. □

Remark. Theorem 5.3.2 gives the so-called *equation in variation* in the form (5.3.16) (with respect to the domain) for the problem (5.3.9). The first paper to put in evidence such a relation is the work of Zolésio [1981]. The reader may also consult Fujii [1986], Simon [1980], and Delfour and Zolésio [2001].

If z and y belong to $H^2(\Omega)$, then integration by parts in (5.3.16) yields the "boundary condition"

$$\int_{\partial\Omega} z\, \frac{\partial \varphi}{\partial n}\, d\sigma = \int_{\partial\Omega} \frac{\partial y}{\partial n}\, [\nabla \varphi \cdot V]\, d\sigma. \tag{5.3.28}$$

In particular, (5.3.28) and z harmonic show that only the values of V on $\partial\Omega$ are important in finding z, that is, in the investigation of the differentiability properties of the mapping $\Omega \mapsto y = y(\Omega)$ defined by (5.3.9).

Assume now that at any point $x \in \partial\Omega$ an outward unit normal vector $\overline{n}(x)$ can be defined, and take $\alpha \in C^2(D)$ with variable sign. We define the perturbed domain $\Omega_{\lambda\alpha}$ by its boundary

$$\partial\Omega_{\lambda\alpha} = \{x + \lambda\, \alpha(x)\, \overline{n}(x) : x \in \partial\Omega\}. \tag{5.3.29}$$

Here, it is assumed that $|\lambda|$ is small enough such that $\partial\Omega_{\lambda\alpha}$ is indeed the boundary of an open bounded set $\Omega_{\lambda\alpha}$ that approximates Ω for $\lambda \to 0$ in the Hausdorff–Pompeiu complementary topology. This is the basic idea of the *boundary variation* technique introduced by Hadamard [1968] in his Mémoire back in 1907.

This method was used by Pironneau [1984, Ch. VI], and by Fujii [1986], in the study of the differentiability properties of the mapping $\Omega \mapsto y(\Omega)$ in linear elliptic problems with various boundary conditions and in some semilinear elliptic equations with Dirichlet conditions. It is also quite popular in numerical applications, where it is sometimes called the *moving grid method* (see Example 5.2.20). The boundary of the unknown domain is approximated by a polygonal curve connecting finitely many points (the nodes). These nodes are allowed to vary and become the true unknowns of the problem.

If the boundary is given by (5.3.29), then $V(x) = \alpha(x)\,\bar{n}(x)$ for any $x \in \partial\Omega$. Then, under regularity hypotheses such as those made for (5.3.28), we get the boundary condition

$$z(x) = \alpha(x)\,\frac{\partial y}{\partial n}(x), \quad x \in \partial\Omega, \tag{5.3.30}$$

since $\frac{\partial \varphi}{\partial n}$ may vary arbitrarily in $H^{1/2}(\partial\Omega)$, and

$$\nabla\varphi \cdot V = \nabla\varphi \cdot (\alpha\,\bar{n}) = \alpha\,\frac{\partial \varphi}{\partial n}\;.$$

Equation (5.3.30) was derived by Pironneau [1984, p. 86] and by Fujii [1986] for a semilinear elliptic operator. Together with the property that z is harmonic in Ω (in our case), it provides a complete characterization of the variation of the solution of (5.3.9) with respect to the domain Ω.

If in the two-dimensional case the admissible domains are assumed to be star-shaped, then a parametrization of their boundaries via polar coordinates is possible. A complete analysis for this situation has been performed in Eppler [2000].

Let us now impose a stronger differentiability property for the mapping $\Omega \mapsto y(\Omega)$, namely,

$$\frac{\tilde{y}_\lambda|_\Omega - y}{\lambda} \to z \text{ as } \lambda \to 0, \quad \text{weakly in } L^2(\Omega), \quad \text{for any } \Omega. \tag{5.3.31}$$

Under stronger regularity conditions on Ω, y, and Ω_λ, y_λ, the validity of (5.3.31) may be proved by the mapping method, for instance. The interested reader may consult Simon [1980], Litvinov [2000], and Delfour and Zolésio [2001].

Our aim is to examine the following model shape optimization problem.

Example 5.3.3

$$\operatorname*{Min}_{\Omega\in\mathcal{O}} \left\{ \int_E |y(x) - y_d(x)|^2\,dx \right\} \tag{5.3.32}$$

with $y = y(\Omega)$ given by (5.3.9), $y_d \in L^2(D)$, and $\mathcal{O}$ being a suitable family of open connected subdomains of D containing the given measurable set E. We assume that (5.3.32), (5.3.9) admits at least one (sufficiently smooth) optimal domain $\Omega^* \in \mathcal{O}$ with corresponding optimal state $y^* \in H_0^1(\Omega^*)$.

Consider variations $\Omega_\lambda^* = (id + \lambda V)^{-1}(\Omega^*)$, and let $y_\lambda^* \in H_0^1(\Omega_\lambda^*)$ be the solution of (5.3.9) in Ω_λ^*, as before. Since $\tilde{y}_\lambda^* \to \tilde{y}^*$ strongly in $L^2(D)$ (at least on a subsequence, which is again indexed by λ), it follows from (5.3.31) that for $\lambda \searrow 0$,

$$\begin{aligned} 0 &\le \frac{1}{\lambda}\left\{\int_E |y_\lambda^* - y_d|^2\,dx - \int_E |y^* - y_d|^2\,dx\right\} \\ &= \int_E \frac{y_\lambda^* - y^*}{\lambda}\,(y_\lambda^* + y^* - 2y_d)\,dx = \int_{\Omega^*} \frac{y_\lambda^* - y^*}{\lambda}\,\chi_E\,(y_\lambda^* + y^* - 2y_d)\,dx \\ &\to 2\int_{\Omega^*} z\,\chi_E\,(y^* - y_d)\,dx. \end{aligned} \tag{5.3.33}$$

We introduce the adjoint system for p^*,

$$-\Delta p^* = 2\,\chi_E\,(y^* - y_d) \quad \text{in } \Omega^*, \tag{5.3.34}$$

$$p^* = 0 \quad \text{on } \partial\Omega^*. \tag{5.3.35}$$

We then have the following result.

Corollary 5.3.4 *Suppose that Ω^* is a sufficiently smooth domain and that $y^* \in H^2(\Omega^*) \cap H_0^1(\Omega^*) \cap L^{r'}(\Omega^*)$. Then*

$$\int_{\partial\Omega^*} \frac{\partial y^*}{\partial n}(\sigma)\cdot\frac{\partial p^*}{\partial V}(\sigma)\,d\sigma \le 0 \quad \textit{for any admissible } V \in W^{1,\infty}(D). \tag{5.3.36}$$

Proof. From (5.3.33)–(5.3.35), (5.3.16), we get $p^* \in W^{2,r'}(\Omega^*) \cap H_0^1(\Omega^*)$, and we infer that

$$0 \le -\int_{\Omega^*} z\Delta p^*\,dx = \int_{\Omega^*} f\,(\nabla p^*\cdot V)\,dx - \int_{\Omega^*} \nabla y^*\cdot\nabla[\nabla p^*\cdot V]\,dx,$$

whence, using integration by parts,

$$0 \le \int_{\Omega^*} f\,(\nabla p^*\cdot V)\,dx + \int_{\Omega^*} \Delta y^*\,(\nabla p^*\cdot V)\,dx - \int_{\partial\Omega^*} \frac{\partial y^*}{\partial n}\,(\nabla p^*\cdot V)\,d\sigma,$$

and (5.3.36) follows since $-\Delta y^* = f$ in Ω^*. □

Remark. The optimality condition (5.3.36) may be compared with the one obtained by Fujii [1986] and Pironneau [1984, p. 87], with $\frac{\partial p^*}{\partial n}$ in place of $\frac{\partial p^*}{\partial V}$. Since these authors use the boundary variation technique, (5.3.30) explains the difference between the two conditions. If V may attain arbitrary signs on $\partial\Omega^*$, then (5.3.36) holds with equality. It is interesting that one of the necessary conditions obtained by Buttazzo and Dal Maso [1991] via a relaxation method also involves $\frac{\partial p^*}{\partial n}$; this is a hint that there might exist a connection to the boundary variation technique. Since p^* vanishes on $\partial\Omega^*$, one can also write

$$\int_{\partial\Omega^*} \frac{\partial y^*}{\partial n} (\nabla p^* \cdot V)\, d\sigma = \int_{\partial\Omega^*} \frac{\partial y^*}{\partial n} \frac{\partial p^*}{\partial n} (\bar{n} \cdot V)\, d\sigma.$$

This corresponds to the type of optimality conditions obtained by Delfour and Zolésio [2001, Ch. 9.2], where more general problems have been discussed.

Remark. Higher-order "directional" derivatives with respect to such variations of the domains have been studied by Simon [1989], Guillaume and Masmoudi [1994], Guillaume [1996], and Eppler [2000].

5.3.2 Topological Asymptotics

The mapping $z \in L^r(\Omega)$ introduced in Theorem 5.3.2 is sometimes called the *shape gradient* of the mapping $\Omega \mapsto y(\Omega)$ defined by (5.3.9). It plays an essential role in the study of the sensitivity of a cost functional when small perturbations of Ω by diffeomorphisms close to the identity operator are considered. The aim of the notion of the *topological gradient*, which will be briefly introduced in this paragraph, is to characterize the variation of a cost functional when small topological-type modifications of the domain have to be taken into account. The creation of a small hole or the filling in of a small hole are the simplest examples of such topological perturbations of open sets. The general case would be the insertion of an arbitrary number of small holes in a given domain; however, such variations have not yet been studied in the literature.

It turns out that the concept of topological gradients provides a mathematical justification of the method of Céa, Gioan, and Michel [1973] discussed above in §5.2.1. In this context, the integrated topology optimization method of Bendsøe [1995], and the homogenization method (Allaire [2001]; see §5.2.2) offer general relaxation procedures that have proven to be very efficient in certain applications.

Let us also mention that in connection with these concepts, one can distinguish in the setting of optimal design theory between *shape optimization* problems, where variations specific to shape gradients are used, and *topological optimization* problems, where the topology of the optimal domain is a priori unknown. A very special optimal design problem is the so-called *size optimization*, which is characterized by a good knowledge of the desired structure, the unknowns to be optimized being just some scalar parameters such as dimensions of cross sections of various parts. Such simplified problems are encountered in many engineering applications. The above classification is discussed in Neittaanmäki [1991], where further details may be found. To complete this brief classification, we also recall the material distribution or *structural optimization* problems (see §2.3.4 in Chapter 2).

The notion of topological gradient is due to Schumacher [1995], but it implicitly appears in the work of Céa et al. [1973]; cf. §5.2.1. The subject is still under very active development, as the references indicated in the beginning of Section 5.3 show.

Let $\Omega \subset \mathbf{R}^d$ be a given bounded domain. If $B(x, r)$ denotes the open ball of radius $r > 0$ centered at x, then for any $x \in \Omega$, $B(x,r) \subset \Omega$ provided that $0 < r < r_x$ with some sufficiently small $r_x > 0$.

Now let $J(\Omega) = j(\Omega, y(\Omega))$ be some cost functional associated with Ω, for instance as in (5.3.8), (5.3.9). The *topological gradient* of J is defined by the limit (if it exists)

$$\nabla_T J(x) = \lim_{r \to 0} \frac{J(\Omega) - J(\Omega \setminus B(x,r))}{\text{meas}(B(x,r))}. \tag{5.3.37}$$

In many applications, the denominator $\text{meas}(B(x,r))$ in this definition has to be replaced by some other function $h(r)$ such that $h(r) \to 0$ as $r \searrow 0$. The appropriate choice of h depends on the actual problem; for instance, one can choose

$$h(r) = \left[\log r^{\frac{1}{2\pi}}\right]^{-1}$$

in the case of the two-dimensional Laplace equation with Dirichlet boundary conditions on $\partial B(x,r)$, cf. Céa et al. [2000]. Usually, Neumann boundary conditions are imposed on $\partial B(x,r)$, since this ensures better convergence properties. Let us also notice that the limit in (5.3.37) may heavily depend on the shape of the hole to be inserted in the domain Ω; the ball is just a special case. In this respect, a general analysis for the Poisson equation with Dirichlet boundary conditions has been performed in Guillaume and Idris [2002].

Again, we consider the state equation (5.3.9), to which we associate the cost functional (5.3.32). The minimization parameter is the domain $\Omega \in \mathcal{O}$, as indicated in Example 5.3.3. However, the variations of Ω taken into account here are defined as in (5.3.37). For some fixed $x \in \text{int}(\Omega \setminus E)$, set $\Omega_r = \Omega \setminus B(x,r)$. Then, for sufficiently small $r > 0$, we have $\Omega_r \in \mathcal{O}$. Let y_r denote the solution of (5.3.9) corresponding to Ω_r. If $r = 0$, then y_0 coincides with $y(\Omega)$ defined in (5.3.9).

The study of the mapping $\Omega_r \mapsto y_r$ from the point of view of its differentiability properties in $r = 0$ encounters the same general difficulty as in the previous paragraph: namely, the functions y_r are defined in Ω_r, which changes when $r \searrow 0$. Moreover, it is not clear how a mapping $\widetilde{T}_r$ satisfying $\Omega_r = \widetilde{T}_r(\Omega)$ (as in (5.3.8)′, (5.3.9)′) can be constructed. In Garreau, Guillaume, and Masmoudi [2001], a *truncation* method is introduced to overcome this difficulty. The method works for any shape ω of the hole to be inserted (in which case the perturbed domain becomes $\Omega\setminus(x+r\omega)$, with $x\in\Omega$ and $r>0$ sufficiently small).

Here, we confine ourselves to the case of a ball. The idea is to fix some $R>0$ such that $B(x,R)\subset\Omega\setminus E$ and to choose $0<r<R$ such that $\Omega\setminus B(x,R)\subset\Omega_r\in\mathcal{O}$. In $\Omega_R=\Omega\setminus B(x,R)$, the following boundary value problem is considered:

$$-\Delta y_r = f \quad \text{in } \Omega_R, \tag{5.3.38}$$

$$y_r = 0 \quad \text{on } \partial\Omega, \tag{5.3.39}$$

$$-\frac{\partial y_r}{\partial n}+T_r y_r = -\frac{\partial y^r}{\partial n} \quad \text{on } \partial B(x,R). \tag{5.3.40}$$

Here, T_r is defined via the relations (with $\alpha\in H^{\frac{1}{2}}(\partial B(x,R))$)

$$-\Delta y_\alpha = 0 \quad \text{in } B(x,R)\setminus B(x,r), \tag{5.3.41}$$

$$y_\alpha = \alpha \quad \text{in } \partial B(x,R), \tag{5.3.42}$$

$$y_\alpha = 0 \quad \text{in } \partial B(x,r), \tag{5.3.43}$$

$$T_r\alpha = \frac{\partial y_\alpha}{\partial n}\Big|_{\partial B(x,R)} \in H^{-\frac{1}{2}}(\partial B(x,R)), \tag{5.3.44}$$

and y^r denotes the solution of (5.3.9) in $B(x,R)\setminus B(x,r)$. Apparently, T_r in (5.3.44) is the Dirichlet-to-Neumann operator, and (5.3.38)–(5.3.40), (5.3.41)–(5.3.43) should be understood in the weak sense; cf. Appendix 2. In particular,

$$\int_{\partial B(x,R)}(T_r\alpha)\,\mu\,d\sigma=\int_{B(x,R)\setminus B(x,r)}\nabla y_\alpha\cdot\nabla y_\mu\,dx$$

for all $\alpha,\mu\in H^{1/2}(\partial B(x,R))$. It follows that y_r is exactly the restriction of $y(\Omega\setminus B(x,r))$, given by (5.3.9), to $\Omega\setminus B(x,R)$.

This reformulation of the equation (5.3.9), when defined in Ω_r, allows one to work in $H^1(\Omega_R)$ for $r<R$, via (5.3.38)–(5.3.40). The proof of the corresponding asymptotic expansion with respect to r necessitates a deep study of the operator T_r and of the boundary value problem (5.3.41)–(5.3.44). The interested reader may consult Herwig [1989], Garreau, Guillaume, and Masmoudi [2001], and Guillaume and Idris [2002].

The computation then follows as in Example 5.3.3, and the topological gradient $\nabla_T J(x)$, $x\in\Omega$, is obtained by (for $r>0$ so small that $E\subset\Omega_r$)

$$\lim_{r\to 0}\frac{1}{h(r)}\left[\int_E(y_0-y_d)^2\,dt-\int_E(y_r-y_d)^2\,dt\right]$$

$$= \lim_{r \to 0} \frac{1}{h(r)} \int_E (y_0 + y_r - 2y_d)(y_0 - y_r)\, dt$$

$$= 2 \int_\Omega \chi_E \, (y_0 - y_d)\, \xi_x \, dt.$$

Here, we have used the weak convergence $y_r \to y_0$ in $H^1(\Omega_R)$, and we have assumed the differentiability property

$$\frac{1}{h(r)}(y_0 - y_r) \to \xi_x$$

in $L^2(\Omega_R)$ (we have specified no restrictions on the dimension of Ω_R, but the results are valid in dimensions 2 and 3). From the adjoint system (5.3.34), (5.3.35) (rewritten without the "*"), we then get that

$$\nabla_T J(x) = -2 \int_\Omega \Delta p \, \xi_x \, dx, \quad \forall x \in \Omega.$$

A particularly simple form for $\nabla_T J(x)$ has been obtained in Céa et al. [2000] (although the proof was very elaborate): if $\Omega \subset \mathbf{R}^3$, and if the hole is a small ball, then

$$\nabla_T J(x) = 4\pi \, y(x) \, p(x), \quad x \in \Omega, \tag{5.3.45}$$

with the remark that in (5.3.37) $h(r) = \text{meas}(B(x,r))$ must be replaced by $h(r) = r$. If zero Neumann boundary conditions are imposed on the boundary of the small hole, then $h(r) = \text{meas}(B(x,r))$ is an appropriate choice in arbitrary dimension (compare Sokołowski and Zochowski [1999]). The right-hand side f has to belong to H^2, respectively in C^1, for these results to be valid.

An important observation is that the adjoint state p appearing in (5.3.45) is independent (as a function) of the point in Ω where the perturbation is produced. This suggests the following algorithm for the solution of the *topological* optimization problem (5.3.8), (5.3.9), which is a generalization of the method introduced in §5.2.1.

Algorithm 5.3.5

(1) Choose $\Omega_0 = D$ and set $k = 0$.

(2) Solve the state equation (5.3.9) in Ω_k.

(3) Solve the adjoint equation (5.3.34), (5.3.35) in Ω_k.

(4) Compute $\nabla_T J_k$ by (5.3.45) in Ω_k (the notation J_k stresses that Ω_k is known and variations of Ω_k as in (5.3.37) are considered).

(5) Set $\Omega_{k+1} = \{x \in \Omega_k : \nabla_T J_k(x) \geq c_{k+1}\}$, with c_{k+1} fixed such that $\text{meas}(\Omega_{k+1}) = m_{k+1}$.

(6) IF "convergence" is achieved THEN STOP! ELSE

(7) $k := k + 1$ and GO TO (2).

The "convergence" in step (6) is usually based on an a priori fixed bound for the expression $|J_k - J_{k-1}|$. The quantities $\{m_k\}$ appearing in step (5) are chosen in advance such that $\text{meas}(D) \geq m_k > m_{k+1}$. It is clear that in step (5) an arbitrary number of "holes" of unknown shapes may be inserted in Ω_k. From this point of view, Algorithm 5.3.5 is far from being mathematically justified. Some numerical experiments have been reported in Céa et al. [2000], Garreau et al. [2001], and Guillaume and Idris [2002].

Example 5.3.6 We conclude this paragraph with an illustrative example. Let $B(0,1)$ denote the unit ball in $\mathbf{R}^2$, and let $\alpha \in L^1(B(0,1))$ be given with $\alpha(x_1, x_2) \geq c > 0$ a.e. in $B(0,1)$. Furthermore, let the function $a \in L^\infty(B(0,1))$ be defined by

$$a(x_1, x_2) = \begin{cases} -4, & \text{if } x_1^2 + x_2^2 < \frac{1}{4}, \\ 4 - 2(x_1^2 + x_2^2)^{-1/2}, & \text{if } \frac{1}{4} \leq x_1^2 + x_2^2 \leq 1. \end{cases} \tag{5.3.46}$$

We then consider the minimization problem

$$\operatorname*{Min}_{\Omega \in \mathcal{O}} \left\{ \int_\Omega \alpha(x_1, x_2)\, y(x_1, x_2)\, dx_1\, dx_2 \right\}, \tag{5.3.47}$$

subject to

$$-\Delta y = a, \quad \text{in } \Omega, \qquad y = 0 \quad \text{on } \partial\Omega. \tag{5.3.48}$$

We assume that the admissible domains $\Omega \in \mathcal{O}$, $\Omega \subset B(0,1)$, are smooth, so that the corresponding solutions $y = y(\Omega)$ belong to $W^{2,p}(\Omega) \cap C^1(\overline{\Omega})$ for any $p > 1$.

Let $\Omega = \Omega_0 = B(0,1)$ initially. Then it is easily verified that the corresponding solution $y_0 = y(\Omega_0)$ of (5.3.48) has the form

$$y_0(x_1, x_2) = \begin{cases} \left(x_1^2 + x_2^2\right) - \frac{1}{2}, & \text{if } x_1^2 + x_2^2 \leq \frac{1}{4}, \\ -\left(\sqrt{x_1^2 + x_2^2} - 1\right)^2, & \text{if } \frac{1}{4} \leq x_1^2 + x_2^2 \leq 1. \end{cases} \tag{5.3.49}$$

Clearly, $y_0 \in C^1(\overline{\Omega}_0)$ and $y_0 < 0$ in Ω_0.

Now let $B_\varepsilon \subset B(0,1)$ be any small ball, and let y_ε be the solution to (5.3.48) in $\Omega_\varepsilon = B(0,1) \setminus B_\varepsilon$. Then

$$-\Delta(y_\varepsilon - y_0) = 0 \quad \text{in } \Omega_\varepsilon, \qquad y_\varepsilon = y_0 \quad \text{on } \partial B(0,1), \qquad y_\varepsilon > y_0 \quad \text{on } \partial B_\varepsilon,$$

and it follows from the maximum principle that $y_\varepsilon > y_0$ in Ω_ε, whence

$$\int_{\Omega_\varepsilon} \alpha(x_1,x_2)\, y_\varepsilon(x_1,x_2)\, dx_1\, dx_2 > \int_{\Omega_\varepsilon} \alpha(x_1,x_2)\, y_0(x_1,x_2)\, dx_1\, dx_2$$
$$> \int_{B(0,1)} \alpha(x_1,x_2)\, y_0(x_1,x_2)\, dx_1\, dx_2; \quad (5.3.50)$$

that is, the insertion of a small hole anywhere in $B(0,1)$ increases the value of the cost functional.

Next, we take $\Omega_1 = B(0,1) \setminus B(0,\frac{1}{2})$. By (5.3.46), we have $a > 0$ in Ω_1, and thus we can infer from the maximum principle that the associated solution y_1 of (5.3.48) is positive in Ω_1. Now let $B_\varepsilon \subset \Omega_1$ be a small hole, and let $\tilde{y}_\varepsilon$ denote the solution of (5.3.48) in $\Omega_1 \setminus B_\varepsilon$. Then it follows again from the maximum principle that $\tilde{y}_\varepsilon < y_1$ in $\Omega_1 \setminus B_\varepsilon$ and thus

$$\int_{\Omega_1\setminus B_\varepsilon} \alpha(x_1,x_2)\, \tilde{y}_\varepsilon(x_1,x_2)\, dx_1\, dx_2 < \int_{\Omega_1\setminus B_\varepsilon} \alpha(x_1,x_2)\, y_1(x_1,x_2)\, dx_1\, dx_2$$
$$< \int_{\Omega_1} \alpha(x_1,x_2)\, y_1(x_1,x_2)\, dx_1\, dx_2\,.$$

Therefore, the insertion of a small hole anywhere in Ω_1 decreases the value of the cost.

The fact that the creation of small circular holes in Ω_0 and Ω_1 has opposite effects on the performance index (although $\Omega_1 \subset \Omega_0$) originates from the dependence of the topological derivative (5.3.37) on the domain on which it is calculated (here, once on Ω_0 and once on Ω_1).

Remark. We see that the topological derivative cannot predict the connectivity properties of the global solution in shape optimization problems. Indeed, it is possible that an optimal design problem admits several global solutions having different connectivity characteristics. Like other methods in shape optimization, the topological optimimization produces, in general, just local solutions (in the Hausdorff–Pompeiu distance) for the given problems. In Sokołowski and Zochowski [2003] a similar example is used to perform a comparison between the topological optimization approach and other boundary variation techniques.

In general, no method can ensure more than to find a local solution depending on the initial guess. However, it is to be noted that the methods discussed in this section are characterized by an a priori assumption (which varies with the method) on the type of variations to be taken into account. It is even possible to combine several types of prescribed variations as in Sokołowski and Zochowski [2003]. From this point of view, the approaches studied in Sections 1 and 2 offer complete "freedom," but the intrinsic nonconvexity of optimal design problems again limits the search just to local minima.

Chapter 6

Optimization of Curved Mechanical Systems

In this chapter, we investigate optimal design problems for "thin" mechanical structures such as planar *arches* or three-dimensional curved *rods* and *shells* in the setting of linear elasticity theory. In contrast to the previously studied case of plates, where the thickness was the natural optimization parameter, the minimization problems in this chapter will be formulated in terms of the (a priori unknown) *shape* of the structure. It seems to us that this is a very important problem from the practical point of view, and therefore thickness problems will not be considered here; moreover, this choice also limits the complexity of the arguments, making the text more accessible to the reader.

Since the modeling of thin curved mechanical structures is still under active development (we just quote the recent monographs of Ciarlet [2000], Trabucho and Viaño [1996], and Antman [1995]), we also introduce new convenient models of a generalized *Naghdi type* for shells and for curved rods. These new models fall into the class of the so-called *polynomial models*. Our method has the advantage that the required regularity hypotheses on the geometry are minimal. More precisely, we will impose only C^2 regularity assumptions, while in the literature usually three derivatives are required. In the case of arches, we will employ the classical *Kirchhoff–Love model*, but our treatment differs completely from the usual approach in the literature: indeed, it relies neither on Korn's inequality nor on the Lax–Milgram lemma, and it requires only Lipschitz continuous parametrizations. Hence, after the standard reparametrization, the theory applies to regular curves that are only absolutely continuous. The key to this achievement is to employ the control variational method that was already introduced in §3.4.2 of Chapter 3.

Let us also mention that all the optimization problems to be analyzed in this chapter are of the control into coefficients type. Since they are highly nonconvex, just the existence of optimal geometries can be proved, while uniqueness cannot be expected in general. We will also provide a number of numerical examples. Some of the observed results have a clear physical interpretation, which is a hint that the models and the optimization methods to be discussed in this chapter are well founded from the viewpoint of physics and have good stability properties.

6.1 Kirchhoff–Love Arches

In this section, we will study *Kirchhoff–Love arches*. We recall that arches are mechanical structures in the two-dimensional plane $\mathbf{R}^2$, which, under the impact of forces acting in that plane, may undergo planar deformations. The arches are described by planar curves parametrized with respect to their arc length s. To fix ideas, we generally assume that $s \in [0,1]$, that is, that the arches under consideration have unit length. We also assume that the arches are *clamped*, i.e., fixed at both endpoints.

As a relevant physical example we mention cylindrical shells (i.e., shells that are constant in one direction) that are clamped along two of their generators. If the forces are constant in the same direction and act in the plane perpendicular to it, then it suffices to study the deformation of a two-dimensional cross section perpendicular to the "constant" direction; clearly, this cross section then forms an arch.

We recall that if $\varphi = (\varphi_1, \varphi_2) : [0,1] \to \mathbf{R}^2$ is the parametrization of a Jordan (no self-intersection) smooth clamped arch with respect to its arc length, then $c : [0,1] \to \mathbf{R}$,

$$c(s) = \varphi_2''(s)\varphi_1'(s) - \varphi_1''(s)\varphi_2'(s),$$

denotes its *curvature*, and we have $\theta'(s) = c(s)$, where

$$\theta(s) = \arctan\left(\frac{\varphi_2'(s)}{\varphi_1'(s)}\right)$$

denotes the angle between the horizontal coordinate axis and the tangent vector $\varphi'(s)$ (with $|\varphi'(s)|_{\mathbf{R}^2} = 1$) to the arch in the point $\varphi(s)$. If φ is smooth, then the classical Kirchhoff–Love model (with normalized mechanical constants) consists in finding $v_1 \in H_0^1(0,1)$ and $v_2 \in H_0^2(0,1)$ such that

$$\int_0^1 \left[\frac{1}{\varepsilon}(v_1' - c\,v_2)(s)\,(u_1' - c\,u_2)(s) + (v_2' + c\,v_1)'(s)\,(u_2' + c\,u_1)'(s)\right] ds$$

$$= \int_0^1 (f_1\,u_1 + f_2\,u_2)(s)\,ds \quad \forall\, u_1 \in H_0^1(0,1),\ \forall\, u_2 \in H_0^2(0,1). \quad (6.1.1)$$

Here, $\sqrt{\varepsilon}$ represents the constant thickness of the arch, $[f_1, f_2] \in L^2(0,1)^2$ are, respectively, the tangential and normal components of the forces (internal and external) loading the clamped arch (assumed to act in its plane), while the tangential component v_1 and the normal component v_2 define a similar representation of the deformation.

The decomposition of forces and deformation is performed with respect to the *local basis* (tangent, normal) of $\mathbf{R}^2$, which is constructed for each individual point on the arch; that is, the local basis varies from point to point on the arch. Observe that owing to the definition of c and in order that (6.1.1) be meaningful, the natural smoothness assumption for the geometry of the arch is that $\varphi \in W^{3,\infty}(0,1)^2$.

A thorough investigation, using Dirichlet's principle, Korn's inequality, and the Lax–Milgram lemma for the solvability of (6.1.1), may be found in Ciarlet [1978, p. 432] under the assumption $\varphi \in C^3[0,1]^2$. We will demonstrate in the next paragraph that this regularity hypothesis may be relaxed considerably. In addition, our approach will lead to explicit integration rules for (6.1.1) and is also applicable to variational inequalities.

The associated shape optimization problems will be discussed in §6.1.2. Numerical examples based on our theory will be presented in both parts of this section.

6.1.1 Application of the Control Variational Method

Throughout this section, we will just assume that $\varphi \in W^{1,\infty}(0,1)^2$. Consequently, $\theta \in L^\infty(0,1)$ and c is a distribution, while (6.1.1) becomes meaningless. Consider (formally) the linear homogeneous system of ordinary differential equations that takes into account the terms appearing in (6.1.1), namely

$$v_1'(s) = c(s)\, v_2(s), \quad v_2'(s) = -c(s)\, v_1(s), \quad \text{for a.e. } s \in [0,1].$$

The corresponding fundamental matrix $W(s)$ is obviously given by

$$W(s) = \begin{pmatrix} \cos(\theta(s)) & \sin(\theta(s)) \\ -\sin(\theta(s)) & \cos(\theta(s)) \end{pmatrix}. \tag{6.1.2}$$

Notice that (6.1.2) is meaningful for a.e. $s \in [0,1]$ if $\theta \in L^\infty(0,1)$.

As we have seen in §3.4.2 of Chapter 3, the control variational method proposes a variational formulation of the considered equation in the form of an optimal control problem. The optimal control problem that we associate with (6.1.1) does not have an intuitive character, due to the effort made to relax the regularity assumptions. We will use in (6.1.4) below the *mild formulation* of the Cauchy problem for inhomogeneous ordinary differential equations (here, this coincides with the variation of constants formula). We first introduce the affine part of the state equation (6.1.4) by defining the functions

$$\begin{bmatrix} l \\ h \end{bmatrix}(t) = -\int_0^t W(t)\, W^{-1}(s) \begin{bmatrix} f_1(s) \\ f_2(s) \end{bmatrix} ds, \quad \text{for a.e. } t \in [0,1],$$

$$g_1 = \varepsilon\, l, \quad -g_2'' = h, \text{ a.e. in } [0,1], \quad g_2(0) = g_2(1) = 0. \tag{6.1.3}$$

Then the optimal control problem associated with (6.1.1) reads

$$\text{(P)} \qquad \text{Min}\left\{ L(u,z) = \frac{1}{2\varepsilon}\int_0^1 u^2(s)\, ds + \frac{1}{2}\int_0^1 (z'(s))^2\, ds \right\},$$

subject to $u \in L^2(0,1)$, $z \in H_0^1(0,1)$, to the state equation

$$\begin{bmatrix} v_1 \\ v_2 \end{bmatrix}(t) = \int_0^t W(t)\,W^{-1}(s) \begin{bmatrix} u(s) + g_1(s) \\ z(s) + g_2(s) \end{bmatrix} ds, \quad \text{for a.e. } t \in [0,1], \tag{6.1.4}$$

and to the constraint

$$\int_0^1 W^{-1}(s) \begin{bmatrix} u(s) + g_1(s) \\ z(s) + g_2(s) \end{bmatrix} ds = \begin{bmatrix} 0 \\ 0 \end{bmatrix}. \tag{6.1.5}$$

Notice that (6.1.5) is nothing but a weak form of the boundary condition $v_1(1) = v_2(1) = 0$; indeed, this results from (6.1.4) under the assumption that $W(1)^{-1}$ exists (which cannot be guaranteed, in general, since the entries of W only belong to $L^\infty(0,1)$).

Obviously, $[u,z] = [-g_1, -g_2]$ is admissible for (P). Since the cost functional $L(u,z)$ defined by (P) satisfies the coercivity condition (2.1.6), Theorem 2.1.2 in Chapter 2 yields the existence of a minimizer $[u_\varepsilon, z_\varepsilon] \in L^2(0,1) \times H_0^1(0,1)$, which, owing to the strict convexity of the cost functional, is unique.

Let us denote by $S \subset L^2(0,1) \times H_0^1(0,1)$ the closed subspace of admissible variations for (P). Obviously, $[\mu, \xi] \in S$ if and only if

$$\int_0^1 W^{-1}(s) \begin{bmatrix} \mu(s) \\ \xi(s) \end{bmatrix} ds = \begin{bmatrix} 0 \\ 0 \end{bmatrix}. \tag{6.1.6}$$

Now, for any $[\mu, \xi] \in S$ and any $\lambda \in \mathbf{R}$, we have

$$L(u_\varepsilon, z_\varepsilon) \le L(u_\varepsilon + \lambda\mu, z_\varepsilon + \lambda\xi).$$

Writing this inequality in explicit form, dividing by $\lambda \neq 0$, and taking the limit as $\lambda \to 0$, we easily conclude that $[u_\varepsilon, z_\varepsilon]$ satisfies

$$\frac{1}{\varepsilon}\int_0^1 u_\varepsilon(s)\,\mu(s)\,ds + \int_0^1 z_\varepsilon'(s)\,\xi'(s)\,ds = 0 \quad \forall\,[\mu,\xi] \in S. \tag{6.1.7}$$

Observe that the left-hand side of (6.1.7) defines a new scalar product $(\cdot\,,\,\cdot)_\varepsilon$ on the space $L^2(0,1) \times H_0^1(0,1)$; that is, for all $[u,z], [\mu,\xi] \in L^2(0,1) \times H_0^1(0,1)$ we take

$$([u,z],[\mu,\xi])_\varepsilon = \frac{1}{\varepsilon}\int_0^1 u(s)\,\mu(s)\,ds + \int_0^1 z'(s)\,\xi'(s)\,ds.$$

Then (6.1.7) means that $[u_\varepsilon, z_\varepsilon] \in S_\varepsilon^\perp$, where $S_\varepsilon^\perp$ denotes the orthogonal complement of S in $L^2(0,1) \times H_0^1(0,1)$ with respect to $(\cdot\,,\,\cdot)_\varepsilon$.

Remark. If $\theta \in W^{1,1}(0,1)$, then $c \in L^1(0,1)$, and (6.1.4) can be rewritten in differential form as

$$v_1' - c\,v_2 = u + g_1 \qquad \text{a.e. in } [0,1], \tag{6.1.8}$$

$$v_2' + c\,v_1 = z + g_2 \qquad \text{a.e. in } [0,1]. \tag{6.1.9}$$

Formula (6.1.4) gives the "mild" solution of (6.1.8), (6.1.9) with zero initial conditions in the sense of semigroup theory (cf. Bénilan [1972], and Barbu [1993]). Relation (6.1.5) is a state constraint. It is expressed directly as a control constraint, since the system (6.1.8), (6.1.9) is solved by (6.1.4), and the matrix $W(t)$ exists and is nonsingular for every $t \in [0,1]$ if $\theta \in W^{1,1}(0,1)$.

In the following, we will denote by $[v_1^\varepsilon, v_2^\varepsilon] \in L^\infty(0,1)^2$ the optimal state of (P), obtained from $[u_\varepsilon, z_\varepsilon]$ via (6.1.4). The following result relates the control problem (P) directly to the original problem (6.1.1).

Theorem 6.1.1 *If $\varphi \in (W^{3,\infty}(0,1))^2$, then $[v_1^\varepsilon, v_2^\varepsilon]$ is the unique solution to (6.1.1).*

Proof. Under the regularity assumption on φ, (6.1.4) can be written in the form (6.1.8), (6.1.9). For any $u_1 \in H_0^1(0,1)$, $u_2 \in H_0^2(0,1)$, we introduce

$$\tilde{\mu} = u_1' - c\,u_2 \in L^2(0,1), \tag{6.1.10}$$

$$\tilde{\xi} = u_2' + c\,u_1 \in H_0^1(0,1). \tag{6.1.11}$$

Obviously, since W is a fundamental system also for (6.1.10), (6.1.11), we have, for every $t \in [0,1]$,

$$\begin{bmatrix} u_1 \\ u_2 \end{bmatrix}(t) = \int_0^t W(t)\,W^{-1}(s) \begin{bmatrix} \tilde{\mu} \\ \tilde{\xi} \end{bmatrix}(s)\,ds. \tag{6.1.12}$$

Since u_1, u_2 vanish at both ends of $[0,1]$, it follows from (6.1.12) and (6.1.6) that $[\tilde{\mu}, \tilde{\xi}] \in S$; hence it may be inserted in (6.1.7). Taking into account that v_1^ε, v_2^ε satisfy (6.1.8), (6.1.9), and invoking (6.1.10), (6.1.11), and (6.1.3), we obtain that

$$\begin{aligned} 0 &= \frac{1}{\varepsilon}\int_0^1 ((v_1^\varepsilon)' - cv_2^\varepsilon - g_1)\,(u_1' - cu_2)\,ds \\ &\quad + \int_0^1 ((v_2^\varepsilon)' + cv_1^\varepsilon - g_2)'\,(u_2' + cu_1)'\,ds \\ &= \frac{1}{\varepsilon}\int_0^1 ((v_1^\varepsilon)' - cv_2^\varepsilon)\,(u_1' - cu_2)\,ds + \int_0^1 ((v_2^\varepsilon)' + cv_1^\varepsilon)'\,(u_2' + cu_1)'\,ds \\ &\quad - \int_0^1 l\,(u_1' - cu_2)\,ds - \int_0^1 h\,(u_2' + cu_1)\,ds. \end{aligned}$$

By the regularity assumption, (6.1.3) can be rewritten in the differential form (6.1.8), (6.1.9), and we can infer that

$$\begin{aligned} &\int_0^1 l\,(u_1' - c\,u_2)\,ds + \int_0^1 h\,(u_2' - c\,u_1)\,ds \\ &\quad = -\int_0^1 u_1\,(l' - c\,h)\,ds - \int_0^1 u_2\,(h' + c\,l)\,ds \\ &\quad = \int_0^1 (f_1\,u_1 + f_2\,u_2)\,ds. \end{aligned}$$

The last two relations imply (6.1.1), which finishes the proof. □

Remark. The approach via the control problem (P) is constructive and uses neither Dirichlet's principle nor Korn's inequality. Since the formulation of (P) is meaningful for $\theta \in L^\infty(0,1)$, this method introduces a notion of weak solution for the arch problem even in nonsmooth situations in which Korn's inequality is not valid (for such cases, we refer to Geymonat and Gilardi [1998]). This notion is a natural extension of the usual one given by (6.1.1). This will be further justified below via an approximation argument (see the second remark following Corollary 6.1.6).

We now introduce the auxiliary mappings $w_1, w_2 \in H^2(0,1) \cap H_0^1(0,1)$ given as the unique solutions to the boundary value problems

$$w_1''(s) = \sin(\theta(s)) \quad \text{for a.e. } s \in [0,1], \quad w_1(0) = w_1(1) = 0, \tag{6.1.13}$$

$$w_2''(s) = -\cos(\theta(s)) \quad \text{for a.e. } s \in [0,1], \quad w_2(0) = w_2(1) = 0. \tag{6.1.14}$$

Taking the form of W into account, the definition (6.1.6) of S can be rewritten as

$$\begin{aligned}
\int_0^1 [\mu(s)\cos(\theta(s)) - \xi(s)\sin(\theta(s))]\, ds &= 0,\\
\int_0^1 [\mu(s)\sin(\theta(s)) + \xi(s)\cos(\theta(s))]\, ds &= 0.
\end{aligned}$$

Replacing the factors multiplying $\xi(s)$ according to (6.1.13), (6.1.14), and integrating once by parts, we find that (6.1.6) may be written in the equivalent form

$$\frac{1}{\varepsilon}\int_0^1 \varepsilon\, \mu(s)\, \cos(\theta(s))\, ds + \int_0^1 w_1'(s)\, \xi'(s)\, ds = 0, \tag{6.1.15}$$

$$\frac{1}{\varepsilon}\int_0^1 \varepsilon\, \mu(s)\, \sin(\theta(s))\, ds + \int_0^1 w_2'(s)\, \xi'(s)\, ds = 0. \tag{6.1.16}$$

From this it follows that the set $\{b_1, b_2\}$, where

$$\begin{aligned}
b_1(s) &= [\varepsilon\, \cos(\theta(s)), w_1(s)], \qquad \text{a.e. in } [0,1],\\
b_2(s) &= [\varepsilon\, \sin(\theta(s)), w_2(s)], \qquad \text{a.e. in } [0,1],
\end{aligned}$$

spans $S_\varepsilon^\perp$; a fortiori, this set forms a basis of $S_\varepsilon^\perp$ (which is therefore two-dimensional). To verify this, assume there are $r_1, r_2 \in \mathbf{R}$ such that

$$\begin{aligned}
0 &= r_1\, \varepsilon\, \cos(\theta(s)) + r_2\, \varepsilon\, \sin(\theta(s)) \quad \text{a.e. in } [0,1],\\
0 &= r_1\, w_1(s) + r_2\, w_2(s) \quad \text{a.e. in } [0,1].
\end{aligned}$$

In view of (6.1.13) and (6.1.14), we obtain from the second equality that

$$0 = r_1\, w_1''(s) + r_2\, w_2''(s) = r_1\, \sin(\theta(s)) - r_2\, \cos(\theta(s)) \quad \text{a.e. in } [0,1],$$

whence, using also the first equality, $r_1 = r_2 = 0$.

Observe now that from the relations (6.1.5) and (6.1.6) we can infer that $[u_\varepsilon + g_1, z_\varepsilon + g_2] \in S$. Consequently, since (6.1.7) yields that $[u_\varepsilon, z_\varepsilon] \in S_\varepsilon^\perp$, the orthogonal decomposition theorem implies that

$$[u_\varepsilon, z_\varepsilon] = -P_{S_\varepsilon^\perp}[g_1, g_2], \tag{6.1.17}$$

where $P_{S_\varepsilon^\perp}$ is the orthogonal projection operator onto $S_\varepsilon^\perp$, with respect to $(\cdot\,,\cdot)_\varepsilon$. Using once more the fact that $[u_\varepsilon, z_\varepsilon] \in S_\varepsilon^\perp$, which has the basis $\{b_1, b_2\}$, we find that

$$[u_\varepsilon, z_\varepsilon] = \lambda_1^\varepsilon\,[\varepsilon\,\cos(\theta), w_1] + \lambda_2^\varepsilon\,[\varepsilon\,\sin(\theta), w_2] \tag{6.1.18}$$

with suitable $\lambda_1^\varepsilon, \lambda_2^\varepsilon \in \mathbf{R}$. By virtue of the definition of the projection operator, and owing to (6.1.17), (6.1.18), we see that $[\lambda_1^\varepsilon, \lambda_2^\varepsilon]$ is the unique minimizer of the unconstrained optimization problem

$$\text{(D)} \qquad \operatorname*{Min}_{[\lambda_1,\lambda_2]\in\mathbf{R}^2} \left\{ \frac{1}{2\varepsilon}\int_0^1 (\lambda_1\,\varepsilon\,\cos(\theta(s)) + \lambda_2\,\varepsilon\,\sin(\theta(s)) + \varepsilon\, l(s))^2\,ds \right.$$
$$\left. + \frac{1}{2}\int_0^1 ((\lambda_1\,w_1 + \lambda_2\,w_2 + g_2)'(s))^2\,ds \right\}.$$

Problem (D) can be explicitly solved by putting the partial derivatives of the quadratic form with respect to λ_1, λ_2, equal to zero, which leads to the linear-algebraic system

$$\varepsilon\,\lambda_1 \int_0^1 \cos^2(\theta(s))\,ds + \lambda_1\,|w_1|^2_{H_0^1(0,1)} + \varepsilon\,\lambda_2 \int_0^1 \cos(\theta(s))\,\sin(\theta(s))\,ds$$
$$+\lambda_2 \int_0^1 w_1'(s)\,w_2'(s)\,ds + \varepsilon \int_0^1 l(s)\,\cos(\theta(s))\,ds + \int_0^1 g_2'(s)\,w_1'(s)\,ds \;=\; 0,$$
$$\varepsilon\,\lambda_1 \int_0^1 \cos(\theta(s))\,\sin(\theta(s))\,ds + \lambda_1 \int_0^1 w_1'(s)\,w_2'(s)\,ds + \varepsilon\,\lambda_2 \int_0^1 \sin^2(\theta(s))\,ds$$
$$+\lambda_2\,|w_2|^2_{H_0^1(0,1)} + \varepsilon \int_0^1 l(s)\,\sin(\theta(s))\,ds + \int_0^1 g_2'(s)\,w_2'(s)\,ds \;=\; 0. \tag{6.1.19}$$

Obviously, the coefficient matrix of this linear-algebraic system is of the form

$$\begin{pmatrix} (b_1, b_1)_\varepsilon & (b_1, b_2)_\varepsilon \\ (b_1, b_2)_\varepsilon & (b_2, b_2)_\varepsilon \end{pmatrix},$$

and thus, since $\{b_1, b_2\}$ is a basis, is invertible with a positive determinant. We have arrived at the following result.

Theorem 6.1.2 *If $\theta \in L^\infty(0,1)$, then the solution of (P) is given by (6.1.18) and (6.1.4), where $[\lambda_1^\varepsilon, \lambda_2^\varepsilon]$ is the unique solution to (6.1.19), and where w_1, w_2, g_1, g_2 are defined by (6.1.2), (6.1.3), (6.1.13), (6.1.14).*

Remark. In terms of optimization theory, (D) is the *dual* problem of (P). Its complete solution is possible in this case since the constraints of (P) are affine and finite-dimensional (compare Barbu and Precupanu [1986]). For simple choices of the given data θ, f_1, f_2, explicit expressions can be given for the deformation $[v_1^\varepsilon, v_2^\varepsilon]$, while in general, numerical quadrature formulas have to be employed to evaluate the occurring integrals.

Remark. If $\tilde{\varphi} : [a,b] \to \mathbf{R}^2$ is an absolutely continuous Jordan arc of unit length satisfying $\tilde{\varphi}' \neq 0$ a.e. in $[a,b]$, then the usual reparametrization via the arc length function $s : [a,b] \to [0,1]$, $s(0) = 0$, $s'(t) = |\tilde{\varphi}'(t)|_{\mathbf{R}^2}$, yields that $\varphi(t) = \tilde{\varphi}(s^{-1}(t))$ satisfies $|\varphi'(t)|_{\mathbf{R}^2} = 1$ for almost every $t \in [0,1]$, that is, φ is Lipschitz continuous, and our results still apply.

Remark. If $\theta \in L^\infty(0,1)$, then $v_1^\varepsilon, v_2^\varepsilon \in L^\infty(0,1)$. However, the representation in global Cartesian coordinates is given by

$$W(t)^{-1} \begin{bmatrix} v_1^\varepsilon(t) \\ v_2^\varepsilon(t) \end{bmatrix} = \int_0^t W^{-1}(s) \begin{bmatrix} u_\varepsilon(s) + g_1(s) \\ z_\varepsilon(s) + g_2(s) \end{bmatrix} ds,$$

and belongs to $W^{1,2}(0,1)^2$. This means that the apparent lack of smoothness is due just to the local coordinates (θ is defined only a.e. and may have jumps), while the constructed deformation is in fact continuous.

The next result brings a characterization of the solution to the problem (P) (or, equivalently, to the problem (D)) as a system of first-order differential equations, which will be used frequently in the following. It implicitly provides a nonstandard decomposition of (6.1.1) in the case of nonsmooth coefficients. Basically, it is given by the first-order necessary conditions for (P), but the form differs from the classical Pontryagin principle.

Theorem 6.1.3 *Suppose that $\theta \in L^\infty(0,1)$. Then the pair $[u_\varepsilon, z_\varepsilon]$ is optimal for (P) if and only if there are $\lambda_1^\varepsilon, \lambda_2^\varepsilon \in \mathbf{R}$ and $p_\varepsilon, q_\varepsilon, v_1^\varepsilon, v_2^\varepsilon \in L^\infty(0,1)$ such that*

$$\begin{bmatrix} v_1^\varepsilon \\ v_2^\varepsilon \end{bmatrix}(t) = \int_0^t W(t)\, W^{-1}(s) \begin{bmatrix} u_\varepsilon(s) + g_1(s) \\ z_\varepsilon(s) + g_2(s) \end{bmatrix} ds, \quad \textit{for a.e. } t \in [0,1], \tag{6.1.20}$$

$$\int_0^1 W^{-1}(s) \begin{bmatrix} u_\varepsilon(s) + g_1(s) \\ z_\varepsilon(s) + g_2(s) \end{bmatrix} ds = \begin{bmatrix} 0 \\ 0 \end{bmatrix}, \tag{6.1.21}$$

$$\begin{bmatrix} p_\varepsilon \\ q_\varepsilon \end{bmatrix}(t) = W(t) \begin{bmatrix} \lambda_1^\varepsilon \\ \lambda_2^\varepsilon \end{bmatrix}, \quad \textit{for a.e. } t \in [0,1], \tag{6.1.22}$$

$$u_\varepsilon = \varepsilon\, p_\varepsilon \quad \textit{a.e. in } [0,1], \tag{6.1.23}$$

$$z_\varepsilon'' = -q_\varepsilon \quad \textit{a.e. in } [0,1], \quad z_\varepsilon(0) = z_\varepsilon(1) = 0. \tag{6.1.24}$$

Proof. Assume first that $[u_\varepsilon, z_\varepsilon]$ satisfies the system (6.1.20)–(6.1.24) with some $\lambda_1^\varepsilon, \lambda_2^\varepsilon \in \mathbf{R}$ and $p_\varepsilon, q_\varepsilon, v_1^\varepsilon, v_2^\varepsilon \in L^\infty(0,1)$. Then clearly, $[u_\varepsilon + g_1, z_\varepsilon + g_2] \in S$, i.e.,

$[u_\varepsilon, z_\varepsilon]$ is admissible for (P). Using (6.1.22)–(6.1.24), the definition of S, and the orthogonality of the matrix $W(t)$, we find that for any $[\mu, \xi] \in S$,

$$\frac{1}{\varepsilon}\int_0^1 u_\varepsilon(s)\,\mu(s)\,ds + \int_0^1 z_\varepsilon'(s)\,\xi'(s)\,ds = \int_0^1 p_\varepsilon(s)\,\mu(s)\,ds + \int_0^1 q_\varepsilon(s)\,\xi(s)\,ds$$

$$= \int_0^1 [\mu(s), \xi(s)]\, W(s) \begin{bmatrix} \lambda_1^\varepsilon \\ \lambda_2^\varepsilon \end{bmatrix} ds = [\lambda_1^\varepsilon, \lambda_2^\varepsilon] \int_0^1 W(s)^{-1} \begin{bmatrix} \mu(s) \\ \xi(s) \end{bmatrix} ds = 0.$$

Consequently, $[u_\varepsilon, z_\varepsilon] \in S_\varepsilon^\perp$, and it follows that $[u_\varepsilon, z_\varepsilon]$ is the unique minimizer of (P). Indeed, if $[\bar{u}, \bar{z}]$ is another admissible control for (P), then $[\bar{u} - u_\varepsilon, \bar{z} - z_\varepsilon] \in S$ is orthogonal to $[u_\varepsilon, z_\varepsilon]$ with respect to the scalar product $(\cdot,\cdot)_\varepsilon$. A direct computation of the cost proves the minimum property for $[u_\varepsilon, z_\varepsilon]$.

Conversely, notice that (6.1.22)–(6.1.24) give a complete description of the two-dimensional space $S_\varepsilon^\perp$ when $\lambda_1, \lambda_2 \in \mathbf{R}$ are arbitrary. By virtue of (6.1.6), the optimal control $[u_\varepsilon, z_\varepsilon]$ belongs to $S_\varepsilon^\perp$. Hence, there are $\lambda_1^\varepsilon, \lambda_2^\varepsilon \in \mathbf{R}$ such that $[u_\varepsilon, z_\varepsilon]$ can be represented via (6.1.22)–(6.1.24) (which is in fact the same representation as in (6.1.18)). Moreover, since $[u_\varepsilon, z_\varepsilon]$ is admissible for (P), also (6.1.20) and (6.1.21) are fulfilled, which ends the proof of the assertion. □

Example 6.1.4 We close this section with three examples for the computation of the deformation of arches having various shapes and thicknesses under the action of given loads. In Figures 1.1–1.3 a Gothic, a Roman, and a closed arch, respectively, are depicted. The forces are constant and act at every point on the considered curve in the normal direction (Figure 1.1), or in the "vertical" direction (Figures 1.2 and 1.3). The calculations are based on Theorem 1.2, where the corresponding solutions have been obtained using the software package Maple (see Richards [2002]).

The parametrization $\varphi = [\varphi_1, \varphi_2]$ of an arch associated with some function θ on a prescribed interval is given by the identities $\varphi_1' = \cos(\theta)$, $\varphi_2' = \sin(\theta)$, together with zero initial conditions. Notice that in the case of the Gothic arch as in Figure 1.1 the function θ is obviously discontinuous, which shows the need of relaxing the regularity assumptions for φ.

Figures 1.2 and 1.3 show the same type of arch with similar loading. The difference in the shapes of the resulting deformations is due to the fact that the first arch is clamped at both ends, while the closed arch is clamped only in the point $(0,0)$. The constant E represents the Young modulus of the material, while $S = \varepsilon^{3/2}$ reflects the influence of the thickness. As an example, we give the explicit form of the deformation $[v_1, v_2]$ corresponding to the situation described in Figure 1.2. We have:

$$\begin{aligned} v_1(t) = {} & (6\varepsilon \sin(t) + 4\sin(t) + 2\pi\varepsilon \sin(t) + \pi\varepsilon^2\, t\sin(t) - 4\varepsilon\, t\cos(t) \\ & -2\varepsilon^2 \sin(t) - 2\varepsilon\, t^2 \sin(t) - \varepsilon^2 t^2 \sin(t) + \pi\, t \sin(t) - 4\, t \cos(t) \\ & -t^2 \sin(t) - 2\pi - 2\varepsilon\pi + 2\pi \cos(t) + 2\pi\varepsilon \cos(t))/[4\,\varepsilon^{3/2}\,(\varepsilon + 1)], \\ v_2(t) = {} & (\varepsilon + 1)\,(2t \sin(t) + \pi\, t \cos(t) - \pi \sin(t) - t^2 \cos(t))/(4\,\varepsilon^{3/2}). \end{aligned}$$

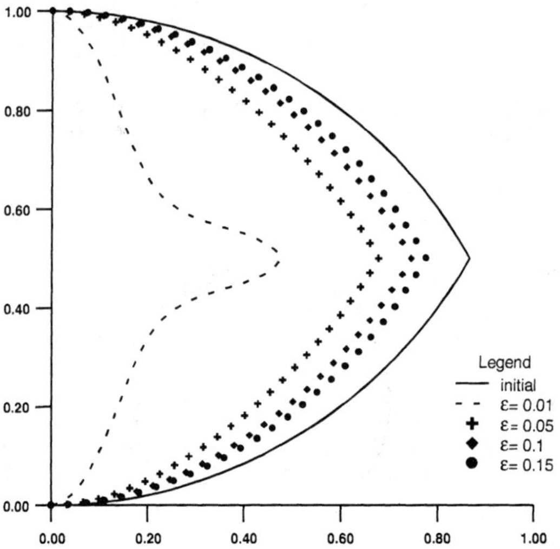

Figure 1.1. $\theta(t) = t,\ t \in \left[0, \frac{\pi}{3}\right],\ \theta(t) = t + \frac{\pi}{3},\ t \in \left[\frac{\pi}{3}, \frac{2\pi}{3}\right],\ f_1(t) = 0,$ $f_2(t) = \frac{1}{SE},\ E = 10,$ © SIAM.

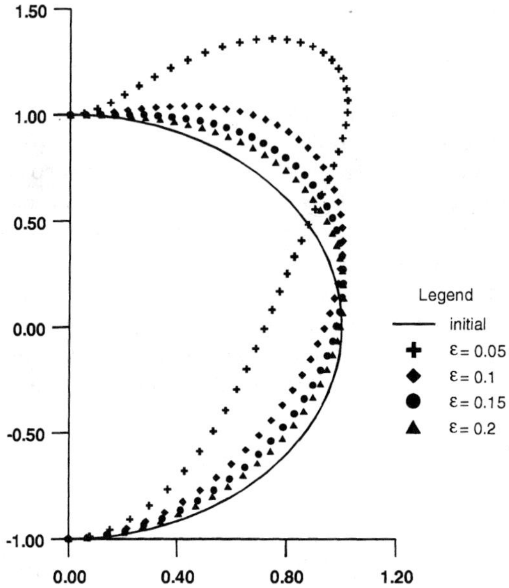

Figure 1.2. $\theta(t) = t,\ t \in [0, \pi],\ f_1(t) = \frac{\sin(t)}{S},\ f_2(t) = \frac{\cos(t)}{S},$ © SIAM.

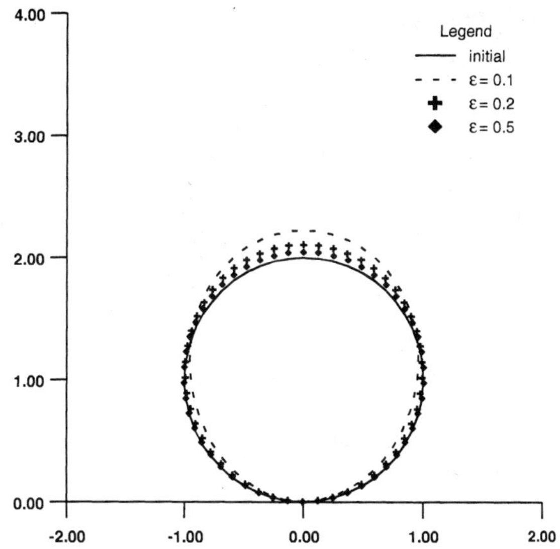

Figure 1.3. $\theta(t) = t,\ f_1(t) = \frac{\sin(t)}{SE},\ f_2(t) = \frac{\cos(t)}{SE},\ t \in [0, 2\pi],\ E = 100,$ © SIAM.

6.1.2 Optimization of Nonsmooth Arches

Assuming that the load $[f_1, f_2]$ is prescribed, the question examined in this paragraph is how to determine the shape of the arch (i.e., its parametrization) that minimizes some given integral cost functional. This shape optimization problem is quite similar to the type of problem studied in Chapter 5. It arises naturally as a control into coefficients problem since the geometry of the arch is completely characterized by the curvature c (the coefficient occurring in (6.1.1)) or, equivalently, by θ, which appears in the formulation of (P) (see (6.1.25)–(6.1.29) below). We study the model problem

$$\text{(Q)} \qquad \operatorname*{Min}_{\theta \in U_{ad}} \left\{ \frac{1}{2} \int_0^1 v_2^2(s)\, ds \right\},$$

where $U_{ad} \subset L^\infty(0,1)$ is a nonempty and closed set, and where v_2 represents the normal deformation of the arch, as in §6.1.1. The cost functional of the optimal design problem (Q) can be motivated by natural safety requirements for the structure. Various other types of cost functionals may be investigated by the technique to be developed below.

The general assumption $U_{ad} \subset L^\infty(0,1)$, that is, $\varphi \in W^{1,\infty}(0,1)^2$, is justified by applications (see Example 6.1.4) and by the remark that many numerical methods lead to this type of condition for the computed approximating arches, which are, in general, just piecewise regular. Then we do not have enough smoothness to use the form (6.1.1) of the Kirchhoff–Love model. Instead, we employ the corresponding characterization (6.1.20)–(6.1.24), which we rewrite in a slightly modified form:

$$\begin{bmatrix} v_1 \\ v_2 \end{bmatrix}(t) = \int_0^t W_\theta(t)\, W_\theta^{-1}(s) \begin{bmatrix} u(s)+g_1(s) \\ z(s)+g_2(s) \end{bmatrix} ds, \quad \text{a.e. in } [0,1], \tag{6.1.25}$$

$$\int_0^1 W_\theta^{-1}(s) \begin{bmatrix} u(s)+g_1(s) \\ z(s)+g_2(s) \end{bmatrix} ds = \begin{bmatrix} 0 \\ 0 \end{bmatrix}, \tag{6.1.26}$$

$$\begin{bmatrix} p \\ q \end{bmatrix}(t) = W_\theta(t) \begin{bmatrix} \lambda_1 \\ \lambda_2 \end{bmatrix} \quad \text{a.e. in } [0,1], \tag{6.1.27}$$

$$u = \varepsilon\, p \quad \text{a.e. in } [0,1], \tag{6.1.28}$$

$$z'' = -q \quad \text{a.e. in } [0,1], \quad z(0) = z(1) = 0. \tag{6.1.29}$$

The new matrix notation W_θ instead of W as in (6.1.2) stresses the dependence on $\theta \in U_{ad}$, which is now the minimization parameter. All quantities appearing in (6.1.25)–(6.1.29) depend on θ, including the "data" $[g_1, g_2]$ that are derived from the prescribed $[f_1, f_2]$ via (6.1.3). The load $[f_1, f_2]$ is fixed; however, the use of the local system of axes in the formulation of (6.1.1) or in (6.1.20)–(6.1.24), (6.1.25)–(6.1.29), makes it necessary to rewrite $[f_1, f_2]$ in terms of the local coordinate frame, which introduces the dependence on θ. In the following, we will indicate this dependence by writing $v_1(\theta)$, $v_2(\theta)$, $\lambda_1(\theta)$, $\lambda_2(\theta)$, and so on.

We notice that the shape optimization problem (Q) is a nonconvex control into coefficients problem and that the mapping $\theta \mapsto v_2(\theta)$ is strongly nonlinear. We have, however, continuity, as the following result shows.

Theorem 6.1.5 *Suppose that a sequence $\{\theta_n\} \subset L^\infty(0,1)$ is given such that $\theta_n \to \theta$ strongly in $L^\infty(0,1)$ and $f_i(\theta_n) \to f_i(\theta)$, $i = 1, 2$, strongly in $L^1(0,1)$. Then*

$$\begin{aligned}
& & W_{\theta_n} \to W_\theta, \quad & \textit{strongly in } L^\infty(0,1)^4,\\
& h(\theta_n) \to h(\theta), & l(\theta_n) \to l(\theta), \quad & g_1(\theta_n) \to g_1(\theta), \quad p(\theta_n) \to p(\theta),\\
& q(\theta_n) \to q(\theta), & u(\theta_n) \to u(\theta), \quad & \textit{all strongly in } L^\infty(0,1),\\
& g_2(\theta_n) \to g_2(\theta), & z(\theta_n) \to z(\theta), \quad & \textit{strongly in } W^{2,\infty}(0,1),\\
& v_1(\theta_n) \to v_1(\theta), & v_2(\theta_n) \to v_2(\theta), \quad & \textit{strongly in } L^\infty(0,1),\\
& \lambda_1(\theta_n) \to \lambda(\theta), & \lambda_2(\theta_n) \to \lambda(\theta). \quad &
\end{aligned}$$

If $\theta_n \to \theta$ strongly in $C[0,1]$, then the above convergences are valid in $C[0,1]$ and $C^2[0,1]$.

Proof. Since $\theta_n \to \theta$ strongly in $L^\infty(0,1)$, we conclude that $\cos(\theta_n) \to \cos(\theta)$ and $\sin(\theta_n) \to \sin(\theta)$ strongly in $L^\infty(0,1)$. Hence, $W_{\theta_n} \to W_\theta$ and $W_{\theta_n}^{-1} \to W_\theta^{-1}$ strongly in $L^\infty(0,1)^4$. Moreover, if $\Delta(\theta_n)$ denotes the determinant associated with the system (6.1.19) (written for θ_n), we obtain by a direct calculation that $\Delta(\theta_n) \to \Delta(\theta)$. In addition, relation (6.1.3) implies that

$$\begin{aligned}
&\left| \begin{bmatrix} l(\theta_n) \\ h(\theta_n) \end{bmatrix} (t) - \begin{bmatrix} l(\theta) \\ h(\theta) \end{bmatrix} (t) \right|_{\mathbf{R}^2} \\
&\quad \le |W_{\theta_n} - W_\theta|_{L^\infty(0,1)^4} \, |W_{\theta_n}^{-1}|_{L^\infty(0,1)^4} \left| \begin{bmatrix} f_1(\theta_n) \\ f_2(\theta_n) \end{bmatrix} \right|_{L^1(0,1)^2} \\
&\qquad + |W_\theta|_{L^\infty(0,1)^4} \, |W_{\theta_n}^{-1} - W_\theta^{-1}|_{L^\infty(0,1)^4} \left| \begin{bmatrix} f_1(\theta_n) \\ f_2(\theta_n) \end{bmatrix} \right|_{L^1(0,1)^2} \\
&\qquad + |W_\theta|_{L^\infty(0,1)^4} \, |W_\theta|_{L^\infty(0,1)^4}^{-1} \left| \begin{bmatrix} f_1(\theta_n) - f_1(\theta) \\ f_2(\theta_n) - f_2(\theta) \end{bmatrix} \right|_{L^1(0,1)^2} .
\end{aligned} \tag{6.1.30}$$

Consequently, $l(\theta_n) \to l(\theta)$ and $h(\theta_n) \to h(\theta)$ strongly in $L^\infty(0,1)$. Using (6.1.3) again, we then conclude that $g_1(\theta_n) \to g_1(\theta)$ strongly in $L^\infty(0,1)$, as well as $g_2(\theta_n) \to g_2(\theta)$ strongly in $W^{2,\infty}(0,1)$. Similarly, $w_1(\theta_n) \to w_1(\theta)$ and $w_2(\theta_n) \to w_2(\theta)$ strongly in $W^{2,\infty}(0,1)$, by (6.1.13) and (6.1.14). Since $\Delta(\theta_n) \to \Delta(\theta)$, one also obtains, by solving (6.1.19), that $\lambda_1(\theta_n) \to \lambda_1(\theta)$ and $\lambda_2(\theta_n) \to \lambda_2(\theta)$.

Next, the equations (6.1.27)–(6.1.29) yield the asserted convergences for the sequences $\{p(\theta_n)\}$, $\{q(\theta_n)\}$, $\{u(\theta_n)\}$, and $\{z(\theta_n)\}$. The proof of the convergences $v_1(\theta_n) \to v_1(\theta)$ and $v_2(\theta_n) \to v_2(\theta)$ is similar to that used in the derivation of (6.1.30). Finally, if $\theta_n \to \theta$ in $C[0,1]$, then the asserted stronger convergences follow with minor modifications by arguing along the same lines.

□

Corollary 6.1.6 *If U_{ad} is compact in $L^\infty(0,1)$, then the shape optimization problem (Q) has at least one solution $\theta^* \in U_{ad}$.*

Proof. Suppose that $\{\theta_n\} \subset U_{ad}$ is a minimizing sequence. Since U_{ad} is compact, there is a subsequence, again indexed by n, such that $\theta_n \to \hat\theta \in U_{ad}$ strongly in $L^\infty(0,1)$. By Theorem 6.1.5, we may pass to the limit as $n \to \infty$ in all expressions, in particular in the constraint (6.1.26) and in the cost functional. It follows that $\theta^* = \hat\theta$ minimizes (Q). □

Remark. In the classical setting (6.1.1) it suffices to assume that the curvature $c = \theta'$ is bounded in $L^r(0,1)$ for some $r > 1$ in order to guarantee that U_{ad} is compact in $C[0,1]$. Our compactness assumption is very weak compared with the standard assumptions imposed in the literature for control into coefficients problems, which usually postulate that also the derivative c' is bounded in $L^r(0,1)$.

Remark. For any $\theta \in L^\infty(0,1)$, we may use Friedrichs mollifiers to construct a sequence $\{\theta_n\} \subset C^\infty(0,1)$ such that $\theta_n \to \theta$ in $L^r(0,1)$ for all $r \geq 1$. Then, keeping $[f_1, f_2] \in L^2(0,1)^2$ fixed, it is possible to modify the proof of Theorem 6.1.5 and to show that $v_1(\theta_n) \to v_1(\theta)$ and $v_2(\theta_n) \to v_2(\theta)$ strongly in $L^r(0,1)$ for all $r \geq 1$. Now observe that the "global" coordinates of $[v_1(\theta_n), v_2(\theta_n)]$ are given by

$$W_{\theta_n}^{-1}(t) \begin{bmatrix} v_1(\theta_n)(t) \\ v_2(\theta_n)(t) \end{bmatrix},$$

and, thanks to (6.1.25), are convergent in $W^{1,r}(0,1)^2$ for all $r \geq 1$. Since for θ_n the corresponding solutions of (P) coincide with the solution of (6.1.1), we conclude from Theorem 6.1.1 that for any $\theta \in L^\infty(0,1)$ the optimal state of (P) may be approximated by usual solutions of (6.1.1). In addition to Theorem 6.1.1, this is another property showing that the optimal state of (P) provides a generalized solution to (6.1.1).

We continue with the sensitivity analysis of the Kirchhoff–Love model with respect to geometrical variations. We proceed in two steps: first, we assume that $c \in L^1(0,1)$ and, consequently, $\theta \in W^{1,1}(0,1)$, and compute the gradient of the cost functional for this case; then, we use an approximation argument to treat the general case $\theta \in L^\infty(0,1)$.

Now let $c \in L^1(0,1)$. Recalling (6.1.2), and the definition of W_θ as a fundamental matrix, we may rewrite the state system (6.1.25)–(6.1.29) of (Q) in the differential form

$$\begin{aligned}
v_1' - c\,v_2 &= u + g_1 \quad \text{a.e. in } [0,1], && (6.1.31)\\
v_2' + c\,v_1 &= z + g_2 \quad \text{a.e. in } [0,1], && (6.1.32)\\
v_1(0) &= v_2(0) = 0, && (6.1.33)\\
v_1(1) &= v_2(1) = 0, && (6.1.34)\\
p' - c\,q &= 0 \quad \text{a.e. in } [0,1], && (6.1.35)\\
q' + c\,p &= 0 \quad \text{a.e. in } [0,1], && (6.1.36)\\
p(0) &= \lambda_1 \cos(\theta(0)) + \lambda_2 \sin(\theta(0)), &&\\
q(0) &= -\lambda_1 \sin(\theta(0)) + \lambda_2 \cos(\theta(0)), && (6.1.37)\\
u &= \varepsilon\, p \quad \text{a.e. in } [0,1], && (6.1.38)\\
z'' &= -q \quad \text{a.e. in } [0,1], && (6.1.39)\\
z(0) &= z(1) = 0. && (6.1.40)
\end{aligned}$$

We denote by $f_i(c)$, $g_i(c)$, $v_i(c)$, $\lambda_i(c)$, $i = 1,2$, and $u(c)$, $z(c)$, $p(c)$, $q(c)$ the dependence on $c \in L^1(0,1)$ of the data $[f_1, f_2, g_1, g_2]$ and of the solution $[v_1, v_2, u, z, p, q, \lambda_1, \lambda_2]$ to (6.1.31)–(6.1.40), which is now considered instead of the related dependence on θ. We aim to show the Gâteaux differentiability of these dependences viewed as mappings from $L^1(0,1)$ into appropriate Banach spaces.

Let $c \in L^1(0,1)$ be fixed. We assume that the mappings $c \mapsto f_i(c)$ are Gâteaux differentiable at c as nonlinear operators from $L^1(0,1)$ into $L^1(0,1)$, with the Gâteaux derivatives $\nabla f_i(c)$, $i = 1, 2$.

First, we analyze the dependences $g_1(c)$, $g_2(c)$, $h(c)$, $l(c)$. For this purpose, notice that we may rewrite (6.1.3) in the differential form

$$g_1(c) = \varepsilon\, l(c), \tag{6.1.41}$$

$$g_2(c)'' = -h(c) \quad \text{a.e. in } [0,1], \tag{6.1.42}$$

$$g_2(c)(0) = g_2(c)(1) = 0, \tag{6.1.43}$$

$$l(c)' - c\, h(c) = -f_1(c) \quad \text{a.e. in } [0,1], \tag{6.1.44}$$

$$h(c)' + c\, l(c) = -f_2(c) \quad \text{a.e. in } [0,1], \tag{6.1.45}$$

$$l(c)(0) = h(c)(0) = 0. \tag{6.1.46}$$

Now let any $d \in L^1(0,1)$ be given. We have, by (6.1.44), (6.1.45),

$$\frac{l\,(c+\delta\, d)' - l(c)'}{\delta} - (c+\delta\, d)\,\frac{h(c+\delta\, d) - h(c)}{\delta} = d\, h(c) - \frac{f_1(c+\delta\, d) - f_1(c)}{\delta}, \tag{6.1.47}$$

$$\frac{h\,(c+\delta\, d)' - h(c)'}{\delta} + (c+\delta\, d)\,\frac{l(c+\delta\, d) - l(c)}{\delta} = -d\, l(c) - \frac{f_2(c+\delta\, d) - f_2(c)}{\delta}. \tag{6.1.48}$$

Multiplying (6.1.47) and (6.1.48) by $\frac{1}{\delta}(l(c+\delta\, d) - l(c))$ and $\frac{1}{\delta}(h(c+\delta\, d) - h(c))$, respectively, and integrating over $[0,t]$, we find that

$$\begin{aligned} &\frac{1}{2}\left(\delta^{-2}\,|l(c+\delta\, d) - l(c)|^2 + \delta^{-2}\,|h(c+\delta\, d) - h(c)|^2\right)(t) \\ &\quad \le \int_0^t \Big(\left|d\, h(c) - \delta^{-1}\,(f_1(c+\delta\, d) - f_1(c))\right| \left|\delta^{-1}\,(l(c+\delta\, d) - l(c))\right| \\ &\qquad + \left|-d\, l(c) - \delta^{-1}\,(f_2(c+\delta\, d) - f_2(c))\right| \left|\delta^{-1}\,(h(c+\delta\, d) - h(c))\right|\Big)\, ds. \end{aligned} \tag{6.1.49}$$

The Brézis [1973] variant of Gronwall's lemma and (6.1.49) imply that the ratio families $\{\frac{1}{\delta}\,(l(c+\delta\, d) - l(c))\}$ and $\{\frac{1}{\delta}\,(h(c+\delta\, d) - h(c))\}$ are bounded in $L^\infty(0,1)$ for $\delta \to 0$. For the reader's convenience, we recall that the inequality (6.1.49) may be written in the form

$$\frac{1}{2}\,\chi(t)^2 \le \int_0^t \rho(s)\,\chi(s)\, ds \le \frac{1}{2}\int_0^t \rho(s)\,\left(1 + \chi(s)^2\right)\, ds,$$

where $\chi(t)$ denotes the norm on the left-hand side and ρ belongs to $L^1(0,1)_+$, so that also the usual Gronwall inequality may be applied.

Owing to (6.1.47) and (6.1.48), we even have boundedness in $W^{1,1}(0,1)$. Moreover, in view of the equiabsolute integrability of $\{\frac{1}{\delta}(f_i(c+\delta\, d)-f_i(c))\}$, $i=1,2$, the above ratio families are equiuniformly continuous. Consequently, there is a sequence $\delta_n \to 0$, $\delta_n \neq 0$ for all $n \in \mathbf{N}$, such that as $n \to \infty$,

$$\delta_n^{-1}\,(l(c+\delta_n\, d)-l(c)) \to \bar{l}, \quad \delta_n^{-1}\,(h(c+\delta_n\, d)-h(c)) \to \bar{h},$$

strongly in $C[0,1]$, and (6.1.47), (6.1.48) imply the strong convergence in the space $W^{1,1}(0,1)$. But then also, thanks to (6.1.41)–(6.1.46),

$$\begin{aligned}
\delta_n^{-1}\,(g_1(c+\delta_n\, d)-g_1(c)) &\to \bar{g}_1, \quad \text{strongly in } L^2(0,1),\\
\delta_n^{-1}\,(g_2(c+\delta_n\, d)-g_2(c)) &\to \bar{g}_2, \quad \text{strongly in } H^2(0,1),
\end{aligned}$$

where

$$\begin{aligned}
\bar{g}_1 &= \varepsilon\,\bar{l}, \quad \bar{g}_2'' = -\bar{h}, \quad \text{a.e. in } [0,1],\\
\bar{g}_2(0) &= \bar{g}_2(1) = 0,\\
\bar{l}' - c\,\bar{h} &= d\,h(c) - \nabla f_1(c)d, \quad \text{a.e. in } [0,1],\\
\bar{h}' + c\,\bar{l} &= -d\,l(c) - \nabla f_2(c)d, \quad \text{a.e. in } [0,1],\\
\bar{l}(0) &= \bar{h}(0) = 0.
\end{aligned}$$

Since this system has a unique solution and the unknowns $\bar{g}_1$, $\bar{g}_2$, $\bar{l}$, $\bar{h}$ depend linearly on d, the above convergences hold generally for $\delta \to 0$, and we have shown the existence of the Gâteaux derivatives

$$\begin{aligned}
\nabla h(c) &\in L(L^1(0,1), L^2(0,1)), \quad \nabla h(c)d = \bar{h},\\
\nabla l(c) &\in L(L^1(0,1), L^2(0,1)), \quad \nabla l(c)d = \bar{l},\\
\nabla g_1(c) &\in L(L^1(0,1), L^2(0,1)), \quad \nabla g_1(c)d = \bar{g}_1,\\
\nabla g_2(c) &\in L(L^1(0,1), H^2(0,1)), \quad \nabla g_2(c)d = \bar{g}_2.
\end{aligned}$$

Next, we examine the Gâteaux differentiability of the auxiliary mappings $w_1 = w_1(c)$, $w_2 = w_2(c)$ introduced in (6.1.13), (6.1.14). Recalling that $\theta(c)' = c$ a.e. in $[0,1]$, we have $w_1(c), w_2(c) \in H^2(0,1) \cap H_0^1(0,1)$ with

$$w_1(c)''(t) = \sin\left(\theta(c)(0) + \int_0^t c(s)\,ds\right), \quad \text{a.e. in } [0,1],$$

$$w_2(c)''(t) = -\cos\left(\theta(c)(0) + \int_0^t c(s)\,ds\right), \quad \text{a.e. in } [0,1].$$

Thus, assuming that the perturbations $(\tilde{\theta}_\delta)' = \theta(c)' + \delta\,\eta(d)' = c + \delta\, d$ satisfy the condition $\tilde{\theta}_\delta(0) = \theta(c)(0) + \delta\,\eta(d)(0)$ with some $\eta(d)(0) \in \mathbf{R}$, we easily conclude from the mean value theorem the existence of the Gâteaux derivative

$$\nabla w_i(c) \in L(L^1(0,1), H^2(0,1) \cap H_0^1(0,1)), \quad i = 1,2,$$

where the directional derivatives $\overline{w}_i = \nabla w_i(c)d$ in the direction d are given by

$$\overline{w}_1''(t) = \Big(\eta(d)(0) + \int_0^t d(s)\,ds\Big) \cos\Big(\theta(c)(0) + \int_0^t c(s)\,ds\Big),$$
$$\overline{w}_1(0) = \overline{w}_1(1) = 0, \tag{6.1.50}$$
$$\overline{w}_2''(t) = \Big(\eta(d)(0) + \int_0^t d(s)\,ds\Big) \sin\Big(\theta(c)(0) + \int_0^t c(s)\,ds\Big),$$
$$\overline{w}_2(0) = \overline{w}_2(1) = 0. \tag{6.1.51}$$

Now recall that by (6.1.19), $\lambda_1(c)$, $\lambda_2(c)$ are the solutions of a linear system whose determinant $\Delta(c)$ is positive and whose coefficient matrix has (cf. (6.1.50), (6.1.51)) Gâteaux differentiable entries. Thus, a lengthy but straightforward calculation yields the Gâteaux differentiability of the mappings $c \mapsto \lambda_i(c)$ from $L^1(0,1)$ into $\mathbf{R}$ and the corresponding Gâteaux derivatives $\nabla\lambda_i(c)$, $i = 1, 2$. Since the resulting expressions are rather involved, we do not give an explicit representation of the Gâteaux derivatives here.

Next, observe that (6.1.35)–(6.1.37) imply that the mappings $c \mapsto q(c)$ and $c \mapsto p(c)$ are Gâteaux differentiable from $L^1(0,1)$ into $L^2(0,1)$. It then follows immediately from (6.1.38), (6.1.39) that $c \mapsto u(c)$ and $c \mapsto z(c)$ are Gâteaux differentiable from $L^1(0,1)$ into $L^2(0,1)$ and $H^2(0,1)$, respectively. Finally, applying to (6.1.31)–(6.1.33) similar arguments as in (6.1.47)–(6.1.49), we also obtain the existence of the Gâteaux derivatives $\nabla v_i(c) \in L(L^1(0,1), L^2(0,1))$.

We thus have established the following result.

Theorem 6.1.7 *Suppose that the mappings $c \mapsto f_i(c)$, $i = 1, 2$, are Gâteaux differentiable in $c \in L^1(0,1)$, and assume that $d \in L^1(0,1)$ and $\eta(d)(0) \in \mathbf{R}$ are given and the perturbations $(\tilde{\theta}_\delta)' = \theta(c)' + \delta\,\eta(d)' = c + \delta\, d$ satisfy the initial condition $\tilde{\theta}_\delta(0) = \theta(c)(0) + \delta\,\eta(d)(0)$. Then the directional derivatives*

$$\overline{v}_i = \nabla v_i(c)d, \quad \overline{g}_i = \nabla g_i(c)d, \quad \overline{\lambda}_i = \nabla\lambda_i(c)d, \quad i = 1, 2,$$
$$\overline{u} = \nabla u(c)d, \quad \overline{z} = \nabla z(c)d, \quad \overline{p} = \nabla p(c)d, \quad \overline{q} = \nabla q(c)d,$$

exist and satisfy the system

$$\overline{v}_1' - c\,\overline{v}_2 = d\,v_2(c) + \overline{u} + \overline{g}_1, \quad a.e. \text{ in } [0,1], \tag{6.1.52}$$
$$\overline{v}_2' + c\,\overline{v}_1 = -d\,v_1(c) + \overline{z} + \overline{g}_2, \quad a.e. \text{ in } [0,1], \tag{6.1.53}$$
$$\overline{v}_1(0) = \overline{v}_2(0) = 0, \tag{6.1.54}$$
$$\overline{v}_1(1) = \overline{v}_2(1) = 0, \tag{6.1.55}$$
$$\overline{p}' - c\,\overline{q} = d\,q(c), \quad a.e. \text{ in } [0,1], \tag{6.1.56}$$
$$\overline{q}' + c\,\overline{p} = -d\,p(c), \quad a.e. \text{ in } [0,1], \tag{6.1.57}$$
$$\overline{p}(0) = \overline{\lambda}_1 \cos(\theta(c)(0)) + \overline{\lambda}_2 \sin(\theta(c)(0))$$
$$+\eta(d)(0)\,[\lambda_2(c) \cos(\theta(c)(0)) - \lambda_1(c) \sin(\theta(c)(0))],$$

$$\overline{q}(0) = -\overline{\lambda}_1 \sin(\theta(c)(0)) + \overline{\lambda}_2 \cos(\theta(c)(0))$$
$$-\eta(d)(0)\,[\lambda_1(c)\cos(\theta(c)(0)) + \lambda_2(c)\sin(\theta(c)(0))], \quad (6.1.58)$$
$$\overline{u} = \varepsilon\,\overline{p}, \quad a.e.\ in\ [0,1], \quad (6.1.59)$$
$$\overline{z}'' = -\overline{q}, \quad a.e.\ in\ [0,1], \quad (6.1.60)$$
$$\overline{z}(0) = \overline{z}(1) = 0. \quad (6.1.61)$$

Remark. The system (6.1.52)–(6.1.61) admits a unique solution, since its homogeneous variant is of the type (6.1.31)–(6.1.40) and may be reformulated in the language of the control problem (P) (in this connection, "homogeneous" means that $\bar{g}_1 = 0,\ \bar{g}_2 = 0,\ d = 0,\ \eta(d)(0) = 0$), and since the corresponding solution of (P) vanishes identically if the inhomogeneous terms equal zero.

Next, we introduce the adjoint system associated with (6.1.52)–(6.1.61):

$$P_1' - c\,P_2 = 0, \quad (6.1.62)$$
$$P_2' + cP_1 = -v_2(c), \quad (6.1.63)$$
$$P_3' - c\,P_4 = R, \quad (6.1.64)$$
$$P_4' + c\,P_3 = Q, \quad (6.1.65)$$
$$Q'' = -P_2, \quad (6.1.66)$$
$$R = \varepsilon\,P_1, \quad (6.1.67)$$
$$Q(0) = Q(1) = P_3(0) = P_3(1) = P_4(0) = P_4(1) = 0. \quad (6.1.68)$$

Proposition 6.1.8 *The system (6.1.62)–(6.1.68) has a unique solution such that* $P_1, P_2, R \in W^{1,1}(0,1)$, $P_3, P_4 \in W_0^{1,1}(0,1)$, *and* $Q \in W^{2,\infty}(0,1) \cap H_0^1(0,1)$.

Proof. Let $\mu_1, \mu_2 \in \mathbf{R}^2$ be some arbitrary initial conditions for (6.1.62), (6.1.63). Then

$$\begin{bmatrix} P_1 \\ P_2 \end{bmatrix}(t) = W_c(t)\begin{bmatrix} \mu_1 \\ \mu_2 \end{bmatrix} + \begin{bmatrix} \gamma_1(t) \\ \gamma_2(t) \end{bmatrix},$$

where

$$\begin{bmatrix} \gamma_1(t) \\ \gamma_2(t) \end{bmatrix} = \int_0^t W_c(t)\,W_c^{-1}(s)\begin{bmatrix} 0 \\ -v_2(c) \end{bmatrix}(s)\,ds,$$

and $P_1, P_2 \in W^{1,1}(0,1)$ if $c \in L^1(0,1)$. Here, the notation W_c indicates the dependence of the matrix W on c. Consequently, $R = \varepsilon\,P_1$ and Q depend in an affine manner on μ_1, μ_2 and belong to $W^{1,1}(0,1)$ and $W^{2,\infty}(0,1) \cap H_0^1(0,1)$, respectively. Then

$$\begin{bmatrix} P_3 \\ P_4 \end{bmatrix}(t) = -\int_t^1 W_c(t)\,W_c^{-1}(s)\begin{bmatrix} R(s) \\ Q(s) \end{bmatrix}ds$$

belongs to $W^{1,1}(0,1)^2$, where we have used the final null conditions. In order to comply with the initial null conditions for P_3, P_4, the constraint

$$\int_0^1 W_c^{-1}(s)\left[\begin{array}{c} R(s) \\ Q(s) \end{array}\right] ds = \left[\begin{array}{c} 0 \\ 0 \end{array}\right] \tag{6.1.69}$$

has to be fulfilled. Writing (6.1.69) explicitly, we obtain a linear system like (6.1.19) for μ_1, μ_2. Since its determinant is positive, it is uniquely solvable. This concludes the proof of the assertion. □

Theorem 6.1.9 *The directional derivative of the cost functional in the problem (Q) at the point $c \in L^1(0,1)$ in the direction $d \in L^1(0,1)$ is given by*

$$\int_0^1 d\,[P_1\,v_2(c) - P_2\,v_1(c) + \nabla g_1(c)^* P_1 + \nabla g_2(c)^* P_2 - P_3\,q(c) + P_4\,p(c)]\,ds, \tag{6.1.70}$$

where $\nabla g_i(c)^* \in L(L^2(0,1), L^\infty(0,1))$ denotes the dual operator of the Gâteaux derivative $\nabla g_i(c)$, $i = 1, 2$.

Proof. Using (6.1.62), (6.1.63), and integration by parts, we have

$$\begin{aligned}
\lim_{\delta\to 0}\frac{1}{2\delta}&\left[\int_0^1 (v_2(c+\delta d)(s))^2\,ds - \int_0^1 (v_2(c)(s))^2\,ds\right] = \int_0^1 v_2(c)(s)\,\overline{v}_2(s)\,ds \\
&= -\int_0^1 (P_2' + c\,P_1)(s)\,\overline{v}_2(s)\,ds - \int_0^1 (P_1' - c\,P_2)(s)\,\overline{v}_1(s)\,ds \\
&= \int_0^1 P_1(s)\,(\overline{v}_1' - c\,\overline{v}_2)(s)\,ds + \int_0^1 P_2(s)(\overline{v}_2' + c\,\overline{v}_1)(s)\,ds \\
&= \int_0^1 d(s)\,(P_1\,v_2(c) - P_2\,v_1(c))(s)\,ds + \int_0^1 P_1(s)\,(\overline{u} + \overline{g}_1)(s)\,ds \\
&\quad + \int_0^1 P_2(s)\,(\overline{z} + \overline{g}_2)(s)\,ds.
\end{aligned}$$

Now recall that $\overline{g}_i = \nabla g_i(c)d$, $i = 1, 2$. Hence, invoking (6.1.64)–(6.1.67) as well, we can infer that

$$\begin{aligned}
&\int_0^1 v_2(c)(s)\,\overline{v}_2(s)\,ds \\
&\quad = \int_0^1 d(s)\,[P_1 v_2(c) - P_2 v_1(c) + \nabla g_1(c)^* P_1 + \nabla g_2(c)^* P_2](s)\,ds \\
&\qquad + \int_0^1 \varepsilon^{-1}\,R(s)\,\overline{u}(s)\,ds - \int_0^1 Q''(s)\,\overline{z}(s)\,ds \\
&\quad = \int_0^1 d(s)\,[P_1 v_2(c) - P_2 v_1(c) + \nabla g_1(c)^* P_1 + \nabla g_2(c)^* P_2](s)\,ds \\
&\qquad + \int_0^1 R(s)\,\overline{p}(s)\,ds + \int_0^1 Q(s)\,\overline{q}(s)\,ds \\
&\quad = \int_0^1 d(s)\,[P_1 v_2(c) - P_2 v_1(c) + \nabla g_1(c)^* P_1 + \nabla g_2(c)^* P_2](s)\,ds \\
&\qquad + \int_0^1 \overline{p}(s)\,(P_3' - c\,P_4)(s)\,ds + \int_0^1 \overline{q}(s)\,(P_4' + c\,P_3)(s)\,ds.
\end{aligned}$$

From this, using (6.1.56), (6.1.57), and integration by parts, we obtain (6.1.70), which finishes the proof of the assertion. □

Next, we shall study the differentiability properties of the problem (Q) in the general case in which only $\theta \in L^\infty(0,1)$. We consider variations of the form $\theta + \sigma\eta$ with $\eta \in L^\infty(0,1)$ and "small" $\sigma \in \mathbf{R}$. We assume that the operators

$$f_i : L^\infty(0,1) \to L^2(0,1), \quad \theta \mapsto f_i(\theta), \quad i = 1, 2,$$

depend directly on θ and are Gâteaux differentiable. A direct calculation, starting from (6.1.3) and taking into account the dependence of $W(t)$ on θ, leads to

$$\begin{aligned} \begin{bmatrix} \overline{l} \\ \overline{h} \end{bmatrix}(t) &= -\int_0^t W_\theta(t)\, W_\theta^{-1}(s) \begin{bmatrix} \overline{f}_1(\theta) \\ \overline{f}_2(\theta) \end{bmatrix} ds - \begin{pmatrix} 0 & \eta(t) \\ -\eta(t) & 0 \end{pmatrix} \begin{bmatrix} l(\theta) \\ h(\theta) \end{bmatrix}(t) \\ &\quad + \int_0^t \begin{pmatrix} 0 & \eta(s) \\ -\eta(s) & 0 \end{pmatrix} W_\theta(t)\, W_\theta^{-1}(s) \begin{bmatrix} f_1(\theta) \\ f_2(\theta) \end{bmatrix}(s)\, ds. \end{aligned} \tag{6.1.71}$$

Moreover, (6.1.3) implies that

$$\overline{g}_1 = \varepsilon\, \overline{l}, \quad -\overline{g}_2'' = \overline{h}, \quad \overline{g}_2(0) = \overline{g}_2(1) = 0. \tag{6.1.72}$$

Comparing (6.1.71) with (6.1.44)–(6.1.46), we see that the integral formulation is more difficult to handle since it involves more products that generate additional terms via differentiation.

For the auxiliary mappings w_1, w_2 defined in (6.1.13), (6.1.14), we write directly the increment ratios corresponding to θ and $\theta + \sigma\eta$, and we compute the limit as $\sigma \to 0$ to find that

$$\overline{w}_1'' = \eta \cos(\theta), \quad \overline{w}_2'' = \eta \sin(\theta), \quad \overline{w}_i(0) = \overline{w}_i(1) = 0, \quad i = 1, 2. \tag{6.1.73}$$

Relations (6.1.71)–(6.1.73) also show the continuous dependence in $L^2(0,1)$ of $\overline{g}_i$, $\overline{w}_i$, $i = 1, 2$, and of $\overline{l}$, $\overline{h}$ with respect to regularizations of η and θ if the same is assumed for f_i, $\overline{f}_i$, $i = 1, 2$. For $\overline{w}_i$, $i = 1, 2$, and for $\overline{g}_2$, the continuous dependence holds even in $H^2(0,1)$. An elementary calculation, starting from (6.1.19), shows that also $\overline{\lambda}_1$, $\overline{\lambda}_2$ depend continuously on regularizations of η and θ.

Next, we conclude from relation (6.1.27) that

$$\begin{bmatrix} \overline{p} \\ \overline{q} \end{bmatrix}(t) = \begin{pmatrix} 0 & \eta(t) \\ -\eta(t) & 0 \end{pmatrix} W_\theta(t) \begin{bmatrix} \lambda_1(\theta) \\ \lambda_2(\theta) \end{bmatrix} + W_\theta(t) \begin{bmatrix} \overline{\lambda}_1 \\ \overline{\lambda}_2 \end{bmatrix}, \tag{6.1.74}$$

with the same continuity property in $L^2(0,1)^2$ with respect to regularizations of η and θ. By virtue of (6.1.28), (6.1.29), this property is preserved for $\overline{u}$ and $\overline{z}$, and we have

$$\overline{u} = \varepsilon\, \overline{p}, \quad \overline{z}'' = -\overline{q}, \quad \overline{z}(0) = \overline{z}(1) = 0. \tag{6.1.75}$$

Finally, (6.1.25) gives

$$\begin{aligned}\begin{bmatrix}\overline{v}_1\\ \overline{v}_2\end{bmatrix}(t) &= \int_0^t W_\theta(t)\,W_\theta^{-1}(s)\begin{bmatrix}\overline{u}(s)+\overline{g}_1(s)\\ \overline{z}(s)+\overline{g}_2(s)\end{bmatrix} ds\\ &\quad -\int_0^t \begin{pmatrix}0 & \eta(s)\\ -\eta(s) & 0\end{pmatrix} W_\theta(t)\,W_\theta^{-1}(s)\begin{bmatrix}u(\theta)+g_1(\theta)\\ z(\theta)+g_2(\theta)\end{bmatrix}(s)\,ds\\ &\quad +\begin{pmatrix}0 & \eta(t)\\ -\eta(t) & 0\end{pmatrix}\begin{bmatrix}v_1(\theta)\\ v_2(\theta)\end{bmatrix}(t), \end{aligned} \tag{6.1.76}$$

with the same conclusion concerning the continuity of the dependence on η and θ.

Let us now explicitly introduce the regularizations of θ and η,

$$\theta_\delta(t) = \int_{\mathbf{R}} \theta(t-\delta y)\,\rho(y)\,dy, \qquad \eta_\delta(t) = \int_{\mathbf{R}} \eta(t-\delta y)\,\rho(y)\,dy, \tag{6.1.77}$$

where θ and η are extended by zero outside the interval $[0,1]$, $\delta > 0$, and where $\rho \in C_0^\infty(\mathbf{R})$ is a Friedrichs mollifier. Then

$$\theta_\delta \to \theta, \quad \eta_\delta \to \eta, \quad \text{strongly in } L^s(0,1), \quad \text{for all } s \geq 1.$$

We also define $d_\delta = \eta_\delta'$ and $c_\delta = \theta_\delta'$, which exist in $L^1(0,1)$ but do not have good convergence properties for $\delta \to 0$. Then, the systems (6.1.31)–(6.1.40), (6.1.52)–(6.1.61), and (6.1.62)–(6.1.68) can be solved for the data c_δ, d_δ. Let us denote the corresponding solutions with a subscript or an exponent δ. We have the following result.

Theorem 6.1.10 *The gradient of the cost functional of the problem (Q) at $\theta \in L^\infty(0,1)$ in the direction $\eta \in L^\infty(0,1)$ is given by*

$$\begin{aligned}&\int_0^1 v_2(\theta)\,\overline{v}_2\,ds\\ &\quad = \int_0^1 \eta\,\Big[\nabla g_1(\theta)^* P_1 + \nabla g_2(\theta)^* P_2 - v_1(\theta)\,v_2(\theta) - P_1(\theta)\,(z(\theta)+g_2(\theta))\\ &\qquad\qquad + P_2(\theta)\,(u(\theta)+g_1(\theta)) + q(\theta)\,R(\theta) - p(\theta)\,Q(\theta)\Big]\,ds\,,\end{aligned} \tag{6.1.78}$$

where $v_1(\theta)$, $v_2(\theta)$, $u(\theta)$, $z(\theta)$, $p(\theta)$, $q(\theta)$ are obtained from (6.1.25)–(6.1.29) with $g_1(\theta)$, $g_2(\theta)$ given by (6.1.3), and where P_1, P_2, P_3, P_4, R, Q are computed via (6.1.62)–(6.1.68) (rewritten in the corresponding obvious integral form).

Proof. Thanks to (6.1.76), (6.1.77), we can write

$$\int_0^1 v_2(\theta)\,\overline{v}_2\,ds = \lim_{\delta\to 0}\int_0^1 v_2^\delta\,\overline{v}_2^\delta\,ds\,. \tag{6.1.79}$$

From Theorem 6.1.9, we obtain that

$$\int_0^1 v_2^\delta \, \overline{v}_2^\delta \, ds = \int_0^1 d_\delta \Big[P_1^\delta \, v_2^\delta - P_2^\delta \, v_1^\delta - P_3^\delta \, q^\delta + P_4^\delta \, p^\delta + P_1^\delta \, \overline{g}_1^\delta + P_2^\delta \, \overline{g}_2^\delta \Big] \, ds \, .$$

Using the boundary conditions and the differentiability properties, we first find that

$$\begin{aligned}
&\int_0^1 d_\delta \left[P_1^\delta \, v_2^\delta - P_2^\delta \, v_1^\delta - P_3^\delta \, q^\delta + P_4^\delta \, p^\delta \right] ds \\
&= - \int_0^1 \eta_\delta \Big[(P_1^\delta)' \, v_2^\delta + P_1^\delta (v_2^\delta)' - (P_2^\delta)' \, v_1^\delta - P_2^\delta \, (v_1^\delta)' - (P_3^\delta)' \, q^\delta \\
&\qquad\qquad - P_3^\delta \, (q^\delta)' + (P_4^\delta)' \, p^\delta + P_4^\delta \, (p^\delta)' \Big] \, ds \\
&= - \int_0^1 \eta_\delta \left[v_1^\delta \, v_2^\delta + P_1^\delta \, (z^\delta + g_2^\delta) - P_2^\delta \, (u^\delta + g_1^\delta) - q^\delta \, R^\delta + p^\delta \, Q^\delta \right] ds \, . \qquad (6.1.80)
\end{aligned}$$

We indicate partially how the last equality in (6.1.80) is established. We have, using (6.1.35), (6.1.36), (6.1.64), and (6.1.65),

$$\begin{aligned}
&(P_4^\delta)' \, p^\delta + P_4^\delta \, (p^\delta)' - (P_3^\delta)' \, q^\delta - P_3^\delta \, (q^\delta)' \\
&\quad = (P_4^\delta)' \, p^\delta + P_4^\delta \, c_\delta \, q^\delta - (P_3^\delta)' \, q^\delta + P_3^\delta \, c_\delta \, p^\delta \\
&\quad = q^\delta [- (P_3^\delta)' + c_\delta \, P_4^\delta] + p^\delta \, [(P_4^\delta)' + c_\delta \, P_3^\delta] \\
&\quad = -q^\delta \, R^\delta + p^\delta \, Q^\delta .
\end{aligned}$$

Moreover,

$$\begin{aligned}
\int_0^1 (P_1^\delta \, \overline{g}_1^\delta + P_2^\delta \, \overline{g}_2^\delta) \, ds &= \int_0^1 (P_1^\delta \, \nabla g_1^\delta(\theta_\delta)\eta_\delta + P_2^\delta \, \nabla g_2^\delta(\theta_\delta)\eta_\delta) \, ds \\
&= \int_0^1 \eta_\delta \big([\nabla g_1^\delta(\theta_\delta)]^* P_1^\delta + [\nabla g_2^\delta(\theta_\delta)]^* P_2^\delta \big) \, ds. \qquad (6.1.81)
\end{aligned}$$

The derivatives of g_1, g_2 may be taken directly with respect to θ. This can easily be seen from (6.1.47)–(6.1.49) (where f_i now depend on θ) without modifying the argument.

We combine (6.1.79)–(6.1.81) and pass to the limit as $\delta \to 0$. The continuity properties with respect to both η_δ and θ_δ have been explained in (6.1.71)–(6.1.76). We remark that continuous dependence for $\delta \to 0$ is also valid for P_1^δ, P_2^δ, P_3^δ, P_4^δ, R^δ, Q^δ, since the system (6.1.62)–(6.1.68) can also be brought into integral (mild) form. We thus arrive at the desired conclusion. □

Example 6.1.11 We have employed the gradient provided by Theorem 6.1.10 in several numerical shape optimization experiments for arches. In order to apply formula (6.1.78), one has to determine numerically the solutions to the state system (6.1.25)–(6.1.29) and to the adjoint system (6.1.62)–(6.1.68) as well as an approximation to the mappings $[\nabla g_1(\theta)]^* P_1$ and $[\nabla g_2(\theta)]^* P_2$. By the nature of the data it is obvious that an explicit calculation is not possible in the optimization routine.

We have chosen an equidistant division of the interval $[0, L]$ (where here the length of the arc was assumed as $L > 0$) into $N_0 \in \mathbf{N}$ subintervals $[t_i, t_{i+1}]$, where $t_i = ih$, $h = \frac{L}{N_0}$. The mapping $\theta \in L^\infty(0, L)$ has been approximated in the different examples by either piecewise linear splines or piecewise constant functions. The integrals have been evaluated accordingly using standard quadrature formulas.

The solution of the ordinary differential systems was obtained using linear finite elements, and the scalars λ_1, λ_2 from (6.1.27) were determined by solving the algebraic system (6.1.19). Similarly, the unknown initial conditions μ_1, μ_2 for (6.1.62), (6.1.63) satisfy a system of the same type as (6.1.19), where the mappings l and g_2 have to be replaced by γ_1 and γ, respectively, with $\gamma'' = -\gamma_2$ and $\gamma(0) = \gamma(L) = 0$; see Proposition 6.1.8 and its proof.

The functions $[\nabla g_1(\theta)]^* P_1$ and $[\nabla g_2(\theta)]^* P_2$ have been approximated as follows:

$$
\begin{aligned}
[\nabla g_k(\theta)]^* P_k(t_i) &\approx \frac{1}{h} \int_{t_i}^{t_{i+1}} [\nabla g_k(\theta)]^* P_k(s)\, ds \\
&= \frac{1}{h} \int_0^L P_k(s) (\bar{g}_k\, \chi_{[t_i, t_{i+1}]})(s)\, ds, \qquad k = 1, 2, \qquad i = \overline{0, N-1}, \\
[\nabla g_k(\theta)]^* P_k(L) &\approx 0, \qquad k = 1, 2.
\end{aligned}
$$

Here, $\bar{g}_k$ has been determined using (6.1.72), and $\chi_{[t_i, t_{i+1}]}$ denotes the characteristic function of the subinterval $[t_i, t_{i+1}]$.

Although the optimization problems under study are nonconvex, adaptations of Rosen's and Uzawa's gradient algorithms with projection (Gruver and Sachs [1981], Arnăutu [2001], Arnăutu and Neittaanmäki [2003]), have been employed. A maximal number of iterations (between 200 and 300) was prescribed, and as minimum solution we chose the approximation giving the best value of the cost functional. The algorithm also stopped for vanishing value of the gradient of the cost functional and of the cost itself (which is positive).

For each example several tests with both algorithms were performed, using various values for the parameters N_0 and α (the parameter from the Rosen algorithm). In general, the Rosen algorithm gave better results than the Uzawa algorithm.

In the optimization problems, we fixed the "thickness" $\varepsilon = 0.1$ and employed a simple *line search* strategy, subdividing $]0, 1]$ into N_1 equal parts and evaluating the cost functional for the values $\frac{i}{N_1}$, $i = \overline{1, N_1}$, of the line search parameter. The parameter yielding the smallest value was chosen to generate the next iterate. With this simple strategy we have avoided, with good numerical results, the usual determination of the line search parameter via a one-dimensional optimization problem (which may be very time-consuming). In addition, in each iteration a projection onto the admissible set was performed.

The optimization problem (Q) searches the shape of an arch having a minimal normal deformation (in some integral sense) under the action of a prescribed force. We have examined purely tangential ($f_2 = 0$) and normal ($f_1 = 0$) forces, since these act in the direction of the axes of the local coordinate system, as well as forces that do not depend on the unknown arch. For instance, forces parallel to the vertical axis are described by $f_1(t) = \sin(\theta(t))/S$ and $f_2(t) = \cos(\theta(t))/S$ in the local system of coordinates, and in converse order for forces parallel to the horizontal axis. It should be noticed that while the force is independent of the arch, its local representation still is dependent via θ.

The constraints for θ were given in terms of subintervals of $[0, \pi]$ as indicated in the figures. This suffices for many applications and avoids a self-intersection of the arches. However, degenerate cases are still possible (see Figure 1.7).

In Figure 1.4, under the action of a tangential force, and starting from a Roman arch as initial iterate, it is seen that the global solution is a beam, which clearly has no normal deflection under such a load. In our representation, two global solutions (beams) are put into evidence, associated with $\theta = 0$ and $\theta = \pi$. The figure depicts some iterates produced by the algorithm and the corresponding values of the cost.

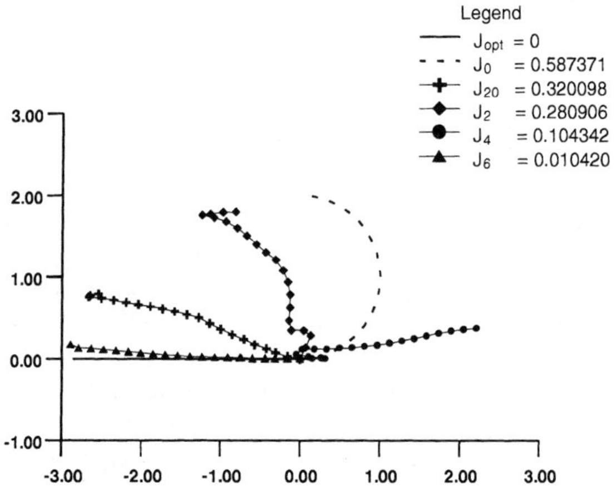

Figure 1.4. $\theta(t) \in [0, \pi]$, $f_1(t) = \frac{1}{S}$, $f_2(t) = 0$, $\theta_0(t) = t$, $t \in [0, \pi]$, © SIAM.

In this experiment, we have used $N_0 = 200$, $N_1 = 10$, $\alpha = 0,75$, and the arch close to the beam was obtained after 24 iterations. Notice that in this example there are infinitely many global solutions (beams of any slope), which shows the difficulty of the numerical computations.

However, from the physical viewpoint all these solutions cannot in fact be distinguished. They arise in the mathematical formulation, since the considered parametrization also takes the slope into account (which in this example is irrelevant). Such apparent situations of nonuniqueness have also been considered by

Eppler [2000]. One idea for their removal is the introduction of supplementary restrictions in the formulation of the problem.

In Figure 1.5 the initial iterate was again a Roman arch, but this time the force was of constant modulus $\frac{1}{S}$ and parallel to the vertical axis. The depicted iterates again demonstrate how the routine finds the (unique, if θ is confined to $[0,\pi]$) global solution given by the vertical beam characterized by $\theta = \frac{\pi}{2}$. In this configuration the prescribed force becomes purely tangential to the arch, and the global solution is a special case of the previous example (but not of the problem as a whole). Here, we took $N_0 = 200$, $N_1 = 10$, $\alpha = 1$, and the global optimum was reached after 139 iterations.

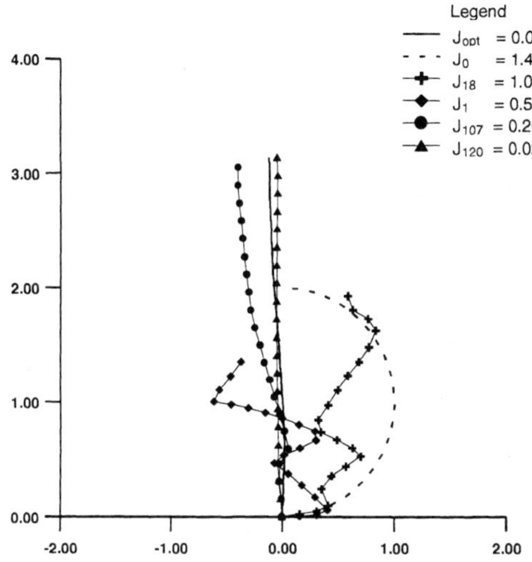

Figure 1.5. $\theta(t) \in [0,\pi]$, $f_1(t) = \frac{\sin(\theta(t))}{S}$, $f_2(t) = \frac{\cos(\theta(t))}{S}$, $\theta_0(t) = t$, $t \in [0,\pi]$, © SIAM.

The numerical results from Figures 1.4 and 1.5 match the physical interpretation perfectly, which is a strong justification for the notion of weak solutions introduced here and also indicates the good stability properties of the approximation methods used.

In Figure 1.6 a more "realistic" example has been studied, namely the construction of a most-resistant roof (made of one piece) subject to a vertical constant load of modulus $\frac{1}{S}$. The reader should note that in this figure we have interchanged the axes to make the representation look more "physical." We took $N_0 = 500$, $N_1 = 100$, $\alpha = 10$. Two experiments are depicted in Figure 1.6, one in which the initial guess θ_{02} was given by a fragment of a Roman arch, and another where the initial configuration θ_{01} consisted of two coupled fragments of Roman arches. In both cases the numerical solutions were obtained in the first iteration and look very similar. In this example the theoretical optimal value is "far" away from zero.

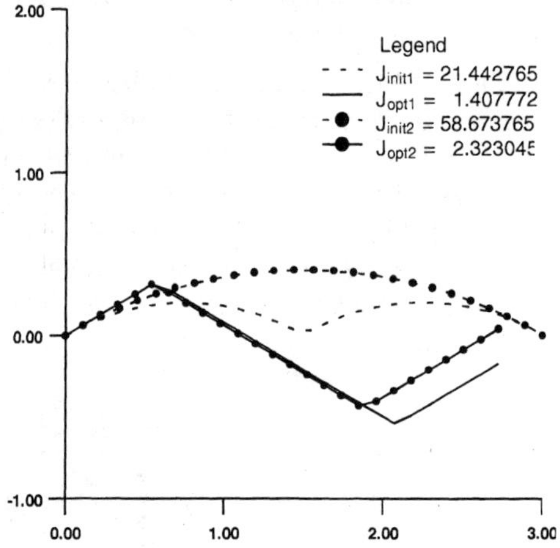

Figure 1.6. $\theta(t) \in [\frac{\pi}{3}, \frac{2\pi}{3}]$, $f_1(t) = \frac{\cos(\theta(t))}{S}$, $f_2(t) = \frac{\sin(\theta(t))}{S}$, $t \in [0, \pi]$, $\theta_{01}(t) = \frac{2t+\pi}{3}$, $t \in [0, \frac{\pi}{2})$, $\theta_{01}(t) = \frac{2t}{3}$, $t \in [\frac{\pi}{2}, \pi]$, $\theta_{02}(t) = \frac{t+\pi}{3}$, $t \in [0, \pi]$, © SIAM.

In Figures 1.7 and 1.8, the case of a purely normal load is discussed, the difference being given by the constraints imposed on θ, namely $\theta(t) \in [0, \pi]$ and $\theta(t) \in [\frac{\pi}{6}, \frac{5\pi}{6}]$, respectively.

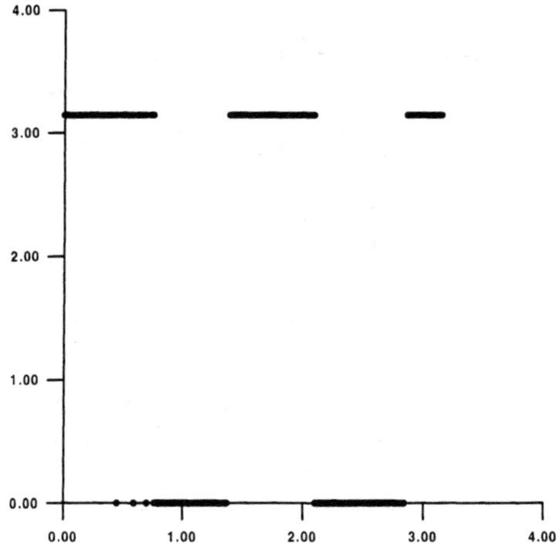

Figure 1.7. $\theta(t) \in [0, \pi]$, $f_1(t) = 0$, $f_2(t) = \frac{1}{S}$, $\theta_0(t) = t$, $t \in [0, \pi]$, $J_{\text{init}} = 82.922993$, $J_{\text{opt}} = 0.0024772$, © SIAM.

In Figure 1.7, the "optimal" found θ is represented, not (as usual) the arch. The computations used $N_0 = 200$, $N_1 = 20$, $\alpha = 1.5$, and 27 iterations. Since the resulting solution is of *bang-bang type* (we have $\theta \in \{0, \pi\}$ for a.e. $t \in [0, \pi]$), the arch degenerates and cannot be represented graphically.

Motivated by the bang-bang character of the optimal solution, we generated a sequence $\{\theta^N\}$, giving N the values listed in Table 1.1 and θ^N the values 0 and π, alternately on successive subintervals of the partition in N parts of $[0, \pi]$, and computed the corresponding costs $J(\theta^N)$, also listed in Table 1.1. The conclusion is that $\{\theta^N\}$ is a very efficient minimizing sequence for this problem that yields for $N \geq 50$ smaller values of the cost functional than the one computed by the complete numerical procedure (although this provides a satisfactory result as well).

N	$J(\theta^N)$
30	0.0095834975
50	0.0012420426
100	0.0000776279
200	0.0000048517
300	0.0000009584
500	0.0000001242
800	0.0000000190
1000	0.0000000078

Table 1.1. $\theta(t) \in \{0, \pi\}$, $f_1(t) = 0$, $f_2(t) = \frac{1}{8}$, $t \in [0, \pi]$, © SIAM.

We stress that the oscillatory nature of the minimizing sequence $\{\theta^N\}$ is related to the noncompactness of the constraints set $\{\theta \in L^\infty : \theta(t) \in [0, \pi]$ for a.e. $t \in (0, \pi)\}$ in $L^\infty(0, \pi)$. This set is only bounded and closed, which is not enough to ensure the existence of an optimal θ as discussed in Theorem 6.1.5 and Corollary 6.1.6. This numerical example indicates that the assumptions of Corollary 6.1.6 seem to be sharp. Notice that the lack of compactness of the admissible set is also true for Figures 1.4 and 1.5, even though global minimum points exist in these examples.

In Figure 1.8 the initial Roman arch and the obtained solution (under normal load) are depicted. The numerical test was performed with $N_0 = 300$, $N_1 = 10$, $\alpha = 1, 5$, and the optimum was achieved after 160 iteration steps. Obviously, the solution is again of bang-bang type (recall that θ is the angle between the tangent to the arch and the horizontal axis). However, Table 1.2 shows that the simple sequence $\{\theta^N\}$, constructed as in the previous example, but with the values $\frac{\pi}{6}$, $\frac{5\pi}{6}$, is no longer a minimizing sequence for this problem. The commuting points for the bang-bang solution are not equidistant in this example.

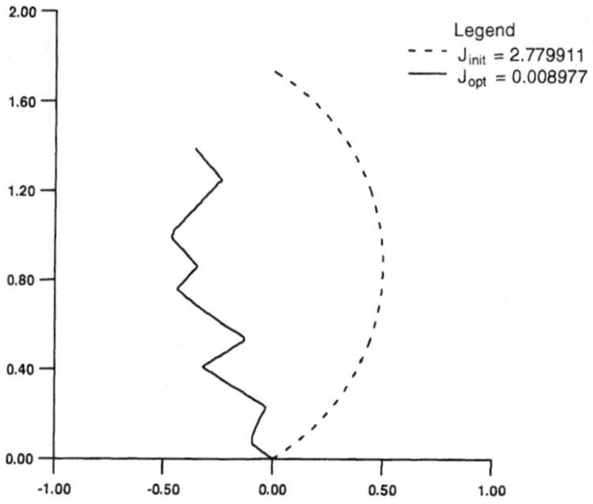

Figure 1.8. $\theta(t) \in [\frac{\pi}{6}, \frac{5\pi}{6}]$, $f_1(t) = 0$, $f_2(t) = \frac{1}{S}$, $\theta_0(t) = t + \frac{\pi}{3}$, $t \in [0, \frac{2\pi}{3}]$, $J_{\text{init}} = 2.779911$, $J_{\text{opt}} = 0.008977$, © SIAM.

N	$J(\theta^N)$
30	0.0141367792
50	0.0247750769
100	0.0303698330
200	0.0318697376
300	0.0321519172
500	0.0322969269
800	0.0323467156
1000	0.0323582113

Table 1.2. $\theta(t) \in [\frac{\pi}{6}, \frac{5\pi}{6}]$, $f_1(t) = 0$, $f_2(t) = \frac{1}{S}$, $t \in [0, \frac{2\pi}{3}]$, © SIAM.

We close this paragraph by stressing the fact that working with low regularity assumptions was essential for the optimization applications, in view of the bang-bang structure of the optimal θ found in many examples. However, in Figure 1.5 the global solution is not bang-bang, which seems to be due just to the force applied. Bang-bang properties are known in thickness optimization problems for plates; compare Corollaries 3.4.5 and 3.4.7 and Proposition 3.4.13 in Chapter 3. We also underline the nonlocal optimization character of our numerical experiments, which is obvious from the line search used and the reported results. More examples may be found in Ignat, Sprekels, and Tiba [2001].

6.1.3 Variational Inequalities for Arches

We analyze a variant of the control problem studied above in §6.1.1, namely

$$\underset{[u,z]\in L^2(0,1)\times V}{\mathrm{Min}} \left\{ \frac{1}{2\varepsilon} \int_0^1 u^2(s)\, ds + \frac{1}{2} \int_0^1 (z'(s))^2\, ds \right\}, \tag{6.1.82}$$

subject to the constraint that the state pair $[v_1, v_2] \in L^\infty(0,1)^2$ given by (6.1.4) satisfy

$$[v_1, v_2] \in C, \tag{6.1.83}$$

where C is a closed and convex subset of $L^\infty(0,1)^2$ whose elements satisfy the null initial conditions imposed in (6.1.4). Notice that we have replaced the final state constraint (6.1.5) by (6.1.83). This describes arches that are clamped just in $t = 0$ and subject to unilateral conditions given by (6.1.83).

The notation used here is the same as in the previous paragraphs of this section; however, the control space for z is $V = \{w \in H^1(0,1) : w(0) = 0\}$, and the definition (6.1.3) of g_1, g_2 is replaced by

$$\begin{aligned} \begin{bmatrix} l \\ h \end{bmatrix}(t) &= \int_t^1 W(t)\, W^{-1}(s) \begin{bmatrix} f_1(s) \\ f_2(s) \end{bmatrix} ds, \quad \text{for a.e. } t \in [0,1], \\ g_1 &= \varepsilon l, \quad -g_2'' = h, \quad g_2(0) = g_2'(1) = 0. \end{aligned} \tag{6.1.84}$$

Since there are no constraints imposed on the controls $[u, z]$, admissibility may be assumed in (6.1.82), (6.1.83), and the existence of a unique optimal pair $[u_\varepsilon, z_\varepsilon] \in L^2(0,1) \times V$ with corresponding optimal state $[v_1^\varepsilon, v_2^\varepsilon]$ belonging to C follows easily using the technique of the proof of Theorem 2.1.2 in Chapter 2.

Now let $[u, z]$ be any admissible control pair. Then the convexity of C yields that also the control variations

$$[u_\varepsilon, z_\varepsilon] + \lambda\, [u - u_\varepsilon, z - z_\varepsilon], \quad \lambda \in [0,1], \tag{6.1.85}$$

are admissible. Thus

$$\begin{aligned} \frac{1}{2\varepsilon} \int_0^1 u_\varepsilon^2(s)\, ds + \frac{1}{2} \int_0^1 (z_\varepsilon'(s))^2\, ds \;\le\; & \frac{1}{2\varepsilon} \int_0^1 (u_\varepsilon + \lambda\,(u - u_\varepsilon))^2(s)\, ds \\ & + \frac{1}{2} \int_0^1 (z_\varepsilon' + \lambda\,(z' - z_\varepsilon'))^2(s)\, ds, \end{aligned}$$

whence, dividing by $\lambda > 0$ and taking the limit as $\lambda \searrow 0$, we easily obtain the Euler inequality for (6.1.82), (6.1.83),

$$0 \le \frac{1}{\varepsilon} \int_0^1 u_\varepsilon(s)\,(u(s) - u_\varepsilon(s))\, ds + \int_0^1 z_\varepsilon'(s)\,(z'(s) - z_\varepsilon'(s))\, ds. \tag{6.1.86}$$

We are now going to show under the assumption $W \in W^{2,\infty}(0,1)^4$ (as in Theorem 6.1.1) that $[v_1^\varepsilon, v_2^\varepsilon]$ satisfies a general variational inequality. This will be another application of the *control variational method* as in Theorem 3.4.17 in §3.4.2.1 of Chapter 3.

To this end, fix any $[w_1, w_2] \in (V \times U) \cap C$, where

$$U = \{w_2 \in H^2(0,1) : \, w_2(0) = w_2'(0) = 0\}.$$

Then w_1, w_2 may be generated via the state system (6.1.4) from an admissible control $[\mu, \xi] \in L^2(0,1) \times V$, where

$$\mu = w_1' - c\,w_2 - g_1, \tag{6.1.87}$$

$$\xi = w_2' + c\,w_1 - g_2. \tag{6.1.88}$$

Here, the assumed regularity $W \in W^{2,\infty}(0,1)^4$ is used essentially. Moreover, v_1^ε, v_2^ε satisfy similar differential equations with $[\mu, \xi]$ replaced by $[u^\varepsilon, z^\varepsilon]$. We insert (6.1.87), (6.1.88), by taking $u = \mu$ and $z = \xi$ (and the corresponding relations for v_1^ε, v_2^ε) in (6.1.86), to obtain that

$$\begin{aligned} 0 \le \frac{1}{\varepsilon}\int_0^1 &(- g_1 + (v_1^\varepsilon)' - c\,v_2^\varepsilon)\,(w_1' - c\,w_2 - (v_1^\varepsilon)' + c\,v_2^\varepsilon)\,ds \\ &+ \int_0^1 (- g_2 + (v_2^\varepsilon)' + c\,v_1^\varepsilon)'\,(w_2' + c\,w_1 - (v_2^\varepsilon)' - c\,v_1^\varepsilon)'\,ds. \end{aligned} \tag{6.1.89}$$

Using integration by parts repeatedly, and invoking (6.1.84), we find that

$$\begin{aligned} -\frac{1}{\varepsilon}\int_0^1 & g_1\,(w_1' - c\,w_2 - (v_1^\varepsilon)' + c\,v_2^\varepsilon)\,ds + \int_0^1 g_2''\,(w_2' + c\,w_1 - (v_2^\varepsilon)' - c\,v_1^\varepsilon)\,ds \\ &= -\int_0^1 l\,(w_1' - c\,w_2 - (v_1^\varepsilon)' + c\,v_2^\varepsilon)\,ds - \int_0^1 h\,(w_2' + c\,w_1 - (v_2^\varepsilon)' - c\,v_1^\varepsilon)\,ds \\ &= -\int_0^1 f_1\,(w_1 - v_1^\varepsilon)\,ds - \int_0^1 f_2\,(w_2 - v_2^\varepsilon)\,ds. \end{aligned} \tag{6.1.90}$$

Combining (6.1.89), (6.1.90), we have proved the following result:

Theorem 6.1.12 *If $W \in W^{2,\infty}(0,1)^4$, then v_1^ε, v_2^ε satisfy*

$$\begin{aligned} \frac{1}{\varepsilon}\int_0^1 &((v_1^\varepsilon)' - c\,v_2^\varepsilon)\,((v_1^\varepsilon)' - c\,v_2^\varepsilon - w_1' + c\,w_2)\,ds \\ &\qquad + \int_0^1 ((v_2^\varepsilon)' + c\,v_1^\varepsilon)'\,((v_2^\varepsilon)' + c\,v_1^\varepsilon - w_2' - c\,w_1)'\,ds \\ &\le \int_0^1 f_1\,(v_1^\varepsilon - w_1)\,ds + \int_0^1 f_2\,(v_1^\varepsilon - w_2)\,ds, \\ &\textit{for all } [w_1, w_2] \in (V \times U) \cap C. \end{aligned} \tag{6.1.91}$$

Remark. If C imposes null conditions at $t = 1$, then (6.1.91) is a variational inequality for a clamped arch.

Remark. In the works of Hlavacek, Bock, and Lovisek [1984], [1985], and Sprekels and Tiba [2000], variational inequalities for fourth-order ordinary differential equations associated with beam models are discussed. A short example has already been mentioned in Example 1.2.7 in Chapter 1. Notice that the formulation (6.1.82), (6.1.83), (6.1.4) may be interpreted as a weak formulation of the variational inequality that is valid even if $W \in L^\infty(0,1)^4$.

Remark. If $C = L^\infty(0,1) \times \{v_2 \in L^\infty(0,1) : \alpha \le v_2 \le \beta \text{ a.e. in } [0,1]\}$, then (6.1.91) constitutes an obstacle-type problem for the normal component of the deflection of the arch. Here, $\alpha, \beta \in C[0,1]$ have to satisfy $\alpha(0) \le 0 \le \beta(0)$ in order that the initial condition for v_2 can be fulfilled.

Remark. Under the smoothness condition $W \in W^{2,\infty}(0,1)^4$ the solution $[v_1^\varepsilon, v_2^\varepsilon]$ belongs to $V \times U$, and we can choose C as a closed convex subset of $V \times U$. For instance, if

$$C = \{[v_1, v_2] \in V \times U : v_1(1) \ge r\}$$

with some fixed $r \in \mathbf{R}$, then (6.1.91) describes an arch that is clamped in $t = 0$ and obeys a unilateral condition for the tangential component in $t = 1$. Other examples of variational inequalities may easily be constructed in this way.

6.2 General Three-Dimensional Curved Rods

In this section, the results for plane arches will be extended to curved rods in $\mathbf{R}^3$, i.e., to Jordan curves of class $W^{2,\infty}(0,L)^3$. Since rods, while being "thin," are fully three-dimensional structures, they are more difficult to describe geometrically, and a theory analyzing their deformation under general mechanical loads is more complex than for arches.

One way to investigate the deformation of curved rods is of course to use the general theory of linear elasticity, which, under the assumption of small deformations, leads to a boundary value problem for a linear elliptic system of three partial differential equations for the three components of the displacement vector $\bar{y}$ in $\mathbf{R}^3$. Under appropriate conditions, existence and uniqueness of a weak solution can be shown using Korn's inequality.

It turns out, however, that this "fully three-dimensional" approach has several disadvantages since it does not take into account the special geometrical feature (namely, the thinness) of rods: the resulting system is overly complex, and one encounters stability difficulties in the numerical approximation of the solutions. In the literature, there are many models that are better adapted to the geometry of rods (cf. Trabucho and Viaño [1996]). We will also propose such an adapted "polynomial" model in this section, leading to a boundary value problem for a linear system of nine ordinary differential equations, which constitutes a considerable reduction of complexity.

The $W^{2,\infty}(0,L)^3$ regularity assumptions needed for this approach, while stronger than those in the previous section, are weaker than those in the available mathematical literature, where at least three derivatives are required for this case. In order to achieve this relaxation, we are going to show in the first paragraph below that local coordinate frames on curves can be defined already for curves of class $C^1[0,L]^3$ in $\mathbf{R}^3$. Additional regularity conditions will have to be imposed later in the discussion of the model to be introduced and its shape optimization. In Tiba and Vodák [2004] an asymptotic model for three-dimensional curved rods is studied under mere Lipschitz assumptions on the parametrization.

6.2.1 A Local Frame Under Low Differentiability Assumptions

We begin by constructing local coordinate frames on curves that are parametrized by vector functions $\bar{\theta} \in C^1[0,L]^3$. This will lead us to a local basis consisting of tangent, normal, and binormal in $C[0,L]^3$. Thus, all basis vectors will have the same regularity, in contrast to the classical Frenet construction (see Cartan [1967]), where the normal and the binormal vectors are less regular than the tangent vector.

We assume again that $\bar{\theta}$ is a unit speed curve, that is, the tangent vector $\bar{t} = \bar{\theta}' \in C[0,L]$ has unit length at any point on the curve. We choose some "sufficiently small" (to be made precise later) $\delta > 0$ and a partition $0 = s_0 < s_1 < \ldots < s_{N+1} = L$, with some $N \in \mathbf{N}$, of the interval $[0,L]$ such that

$$|\bar{t}(s_i) - \bar{t}(s_{i+1})|_{\mathbf{R}^3} < \delta \quad \forall i = \overline{0,N}. \tag{6.2.1}$$

Observe that (6.2.1) can always be achieved for sufficiently large N since $\bar{t}$ is uniformly continuous in $[0,L]$.

For $0 \le i \le N+1$ we define the vectors

$$\tilde{n}(s_i) = \begin{cases} \dfrac{(0, -\theta_3'(s_i), \theta_2'(s_i))}{\sqrt{\theta_3'(s_i)^2 + \theta_2'(s_i)^2}}, & \text{if } (\theta_2'(s_i), \theta_3'(s_i)) \neq (0,0), \\[2ex] \dfrac{(0, \theta_1'(s_i), 0)}{|\theta_1'(s_i)|}, & \text{otherwise.} \end{cases} \tag{6.2.2}$$

Then

$$|\tilde{n}(s_i)|_{\mathbf{R}^3} = 1, \quad \tilde{n}(s_i) \perp \bar{t}(s_i), \quad 0 \le i \le N+1. \tag{6.2.3}$$

We can also assume that the angle between $\tilde{n}(s_i)$ and $\tilde{n}(s_{i+1})$ is acute, i.e., that

$$(\tilde{n}(s_i), \tilde{n}(s_{i+1}))_{\mathbf{R}^3} \ge 0, \quad 0 \le i \le N. \tag{6.2.4}$$

This can be achieved by simply replacing $\tilde{n}(s_{i+1})$ by $-\tilde{n}(s_{i+1})$ if necessary and proceeding inductively from s_1 to s_{N+1}. We denote by $\bar{n}(s_i)$ the vectors thus obtained.

On each interval $[s_i, s_{i+1}]$, we define the linear interpolate

$$\bar{m}_i(s) = \bar{n}(s_i) + \frac{s - s_i}{s_{i+1} - s_i}\,(\bar{n}(s_{i+1}) - \bar{n}(s_i)), \quad i = \overline{0, N}. \tag{6.2.5}$$

Clearly, $\bar{m}_i(s_i) = \bar{n}_i(s_i)$, $\bar{m}_i(s_{i+1}) = \bar{n}_i(s_{i+1})$, and the function $\bar{m}(s) = \bar{m}_i(s)$, $s \in [s_i, s_{i+1}]$, $0 \le i \le N$, is Lipschitz continuous in $[0, L]$. Moreover,

$$|\bar{m}_i(s)|_{\mathbf{R}^3} \ge \frac{\sqrt{2}}{2}, \quad \forall\, s \in [0, L], \quad \forall\, i = \overline{0, N}. \tag{6.2.6}$$

Inequality (6.2.6) is a consequence of elementary geometric arguments in the triangle having the two unit edges $\bar{n}(s_i)$ and $\bar{n}(s_{i+1})$ and an acute angle, using (6.2.4). We note only that $|\bar{m}_i(s_i)|_{\mathbf{R}^3}$ is just the length of the line segment connecting arbitrary points on the "basis" of this triangle with its "top" point, whence (6.2.6) easily follows.

Now let us assume for the moment that $\bar{m}(\hat{s})$ is collinear with $\bar{t}(\hat{s})$ for some $\hat{s} \in [0, L]$. Since $|\bar{t}(\hat{s})|_{\mathbf{R}^3} = 1$, we can infer from (6.2.5) that

$$\bar{n}(s_i) + \frac{\hat{s} - s_i}{s_{i+1} - s_i}\,(\bar{n}(s_{i+1}) - \bar{n}(s_i)) = \pm\,|\bar{m}_i(\hat{s})|_{\mathbf{R}^3}\,\bar{t}(\hat{s}), \tag{6.2.7}$$

where i is fixed such that $\hat{s} \in [s_i, s_{i+1}]$. We multiply (6.2.7) by $\bar{t}(\hat{s})$ and add and subtract $\bar{t}(s_i)$ or $\bar{t}(s_{i+1})$, apply (6.2.6), the triangle inequality, and the Schwarz inequality, to obtain

$$\begin{aligned} \frac{\sqrt{2}}{2} \le |\bar{m}_i(\hat{s})|_{\mathbf{R}^3} \le & \left(1 - \frac{\hat{s} - s_i}{s_{i+1} - s_i}\right) |\bar{t}(\hat{s}) - \bar{t}(s_i)|_{\mathbf{R}^3} \\ & + \frac{\hat{s} - s_i}{s_{i+1} - s_i}\, |\bar{t}(\hat{s}) - \bar{t}(s_{i+1})|_{\mathbf{R}^3} \le \delta. \end{aligned} \tag{6.2.8}$$

Here, we have also used (6.2.1) and (6.2.3).

Clearly, (6.2.8) leads to a contradiction for $\delta < \frac{1}{\sqrt{2}}$. For such a choice of $\delta > 0$ (which will be assumed henceforth), the vectors $\bar{m}(s)$ and $\bar{t}(s)$ are linearly independent for any $s \in [0, L]$. We thus can apply the Schmidt orthogonalization process and define, for any $s \in [0, L]$,

$$\hat{n}(s) = \bar{m}(s) - (\bar{m}(s), \bar{t}(s))_{\mathbf{R}^3}\,\bar{t}(s)\,, \tag{6.2.9}$$

$$\bar{n}(s) = \hat{n}(s)/|\hat{n}(s)|_{\mathbf{R}^3}\,. \tag{6.2.10}$$

Notice that $\hat{n}(s) \ne (0, 0, 0)$ by the linear independence of $\bar{m}(s)$ and $\bar{t}(s)$; moreover, $|\hat{n}(s)|_{\mathbf{R}^3} \ge \hat{c} > 0$ for all $s \in [0, L]$ with some $\hat{c} > 0$, since $\hat{n}$ is continuous on the compact set $[0, L]$. In addition, since $\bar{m}$ is Lipschitz continuous, it follows from (6.2.9) and (6.2.10) that $\bar{n}$ can be at most a $W^{1,\infty}(0, L)^3$ function; on the other hand, if $\bar{t}$ belongs to $W^{1,\infty}(0, L)^3$, then so does $\bar{n}$. We thus have proved the following result.

Theorem 6.2.1 *If* $\bar{\theta} \in C^1[0,L]^3$ *(or* $\bar{\theta} \in W^{2,\infty}(0,L)^3$*) is a unit speed curve then* $\bar{t}, \bar{n} \in C[0,L]^3$ *(or* $\bar{t}, \bar{n} \in W^{1,\infty}(0,L)^3$*), and*

$$|\bar{n}(s)|_{\mathbf{R}^3} = 1, \quad (\bar{n}(s), \bar{t}(s))_{\mathbf{R}^3} = 0, \quad \textit{for every } s \in [0,L].$$

Moreover, the vector function $\bar{b} = \bar{t} \wedge \bar{n}$ *completing the local frame has the same regularity properties.*

For the reader's convenience we recall that the vector product $\bar{b} = \bar{t} \wedge \bar{n}$ of two unit vectors $\bar{t} = (t_1, t_2, t_3)$, $\bar{n} = (n_1, n_2, n_3) \in \mathbf{R}^3$, defined by

$$\bar{b} = (t_2 n_3 - t_3 n_2, t_3 n_1 - t_1 n_3, t_1 n_2 - t_2 n_1),$$

is a unit vector that is orthogonal to both $\bar{t}$ and $\bar{n}$ and is such that the coordinate system $\{\bar{t}, \bar{n}, \bar{b}\}$ has a positive orientation.

Remark. The construction given above is particularly well suited for numerical simulations. Notice that $\bar{m}'$ is piecewise constant. If $\bar{\theta} \in W^{2,\infty}(0,L)^3$ has $\bar{\theta}''$ piecewise continuous, then $\bar{t}'$, $\bar{n}'$, $\bar{b}'$ are piecewise continuous as well.

Remark. In Cartan [1967] the so-called *Darboux local frame* also is discussed. In this case, the normal $\bar{n}$ is constructed under the assumption that the curve $\bar{\theta}$ lies on some differentiable surface $\mathcal{S}$; namely, $\bar{n}(s)$ is just a unit normal to $\mathcal{S}$ in the corresponding points $\bar{\theta}(s) \in \mathcal{S}$. The local frame is again completed by the tangent vector $\bar{t}$ and the vector $\bar{b} = \bar{t} \wedge \bar{n}$. Also, this construction requires less regularity than the Frenet frame, but it uses another geometric assumption (for $\mathcal{S}$), which seems to be unnecessary, as Theorem 6.2.1 shows.

6.2.2 A Generalized Naghdi-Type Model

We now introduce a model of so-called *polynomial type* for the deformation of a "thin" elastic curved rod occupying the open domain $\tilde{\Omega} \subset \mathbf{R}^3$. For a general discussion of various models that may be considered in this setting, we recommend to the reader the monograph of Trabucho and Viaño [1996]. The present model was studied in detail in the work of Ignat, Sprekels, and Tiba [2002]. Motivated by the analogy to a corresponding model for shells investigated below in §3.1, we call it a *generalized Naghdi-type model.* More precisely, the subsequent conditions (6.2.17) and (6.3.7) are very similar in nature, where (6.3.7) is a slight generalization of the classical *Naghdi assumption* for shells.

To begin with, let $L > 0$ be given, and suppose that $\omega(x_3) \subset \mathbf{R}^2$ is a bounded domain, not necessarily simply connected, for any $x_3 \in [0,L]$. We assume that there exists a given open set $\omega \subset \mathbf{R}^2$ with $\omega(x_3) \supset \omega$ for every $x_3 \in [0,L]$ and such that

$$0 = \int_\omega x_1 \, dx_1 \, dx_2 = \int_\omega x_2 \, dx_1 \, dx_2 = \int_\omega x_1 x_2 \, dx_1 \, dx_2. \tag{6.2.11}$$

Condition (6.2.11) generalizes similar hypotheses used by Murat and Sili [1999] and is related to the choice of the global system of axes in $\mathbf{R}^2 \supset \omega$.

Introducing the open set

$$\Omega = \bigcup_{x_3 \in]0,L[} (\omega(x_3) \times \{x_3\}) \subset \mathbf{R}^3, \tag{6.2.12}$$

we then define the curved rod $\tilde{\Omega} \subset \mathbf{R}^3$ as the image of Ω under a geometric transformation of special form (to be justified below, see (6.2.20)), namely

$$\begin{aligned} (x_1, x_2, x_3) &= \bar{x} \in \Omega \;\mapsto\; F\bar{x} = \tilde{x} = (\tilde{x}_1, \tilde{x}_2, \tilde{x}_3) \in \tilde{\Omega}, \\ (\tilde{x}_1, \tilde{x}_2, \tilde{x}_3) &= \bar{\theta}(x_3) + x_1\, \bar{n}(x_3) + x_2\, \bar{b}(x_3) \in \tilde{\Omega} \quad \forall \bar{x} \in \Omega, \qquad (6.2.13) \\ \tilde{\Omega} &= \{\tilde{x} = F\bar{x} : \bar{x} \in \Omega\}. \qquad (6.2.14) \end{aligned}$$

Here, $\bar{\theta} \in W^{2,\infty}(0,L)^3$, where $\bar{\theta}''$ is assumed piecewise continuous, represents a three-dimensional unit speed curve (called the *line of centroids*) that is parametrized with respect to its arc length, and the vectors $\bar{n}(x_3)$, $\bar{b}(x_3) \in \mathbf{R}^3$ are as defined in Theorem 6.2.1.

Since $\bar{t}$, $\bar{n}$, $\bar{b}$ are differentiable and

$$|\bar{t}(x_3)|^2_{\mathbf{R}^3} = |\bar{n}(x_3)|^2_{\mathbf{R}^3} = |\bar{b}(x_3)|^2_{\mathbf{R}^3} = 1 \quad \forall x_3 \in [0, L],$$

it follows that for almost all $x_3 \in [0, L]$,

$$(\bar{t}(x_3), \bar{t}'(x_3))_{\mathbf{R}^3} = (\bar{n}(x_3), \bar{n}'(x_3))_{\mathbf{R}^3} = (\bar{b}(x_3), \bar{b}'(x_3))_{\mathbf{R}^3} = 0. \tag{6.2.15}$$

The orthogonality relations (6.2.15) give the "equations of motion" of the local frame,

$$\begin{aligned} \bar{t}'(x_3) &= a(x_3)\, \bar{b}(x_3) + \beta(x_3)\, \bar{n}(x_3), \\ \bar{b}'(x_3) &= -a(x_3)\, \bar{t}(x_3) + c(x_3)\, \bar{n}(x_3), \\ \bar{n}'(x_3) &= -\beta(x_3)\, \bar{t}(x_3) - c(x_3)\, \bar{b}(x_3), \end{aligned} \tag{6.2.16}$$

with suitable piecewise continuous functions a, β, c that have a similar meaning as the curvatures in the standard Frenet frame.

The curved rod $\tilde{\Omega}$ is assumed to be clamped at both ends and subjected to body forces $\tilde{f}$ acting in $\tilde{\Omega}$ (weight, electromagnetic field, etc.) and to surface tractions $\tilde{g}$ on the lateral surface denoted by $\tilde{\Sigma}$. On the "inside" lateral face of $\tilde{\Omega}$ (corresponding to possible holes in the cross section), we assume that $\tilde{g} \equiv 0$.

We denote by $\bar{y} : \tilde{\Omega} \to \mathbf{R}^3$ the corresponding displacement of each point in $\tilde{\Omega}$ under the action of the given forces. Our "mechanical" assumption is that $\bar{y}$ has the special form

$$\bar{y}(\tilde{x}) = \bar{r}(x_3) + x_1\, \bar{N}(x_3) + x_2\, \bar{B}(x_3) \quad \forall \tilde{x} \in \tilde{\Omega}, \tag{6.2.17}$$

with $\bar{x} = (x_1, x_2, x_3) = F^{-1}(\tilde{x}) \in \Omega$, and where $\bar{\tau}$, $\bar{N}$, $\bar{B} \in H_0^1(0,L)^3$ are unknown functions.

Relation (6.2.17) means that transverse sections of $\tilde{\Omega}$ (i.e., those that are perpendicular to $\bar{t}(x_3)$, $x_3 \in [0, L]$) may deform but remain plane or degenerate, that is, become one-dimensional, after the deformation. However, taking also into account the small-deformations assumption of linear elasticity theory, we see that degeneracy is in fact not possible, and our comment just shows that (6.2.17) is a very weak hypothesis. The vector $\bar{\tau}(x_3)$ describes the translation of the points on the line of centroids parametrized by $\bar{\theta}$, and the vectors $\bar{N}(x_3) + \bar{n}(x_3)$, $\bar{B}(x_3) + \bar{b}(x_3)$ show the deformation of the orthogonal frame in the cross section (which does not necessarily remain orthogonal to the tangent of the new centroid line, i.e., to $\bar{\theta}'(x_3) + \bar{\tau}'(x_3)$). This allows for shear and for length or volume changes after the deformation. The classical Timoshenko condition is fulfilled (namely that the cross section remains plane), but not the Navier condition, since the cross section may deform within its own plane.

One can easily compute the Jacobian of F, denoted by $J(\bar{x}) = \nabla F(\bar{x})$, its determinant, and its inverse (using (6.2.16)). We find that for all $\bar{x} = (x_1, x_2, x_3) \in \Omega$,

$$J(\bar{x}) = \begin{bmatrix} n_1(x_3) & b_1(x_3) & t_1(x_3) + x_1\, n_1'(x_3) + x_2\, b_1'(x_3) \\ n_2(x_3) & b_2(x_3) & t_2(x_3) + x_1\, n_2'(x_3) + x_2\, b_2'(x_3) \\ n_3(x_3) & b_3(x_3) & t_3(x_3) + x_1\, n_3'(x_3) + x_2\, b_3'(x_3) \end{bmatrix} \tag{6.2.18}$$

$$J(\bar{x})^{-1} = \begin{bmatrix} n_1 - \dfrac{ct_1 x_2}{1-\beta x_1 - a x_2} & n_2 - \dfrac{ct_2 x_2}{1-\beta x_1 - a x_2} & n_3 - \dfrac{ct_3 x_2}{1-\beta x_1 - a x_2} \\ b_1 + \dfrac{ct_1 x_1}{1-\beta x_1 - a x_2} & b_2 + \dfrac{ct_2 x_1}{1-\beta x_1 - a x_2} & b_3 + \dfrac{ct_3 x_1}{1-\beta x_1 - a x_2} \\ \dfrac{t_1}{1-\beta x_1 - a x_2} & \dfrac{t_2}{1-\beta x_1 - a x_2} & \dfrac{t_3}{1-\beta x_1 - a x_2} \end{bmatrix}, \tag{6.2.19}$$

$$\det J(\bar{x}) = 1 - \beta(x_3)\, x_1 - a(x_3)\, x_2 \,. \tag{6.2.20}$$

If $\omega(x_3)$ is contained in a ball that has a sufficiently small radius with respect to the bounded "curvatures" a, β, c, then

$$\det J(\bar{x}) \geq c > 0 \quad \forall \bar{x} \in \overline{\Omega}, \quad \text{with some } c > 0,$$

which will henceforth be assumed. This assumption, which is natural in the modeling of curved rods, postulates that either the "curvatures" or the diameters of the cross sections $\omega(x_3)$ of Ω have to be sufficiently small. It ensures that F is a one-to-one transformation and justifies the definition (6.2.13), (6.2.14) of the curved rod $\tilde{\Omega}$.

We now introduce the mapping $\bar{w} : \Omega \to \mathbf{R}^3$ by

$$\bar{w}(\bar{x}) = \bar{\tau}(x_3) + x_1\,\bar{N}(x_3) + x_2\,\bar{B}(x_3) \quad \forall \bar{x} \in \Omega. \tag{6.2.21}$$

Clearly,

$$\bar{y}(\tilde{x}) = \bar{w}\left(F^{-1}(\tilde{x})\right) \quad \forall \tilde{x} \in \tilde{\Omega}. \tag{6.2.22}$$

Using the chain rule, we find that for all $\tilde{x} \in \tilde{\Omega}$,

$$\nabla\bar{y}(\tilde{x}) = \nabla\bar{w}(F^{-1}(\tilde{x}))\,\nabla F^{-1}(\tilde{x}) = \nabla\bar{w}(F^{-1}(\tilde{x}))\,J(F^{-1}(\tilde{x}))^{-1}. \tag{6.2.23}$$

To simplify the exposition, we define

$$\nabla F^{-1}(\tilde{x}) = \left(d_{ij}(\tilde{x})\right)_{i,j=\overline{1,3}}, \quad J(\bar{x})^{-1} = \left(h_{ij}(\bar{x})\right)_{i,j=\overline{1,3}}. \tag{6.2.24}$$

Invoking (6.2.18)–(6.2.24), we can compute the components e_{ij} of the linearized strain tensor in terms of the displacement vector $\bar{y}(\tilde{x})$. We have

$$\begin{aligned} \frac{\partial y_i}{\partial \tilde{x}_j}(\tilde{x}) &= N_i(x_3(\tilde{x}))\,d_{1j}(\tilde{x}) + B_i(x_3(\tilde{x}))\,d_{2j}(\tilde{x}) + [\tau_i'(x_3(\tilde{x})) \\ &\quad + x_1(\tilde{x})\,N_i'(x_3(\tilde{x})) + x_2(\tilde{x})\,B_i'(x_3(\tilde{x}))]\,d_{3j}(\tilde{x}), \quad i,j=\overline{1,3}, \qquad (6.2.25)\\ e_{ij}(\tilde{x}) &= \frac{1}{2}\left(\frac{\partial y_i}{\partial \tilde{x}_j} + \frac{\partial y_j}{\partial \tilde{x}_i}\right)(\tilde{x}), \quad i,j=\overline{1,3}. \qquad (6.2.26) \end{aligned}$$

Consequently, we get (without summation convention)

$$\begin{aligned} e_{ii}(\tilde{x}) &= \frac{\partial y_i}{\partial \tilde{x}_i}(\tilde{x}), \quad i=\overline{1,3}, \qquad (6.2.27)\\ e_{ij}(\tilde{x}) &= e_{ji}(\tilde{x}) \\ &= \frac{1}{2}\Big\{N_i(x_3(\tilde{x}))\,d_{1j}(\tilde{x}) + B_i(x_3(\tilde{x}))\,d_{2j}(\tilde{x}) \\ &\quad + \left[\tau_i'(x_3(\tilde{x})) + x_1(\tilde{x})\,N_i'(x_3(\tilde{x})) + x_2(\tilde{x})\,B_i'(x_3(\tilde{x}))\right]d_{3j}(\tilde{x}) \\ &\quad + N_j(x_3(\tilde{x}))\,d_{1i}(\tilde{x}) + B_j(x_3(\tilde{x}))\,d_{2i}(\tilde{x}) + \left[\tau_j'(x_3(\tilde{x}))\right. \\ &\quad \left. + x_1(\tilde{x})\,N_j'(x_3(\tilde{x})) + x_2(\tilde{x})\,B_j'(x_3(\tilde{x}))\right]d_{3i}(\tilde{x})\Big\}, \quad i,j=\overline{1,3}. \end{aligned} \tag{6.2.28}$$

Now let us assume that the rod consists of an isotropic material having the Lamé constants $\tilde{\lambda} > 0$ and $\tilde{\mu} > 0$. Then the equations of linear elasticity for the displacement $\hat{y}$ read (see (6.6.29) below)

$$\mathcal{B}_0(\hat{y}, \bar{v}) = \mathcal{G}(\bar{v}),$$

where the bilinear form $\mathcal{B}_0$ and the linear functional $\mathcal{G}$ are defined on $\tilde{Z} \times \tilde{Z}$ and on $\tilde{Z}$, respectively, where $\tilde{Z} \subset H^1(\tilde{\Omega})^3$ denotes the subspace of functions having zero traces on the "bases" of the rod $\tilde{\Omega}$, i.e., on $F(\omega(0)) \cup F(\omega(L))$. Using the summation convention, the linear elasticity system reads (cf. Ciarlet [2000])

$$\begin{aligned}\mathcal{B}_0(\hat{y}, \bar{v}) &= \int_{\tilde{\Omega}} \left[\tilde{\lambda}\, e_{pp}(\hat{y}(\tilde{x}))\, e_{qq}(\bar{v}(\tilde{x})) + 2\,\tilde{\mu}\, e_{ij}(\hat{y}(\tilde{x})) e_{ij}(\bar{v}(\tilde{x}))\right] d\tilde{x} \\ &= \mathcal{G}(\bar{v}) = \int_{\tilde{\Omega}} \tilde{f}_i(\tilde{x})\, v_i(\tilde{x})\, d\tilde{x} + \int_{\tilde{\Sigma}} \tilde{g}_i(\tilde{\sigma})\, v_i(\tilde{\sigma})\, d\tilde{\sigma}. \qquad (6.2.29)\end{aligned}$$

Notice that $\bar{y}$ as given by (6.2.17) belongs to $\tilde{Z}$. We also restrict the class of admissible test functions in (6.2.29) by admitting only test functions $\bar{v} = (v_1, v_2, v_3) \in \tilde{Z}$ of the special form

$$\bar{v}(\tilde{x}) = \bar{\mu}(x_3) + x_1\, \bar{M}(x_3) + x_2\, \bar{D}(x_3),$$

where the components M_i, D_i, μ_i, $i = 1, 2, 3$, all belong to $H_0^1(0, L)$. In this way, we project the original fully three-dimensional problem onto the (still infinite dimensional) subspace $Z \subset \tilde{Z}$ of all mappings $\bar{y}$ of the form (6.2.17) with $\bar{\tau}, \bar{N}, \bar{B} \in H_0^1(0, L)^3$, thereby reducing the original problem to a linear system of nine ordinary differential equations in the variable x_3.

Observe that this projection follows essentially the same procedure as in the finite element method (where the subspaces are finite-dimensional). Some convergence and approximation results for such projected problems have been proved by Trabucho and Viaño [1996] in a different setting. Moreover, owing to the above choice, the space Z can be identified with $H_0^1(0, L)^9$ which will be done in the following. We denote by $\mathcal{B}$ the restriction of $\mathcal{B}_0$ to the subspace $Z \times Z$ of $\tilde{Z} \times \tilde{Z}$ or, equivalently, to $H_0^1(0, L)^9 \times H_0^1(0, L)^9$.

We also perform in (6.2.29) the change of variables $\tilde{x} = F\bar{x}$, and invoking (6.2.17) and (6.2.24), we find that in this case $\mathcal{B}$ can be written in the form

$$\begin{aligned}&\mathcal{B}(\bar{y}, \bar{v}) \\ &= \int_{F(\Omega)} \left[\tilde{\lambda}\, e_{pp}(\bar{y}(F\bar{x}))\, e_{qq}(\bar{v}(F\bar{x})) + 2\,\tilde{\mu}\, e_{ij}(\bar{y}(F\bar{x}))\, e_{ij}(\bar{v}(F\bar{x}))\right] d\tilde{x} \\ &= \tilde{\lambda} \int_{\Omega} \sum_{i,j=1}^{3} \Big[N_i(x_3)\, h_{1i}(\bar{x}) + B_i(x_3)\, h_{2i}(\bar{x}) + (\tau_i'(x_3) + x_1\, N_i'(x_3) \\ &\qquad + x_2\, B_i'(x_3))\, h_{3i}(\bar{x})\Big] \Big[M_j(x_3)\, h_{1j}(\bar{x}) + D_j(x_3)\, h_{2j}(\bar{x}) \\ &\qquad + (\mu_j'(x_3) + x_1\, M_j'(x_3) + x_2\, D_j'(x_3))\, h_{3j}(\bar{x})\Big]\, |\det J(\bar{x})|\, d\bar{x} \\ &\quad + \tilde{\mu} \int_{\Omega} \sum_{i<j} \Big[N_i(x_3)\, h_{1j}(\bar{x}) + B_i(x_3)\, h_{2j}(\bar{x}) + (\tau_i'(x_3) + x_1\, N_i'(x_3) \\ &\qquad + x_2\, B_i'(x_3))\, h_{3j}(\bar{x}) + N_j(x_3)\, h_{1i}(\bar{x}) + B_j(x_3)\, h_{2i}(\bar{x}) \\ &\qquad + (\tau_j'(x_3) + x_1\, N_j'(x_3) + x_2\, B_j'(x_3))\, h_{3i}(\bar{x})\Big]\end{aligned}$$

$$\cdot \Big[M_i(x_3)\, h_{1j}(\bar{x}) + D_i(x_3)\, h_{2j}(\bar{x}) + (\mu_i'(x_3) + x_1\, M_i'(x_3) + x_2\, D_i'(x_3))\, h_{3j}(\bar{x}) + M_j(x_3)\, h_{1i}(\bar{x}) + D_j(x_3)\, h_{2i}(\bar{x}) + (\mu_j'(x_3) + x_1\, M_j'(x_3) + x_2\, D_j'(x_3))\, h_{3i}(\bar{x})\Big] \, |\det J(\bar{x})|\, d\bar{x}$$

$$+ 2\,\tilde{\mu} \int_\Omega \sum_{i=1}^{3} \Big[N_i(x_3)\, h_{1i}(\bar{x}) + B_i(x_3)\, h_{2i}(\bar{x}) + (\tau_i'(x_3) + x_1\, N_i'(x_3) + x_2\, B_i'(x_3))\, h_{3i}(\bar{x})\Big] \Big[M_i(x_3) h_{1i}(\bar{x}) + D_i(x_3)\, h_{2i}(\bar{x}) + (\mu_i'(x_3) + x_1\, M_i'(x_3) + x_2\, D_i'(x_3))\, h_{3i}(\bar{x})\Big] \, |\det J(\bar{x})|\, d\bar{x}\,. \tag{6.2.30}$$

Notice, however, that in general the assumption (6.2.17) is not satisfied by the solution to (6.2.29). Therefore, a direct analysis of the bilinear form (6.2.30) is necessary. We have the following result.

Theorem 6.2.2 *Under the above assumptions, there are constants* c, m *such that*

$$0 < c \le 1 - \beta(x_3)\, x_1 - a(x_3)\, x_2 \le m \quad \forall \bar{x} \in \Omega,$$

and the bilinear form $\mathcal{B}$ *is coercive and bounded on* $H_0^1(0,L)^9$.

We begin the proof with the following lemma.

Lemma 6.2.3 *There are* $c_1 > 0,\ c_2 > 0$ *such that*

$$\mathcal{B}(\bar{y}, \bar{y}) \ge c_1 |\bar{y}|^2_{H_0^1(0,L)^9} - c_2 |\bar{y}|^2_{L^2(0,L)^9}, \tag{6.2.31}$$

where $\bar{y}$, *given by (6.2.17), is identified with the vector* $(\tau_1, \tau_2, \tau_3, N_1, N_2, N_3, B_1, B_2, B_3) \in H_0^1(0,L)^9$.

Proof. By virtue of (6.2.30), (6.2.20), and since $\omega(x_3) \supset \omega$ for all $x_3 \in [0, L]$, we have

$$\mathcal{B}(\bar{y}, \bar{y}) \ge \tilde{\mu}\, c \int_{\omega \times [0,L]} \sum_{i<j} \Big[N_i(x_3) h_{1j}(\bar{x}) + B_i(x_3) h_{2j}(\bar{x}) + (\tau_i'(x_3) + x_1 N_i'(x_3) + x_2 B_i'(x_3)) h_{3j}(\bar{x}) + N_j(x_3) h_{1i}(\bar{x}) + B_j(x_3) h_{2i}(\bar{x}) + (\tau_j'(x_3) + x_1 N_j'(x_3) + x_2 B_j'(x_3)) h_{3i}(\bar{x})\Big]^2 d\bar{x}$$

$$+ 2\,\tilde{\mu}\, c \int_{\omega \times [0,L]} \sum_{i=1}^{3} \Big[N_i(x_3) h_{1i}(\bar{x}) + B_i(x_3) h_{2i}(\bar{x}) + (\tau_i'(x_3) + x_1 N_i'(x_3) + x_2 B_i'(x_3)) h_{3i}(\bar{x})\Big]^2 d\bar{x}.$$

Consequently, usual binomial inequalities imply that

$$\begin{aligned}\frac{1}{\tilde{\mu}\,c}\,\mathcal{B}(\bar{y},\bar{y}) \ \geq\ & \frac{1}{2}\int_{\omega\times[0,L]}\sum_{i<j}\Big[(\tau_i'(x_3)+x_1N_i'(x_3)+x_2B_i'(x_3))h_{3j}(\bar{x})\\ &\qquad +(\tau_j'(x_3)+x_1N_j'(x_3)+x_2B_j'(x_3))h_{3i}(\bar{x})\Big]^2\,d\bar{x}\\ &+\int_{\omega\times[0,L]}\sum_{i=1}^{3}\Big[(\tau_i'(x_3)+x_1N_i'(x_3)+x_2B_i'(x_3))h_{3i}(\bar{x})\Big]^2\,d\bar{x}\\ &-C\,|\bar{y}|^2_{L^2(0,L)^9}, \end{aligned} \tag{6.2.32}$$

where we have used that $h_{ij}\in L^\infty(\Omega)$, $i,j=\overline{1,3}$, which follows from (6.2.18)–(6.2.20), (6.2.24), and from the regularity of the parametrization $\bar{\theta}$ and of the local frame.

Putting $z_i=\tau_i'+x_1N_i'+x_2B_i'$, $i=\overline{1,3}$, we notice the algebraic transformation

$$\begin{aligned}&\frac{1}{2}[(z_1h_{32}+z_2h_{31})^2+(z_2h_{33}+z_3h_{32})^2+(z_1h_{33}+z_3h_{31})^2]\\ &\quad+\frac{3}{2}(z_1^2h_{31}^2+z_2^2h_{32}^2+z_3^2h_{33}^2)\\ &=\frac{1}{2}z_1^2h_{32}^2+\frac{1}{2}z_2^2h_{31}^2+\frac{1}{2}z_2^2h_{33}^2+\frac{1}{2}z_3^2h_{32}^2+\frac{1}{2}z_1^2h_{33}^2+\frac{1}{2}z_3^2h_{31}^2+z_1z_2h_{31}h_{32}\\ &\quad+z_2z_3h_{32}h_{33}+z_1z_3h_{31}h_{33}+\frac{3}{2}(z_1^2h_{31}^2+z_2^2h_{32}^2+z_3^2h_{33}^2)\\ &=\frac{1}{2}(z_1^2+z_2^2+z_3^2)(h_{31}^2+h_{32}^2+h_{33}^2)+\frac{1}{2}(z_1h_{31}+z_2h_{32})^2\\ &\quad+\frac{1}{2}(z_1h_{31}+z_3h_{33})^2+\frac{1}{2}(z_2h_{32}+z_3h_{33})^2.\end{aligned}$$

Hence, we can infer from (6.2.32) that

$$\begin{aligned}\frac{1}{\tilde{\mu}\,c}\,\mathcal{B}(\bar{y},\bar{y}) \ \geq\ & \frac{1}{4}\int_{\omega\times[0,L]}\sum_{i=1}^{3}(\tau_i'(x_3)+x_1N_i'(x_3)+x_2B_i'(x_3))^2\sum_{i=1}^{3}h_{3i}^2(\bar{x})\,d\bar{x}\\ &-C\,|\bar{y}|^2_{L^2(0,L)^9}.\end{aligned} \tag{6.2.33}$$

Examining (6.2.19), we see that under the assumptions of the theorem,

$$\begin{aligned}\sum_{i=1}^{3}h_{3i}^2 &= \frac{1}{(1-\beta(x_3)x_1-a(x_3)x_2)^2}\sum_{i=1}^{3}t_i(x_3)^2\\ &=\frac{1}{(1-\beta(x_3)x_1-a(x_3)x_2)^2}\geq\kappa>0\,,\end{aligned} \tag{6.2.34}$$

for some $\kappa>0$, since $|\bar{t}|_{\mathbf{R}^3}=1$. Moreover, by (6.2.11) we have, for $i=\overline{1,3}$,

$$\int_{\omega\times[0,L]} (\tau_i'(x_3)+x_1N_i'(x_3)+x_2B_i'(x_3))^2\,d\bar{x} = \text{meas}(\omega)\,|\tau_i|^2_{H_0^1(0,L)}$$
$$+\int_\omega x_1^2\,dx_1\,dx_2\,|N_i|^2_{H_0^1(0,L)}+\int_\omega x_2^2\,dx_1\,dx_2\,|B_i|^2_{H_0^1(0,L)}, \quad (6.2.35)$$

where meas(ω) is the Lebesgue measure in $\mathbf{R}^2$ of the domain ω. Combining (6.2.33)–(6.2.35), we obtain the desired result. □

Lemma 6.2.4 *If $\mathcal{B}(\bar{y},\bar{y})=0$, then $\bar{y}=0$.*

Proof. Let $\mathcal{B}(\bar{y},\bar{y})=0$. Defining z_i as in the previous proof, we have, for a.e. $\bar{x}\in\Omega$,

$$0 = N_i(x_3)\,h_{1i}(\bar{x})+B_i(x_3)\,h_{2i}(\bar{x})+z_i\,h_{3i}(\bar{x}), \quad i=\overline{1,3}, \quad (6.2.36)$$
$$0 = N_i(x_3)\,h_{1j}(\bar{x})+B_i(x_3)\,h_{2j}(\bar{x})+N_j(x_3)\,h_{1i}(\bar{x})$$
$$+B_j(x_3)\,h_{2i}(\bar{x})+z_i\,h_{3j}(\bar{x})+z_j\,h_{3i}, \quad i,j=\overline{1,3}, \quad i\neq j. \quad (6.2.37)$$

We first multiply (6.2.37) by $h_{3i}(\bar{x})$ and (6.2.36) by $h_{3j}(\bar{x})$ (j is fixed!) and subtract. Then multiply (6.2.36) by $h_{3i}(\bar{x})$ for $i=j$, and add to the previous results, obtained for $i\neq j$. We then obtain that

$$\tilde{\Gamma}(\bar{N},\bar{B})+z_j\sum_{i=1}^{3}h_{3i}^2(\bar{x})=0, \quad \forall j=\overline{1,3}, \quad (6.2.38)$$

where $\tilde{\Gamma}(\bar{N},\bar{B})$ is some expression that is linear with respect to the vector functions $\bar{N},\bar{B}$. Taking into account (6.2.19) and $|\bar{t}|_{\mathbf{R}^3}=1$, we infer from (6.2.38) that

$$\tau_i'(x_3)+x_1\,N_i'(x_3)+x_2\,B_i'(x_3)=\Gamma(\bar{N},\bar{B}), \quad \forall i=\overline{1,3}, \quad (6.2.39)$$

with an obvious modification Γ of the expression $\tilde{\Gamma}$.

For any i, we choose three different pairs $(x_1,x_2)\in\omega$, obtaining a linear differential system in normalized form with zero initial conditions, since $\tau_i,N_i,B_i\in H_0^1(0,L)$ for all $i=\overline{1,3}$. Hence, the unique solution must vanish identically, which implies that $\bar{y}=0$ in Ω, as claimed. Here, we have again used that the coefficients h_{ij} are bounded. □

Proof of Theorem 6.2.2. We argue by contradiction and assume that there are sequences $\varepsilon_k\searrow 0$ and $y_{\varepsilon_k}=(\tau_1^{\varepsilon_k},\tau_2^{\varepsilon_k},\tau_3^{\varepsilon_k},N_1^{\varepsilon_k},N_2^{\varepsilon_k},N_3^{\varepsilon_k},B_1^{\varepsilon_k},B_2^{\varepsilon_k},B_3^{\varepsilon_k})\in H_0^1(0,L)^9$ such that

$$0\le \mathcal{B}(y_{\varepsilon_k},y_{\varepsilon_k})\le \varepsilon_k\,|y_{\varepsilon_k}|^2_{H_0^1(0,L)^9}.$$

Without loss of generality, we may assume that $|y_{\varepsilon_k}|_{H_0^1(0,L)^9}=1$ for all $k\in\mathbf{N}$, and that $y_{\varepsilon_k}\to\hat{y}$ weakly in $H_0^1(0,L)^9$ as $k\to\infty$. The weak lower semicontinuity of the quadratic form then implies that $0 \leq \mathcal{B}(\hat{y},\hat{y}) \leq 0$, that is, $\mathcal{B}(\hat{y},\hat{y}) = 0$, and we can infer from Lemma 6.2.4 that $\hat{y} = 0$.

Consequently, $y_{\varepsilon_k} \to 0$ weakly in $H_0^1(0,L)^9$ and, by compact embedding, strongly in $L^2(0,L)^9$. Then, applying inequality (6.2.31) in Lemma 6.2.3 to y_{ε_k} for every $k \in \mathbf{N}$, we find that

$$\varepsilon_k \geq \mathcal{B}(y_{\varepsilon_k}, y_{\varepsilon_k}) \geq c_1 \, |y_{\varepsilon_k}|^2_{H_0^1(0,L)^9} - c_2 \, |y_{\varepsilon_k}|^2_{L^2(0,L)^9} = c_1 - c_2 \, |y_{\varepsilon_k}|^2_{L^2(0,L)^9},$$

and passage to the limit as $k \to \infty$ leads to the contradiction $0 \geq c_1$. This concludes the proof of the assertion. □

Having determined the form (cf. (6.2.29), (6.2.30)) and the continuity and coercivity properties of the bilinear form $\mathcal{B}$ for the displacement $\bar{y}$ defined by (6.2.17), we focus our attention on the linear functional $\mathcal{G}$ that collects the contributions of the volume forces and surface tractions and forms the right-hand side of the system of linear elasticity. Performing again the change of variables $\tilde{x} = F\bar{x}$, and using the summation convention, we find that $\mathcal{G}$ can be expressed in the form

$$\begin{aligned}\mathcal{G}(\bar{v}) &= \int_\Omega f_i(\bar{x}) \, (\mu_i(x_3) + x_1 \, M_i(x_3) + x_2 \, D_i(x_3)) \, |\det J(\bar{x})| \, d\bar{x} \\ &\quad + \int_{\partial\Omega} g_i(\bar{\sigma}) \, (\mu_i(\sigma_3) + \sigma_1 \, M_i(\sigma_3) + \sigma_2 \, D_i(\sigma_3)) \, |\det J(\bar{\sigma})| \\ &\quad \cdot \sqrt{\nu_i(\bar{\sigma}) g^{ij}(\bar{\sigma}) \nu_j(\bar{\sigma})} \, d\bar{\sigma}, \end{aligned} \tag{6.2.40}$$

where $(\nu_i)_{i=\overline{1,3}}$ denotes the outward unit normal to $\partial\Omega$. We recall that $(\tilde{f}_i)_{i=\overline{1,3}}$ and $(\tilde{g}_i)_{i=\overline{1,3}}$ are the body forces and surface tractions, respectively, acting on the curved rod $\tilde{\Omega}$, and we have denoted by $\bar{f} = (f_i)_{i=\overline{1,3}}$ and $\bar{g} = (g_i)_{i=\overline{1,3}}$ the quantities

$$\bar{f}(\bar{x}) = \tilde{f}(F\bar{x}), \quad \bar{g}(\bar{x}) = \tilde{g}(F\bar{x}). \tag{6.2.41}$$

The coefficients $g^{ij}(\bar{\sigma})$ are obtained as (cf. Ciarlet [2000])

$$(g_{ij}(\bar{\sigma}))_{i,j=\overline{1,3}} = J(\bar{\sigma})^T J(\bar{\sigma}), \tag{6.2.42}$$

$$(g^{ij}(\bar{\sigma}))_{i,j=\overline{1,3}} = (g_{ij}(\bar{\sigma}))^{-1}_{i,j=1,3}. \tag{6.2.43}$$

Observe that in order that the right-hand side of (6.2.40) be meaningful, at least Lipschitz regularity is needed for that part of $\partial\tilde{\Omega}$ where the tractions are nonzero. This constitutes a regularity condition for the "variation" of the mapping $x_3 \mapsto \omega(x_3)$. In particular, (6.2.40) is fully justified if the mapping $x_3 \mapsto \omega(x_3)$ is "constant" and if $\omega(x_3)$ has a smooth boundary in $\mathbf{R}^2$.

Noticing that $\mathcal{G}$ is under our general assumptions a bounded linear functional on Z (which is identified with $H_0^1(0,L)^9$), and summarizing the above results, we conclude from the Lax–Milgram lemma the following existence and uniqueness result:

Corollary 6.2.5 *Under the above assumptions, the variational equation*

$$\mathcal{B}(\bar{y}, \bar{v}) = \mathcal{G}(\bar{v}) \quad \forall\, \bar{v} \in H_0^1(0, L)^9,$$

with $\mathcal{B}$ *and* $\mathcal{G}$ *given by (6.2.30) and (6.2.40), respectively, has a unique solution* $\bar{y} \in H_0^1(0, L)^9$.

Example 6.2.6 We have determined numerically the deformation of a large class of rods having various cross sections, under different types of applied forces. As three-dimensional curves representing the lines of centroids of the rods, we took the spirals parametrized by

$$\bar{\theta}(t) = \left(\cos\left(\tfrac{t}{\sqrt{2}}\right), \sin\left(\tfrac{t}{\sqrt{2}}\right), \tfrac{t}{\sqrt{2}}\right), \quad t \in [T_1, T_2], \tag{6.2.44}$$

where we considered the cases

$$T_1 = -\tfrac{\pi}{\sqrt{2}}, \quad T_2 = \tfrac{\pi}{\sqrt{2}}, \tag{6.2.45}$$

$$T_1 = 0, \quad T_2 = 8\pi\sqrt{2}. \tag{6.2.46}$$

All these spirals lie on the cylindrical surface

$$\mathcal{S} = \left\{(x_1, x_2, x_3) \in \mathbf{R}^3 : x_1^2 + x_2^2 = 1, \quad \tfrac{T_1}{\sqrt{2}} \le x_3 \le \tfrac{T_2}{\sqrt{2}}\right\}$$

and were assumed to be clamped at their endpoints.

Examples (6.2.44), (6.2.45) have also been discussed by Arunakirinathar and Reddy [1993], and by Chapelle [1997], using the Frenet frame, while our numerical calculations were based on the Darboux frame (for Figures 2.1–2.8 and 2.10–2.11) or on the new local frame introduced in §6.2.1 (for Figure 2.9).

Another form of the line of centroids (a "decreasing" spiral) that has been used is given by

$$\bar{\theta}(t) = \left(-\tfrac{(6\pi - t)^2}{20}\cos(t), -\tfrac{(6\pi - t)^2}{20}\sin(t), t\right), \quad t \in [0, 4\pi], \tag{6.2.47}$$

which is inspired by the spiraling walkway accessing the new dome-shaped cupola of the Reichstag in Berlin. Notice that the parametrization (6.2.47) is not with respect to the arc length, and the standard reparametrization is given by $s(0) = 0$, $s' = |\bar{\theta}'|_{\mathbf{R}^3}$, $\bar{\theta}_1(t) = \bar{\theta}(s^{-1}(t))$. While this relation is difficult to integrate numerically, we remark that in the direct approach of §6.2.1 just $\bar{\theta}_1' = \bar{\theta}'/|\bar{\theta}'|_{\mathbf{R}^3}$ is needed, which is easy to obtain.

Our choices of cross sections included the following cases:

(a) Disk of radius 0.3 centered at the points on $\bar{\theta}$: $x_1^2 + x_2^2 \le 0.3^2$.

(b) Elliptical crown, centered at the points on $\bar{\theta}$: $0.3^2 \le x_1^2 + \frac{x_2^2}{16} \le 0.5^2$.

(c) Rectangle of dimensions 0.6×0.2 centered at the points on $\bar{\theta}$.

The curved rods described above were "subjected" to distributed forces of the following types:

- torsional force

$$f(x_1, x_2, x_3) = (-x_2, x_1, 0), \quad x_3 \in \left[-\tfrac{\pi}{\sqrt{2}}, \tfrac{\pi}{\sqrt{2}}\right]; \tag{6.2.48}$$

- torsional force in two opposite directions

$$f(x_1, x_2, x_3) = \begin{cases} (-x_2, x_1, 0), & x_3 \in \left[-\tfrac{\pi}{\sqrt{2}}, 0\right], \\ (x_2, -x_1, 0), & x_3 \in \left[0, \tfrac{\pi}{\sqrt{2}}\right]; \end{cases} \tag{6.2.49}$$

- pushing in one or two opposite given directions

$$f(x_1, x_2, x_3) = (0,0,1), \quad x_3 \in \left[-\tfrac{\pi}{\sqrt{2}}, \tfrac{\pi}{\sqrt{2}}\right], \tag{6.2.50}$$

$$f(x_1, x_2, x_3) = \begin{cases} (0,0,1), & x_3 \in \left[T_1, \tfrac{T_1+T_2}{2}\right], \\ (0,0,-1), & x_3 \in \left[\tfrac{T_1+T_2}{2}, T_2\right], \end{cases} \tag{6.2.51}$$

$$f(x_1, x_2, x_3) = \begin{cases} (0,0,-1000), & x_3 \in [2\pi, 3\pi], \\ (0,0,0), & \text{otherwise}; \end{cases} \tag{6.2.52}$$

- tangential forces

$$f(x_1, x_2, x_3) = \begin{cases} \bar{\theta}'(x_3), & x_3 \in \left[T_1, \tfrac{T_1+T_2}{2}\right], \\ -\bar{\theta}'(x_3), & x_3 \in \left[\tfrac{T_1+T_2}{2}, T_2\right]; \end{cases} \tag{6.2.53}$$

- normal forces

$$f(x_1, x_2, x_3) = \bar{n}(x_3), \tag{6.2.54}$$

where $\bar{n}$ is determined from the Darboux frame, i.e., is given by the normal to the cylindrical surface $\mathcal{S}$ in the points of the curve $\bar{\theta}$. We also mention that in all cases where the Darboux frame was applied, the computation of $\bar{t}$, $\bar{n}$, $\bar{b}$ and of the quantities $\det J(\bar{x})$, $h_{ij}(\bar{x})$, $a(x_3)$, $\beta(x_3)$, $c(x_3)$ was performed exactly, using the software package *Mathematica*. We also employed *Mathematica* for the graphical representations. For the method developed in §6.2.1, direct numerical calculations were performed to obtain the values of the above parameters in the discretization points.

In Table 2.1, we have collected details about the geometrical and mechanical data assumed in the examples (Figures 2.1–2.9).

Example	Interval	Force	Section	Curve	Scaling
Figure 2.1	(6.2.45)	(6.2.49)	(a)	(6.2.44)	500
Figure 2.2	(6.2.46)	(6.2.51)	(a)	(6.2.44)	0.2
Figure 2.3	(6.2.45)	(6.2.51)	(a)	(6.2.44)	50
Figure 2.4	(6.2.45)	(6.2.50)	(b)	(6.2.44)	50
Figure 2.5	(6.2.45)	(6.2.48)	(b)	(6.2.44)	50
Figure 2.6	(6.2.46)	(6.2.53)	(a)	(6.2.44)	0.05
Figure 2.7	(6.2.45)	(6.2.53)	(a)	(6.2.44)	50
Figure 2.8	(6.2.46)	(6.2.54)	(a)	(6.2.44)	30
Figure 2.9	(6.2.47)	(6.2.52)	(c)	(6.2.47)	300

Table 2.1.

The Lamé constants were in all examples fixed as $\tilde{\lambda} = 50$, $\tilde{\mu} = 100$. For the case of Figure 2.5, we have also depicted in Figures 2.10 and 2.11 the deformation of the cross section in the two points that correspond to the division points with the indices $i = 50$ and $i = 100$. In all examples, the interval $[T_1, T_2]$ has been subdivided equidistantly into 200 subintervals; i.e., the indices $i = 50$ and $i = 100$ correspond to the first quarter and to the middle of the curved rod.

The scalings used for the various figures (cf. Table 2.1) were chosen to obtain a clear graphical representation. In view of the linearity of the equations, this amounts to multiplying the force by the given factor. In Figures 2.10 and 2.11 the scaling factors were equal to 20 and 10, respectively. We also notice that the shear and the torsion effects are not represented, since this would require three-dimensional figures.

Our aim in the numerical simulation was to test the modeling approach, and all experiments produced results that are meaningful from the viewpoint of mechanics. We have avoided the well-known *locking problem* appearing in the numerical approximation of arches, rods, and shells (see Chenais and Paumier [1994], Arunakirinathar and Reddy [1993], Chapelle [1997], and Pitkäranta and Leino [1994]), by choosing a very fine division of $[T_1, T_2]$ in comparison with the dimension of the cross sections, that is, by taking "large" cross sections.

Consequently, we could use standard piecewise linear finite elements. If V_m, $m = 200$, denotes the discrete subspace of $H_0^1(T_1, T_2)$, then V_m^9 is associated with $H_0^1(T_1, T_2)^9$. Its finite element basis is constructed in a canonical way. The rigidity matrix is sparse; the number of unknowns in the obtained linear algebraic system was equal to 1791. The integrals over cross sections that appear in the coefficients were computed using a change of variables to polar coordinates, which reduced all the cases to various rectangles, and then classical interpolation methods. Since the dimension was not too big, the algebraic linear system could be stably solved by Gauss elimination. In all examples, we have observed that the vectors $\bar{N}^m\left(\frac{T_1+T_2}{2}\right)$ and $\bar{B}^m\left(\frac{T_1+T_2}{2}\right)$ (the deformations of the "normal" and "binormal" vectors in the middle of the rod) are orthogonal.

Remark. We finally point out that the theoretical results allow for variable cross sections, nonzero surface tractions, etc. We did not implement such cases numerically in order to keep the complexity at a reasonable level.

(Figures 2.1–2.11 © John Wiley & Sons Ltd.)

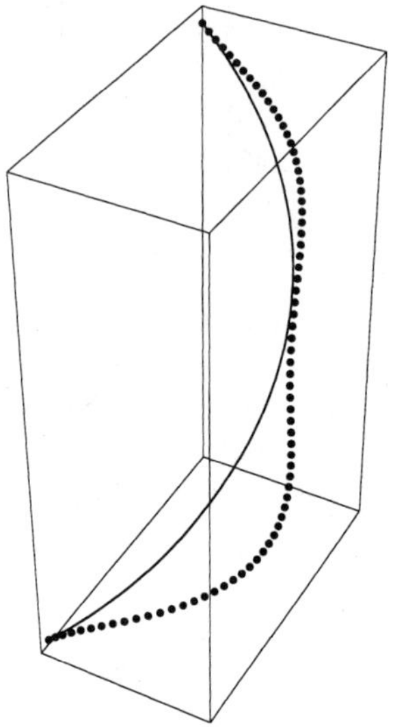

Figure 2.1.

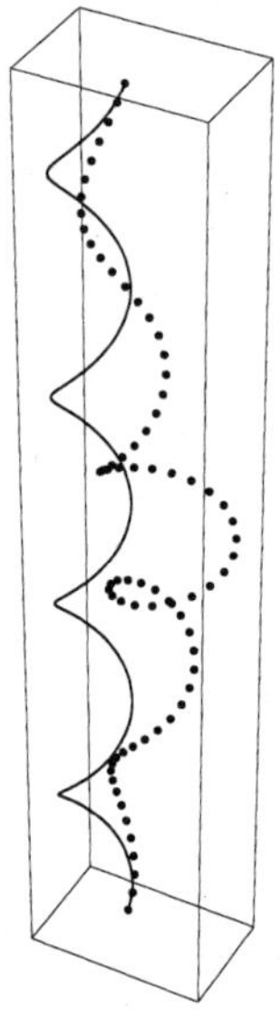

Figure 2.2.

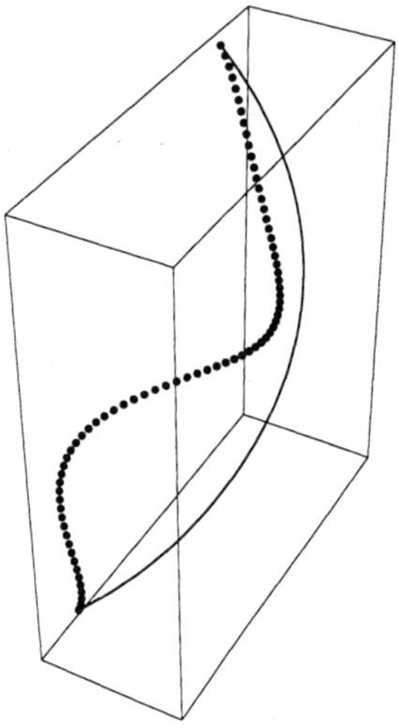

Figure 2.3.

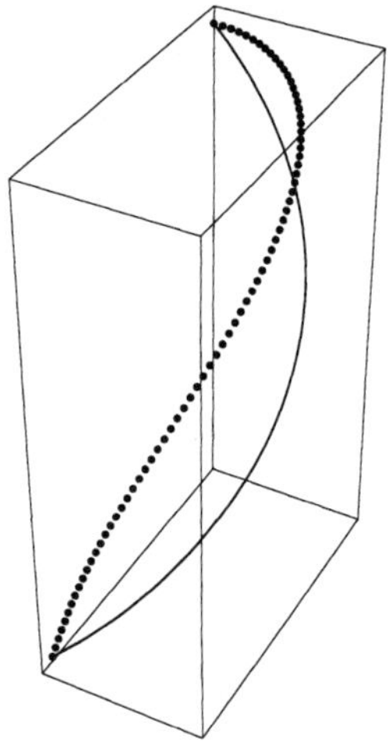

Figure 2.4.

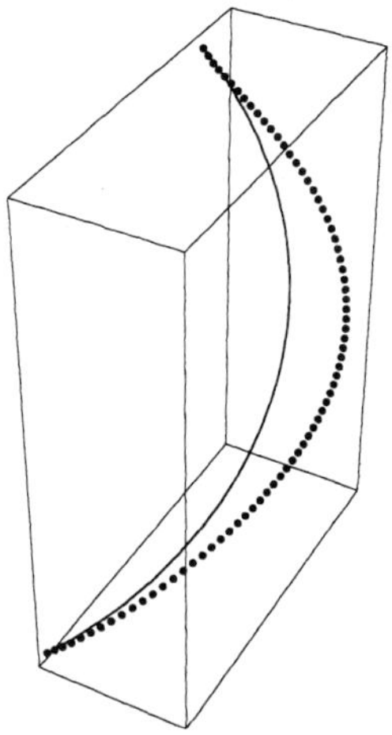

Figure 2.5.

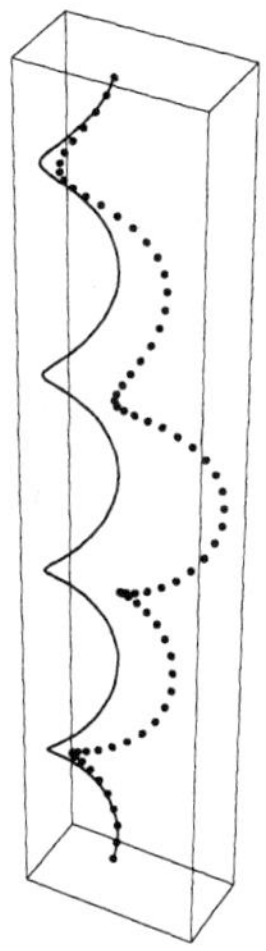

Figure 2.6.

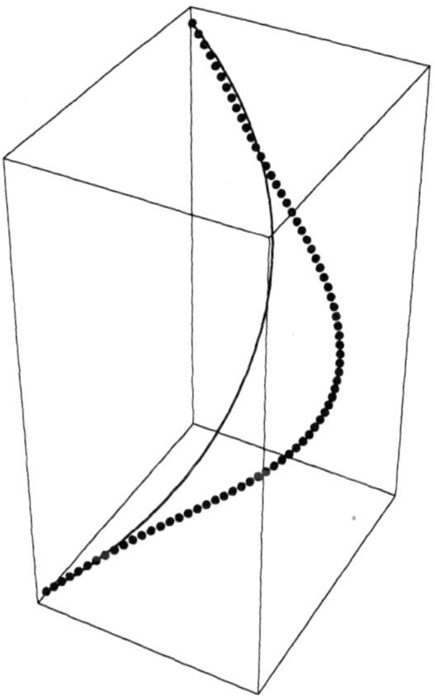

Figure 2.7.

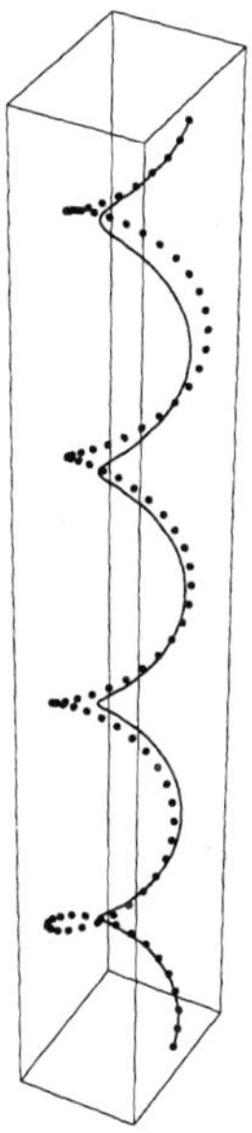

Figure 2.8.

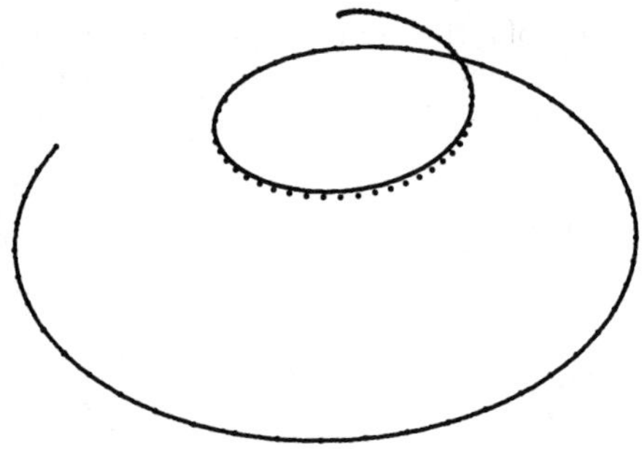

Figure 2.9.

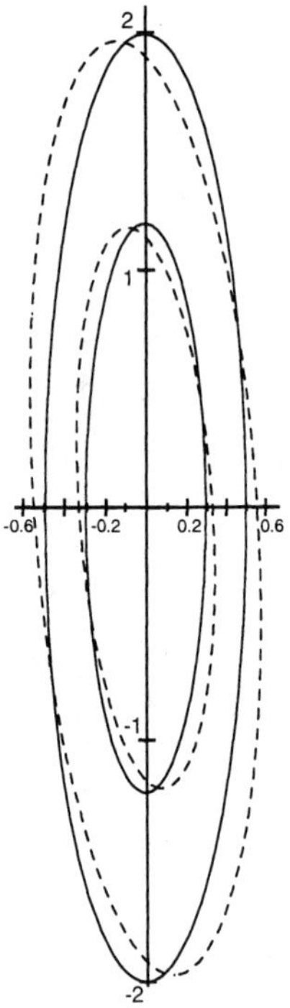

Figure 2.10.

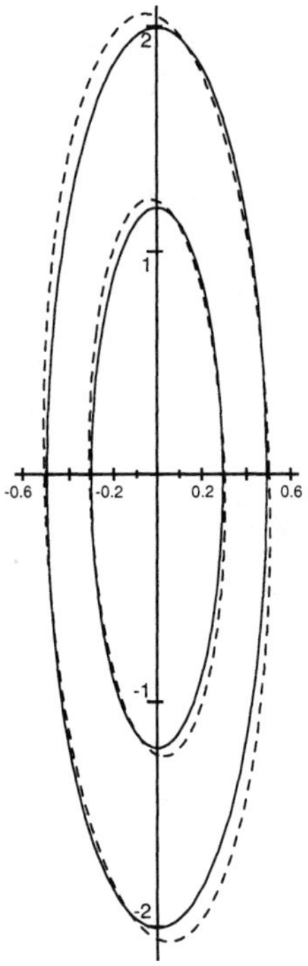

Figure 2.11.

6.2.3 Shape Optimization

Finding the "best" form of a three-dimensional curved rod with respect to some given performance index relies, as usual, on the continuity and differentiability properties of the solution $\bar{y}$ of the variational equation given by (6.2.30), (6.2.40) with respect to variations of the geometry. We assume that the volume forces $\bar{f}$ and the surface tractions $\bar{g}$ are given for any point in a sufficiently large domain in $\mathbf{R}^3$.

In dealing with these questions, we adopt a new parametrization idea that is more advantageous in this setting. Namely, instead of using a parametrization $\bar{\theta}$ of the line of centroids as in Theorem 6.2.1, we are now going to parametrize the tangent vector field $\bar{t}$ directly. Since $\bar{t}$ is always assumed to be nonzero, the standard reparametrization will make $\bar{\theta}$ a unit speed curve, i.e., we will have $|\bar{t}(x_3)|_{\mathbf{R}^3} = 1$ for all $x_3 \in [0, L]$. Therefore, we can always find functions φ, ψ that represent the spherical coordinates of $\bar{t}(x_3)$; that is, we have

$$\bar{t}(x_3) = (\sin(\varphi(x_3))\cos(\psi(x_3)),\ \sin(\varphi(x_3))\sin(\psi(x_3)),\ \cos(\varphi(x_3))), \quad (6.2.55)$$

$$\bar{\theta}(x_3) = \int_0^{x_3} \bar{t}(\tau)\, d\tau, \quad x_3 \in [0, L]. \quad (6.2.56)$$

In the following, we will always assume that $[\varphi, \psi] \in C^1[0, L]^2$. Then it follows from (6.2.55) and (6.2.56) that $\bar{t} \in C^1[0, L]$ and $\bar{\theta} \in C^2[0, L]$.

One advantage of this approach is that a local frame may be defined by purely algebraic means, namely, we can in every point on the line of centroids choose (with obvious notation)

$$\bar{n} = (\cos(\varphi)\cos(\psi),\ \cos(\varphi)\sin(\psi),\ -\sin(\varphi)), \quad (6.2.57)$$

$$\bar{b} = (-\sin(\psi),\ \cos(\psi), 0). \quad (6.2.58)$$

Moreover, the line of centroids is automatically parametrized with respect to the arc length, and its length is fixed and equal to $L > 0$. This is particularly convenient in optimization problems, since the cost functional may depend not only on the "shape" but also on the length, and so a length dependence in the parametrization should be avoided. We also remark that while the spherical coordinates may not be uniquely determined in certain cases, the relations (6.2.55), (6.2.56) with arbitrary $[\varphi, \psi]$ already generate a rich class of unit speed curves in $C^2[0, L]$ that suffices for many applications.

Taking into account (6.2.55)–(6.2.58), the whole machinery developed in §6.2.2 can be applied correspondingly. We begin our analysis by proving a continuous dependence result.

Theorem 6.2.7 *Suppose that $\varphi_k \to \varphi$ and $\psi_k \to \psi$ strongly in $C^1[0, L]$. If $\bar{\theta}_k$ is obtained by (6.2.55), (6.2.56) from φ_k, ψ_k, and $\bar{y}_k$, $\bar{y}$ denote the corresponding solutions of the associated variational equations*

$$\mathcal{B}_k(\bar{y}_k, \bar{v}) = \mathcal{G}_k(\bar{v}) \quad \forall\, \bar{v} \in H_0^1(0, L)^9,$$

where $\mathcal{B}_k$ *and* $\mathcal{G}_k$ *denote the bilinear form and the linear functional corresponding to* $\bar{\theta}_k$ *by (6.2.30) and (6.2.40), respectively, for* $k \in \mathbf{N}$*, then*

$$\bar{y}_k \to \bar{y} \quad \textit{strongly in } H_0^1(0,L)^9. \tag{6.2.59}$$

Proof. Clearly, we have

$$\begin{aligned} \bar{t}_k &= (\cos(\psi_k)\sin(\varphi_k), \sin(\psi_k)\sin(\varphi_k), \cos(\varphi_k)) \\ \to \bar{t} &= (\cos(\psi)\sin(\varphi), \sin(\psi)\sin(\varphi), \cos(\varphi)) \end{aligned} \tag{6.2.60}$$

strongly in $C^1[0,L]^3$. Then, in view of (6.2.56), $\bar{\theta}_k \to \bar{\theta}$ strongly in $C^2[0,L]^3$, and it follows from (6.2.57), (6.2.58) with obvious notation that $\bar{n}_k \to \bar{n}$ and $\bar{b}_k \to \bar{b}$ strongly in $C^1[0,L]^3$. Moreover, adapting (6.2.16) to the present situation, it is easy to infer that

$$a_k = (\bar{t}_k', \bar{b}_k)_{\mathbf{R}^3} \to a = (\bar{t}', \bar{b})_{\mathbf{R}^3} \quad \text{strongly in } C[0,L]. \tag{6.2.61}$$

Likewise, we have $\beta_k \to \beta$ and $c_k \to c$ strongly in $C[0,L]$. Similarly, the relations (6.2.20), (6.2.61) show that

$$\det J_k(\bar{x}) \to \det J(\bar{x}) \quad \text{strongly in } C(\overline{\Omega}), \tag{6.2.62}$$

and thus we can without loss of generality assume that $\{\det J_k(\bar{x})\}_{k\in\mathbf{N}}$ is bounded from below by some positive constant $c > 0$. In addition, (6.2.18) yields that $J_k(\cdot) \to J(\cdot)$ strongly in $C(\overline{\Omega})^9$ and, likewise, that $J_k^{-1}(\bar{x}) \to J^{-1}(\bar{x})$ uniformly in $\overline{\Omega}$, thanks to (6.2.19) and (6.2.62). In particular,

$$h_{ij}^k(\cdot) \to h_{ij}(\cdot) \quad \text{strongly in } C(\overline{\Omega}). \tag{6.2.63}$$

Now recall that $\mathcal{B}_k$ denotes the bilinear functional (6.2.30) associated with the coefficients h_{ij}^k and $\det J_k$, $k \in \mathbf{N}$. We first show that the statement of Lemma 6.2.3 persists uniformly with respect to $k \in \mathbf{N}$.

Lemma 6.2.8 *There are constants* $c_1 > 0$ *and* $c_2 > 0$ *such that*

$$\mathcal{B}_k(\bar{y},\bar{y}) \ge c_1 |\bar{y}|^2_{H_0^1(0,L)^9} - c_2 |\bar{y}|^2_{L^2(0,L)^9}, \quad \forall \bar{y} \in H_0^1(0,L)^9, \quad \forall k \in \mathbf{N}. \tag{6.2.64}$$

Proof. Since $\det J_k(\bar{x}) \ge c > 0$ for all $k \in \mathbf{N}$, we have (see the proof of Lemma 6.2.3)

$$\begin{aligned} \mathcal{B}_k(\bar{y},\bar{y}) \ge \tilde{\mu}\, c \int_\Omega \sum_{i<j} \Big[& N_i(x_3)\, h_{1j}^k + B_i(x_3)\, h_{2j}^k + (\tau_i'(x_3) + x_1\, N_i'(x_3) \\ & + x_2\, B_i'(x_3))\, h_{3j}^k \; + \; N_j(x_3)\, h_{1i}^k + B_j(x_3)\, h_{2i}^k \\ & + (\tau_j'(x_3) + x_1\, N_j'(x_3) + x_2\, B_j'(x_3))\, h_{3i}^k \Big]^2 \, d\bar{x} \\ + 2\,\tilde{\mu}\, c \int_\Omega \sum_{i=1}^3 \Big[& N_i(x_3)\, h_{1i}^k + B_i(x_3)\, h_{2i}^k + (\tau_i'(x_3) + x_1\, N_i'(x_3) \\ & + x_2\, B_i'(x_3))\, h_{3i}^k \Big]^2 \, d\bar{x}. \end{aligned}$$

Thanks to (6.2.63), the coefficients are uniformly bounded, and invoking standard binomial inequalities, we find that

$$\frac{1}{\tilde{\mu}c}\,\mathcal{B}_k(\bar{y},\bar{y}) \;\geq\; \frac{1}{2}\int_\Omega \sum_{i<j}\Big[(\tau_i' + x_1\,N_i' + x_2\,B_i')\,h_{3j}^k + (\tau_j' + x_1\,N_j' + x_2\,B_j')\,h_{3i}^k\Big]^2\,d\bar{x}$$

$$+\int_\Omega \sum_{i=1}^{3}\Big[(\tau_i' + x_1\,N_i' + x_2\,B_i')\,h_{3i}^k\Big]^2\,d\bar{x} - C\,|\bar{y}|^2_{L^2(0,L)^9},$$

with some constant $C > 0$ that does not depend on $k \in \mathbf{N}$. Using again the algebraic identity from the proof of Lemma 6.2.3 with $z_i = \tau_i' + x_1 N_i' + x_2 B_i'$, $i = \overline{1,3}$, we conclude that

$$\frac{1}{\tilde{\mu}c}\,\mathcal{B}_k(\bar{y},\bar{y}) \geq \frac{1}{4}\int_\Omega \sum_{i=1}^{3}(\tau_i' + x_1\,N_i' + x_2\,B_i')^2\,\sum_{i=1}^{3}(h_{3i}^k)^2\,d\bar{x} - C\,|\bar{y}|^2_{L^2(0,L)^9}. \quad (6.2.65)$$

By a lengthy but straightforward calculation, which is omitted here, we can determine h_{ij}^k as in (6.2.19) and check that with some $\tilde{c} > 0$,

$$\sum_{i=1}^{3}(h_{3i}^k)^2 = [\det J_k]^{-2}\,\sum_{i=1}^{3}(t_i^k)^2 = [\det J_k]^{-2} \geq \tilde{c} > 0, \quad (6.2.66)$$

where we have used that $|\bar{t}_k|_{\mathbf{R}^3} = 1$. From (6.2.65), (6.2.66) it follows that

$$\frac{1}{\tilde{\mu}c}\mathcal{B}_k(\bar{y},\bar{y}) \geq \frac{\tilde{c}}{4}\int_\Omega \sum_{i=1}^{3}(\tau_i' + x_1 N_i' + x_2 B_i')^2\,d\bar{x} - C\,|\bar{y}|^2_{L^2(0,L)^9}.$$

Performing the computations on the right-hand side, and integrating with respect to x_1, x_2, we obtain the asserted inequality (6.2.64) by means of (6.2.11).

□

Proof of Theorem 6.2.7 (continued). We argue by contradiction to show that the sequence $\mathcal{B}_k$ of functionals is uniformly coercive. Assume there are $\varepsilon_k \searrow 0$ and $\tilde{y}_k \in H_0^1(0,L)^9$ with $|\tilde{y}_k|_{H_0^1(0,L)^9} = 1$, $k \in \mathbf{N}$, such that

$$0 \leq \mathcal{B}_k(\tilde{y}_k,\tilde{y}_k) \leq \varepsilon_k\,|\tilde{y}_k|^2_{H_0^1(0,L)^9} = \varepsilon_k. \quad (6.2.67)$$

Without loss of generality, we may assume that $\tilde{y}_k = (\tilde{\tau}^k, \tilde{N}^k, \tilde{B}^k) \to \hat{y} = (\hat{\tau}, \hat{N}, \hat{B})$ weakly in $H_0^1(0,L)^9$ and strongly in $L^2(0,L)^9$.

We give a detailed computation only for the last integral occurring in the definition of the expression $\mathcal{B}_k(\tilde{y}_k,\tilde{y}_k)$. We have

$$I_k = 2\,\tilde{\mu}\int_\Omega \sum_{i=1}^{3}\Big[\tilde{N}_i^k\,h_{1i}^k + \tilde{B}_i^k\,h_{2i}^k + ((\tilde{\tau}_i^k)' + x_1\,(\tilde{N}_i^k)' + x_2\,(\tilde{B}_i^k)')\,h_{3i}^k\Big]^2\,\det J^k\,d\bar{x}$$

$$= 2\,\tilde{\mu}\int_\Omega \sum_{i=1}^{3}\Big[\tilde{N}_i^k\,h_{1i} + \tilde{B}_i^k\,h_{2i} + ((\tilde{\tau}_i^k)' + x_1\,(\tilde{N}_i^k)' + x_2\,(\tilde{B}_i^k)')\,h_{3i}\Big]^2\,\det J\,d\bar{x}$$

$$
\begin{aligned}
+2\,\tilde{\mu} \int_{\Omega} \sum_{i=1}^{3} \Big[& \tilde{N}_i^k \,(h_{1i}^k \sqrt{\det J^k} \;-\; h_{1i} \sqrt{\det J}) + \tilde{B}_i^k \,(h_{2i}^k \sqrt{\det J^k} \\
& - h_{2i}\sqrt{\det J}) + ((\tilde{\tau}_i^k)' + x_1\,(\tilde{N}_i^k)' + x_2\,(\tilde{B}_i^k)') \\
& \cdot (h_{3i}^k \sqrt{\det J^k} - h_{3i} \sqrt{\det J})\Big] \\
& \cdot \Big[\tilde{N}_i^k \,(h_{1i}^k \sqrt{\det J^k} + h_{1i} \sqrt{\det J}) + \tilde{B}_i^k \,(h_{2i}^k \sqrt{\det J^k} \\
& + h_{2i} \sqrt{\det J}) + ((\tilde{\tau}_i^k)' + x_1\,(\tilde{N}_i^k)' + x_2\,(\tilde{B}_i^k)') \\
& \cdot (h_{3i}^k \sqrt{\det J^k} + h_{3i} \sqrt{\det J})\Big]\, d\bar{x}.
\end{aligned}
$$

The boundedness of $\left\{|\tilde{y}_k|_{H_0^1(0,L)^9}\right\}_{k\in\mathbf{N}}$ and the uniform convergence of the coefficients (see (6.2.62), (6.2.63)) imply that the last integral converges to zero as $k \to \infty$. Moreover, by the weak lower semicontinuity of quadratic forms,

$$
\begin{aligned}
\liminf_{k\to\infty} I_k \;&=\; 2\,\tilde{\mu}\, \liminf_{k\to\infty} \int_{\Omega} \sum_{i=1}^{3} \Big[\tilde{N}_i^k\, h_{1i} + \tilde{B}_i^k\, h_{2i} + ((\tilde{\tau}_i^k)' + x_1\,(\tilde{N}_i^k)' \\
&\qquad\qquad + x_2\,(\tilde{B}^k)')\, h_{3i}\Big]^2 \det J\, d\bar{x} \\
&\geq\; 2\tilde{\mu} \int_{\Omega} \sum_{i=1}^{3} \Big[\hat{N}_i\, h_{1i} + \hat{B}_i\, h_{2i} + (\hat{\tau}_i' + x_1\,\hat{N}_i' + x_2\,\hat{B}_i')\, h_{3i}\Big]^2 \det J\, d\bar{x}.
\end{aligned}
$$

Performing analogous computations for the other terms occurring in $\mathcal{B}_k(\tilde{y}_k, \tilde{y}_k)$, we find that

$$
\begin{aligned}
& \liminf_{k\to\infty} \mathcal{B}_k(\tilde{y}_k, \tilde{y}_k) \\
& \geq 2\,\tilde{\mu} \int_{\Omega} \sum_{i=1}^{3} \Big[\hat{N}_i\, h_{1i} + \hat{B}_i\, h_{2i} + (\hat{\tau}_i' + x_1\,\hat{N}_i' + x_2\,\hat{B}_i')\, h_{3i}\Big]^2 \det J\, d\bar{x} \\
& \quad + \tilde{\lambda} \int_{\Omega} \sum_{i,j=1}^{3} \Big[\hat{N}_i\, h_{1i} + \hat{B}_i\, h_{2i} + (\hat{\tau}_i' + x_1\,\hat{N}_i' + x_2\,\hat{B}_i')\, h_{3i}\Big] \\
& \qquad\qquad \cdot \Big[\hat{N}_j\, h_{1j} + \hat{B}_j\, h_{2j} + (\hat{\tau}_j' + x_1\,\hat{N}_j' + x_2\,\hat{B}_j')\, h_{3j}\Big] \det J\, d\bar{x} \\
& \quad + \tilde{\mu} \int_{\Omega} \sum_{i<j} \Big[\hat{N}_i\, h_{1j} + \hat{B}_i\, h_{2j} + (\hat{\tau}_i' + x_1\,\hat{N}_i' + x_2\,\hat{B}_i')\, h_{3j} \\
& \qquad\qquad + \hat{N}_j\, h_{1i} + \hat{B}_j\, h_{2i} + (\hat{\tau}_j' + x_1\,\hat{N}_j' + x_2\,\hat{B}_j') h_{3i}\Big]^2 \det J\, d\bar{x} \\
& = \mathcal{B}(\hat{y}, \hat{y}).
\end{aligned}
\tag{6.2.68}
$$

By assumption (6.2.67), and by (6.2.68), we have $\mathcal{B}(\hat{y}, \hat{y}) = 0$, and thus, thanks to Lemma 6.2.3, $\hat{y} = 0$. Using (6.2.67) together with Lemma 6.2.8, we find that

$$
\varepsilon_k \;\geq\; \mathcal{B}_k(\tilde{y}_k, \tilde{y}_k) \;\geq\; c_1 - c_2 |\tilde{y}_k|^2_{L^2(0,L)^9}, \tag{6.2.69}
$$

since $|\tilde{y}_k|_{H_0^1(0,L)^9} = 1$. Thus, since $\tilde{y}_k \to \hat{y} = 0$ strongly in $L^2(0,L)^9$, we arrive at the contradiction $0 \geq c_1$. Therefore, there is some $\delta > 0$ such that

$$\mathcal{B}_k(\bar{y},\bar{y}) \geq \delta\, |\bar{y}|^2_{H_0^1(0,L)^9}, \quad \forall\, \bar{y} \in H_0^1(0,L)^9, \quad \forall\, k \in \mathbf{N}. \tag{6.2.70}$$

Now let us insert $\bar{v} = \bar{y}_k$ in the state equation (6.2.40) corresponding to $\mathcal{B}_k$, $k \in \mathbf{N}$. Taking (6.2.70) into account, we immediately obtain that $\{\bar{y}_k\}$ is bounded in $H_0^1(0,L)^9$. We may thus select a subsequence, again indexed by k, such that $\bar{y}_k \to \check{y}$ weakly in $H_0^1(0,L)^9$. By virtue of the uniform convergence of the coefficients h_{ij}^k, $\det J_k$, and g_k^{ij}, one may pass to the limit as $k \to \infty$ in (6.2.40) to see that $\check{y}$ is in fact the solution of (6.2.40) associated with $[\varphi,\psi]$. The uniqueness of the limit point entails that the convergence holds for the entire sequence.

The last step of the proof is to show that the convergence is valid in the strong topology of $H_0^1(0,L)^9$. To this end, we subtract the equations corresponding to (τ^k, N^k, B^k), respectively to $(\check{\tau}, \check{N}, \check{B})$, add and subtract advantageous terms, and we take as test functions $(\tau^k, N^k, B^k) - (\check{\tau}, \check{N}, \check{B}) \in H_0^1(0,L)^9$.

The resulting expression is rather complex, and we confine ourselves to write in detail just the simplest term (coming from the last integral). We obtain

$$\begin{aligned}
&2\tilde{\mu} \int_\Omega \sum_{i=1}^3 \Big[N_i^k\, h_{1i}^k + B_1^k\, h_{2i}^k + ((\tau_i^k)' + x_1\,(N_i^k)' + x_2\,(B_i^k)')\, h_{3i}^k\Big]\, \Big[(N_i^k - \check{N}_i)\, h_{1i}^k \\
&\qquad + (B_i^k - \check{B}_i)\, h_{2i}^k + ((\tau_i^k)' - \check{\tau}_i' + x_1((N_i^k)' - \check{N}_i') + x_2((B_i^k)' - \check{B}_i'))\, h_{3i}^k\Big] \\
&\qquad\qquad \cdot \det J^k\, d\bar{x} \\
&-2\tilde{\mu} \int_\Omega \sum_{i=1}^3 \Big[\check{N}_i\, h_{1i} + \check{B}_i\, h_{2i} + (\check{\tau}_i' + x_1\, \check{N}_i' + x_2\, \check{B}_i')\, h_{3i}\Big]\, \Big[(N_i^k - \check{N}_i)\, h_{1i} \\
&\qquad + (B_i^k - \check{B}_i)\, h_{2i} + ((\tau_i^k)' - \check{\tau}_i' + x_1((N_i^k)' - \check{N}_i') + x_2((B_i^k)' - \check{B}_i'))\, h_{3i}\Big] \\
&\qquad\qquad \cdot \det J\, d\bar{x} \\
&= 2\tilde{\mu} \int_\Omega \sum_{i=1}^3 \Big[(N_i^k - \check{N}_i)\, h_{1i} + (B_i^k - \check{B}_i)\, h_{2i} + ((\tau_i^k)' - \check{\tau}_i' + x_1\,((N_i^k)' - \check{N}_i') \\
&\qquad\qquad + x_2\,((B_i^k)' - \check{B}_i'))\, h_{3i}\Big]^2 \det J\, d\bar{x} \\
&+2\tilde{\mu} \int_\Omega \sum_{i=1}^3 \Big[N_i^k\, h_{1i} + B_i^k\, h_{2i} + ((\tau_i^k)' + x_1\,(N_i^k)' + x_2\,(B_i^k)')\, h_{3i}\Big] \\
&\qquad \cdot \Big[(N_i^k - \check{N}_i)\,(h_{1i}^k - h_{1i}) + (B_i^k - \check{B}_i)\,(h_{2i}^k - h_{2i}) + ((\tau_i^k)' - \check{\tau}_i' \\
&\qquad\qquad + x_1\,((N_i^k)' - \check{N}_i') + x_2\,((B_i^k)' - \check{B}_i'))\,(h_{3i}^k - h_{3i})\Big] \det J\, d\bar{x}
\end{aligned}$$

$$+2\tilde{\mu}\int_{\Omega}\sum_{i=1}^{3}\left[N_i^k\,h_{1i}+B_i^k\,h_{2i}+((\tau_i^k)'+x_1\,(N_i^k)'+x_2\,(B_i^k)')\,h_{3i}\right]$$

$$\cdot\left[(N_i^k-\check{N}_i)\,h_{1i}^k+(B_i^k-\check{B}_i)\,h_{2i}^k+((\tau_i^k)'-\check{\tau}_i'+x_1\,((N_i^k)'-\check{N}_i')\right.$$

$$\left.+\,x_2\,((B_i^k)'-\check{B}_i'))\,h_{3i}^k\right]\,[\det J^k-\det J]\,d\bar{x}$$

$$+2\tilde{\mu}\int_{\Omega}\sum_{i=1}^{3}\left[N_i^k\,(h_{1i}^k-h_{1i})+B_i^k\,(h_{2i}^k-h_{2i})+((\tau_i^k)'+x_1\,(N_i^k)'+x_2\,(B_i^k)')\right.$$

$$\left.\cdot\,(h_{3i}^k-h_{3i})\right]\,\left[(N_i^k-\check{N}_i)\,h_{1i}^k+(B_i^n-\check{B}_i)\,h_{2i}^k\right.$$

$$\left.+((\tau_i^k)'-\check{\tau}_i'+x_1\,((N_i^k)'-\check{N}_i')+x_2\,((B_i^k)'-\check{B}_i'))\,h_{31}^k\right]\,\det J^k\,d\bar{x}.$$

All the terms above, except the first one after the equality sign (the quadratic one), converge to zero in view of the weak convergence of $\{(\tau^k, N^k, B^k)\}$ and of the uniform convergence of the coefficients. Similar computations may be performed for all the integrals in the variational equations, and we may conclude that

$$\lim_{k\to\infty}\mathcal{B}(\bar{y}_k-\check{y},\bar{y}_k-\check{y})=0, \tag{6.2.71}$$

which concludes the proof of the assertion. □

Next, we will study the differentiability properties of the "solution operator" that assigns to $[\varphi,\psi]\in C^1[0,L]^2$ the corresponding solution $\bar{y}\in H_0^1(0,L)^9$ of the variational equality defined by the expressions (6.2.30) and (6.2.40). To this end, we consider admissible variations around $[\varphi,\psi]$ of the form $[\varphi_\lambda,\psi_\lambda]=[\varphi+\lambda\gamma,\psi+\lambda\xi]$ with $[\gamma,\xi]\in C^1[0,1]^2$ and $\lambda\in\mathbf{R}$. Let $\bar{y}_\lambda=(\bar{\tau}_\lambda,\bar{N}_\lambda,\bar{B}_\lambda)\in H_0^1(0,L)^9$ denote the solutions to the corresponding variational problems. Likewise, we denote by $\bar{t}_\lambda$, $\bar{\theta}_\lambda$, $\bar{n}_\lambda$, $\bar{b}_\lambda$, a_λ, β_λ, c_λ, J_λ, h_{ij}^λ, g_λ^{ij}, the related quantities defined previously, beginning from $[\varphi_\lambda,\psi_\lambda]$.

Notice that by construction, the perturbed curved rods described by $\bar{\theta}_\lambda$ have length L and are parametrized with respect to their arc length; i.e., we still have $|\bar{t}_\lambda|_{\mathbf{R}^3}=1$. We also notice that "admissibility" means, in particular, that

$$\det J_\lambda(\bar{x})\geq c\quad\forall\,\bar{x}\in\overline{\Omega},$$

with some fixed constant $c>0$. This condition can be satisfied for "sufficiently small" perturbations $[\lambda\,\gamma,\lambda\,\xi]$, since the mapping $[\varphi,\psi]\mapsto\det J$ is continuous. An elementary, though rather lengthy and tedious, calculation (which must be omitted here) shows that the limits and operators listed below exist and are defined as mappings between the indicated function spaces:

$$\tilde{t}:C^1[0,L]^2\to C^1[0,L]^3;\qquad \tilde{t}([\gamma,\xi])=\lim_{\lambda\to 0}\frac{\bar{t}_\lambda-\bar{t}}{\lambda}, \tag{6.2.72}$$

$$\tilde{\theta} : C^1[0,L]^2 \to C^2[0,L]^3\,; \qquad \tilde{\theta}([\gamma,\xi]) = \lim_{\lambda\to 0} \frac{\bar{\theta}_\lambda - \bar{\theta}}{\lambda}\,, \tag{6.2.73}$$

$$\tilde{n} : C^1[0,L]^2 \to C^1[0,L]^3\,; \qquad \tilde{n}([\gamma,\xi]) = \lim_{\lambda\to 0} \frac{\bar{n}_\lambda - \bar{n}}{\lambda}\,, \tag{6.2.74}$$

$$\tilde{b} : C^1[0,L]^2 \to C^1[0,L]^3\,; \qquad \tilde{b}([\gamma,\xi]) = \lim_{\lambda\to 0} \frac{\bar{b}_\lambda - \bar{b}}{\lambda}\,, \tag{6.2.75}$$

$$\tilde{a} : C^1[0,L]^2 \to C[0,L]\,; \qquad \tilde{a}([\gamma,\xi]) = \lim_{\lambda\to 0} \frac{a_\lambda - a}{\lambda}\,, \tag{6.2.76}$$

$$\tilde{\beta} : C^1[0,L]^2 \to C[0,L]\,; \qquad \tilde{\beta}([\gamma,\xi]) = \lim_{\lambda\to 0} \frac{\beta_\lambda - \beta}{\lambda}\,, \tag{6.2.77}$$

$$\tilde{c} : C^1[0,L]^2 \to C[0,L]\,; \qquad \tilde{c}([\gamma,\xi]) = \lim_{\lambda\to 0} \frac{c_\lambda - c}{\lambda}\,, \tag{6.2.78}$$

$$\tilde{\mathcal{D}} : C^1[0,L]^2 \to C(\overline{\Omega})\,; \qquad \tilde{\mathcal{D}}([\gamma,\xi]) = \lim_{\lambda\to 0} \frac{\det J_\lambda - \det J}{\lambda}\,, \tag{6.2.79}$$

$$\tilde{J} : C^1[0,L]^2 \to C(\overline{\Omega})^9\,; \qquad \tilde{J}([\gamma,\xi]) = \lim_{\lambda\to 0} \frac{J_\lambda - J}{\lambda}\,, \tag{6.2.80}$$

$$\tilde{I} : C^1[0,L]^2 \to C(\overline{\Omega})^9\,; \qquad \tilde{I}([\gamma,\xi]) = \lim_{\lambda\to 0} \frac{J_\lambda^{-1} - J^{-1}}{\lambda}\,, \tag{6.2.81}$$

$$\tilde{h}_{ij} : C^1[0,L]^2 \to C(\overline{\Omega})\,; \qquad \tilde{h}_{ij}([\gamma,\xi]) = \lim_{\lambda\to 0} \frac{h_{ij}^\lambda - h_{ij}}{\lambda}\,, \tag{6.2.82}$$

$$\tilde{g}^{ij} : C^1[0,L]^2 \to C(\overline{\Omega})\,; \qquad \tilde{g}^{ij}([\gamma,\xi]) = \lim_{\lambda\to 0} \frac{g_\lambda^{ij} - g^{ij}}{\lambda}\,. \tag{6.2.83}$$

All these operators are linear and bounded with respect to the indicated spaces, and they are the Gâteaux derivatives at $[\varphi,\psi]$ of the associated mappings. For instance, we have $\tilde{t}([\gamma,\xi]) = \nabla\bar{t}([\varphi,\psi])([\gamma,\xi])$, and the other expressions have analogous meanings. The explicit forms of these derivatives are rather involved and cannot be given here in order to keep the exposition at a reasonable length; we trust that the reader will be able to carry out the corresponding lengthy, but elementary, calculations.

Now notice that by Theorem 6.2.7 we also have

$$\bar{y}_\lambda \to \bar{y} \text{ strongly in } H_0^1(0,L)^9. \tag{6.2.84}$$

In order to prove Gâteaux differentiability of the mapping $[\varphi,\psi] \mapsto \bar{y}$, we subtract the equations associated with $\bar{y}_\lambda$ and $\bar{y}$ (cf. (6.2.29), (6.2.30), (6.2.40)), then divide by λ, and add and subtract advantageous terms. In the resulting equation we insert as test function $\bar{v} = \lambda^{-1}(\bar{y}_\lambda - \bar{y}) \in H_0^1(0,L)^9$.

We first examine the right-hand side of the equation resulting from (6.2.40). Invoking the convergence results established above, we can pass to the limit as

$\lambda \to 0$ to obtain that

$$\lim_{\lambda\to 0}\Big\{\sum_{l=1}^{3}\int_\Omega f_l(\bar{x})\,(\mu_l(x_3)+x_1\,M_l(x_3)+x_2\,D_l(x_3))\,\frac{\det J_\lambda-\det J}{\lambda}\,d\bar{x}$$

$$+\sum_{i,j=1}^{3}\sum_{l=1}^{3}\int_{\partial\Omega} g_l(\bar{\sigma})\,(\mu_l(\sigma_3)+\sigma_1\,M_l(\sigma_3)+\sigma_2\,D_l(\bar{\sigma}_3))$$

$$\cdot\frac{\det J_\lambda\sqrt{\nu_i(\bar{\sigma})\,g_\lambda^{ij}\,\nu_j(\bar{\sigma})}-\det J\sqrt{\nu_i(\bar{\sigma})\,g^{ij}\,\nu_j(\bar{\sigma})}}{\lambda}\,d\bar{\sigma}\Big\}$$

$$=\sum_{l=1}^{3}\int_\Omega f_l(\bar{x})\,(\mu_l(x_3)+x_1\,M_l(x_3)+x_2\,D_l(x_3))\,\tilde{\mathcal{D}}([\gamma,\xi])(\bar{x})\,d\bar{x}$$

$$+\sum_{i,j=1}^{3}\sum_{l=1}^{3}\int_{\partial\Omega} g_l(\bar{\sigma})\,(\mu_l(\sigma_3)+\sigma_1\,M_l(\sigma_3)+\sigma_2\,D_l(\sigma_3))$$

$$\cdot\left[\tilde{\mathcal{D}}([\gamma,\xi])\sqrt{\nu_i\,g^{ij}\,\nu_j}+\det J\,\frac{\nu_i\,\tilde{g}^{ij}([\gamma,\xi])\,\nu_j}{2\sqrt{\nu_i\,g^{ij}\,\nu_j}}\right](\bar{\sigma})\,d\bar{\sigma}. \qquad (6.2.85)$$

We also write explicitly the corresponding transformation of the simplest term that occurs in the bilinear form $\mathcal{B}_\lambda$ given by (6.2.30) for the pair $[\varphi_\lambda,\psi_\lambda]$:

$$\frac{1}{\lambda}\Big\{2\tilde{\mu}\int_\Omega\sum_{i=1}^{3}\left[N_i^\lambda\,h_{1i}^\lambda+B_i^\lambda\,h_{2i}^\lambda+((\tau_i^\lambda)'+x_1\,(N_i^\lambda)'+x_2\,(B_i^\lambda)')\,h_{3i}^\lambda\right]$$

$$\cdot\left[M_i\,h_{1i}^\lambda+D_i\,h_{2i}^\lambda+(\mu_i'+x_1\,M_i'+x_2\,D_i')\,h_{3i}^\lambda\right]\det J_\lambda\,d\bar{x}$$

$$-2\,\tilde{\mu}\int_\Omega\sum_{i=1}^{3}[N_i\,h_{1i}+B_i\,h_{2i}+(\tau_i'+x_1\,N_i'+x_2\,B_i')\,h_{3i}]$$

$$\cdot\,[M_i\,h_{1i}+D_i\,h_{2i}+(\mu_i'+x_1\,M_i'+x_2\,D_i')h_{3i}]\det J\,d\bar{x}\Big\}$$

$$=2\,\tilde{\mu}\int_\Omega\sum_{i=1}^{3}\Big[\frac{N_i^\lambda-N_i}{\lambda}\,h_{1i}+\frac{B_i^\lambda-B_i}{\lambda}\,h_{2i}$$

$$+\Big(\frac{(\tau_i^\lambda)'-\tau_i'}{\lambda}+x_1\,\frac{(N_i^\lambda)'-N_i'}{\lambda}+x_2\,\frac{(B_i^\lambda)'-B_i'}{\lambda}\Big)\,h_{3i}\Big]$$

$$\cdot\,[M_i\,h_{1i}+D_i\,h_{2i}+(\mu_i'+x_1\,M_i'+x_2\,D_i')\,h_{3i}]\det J\,d\bar{x}$$

$$+2\,\tilde{\mu}\int_\Omega\sum_{i=1}^{3}\left[N_i^\lambda\,h_{1i}+B_i^\lambda\,h_{2i}+((\tau_i^\lambda)'+x_1\,(N_i^\lambda)'+x_2\,(B_i^\lambda)')\,h_{3i}\right]\det J$$

$$\cdot\left[M_i\,\frac{h_{1i}^\lambda-h_{1i}}{\lambda}+D_i\,\frac{h_{2i}^\lambda-h_{2i}}{\lambda}+(\mu_i'+x_1\,M_i'+x_2\,D_i')\,\frac{h_{3i}^\lambda-h_{3i}}{\lambda}\right]d\bar{x}$$

$$+2\,\tilde{\mu}\int_{\Omega}\sum_{i=1}^{3}\Bigg[N_i^{\lambda}\,\frac{h_{1i}^{\lambda}\det J_{\lambda}-h_{1i}\det J}{\lambda}+B_i^{\lambda}\,\frac{h_{2i}^{\lambda}\det J_{\lambda}-h_{2i}\det J}{\lambda}$$

$$+((\tau_i^{\lambda})'+x_1\,(N_i^{\lambda})'+x_2\,(B_i^{\lambda})')\,\frac{h_{31}^{\lambda}\det J_{\lambda}-h_{3i}\det J}{\lambda}\Bigg]$$

$$\cdot\Big[M_i\,h_{1i}^{\lambda}+D_i\,h_{2i}^{\lambda}+(\mu_i'+x_1\,M_i'+x_2\,D_i')\,h_{31}^{\lambda}\Big]\,d\bar{x}. \qquad (6.2.86)$$

The important term in (6.2.86) is the first term after the equality sign. Taking (6.2.85) into account, and performing similar transformations in the other integrals defining $\mathcal{B}_{\lambda}$, we obtain a relation of the form

$$\mathcal{B}\left(\frac{\bar{y}_{\lambda}-\bar{y}}{\lambda},\bar{v}\right)=Z_{\lambda}(\bar{v}),\quad \forall\,\bar{v}=(\bar{\mu},\bar{M},\bar{D})\in H_0^1(0,L)^9,\quad \forall\,\lambda\neq 0, \qquad (6.2.87)$$

with suitable linear and bounded functionals $Z_{\lambda}: H_0^1(0,L)^9 \to \mathbf{R}$. From the relations (6.2.85) and (6.2.86), invoking the differentiability properties of the coefficients expressed in (6.2.72)–(6.2.83), as well as the convergence of $\bar{y}_{\lambda}$ stated in (6.2.84), we can infer that there is some constant $C>0$, independent of λ, such that

$$|Z_{\lambda}(\bar{v})|\,\leq\,C\,|\bar{v}|_{H_0^1(0,L)^9}\quad \forall\,\bar{v}\in H_0^1(0,L)^9. \qquad (6.2.88)$$

Hence, $\{Z_{\lambda}\}_{\lambda}$ is bounded in the Hilbert space $H^{-1}(0,L)^9$. We may thus select a subsequence $\lambda_n \to 0$ such that with some $Z\in H^{-1}(0,L)^9$,

$$Z_{\lambda_n}(\bar{v})\to Z(\bar{v})\quad \forall\,\bar{v}\in H_0^1(0,L)^9.$$

Actually, the weak limit point $Z\in H^{-1}(0,L)^9$ is uniquely determined so that the above convergence holds generally for $\lambda\to 0$ and not only for the selected subsequence: indeed, the terms forming Z_{λ} are obtained from the term inside the curly brackets on the left-hand side of (6.2.85) (whose convergence to the expression on the right-hand side of (6.2.85) is already known), from the last two terms on the right of (6.2.86), and from further terms coming from the other similar parts. Owing to (6.2.84), and thanks to the differentiability properties (6.2.72)–(6.2.83), all these remaining terms have (unique) limits as $\lambda\to 0$. Observe that the mapping $[\gamma,\xi]\mapsto Z$ is linear and bounded from $C^1[0,L]^2$ to $H^{-1}(0,L)^9$.

Next, we insert $\bar{v}=\lambda^{-1}(\bar{y}_{\lambda}-\bar{y})$ in (6.2.87). Using (6.2.88) and the coercivity of $\mathcal{B}$, we conclude that $\{\lambda^{-1}(\bar{y}_{\lambda}-\bar{y})\}_{\lambda}$ is bounded in $H_0^1(0,L)^9$. Hence there is a sequence $\lambda_n\to 0$ such that

$$\frac{\bar{y}_{\lambda_n}-\bar{y}}{\lambda_n}\to\hat{y}\quad\text{weakly in } H_0^1(0,L)^9. \qquad (6.2.89)$$

As in the previous section, one can conclude that the convergence is valid even in the strong topology of $H_0^1(0,L)^9$. The equation in variations has the form

$$\mathcal{B}(\hat{y},\bar{v})=Z(\bar{v})\quad\forall\,\bar{v}\in H_0^1(0,L)^9, \qquad (6.2.90)$$

and admits a unique solution by the Lax–Milgram lemma. Thus, the limit point $\hat{y}$ is uniquely determined, and the convergence in (6.2.89) holds generally for $\lambda \to 0$. We thus have proved the following result.

Proposition 6.2.9 *The mapping* $[\varphi, \psi] \in C^1[0,L]^2 \mapsto \bar{y} \in H_0^1(0,L)^9$ *is Gâteaux differentiable, and its directional derivative* $\hat{y}$, *at* $[\varphi, \psi]$ *in the direction* $[\gamma, \xi] \in C^1[0,L]^2$, *satisfies (6.2.90).*

We now formulate a general shape optimization problem associated with the variational equation defined by the bilinear form (6.2.30) and by the linear functional (6.2.40). To this end, let $\bar{f}$, $\bar{g}$ be given in some sufficiently large region in $\mathbf{R}^3$, and let K denote a bounded and closed subset of $C^2[0,L]^3$. We consider the optimal design problem

$$\operatorname*{Min}_{[\varphi,\psi]} \left\{ \Pi([\varphi,\psi]) = j(\bar{\theta}, \bar{y}) \right\}, \tag{6.2.91}$$

subject to $\bar{\theta} \in K$, where $\bar{y} \in H_0^1(0,L)^9$ denotes the unique solution to the variational equation

$$\mathcal{B}(\bar{y}, \bar{v}) = \mathcal{G}(\bar{v}) \quad \forall \bar{v} \in H_0^1(0,L)^9$$

associated with $\bar{\theta}$.

From Theorem 6.2.7 we immediately deduce the following result.

Corollary 6.2.10 *Suppose that* $K \subset C^2[0,L]^3$ *is generated by a compact subset of* $[\varphi,\psi] \in C^1[0,L]^2$, *and suppose that the cost functional* $j : C^2[0,L]^3 \times H_0^1(0,L)^9 \to \mathbf{R}$ *is lower semicontinuous. Then the shape optimization problem (6.2.91) admits at least one optimal solution* $\bar{\theta}^* \in K$.

It should be clear that in general the global optimum for (6.2.91) will not be unique, and that there may be many local optimum "points." A standard example for the functional j in (6.2.91) is the quadratic case, for instance

$$j(\bar{\theta}, \bar{y}) = |\tau_1|^2_{H_0^1(0,L)} + |\tau_2|^2_{H_0^1(0,L)} + |\tau_3|^2_{H_0^1(0,L)}. \tag{6.2.92}$$

Concerning the constraints to which the curved rod may be subjected, we underline again that our formalism automatically ensures a prescribed length $L > 0$. This eliminates possible trivial cases such as when $\bar{\theta}$ is constant in $[0,L]$, i.e., the curved rod degenerates to a point. A simple sufficient condition under which $\bar{\theta}$ does not have multiple points is

$$0 \le \varphi(x_3) \le \frac{\pi}{2} - \varepsilon, \quad x_3 \in [0,L], \tag{6.2.93}$$

with some "small" $\varepsilon > 0$. Indeed, then $t_3 > 0$, and θ_3 is strictly increasing in $[0, L]$. Condition (6.2.93) may be employed in problems concerning the optimization of strings where the periodicity condition

$$\int_0^L t_1(s)\, ds = \int_0^L t_2(s)\, ds = 0$$

is also important.

Remark. An alternative parametrization of the tangent vector is given by $\bar{t} = (u_1, u_2, (1 - u_1^2 - u_2^2)^{1/2})$, but this already assumes positivity of t_3. However, in this case the above periodicity condition becomes linear, which may be useful in certain applications.

Next, we introduce the adjoint system, with the unknowns $\bar{T} = (\bar{R}, \bar{P}, \bar{Q}) \in H_0^1(0, L)^9$. It reads

$$\mathcal{B}(\bar{T}, \bar{v}) = \nabla_2 j([\bar{\theta}, \bar{y}])(\bar{v}), \quad \forall \bar{v} \in H_0^1(0, L)^9. \tag{6.2.94}$$

In (6.2.94), we have assumed that $j : C^2[0, L]^3 \times H_0^1(0, L)^9 \to \mathbf{R}$ is Fréchet differentiable, and that $\nabla_2 j$ denotes the second component of ∇j or, equivalently, the partial Fréchet differential with respect to $\bar{y}$. Existence and uniqueness of a solution $\bar{T} \in H_0^1(0, L)^9$ to (6.2.94) follows again from the Lax–Milgram lemma. We have the following result.

Proposition 6.2.11 *Suppose that j is Fréchet differentiable. Then the directional derivative of the cost functional Π in the problem (6.2.91) at the point $[\varphi, \psi] \in C^1[0, L]^2$ in the direction $[\gamma, \xi] \in C^1[0, L]^2$ is given by the expression*

$$\begin{aligned}
&\nabla \Pi([\varphi, \psi])([\gamma, \xi]) \\
&\quad = \nabla_1 j([\bar{\theta}, \bar{y}])\tilde{\theta}([\gamma, \xi]) + \sum_{l=1}^{3} \int_\Omega f_l(\bar{x})\, (R_l(x_3) + x_1\, P_l(x_3) + x_2\, Q_l(x_3)) \\
&\qquad\qquad \cdot \tilde{\mathcal{D}}([\gamma, \xi])(\bar{x})\, d\bar{x} \\
&\qquad + \sum_{i,j=1}^{3} \sum_{l=1}^{3} \int_{\partial\Omega} g_l(\bar{\sigma})\, (R_l(\sigma_3) + \sigma_1\, P_l(\sigma_3) + \sigma_2\, Q_l(\sigma_3))\, \tilde{\mathcal{D}}([\gamma, \xi])(\bar{\sigma}) \\
&\qquad\qquad \cdot \sqrt{\nu_i\, g^{ij}\, \nu_j}\, d\bar{\sigma} \\
&\qquad + \sum_{i,j=1}^{3} \sum_{l=1}^{3} \int_{\partial\Omega} g_l(\bar{\sigma})\, (R_l(\sigma_3) + \sigma_1\, P_l(\sigma_3) + \sigma_2\, Q_l(\sigma_3)) \\
&\qquad\qquad \cdot \det J(\bar{\sigma})\, \frac{1}{\sqrt{\nu_i\, g^{ij}\, \nu_j}}\, \nu_i\, \tilde{g}^{ij}([\gamma, \xi])(\bar{\sigma})\, \nu_j\, d\bar{\sigma}
\end{aligned}$$

$$-2\,\tilde{\mu}\int_\Omega \sum_{i=1}^{3}\Big[N_i\,h_{1i}+B_i\,h_{2i}+(\tau_i'+x_1\,N_i'+x_2\,B_i')\,h_{3i}\Big]\det J$$

$$\cdot\Big[P_i\,\tilde{h}_{1i}([\gamma,\xi])+Q_i\,\tilde{h}_{2i}([\gamma,\xi])+(R_i'+x_1\,P_i'+x_2\,Q_i')\,\tilde{h}_{3i}([\gamma,\xi])\Big]\,d\bar{x}$$

$$-2\,\tilde{\mu}\int_\Omega \sum_{i=1}^{3}\Big[N_i\,\tilde{h}_{1i}([\gamma,\xi])+B_i\,\tilde{h}_{2i}([\gamma,\xi])+(\tau_i'+x_1\,N_i'+x_2\,B_i')\,\tilde{h}_{3i}([\gamma,\xi])\Big]$$

$$\cdot\,[P_i\,h_{1i}+Q_i\,h_{2i}+(R_i'+x_1\,P_i'+x_2\,Q_i')\,h_{3i}]\det J\,d\bar{x}$$

$$-2\,\tilde{\mu}\int_\Omega \sum_{i=1}^{3}[N_i\,h_{1i}+B_i\,h_{2i}+(\tau_i'+x_1\,N_i'+x_2\,B_i')\,h_{3i}]$$

$$\cdot\,[P_i\,h_{1i}+Q_i\,h_{2i}+(R_i'+x_1\,P_i'+x_2\,Q_i')\,h_{3i}]\;\tilde{\mathcal{D}}([\gamma,\xi])\,d\bar{x}$$

$$-\,\tilde{\lambda}\int_\Omega \sum_{i,j=1}^{3}[N_i\,h_{1i}+B_i\,h_{2i}+(\tau_i'+x_1\,N_i'+x_2\,B_i')\,h_{3i}]\det J$$

$$\cdot\Big[P_j\,\tilde{h}_{1j}([\gamma,\xi])+Q_j\,\tilde{h}_{2j}([\gamma,\xi])+(R_j'+x_1P_j'+x_2Q_j')\,\tilde{h}_{3j}([\gamma,\xi])\Big]\,d\bar{x}$$

$$-\tilde{\lambda}\int_\Omega \sum_{i,j=1}^{3}\Big[N_i\,\tilde{h}_{1i}([\gamma,\xi])+B_i\,\tilde{h}_{2i}([\gamma,\xi])+(\tau_i'+x_1\,N_i'+x_2\,B_i')\,\tilde{h}_{3i}([\gamma,\xi])\Big]$$

$$\cdot\Big[P_j\,h_{1j}+Q_j\,h_{2j}+(R_j'+x_1\,P_j'+x_2\,Q_j')\,h_{3j}\Big]\det J\,d\bar{x}$$

$$-\tilde{\lambda}\int_\Omega \sum_{i,j=1}^{3}[N_i\,h_{1i}+B_i\,h_{2i}+(\tau_i'+x_1\,N_i'+x_2\,B_i')\,h_{3i}]$$

$$\cdot\Big[P_j\,h_{1j}+Q_j\,h_{2j}+(R_j'+x_1\,P_j'+x_2\,Q_j')\,h_{3j}\Big]\;\tilde{\mathcal{D}}([\gamma,\xi])\,d\bar{x}$$

$$-\tilde{\mu}\int_\Omega \sum_{i<j}[N_i\,h_{1j}+B_i\,h_{2j}+(\tau_i'+x_1\,N_i'+x_2\,B_i')\,h_{3j}+N_j\,h_{1i}+B_j\,h_{2i}$$

$$+(\tau_j'+x_1\,N_j'+x_2\,B_j')\,h_{3i}\Big]\det J$$

$$\cdot\Big[P_i\,\tilde{h}_{1j}([\gamma,\xi])+Q_i\,\tilde{h}_{2j}([\gamma,\xi])+(R_i'+x_1\,P_i'+x_2\,Q_i')\,\tilde{h}_{3j}([\gamma,\xi])$$

$$+P_j\tilde{h}_{1i}([\gamma,\xi])+Q_j\tilde{h}_{2i}([\gamma,\xi])+(R_j'+x_1P_j'+x_2Q_j')\,\tilde{h}_{3i}([\gamma,\xi])\Big]\,d\bar{x}$$

$$-\tilde{\mu}\int_\Omega \sum_{i<j}\Big[N_i\,\tilde{h}_{1j}([\gamma,\xi])+B_i\,\tilde{h}_{2j}([\gamma,\xi])+(\tau_i'+x_1\,N_i'+x_2\,B_i')\,\tilde{h}_{3j}([\gamma,\xi])$$

$$+N_j\,\tilde{h}_{1i}([\gamma,\xi])+B_j\,\tilde{h}_{2i}([\gamma,\xi])+(\tau_j'+x_1\,N_j'+x_2\,B_j')\,\tilde{h}_{3i}([\gamma,\xi])\Big]$$

$$\cdot\,[P_i\,h_{1j}+Q_i\,h_{2j}+(R_i'+x_1\,P_i'+x_2\,Q_i')\,h_{3j}+P_j\,h_{1i}+Q_j\,h_{2i}$$

$$+(R_j'+x_1\,P_j'+x_2\,Q_j')\,h_{3i}\Big]\det J\,d\bar{x}$$

$$-\tilde{\mu}\int_\Omega \sum_{i<j}\big[N_i\,h_{1j}+B_i\,h_{2j}+(\tau_i'+x_1\,N_i'+x_2\,B_i')\,h_{3j}+N_j\,h_{1i}+B_j\,h_{2i}$$
$$+(\tau_j'+x_1\,N_j'+x_2\,B_j')\,h_{3i}\big]$$
$$\cdot\,\big[P_i\,h_{1j}+Q_i\,h_{2j}+(R_i'+x_1\,P_i'+x_2\,Q_i')\,h_{3j}+P_j\,h_{1i}+Q_j\,h_{2i}$$
$$+(R_j'+x_1\,P_j'+x_2\,Q_j')\,h_{3i}\big]\;\tilde{\mathcal{D}}([\gamma,\xi])\,d\bar{x}. \tag{6.2.95}$$

Remark. In order to determine $\nabla\Pi([\varphi,\psi])([\gamma,\xi])$ for $[\varphi,\psi],[\gamma,\xi]\in C^1[0,L]^2$ from (6.2.95), one has to perform the following steps: first, determine $\bar{\theta}\in C^2[0,L]^3$ using (6.2.55), (6.2.56); then calculate the corresponding solution $\bar{y}=(\bar{\tau},\bar{N},\bar{B})\in H_0^1(0,L)^9$ of the state equation (using (6.2.30), (6.2.40)); then, solve the adjoint state equation (6.2.94) to obtain $\bar{T}=(\bar{R},\bar{P},\bar{Q})\in H_0^1(0,L)^9$, and finally evaluate (6.2.95) invoking (6.2.72)–(6.2.83). This procedure was applied in the numerical experiments reported below.

Since we are working in spaces of continuous functions, it is not advantageous to rewrite (6.2.95) using adjoint operators. It is also to be noticed that the above argument carries over to the case that φ,ψ,γ,ξ are only piecewise continuously differentiable in $[0,L]$, which is important for the numerical simulations.

Remark. Assuming that the cross section of the rod is not constant, one can also study optimization problems with respect to the cross section, under appropriate regularity conditions.

We will now state the standard first-order optimality conditions for the problem (6.2.91) (cf. Tröltzsch [1984]). To this end, let

$$\mathcal{C}=\Big\{[\varphi,\psi]\in C^1[0,L]^2:\ \bar{\theta}([\varphi,\psi])\in K\Big\},$$

and let $u_0=[\varphi_0,\psi_0]\in\mathcal{C}$ be arbitrarily fixed. We denote by

$$T(\mathcal{C};u_0)=\Big\{u\in C^1[0,L]^2:\ u=\lim_{n\to\infty}\lambda_n(u_n-u_0),\,\lambda_n\geq 0,\,u_n\in\mathcal{C},\quad u_n\to u_0\Big\}$$

the tangent cone to $\mathcal{C}$ at u_0 (see Barbu and Precupanu [1986]). It is known that if $\mathcal{C}$ is convex (see the examples following Corollary 6.2.10), then $T(\mathcal{C};u_0)=\overline{\bigcup_{\lambda>0}\lambda(\mathcal{C}-u_0)}$. We have the following result.

Corollary 6.2.12 *Assume that $u^*=[\varphi^*,\psi^*]$ is a (local) optimum point for (6.2.91). Then the following statements are valid:*

(i) *If Π is Fréchet differentiable on $C^1[0,L]^2$, then*

$$\nabla\Pi([\varphi^*,\psi^*])([\gamma,\xi])\geq 0\quad\forall\,[\gamma,\xi]\in T(\mathcal{C};u^*).$$

(ii) *If* $\mathcal{C}$ *is convex, then the directional derivative of* Π *satisfies*

$$\nabla\Pi([\varphi^*,\psi^*])([\gamma,\xi]) \geq 0 \quad \forall\,[\gamma,\xi] \in \mathcal{C} - u^*.$$

Example 6.2.13 We briefly comment on numerical examples related to the methods developed in this paragraph. We took the interval $[0,L]$ with $L = 4\pi\sqrt{2}$ and divided it in 100 equal parts. As cross section of the curved rod we always took a disk of radius $R = 0.3$. To evaluate the integrals over the cross section, the usual change of variables to polar coordinates was performed, which leads to an integration over the rectangle $[0,R]\times[0,2\pi]$, making possible the use of simple quadrature formulas. Here, we employed Simpson's iterative formula.

As initial iterate for the optimization algorithm, we have always considered a spiral lying on the cylindrical surface $\{(x_1,x_2,x_3) : x_1^2 + x_2^2 = 1\}$, namely

$$\varphi^0(x_3) = \frac{\pi}{4}, \quad \psi^0(x_3) = \frac{\pi}{2} + \frac{x_3}{\sqrt{2}}, \quad x_3 \in [0,L]. \tag{6.2.96}$$

A simple calculation shows that the rod parametrization corresponding to (6.2.96) is

$$\bar{\theta}(x_3) = \left(\cos\left(\tfrac{x_3}{\sqrt{2}}\right), \sin\left(\tfrac{x_3}{\sqrt{2}}\right), \tfrac{x_3}{\sqrt{2}}\right), \quad x_3 \in [0,L]. \tag{6.2.97}$$

The Lamé constants were chosen as $\tilde{\lambda} = 50$, $\tilde{\mu} = 100$. The solution of the state system in the Sobolev space $H_0^1(0,L)^9$ has been approximated using linear splines in V_h^9, where $h > 0$ is the norm of the subdivision of $[0,L]$, and where

$$V_h = \{v_h \in C[0,L] : \ v_h(0) = v_h(L) = 0, \ v_h \text{ is piecewise linear in } [0,L]\}. \tag{6.2.98}$$

The same matrix governs both the state and the adjoint discretized systems. It has to be recomputed in each iteration step which is the most time-consuming part of the algorithm. This is due to the three-dimensional character of the objects under study. The variational equation defined by (6.2.30), (6.2.40) provides a dimension reduction to ordinary differential equations that is reflected in the fact that we have to evaluate many integrals over the cross section to obtain the coefficients h_{ij}. One can compute the gradient of the cost functional and use descent algorithms as explained in Proposition 6.2.11. Here, we have used the Uzawa algorithm combined with the Armijo line search rule (compare Arnăutu and Neittaanmäki [2003]).

In a series of examples, we used the cost functional

$$\Pi([\varphi,\psi]) = \frac{1}{2}\,|\tau_3|^2_{L^2(0,L)},$$

and considered forces of the form $\bar{f} = (0,0,f_3)$ with

$$f_3(x_3) = \begin{cases} 10, & x_3 \in \left[0,\frac{L}{2}\right], \\ -10, & x_3 \in \left]\frac{L}{2},L\right], \end{cases} \tag{6.2.99}$$

and several variants thereof. We have also imposed the constraint (6.2.93) with $\varepsilon = \frac{\pi}{8}$ in order to prevent self-intersections. We have neglected the requirement that $\det J$ be strictly positive, but in all examples it could be verified a posteriori that $\det J \neq 0$.

In all considered cases, the vertical column, which corresponds to $\varphi \equiv 0$, was the optimal geometric solution of the problem. Indeed, the vertical column is the most resistant structure against vertical forces of the above form. In accordance with this physical expectation, the lateral displacements τ_1, τ_2 of the optimal state were several orders of magnitude smaller than the corresponding optimal vertical displacement.

Figure 2.12 shows the initial and the final geometries, obtained in one or two iteration steps. We underline the very good correspondence between the results of the numerical experiments and the physical interpretation. This is an argument showing the validity of the model developed in this section and the stability of the employed approximations.

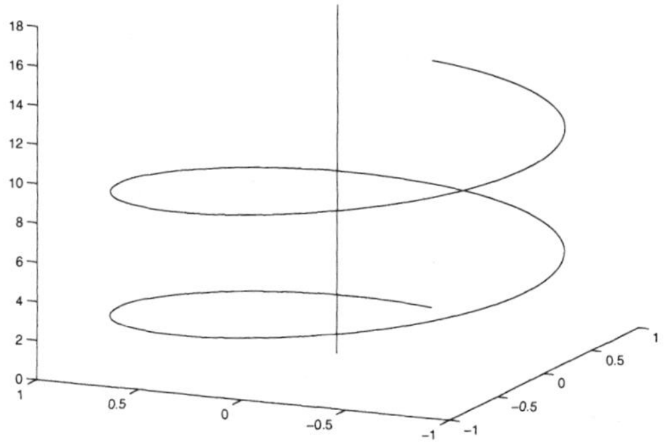

Figure 2.12, © SIAM.

In another set of numerical tests, we considered the force $\bar{f} = 10\,\bar{b}$ with $\bar{b}$ given by (6.2.58). The cost functional was the same as above, and the initial guess was again given by (6.2.96) or by a perturbation thereof, namely by

$$\varphi^0(x_3) = \begin{cases} \frac{\pi}{4} + \frac{1}{10}, & x_3 \in \left[0, \frac{L}{2}\right], \\ \frac{\pi}{4} - \frac{1}{10}, & x_3 \in \left]\frac{L}{2}, L\right]. \end{cases} \tag{6.2.100}$$

Notice that under our parametrization it is very simple to change the initial guess, which is an important advantage in nonconvex optimization problems. The main property of the above choice of the force $\bar{f}$ is that it acts always in the horizontal plane, although with varying directions.

For the constraints, we took $\varepsilon = 0$ in (6.2.93). This also allows for horizontal curves, but self-intersections may occur (which indeed was the case). In all computed examples a clear tendency to produce a horizontal curve as optimal solution was observed. Even if self-intersections are present, horizontal curves will deform just in the horizontal plane under the action of $\bar{f} = 10\,\bar{b}$, which corresponds to a null cost. That is, a mechanical interpretation is still possible (and, due to this requirement, it was necessary to allow $\varepsilon = 0$ in (6.2.93)).

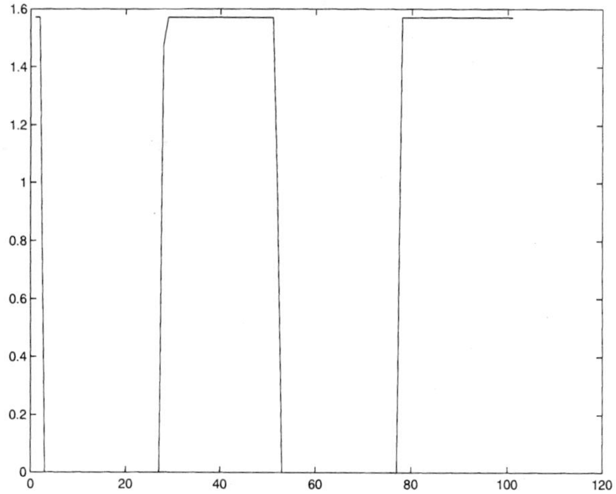

Figure 2.13, © SIAM.

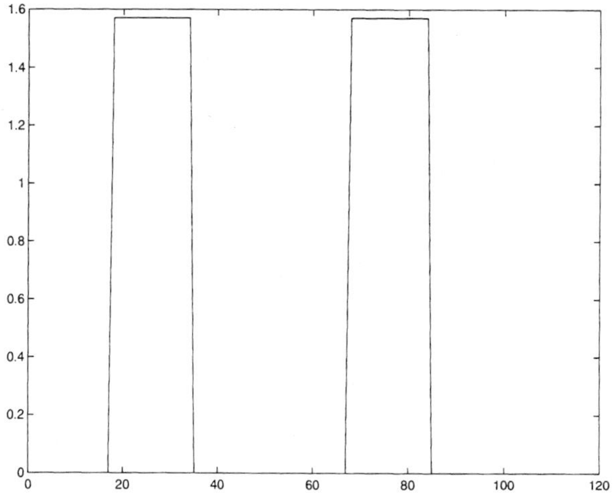

Figure 2.14, © SIAM.

An interesting feature of this type of experiment was that the optimal φ was bang-bang. Figures 2.13 and 2.14 show that this is the case if the initial guess is given by (6.2.100) and (6.2.96), respectively.

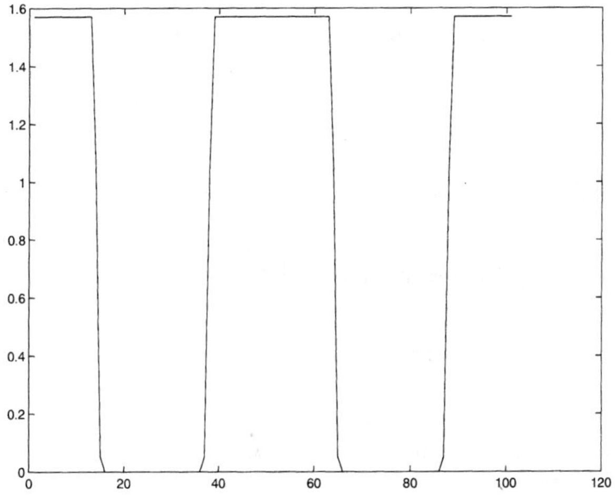

Figure 2.15.

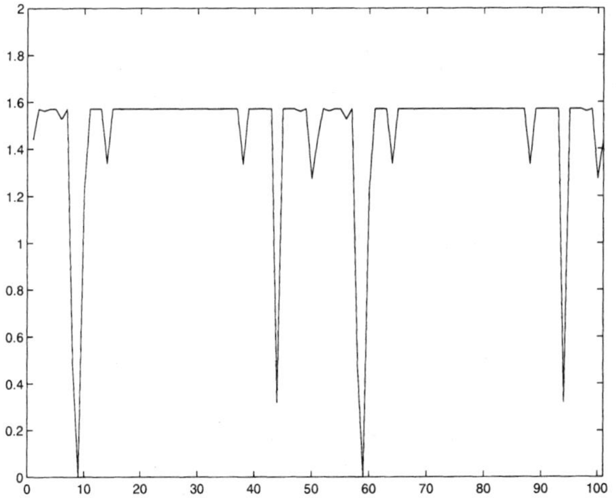

Figure 2.16.

In Figures 2.15 and 2.16 the last two iterates of another example are depicted, with initial guess (6.2.96) and cost functional

$$\Pi([\varphi,\psi]) = \frac{1}{2}\,|\tau_2|^2_{L^2(0,L)}\;.$$

Since the optimal φ is very close to $\frac{\pi}{2}$, the resulting (self-intersecting) structure is almost horizontal. In this example, the values of the optimal τ_3 are also very small, which corresponds well to the mechanical interpretation.

6.3 Applications to Shells

6.3.1 A Generalized Naghdi Shell Model

In this paragraph we will introduce in a direct analytic way (somewhat similar to §6.2.2) a shell model that constitutes a slight generalization of the classical *Naghdi model* (cf. Ciarlet [2000]). For the subsequent applications to optimization problems, we simplify our setting by assuming that the midsurface of the shell is given by the graph of a function. This assumption is not too restrictive and sufficient for many practical situations. Concerning the regularity of this surface, our hypotheses require just a piecewise C^2 representation, similar to the works of Geymonat and Sanchez-Palencia [1995] and of Blouza [1997]. We also underline that our argument is original, following Sprekels and Tiba [2002] and Arnăutu, Sprekels, and Tiba [2003], and relatively simple. It is based on a perturbation analysis and on certain symmetries of the involved differential operators.

To begin with, let $\omega \subset \mathbf{R}^2$ be an open, bounded, and connected set, not necessarily simply connected, with Lipschitz boundary $\partial\omega$. For $\varepsilon > 0$, we define the set

$$\Omega = \omega \times]-\varepsilon,\varepsilon[\subset \mathbf{R}^3,$$

which is obviously open, bounded, and connected in $\mathbf{R}^3$. We denote the independent variables by

$$(x_1,x_2)\in\omega,\quad x_3\in\,]-\varepsilon,\varepsilon[\,,\quad \bar{x}=(x_1,x_2,x_3)\in\Omega.$$

Let $p:\omega\to\mathbf{R}$ denote a piecewise $C^2(\bar\omega)$ mapping whose graph represents the midsurface S of the shell. We consider the geometric transformation $F:\Omega\to\mathbf{R}^3$,

$$F(\bar{x}) = \bar{\pi}(x_1,x_2) + x_3\,\bar{n}(x_1,x_2), \tag{6.3.1}$$

where

$$\bar{\pi} = (\pi_1,\pi_2,\pi_3) = (x_1,x_2,p(x_1,x_2)),$$

and where $\bar{n} = (n_1, n_2, n_3)$ denotes a normal vector to $\mathcal{S}$ in the point $\bar{\pi}(x_1, x_2) \in \mathcal{S}$. Notice that the tangent vectors

$$\frac{\partial \bar{\pi}}{\partial x_1} = (1, 0, p_1), \quad \frac{\partial \bar{\pi}}{\partial x_2} = (0, 1, p_2), \quad \text{where } p_1 = \frac{\partial p}{\partial x_1} \quad \text{and } p_2 = \frac{\partial p}{\partial x_2},$$

are always linearly independent; consequently, we have

$$\bar{n} = \frac{\dfrac{\partial \bar{\pi}}{\partial x_1} \wedge \dfrac{\partial \bar{\pi}}{\partial x_2}}{\left|\dfrac{\partial \bar{\pi}}{\partial x_1} \wedge \dfrac{\partial \bar{\pi}}{\partial x_2}\right|_{\mathbf{R}^3}} = \left(-\frac{p_1}{\sqrt{1 + p_1^2 + p_2^2}}, -\frac{p_2}{\sqrt{1 + p_1^2 + p_2^2}}, \frac{1}{\sqrt{1 + p_1^2 + p_2^2}} \right). \tag{6.3.2}$$

Assume that $\partial\omega$ is divided into two nonoverlapping (relatively) open parts γ_0, γ_1. We introduce the notation $\Gamma_0 = \gamma_0 \times]-\varepsilon, \varepsilon[$, $\Gamma_1 := \partial\Omega \setminus \Gamma_0$, as well as

$$\hat{\Omega} = F(\Omega), \quad \hat{\Gamma}_0 = F(\Gamma_0), \quad \hat{\Gamma}_1 = F(\Gamma_1). \tag{6.3.3}$$

Under the assumptions made below (see (6.3.15)), F is a homeomorphism, and $\hat{\Omega}$ is an open, bounded, and connected set in $\mathbf{R}^3$, representing the shell, that has the Lipschitz boundary $\partial\hat{\Omega} = \overline{\hat{\Gamma}_0} \cup \overline{\hat{\Gamma}_1}$. For $\hat{\Omega}$, we introduce the Hilbert space

$$V(\hat{\Omega}) = \left\{ \hat{y} \in H^1(\hat{\Omega})^3 : \hat{y}|_{\hat{\Gamma}_0} = 0 \right\}, \tag{6.3.4}$$

and recall the linear elasticity system in weak formulation (cf. Ciarlet [2000]),

$$\int_{\hat{\Omega}} \left[\tilde{\lambda}\, \hat{e}_{pp}(\hat{y})\, \hat{e}_{qq}(\hat{v}) + 2\,\tilde{\mu}\, \hat{e}_{ij}(\hat{y})\, \hat{e}_{ij}(\hat{v}) \right] d\hat{x} = \int_{\hat{\Omega}} \hat{f}_i\, \hat{v}_i\, d\hat{x} + \int_{\hat{\Gamma}_1} \hat{h}_i\, \hat{v}_i\, d\hat{\sigma} \quad \forall\, \hat{v} \in V(\hat{\Omega}). \tag{6.3.5}$$

Here, $\tilde{\lambda} \geq 0$, $\tilde{\mu} > 0$ are the Lamé constants of the material under consideration, $\hat{f}_i \in L^2(\hat{\Omega})$ are the body forces, $\hat{h}_i \in L^2(\hat{\Gamma}_1)$ are the surface tractions, and the summation convention is used. The components of the linearized strain or change of metric tensor are given by

$$\hat{e}_{ij}(\hat{y}) = \frac{1}{2} \left(\frac{\partial \hat{y}_i}{\partial \hat{x}_j} + \frac{\partial \hat{y}_j}{\partial \hat{x}_i} \right), \quad i, j = \overline{1, 3}. \tag{6.3.6}$$

Our main *geometric assumption* is that the displacement $\hat{y} \in V(\hat{\Omega})$ has the form

$$\hat{y}(\hat{x}) = \bar{u}(x_1, x_2) + x_3\, \bar{r}(x_1, x_2), \quad \hat{x} \in \hat{\Omega}, \tag{6.3.7}$$

with $\bar{x} = (x_1, x_2, x_3) = F^{-1}(\hat{x})$, and where $\bar{u} = (u_1, u_2, u_3)$ and $\bar{r} = (r_1, r_2, r_3)$ belong to the space

$$V(\omega) = \{\bar{v} = (v_1, v_2, v_3) \in H^1(\omega)^3 : \bar{v}|_{\gamma_0} = 0\}. \tag{6.3.8}$$

This means that we are looking for solutions in the infinite-dimensional subspace

$$\tilde{V}(\hat{\Omega}) = \{\hat{y} \in V(\hat{\Omega}) : \hat{y} \quad \text{is of the form (6.3.7)}\}. \tag{6.3.9}$$

Note that $\tilde{V}(\hat{\Omega})$ can be identified through the relation (6.3.7) with the product space $V(\omega)^2 := V(\omega) \times V(\omega)$. Therefore, instead of working in the space $\tilde{V}(\hat{\Omega})$, we can always work in $V(\omega)^2$. We will do this repeatedly later in this section.

From the geometrical point of view, it should be clear that $\bar{u}$ represents the displacement of the midsurface $\mathcal{S}$ of the shell, while $\bar{r}$ is the modification of the points along the normal $\bar{n}(x_1, x_2)$, which are assumed to remain on a line. However, it is not required that this line (the deformation of the normal) remain perpendicular to the deformation of the midsurface $\mathcal{S}$. Note also that the form (6.3.7) allows for both dilation and contraction of the elastic material, and that it constitutes a generalization of the standard assumptions associated with the so-called *Naghdi model* (cf. Ciarlet [2000], Blouza [1997]). Assumption (6.3.7) is similar to (6.2.17) used in §6.2.2.

Let us now collect some properties of the transformation F. The Jacobian $J = \nabla F$ of F is given by

$$J(\bar{x}) = \begin{bmatrix} 1 + x_3 \dfrac{\partial n_1}{\partial x_1} & x_3 \dfrac{\partial n_1}{\partial x_2} & n_1 \\ x_3 \dfrac{\partial n_2}{\partial x_1} & 1 + x_3 \dfrac{\partial n_2}{\partial x_2} & n_2 \\ p_1 + x_3 \dfrac{\partial n_3}{\partial x_1} & p_2 + x_3 \dfrac{\partial n_3}{\partial x_2} & n_3 \end{bmatrix}. \tag{6.3.10}$$

We recall the relations

$$n_1 = -n_3\, p_1, \quad n_2 = -n_3\, p_2, \tag{6.3.11}$$

$$\frac{\partial \bar{n}}{\partial x_1}(x_1, x_2) = \frac{\partial n_1}{\partial x_1}\frac{\partial \bar{\pi}}{\partial x_1} + \frac{\partial n_2}{\partial x_1}\frac{\partial \bar{\pi}}{\partial x_2}, \tag{6.3.12}$$

$$\frac{\partial \bar{n}}{\partial x_2}(x_1, x_2) = \frac{\partial n_1}{\partial x_2}\frac{\partial \bar{\pi}}{\partial x_1} + \frac{\partial n_2}{\partial x_2}\frac{\partial \bar{\pi}}{\partial x_2}, \tag{6.3.13}$$

which are easy consequences of (6.3.2), and of $|\bar{n}|_{\mathbf{R}^3} = 1$, which implies that

$$\left(\bar{n}, \frac{\partial \bar{n}}{\partial x_i}\right)_{\mathbf{R}^3} = 0, \quad i = 1, 2.$$

Hence, $\frac{\partial \bar{n}}{\partial x_i}$ is orthogonal to $\bar{n}$ and may be generated from $\frac{\partial \bar{\pi}}{\partial x_1}$ and $\frac{\partial \bar{\pi}}{\partial x_2}$, $i = 1, 2$.

Notice that (6.3.12), (6.3.13) are special cases of the equations of movement of the local frame on the surface $\mathcal{S}$; see Cartan [1967]. The coefficients

$$\frac{\partial n_i}{\partial x_\alpha}, \quad i = 1, 2, 3, \quad \alpha = 1, 2,$$

may be interpreted as various curvatures of $\mathcal{S}$. Moreover, using (6.3.10)–(6.3.13), one can easily check that

$$\begin{aligned} \det J(\bar{x}) &= \left[1 + x_3 \left(\frac{\partial n_1}{\partial x_1} + \frac{\partial n_2}{\partial x_2}\right) + x_3^2 \left(\frac{\partial n_1}{\partial x_1}\frac{\partial n_2}{\partial x_2} - \frac{\partial n_1}{\partial x_2}\frac{\partial n_2}{\partial x_1}\right)\right] \\ &\quad \cdot \sqrt{1 + p_1^2 + p_2^2} \; . \end{aligned} \tag{6.3.14}$$

Since $p \in W^{2,\infty}(\omega)$, it follows from (6.3.14) that if $\varepsilon > 0$ is assumed to be "small," then

$$\det J(\bar{x}) \geq c > 0 \quad \forall \bar{x} \in \overline{\Omega}, \tag{6.3.15}$$

with some constant $c > 0$. This justifies the definition (6.3.3) of the shell $\hat{\Omega}$ via the bijective geometric transformation F from (6.3.1). From now on, we will always assume that $0 < \varepsilon < 1$ is small enough to guarantee the validity of (6.3.15).

In the following, the inverse of J and the Jacobian of F^{-1} will be needed. We denote them by

$$J(\bar{x})^{-1} = (h_{ij}(\bar{x}))_{i,j=\overline{1,3}}, \quad \nabla F^{-1}(\hat{x}) = (d_{ij}(\hat{x}))_{i,j=\overline{1,3}}. \tag{6.3.16}$$

Their calculation is tedious (but straightforward), and we just list some elements of the matrix $J(\bar{x})^{-1} \cdot \det J(\bar{x})$ (with obvious notation):

$$\tilde{h}_{11} = n_3 \left(1 + x_3 \frac{\partial n_2}{\partial x_2}\right) - n_2 \left(p_2 + x_3 \frac{\partial n_3}{\partial x_2}\right), \tag{6.3.17}$$

$$\tilde{h}_{21} = n_2 \left(p_1 + x_3 \frac{\partial n_3}{\partial x_1}\right) - n_3 \, x_3 \frac{\partial n_2}{\partial x_1}, \tag{6.3.18}$$

$$\tilde{h}_{31} = x_3 \frac{\partial n_2}{\partial x_1} \left(p_2 + x_3 \frac{\partial n_3}{\partial x_2}\right) - \left(1 + x_3 \frac{\partial n_2}{\partial x_2}\right) \left(p_1 + x_3 \frac{\partial n_3}{\partial x_1}\right), \tag{6.3.19}$$

$$\tilde{h}_{32} = x_3 \frac{\partial n_1}{\partial x_2} \left(p_1 + x_3 \frac{\partial n_3}{\partial x_1}\right) - \left(1 + x_3 \frac{\partial n_1}{\partial x_1}\right) \left(p_2 + x_3 \frac{\partial n_3}{\partial x_2}\right). \tag{6.3.20}$$

We introduce the vectorial mapping $\bar{w} : \Omega \to \mathbf{R}^3$ by

$$\bar{w}(\bar{x}) = \bar{u}(x_1, x_2) + x_3\, \bar{r}(x_1, x_2), \quad \bar{x} \in \Omega, \tag{6.3.21}$$

so that

$$\hat{y}(\hat{x}) = \bar{w}(F^{-1}(\hat{x})), \quad \hat{x} \in \hat{\Omega}. \tag{6.3.22}$$

The Jacobian of $\bar{w}$ is given by

$$\nabla\bar{w}(\bar{x}) = \begin{bmatrix} \frac{\partial u_1}{\partial x_1} + x_3 \frac{\partial r_1}{\partial x_1} & \frac{\partial u_1}{\partial x_2} + x_3 \frac{\partial r_1}{\partial x_2} & r_1 \\ \frac{\partial u_2}{\partial x_1} + x_3 \frac{\partial r_2}{\partial x_1} & \frac{\partial u_2}{\partial x_2} + x_3 \frac{\partial r_2}{\partial x_2} & r_2 \\ \frac{\partial u_3}{\partial x_1} + x_3 \frac{\partial r_3}{\partial x_1} & \frac{\partial u_3}{\partial x_2} + x_3 \frac{\partial r_3}{\partial x_2} & r_3 \end{bmatrix}. \tag{6.3.23}$$

We infer that for $\bar{x} = F^{-1}(\hat{x})$,

$$\begin{aligned} \nabla\hat{y}(\hat{x}) &= \nabla\bar{w}\left(F^{-1}(\hat{x})\right) \nabla F^{-1}(\hat{x}) = \nabla\bar{w}(\bar{x}) \cdot (d_{ij}(\hat{x}))_{i,j=\overline{1,3}} \\ &= \nabla\bar{w}\left(F^{-1}(\hat{x})\right) J\left(F^{-1}(\hat{x})\right)^{-1} = \nabla\bar{w}(\bar{x}) J(\bar{x})^{-1} \\ &= \nabla\bar{w}(\bar{x}) \cdot (h_{ij}(\bar{x}))_{i,j=\overline{1,3}}\,. \end{aligned} \tag{6.3.24}$$

Consequently, we have (again, $\bar{x} = F^{-1}(\hat{x})$)

$$\frac{\partial\hat{y}_i}{\partial\hat{x}_j}(\hat{x}) = \left(\left(\frac{\partial u_i}{\partial x_1} + x_3 \frac{\partial r_i}{\partial x_1}, \frac{\partial u_i}{\partial x_2} + x_3 \frac{\partial r_i}{\partial x_2}, r_i\right), (d_{1j}(\hat{x}), d_{2j}(\hat{x}), d_{3j}(\hat{x}))\right)_{\mathbf{R}^3}. \tag{6.3.25}$$

To arrive at our model, we now restrict the set of admissible test functions $\hat{v} \in V(\hat{\Omega})$. In accordance with the imposed special form (6.2.7) of the displacement, we consider test functions $\hat{v} \in \tilde{V}(\hat{\Omega})$,

$$\hat{v}(\hat{x}) = \bar{\mu}(x_1, x_2) + x_3\, \bar{\rho}(x_1, x_2), \quad \hat{x} \in \hat{\Omega}, \tag{6.3.26}$$

where $\bar{x} = F^{-1}(\hat{x})$ and $\bar{\mu} = (\mu_1, \mu_2, \mu_3)$, $\bar{\rho} = (\rho_1, \rho_2, \rho_3) \in V(\omega)$. Inserting $\hat{y}, \hat{v} \in V(\hat{\Omega})$ in (6.3.5), we obtain the bilinear form governing our generalized Naghdi model,

$$B(\hat{y}, \hat{v}) = \tilde{\lambda} \int_{\hat{\Omega}} \Big\{ \sum_{i=1}^{3} \Big[\Big(\frac{\partial u_i}{\partial x_1} + x_3 \frac{\partial r_i}{\partial x_1}\Big) d_{1i} + \Big(\frac{\partial u_i}{\partial x_2} + x_3 \frac{\partial r_i}{\partial x_2}\Big) d_{2i} + r_i d_{3i} \Big] \Big\}$$

$$\cdot\Big\{\sum_{j=1}^{3}\Big[\Big(\frac{\partial \mu_j}{\partial x_1}+x_3\frac{\partial \rho_j}{\partial x_1}\Big)d_{1j}+\Big(\frac{\partial \mu_j}{\partial x_2}+x_3\frac{\partial \rho_j}{\partial x_2}\Big)d_{2j}+\rho_j d_{3j}\Big]\Big\}\,d\hat{x}$$

$$+\,2\,\tilde{\mu}\int\limits_{\hat{\Omega}}\sum_{i=1}^{3}\Big[\Big(\frac{\partial u_i}{\partial x_1}+x_3\frac{\partial r_i}{\partial x_1}\Big)d_{1i}+\Big(\frac{\partial u_i}{\partial x_2}+x_3\frac{\partial r_i}{\partial x_2}\Big)d_{2i}+r_i d_{3i}\Big]$$

$$\cdot\Big[\Big(\frac{\partial \mu_i}{\partial x_1}+x_3\frac{\partial \rho_i}{\partial x_1}\Big)d_{1i}+\Big(\frac{\partial \mu_i}{\partial x_2}+x_3\frac{\partial \rho_i}{\partial x_2}\Big)d_{2i}+\rho_i d_{3i}\Big]\,d\hat{x}$$

$$+\,\tilde{\mu}\int\limits_{\hat{\Omega}}\Big\{\Big[\Big(\frac{\partial u_1}{\partial x_1}+x_3\frac{\partial r_1}{\partial x_1}\Big)d_{12}+\Big(\frac{\partial u_1}{\partial x_2}+x_3\frac{\partial r_1}{\partial x_2}\Big)d_{22}+r_1 d_{32}$$

$$+\Big(\frac{\partial u_2}{\partial x_1}+x_3\frac{\partial r_2}{\partial x_1}\Big)d_{11}+\Big(\frac{\partial u_2}{\partial x_2}+x_3\frac{\partial r_2}{\partial x_2}\Big)d_{21}+r_2 d_{31}\Big]$$

$$\cdot\Big[\Big(\frac{\partial \mu_1}{\partial x_1}+x_3\frac{\partial \rho_1}{\partial x_1}\Big)d_{12}+\Big(\frac{\partial \mu_1}{\partial x_2}+x_3\frac{\partial \rho_1}{\partial x_2}\Big)d_{22}+\rho_1 d_{32}$$

$$+\Big(\frac{\partial \mu_2}{\partial x_1}+x_3\frac{\partial \rho_2}{\partial x_1}\Big)d_{11}+\Big(\frac{\partial \mu_2}{\partial x_2}+x_3\frac{\partial \rho_2}{\partial x_2}\Big)d_{21}+\rho_2 d_{31}\Big]$$

$$+\Big[\Big(\frac{\partial u_1}{\partial x_1}+x_3\frac{\partial r_1}{\partial x_1}\Big)d_{13}+\Big(\frac{\partial u_1}{\partial x_2}+x_3\frac{\partial r_1}{\partial x_2}\Big)d_{23}+r_1 d_{33}$$

$$+\Big(\frac{\partial u_3}{\partial x_1}+x_3\frac{\partial r_3}{\partial x_1}\Big)d_{11}+\Big(\frac{\partial u_3}{\partial x_2}+x_3\frac{\partial r_3}{\partial x_2}\Big)d_{21}+r_3 d_{31}\Big]$$

$$\cdot\Big[\Big(\frac{\partial \mu_1}{\partial x_1}+x_3\frac{\partial \rho_1}{\partial x_1}\Big)d_{13}+\Big(\frac{\partial \mu_1}{\partial x_2}+x_3\frac{\partial \rho_1}{\partial x_2}\Big)d_{23}+\rho_1 d_{33}$$

$$+\Big(\frac{\partial \mu_3}{\partial x_1}+x_3\frac{\partial \rho_3}{\partial x_1}\Big)d_{11}+\Big(\frac{\partial \mu_3}{\partial x_2}+x_3\frac{\partial \rho_3}{\partial x_2}\Big)d_{21}+\rho_3 d_{31}\Big]$$

$$+\Big[\Big(\frac{\partial u_2}{\partial x_1}+x_3\frac{\partial r_2}{\partial x_2}\Big)d_{13}+\Big(\frac{\partial u_2}{\partial x_2}+x_3\frac{\partial r_2}{\partial x_2}\Big)d_{23}+r_2 d_{33}$$

$$+\Big(\frac{\partial u_3}{\partial x_1}+x_3\frac{\partial r_3}{\partial x_2}\Big)d_{12}+\Big(\frac{\partial u_3}{\partial x_2}+x_3\frac{\partial r_3}{\partial x_2}\Big)d_{22}+r_3 d_{32}\Big]$$

$$\cdot\Big[\Big(\frac{\partial \mu_2}{\partial x_1}+x_3\frac{\partial \rho_2}{\partial x_1}\Big)d_{13}+\Big(\frac{\partial \mu_2}{\partial x_2}+x_3\frac{\partial \rho_2}{\partial x_2}\Big)d_{23}+\rho_2 d_{33}$$

$$+\Big(\frac{\partial \mu_3}{\partial x_1}+x_3\frac{\partial \rho_3}{\partial x_1}\Big)d_{12}+\Big(\frac{\partial \mu_3}{\partial x_2}+x_3\frac{\partial \rho_3}{\partial x_2}\Big)d_{22}+\rho_3 d_{32}\Big]\Big\}\,d\hat{x}. \tag{6.3.27}$$

The generalized Naghdi model of a partially clamped shell is now finally obtained by (6.3.7), (6.3.26), (6.3.27), and by the variational equation

$$B(\hat{y},\hat{v})=\int\limits_{\hat{\Omega}}\hat{f}_i(\hat{x})\,\hat{v}_i(\hat{x})\,d\hat{x}+\int\limits_{\hat{\Gamma}_1}\hat{h}_i(\hat{\sigma})\,\hat{v}_i(\hat{\sigma})\,d\hat{\sigma}\quad\forall\,\hat{v}\in\tilde{V}(\hat{\Omega}). \tag{6.3.28}$$

We underline that (6.3.28) constitutes a projection of the general elasticity system (6.3.5) from $V(\hat{\Omega})$ onto the infinite-dimensional subspace $\tilde{V}(\hat{\Omega})$. This process should be interpreted as an approximation of the solution of the linear elasticity system by expressions of the type (6.3.7). It is reminiscent of the finite element approximation method where the projection subspaces are however only finite-dimensional. We also note that with the bilinear form B acting on $\tilde{V}(\hat{\Omega}) \times \tilde{V}(\hat{\Omega})$ we can associate a bilinear form $\mathcal{B}$ acting on $V(\omega)^2 \times V(\omega)^2$ through the identity

$$\mathcal{B}([\bar{u}, \bar{r}], [\bar{\mu}, \bar{\rho}]) = B(\hat{y}, \hat{v}). \tag{6.3.29}$$

In what follows, we will mainly work with the bilinear form B even if $\mathcal{B}$ is actually meant. From this no confusion will arise. After a standard change of variables, also using (6.3.22), we can rewrite the bilinear forms B and $\mathcal{B}$, respectively, as

$$\begin{aligned}
&B(\hat{y}, \hat{v}) \\
&= \mathcal{B}([\bar{u}, \bar{r}], [\bar{\mu}, \bar{\rho}]) \\
&= \tilde{\lambda} \int_{\Omega} \Big\{ \sum_{i=1}^{3} \Big[\Big(\frac{\partial u_i}{\partial x_1} + x_3 \frac{\partial r_i}{\partial x_1} \Big) h_{1i} + \Big(\frac{\partial u_i}{\partial x_2} + x_3 \frac{\partial r_i}{\partial x_2} \Big) h_{2i} + r_i h_{3i} \Big] \Big\} \\
&\quad \cdot \Big\{ \sum_{j=1}^{3} \Big[\Big(\frac{\partial \mu_j}{\partial x_1} + x_3 \frac{\partial \rho_j}{\partial x_1} \Big) h_{1j} + \Big(\frac{\partial \mu_j}{\partial x_2} + x_3 \frac{\partial \rho_j}{\partial x_2} \Big) h_{2j} + \rho_j h_{3j} \Big] \Big\} \, |\det J(\bar{x})| \, d\bar{x} \\
&\quad + 2\tilde{\mu} \int_{\Omega} \sum_{i=1}^{3} \Big[\Big(\frac{\partial u_i}{\partial x_1} + x_3 \frac{\partial r_i}{\partial x_1} \Big) h_{1i} + \Big(\frac{\partial u_i}{\partial x_2} + x_3 \frac{\partial r_i}{\partial x_2} \Big) h_{2i} + r_i h_{3i} \Big] \\
&\quad \cdot \Big[\Big(\frac{\partial \mu_i}{\partial x_1} + x_3 \frac{\partial \rho_i}{\partial x_1} \Big) h_{1i} + \Big(\frac{\partial \mu_i}{\partial x_2} + x_3 \frac{\partial \rho_i}{\partial x_2} \Big) h_{2i} + \rho_i h_{3i} \Big] \, |\det J(\bar{x})| \, d\bar{x} \\
&\quad + \tilde{\mu} \int_{\Omega} \Big\{ \Big[\Big(\frac{\partial u_1}{\partial x_1} + x_3 \frac{\partial r_1}{\partial x_1} \Big) h_{12} + \Big(\frac{\partial u_1}{\partial x_2} + x_3 \frac{\partial r_1}{\partial x_2} \Big) h_{22} + r_1 h_{32} \\
&\qquad + \Big(\frac{\partial u_2}{\partial x_1} + x_3 \frac{\partial r_2}{\partial x_1} \Big) h_{11} + \Big(\frac{\partial u_2}{\partial x_2} + x_3 \frac{\partial r_2}{\partial x_2} \Big) h_{21} + r_2 h_{31} \Big] \\
&\qquad \cdot \Big[\Big(\frac{\partial \mu_1}{\partial x_1} + x_3 \frac{\partial \rho_1}{\partial x_1} \Big) h_{12} + \Big(\frac{\partial \mu_1}{\partial x_2} + x_3 \frac{\partial \rho_1}{\partial x_2} \Big) h_{22} + \rho_1 h_{32} \\
&\qquad + \Big(\frac{\partial \mu_2}{\partial x_1} + x_3 \frac{\partial \rho_2}{\partial x_1} \Big) h_{11} + \Big(\frac{\partial \mu_2}{\partial x_2} + x_3 \frac{\partial \rho_2}{\partial x_2} \Big) h_{21} + \rho_2 h_{31} \Big] \\
&\qquad + \Big[\Big(\frac{\partial u_1}{\partial x_1} + x_3 \frac{\partial r_1}{\partial x_1} \Big) h_{13} + \Big(\frac{\partial u_1}{\partial x_2} + x_3 \frac{\partial r_1}{\partial x_2} \Big) h_{23} + r_1 h_{33}
\end{aligned}$$

$$+\Big(\frac{\partial u_3}{\partial x_1}+x_3\frac{\partial r_3}{\partial x_1}\Big)h_{11}+\Big(\frac{\partial u_3}{\partial x_2}+x_3\frac{\partial r_3}{\partial x_2}\Big)h_{21}+r_3h_{31}\Big]$$

$$\cdot\Big[\Big(\frac{\partial \mu_1}{\partial x_1}+x_3\frac{\partial \rho_1}{\partial x_1}\Big)h_{13}+\Big(\frac{\partial \mu_1}{\partial x_2}+x_3\frac{\partial \rho_1}{\partial x_2}\Big)h_{23}+\rho_1h_{33}$$

$$+\Big(\frac{\partial \mu_3}{\partial x_1}+x_3\frac{\partial \rho_3}{\partial x_1}\Big)h_{11}+\Big(\frac{\partial \mu_3}{\partial x_2}+x_3\frac{\partial \rho_3}{\partial x_2}\Big)h_{21}+\rho_3h_{31}\Big]$$

$$+\Big[\Big(\frac{\partial u_2}{\partial x_1}+x_3\frac{\partial r_2}{\partial x_1}\Big)h_{13}+\Big(\frac{\partial u_2}{\partial x_2}+x_3\frac{\partial r_2}{\partial x_2}\Big)h_{23}+r_2h_{33}$$

$$+\Big(\frac{\partial u_3}{\partial x_1}+x_3\frac{\partial r_3}{\partial x_1}\Big)h_{12}+\Big(\frac{\partial u_3}{\partial x_2}+x_3\frac{\partial r_3}{\partial x_2}\Big)h_{22}+r_3h_{32}\Big]$$

$$\cdot\Big[\Big(\frac{\partial \mu_3}{\partial x_1}+x_3\frac{\partial \rho_3}{\partial x_1}\Big)h_{12}+\Big(\frac{\partial \mu_3}{\partial x_2}+x_3\frac{\partial \rho_3}{\partial x_2}\Big)h_{22}+\rho_3h_{32}$$

$$+\Big(\frac{\partial \mu_2}{\partial x_1}+x_3\frac{\partial \rho_2}{\partial x_1}\Big)h_{13}+\Big(\frac{\partial \mu_2}{\partial x_2}+x_3\frac{\partial \rho_2}{\partial x_2}\Big)h_{23}+\rho_2h_{33}\Big]\Big\}$$

$$\cdot\,|\det J(\bar{x})|\;d\bar{x}\,. \tag{6.3.30}$$

Remark. It is here that the piecewise $C^2(\bar{\omega})$ regularity of p is in fact used. However, this assumption may be slightly relaxed using more refined change of variables theorems (see, for instance, Rudin [1987, p. 153]).

By performing a similar change of variables in the right-hand side of (6.3.28), the generalized Naghdi model can be expressed directly on the domain Ω. The computations are rather tedious, and for the sake of brevity, we do not give them in detail here. The reader may get a hint in this direction from the arguments developed in the next section. We close this part by stating the main result of this paragraph.

Theorem 6.3.1 *If $\varepsilon > 0$ is sufficiently small, then the generalized Naghdi model (6.3.28) has a unique solution of the form $\hat{y}(\hat{x}) = \bar{u}(x_1,x_2)+x_3\,\bar{r}(x_1,x_2)$ with $[\bar{u},\bar{r}] \in V(\omega)^2$ and $\bar{x}=(x_1,x_2,x_3)=F^{-1}(\hat{x})$.*

This result will be a consequence of the Lax–Milgram lemma applied to the bilinear form (6.3.30). To this end, we have to show its coercivity, which will be done in the next paragraph.

6.3.2 Proof of Coercivity

In the following we will fix $\tilde{\lambda}=0$, $\tilde{\mu}=\frac{1}{2}$, without loss of generality. The classical Korn's inequality with boundary conditions (cf. Ciarlet [2000]) yields that there is some constant $c_0 > 0$ such that

$$B(\hat{y},\hat{y}) \geq \int_{\hat{\Omega}} \sum_{i,j=1}^{3} |\hat{e}_{ij}(\hat{y})(\hat{x})|^2 \, d\hat{x} \geq c_0 \, |\hat{y}|^2_{H^1(\hat{\Omega})} \quad \forall \hat{y} \in V(\hat{\Omega}). \tag{6.3.31}$$

Since $\hat{y}|_{\hat{\Gamma}_0} = 0$, we may replace the standard norm $|\hat{y}|_{H^1(\hat{\Omega})}$ by the equivalent norm

$$||\hat{y}||^2_{H^1(\hat{\Omega})} = \sum_{i,j=1}^{3} \int_{\hat{\Omega}} \left| \frac{\partial \hat{y}_i}{\partial \hat{x}_j}(\hat{x}) \right|^2 d\hat{x}. \tag{6.3.32}$$

Lemma 6.3.2 *If $\hat{y}$ has the form (6.3.7) then*

$$||\hat{y}||^2_{H^1(\hat{\Omega})} = \int_{\Omega} \sum_{i,j=1}^{3} \Big[\Big(\frac{\partial u_i}{\partial x_1} + x_3 \frac{\partial r_i}{\partial x_1} \Big) h_{1j}(\bar{x}) + \Big(\frac{\partial u_i}{\partial x_2} + x_3 \frac{\partial r_i}{\partial x_2} \Big) h_{2j}(\bar{x})$$
$$+ r_i(\bar{x}) h_{3j}(\bar{x}) \Big]^2 \, |\det J(\bar{x})| \, d\bar{x} \, . \tag{6.3.33}$$

Proof. This is the consequence of (6.3.32) and of the change of variables in the integral, similar to that performed in (6.3.27), (6.3.28). □

Our aim is to obtain an estimate directly involving the norms of $\bar{u}, \bar{r} \in V(\omega)$. While Korn's inequality estimates the symmetrized gradients $\bar{e}_{ij}$ in terms of the $H^1(\hat{\Omega})$ norm, our task is more complicated owing to the presence of the nonconstant coefficients h_{ij} appearing in (6.3.3). In the literature, such inequalities are called *Korn's inequalities in curvilinear coordinates*; see Ciarlet [2000]. Here, we indicate a direct approach based on a special approximation of the coefficients h_{ij}.

To this end, recall (6.3.2) and the fact that $\left(\bar{n}, \frac{\partial \bar{n}}{\partial x_i}\right)_{\mathbf{R}^3} = 0$ for $i = 1,2$. Hence, we can verify by a direct calculation that

$$J(\bar{x}) = \begin{bmatrix} 1 & 0 & n_1 \\ 0 & 1 & n_2 \\ p_1 & p_2 & n_3 \end{bmatrix} \begin{bmatrix} 1 + x_3 \dfrac{\partial n_1}{\partial x_1} & x_3 \dfrac{\partial n_1}{\partial x_2} & 0 \\ x_3 \dfrac{\partial n_2}{\partial x_1} & 1 + x_3 \dfrac{\partial n_2}{\partial x_2} & 0 \\ 0 & 0 & 1 \end{bmatrix} = S\,R = S\,(id + x_3\,M), \tag{6.3.34}$$

with obvious meanings of the matrices R, S, M. Apparently, the matrix S does not depend on x_3, while R is a perturbation of the identity matrix id for small values of $|x_3|$. By virtue of the relations (6.3.9), we also have

$$S^{-1} = \frac{1}{\sqrt{1+p_1^2+p_2^2}} \begin{bmatrix} n_3 - n_2 p_2 & n_1 p_2 & -n_1 \\ n_2 p_1 & n_3 - n_1 p_1 & -n_2 \\ -p_1 & -p_2 & 1 \end{bmatrix}. \qquad (6.3.35)$$

The relations (6.3.34), (6.3.35) show that for sufficiently small $|x_3|$ we can approximate the coefficients h_{ij}, $i,j = \overline{1,3}$, by the elements of the matrix $H = (h^0_{ij})_{i,j=\overline{1,3}}$, which is defined by the right-hand side of equation (6.3.35). From (6.3.2) and (6.3.35), we obtain that

$$H = \frac{1}{\sqrt{1+p_1^2+p_2^2}} \begin{bmatrix} \dfrac{1+p_2^2}{\sqrt{1+p_1^2+p_2^2}} & \dfrac{-p_1 p_2}{\sqrt{1+p_1^2+p_2^2}} & \dfrac{p_1}{\sqrt{1+p_1^2+p_2^2}} \\ \dfrac{-p_1 p_2}{\sqrt{1+p_1^2+p_2^2}} & \dfrac{1+p_1^2}{\sqrt{1+p_1^2+p_2^2}} & \dfrac{p_2}{\sqrt{1+p_1^2+p_2^2}} \\ -p_1 & -p_2 & 1 \end{bmatrix}. \qquad (6.3.36)$$

Obviously, $\det H = \sqrt{1+p_1^2+p_2^2}$, and therefore the quadratic form

$$\mathcal{K}([\bar{u},\bar{r}]) = \int_\Omega \sum_{i,j=1}^{3} \left[\left(\frac{\partial u_i}{\partial x_1} + x_3 \frac{\partial r_i}{\partial x_1} \right) h^0_{1j} + \left(\frac{\partial u_i}{\partial x_2} + x_3 \frac{\partial r_i}{\partial x_2} \right) h^0_{2j} + r_i\, h^0_{3j} \right]^2$$
$$\cdot \sqrt{1+p_1^2+p_2^2}\; d\bar{x}, \qquad (6.3.37)$$

where $[\bar{u},\bar{r}] \in V(\omega)^2$, constitutes an approximation to the one given in (6.3.33). It thus makes sense to study this form instead of (6.3.33) first.

Taking into account that all the functions appearing in (6.3.37) are independent of x_3, we can perform the integration with respect to x_3 to obtain

$$\mathcal{K}([\bar{u},\bar{r}]) = 2\varepsilon \int_\omega \sum_{i,j=1}^{3} \left(\frac{\partial u_i}{\partial x_1} h^0_{1j} + \frac{\partial u_i}{\partial x_2} h^0_{2j} + r_i h^0_{3j} \right)^2 \sqrt{1+p_1^2+p_2^2}\; dx_1\, dx_2$$
$$+ \frac{2\varepsilon^3}{3} \int_\omega \sum_{i,j=1}^{3} \left(\frac{\partial r_i}{\partial x_1} h^0_{1j} + \frac{\partial r_i}{\partial x_2} h^0_{2j} \right)^2 \sqrt{1+p_1^2+p_2^2}\; dx_1\, dx_2 . \quad (6.3.38)$$

Lemma 6.3.3 *The quadratic form $\mathcal{K}$ defines a norm on $V(\omega)^2$ through the identity $|||[\bar{u},\bar{r}]||| = \sqrt{\mathcal{K}([\bar{u},\bar{r}])}$, for $[\bar{u},\bar{r}] \in V(\omega)^2$.*

Proof. Due to the quadratic structure of $\mathcal{K}$, we only need to show that it follows from $\mathcal{K}([\bar{u},\bar{r}]) = 0$ that $[\bar{u},\bar{r}] = [0,0]$ almost everywhere in ω.

We prove just that $\bar{r} = 0$; the argument for $\bar{u}$ is similar. We have

$$\frac{\partial r_i}{\partial x_1} h_{1j}^0 + \frac{\partial r_i}{\partial x_2} h_{2j}^0 = 0, \quad i,j = \overline{1,3}\,, \quad \text{a.e. in} \quad \omega. \tag{6.3.39}$$

Let i be fixed. Multiplying (6.3.39) by $-p_1$ for $j = 3$, and adding the result to relation (6.3.39) for $j = 1$, we obtain from (6.3.36) that $\frac{\partial r_i}{\partial x_1} = 0$ a.e. in ω. Likewise, multiplication of (6.3.39) by $-p_2$ for $j = 3$, and addition to relation (6.3.39) for $j = 2$, yield that $\frac{\partial r_i}{\partial x_2} = 0$ a.e. in ω. Since $r_i|_{\gamma_0} = 0$, we conclude that $r_i = 0$ a.e. in ω. □

Lemma 6.3.4 *There is some $\hat{c} > 0$ such that*

$$\int_\omega \sum_{i,j=1}^{3} \left(\frac{\partial v_i}{\partial x_1} h_{1j}^0 + \frac{\partial v_i}{\partial x_2} h_{2j}^0 \right)^2 dx_1\, dx_2 \geq \hat{c}\, ||\bar{v}||^2_{H^1(\omega)^3} \quad \forall\, \bar{v} \in V(\omega), \tag{6.3.40}$$

where

$$||\bar{v}||^2_{H^1(\omega)^3} = \sum_{i=1}^{3} \int_\omega |\nabla v_i(x_1,x_2)|^2\, dx_1\, dx_2. \tag{6.3.41}$$

Proof. Notice first that owing to the zero boundary conditions on γ_0, the norm $||\cdot||_{H^1(\omega)^3}$ is on $V(\omega)$ equivalent to the usual norm of $H^1(\omega)^3$. We consider the linear space

$$W = \left\{ \bar{v} \in L^2(\omega)^3 : \frac{\partial v_i}{\partial x_1} h_{1j}^0 + \frac{\partial v_i}{\partial x_2} h_{2j}^0 \in L^2(\omega),\ \ i,j = \overline{1,3},\ \ \bar{v}|\gamma_0 = 0 \right\}. \tag{6.3.42}$$

Arguing as in the proof of Lemma 6.3.3, we can infer that

$$|\bar{v}|_W = \left(\int_\omega \sum_{i,j=1}^{3} \left(\frac{\partial v_i}{\partial x_1} h_{1j}^0 + \frac{\partial v_i}{\partial x_2} h_{2j}^0 \right)^2 dx_1\, dx_2 \right)^{1/2} \tag{6.3.43}$$

defines a norm on W. Clearly, we have $V(\omega) \subset W$, and for every $\bar{v} \in V(\omega)$,

$$|\bar{v}|_W \leq M\, ||\bar{v}||_{H^1(\omega)^3}, \tag{6.3.44}$$

with some fixed $M > 0$. We now show that also $W \subset V(\omega)$, i.e., that $W = V(\omega)$. To this end, suppose that $\bar{v} \in W$, and let

$$f_{ij} = \frac{\partial v_i}{\partial x_1} h^0_{1j} + \frac{\partial v_i}{\partial x_2} h^0_{2j}, \quad i,j = \overline{1,3}. \tag{6.3.45}$$

Then $f_{ij} \in L^2(\omega)$, $i,j = \overline{1,3}$. Now let i be fixed. As in the proof of Lemma 6.3.3, we multiply (6.3.45) by $-p_1$ for $j = 3$ and add the result to (6.3.45) for $j = 1$ to find that

$$\frac{\partial v_i}{\partial x_1} = \frac{-p_1 f_{i3} + f_{i1}}{\sqrt{1 + p_1^2 + p_2^2}} \in L^2(\omega). \tag{6.3.46}$$

Similarly, we prove that also $\frac{\partial v_i}{\partial x_2} \in L^2(\omega)$. In conclusion, $v_i \in H^1(\omega)$ (which also makes the boundary condition $\bar{v}_{|\gamma_0} = 0$ meaningful), and thus $\bar{v} \in V(\omega)$.

We now consider the identity mapping id acting between the Banach space $(V(\omega), \|\cdot\|_{H^1(\omega)^3})$ and the normed space $(W, |\cdot|_W)$. Clearly, id is linear and injective, and we have just shown its surjectivity. Besides, (6.3.44) implies that id is continuous. Therefore, if $(W, |\cdot|_W)$ is also complete, i.e., a Banach space, then it follows from the open mapping theorem (cf. Yosida [1980]) that also the inverse id^{-1} is continuous, which then proves (6.3.40).

To prove the completeness, take any Cauchy sequence $\{\bar{v}^n\} \subset W$ with respect to the norm $|\cdot|_W$. Then, for $i,j = \overline{1,3}$,

$$q^{n,m}_{ij} = \left(\frac{\partial v^n_i}{\partial x_1} - \frac{\partial v^m_i}{\partial x_1}\right) h^0_{1j} + \left(\frac{\partial v^n_i}{\partial x_2} - \frac{\partial v^m_i}{\partial x_2}\right) h^0_{2j} \to 0, \quad n,m \to \infty, \tag{6.3.47}$$

strongly in $L^2(\omega)$. Using the same argument as in the derivation of (6.3.45), we have, for $i = \overline{1,3}$,

$$\frac{\partial(v^n_i - v^m_i)}{\partial x_1} = \frac{-p_1\, q^{n,m}_{i3} + q^{n,m}_{i1}}{\sqrt{1 + p_1^2 + p_2^2}}, \tag{6.3.48}$$

which converges strongly to 0 in $L^2(\omega)$ as n, $m \to \infty$. Arguing similarly for $\frac{\partial(v^n_i - v^m_i)}{\partial x_2}$, we conclude that $\{\bar{v}^n\}$ is a Cauchy sequence in $(V(\omega), \|\cdot\|_{H^1(\omega)^3})$, hence convergent to some $\bar{v} \in V(\omega)$. By (6.3.44), $|\bar{v}_n - \bar{v}|_W \to 0$, which concludes the proof of the assertion. □

Lemma 6.3.5 *$\mathcal{K}$ is coercive on $V(\omega)^2$, with $V(\omega)$ equipped with the norm (6.3.41).*

Proof. Let $[\bar{u}, \bar{r}] \in V(\omega)$. Using Young's inequality and Lemma 6.3.4, we infer that with some $\hat{C} > 0$,

$$
\begin{aligned}
\mathcal{K}([\bar{u},\bar{r}]) \geq\ & \varepsilon \int_\omega \sum_{i,j=1}^{3} \Big(\frac{\partial u_i}{\partial x_1} h_{1j}^0 + \frac{\partial u_i}{\partial x_2} h_{2j}^0\Big)^2 \sqrt{1+p_1^2+p_2^2}\, dx_1\, dx_2 \\
& -\hat{C}\,\varepsilon \int_\omega \sum_{i,j=1}^{3} r_i^2\, (h_{3j}^0)^2 \sqrt{1+p_1^2+p_2^2}\, dx_1\, dx_2 \\
& +\frac{2\,\varepsilon^3}{3} \int_\omega \sum_{i,j=1}^{3} \Big(\frac{\partial r_i}{\partial x_1} h_{1j}^0 + \frac{\partial r_i}{\partial x_2} h_{2j}^0\Big)^2 \sqrt{1+p_1^2+p_2^2}\, dx_1\, dx_2 \\
\geq\ & \hat{c}\,\varepsilon \int_\omega \sum_{i=1}^{3} \Big(\Big|\frac{\partial u_i}{\partial x_1}\Big|^2 + \Big|\frac{\partial u_i}{\partial x_2}\Big|^2\Big)\, dx_1\, dx_2 \\
& +\frac{2\,\hat{c}\,\varepsilon^3}{3} \int_\omega \sum_{i=1}^{3} \Big(\Big|\frac{\partial r_i}{\partial x_1}\Big|^2 + \Big|\frac{\partial r_i}{\partial x_2}\Big|^2\Big)\, dx_1\, dx_2 \\
& -\hat{C}\,\varepsilon \int_\omega \sum_{i,j=1}^{3} r_i^2\, (h_{3j}^0)^2 \sqrt{1+p_1^2+p_2^2}\, dx_1\, dx_2\,.
\end{aligned}
\tag{6.3.49}
$$

Arguing by contradiction, let us assume now that $\mathcal{K}$ is not coercive with respect to the norm (6.3.41). Then there exists a sequence $\{[\bar{u}^n,\bar{r}^n]\} \subset V(\omega)^2$ satisfying

$$
\sum_{i=1}^{3} \int_\omega \Big(\Big|\frac{\partial u_i^n}{\partial x_1}\Big|^2 + \Big|\frac{\partial u_i^n}{\partial x_2}\Big|^2 + \Big|\frac{\partial r_i^n}{\partial x_1}\Big|^2 + \Big|\frac{\partial r_i^n}{\partial x_2}\Big|^2\Big)\, dx_1\, dx_2 = 1 \quad \forall n \in \mathbf{N} \tag{6.3.50}
$$

such that

$$
\mathcal{K}([\bar{u}^n,\bar{r}^n]) \to 0 \quad \text{for} \quad n \to \infty. \tag{6.3.51}
$$

In view of (6.3.50), we can without loss of generality assume that $\bar{u}^n \to \bar{u}$ and $\bar{r}^n \to \bar{r}$ weakly in $V(\omega)$ and, by compact embedding, strongly in $L^2(\omega)^3$. The weak lower semicontinuity of the quadratic form yields that

$$
\lim_{n\to\infty} \mathcal{K}([\bar{u}^n,\bar{r}^n]) \geq \mathcal{K}([\bar{u},\bar{r}]) \geq 0, \tag{6.3.52}
$$

and we get from (6.3.52), (6.3.51), and Lemma 6.3.3 that $u_i = 0$, $r_i = 0$, a.e. in ω, $i = \overline{1,3}$. However, from (6.3.49) and (6.3.50), and since $0 < \varepsilon < 1$, we can infer that

$$
\mathcal{K}([\bar{u}^n,\bar{r}^n]) \geq \frac{2\,\hat{c}\,\varepsilon^3}{3} \int_\omega \sum_{i=1}^{3} \Big\{\Big|\frac{\partial u_i^n}{\partial x_1}\Big|^2 + \Big|\frac{\partial u_i^n}{\partial x_2}\Big|^2 + \Big|\frac{\partial r_i^n}{\partial x_1}\Big|^2 + \Big|\frac{\partial r_i^n}{\partial x_2}\Big|^2\Big\}\, dx_1\, dx_2
$$

$$-\hat{C}\,\varepsilon \int_\omega \sum_{i,j=1}^{3} (r_{3j}^n)^2\,(h_{3j}^0)^2\,\sqrt{1+p_1^2+p_2^2}\;dx_1\,dx_2$$

$$= \frac{2\,\hat{c}\,\varepsilon^3}{3} - \hat{C}\,\varepsilon \int_\omega \sum_{i,j=1}^{3} (r_i^n)^2\,(h_{3j}^0)^2\,\sqrt{1+p_1^2+p_2^2}\;dx_1\,dx_2. \qquad (6.3.53)$$

By the strong convergence of $\{\bar{r}^n\}$ in $L^2(\omega)^3$, we may pass to the limit as $n \to \infty$ in (6.3.53), arriving at a contradiction. This concludes the proof of the lemma. □

Remark. The coercivity constant of $\mathcal{K}$ can be read off from (6.3.49), with the last term (the one containing the r_i, $i=\overline{1,3}$) just neglected.

Proof of Theorem 6.3.1. We use the form (6.3.30) of $B(\hat{y},\hat{v})$ and (6.3.31), (6.3.33). We estimate the expression (the difference between two corresponding terms in $B(\hat{y},\hat{y})$ and $\mathcal{K}(\hat{y},\hat{y})$):

$$A = \int_\Omega \Big[\Big(\frac{\partial u_1}{\partial x_1} + x_3\frac{\partial r_1}{\partial x_1}\Big)h_{11} + \Big(\frac{\partial u_1}{\partial x_2} + x_3\frac{\partial r_1}{\partial x_2}\Big)h_{21} + r_1 h_{31}\Big]^2\,|\det J(\bar{x})|\,d\bar{x}$$

$$-\int_\Omega \Big[\Big(\frac{\partial u_1}{\partial x_1} + x_3\frac{\partial r_1}{\partial x_1}\Big)h_{11}^0 + \Big(\frac{\partial u_1}{\partial x_2} + x_3\frac{\partial r_1}{\partial x_2}\Big)h_{21}^0 + r_1 h_{31}^0\Big]^2\sqrt{1+p_1^2+p_2^2}\;d\bar{x}\,.$$

From the way that we will prove an advantageous estimate for A, it will become clear that similar estimates can be obtained for all the other terms occurring in $B(\hat{y},\hat{y})$, and therefore we will be able to employ Lemma 6.3.5 to get the desired coercivity property of B.

We define the quantities

$$M = \Big(\frac{\partial u_1}{\partial x_1} + x_3\frac{\partial r_1}{\partial x_1}\Big)h_{11} + \Big(\frac{\partial u_1}{\partial x_2} + x_3\frac{\partial r_1}{\partial x_2}\Big)h_{21} + r_1 h_{31},$$

$$\tilde{M} = \Big(\frac{\partial u_1}{\partial x_1} + x_3\frac{\partial r_1}{\partial x_1}\Big)\tilde{h}_{11} + \Big(\frac{\partial u_1}{\partial x_2} + x_3\frac{\partial r_1}{\partial x_2}\Big)\tilde{h}_{21} + r_1 \tilde{h}_{31},$$

$$M_0 = \Big(\frac{\partial u_1}{\partial x_1} + x_3\frac{\partial r_1}{\partial x_1}\Big)h_{11}^0 + \Big(\frac{\partial u_1}{\partial x_2} + x_3\frac{\partial r_1}{\partial x_2}\Big)h_{21}^0 + r_1 h_{31}^0. \qquad (6.3.54)$$

Then

$$A = \int_\Omega M^2\,|\det J(\bar{x})|\,d\bar{x} - \int_\Omega M_0^2\,\sqrt{1+p_1^2+p_2^2}\;d\bar{x}$$

$$= \left(\int_\Omega M^2\,|\det J(\bar x)|\,d\bar x - \int_\Omega \frac{\tilde M^2}{\sqrt{1+p_1^2+p_2^2}}\,d\bar x\right)$$

$$+\left(\int_\Omega \frac{\tilde M^2}{\sqrt{1+p_1^2+p_2^2}}\,d\bar x - \int_\Omega M_0^2\sqrt{1+p_1^2+p_2^2}\,d\bar x\right)$$

$$= A_1 + A_2, \tag{6.3.55}$$

with obvious meaning of $A_1,\ A_2$. We have, by (6.3.17)–(6.3.20),

$$\frac{\tilde h_{11}}{\sqrt{1+p_1^2+p_2^2}} - h_{11}^0 = \frac{x_3}{1+p_1^2+p_2^2}\left(\frac{\partial n_2}{\partial x_2} + p_2\frac{\partial n_3}{\partial x_2}\right),$$

$$\frac{\tilde h_{21}}{\sqrt{1+p_1^2+p_2^2}} - h_{21}^0 = \frac{-x_3}{1+p_1^2+p_2^2}\left(\frac{\partial n_2}{\partial x_1} + p_2\frac{\partial n_3}{\partial x_1}\right),$$

$$\frac{\tilde h_{31}}{\sqrt{1+p_1^2+p_2^2}} - h_{31}^0 = \frac{x_3}{\sqrt{1+p_1^2+p_2^2}}\left[\frac{\partial n_2}{\partial x_1}\left(p_2 + x_3\frac{\partial n_3}{\partial x_2}\right)\right.$$
$$\left.-\frac{\partial n_2}{\partial x_2}\left(p_1 + x_3\frac{\partial n_3}{\partial x_1}\right) - \frac{\partial n_3}{\partial x_1}\right]. \tag{6.3.56}$$

Invoking (6.3.54) and (6.3.56), we find that

$$A_2 = \int_\Omega \left\{\left(\frac{\partial u_1}{\partial x_1} + x_3\frac{\partial r_1}{\partial x_1}\right)\left(\frac{\partial n_2}{\partial x_2} + p_2\frac{\partial n_3}{\partial x_2}\right) - \left(\frac{\partial u_1}{\partial x_2} + x_3\frac{\partial r_1}{\partial x_2}\right)\left(p_2\frac{\partial n_3}{\partial x_1} + \frac{\partial n_2}{\partial x_1}\right)\right.$$
$$\left.+r_1\sqrt{1+p_1^2+p_2^2}\left[\frac{\partial n_2}{\partial x_1}\left(p_2 + x_3\frac{\partial n_3}{\partial x_2}\right) - \frac{\partial n_2}{\partial x_2}\left(p_1 + x_3\frac{\partial n_3}{\partial x_1}\right) - \frac{\partial n_3}{\partial x_1}\right]\right\}$$
$$\cdot\frac{x_3}{1+p_1^2+p_2^2}\left(\tilde M + M_0\sqrt{1+p_1^2+p_2^2}\right)d\bar x. \tag{6.3.57}$$

From this expression, and from the definitions of $\tilde M$ and of M_0, it is clear that A_2 is of the form

$$A_2 = \int_\Omega \left[x_3\,X(x_1,x_2) + x_3^2\,Y(x_1,x_2) + x_3^3\,Z(x_1,x_2)\right]d\bar x, \tag{6.3.58}$$

where $X,\ Y,\ Z$ are quadratic polynomials of the variables $\frac{\partial u_1}{\partial x_1}, \frac{\partial u_1}{\partial x_2}, \frac{\partial r_1}{\partial x_1}, \frac{\partial r_1}{\partial x_2}$, and r_1, whose coefficients all belong to $L^\infty(\omega)$ since $p \in W^{2,\infty}(\omega)$. The terms with odd powers of x_3 vanish after integration with respect to x_3, and thus we have only to examine the expression

$$L = \int_{\Omega} x_3^2 \, Y(x_1, x_2) \, d\bar{x} = \frac{2\,\varepsilon^3}{3} \int_{\omega} Y(x_1, x_2) \, dx_1 \, dx_2 \,. \tag{6.3.59}$$

It is clear that $Y(x_1, x_2)$ is formed from the summation of terms that appear when terms in A_2 without the factor x_3 are multiplied by terms having the factor x_3. From the definitions of $\tilde{M}$ and M_0, and from inspecting (6.3.57), we find that

$$\begin{aligned} L \; = \; & \frac{2\,\varepsilon^3}{3} \int_{\omega} \Big\{ r_1^2 \, y^{(1)}(x_1, x_2) + \sum_{i,j=1}^{2} \frac{\partial u_1}{\partial x_i} \frac{\partial r_1}{\partial x_j} \, y_{ij}^{(2)}(x_1, x_2) \\ & + \sum_{i=1}^{2} r_1 \left(\frac{\partial u_1}{\partial x_i} \, y_i^{(3)}(x_1, x_2) + \frac{\partial r_1}{\partial x_i} \, y_i^{(4)}(x_1, x_2) \right) \Big\} \, dx_1 \, dx_2, \end{aligned} \tag{6.3.60}$$

where all the coefficient functions $y^{(1)}$, $y_{ij}^{(2)}$, $y_i^{(3)}$, and $y_i^{(4)}$ are known to be bounded in $L^\infty(\omega)$, since $p \in W^{2,\infty}(\omega)$. We thus can estimate, using Young's inequality and the fact that $0 < \varepsilon < 1$,

$$\begin{aligned} |L| \; \le \; & \frac{2\hat{C}_1 \varepsilon^3}{3} \int_{\omega} \Big\{ r_1^2 + \sum_{i,j=1}^{2} \left| \frac{\partial u_1}{\partial x_i} \right| \left| \frac{\partial r_1}{\partial x_j} \right| + \sum_{i=1}^{2} |r_1| \left(\left| \frac{\partial u_1}{\partial x_i} \right| + \left| \frac{\partial r_1}{\partial x_i} \right| \right) \Big\} \, dx_1 \, dx_2 \\ \le \; & \hat{C}_2 \, \varepsilon^2 \sum_{i=1}^{3} \|u_i\|_{H^1(\omega)}^2 + \hat{C}_2 \, \varepsilon^4 \sum_{i=1}^{3} \|r_i\|_{H^1(\omega)}^2 + \hat{C}_2 \, \varepsilon^2 \sum_{i=1}^{3} \|r_i\|_{L^2(\omega)}^2 \,, \end{aligned} \tag{6.3.61}$$

with constants $\hat{C}_1 > 0$, $\hat{C}_2 > 0$ that depend only on the $L^\infty(\omega)$ norms of the functions $y^{(1)}$, $y_{ij}^{(2)}$, $y_i^{(3)}$, and $y_i^{(4)}$.

By comparing this inequality with (6.3.49), which provides the form of the coercivity constants of $\mathcal{K}$, we see that L is dominated by $\mathcal{K}([\bar{u}, \bar{r}])$ provided that $\varepsilon > 0$ is sufficiently small in comparison with the (a priori known) constant $\hat{C}_2$.

It remains to estimate A_1. Note that owing to (6.3.54), and in view of (6.3.17) to (6.3.20), we have $\tilde{M} = M \cdot \det J(\bar{x})$, and hence it follows from (6.3.14), (6.3.15) that

$$\begin{aligned} A_1 \; & = \; \int_{\Omega} \tilde{M}^2 \left(\frac{1}{\det J(\bar{x})} - \frac{1}{\sqrt{1 + p_1^2 + p_2^2}} \right) d\bar{x} \\ & = \; - \int_{\Omega} \tilde{M}^2 x_3 \, \frac{\dfrac{\partial n_1}{\partial x_1} + \dfrac{\partial n_2}{\partial x_2} + x_3 \left(\dfrac{\partial n_1}{\partial x_1} \dfrac{\partial n_2}{\partial x_2} - \dfrac{\partial n_1}{\partial x_2} \dfrac{\partial n_2}{\partial x_1} \right)}{\det J(\bar{x})} \, d\bar{x} \,. \end{aligned} \tag{6.3.62}$$

Next, we perform a Taylor expansion of the function

$$\varphi(x_3) = \frac{1}{\det J(x_1, x_2, x_3)}$$

at $x_3 = 0$. We easily find that

$$\frac{1}{\det J(x_1, x_2, x_3)} = \frac{1 - x_3\left(\frac{\partial n_1}{\partial x_1}(x_1, x_2) + \frac{\partial n_2}{\partial x_2}(x_1, x_2)\right) + x_3^2\, \alpha(x_1, x_2, x_3)}{\sqrt{1 + p_1^2(x_1, x_2) + p_2^2(x_1, x_2)}} \tag{6.3.63}$$

with some function $\alpha \in L^\infty(\Omega)$ whose $L^\infty(\Omega)$ norm is bounded from above by a constant that depends only on $|p|_{W^{2,\infty}(\omega)}$.

We now can argue as follows: the first two terms on the right-hand side of (6.3.63) can be combined with the remaining ones occurring in A_1, and we can explicitly integrate and estimate them as we did in the case of L. Again, they are dominated by $\mathcal{K}([\bar{u}, \bar{r}])$ provided that $\varepsilon > 0$ is sufficiently small. The remaining term on the right of (6.3.63), which depends in a complicated way on x_1, x_2, x_3, is of order x_3^2, and direct estimates can be performed in combination with the other factors in A_1 to see that it is also dominated by $\mathcal{K}([\bar{u}, \bar{r}])$.

We are now in a position to conclude the proof of the assertion: indeed, from the method of estimation used above for A it is apparent that similar computations and estimates can be carried out for all the other terms occurring in $B(\hat{y}, \hat{y})$. Since these estimations are straightforward (while quite lengthy), we do not present them in detail, here. It turns out that all the occurring differences are dominated by $\mathcal{K}([\bar{u}, \bar{r}])$ provided that $\varepsilon > 0$ is sufficiently small. Consequently, $B(\hat{y}, \hat{y})$ inherits the coercivity from $\mathcal{K}$. This concludes the proof of the theorem. □

Remark. Theorem 6.3.1 and its proof remain valid if the shell $\hat{\Omega}$ is of nonconstant thickness, as long as the thickness remains bounded from below by $\varepsilon > 0$. Adequate regularity assumptions on $\partial\hat{\Omega}$ have then to be imposed.

Remark. It is obvious from the proof of Theorem 6.3.1 that the coercivity constant of the bilinear form B is of the order ε^3, and ε must be small for its validity. This explains the well-known instability appearing in numerical computations for shells.

6.3.3 Shell Optimal Design

Let the body forces $\hat{f}$, the surface tractions $\hat{h}$, and the "thickness" $\varepsilon > 0$ be given, and let U_{ad} denote a closed and bounded subset of $C^2(\bar{\omega})$. In the following, we will indicate by a subscript p the dependence of the involved quantities on the "controls" $p \in U_{ad}$.

With $\Omega = \omega \times]-\varepsilon, \varepsilon[$, we can for given $p \in U_{ad}$ define by (6.3.1) the mapping $F_p : \Omega \to \hat{\Omega}_p$ from the reference domain Ω to the shell $\hat{\Omega}_p$ associated with p, where the normal vector $\bar{n}_p$ to the midsurface $\mathcal{S}_p$ at the point $\bar{\pi}_p(x_1, x_2) = (x_1, x_2, p(x_1, x_2)) \in \mathcal{S}_p$ is defined by (6.3.2). Introducing the Jacobian $J_p = \nabla F_p$ as in (6.3.10), we can determine $\det J_p$ using (6.3.14). It then follows from the boundedness of U_{ad} in $C^2(\bar{\omega})$ that there is some $\hat{\varepsilon}_0 = \hat{\varepsilon}_0(U_{ad}) > 0$ such that for any $0 < \varepsilon < \hat{\varepsilon}_0$ the "controls" $p \in U_{ad}$ obey the condition (6.3.15) with a fixed $\hat{c} > 0$, which is independent of both $\varepsilon \in]0, \hat{\varepsilon}_0[$ and $p \in U_{ad}$; in particular, F_p is a bijection for any $p \in U_{ad}$.

We consider the following general shape optimization problem associated with (6.3.28):

$$\text{(P)} \qquad \underset{p \in U_{ad}}{\text{Min}} \left\{ \Pi(p) = j([\bar{y}_p, p]) \right\},$$

where $\bar{y}_p = [\bar{u}_p, \bar{r}_p] \in V(\omega)^2$ is the unique solution to (6.3.28) associated with $p \in U_{ad}$ (which, by Theorem 6.3.1, is known to exist provided that $\varepsilon > 0$ is sufficiently small). Clearly, both the right-hand side of (6.3.28) and the bilinear form B depend on p; we indicate this by writing B_p. Notice that the dependence on p enters the expression (6.3.30) for $B = B_p$ via the coefficient functions $h_{ij} = h_{ij}^p$ and $\det J_p$.

The regularity properties of the mapping $j : V(\omega)^2 \times C^2(\bar{\omega}) \to \mathbf{R}$ will be specified below. A standard example is again the quadratic functional

$$j([\bar{y}, p]) = |u_1|^2_{H^1(\omega)} + |u_2|^2_{H^1(\omega)} + |u_3|^2_{H^1(\omega)} \,. \tag{6.3.64}$$

Then (P) aims at finding the shape of the shell (i.e., the unknown surface $\mathcal{S}$) that minimizes the displacement of the midsurface under the prescribed body forces and surface tractions.

There is a large variety of possible constraints $p \in U_{ad}$ to which the shell itself may be submitted. Some examples are

$$0 \le p(x_1, x_2) \quad \forall\, (x_1, x_2) \in \omega \quad \text{(pointwise constraints)}, \tag{6.3.65}$$

$$\int_\omega p(x_1, x_2)\, dx_1\, dx_2 \ \ge\ \alpha \quad \text{(integral constraints)}, \tag{6.3.66}$$

$$\int_\omega \sqrt{1 + p_1^2 + p_2^2}\, dx_1\, dx_2 \ \le\ \beta. \tag{6.3.67}$$

The last of these constraints means that an upper bound for the area of $\mathcal{S}$ is to be prescribed.

Although the constraints (6.3.64)–(6.3.67) are convex, the shape optimization problem (P) is strongly nonconvex, since the dependence $p \mapsto \bar{y}_p$ is nonlinear. (P) is a control into coefficients problem. We first prove the following continuous dependence result:

Theorem 6.3.6 *Assume that $\{p_k\}_{k\in\mathbf{N}} \subset U_{ad}$ converges strongly in $C^2(\bar{\omega})$ to $p \in U_{ad}$. If $\bar{y}_k = [\bar{u}_k, \bar{r}_k]$ and $\bar{y} = [\bar{u}, \bar{r}]$ denote the solutions to (6.3.28) associated with p_k and p, respectively, then $\bar{y}_k \to \bar{y}$ strongly in $V(\omega)^2$ provided that $\varepsilon \in]0, \hat{\varepsilon}_0[$ is sufficiently small.*

Proof. The relations (6.3.1), (6.3.2), and (6.3.11)–(6.3.14) yield, with obvious notation,

$$\bar{n}_k \to \bar{n} \quad \text{in } C^1(\bar{\omega})^3, \tag{6.3.68}$$

$$F_k = \bar{\pi}_k + x_3\,\bar{n}_k \to F = \bar{\pi} + x_3\,\bar{n} \quad \text{in } C^1(\overline{\Omega})^3, \tag{6.3.69}$$

$$J_k \to J \quad \text{in } C(\overline{\Omega})^9, \tag{6.3.70}$$

$$\det J_k \to \det J \quad \text{in } C(\overline{\Omega}). \tag{6.3.71}$$

Similarly to (6.3.34), we have (with obvious new matrix notation)

$$J_k = S_k\,R_k = S_k\,(id + x_3\,M_k), \quad k \in \mathbf{N}. \tag{6.3.72}$$

A simple calculation gives

$$\begin{aligned} S_k^{-1} &= \frac{1}{\sqrt{1 + (p_1^k)^2 + (p_2^k)^2}} \begin{bmatrix} n_3^k - n_2^k\,p_2^k & n_1^k\,p_2^k & -n_1^k \\ n_2^k\,p_1^k & n_3^k - n_1^k\,p_1^k & -n_2^k \\ -p_1^k & -p_2^k & 1 \end{bmatrix} \\ &\to \frac{1}{\sqrt{1 + p_1^2 + p_2^2}} \begin{bmatrix} n_3 - n_2\,p_2 & n_1\,p_2 & -n_1 \\ n_2\,p_1 & n_3 - n_1\,p_1 & -n_2 \\ -p_1 & -p_2 & 1 \end{bmatrix} = S^{-1}, \end{aligned}$$

strongly in $C^1(\bar{\omega})^9$. Moreover, for sufficiently small $\varepsilon > 0$,

$$\begin{aligned} R_k^{-1} &= (id + x_3\,M_k)^{-1} = id - x_3\,M_k + x_3^2\,M_k^2 - x_3^3\,M_k^3 + \ldots \\ &= \sum_{l=0}^{\infty}(-x_3\,M_k)^l. \end{aligned} \tag{6.3.73}$$

Clearly,

$$M_k = \begin{bmatrix} \dfrac{\partial n_1^k}{\partial x_1} & \dfrac{\partial n_1^k}{\partial x_2} & 0 \\ \dfrac{\partial n_2^k}{\partial x_1} & \dfrac{\partial n_2^k}{\partial x_2} & 0 \\ 0 & 0 & 0 \end{bmatrix} \to M = \begin{bmatrix} \dfrac{\partial n_1}{\partial x_1} & \dfrac{\partial n_1}{\partial x_2} & 0 \\ \dfrac{\partial n_2}{\partial x_1} & \dfrac{\partial n_2}{\partial x_2} & 0 \\ 0 & 0 & 0 \end{bmatrix}, \tag{6.3.74}$$

strongly in $C(\bar{\omega})^9$. The relations (6.3.73), (6.3.74) yield, passing to the limit as $k \to \infty$ in the infinite sum, that $R_k^{-1} \to R^{-1}$ strongly in $C(\overline{\Omega})^9$, provided that $\varepsilon > 0$ is small enough. Then, (6.3.72) and the above argument imply that

$$J_k^{-1} \to J^{-1} \quad \text{strongly in } C(\overline{\Omega})^9. \tag{6.3.75}$$

In particular, invoking (6.3.16), (6.2.43), (6.3.70), and (6.3.71),

$$h_{ij}^k \to h_{ij} \quad \text{strongly in } C(\overline{\Omega}) \quad \forall\, i,j = \overline{1,3}, \tag{6.3.76}$$

$$g_k^{ij} \to g^{ij} \quad \text{strongly in } C(\overline{\Omega}) \quad \forall\, i,j = \overline{1,3}. \tag{6.3.77}$$

The use of (6.3.77), (6.2.43) is needed in order to perform the necessary change of variables in the right-hand side of (6.3.28) as in (6.3.30).

Now let B_k denote the bilinear form B from (6.3.30) corresponding to the coefficient functions h_{ij}^k and $\det J_k$, $k \in \mathbf{N}$. We are now going to show that if $\varepsilon > 0$ is sufficiently small, then the family $\{B_k\}$ has a common coercivity constant $c_\varepsilon > 0$ that does not depend on $k \in \mathbf{N}$; this will entail, in particular, that the problem (6.3.28) corresponding to p_k, $k \in N$, has a unique solution $\bar{y}_k = [\bar{u}_k, \bar{r}_k] \in V(\omega)^2$. We will establish this claim in the next three auxiliary results.

Proposition 6.3.7 *There are constants $\hat{c}_0 = \hat{c}_0(U_{ad}) > 0$, $\hat{m}_0 = \hat{m}_0(U_{ad}) > 0$, and $\hat{\varepsilon} \in]0, \hat{\varepsilon}_0[$ such that for every $\varepsilon \in]0, \hat{\varepsilon}[$ there is some $\delta = \delta(\varepsilon) > 0$ such that $0 < \varepsilon + \delta < \hat{\varepsilon}_0$ and such that for every $p \in U_{ad}$,*

$$B_p(\hat{y}, \hat{y}) \geq \hat{c}_0 \left[\varepsilon\, |\bar{u}|^2_{V(\omega)} + \varepsilon^3\, |\bar{r}|^2_{V(\omega)}\right] - \frac{\hat{m}_0}{\delta} \left[|\bar{r}|^2_{L^2(\omega)^3} + |\bar{u}|^2_{L^2(\omega)^3}\right],$$

$$\forall\, \hat{y} \in \tilde{V}(\hat{\Omega}_p), \quad \hat{y}(\hat{x}) = \bar{u}(x_1, x_2) + x_3\, \bar{r}(x_1, x_2), \quad \textit{with } [\bar{u}, \bar{r}] \in V(\omega)^2. \tag{6.3.78}$$

Proof. Let $0 < \varepsilon < \hat{\varepsilon}_0$, and let $\delta > 0$ be such that $0 < \varepsilon + \delta < \hat{\varepsilon}_0$. Moreover, let $p \in U_{ad}$ be arbitrary, and let $\hat{y} \in \tilde{V}(\hat{\Omega}_p)$ be as in (6.3.78). We confine ourselves to the special case $\bar{u}, \bar{r} \in H_0^1(\omega)$ (that is, $\partial\omega = \gamma_0$) in order to avoid the (only technical) additional complications that would arise in the extension argument below.

We define the mapping $\bar{w} \in H^1(\Omega)^3$ given by (compare (6.3.21))

$$\bar{w}(x_1, x_2, x_3) = \bar{u}(x_1, x_2) + x_3\, \bar{r}(x_1, x_2), \tag{6.3.79}$$

so that $\hat{y}(\hat{x}) = \bar{w}(F_p^{-1}\hat{x})$, $\hat{x} \in \hat{\Omega}_p$, $\bar{x} = F_p^{-1}\hat{x} \in \Omega$. Let

$$S^+ = \bar{\omega} \times [\varepsilon, \varepsilon + \delta], \quad S^- = \bar{\omega} \times [-\varepsilon - \delta, -\varepsilon]. \tag{6.3.80}$$

We extend $\bar{w}$ to $\Omega \cup S^+ \cup S^-$ by putting $\tilde{w}|_\Omega = \bar{w}$ and

$$\tilde{w}(\bar{x}) = \delta^{-1}\left\{(\varepsilon + \delta - x_3)\,\bar{u}(x_1, x_2) + \varepsilon(\varepsilon + \delta - x_3)\,\bar{r}(x_1, x_2)\right\}, \quad x \in S^+, \tag{6.3.81}$$

$$\tilde{w}(\bar{x}) = \delta^{-1}\left\{(\varepsilon + \delta + x_3)\,\bar{u}(x_1, x_2) - \varepsilon(\varepsilon + \delta + x_3)\,\bar{r}(x_1, x_2)\right\}, \quad x \in S^-. \tag{6.3.82}$$

Then, since $\bar{u}, \bar{r} \in H_0^1(\omega)^3$, we may extend $\tilde{w}$ by zero to the whole space $\mathbf{R}^3$ so that $\tilde{w} \in H_0^1(\mathbf{R}^3)$. Notice that in the general case of a partially clamped shell one also can employ an extension procedure around $\omega \subset \mathbf{R}^2$. We may, for instance, use the Calderon extension (cf. Adams [1975]) since $\partial\omega$ is assumed Lipschitzian.

Since $0 < \varepsilon + \delta < \hat{\varepsilon}_0$, the transformation F_p associated with $p \in U_{ad}$ is a bijection on $\Omega \cup S^+ \cup S^-$. We define

$$\Sigma_p^+ = F_p(S^+), \quad \Sigma_p^- = F_p(S^-), \tag{6.3.83}$$

and introduce the extension of $\hat{y} \in \tilde{V}(\hat{\Omega}_p)$ by putting

$$\tilde{y}(\hat{x}) = \tilde{w}\big(F_p^{-1}(\hat{x})\big). \tag{6.3.84}$$

Then, clearly, $\tilde{y} \in H_0^1(\hat{\Omega}_p \cup \Sigma_p^+ \cup \Sigma_p^-)$.

Since $0 < \varepsilon + \delta < \hat{\varepsilon}_0$, there is some open ball $O \subset \mathbf{R}^3$ such that $O \supset (\hat{\Omega}_p \cup \Sigma_p^+ \cup \Sigma_p^-)$ for every $p \in U_{ad}$. We may extend $\tilde{y}$ by zero to O so that $\tilde{y} \in H_0^1(O)$. Since $\tilde{\lambda} \geq 0$, $\tilde{\mu} > 0$, it follows that

$$B_p(\hat{y}, \hat{y}) + \tilde{\mu} \int_{\Sigma_p^+ \cup \Sigma_p^-} \sum_{i,j=1}^{3} |\hat{e}_{ij}(\tilde{y}(\hat{x}))|^2 \, d\hat{x} \geq \tilde{\mu} \int_O \sum_{i,j=1}^{3} |\hat{e}_{ij}(\tilde{y}(\hat{x}))|^2 \, d\hat{x}.$$

Korn's inequality, applied to the last integral, yields that there is some $\bar{c} > 0$, depending only on O (and thus not on $p \in U_{ad}$), such that

$$\begin{aligned} B_p(\hat{y}, \hat{y}) &\geq \bar{c}\,|\tilde{y}|^2_{H_0^1(O)} - \tilde{\mu} \int_{\Sigma_p^+ \cup \Sigma_p^-} \sum_{i,j=1}^{3} \left|\hat{e}_{ij}(\tilde{y}(\hat{x}))\right|^2 d\hat{x} \\ &\geq \bar{c}\,|\hat{y}|^2_{H^1(\hat{\Omega}_p)} - \tilde{\mu} \int_{\Sigma_p^+ \cup \Sigma_p^-} \sum_{i,j=1}^{3} \left|\hat{e}_{ij}(\tilde{y}(\bar{x}))\right|^2 d\hat{x}. \end{aligned} \tag{6.3.85}$$

We have to estimate the last term in (6.3.85). To this end, we compute

$$\int_{\Sigma_p^+\cup\Sigma_p^-} \left|\frac{\partial \tilde{y}_i}{\partial \hat{x}_j}(\hat{x})\right|^2 d\hat{x}$$

$$= \int_{\Sigma_p^+\cup\Sigma_p^-} \left(\left(\frac{\partial \tilde{w}_i}{\partial x_1}(\bar{x}(\hat{x})), \frac{\partial \tilde{w}_i}{\partial x_2}(\bar{x}(\hat{x})), \frac{\partial \tilde{w}_i}{\partial x_3}(\bar{x}(\hat{x}))\right), (d^p_{1j}(\hat{x}), d^p_{2j}(\hat{x}), d^p_{3j}(\hat{x}))\right)^2_{\mathbf{R}^3} d\hat{x}$$

$$= \int_{S^+\cup S^-} \left(\left(\frac{\partial \tilde{w}_i}{\partial x_1}, \frac{\partial \tilde{w}_i}{\partial x_2}, \frac{\partial \tilde{w}_i}{\partial x_3}\right), (h^p_{1j}, h^p_{2j}, h^p_{3j})\right)^2_{\mathbf{R}^3} |\det J_p|\, d\bar{x},$$

where

$$(d^p_{ij})_{i,j=\overline{1,3}} = \nabla F_p^{-1}(\hat{x}), \quad (h^p_{ij})_{i,j=\overline{1,3}} = J_p^{-1}(\bar{x}),$$

and where we have performed the change of variables $\hat{x} = F_p(\bar{x})$ in the integral (see §6.3.2 for a detailed calculation). Observe that the extension of h^p_{ij} to $S^+ \cup S^-$ is obvious thanks to the explicit form (6.3.11).

Since the L^∞ norms of $\det J_p$ and of h^p_{ij} stay uniformly bounded when p varies over U_{ad}, it remains to estimate the gradient of $\tilde{w}$ in $L^2(S^+ \cup S^-)$. We just compute it in S^+. We have

$$\frac{\partial \tilde{w}}{\partial x_\alpha} = \delta^{-1}\Big[(\varepsilon + \delta - x_3)\,\frac{\partial \bar{u}}{\partial x_\alpha} + \varepsilon(\varepsilon + \delta - x_3)\,\frac{\partial \bar{r}}{\partial x_\alpha}\Big], \quad \alpha = 1, 2, \tag{6.3.86}$$

$$\frac{\partial \tilde{w}}{\partial x_3} = -\delta^{-1}(\bar{u} + \varepsilon\,\bar{r}). \tag{6.3.87}$$

Thus,

$$\left|\frac{\partial \tilde{w}}{\partial x_3}\right|_{L^2(S^+\cup S^-)^3} \le \sqrt{2}\,\delta^{-\frac{1}{2}}\,|\bar{u}|_{L^2(\omega)^3} + \sqrt{2}\,\varepsilon\,\delta^{-\frac{1}{2}}\,|\bar{r}|_{L^2(\omega)^3},$$

and, for $\alpha = 1, 2$,

$$\left|\frac{\partial \tilde{w}}{\partial x_\alpha}\right|_{L^2(S^+\cup S^-)^3} \le \frac{\sqrt{2}}{\sqrt{3}}\,\delta^{\frac{1}{2}}\left|\frac{\partial \bar{u}}{\partial x_\alpha}\right|_{L^2(\omega)^3} + \frac{\sqrt{2}}{\sqrt{3}}\,\varepsilon\,\delta^{\frac{1}{2}}\left|\frac{\partial \bar{r}}{\partial x_\alpha}\right|_{L^2(\omega)^3}. \tag{6.3.88}$$

Consequently, there is some constant $c_1 > 0$, which is independent of $p \in U_{ad}$, such that

$$B_p(\hat{y}, \hat{y}) \ge \bar{c}\,|\hat{y}|^2_{H^1(\hat{\Omega}_p)} - c_1\Big[\delta\,|\bar{u}|^2_{V(\omega)} + \varepsilon^2\delta\,|\bar{r}|^2_{V(\omega)} + \delta^{-1}\,|\bar{u}|^2_{L^2(\omega)^3}$$
$$+\,\varepsilon^2\delta^{-1}\,|\bar{r}|^2_{L^2(\omega)^3}\Big]. \tag{6.3.89}$$

To conclude the proof of Proposition 6.3.7, we will need the following result, which we prove first.

Lemma 6.3.8 *There are constants $c_2 > 0$, $c_3 > 0$, and $\hat{\varepsilon}_1 \in]0, \hat{\varepsilon}_0[$ such that for all $p \in U_{ad}$ and for all $\varepsilon \in]0, \hat{\varepsilon}_1[$,*

$$|\hat{y}|^2_{H^1(\hat{\Omega}_p)} \geq c_2 \left[\varepsilon \, |\bar{u}|^2_{V(\omega)} + \varepsilon^3 \, |\bar{r}|^2_{V(\omega)}\right] - c_3 \, \varepsilon \, |\bar{r}|^2_{L^2(\omega)^3},$$

$$\forall \, \hat{y} \in \tilde{V}(\hat{\Omega}_p), \quad \hat{y}(\hat{x}) = \bar{u}(x_1, x_2) + x_3 \, \bar{r}(x_1, x_2), \quad \textit{with } [\bar{u}, \bar{r}] \in V(\omega)^2 .$$

Proof. The proof is quite technical and can only be sketched here. The main point is that it turns out to be possible to verify that all the constants appearing in §6.3.2 can be chosen independently of $p \in U_{ad}$, since U_{ad} is bounded in $C^2(\overline{\Omega})$. We indicate here a precise quantitative argument that replaces the qualitative proof of Lemma 6.3.4, in order to show how to keep control over the constants. We have

$$|\hat{y}|^2_{H^1(\hat{\Omega}_p)} = \int_\Omega \sum_{i,j=1}^{3} \Big[\left(\frac{\partial u_i}{\partial x_1} + x_3 \frac{\partial r_i}{\partial x_1} \right) h^p_{1j}(\bar{x}) + \left(\frac{\partial u_i}{\partial x_2} + x_3 \frac{\partial r_i}{\partial x_2} \right) h^p_{2j}(\bar{x})$$
$$+ \, r_i(\bar{x}) \, h^p_{3j}(\bar{x}) \Big]^2 \, |\det J_p(\bar{x})| \, d\bar{x}, \tag{6.3.90}$$

after the change of variables $\hat{x} = F_p(\bar{x})$. Following the lines of §6.3.2, we define the quadratic form

$$\mathcal{K}_p([\bar{u}, \bar{r}]) = 2\,\varepsilon \int_\omega \sum_{i,j=1}^{3} \left(\frac{\partial u_i}{\partial x_1} h^{p,0}_{1j} + \frac{\partial u_i}{\partial x_2} h^{p,0}_{2j} + r_1 h^{p,0}_{3j} \right)^2 \sqrt{1 + p_1^2 + p_2^2} \, dx_1 \, dx_2$$
$$+ \frac{2\,\varepsilon^3}{3} \int_\omega \sum_{i,j=1}^{3} \left(\frac{\partial r_i}{\partial x_1} h^{p,0}_{1j} + \frac{\partial r_i}{\partial x_2} h^{p,0}_{2j} \right)^2 \sqrt{1 + p_1^2 + p_2^2} \, dx_1 \, dx_2, \tag{6.3.91}$$

which is estimated first. Here, the functions $h^{p,0}_{ij}$, $i, j = \overline{1,3}$, are the elements of the matrix S_p^{-1} (see (6.3.35)) and thus constitute an approximation to h^p_{ij}. Taking into account the structure of S_p^{-1}, we find that

$$\frac{\partial r_i}{\partial x_1} = \frac{p_1}{\sqrt{1 + p_1^2 + p_2^2}} \left(\frac{\partial r_i}{\partial x_1} \, h^{p,0}_{13} + \frac{\partial r_i}{\partial x_2} \, h^{p,0}_{23} \right)$$
$$+ \frac{1}{\sqrt{1 + p_1^2 + p_2^2}} \left(\frac{\partial r_i}{\partial x_1} \, h^{p,0}_{11} + \frac{\partial r_i}{\partial x_2} \, h^{p,0}_{21} \right), \tag{6.3.92}$$

and similar expressions for $\frac{\partial r_i}{\partial x_2}$, $\frac{\partial u_i}{\partial x_\alpha}$, $i = \overline{1,3}$, $\alpha = 1, 2$.

Simple algebraic manipulations in (6.3.91), (6.3.92), involving the triangle inequality and the fact that the coefficients of the expressions in parentheses in the right-hand side of (6.3.92) have a modulus less or equal one, then lead to the conclusion that there is some constant $c > 0$, which is independent of $p \in U_{ad}$, such that for all $\varepsilon \in]0, \hat{\varepsilon}_0[$,

$$\mathcal{K}_p([\bar{u}, \bar{r}]) \geq c \left(\varepsilon \, |\bar{u}|^2_{V(\omega)} + \varepsilon^3 \, |\bar{r}|^2_{V(\omega)} - \varepsilon \, |\bar{r}|^2_{L^2(\omega)^3} \right). \tag{6.3.93}$$

Now take the difference between the expressions on the right-hand sides of (6.3.90) and (6.3.91), and perform the same type of estimates as in §6.3.2 above. As there, it turns out that the modulus of the difference is dominated by the right-hand side of (6.3.93), provided that $\varepsilon > 0$ is sufficiently small. This concludes the proof of Lemma 6.3.8. □

Conclusion of the proof of Proposition 6.3.7. Assume that $0 < \varepsilon < \hat{\varepsilon}_1$. We combine (6.3.89) with Lemma 6.3.8 to conclude that

$$\begin{aligned} B_p(\hat{y}, \hat{y}) \;\geq\; & (\bar{c}\, c_2\, \varepsilon - c_1\, \delta) \, |\bar{u}|^2_{V(\omega)} + (\bar{c}\, c_2\, \varepsilon^3 - c_1\, \varepsilon^2\, \delta) \, |\bar{r}|^2_{V(\omega)} - c_1\, \delta^{-1} \, |\bar{u}|^2_{L^2(\omega)^3} \\ & -(\bar{c}\, c_3\, \varepsilon + c_1\, \varepsilon^2\, \delta^{-1}) \, |\bar{r}|^2_{L^2(\omega)^3}. \end{aligned}$$

We now choose

$$\hat{\varepsilon} = \frac{1}{2} \min \{1, \hat{\varepsilon}_1, \bar{c}\, c_2\}, \quad \delta = \delta(\varepsilon) = \varepsilon^2, \quad \text{for } 0 < \varepsilon < \hat{\varepsilon}.$$

The assertion then follows with $\hat{c}_0 = \frac{\bar{c}\, c_2}{2}$ and $\hat{m}_0 = 2 \max \{c_1, \bar{c}\, c_3\, \hat{\varepsilon}^3 + c_1 \hat{\varepsilon}^2\}$. □

Proposition 6.3.9 *Let $\tilde{U}_{ad}$ be a compact subset of U_{ad} in $C^2(\bar{\omega})$, and let $\hat{\varepsilon} > 0$ be the constant in Proposition 6.3.7. Then there exists for every $\varepsilon \in]0, \hat{\varepsilon}[$ a constant $c_\varepsilon > 0$ such that for every $p \in \tilde{U}_{ad}$,*

$$B_p(\hat{y}, \hat{y}) \geq c_\varepsilon \left[|\bar{u}|^2_{V(\omega)} + |\bar{r}|^2_{V(\omega)} \right], \quad \textit{for all } \hat{y} \textit{ as in } (6.3.78). \tag{6.3.94}$$

Proof. Let $0 < \varepsilon < \hat{\varepsilon}$ be fixed, and let $\delta = \delta(\varepsilon) > 0$ be chosen as in Proposition 6.3.7. Arguing by contradiction, we assume that the assertion is false. Then there exist sequences $\gamma_n \searrow 0$ and $\{p_n\} \subset U_{ad}$ such there are $\hat{y}_n$ and corresponding $\bar{u}_n, \bar{r}_n \in V(\omega)$, $n \in \mathbf{N}$, such that

$$0 \leq B_{p_n}(\hat{y}_n, \hat{y}_n) \leq \gamma_n \left[|\bar{u}_n|^2_{V(\omega)} + |\bar{r}_n|^2_{V(\omega)} \right] \quad \forall n \in \mathbf{N}. \tag{6.3.95}$$

In (6.3.95), we can assume that $\|[\bar{u}_n, \bar{r}_n]\|_{V(\omega)^2} = 1$, $n \in \mathbf{N}$, so that $B_{p_n}(\hat{y}_n, \hat{y}_n) \to 0$ as $n \to \infty$. Moreover, we can without loss of generality suppose that $\bar{u}_n \to \bar{u}$, $\bar{r}_n \to \bar{r}$, both weakly in $V(\omega)$, and since $\tilde{U}_{ad}$ is compact, that $p_n \to \bar{p} \in \tilde{U}_{ad}$ strongly in $C^2(\bar{\omega})$. In particular, we have $h^n_{ij} \to \bar{h}_{ij}$ strongly in $C(\overline{\Omega})$, where $(h^n_{ij})_{i,j=1,3} = J^{-1}_{p_n}$ and $(\bar{h}_{ij})_{i,j=\overline{1,3}} = J^{-1}_{\bar{p}}$.

Thanks to the uniform convergence of the coefficients h^n_{ij}, it is easily verified (see (6.3.27)) that

$$B_{p_n}(\hat{y}_n, \hat{y}_n) - B_{\bar{p}}(\hat{y}_n, \hat{y}_n) \to 0 \quad \text{as } n \to \infty. \tag{6.3.96}$$

The weak lower semicontinuity of $B_{\bar{p}}$ in $H^1(\omega)^3 \times H^1(\omega)^3$, and (6.3.95), (6.3.96), show that

$$0 \geq \liminf_{n\to\infty} B_{p_n}(\hat{y}_n, \hat{y}_n) = \liminf_{n\to\infty} B_{\bar{p}}(\hat{y}_n, \hat{y}_n) \geq \mathcal{B}_{\bar{p}}([\bar{u}, \bar{r}], [\bar{u}, \bar{r}]) \geq 0. \tag{6.3.97}$$

Clearly, (6.3.97) shows that $\mathcal{B}_{\bar{p}}([\bar{u}, \bar{r}], [\bar{u}, \bar{r}]) = 0$, and the coercivity of $B_{\bar{p}}$ implies that $\bar{u} = \bar{r} = 0$, according to §6.3.2. We conclude that $\bar{u}_n \to 0$, $\bar{r}_n \to 0$, both weakly in $V(\omega)$ and strongly in $L^2(\omega)^3$.

Now we combine (6.3.95) and (6.3.78) to obtain that

$$\begin{aligned} \gamma_n &\geq \hat{c}_0 \left[\varepsilon \, |\bar{u}_n|^2_{V(\omega)} + \varepsilon^3 \, |\bar{r}_n|^2_{V(\omega)}\right] - \frac{\hat{m}_0}{\delta} \left[|\bar{r}_n|^2_{L^2(\omega)^3} + |\bar{u}_n|^2_{L^2(\omega)^3}\right] \\ &\geq \hat{c}_0 \, \varepsilon^3 - \frac{\hat{m}_0}{\delta} \left[|\bar{r}_n|^2_{L^2(\omega)^3} + |\bar{u}_n|^2_{L^2(\omega)^3}\right]. \end{aligned}$$

Taking the limit as $n \to \infty$, we arrive at the contradiction $0 \geq \hat{c}_0 \, \varepsilon^3$, which ends the proof of Proposition 6.3.9. □

Conclusion of the proof of Theorem 6.3.6. We first notice that $\{p_k : k \in \mathbf{N}\} \cup \{p\}$ forms a compact subset of U_{ad}. Now let $0 < \varepsilon < \hat{\varepsilon}$ be fixed. It then follows from Proposition 6.3.9 that (6.3.94) is valid for p_k, for all $k \in \mathbf{N}$.

Therefore, if we fix the test functions $[\bar{\mu}, \bar{\rho}] = \bar{y}_k = [\bar{u}_k, \bar{r}_k])$ in (6.3.28) with $p = p_k$, we immediately get that $\{\bar{y}_k\}_{k\in\mathbf{N}}$ is bounded in $V(\omega)^2$. Hence there is a subsequence, again indexed by k, such that $\bar{u}_k \to \bar{u}$, $\bar{r}_k \to \bar{r}$, both weakly in $V(\omega)$. Due to the uniform convergence of the coefficients, one may pass to the limit as $k \to \infty$ in (6.3.28) to see that $\bar{y} = [\bar{u}, \bar{r}]$ is in fact the (unique) solution of (6.3.28) associated with p.

We show that the convergence is valid in the strong topology of $V(\omega)^2$. To this end, we subtract the equations corresponding to $\bar{y}_k$ and $\bar{y}$, add and subtract advantageous terms (compare the last step in the proof of Theorem 6.2.7), and finally, take test functions of the form $\bar{y}_k - \bar{y} \in V(\omega)^2$. Since the difference of the corresponding right-hand sides converges to zero by the weak convergence property, a detailed calculation (which is omitted here) leads to the conclusion that

$$\lim_{k\to\infty} B_p(\bar{y}_k - \bar{y}, \bar{y}_k - \bar{y}) = 0, \tag{6.3.98}$$

and the assertion follows from (6.3.94). The uniqueness of the limit point entails the strong convergence of the entire sequence. □

The following existence result is an immediate consequence of the above results.

Corollary 6.3.10 *If U_{ad} is compact in $C^2(\bar{\omega})$ and $j : V(\omega)^2 \times C^2(\bar{\omega}) \to \mathbf{R}$ is lower semicontinuous, then the shape optimization problem (P) admits at least one optimal solution $p \in U_{ad}$.*

Next, we examine the differentiability properties of the mapping $p \in C^2(\bar{\omega}) \mapsto \bar{y} = \bar{y}_p \in V(\omega)^2$ defined by (6.3.28). Since the expressions occurring in this investigation are very lengthy and involved, we can only sketch the main arguments without going into much detail. The interested reader can fill in the gaps in the following exposition without difficulty.

To begin with, assume that $p \in C^2(\bar{\omega})$ and $\varepsilon > 0$ are such that the condition (6.3.15) is satisfied with some $\hat{c} > 0$. We consider a perturbation of the form $p_\lambda = p + \lambda q$ with $\lambda \in \mathbf{R} \setminus \{0\}$ and $q \in C^2(\bar{\omega})$, where $|q|_{C^2(\bar{\omega})} = 1$. As in §6.3.1, we then can introduce the quantities $\bar{n}_\lambda \in C^1(\bar{\omega})^3$, $F_\lambda \in C^1(\overline{\Omega})^3$, $J_\lambda \in C(\overline{\Omega})^9$, $h^\lambda_{ij} \in C(\overline{\Omega})$, $g^{ij}_\lambda \in C(\overline{\Omega})$, B_λ, etc., which are associated with p_λ. For $\lambda = 0$, we will simply write B instead of B_p.

From the continuity results established above it follows that there is some $\lambda_0 > 0$ such that for $0 < |\lambda| \le \lambda_0$ the inequality

$$\det J_\lambda(\bar{x}) \ge \frac{\hat{c}}{2} > 0 \quad \forall \bar{x} \in \overline{\Omega}$$

holds and (6.3.28) has a unique solution for $p = p_\lambda$. We denote this solution by $\bar{y}_\lambda = [\bar{u}^\lambda, \bar{r}^\lambda] \in V(\omega)^2$.

A lengthy and tedious, but elementary and straightforward, calculation (which has to be omitted here) shows that the following limits and linear operators exist in the indicated spaces:

$$\tilde{n} : C^2(\bar{\omega}) \to C^1(\bar{\omega})^3; \qquad \tilde{n}(q) = \lim_{\lambda\to 0} \frac{\bar{n}_\lambda - \bar{n}}{\lambda}, \tag{6.3.99}$$

$$\tilde{J} : C^2(\bar{\omega}) \to C(\overline{\Omega})^9; \qquad \tilde{J}(q) = \lim_{\lambda\to 0} \frac{J_\lambda - J}{\lambda}, \tag{6.3.100}$$

$$\tilde{I} : C^2(\bar{\omega}) \to C(\overline{\Omega})^9; \qquad \tilde{I}(q) = \lim_{\lambda\to 0} \frac{J_\lambda^{-1} - J^{-1}}{\lambda}, \tag{6.3.101}$$

$$\tilde{h}_{ij} : C^2(\bar{\omega}) \to C(\overline{\Omega}); \qquad \tilde{h}_{ij}(q) = \lim_{\lambda\to 0} \frac{h_{ij}^\lambda - h_{ij}}{\lambda}, \tag{6.3.102}$$

$$\mathcal{D} : C^2(\bar{\omega}) \to C(\overline{\Omega}); \qquad \mathcal{D}(q) = \lim_{\lambda\to 0} \frac{\det J_\lambda - \det J}{\lambda}, \tag{6.3.103}$$

$$\tilde{g}^{ij} : C^2(\bar{\omega}) \to C(\overline{\Omega}); \qquad \tilde{g}^{ij}(q) = \lim_{\lambda\to 0} \frac{g_\lambda^{ij} - g^{ij}}{\lambda}. \tag{6.3.104}$$

By Theorem 6.3.6, we also know that

$$\bar{y}_\lambda \to \bar{y} \quad \text{strongly in } V(\omega)^2. \tag{6.3.105}$$

Next, we subtract the equations for $\bar{y}_\lambda$ and for $\bar{y}$ and divide the result by $\lambda \neq 0$. We are going to show that it is possible to take the limit $\lambda \to 0$. On the right-hand side we have, with an arbitrary test function $\bar{v} = [\bar{\mu}, \bar{\rho}] \in V(\omega)^2$,

$$\begin{aligned}
&\lim_{\lambda\to 0} \Bigg[\int_\Omega \sum_{l=1}^3 f_l\,(\mu_l + x_3\,\rho_l)\, \frac{\det J_\lambda - \det J}{\lambda}\, d\bar{x} \\
&\qquad + \int_{\Gamma_1} \sum_{l=1}^3 \sum_{i,j=1}^3 g_l\,(\mu_l + x_3\,\rho_l)\, \frac{\det J_\lambda \sqrt{\nu_i\, g_\lambda^{ij}\, \nu_j} - \det J \sqrt{\nu_i\, g^{ij}\, \nu_j}}{\lambda}\, d\bar{\sigma} \Bigg] \\
&= \sum_{l=1}^3 \int_\Omega f_l\,(\mu_l + x_3\,\rho_l)\, \mathcal{D}(q)\, d\bar{x} \\
&\qquad + \sum_{l=1}^3 \sum_{i,j=1}^3 \int_{\Gamma_1} g_l\,(\mu_l + x_3\,\rho_l) \left[\mathcal{D}(q) \sqrt{\nu_i\, g^{ij}\, \nu_j} + \det J\, \frac{\nu_i\, \tilde{g}^{ij}(q)\, \nu_j}{2\sqrt{\nu_i\, g^{ij}\, \nu_j}} \right] d\bar{\sigma}.
\end{aligned} \tag{6.3.106}$$

The expression for $\frac{1}{\lambda}(B_\lambda - B)$ is rather lengthy, and we confine ourselves to write explicitly just the terms that originate from its part associated with the coefficient $2\tilde{\mu}$, namely

$$\begin{aligned}
&\frac{1}{\lambda} \Bigg\{ \int_\Omega \sum_{i=1}^3 \left[\Big(\frac{\partial u_i^\lambda}{\partial x_1} + x_3 \frac{\partial r_i^\lambda}{\partial x_1} \Big) h_{1i}^\lambda + \Big(\frac{\partial u_i^\lambda}{\partial x_2} + x_3 \frac{\partial r_i^\lambda}{\partial x_2} \Big) h_{2i}^\lambda + r_i^\lambda\, h_{3i}^\lambda \right] \\
&\qquad \cdot \left[\Big(\frac{\partial \mu_i}{\partial x_1} + x_3 \frac{\partial \rho_i}{\partial x_1} \Big) h_{1i}^\lambda + \Big(\frac{\partial \mu_i}{\partial x_2} + x_3 \frac{\partial \rho_i}{\partial x_2} \Big) h_{2i}^\lambda + \rho_i\, h_{3i}^\lambda \right] |\det J_\lambda|\, d\bar{x} \\
&\quad - \int_\Omega \left[\Big(\frac{\partial u_i}{\partial x_1} + x_3 \frac{\partial r_i}{\partial x_1} \Big) h_{1i} + \Big(\frac{\partial u_i}{\partial x_2} + x_3 \frac{\partial r_i}{\partial x_2} \Big) h_{2i} + r_i\, h_{3i} \right]
\end{aligned}$$

$$\cdot \left[\left(\frac{\partial \mu_i}{\partial x_1} + x_3 \frac{\partial \rho_i}{\partial x_1} \right) h_{1i} + \left(\frac{\partial \mu_i}{\partial x_2} + x_3 \frac{\partial \rho_i}{\partial x_2} \right) h_{2i} + \rho_i \, h_{3i} \right] |\det J| \, d\bar{x} \Bigg\}$$

$$= \int_\Omega \sum_{i=1}^{3} \Bigg[\left(\frac{\partial}{\partial x_1} \left(\frac{u_i^\lambda - u_i}{\lambda} \right) + x_3 \frac{\partial}{\partial x_1} \left(\frac{r_i^\lambda - r_i}{\lambda} \right) \right) h_{1i} + \frac{r_i^\lambda - r_i}{\lambda} h_{3i}$$

$$+ \left(\frac{\partial}{\partial x_2} \left(\frac{u_i^\lambda - u_i}{\lambda} \right) + x_3 \frac{\partial}{\partial x_2} \left(\frac{r_i^\lambda - r_i}{\lambda} \right) \right) h_{2i} \Bigg]$$

$$\cdot \left[\left(\frac{\partial \mu_i}{\partial x_1} + x_3 \frac{\partial \rho_i}{\partial x_1} \right) h_{1i} + \left(\frac{\partial \mu_i}{\partial x_2} + x_3 \frac{\partial \rho_i}{\partial x_2} \right) h_{2i} + \rho_i \, h_{3i} \right] |\det J| \, d\bar{x}$$

$$+ \int_\Omega \sum_{i=1}^{3} \Bigg[\left(\frac{\partial u_i^\lambda}{\partial x_1} + x_3 \frac{\partial r_i^\lambda}{\partial x_1} \right) \frac{h_{1i}^\lambda \det J_\lambda - h_{1i} \det J}{\lambda} + \left(\frac{\partial u_i^\lambda}{\partial x_2} + x_3 \frac{\partial r_i^\lambda}{\partial x_2} \right)$$

$$\cdot \frac{h_{2i}^\lambda \det J_\lambda - h_{2i} \det J}{\lambda} + r_i^\lambda \, \frac{h_{3i}^\lambda \det J_\lambda - h_{3i} \det J}{\lambda} \Bigg]$$

$$\cdot \left[\left(\frac{\partial \mu_i}{\partial x_1} + x_3 \frac{\partial \rho_i}{\partial x_1} \right) h_{1i}^\lambda + \left(\frac{\partial \mu_i}{\partial x_2} + x_3 \frac{\partial \rho_i}{\partial x_2} \right) h_{2i}^\lambda + \rho_i \, h_{3i}^\lambda \right] d\bar{x}$$

$$+ \int_\Omega \sum_{i=1}^{3} \left[\left(\frac{\partial u_i^\lambda}{\partial x_1} + x_3 \frac{\partial r_i^\lambda}{\partial x_1} \right) h_{1i} + \left(\frac{\partial u_i^\lambda}{\partial x_2} + x_3 \frac{\partial r_i^\lambda}{\partial x_2} \right) h_{2i} + r_i^\lambda h_{3i} \right]$$

$$\cdot \Bigg[\left(\frac{\partial \mu_i}{\partial x_1} + x_3 \frac{\partial \rho_i}{\partial x_1} \right) \frac{h_{1i}^\lambda - h_{1i}}{\lambda} + \left(\frac{\partial \mu_i}{\partial x_2} + x_3 \frac{\partial \rho_i}{\partial x_2} \right) \frac{h_{2i}^\lambda - h_{2i}}{\lambda}$$

$$+ \rho_i \frac{h_{3i}^\lambda - h_{3i}}{\lambda} \Bigg] |\det J| \, d\bar{x} \,. \tag{6.3.107}$$

According to (6.3.102), (6.3.103), and (6.3.76), the last two integrals are of the form $Z_\lambda(\bar{y}_\lambda, \bar{v})$, and there is a constant $C > 0$, independent of λ, such that the bilinear forms Z_λ satisfy

$$|Z_\lambda(\bar{y}_\lambda, \bar{v})| \leq C \, |\bar{y}_\lambda|_{V(\omega)^2} \, |\bar{v}|_{V(\omega)^2} \quad \forall \, \bar{v} \in V(\omega)^2, \quad |\lambda| \leq \lambda_0. \tag{6.3.108}$$

Applying the same technique also to the other terms of $B_\lambda - B$, we obtain from (6.3.106)–(6.3.108) that

$$B\Big(\frac{\bar{y}_\lambda - \bar{y}}{\lambda}, \bar{v}\Big) = \tilde{Z}_\lambda(\bar{y}_\lambda, \bar{v}) \quad \forall \, \bar{v} \in V(\omega)^2, \tag{6.3.109}$$

where $\tilde{Z}_\lambda$ results from adding together all the corresponding terms occurring in (6.3.106)–(6.3.108). Using the continuity properties established above, as well as (6.3.105) and the differentiability properties (6.3.99)–(6.3.104), we find that the limit

$$Z(\bar{v}) = \lim_{\lambda \to 0} \tilde{Z}_\lambda(\bar{y}_\lambda, \bar{v})$$

exists for every $\bar{v} \in V(\omega)^2$ without taking subsequences. In fact, we have $Z \in (V(\omega)^2)^*$.

Now insert $\bar{v} = \lambda^{-1}(\bar{y}_\lambda - \bar{y})$, $\lambda \neq 0$, in (6.3.109). Taking (6.3.108) into account, as well as (6.3.105), we see that $\{\lambda^{-1}(\bar{y}_\lambda - \bar{y})\}_\lambda$ is bounded in $V(\omega)^2$, due to Proposition 6.3.9. Hence, there is a sequence $\lambda_n \to 0$ such that

$$\frac{\bar{y}_{\lambda_n} - \bar{y}}{\lambda_n} \to \tilde{y} \quad \text{weakly in } V(\omega)^2 \quad \text{as } n \to \infty. \tag{6.3.110}$$

Passage to the limit as $n \to \infty$ in (6.3.109) yields the equation in variations

$$B(\tilde{y}, \bar{v}) = Z(\bar{v}) \quad \forall \bar{v} \in V(\omega)^2, \tag{6.3.111}$$

which, thanks to (6.3.94) and the Lax–Milgram lemma, has a unique solution. This entails, in particular, that the convergence (6.3.110) holds generally for $\lambda \to 0$ and not only for the selected sequence. We thus have proved the following result.

Proposition 6.3.11 *The mapping $p \in C^2(\bar{\omega}) \mapsto \bar{y} \in V(\omega)^2$ given by (6.3.28) is Gâteaux differentiable on U_{ad}, and the directional derivative $\tilde{y}$ at p in the direction $q \in C^2(\bar{\omega})$ satisfies (6.3.111).*

Next, we introduce the adjoint system with the unknown adjoint state $\bar{s} = [\bar{a}, \bar{b}] \in V(\omega)^2$:

$$B(\bar{s}, \bar{v}) = \nabla_1 j([\bar{y}, p])(\bar{v}) \quad \forall \bar{v} \in V(\omega)^2. \tag{6.3.112}$$

Existence and uniqueness of the solution to (6.3.112) follow again from the Lax–Milgram lemma. Notice that we have assumed that j is Fréchet differentiable on $V(\omega)^2 \times C^2(\bar{\omega})$, and we have denoted by $\nabla_1 j$, $\nabla_2 j$ the partial differentials with respect to $\bar{y}$, p.

Proposition 6.3.12 *If j is Fréchet differentiable, then the directional derivative of the cost functional Π of the problem (P) at the point $p \in C^2(\bar{\omega})$ in the direction $q \in C^2(\bar{\omega})$ is given by*

$$\nabla \Pi(p) q = \nabla_2 j([\bar{y}, p]) q + Z(\bar{s}). \tag{6.3.113}$$

Proof. We have

$$\lim_{\lambda \to 0} \frac{\Pi(p + \lambda q) - \Pi(p)}{\lambda} = \nabla_2 j([\bar{y}, p]) q + \nabla_1 j([\bar{y}, p]) \tilde{y},$$

by the chain rule and by Proposition 6.3.11. Moreover, by virtue of (6.3.112) and (6.3.111) we have

$$\nabla_1 j([\bar{y}, p]) \tilde{y} = B(\bar{s}, \tilde{y}) = B(\tilde{y}, \bar{s}) = Z(\bar{s}).$$

The assertion is proved. □

Remark. In order to evaluate the directional derivative for $p, q \in C^2(\bar{\omega})$ using (6.3.113), one has to determine the state $\bar{y}$ by solving (6.3.28), then the adjoint state $\bar{s}$ by solving (6.3.112), and finally Z from (6.3.111). The computation of Z is standard (see (6.3.107), (6.3.106)), but rather tedious.

We conclude by formulating the first-order necessary optimality conditions:

Corollary 6.3.13 *Suppose that p^* is a (local) minimizer for (P), and let $\bar{y}^*$ denote the associated deformation. Moreover, let all the above assumptions be fulfilled. Then the following hold:*

(i) *If $U_{ad} \subset C^2(\bar{\omega})$ is convex, then*

$$\nabla_2 j([\bar{y}^*, p^*])q + Z(\bar{s}) \geq 0 \quad \forall q \in U_{ad} - p^*.$$

(ii) *If U_{ad} is nonconvex, then*

$$\nabla_2 j([\bar{y}^*, p^*])q + Z(\bar{s}) \geq 0 \quad \forall q \in T(U_{ad}, p^*),$$

where $T(U_{ad}, p^*)$ is the tangent cone to U_{ad} in p^*.

For the definition of the tangent cone, we refer to Corollary 6.2.12.

Appendix 1

Convex Mappings and Monotone Operators

In this section we collect some well-known concepts and results from convex analysis and from the theory of monotone operators that are used repeatedly throughout this book and can be found in the classical monographs of Rockafellar [1970], Brézis [1973], Ekeland and Temam [1974], Barbu and Precupanu [1986], and Zeidler [1990]. We assume that the reader is already familiar with the basic notions of linear functional analysis. In particular, we make free use of such fundamental concepts as Banach and Hilbert spaces, dual spaces, reflexivity, convexity, compactness, strong convergence, weak and weak* convergence, linear continuous operators and their dual operators, and so on. Some brief explanations on function spaces will be given at the beginning of Appendix 2 below.

Let X be a Banach space with dual space X^*, and let $(\cdot,\cdot)_{X^*\times X}$ denote the dual pairing between elements of X^* and of X. We call a function $\varphi : X \to [-\infty,+\infty]$ *convex* if

$$\varphi(\lambda\, x + (1-\lambda)\, y) \le \lambda\, \varphi(x) + (1-\lambda)\, \varphi(y)$$

for all $\lambda \in [0,1]$ and $x, y \in X$ such that the right-hand side is well-defined. The function φ is called *strictly convex* if the inequality is strict for any $\lambda \in]0,1[$ and $x \neq y$. We define the *(effective) domain* $\mathrm{dom}(\varphi)$ of φ by

$$\mathrm{dom}(\varphi) = \{x \in X : \varphi(x) < +\infty\},$$

and call φ *proper* if $\mathrm{dom}(\varphi) \neq \emptyset$ and $\varphi(x) > -\infty$ for all $x \in X$.

A function $\varphi : X \to]-\infty,+\infty]$ is termed *(weakly) lower semicontinuous* if all the sets $\{x \in X : \varphi(x) \le \lambda\}$, $\lambda \in \mathbf{R}$, are (weakly) closed. The *closure* $\mathrm{cl}\varphi$ of φ is the lower semicontinuous hull of φ:

$$(\mathrm{cl}\varphi)(x) = \begin{cases} \liminf_{y\to x} \varphi(y), & \text{if } \varphi(x) > -\infty \text{ for all } x \in X, \\ -\infty, & \text{otherwise.} \end{cases}$$

If $\mathrm{cl}\varphi = \varphi$, then φ is said to be *closed*. For proper and convex functions the closedness is equivalent to lower semicontinuity.

The *subdifferential* $\partial\varphi(x)$ of φ at a point x is the set

$$\partial\varphi(x) = \{w \in X^* : \varphi(x) - \varphi(v) \leq (w, x - v)_{X^* \times X} \quad \forall v \in X\}.$$

It is defined on $\mathrm{dom}(\partial\varphi)$, which is given by all of $x \in X$ for which this set is nonempty. We call the set-valued mapping $x \mapsto \partial\varphi(x)$ the subdifferential of φ. The subdifferential is a generalization of the classical concepts of derivative and tangent.

If φ is convex and is finite at a point x, then for every $h \in X$ the difference quotient

$$\lambda^{-1}[\varphi(x + \lambda h) - \varphi(x)]$$

is a nondecreasing function of λ in $]0, +\infty]$. Thus, the *directional derivative* at x in the direction h,

$$\varphi'(x, h) \;=\; \lim_{\lambda \searrow 0} \lambda^{-1}[\varphi(x + \lambda h) - \varphi(x)] \;=\; \inf_{\lambda > 0} \lambda^{-1}[\varphi(x + \lambda h) - \varphi(x)],$$

exists for every $h \in X$. By the *directional differential* of φ at x we mean the mapping $h \mapsto \varphi'(x, h)$. It is a positively homogeneous and subadditive functional on X. If it is linear and continuous on X, then φ is said to be *Gâteaux differentiable* at x.

Let $\nabla\varphi(x) \in X^*$ denote this functional. Then

$$-\varphi'(x, -h) = \varphi'(x, h) = (\nabla\varphi(x), h)_{X^* \times X} \quad \forall h \in X.$$

We say that φ is *Fréchet differentiable* at x with the Fréchet derivative $\nabla\varphi(x) \in X^*$ if

$$\varphi(x + h) = \varphi(x) + (\nabla\varphi(x), h)_{X^* \times X} + o(|h|_X) \text{ where } \lambda^{-1}o(\lambda) \to 0 \text{ for } \lambda \to 0\,.$$

It is generally the case that

$$\partial\varphi(x_0) = \{x^* \in X^* : \; (x^*, h)_{X^* \times X} \leq \varphi'(x_0, h) \quad \forall h \in X\}.$$

Conversely, if φ is finite and continuous at x_0, then

$$\varphi'(x_0, h) = \sup\{(x^*, h)_{X^* \times X} : \; x^* \in \partial\varphi(x_0)\}.$$

The definitions of Gâteaux and Fréchet differentiability may be straightforwardly extended to mappings not necessarily convex and with values in infinite-dimensional spaces (see Zeidler [1990]).

If $K \subset X$ is a nonempty, closed, and convex set, then the *indicator function* $I_K : X \to]-\infty, +\infty]$ of K,

$$I_K(x) = \begin{cases} 0, & \text{if } x \in K, \\ +\infty, & \text{otherwise,} \end{cases}$$

is a proper, convex, and lower semicontinuous mapping. Its subdifferential at x,

$$\partial I_K(x) = \{x^* \in X^* : (x^*, x-u)_{X^* \times X} \geq 0 \quad \forall u \in K\},$$

is a closed and convex cone in X^* with vertex at the origin. It is called the *normal cone* to K at x.

The subdifferential of the mapping $f(x) = \frac{1}{2}\,|x|_X^2$ is called the *duality mapping* of X. It can be equivalently defined by

$$F(x) = \partial f(x) = \left\{x^* \in X^* : (x^*, x)_{X^* \times X} = |x|_X^2 = |x^*|_{X^*}^2\right\}.$$

We notice that (cf. Pavel [1984]) the norm of X is Gâteaux differentiable at $x \neq 0$ if and only if the duality mapping F is single-valued. Moreover, in this case, for any $x, y \in X$, we have

$$\lim_{t \to 0} \frac{|x+ty|_X - |x|_X}{t} = \frac{1}{|x|_X}(F(x), y)_{X^* \times X}.$$

We now consider the general case of *convex integrands*. To this end, let $\Omega \subset \mathbf{R}^d$ be Lebesgue measurable, and suppose that the mapping $g : \Omega \times \mathbf{R}^m \to]-\infty, +\infty]$, $m \in \mathbf{N}$, satisfies the following conditions:

(i) $g(x,\cdot) : \mathbf{R}^m \to]-\infty, +\infty]$ is proper, convex, and lower semicontinuous for a.e. $x \in \Omega$.

(ii) g is measurable with respect to the σ-field of sets generated by the products between Lebesgue sets in Ω and Borel sets in $\mathbf{R}^m$.

(iii) $g(x,y) \geq (\alpha(x), y)_{\mathbf{R}^m} + \beta(x)$ on $\Omega \times \mathbf{R}^m$, with given functions $\alpha \in L^{p'}(\Omega)^m$, for some $p' > 1$, and $\beta \in L^1(\Omega)$.

(iv) There is some vector function $y_0 \in L^p(\Omega)^m$, where $\frac{1}{p} + \frac{1}{p'} = 1$, such that $g(\cdot, y_0(\cdot)) \in L^1(\Omega)$.

Then, the integral functional $I_g : L^p(\Omega)^m \to]-\infty, +\infty]$,

$$I_g(y) = \begin{cases} \displaystyle\int_\Omega g(x, y(x))\, dx, & \text{if } g(\cdot, y(\cdot)) \in L^1(\Omega), \\ +\infty, & \text{otherwise,} \end{cases}$$

has the following properties (cf. Barbu and Precupanu [1986, p. 117]):

Proposition A1.1 *Under the conditions (i)–(iv) the mapping I_g is proper, convex, and lower semicontinuous on $L^p(\Omega)^m$. For every $y \in L^p(\Omega)^m$ such that the subdifferential $\partial I_g(y)$ exists, we have*

$$\partial I_g(y) = \left\{w \in L^{p'}(\Omega)^m : w(x) \in \partial g(x, y(x)) \text{ for a.e. } x \in \Omega\right\}.$$

Next, we turn to set-valued mappings. We denote by $X \times Y$ the Cartesian product of two sets X and Y, and by $[x,y]$ an ordered pair in $X \times Y$.

We call any subset $A \subset X \times Y$ a *set-valued* or *multivalued operator* defined in X with values in Y. Throughout this monograph, we will use the following convenient notational convention for set-valued operators: instead of $A \subset X \times Y$, we will also write $A : X \to Y$. This will not give rise to confusion. We also define

$$\begin{aligned} Ax &= \{y \in Y : [x,y] \in A\}, \quad \text{for } x \in X, \\ \mathrm{dom}(A) &= \{x \in X : Ax \neq \emptyset\} \subset X, \\ R(A) &= \bigcup_{x \in X} Ax \subset Y, \\ A^{-1} &= \{[y,x] : [x,y] \in A\} \subset Y \times X. \end{aligned}$$

Now let X be a Banach space. We call a set-valued operator $A : X \to X^*$ *monotone* if

$$(y_1 - y_2, x_1 - x_2)_{X^* \times X} \geq 0 \quad \forall\, [x_1,y_1], [x_2,y_2] \in A,$$

and *strictly monotone* if the inequality is strict whenever $x_1, x_2 \in \mathrm{dom}(A)$ and $x_1 \neq x_2$. The operator is called *strongly monotone* if there is some constant $\alpha > 0$ such that

$$(y_1 - y_2, x_1 - x_2)_{X^* \times X} \geq \alpha\, |x_1 - x_2|_X^2 \quad \forall\, [x_1,y_1], [x_2,y_2] \in A.$$

A monotone operator $A : X \to X^*$ is called *maximal monotone* if its graph as a subset in $X \times X^*$ is maximal in the sense that it cannot be strictly included in any other monotone graph in $X \times X^*$. An important class of monotone operators is given by just the subdifferentials of convex functions, as the following result shows.

Theorem A1.2 *Let φ be a proper, convex, and lower semicontinuous mapping in a Banach space X. Then the subdifferential $\partial\varphi$ is maximal monotone.*

Other examples of maximal monotone operators can be obtained as follows:

- If $A : X \to X$ is maximal monotone and X is a Hilbert space (identified with its dual), then also $A^{-1} : X \to X$ is maximal monotone.

- Let $\Omega \subset \mathbf{R}^d$ be a bounded and measurable set. If $A : X \to X$ is maximal monotone, then $\widetilde{A} : L^2(\Omega, X) \to L^2(\Omega, X)$, defined by $v \in \widetilde{A}u$ if $v(x) \in Au(x)$ for a.e. $x \in \Omega$, is maximal monotone.

- Let X be a Banach space and $A : X \to X^*$ a monotone single-valued mapping with $\mathrm{dom}(A) = X$. If A is *hemicontinuous* on X, that is, if for any $x, y \in X$ it follows that $A((1-t)x + ty) \to Ax$ weakly in X as $t \to 0$, then A is maximal monotone.

A fundamental tool in the study of maximal monotonicity is the *Minty characterization*:

Theorem A1.3 *Let X and X^* be reflexive and strictly convex Banach spaces, and let $F : X \to X^*$ be the duality mapping. Then a monotone mapping $A : X \to X^*$ is maximal monotone if and only if $R(A+F) = X^*$.*

A direct consequence of Theorem A1.3 is that for every $\lambda > 0$ and every $x \in X$ the equation

$$F(x_\lambda - x) + \lambda A x_\lambda \ni 0$$

has a unique solution $x_\lambda \in X$. We may therefore define the mappings

$$J_\lambda : X \to X, \qquad J_\lambda x = x_\lambda, \qquad A_\lambda : X \to X^*, \qquad A_\lambda x = -\lambda^{-1} F(J_\lambda x - x),$$

which are called the *resolvent*, respectively the *Yosida approximation*, of A. In Hilbert spaces, they simply take the form

$$J_\lambda x = (id + \lambda A)^{-1} x, \quad A_\lambda x = \lambda^{-1}(id - J_\lambda)x,$$

where $id : X \to X$ denotes the identity mapping. We summarize some properties of J_λ, A_λ.

Proposition A1.4 *Let X and X^* be strictly convex and reflexive Banach spaces, and let $A : X \to X^*$ be maximal monotone. Then the following assertions hold:*

(i) *A_λ is a single-valued, monotone and demicontinuous mapping (that is, whenever $x_n \to x$ strongly in X then $A_\lambda x_n \to A_\lambda x$ weakly in X^*). In particular, A_λ is hemicontinuous and thus maximal monotone.*

(ii) *A_λ and J_λ are bounded operators, i.e., bounded on bounded sets. If X is a Hilbert space, then J_λ is nonexpansive, and A_λ is Lipschitz continuous with Lipschitz constant $\frac{1}{\lambda}$.*

(iii) *$\lim_{\lambda\to 0} J_\lambda x = x$ for every $x \in \overline{\mathrm{conv}(\mathrm{dom}(A))}$, and $|A_\lambda x|_{X^*} \le |A^0 x|_{X^*}$ for every $x \in \mathrm{dom}(A)$, where A^0 is the section of minimal norm of A. If X is a Hilbert space, then $\lim_{\lambda\to 0} A_\lambda x = A^0 x$ for every $x \in \mathrm{dom}(A)$. Moreover, $A_\lambda x \in A J_\lambda x$ for all $x \in X$.*

(iv) *If $\lambda_n \to 0$, $x_n \to x$ weakly in X, $A_{\lambda_n} x_n \to x^*$ weakly in X^*, and*

$$\limsup_{n,m\to\infty} (A_{\lambda_n} x_n - A_{\lambda_m} x_m, x_n - x_m)_{X^* \times X} \le 0,$$

then $x^* \in Ax$ and

$$\lim_{n,m\to\infty} (A_{\lambda_n} x_n - A_{\lambda_m} x_m, x_n - x_m)_{X^* \times X} = 0.$$

Property (iv) is a more general form of the *demiclosedness* of maximal monotone operators A: if $y_n \in Ax_n$, $x_n \to x$ strongly in X, and $y_n \to y$ weakly in X^*, then $y \in Ax$.

In the case of the subdifferentials, Proposition A1.4 can be further sharpened. To this end, we define the *Yosida–Moreau regularization* of the convex mapping φ by

$$\varphi_\lambda(x) = \inf_{y \in X} \left\{ \frac{1}{2\lambda} |x - y|_X^2 + \varphi(y) \right\}, \quad \text{for } \lambda > 0,\ x \in X.$$

Theorem A1.5 *Let X be a reflexive and strictly convex Banach space. Then the function φ_λ is convex, everywhere finite, and Gâteaux differentiable on X. Moreover, with $A = \partial\varphi$, $A_\lambda = \partial\varphi_\lambda$. In addition, we have the following:*

(i) $\varphi(J_\lambda x) \le \varphi_\lambda(x) \le \varphi(x)$ *for all* $x \in X$ *and* $\lambda > 0$.

(ii) $\varphi_\lambda(x) = \varphi(J_\lambda x) + \frac{1}{2\lambda} |x - J_\lambda x|_X^2$ *for all* $x \in X$ *and* $\lambda > 0$.

(iii) $\lim_{\lambda \to 0} \varphi_\lambda(x) = \varphi(x)$ *for all* $x \in X$.

Finally, if X is a Hilbert space, then φ_λ is Fréchet differentiable on X, and $\partial\varphi_\lambda = A_\lambda$ is Lipschitzian.

Property (ii) asserts that the infimum in the definition of φ_λ is attained at $y = J_\lambda x$. The question of the existence of the minimizers and their characterization are fundamental in the framework of convex functions. We have the following result.

Theorem A1.6 *Let X denote a reflexive Banach space. Any proper, closed, and convex function on X is bounded from below by an affine mapping and attains a minimum value on every nonempty, closed, convex, and bounded subset of X. The minimum point is unique if φ is strictly convex. A proper and convex function on X is lower semicontinuous on X if and only if it is weakly lower semicontinuous. A point $x \in X$ is a minimizer of φ on X if and only if $0 \in \partial\varphi(x)$.*

Remark. The boundedness assumption in Theorem A1.6 may be replaced by the *coercivity condition* $\lim_{|x|_X \to \infty} \varphi(x) = +\infty$.

Finding minimum points of various functions plays a crucial role in the context of optimal control theory and in shape optimization problems, which is the subject of this book. It is also very important for the solution of various (elliptic) boundary value problems when variational methods are applied (see Appendix 2 and Chapter 1).

If φ is Fréchet differentiable (but not necessarily convex) and X is a Hilbert space identified with its dual space, then the so-called *gradient algorithms* provide general iterative procedures for the search for (local) minimum points. Given the iterate z_n, the next iterate is constructed in the form

$$z_{n+1} = z_n - \rho \nabla \varphi(z_n), \quad \text{with some } \rho > 0.$$

This iterative procedure is based on the definition of Fréchet differentiability: indeed, we have

$$\varphi(z_n - \rho \nabla \varphi(z_n)) = \varphi(z_n) - \rho \, |\nabla \varphi(z_n)|_X^2 + o(\rho \, |\nabla \varphi(z_n)|_X),$$

which ensures the *descent property* of the algorithm, that is, $\varphi(z_{n+1}) < \varphi(z_n)$ for sufficiently small $\rho > 0$. Gradient algorithms may be applied in a more general setting.

There also is a simple geometrical interpretation: if the level hypersurfaces of φ are smooth, then $\nabla \varphi(x)$ coincides with a normal vector to the level set passing through x, since the tangential component of the gradient vanishes along the level sets. This property motivates the name *method of steepest descent* for the above iterative process. Many variants and convergence results for it are discussed in the literature, for instance, in Céa [1971], Pironneau [1984] (with applications to shape optimization), and Hiriart-Urruty and Lemaréchal [1993]. Using subdifferentials, the approach can be extended to the nonsmooth case, leading to the so-called *bundle* methods (cf. Mäkelä and Neittaanmäki [1992]).

Various *coercivity conditions* play a central role in connection with the surjectivity of maximal monotone operators. We have the following result.

Theorem A1.7 *Let X be a reflexive Banach space. If $A : X \to X^*$ is maximal monotone and coercive, that is, if*

$$\lim_{|x|_X \to \infty} \frac{(x^*, x)_{X^* \times X}}{|x|_X} = +\infty \quad \textit{whenever } [x, x^*] \in A,$$

then A is surjective.

In general, the following characterization of surjectivity is valid.

Theorem A1.8 *Let X be a reflexive Banach space. A maximal monotone operator $A : X \to X^*$ is surjective if and only if A^{-1} is locally bounded on X^* in the following sense: for any $x_0^* \in X^*$ there exists an open neighborhood V of x_0^* such that the set*

$$A^{-1}(V) = \bigcup_{x^* \in V} A^{-1}(x^*)$$

is bounded in X. *If* $A = \partial\varphi$, *then the surjectivity of* A *is equivalent to the boundedness of* A^{-1} *on* X^* *(i.e.,* A^{-1} *maps bounded sets into bounded sets) and with the coercivity condition*

$$\lim_{|x|_X \to \infty} \frac{\varphi(x)}{|x|_X} = +\infty.$$

Let us notice that any maximal monotone operator is *locally bounded* on the interior of its domain (if it is nonempty). In particular, this is valid for subdifferential operators, yielding the following result.

Theorem A1.9 *Let* X *be a Banach space. Then any proper, convex, and lower semicontinuous mapping* $\varphi : X \to]-\infty, +\infty]$ *is locally Lipschitz in the interior of its domain. Moreover,*

$$\mathrm{dom}(\partial\varphi) \subset \mathrm{dom}(\varphi), \quad \overline{\mathrm{dom}(\partial\varphi)} = \overline{\mathrm{dom}(\varphi)}, \quad \mathrm{int}\,(\mathrm{dom}(\partial\varphi)) = \mathrm{int}\,(\mathrm{dom}(\varphi)).$$

We now concentrate on various operations that can be performed with convex functions or monotone operators, while preserving certain basic properties.

Theorem A1.10 *Let* X *be a reflexive Banach space, and let* A *and* B *be maximal monotone operators in* $X \times X^*$ *such that* $\mathrm{int}\,(\mathrm{dom}(A)) \cap \mathrm{dom}(B) \neq \emptyset$. *Then* $A + B$ *is maximal monotone in* $X \times X^*$.

Theorem A1.11 *Let* X *be a reflexive Banach space,* $A : X \to X^*$ *maximal monotone, and* $\varphi : X \to]-\infty, +\infty]$ *proper, convex, and lower semicontinuous. Then* $A + \partial\varphi$ *is maximal monotone in* $X \times X^*$ *provided that one of the following conditions is satisfied:*

(a) $\mathrm{dom}(A) \cap \mathrm{int}\,(\mathrm{dom}(\varphi)) \neq \emptyset$.
(b) $\mathrm{dom}(\varphi) \cap \mathrm{int}\,(\mathrm{dom}(A)) \neq \emptyset$.

Remark. In the special case $A = \partial I_K$, for some nonempty, closed, and convex set $K \subset X$, we have $\partial(\varphi + I_K) = \partial\varphi + \partial I_K$ whenever φ is continuous in some $x_0 \in K$ or $\mathrm{dom}(\varphi) \cap \mathrm{int}(K) \neq \emptyset$.

Another operation frequently met in applications is the composition of mappings and operators. If $\varphi : X \to]-\infty, +\infty]$ is proper, convex, and closed, and if $A : X \to X$ is a linear and bounded operator having the dual operator A^*, then the composed mapping

$$\psi : X \to]-\infty, +\infty], \quad \psi(x) = \varphi(Ax), \quad x \in X,$$

is proper, convex, and lower semicontinuous provided that $R(A) \cap \mathrm{dom}(\varphi) \neq \emptyset$. The following *chain rule* is due to Tiba [1977] and has been studied in more general spaces by Zălinescu [1980].

Theorem A1.12 *Assume that the reflexive Banach space* X *can be decomposed into the direct sum* $X = X_1 \oplus X_2$ *in such a way that* $A^*|_{X_1^0} : X_1^0 \to X^*$ *has a bounded inverse and*

$$R(A) \cap \mathrm{int}_1 [\mathrm{dom}(\varphi) \cap X_1] \neq \emptyset,$$

where int_1 *denotes the relative interior with respect to* X_1. *Then*

$$\partial\psi(x) = A^*(\partial\varphi(Ax)) \quad \forall\, x \in X.$$

Here, we have denoted by $X_1^0 = \{x^* \in X^* : (x^*, x)_{X^* \times X} = 0 \quad \forall\, x \in X\}$ the polar subspace of X_1. Notice that in Hilbert spaces the polar set X_1^0 coincides with the usual orthogonal complement $X_1^\perp$.

We stop briefly to indicate some applications and examples.

Example A1.13 Any maximal monotone graph $\beta \subset \mathbf{R} \times \mathbf{R}$ is a subdifferential; more precisely, there is some proper, convex, and lower semicontinuous mapping $j : \mathbf{R} \to]-\infty, +\infty]$ satisfying $\beta = \partial j$.

If β^0 denotes the minimal section of β and $]a, b[$ is an interval such that $]a, b[\subset \mathrm{dom}(\beta) \subset \mathrm{dom}(j) \subset [a, b]$ (where a and/or b may be infinite), then β^0 is a nondecreasing function on $]a, b[$, and $\beta(x) = [\beta^0(x-), \beta^0(x+)]$ for all $x \in]a, b[$. Then, β is single-valued almost everywhere in its domain. Moreover, if $a \in \mathrm{dom}(\beta)$ (respectively $b \in \mathrm{dom}(\beta)$), then $\beta(a) =]-\infty, \beta^0(a+)]$ (respectively $\beta(b) = [\beta^0(b-), +\infty[$). Here $\beta^0(a+), (\beta^0(a-)$ denote, respectively, the right/left lateral limits of β^0 in a.

We also define a regularization of β, denoted by β^ε and called the *ε-uniform approximation* of β, which differs from the Yosida approximation of β. This regularization is useful for obtaining uniform approximations to the solutions of certain variational inequalities (see Theorem A2.10 in Appendix 2). It is characterized by the following properties:

(i) $\beta(s + \varepsilon) \geq \beta^\varepsilon(s) \geq \beta(s - \varepsilon)$, whenever these expressions are defined.

(ii) $\mathrm{dom}(\beta) \subset \mathrm{dom}(\beta^\varepsilon)$.

It is important to notice that there even exists a continuously differentiable ε-uniform approximation β^ε. To construct it, consider a Friedrichs regularizing kernel $\rho \in C_0^\infty(\mathbf{R})$ such that $\mathrm{supp}\, \rho \subset [0, 1]$, $\rho(r) \geq 0$, and $\int_0^1 \rho(s)\, ds = 1$. The approximation β^ε is described in five basic cases:

Case 1. dom$(\beta) = \mathbf{R}$. Then

$$\beta^\varepsilon(s) = \int_0^1 \beta(s+\varepsilon\sigma)\rho(\sigma)\,d\sigma.$$

Case 2. $0 \in \beta(s)$ for all $s \in]-\infty, s_0] = \mathrm{dom}(\beta)$. Then

$$\beta^\varepsilon(s) = \begin{cases} 0, & \text{if } s < s_0, \\ \tan\left(\dfrac{\pi(s-s_0)}{2\varepsilon}\right) - \dfrac{\pi(s-s_0)}{2\varepsilon}, & \text{if } s \in [s_0, s_0+\varepsilon[. \end{cases}$$

Case 3. $0 \in \beta(s)$ for all $s \in [s_0, +\infty[= \mathrm{dom}(\beta)$. Then

$$\beta^\varepsilon(s) = \begin{cases} \tan\left(\dfrac{\pi(s-s_0)}{2\varepsilon}\right) - \dfrac{\pi(s-s_0)}{2\varepsilon}, & \text{if } s \in]s_0-\varepsilon, s_0], \\ 0, & \text{if } s > s_0. \end{cases}$$

Case 4. dom$(\beta) =]-\infty, s_0[$. Then $\beta^0(s) \to +\infty$ as $s \to s_0$. We put

$$\beta^\varepsilon(s) = \begin{cases} \displaystyle\int_0^1 \beta(s-\varepsilon\sigma)\rho(\sigma)\,d\sigma, & \text{if } s \le s_0 - \varepsilon, \\ \displaystyle\int_0^1 \beta(s-\varepsilon\sigma)\rho(\sigma)\,d\sigma & \\ \quad + \tan\left(\dfrac{\pi(s-s_0+\varepsilon)}{2\varepsilon}\right) - \dfrac{\pi(s-s_0+\varepsilon)}{2\varepsilon}, & \text{if } s \in]s_0-\varepsilon, s_0[. \end{cases}$$

Case 5. dom$(\beta) =]s_0, +\infty[$. Similarly to Case 4, we define

$$\beta^\varepsilon(s) = \begin{cases} \displaystyle\int_0^1 \beta(s+\varepsilon\sigma)\rho(\sigma)\,d\sigma & \\ \quad + \tan\left(\dfrac{\pi(s-s_0-\varepsilon)}{2\varepsilon}\right) - \dfrac{\pi(s-s_0-\varepsilon)}{2\varepsilon}, & \text{if } s \in]s_0, s_0+\varepsilon[, \\ \displaystyle\int_0^1 \beta(s+\varepsilon\sigma)\rho(\sigma)\,d\sigma, & \text{if } s_0+\varepsilon < s. \end{cases}$$

The general case may be obtained using the following observation: if β_1 and β_2 are maximal monotone graphs in $\mathbf{R} \times \mathbf{R}$, as well as their sum $\beta = \beta_1 + \beta_2$, and if β_1^ε, β_2^ε are ε-uniform approximations of β_1 and β_2, respectively, then $\beta_1^\varepsilon + \beta_2^\varepsilon$ is an ε-uniform approximation of β.

We also note that the following decomposition of maximal monotone graphs in $\mathbf{R} \times \mathbf{R}$ is possible: if $s_1 \in \mathrm{int}\,(\mathrm{dom}(\beta))$ with $\beta(s_1)$ a singleton (we exclude the trivial case $\mathrm{dom}(\beta)$ has one element), we may write $\beta = \beta_1 + \beta_2$, where β_1 and β_2 are maximal monotone, and where $\mathrm{dom}(\beta_1) \supset]-\infty, s_1]$ and $\mathrm{dom}(\beta_2) \supset [s_1, +\infty[$. Indeed, we can put

$$\beta_1(s) = \begin{cases} \beta(s_1), & \text{if } s \le s_1, \\ \beta(s), & \text{if } s \in \mathrm{dom}(\beta) \cap [s_1, +\infty[, \\ \emptyset, & \text{otherwise,} \end{cases}$$

$$\beta_2(s) = \begin{cases} \{0\}, & \text{if } s \ge s_1, \\ \beta(s) - \beta(s_1), & \text{if } s \in \mathrm{dom}(\beta) \cap]-\infty, s_1], \\ \emptyset, & \text{otherwise.} \end{cases}$$

If $\mathrm{dom}(\beta_1)$ has the form $]-\infty, s_0[$, we use Case 4; if $\mathrm{dom}(\beta_1)$ has the form $]-\infty, s_0]$, then we put

$$\beta_{1,a}(s) = \begin{cases} \beta_1(s), & \text{if } s \le s_0, \\ \beta_1^0(s_0), & \text{if } s \ge s_0, \end{cases}$$

$$\beta_{1,b}(s) = \begin{cases} \{0\}, & \text{if } s < s_0, \\ [0, +\infty[, & \text{if } s = s_0, \\ \emptyset, & \text{otherwise.} \end{cases}$$

Then $\beta_1 = \beta_{1,a} + \beta_{1,b}$, and the approximation is discussed in Case 1 and Case 2. Using the decomposition property, we obtain the desired approximation of β_1 and, similarly, of β_2.

Example A1.14 As our next example we consider the *Leray–Lions operator* or *generalized divergence operator* of the calculus of variations (see also Example 1.2.8 in Chapter 1 and §2.3.2 in Chapter 2). To this end, let $\Omega \subset \mathbf{R}^d$ denote an open and bounded domain having a regular boundary $\partial\Omega$. We consider in Ω the nonlinear partial differential operator

$$A(y) = \sum_{|\alpha| \le l} (-1)^{|\alpha|} D^\alpha A_\alpha(x, y, Dy, \ldots, D^l y), \quad x \in \Omega,$$

for $y \in W^{l,p}(\Omega)$, $l \in \mathbf{N}$, and $p \ge 2$. Here, α is a multi-index of length $|\alpha| \le l$. The coefficient functions $A_\alpha : \Omega \times \mathbf{R}^T \to \mathbf{R}$, where T denotes the total number of partial derivatives in $\mathbf{R}^d$ from order 0 up to order l (which are collected in the vector $\xi \in \mathbf{R}^T$), are assumed to satisfy the following conditions:

(i) $A_\alpha(\cdot, \xi)$ is measurable in Ω for all $\xi \in \mathbf{R}^T$, and $A_\alpha(x, \cdot)$ is continuous on $\mathbf{R}^T$ for a.e. $x \in \Omega$.

(ii) $|A_\alpha(x,\xi)| \le C\left(|\xi|_{\mathbf{R}^T}^{p-1} + \mu(x)\right)$ for all $(x,\xi) \in \Omega \times \mathbf{R}^T$, with given $C>0$ and $\mu \in L^q(\Omega)$, where $q^{-1}+p^{-1}=1$.

Under these assumptions, A is a well-defined operator from $W_0^{l,p}(\Omega)$ into $W^{-l,q}(\Omega)$ (see Lions [1969, p. 182]). If, in addition,

$$\sum_{|\alpha|\le l} (A_\alpha(x,\xi) - A_\alpha(x,\eta))(\xi_\alpha - \eta_\alpha) \ge 0 \quad \text{for a.e. } x \in \Omega,$$

for any $\xi = (\xi_\alpha) \in \mathbf{R}^T$, $\eta = (\eta_\alpha) \in \mathbf{R}^T$, then A is monotone and hemicontinuous, hence, as seen above, maximal monotone. In this context, it is also of interest to define the *realization* A_H of A in the space $H = L^2(\Omega)$:

$$A_H(y) = A(y) \quad \forall\, y \in \operatorname{dom}(A_H) = \left\{ y \in W_0^{l,p}(\Omega) : A(y) \in H \right\}.$$

If A is *coercive* in the sense that there are $c_1 > 0$ and $c_2 \in \mathbf{R}$ such that

$$(A(y), y)_{W^{-l,q}(\Omega)\times W_0^{l,p}(\Omega)} \ge c_1\, |y|^p_{W_0^{l,p}(\Omega)} + c_2, \quad \forall\, y \in W_0^{l,p}(\Omega),$$

then A_H is maximal monotone in $H \times H$. This follows from Theorem A1.7, which shows that $i+A$ is a surjective mapping onto $W^{-l,q}(\Omega)$ (i is the natural embedding of $W^{l,p}(\Omega)$ into $W^{-l,q}(\Omega)$), and from Theorem A1.3, which proves that this is sufficient for the maximal monotonicity of A_H in $H \times H$.

Next, we introduce the *Fenchel–Moreau conjugate* φ^* of a convex function $\varphi : X \to]-\infty,+\infty]$:

$$\varphi^* : X^* \to]-\infty,+\infty], \quad \varphi^*(x^*) = \sup\{(x,x^*)_{X\times X^*} - \varphi(x) : x \in X\}.$$

Obviously, by definition, *Young's inequality*,

$$\varphi(x) + \varphi^*(x^*) \ge (x,x^*)_{X\times X^*} \quad \forall\, [x,x^*] \in X \times X^*$$

is fulfilled. For proper, convex, and lower semicontinuous functions φ equality occurs if and only if $x^* \in \partial\varphi(x)$ or equivalently, $x \in \partial\varphi^*(x^*)$.

The conjugate of the indicator function I_K of a nonempty, closed, and convex set $K \subset X$ is given by

$$I_K^*(x^*) = \sup\{(x,x^*)_{X\times X^*} : x \in K\}.$$

It is called the *support function* of K. If K is a cone, then its associated *polar* set K^0 is again a cone with vertex at the origin, given by

$$K^0 = \{x^* \in X^* : (x,x^*)_{X\times X^*} \le 0 \ \ \forall\, x \in K\} = \{x^* \in X^* : I_K^*(x^*) \le 0\}.$$

We close this section with a variant of the so-called *Ekeland variational principle* (Ekeland [1979]), which may be viewed as an extension of the classical Fermat theorem.

Theorem A1.15 *Let* (E,d) *be a complete metric space, and let* $F : E \to]-\infty,+\infty]$ *be a lower semicontinuous mapping that is bounded from below. For any* $\varepsilon > 0$, *let* $e_\varepsilon \in E$ *be such that*

$$F(e_\varepsilon) \le \inf\{F(e) : e \in E\} + \varepsilon^2.$$

Then there is some $e \in E$ *such that*

$$F(e) \le F(e_\varepsilon), \quad d(e_\varepsilon, e) \le \varepsilon, \quad F(e) \le F(u) + \varepsilon\, d(e,u) \quad \forall u \in E.$$

Remark. In applications of Theorem A1.15 to optimal control problems, the following situation is typical: Let

$$E = U_{ad} = \{u \in L^\infty(\Omega) : u \text{ is feasible}\}$$

denote a set of admissible controls, where Ω is some measurable subset of $\mathbf{R}^d$. We endow E with the so-called *Ekeland metric*

$$d_E(u,v) = \text{meas}(\{x \in \Omega : u(x) \ne v(x)\}).$$

Then (E, d_E) becomes a complete metric space.

Appendix 2

Elliptic Equations and Variational Inequalities

In this section, we collect some preparatory material concerning elliptic equations and variational inequalities that is needed in this monograph. We begin by recalling the most important facts about some classes of function spaces. For a deeper study of these and other related classes of function spaces, we refer the reader to the textbooks of Alt [1985], Adams [1975], Hewitt and Stromberg [1965], Vulikh [1976], Friedman [1982], and Yosida [1980].

Assume that $\Omega \subset \mathbf{R}^d$, $d \in \mathbf{N}$, is some given measurable and bounded set. We then denote by $L^p(\Omega)$, $1 \le p < +\infty$, the Banach spaces of equivalence classes of Lebesgue-measurable functions having a finite norm

$$|y|_{L^p(\Omega)} = \left(\int_\Omega |y(x)|^p \, dx \right)^{1/p}.$$

For $p = +\infty$ we have the Banach space $L^\infty(\Omega)$ of essentially bounded measurable functions with the usual modification of the norm.

We recall that the dual space of $L^p(\Omega)$ for $1 \le p < \infty$ can be identified with $L^q(\Omega)$, where $\frac{1}{p} + \frac{1}{q} = 1$ (with the convention that $q = +\infty$ if $p = 1$). In this connection, if $y \in L^p(\Omega)$, $1 \le p < +\infty$, and $z \in L^q(\Omega)$, then $y\,z \in L^1(\Omega)$, and we have Hölder's inequality

$$|y\,z|_{L^1(\Omega)} \le |y|_{L^p(\Omega)} \, |z|_{L^q(\Omega)}.$$

If Ω is open, we denote by $C^m(\Omega)$ the set of all functions $y : \Omega \to \mathbf{R}$ such that all partial derivatives up to order $m \in \mathbf{N}$,

$$D^\alpha y = \frac{\partial^{\alpha_1 + \ldots + \alpha_d}}{\partial x_1^{\alpha_1} \ldots \partial x_d^{\alpha_d}} \, y,$$

are continuous in Ω, where $\alpha = (\alpha_1, \ldots, \alpha_d) \in \mathbf{N}_0^d$ denotes any multi-index such that its length $|\alpha|$ satisfies $|\alpha| = \alpha_1 + \ldots + \alpha_d \le m$. We denote by $C^\infty(\Omega)$

the set of all functions belonging to $C^m(\Omega)$ for any $m \in \mathbf{N}$. The functions in $C^m(\Omega)$ having compact support in Ω form a subspace, which is denoted by $C_0^m(\Omega)$. The space of *test functions*, i.e., of infinitely differentiable functions with compact support in Ω, is denoted by $C_0^\infty(\Omega)$ or by $\mathcal{D}(\Omega)$. Its dual space is the space $\mathcal{D}'(\Omega)$ of distributions on Ω (cf. Yosida [1980]). Recall that $C_0^\infty(\Omega)$ is dense in $L^p(\Omega)$ if $1 \le p < \infty$.

The space $C^m(\overline{\Omega})$ consists of all those elements of $C^m(\Omega)$ that together with all their partial derivatives up to order m can be continuously extended onto $\overline{\Omega}$. The space of continuous functions on $\overline{\Omega}$ is simply denoted by $C(\overline{\Omega})$. With the norm

$$|y|_{C^m(\overline{\Omega})} = \max_{0\le|\alpha|\le m} \max_{x\in\overline{\Omega}} |D^\alpha y(x)|,$$

$C^m(\overline{\Omega})$ becomes a separable Banach space. We denote by $C^\infty(\overline{\Omega})$ the set of all functions belonging to $C^m(\overline{\Omega})$ for any $m \in \mathbf{N}$.

For $l \in \mathbf{N}$ and $1 \le p \le +\infty$, the *Sobolev space* $W^{l,p}(\Omega)$ is defined as the space of all functions $y \in L^p(\Omega)$ having generalized partial derivatives $D^\alpha y \in L^p(\Omega)$, for every multi-index α satisfying $0 \le |\alpha| \le l$. Endowed with the norm

$$|y|_{W^{l,p}(\Omega)} = \Big(\sum_{0\le|\alpha|\le l} \int_\Omega |D^\alpha y(x)|^p \, dx \Big)^{1/p},$$

$W^{l,p}(\Omega)$ becomes a Banach space that is separable and reflexive for $1 < p < +\infty$. In the case $p = 2$, we use the notation $H^l(\Omega) = W^{l,2}(\Omega)$ and $|\cdot|_{H^l(\Omega)} = |\cdot|_{W^{l,2}(\Omega)}$. The space $H^l(\Omega)$, $l \in \mathbf{N}$, is a Hilbert space with the inner product

$$(y, z)_{H^l(\Omega)} = \sum_{0\le|\alpha|\le l} \int_\Omega D^\alpha y(x) \, D^\alpha z(x) \, dx \, .$$

The closures of $C_0^\infty(\Omega)$ in $W^{l,p}(\Omega)$ and in $H^l(\Omega)$ are denoted by $W_0^{l,p}(\Omega)$ and $H_0^l(\Omega)$, respectively. By $W^{-l,q}(\Omega)$ and $H^{-l}(\Omega)$ we denote the dual spaces of $W_0^{l,p}(\Omega)$ and of $H_0^l(\Omega)$, respectively, where $l \in \mathbf{N}$ and $1 \le q < +\infty$ with $\frac{1}{p} + \frac{1}{q} = 1$.

Finally, we recall that it is possible to define *fractional-order Sobolev spaces* $W^{s,p}(\Omega)$ and *trace spaces* $W^{s,p}(\partial\Omega)$ for nonintegral $s \in \mathbf{R}$. The spaces $W^{s,2}(\Omega)$ and $W^{s,2}(\partial\Omega)$, which are usually denoted by $H^s(\Omega)$ and $H^s(\partial\Omega)$, respectively, turn out to be Hilbert spaces. For a precise definition of fractional-order spaces, the reader is referred to the standard textbooks by Necas [1967], Lions and Magenes [1968], and Adams [1975].

We list some fundamental properties of Sobolev spaces.

Theorem A2.1 *Let $\Omega \subset \mathbf{R}^d$ denote an open Lipschitz domain. Then the following assertions hold:*

(i) *Let $1 \le p < +\infty$. Then there exists a unique linear and continuous surjective mapping $\gamma_0 : W^{1,p}(\Omega) \to W^{1-\frac{1}{p},p}(\partial\Omega)$ such that $\gamma_0(y) = y|_{\partial\Omega}$*

for all $y \in W^{1,p}(\Omega) \cap C(\overline{\Omega})$. *The mapping* γ_0 *is completely continuous from* $W^{1,p}(\Omega)$ *into* $L^q(\partial\Omega)$ *whenever* $q^{-1} > p^{-1} - (p-1)p^{-1}(d-1)^{-1}$.

(ii) *Suppose that* $m \in \mathbf{N}$, $m \geq 2$, $1 < p < +\infty$, *and that* Ω *belongs to the class* $C^{m-1,1}$, *that is,* Ω *belongs to the class* C^{m-1} *and the derivatives of order* $m-1$ *of all local charts in* $\mathcal{F}_\Omega$ *are Lipschitz continuous (see Definition A3.1 in Appendix 3). Then the mappings* γ_j *that assign to each* $y \in C^m(\overline{\Omega})$ *the* j*th (outward) normal derivative on* $\partial\Omega$,

$$\gamma_j(y) = \left.\frac{\partial^j y}{\partial n^j}\right|_{\partial\Omega}, \quad 1 \leq j \leq m-1,$$

can be uniquely extended to linear, continuous, and surjective mappings from $W^{m,p}(\Omega)$ *to* $W^{m-j-\frac{1}{p},p}(\partial\Omega)$.

Remark. The function $\gamma_0(y)$ is referred to as the *trace* of y on $\partial\Omega$. It is customary to denote it also by y. Likewise, we write $\frac{\partial^j y}{\partial n^j}$ instead of $\gamma_j(y)$.

Theorem A2.2 *Let* $\Omega \subset \mathbf{R}^d$, $d \in \mathbf{N}$, *denote some open and bounded domain. Then the following assertions hold:*

(i) *Let* $l_1, l_2 \in \mathbf{N}_0$ *with* $l_1 > l_2$ *and* $p_1, p_2 \in [1, +\infty[$. *If* $l_1 - \frac{d}{p_1} \geq l_2 - \frac{d}{p_2}$, *then* $W_0^{l_1,p_1}(\Omega)$ *is continuously embedded in* $W_0^{l_2,p_2}(\Omega)$. *If the inequality is strict, then the embedding is compact.*

(ii) *If* Ω *is a Lipschitz domain, then the assertions of (i) also hold if the spaces* $W_0^{l_i,p_i}(\Omega)$ *are replaced by the spaces* $W^{l_i,p_i}(\Omega)$, $i = 1, 2$.

(iii) *Let* $l \in \mathbf{N}$, $1 \leq p < +\infty$, *and* $m \in \mathbf{N} \cup \{0\}$. *If* $l - \frac{d}{p} > m$, *then* $W_0^{l,p}(\Omega)$ *is continuously and compactly embedded in* $C^m(\overline{\Omega})$.

(iv) *If* Ω *is a Lipschitz domain, then the assertions of (iii) also hold if the space* $W_0^{l,p}(\Omega)$ *is replaced by the space* $W^{l,p}(\Omega)$.

(v) *If* Ω *is a Lipschitz domain and* $t, s \in \mathbf{R}$ *satisfy* $t > s$, *then* $H^t(\Omega)$ *is continuously and compactly embedded in* $H^s(\Omega)$ *(Rellich's theorem).*

We now discuss abstract variational inequalities. We begin with a fundamental result due to Lions and Stampacchia [1967] that generalizes the classical Lax–Milgram lemma (cf. Yosida [1980]):

Theorem A2.3 *Let* $a : V \times V \to \mathbf{R}$ *be a continuous (not necessarily symmetric) bilinear form on the real Hilbert space* V *that is coercive in the sense that there is some* $\gamma > 0$ *such that* $a(v,v) \geq \gamma\,|v|_V^2$ *for all* $v \in V$. *Moreover,* *let $K \subset V$ be nonempty, closed, and convex, and let $f \in V^*$ be given. Then the abstract variational inequality*

$$y \in K: \quad a(y, z-y) \geq (f, z-y)_{V^* \times V} \quad \forall z \in K$$

has a unique solution $y \in K$, and the mapping $f \mapsto y$ is Lipschitz continuous from V^* to V. If K is a subspace of V, then the mapping $f \mapsto y$ is linear and bounded.

One far-reaching extension of Theorem A2.3 is the following.

Theorem A2.4 *Let $A : V \to V^*$ be a monotone and demicontinuous operator on the reflexive Banach space V, and let $\varphi : V \to]-\infty, +\infty]$ denote a proper, convex, and lower semicontinuous function. In addition, suppose that there exists some $y_0 \in \mathrm{dom}(\varphi)$ such that the coercivity condition*

$$\lim_{|y|_V \to \infty} \frac{(Ay, y - y_0)_{V^* \times V} + \varphi(y)}{|y|_V} = +\infty$$

is fulfilled. Then the abstract variational inequality

$$(Ay, y-z)_{V^* \times V} + \varphi(y) - \varphi(z) \leq (f, y-z)_{V^* \times V} \quad \forall z \in V$$

admits at least one solution $y \in V$. If A is strictly monotone, then the solution is unique.

The variational inequality can be equivalently expressed as

$$Ay + \partial\varphi(y) \ni f,$$

and Theorem A2.4 is a consequence of perturbation and surjectivity results for maximal monotone operators (compare Appendix 1).

Let us remark that Theorem A2.3 corresponds to the special case that $\varphi = I_K$ is an indicator function and $A \in L(V, V^*)$ is the operator defined by

$$a(y, z) = (Ay, z)_{V^* \times V}, \quad \forall y, z \in V.$$

If some compatibility condition is valid between a and K, then more regularity can be expected for the solution to the variational inequality. To this end, we assume that H is another Hilbert space, identified with its dual, such that $V \subset H \subset V^*$, algebraically and topologically. Let $A : V \to V^*$ be the linear and bounded operator generated by a, and let A_H be its realization in H, that is,

$$A_H : H \to H, \quad A_H y = Ay, \quad \forall y \in \mathrm{dom}(A_H) = \{u \in V : Au \in H\}.$$

We then have the following result.

Theorem A2.5 *Let the assumptions of Theorem A2.3 be fulfilled. In addition, assume that* $f \in H$ *and that there is* $h \in H$ *such that*

$$(id + \lambda A_H)^{-1}(y + \lambda h) \in K, \quad \textit{for any } \lambda > 0,\ y \in K.$$

Then the solution to the variational inequality belongs to $\mathrm{dom}(A_H)$*, and there is some constant* $C > 0$ *such that we have the estimate*

$$|Ay|_H \le C(1 + |f|_H) \quad \forall f \in H.$$

For further details in this respect, we quote the monograph by Barbu [1993].

Now let a bounded and smooth domain $\Omega \subset \mathbf{R}^d$ be given, and consider functions $a_{ij} \in C^1(\overline{\Omega})$, $i, j = \overline{1, d}$, and $a_0 \in C(\overline{\Omega})_+$ such that with some constant $a > 0$ the *ellipticity condition*

$$\sum_{i,j=1}^{d} a_{ij}(x)\, \xi_i\, \xi_j \ge a\, |\xi|^2, \quad \forall x \in \Omega, \quad \forall \xi \in \mathbf{R}^d,$$

is fulfilled. For a given right-hand side $f \in C(\overline{\Omega})$, we consider the second-order partial differential equation

$$-\sum_{i,j=1}^{d} \frac{\partial}{\partial x_i}\Big(a_{ij}(x)\, \frac{\partial y}{\partial x_j}\Big) + a_0\, y = f \quad \text{in } \Omega,$$

to which various types of boundary conditions may be associated, characterizing, respectively, the Dirichlet, Neumann, or third boundary value problems:

$$\begin{aligned}
&\text{(a)} \qquad && y = \overline{y} \quad \text{on } \partial\Omega,\\
&\text{(b)} \qquad && \frac{\partial y}{\partial n_A} = \sum_{i,j=1}^{d} a_{ij}\, \frac{\partial y}{\partial x_j} n_i = g \quad \text{on } \partial\Omega,\\
&\text{(c)} \qquad && cy + \frac{\partial y}{\partial n_A} = g \quad \text{on } \partial\Omega.
\end{aligned}$$

Here, $\frac{\partial y}{\partial n_A}$ is the *outer conormal derivative* of y, where n_i are the components of the outward unit normal n to $\partial\Omega$, and we have $\overline{y}, g, c \in C(\partial\Omega), c$ positive.

A function $y \in C^2(\overline{\Omega})$ is a *classical* solution to one of the above boundary value problems if it, together with its derivatives, satisfies the equation and the corresponding boundary conditions pointwise. If $y \in W^{2,p}(\Omega)$, its derivatives in the equation are understood in the sense of distributions, and if the boundary conditions are interpreted in the sense of traces, then y is called a *strong* solution.

In many practically important applications the regularity properties of the coefficient functions or of the data are too weak to ensure the existence of such solutions. Therefore, various notions of *weak* solutions have been introduced that are relevant for broad classes of applications. We give two examples of this type to clarify this idea.

Example A2.6 Assume that $\partial\Omega = \Gamma_0 \cup \Gamma_1 \cup \Gamma_2 \cup \Gamma_3$, where Γ_0 denotes the set of points on $\partial\Omega$ at which one type of the boundary conditions (a), (b), (c), changes into another, where $\text{meas}(\Gamma_0) = 0$ with respect to the superficial measure, and where Γ_i, $i = \overline{1,3}$, are mutually disjoint (relatively) open parts of $\partial\Omega$, with one or two of them possibly not occurring.

We study the *mixed boundary value problem*, which results if the boundary conditions (a), (b), (c) are imposed respectively on Γ_1, Γ_2, and Γ_3. We assume that the boundary data $\overline{y}$ can be extended to a function $\overline{y} \in H^1(\Omega)$, and we look for a solution in the form $y = y_0 + \overline{y}$, where

$$y_0 \in V = \left\{ y \in H^1(\Omega) : y = 0 \text{ on } \Gamma_1 \right\}.$$

We define on the Hilbert space V a bilinear and a linear form by putting

$$\begin{aligned} a(y,\varphi) &= \int_\Omega \sum_{i,j} a_{ij} \frac{\partial y}{\partial x_i} \frac{\partial \varphi}{\partial x_j}\, dx + \int_\Omega a_0\, y\, \varphi\, dx + \int_{\Gamma_3} c\, y\, \varphi\, dx \quad \forall\, y, \varphi \in V, \\ F(\varphi) &= \int_\Omega f\, \varphi\, dx + \int_{\Gamma_2 \cup \Gamma_3} g\, \varphi\, dx \;-\; a(\overline{y}, \varphi), \quad \forall\, \varphi \in V. \end{aligned}$$

Then, the corresponding weak (or variational) formulation of the mixed boundary value problem is

$$a(y_0, \varphi) = F(\varphi) \quad \forall\, \varphi \in V.$$

If $a(\cdot,\cdot)$ is symmetric, then it is known that the above definition is equivalent to the minimization of the quadratic functional $\frac{1}{2}\, a(y,y) - F(y)$ on V.

Let us notice that this variational problem is meaningful under much weaker assumptions on the data of the system; namely, it suffices to postulate that

$$g \in L^2(\Gamma_2 \cup \Gamma_3), \quad c \in L^\infty(\Gamma_3), \quad a_{ij}, a_0 \in L^\infty(\Omega), \quad \text{and } f \in L^2(\Omega).$$

The existence of a (unique) weak solution in $V + \overline{y}$ is then guaranteed if $(a_{ij})_{i,j=\overline{1,d}}$ is elliptic, $c \geq 0$, and $a_0 \geq 0$, which ensures the coercivity of the bilinear form $a(\cdot\,,\cdot)$. Consequently, one important question related to weak solutions concerns their regularity.

A standard *global regularity* result for the Dirichlet boundary value problem (that is, if $\Gamma_2 = \Gamma_3 = \emptyset$) states the following: if Ω is regular, $f \in L^p(\Omega)$ and $\overline{y} \in W^{2,p}(\Omega)$ for some $p > 1$, and if all a_{ij} are Lipschitz continuous and $a_0 \in L^\infty(\Omega)$, then the weak solution y_0 belongs to $W^{2,p}(\Omega)$, and there is a positive constant C, which depends neither on f nor on $\overline{y}$, such that

$$|y|_{W^{2,p}(\Omega)} \leq C \left(|f|_{L^p(\Omega)} + |\overline{y}|_{W^{2,p}(\Omega)} \right).$$

The interested reader may consult the monograph by Grisvard [1985] and the references therein. Such regularity results are also valid if Ω is only a Lipschitz domain (i.e., may have corners) that is convex.

Example A2.7 Next, we consider a technique that is useful when the regularity properties of the data are very weak. To fix things, we consider the Dirichlet problem

$$-\Delta y = f \quad \text{in } \Omega, \qquad y = g \quad \text{on } \partial\Omega,$$

with $f \in L^2(\Omega)$ and $g \in L^2(\partial\Omega)$. We define a very weak solution $y \in L^2(\Omega)$ to this problem through the relation

$$\int_\Omega y(x)\,\Delta\varphi(x)\,dx = -\int_\Omega f(x)\,\varphi(x)\,dx + \int_{\partial\Omega} g(\sigma)\,\frac{\partial\varphi}{\partial\nu}(\sigma)\,d\sigma \quad \forall\varphi \in H_0^1(\Omega) \cap H^2(\Omega).$$

The existence follows from the classical Riesz representation theorem.

There is an isomorphism T between $L^2(\Omega)$ and $H_0^1(\Omega) \cap H^2(\Omega)$, defined by $\varphi = T\psi$, where

$$-\Delta\varphi = \psi \quad \text{in } \Omega, \qquad \varphi = 0 \quad \text{on } \partial\Omega.$$

Moreover, consider the linear and bounded functional θ on $L^2(\Omega)$ given by

$$-\theta(\psi) = -\int_\Omega f(x)\,(T\psi)(x)\,dx + \int_{\partial\Omega} g(\sigma)\,\frac{\partial(T\psi)}{\partial\nu}(\sigma)\,d\sigma,$$

which, with some uniquely determined $y \in L^2(\Omega)$, admits a Riesz representation as

$$\theta(\psi) = (\psi, y)_{L^2(\Omega)} = \int_\Omega \psi(x)\,y(x)\,dx = -\int_\Omega y(x)\,\Delta\varphi(x)\,dx.$$

Formally, the integral equalities characterizing weak solutions of second-order elliptic equations are obtained by multiplying the original equation by a test function and then integrating once by parts (Green's formula), while the relation characterizing the very weak solution is obtained by integrating by parts twice. This approach is also called the *transposition method*, since the original equation is "transposed" on the test function.

A consequence of this argument is that classical and strong solutions are also weak solutions, since their regularity allows the integration by parts to be performed rigorously.

Another useful property of elliptic equations is the *maximum principle*. To formulate one important variant of it, let us consider the general second-order elliptic equation

$$-\sum_{i,j=1}^{d} \frac{\partial}{\partial x_j}\Big(a_{ij}\,\frac{\partial y}{\partial x_i}\Big) + \sum_{i=1}^{d} a_i\,\frac{\partial y}{\partial x_i} + a_0\,y = f.$$

We have the following result.

Theorem A2.8 *Suppose that the coefficient functions a_{ij}, $i,j = \overline{1,d}$, satisfy the ellipticity condition (here a.e. in Ω), and assume that $a_{ij}, a_i, a_0 \in L^\infty(\Omega)$*

with $a_0 \geq 0$ *a.e. in* Ω*. If* $f \in L^2(\Omega)$ *and* $y \in H^1(\Omega) \cap C(\overline{\Omega})$ *satisfy* $y \geq 0$ *on* $\partial\Omega$*,* $f \geq 0$ *in* Ω*, and*

$$\int_\Omega \sum_{i,j=1}^d a_{ij} \frac{\partial y}{\partial x_i} \frac{\partial \varphi}{\partial x_j}\, dx + \int_\Omega \sum_{i=1}^d a_i \frac{\partial y}{\partial x_i} \varphi\, dx + \int_\Omega a_0\, y\, \varphi\, dx = \int_\Omega f\, \varphi\, dx \quad \forall \varphi \in H_0^1(\Omega),$$

then $y \geq 0$ *in* Ω*.*

This result, together with other variants and references, may be found in Brézis [1987].

Let us finally mention that the regularity theory extends to the case of elliptic variational inequalities. In particular, Theorem A2.5 admits partial extensions to the non-Hilbertian case, as we will see in the following. To this end, let $\beta \subset \mathbf{R} \times \mathbf{R}$ be a maximal monotone graph. We associate with it the variational inequality (in subdifferential form)

$$-\sum_{i,j=1}^d \frac{\partial}{\partial x_i}\Big(a_{ij}(x) \frac{\partial}{\partial x_j} y(x)\Big) + \beta(y(x)) \ni f(x) \quad \text{in } \Omega, \qquad y = 0 \quad \text{on } \partial\Omega.$$

We have the following result.

Theorem A2.9 *Suppose that the coefficient functions* $a_{ij} \in C^1(\overline{\Omega})$*,* $i, j = \overline{1, d}$*, satisfy the ellipticity condition, and assume that* $f \in L^p(\Omega)$ *for some* $p \geq 2$ *and* $0 \in \beta(0)$*. Then the above variational inequality has a unique solution* $y \in W^{2,p}(\Omega) \cap H_0^1(\Omega)$*, and there exists some* $C > 0$ *that depends neither on* β *nor on* f *such that*

$$|y|_{W^{2,p}(\Omega)} \leq C\, |f|_{L^p(\Omega)}.$$

For a sketch of the proof and further references, we quote Bonnans and Tiba [1991].

Finally, let us consider the approximation of the variational inequality via an ε-uniform approximation β^ε of class C^1 (see Appendix 1). Theorem A2.9 may be complemented as follows.

Theorem A2.10 *Both the variational inequality and its* ε*-uniform approximation have unique solutions* $y, y_\varepsilon \in W^{2,p}(\Omega) \cap H_0^1(\Omega) \cap L^\infty(\Omega)$*, and we have the estimate* $|y_\varepsilon - y|_{L^\infty(\Omega)} \leq \varepsilon$*.*

This result still remains valid when the variational inequality is perturbed by a smooth monotone mapping $\varphi(x, y, u)$, where u may be some control parameter. For details, see Tiba [1990, Chapter III.5], and Bonnans and Tiba [1991].

Appendix 3

Domain Convergence

There are many types of problems (some of which are studied in this monograph) in which unknown or variable subsets arise quite naturally: free boundary problems, shape optimization problems, homogenization, discretization of infinite-dimensional sets and its convergence properties, the perturbation of optimization problems and its study via the corresponding family of perturbed epigraphs, and so on.

In this general setting, various notions of solution are used, and their basic properties concerning existence, uniqueness, characterization, computation, etc., strongly depend on certain mappings defined on the class of "admissible" subsets that associate to them functions or real numbers. The continuity or differentiability properties (in a generalized sense) of such "operators" set $\mapsto$ function or "functionals" set $\mapsto$ real number are fundamental for the analysis of variable-domain problems. Such questions have been systematically studied ever since the beginning of the last century; we quote only the pioneering contributions of Hadamard [1968] (originally published in 1907), Courant and Hilbert [1962], and Necas [1967]. Let us also mention that already the first problem in the calculus of variations, Bernoulli's brachistochrone (1696), is a shape optimization problem.

The material of this appendix gives a brief account of the mathematical theory connected with the above-mentioned continuity and convergence questions. It is based on well-known monographs and survey articles including Pironneau [1984], Adams [1975], Kuratowski [1962], Hewitt and Stromberg [1965], Attouch [1984], Azé [1997], Sokołowski and Zolésio [1992], Henrot [1994], Simon [1980], and Delfour and Zolésio [2001].

We begin with the definition of several classes of domains in the Euclidean space $\mathbf{R}^d$ that are used throughout this monograph.

Definition A3.1 *We say that a bounded open set* $\Omega \subset \mathbf{R}^d$ is of class C *(or* has continuous boundary*) if there exists a family* $\mathcal{F}_\Omega$ *of continuous functions* $g : \tilde{B}(0, k_\Omega) \to \mathbf{R}$, *where* $\tilde{B}(0, k_\Omega) \subset \mathbf{R}^{d-1}$ *denotes the open ball of radius*

$k_\Omega > 0$ centered at the origin of $\mathbf{R}^{d-1}$, such that

$$\partial\Omega = \bigcup_{g\in\mathcal{F}_\Omega} \{R_g(\tilde{s},0) + o_g + g(\tilde{s})\, y_g \;:\; \tilde{s} \in \tilde{B}(0,k_\Omega)\},$$

with $y_g = R_g(0,\dots,0,1)$ for some rotation R_g of $\mathbf{R}^d$, and with some $o_g \in \mathbf{R}^d$.

The geometrical interpretation of this definition should be clear: the set $\{(\tilde{s},0) : \tilde{s} \in \tilde{B}(0,k_\Omega)\}$, which forms a $(d-1)$-dimensional ball in $\mathbf{R}^d$, is through the coordinate transformation

$$(\tilde{s},0) \mapsto (s,0) = R_g(\tilde{s},0) + o_g\,,$$

mapped onto another $(d-1)$-dimensional ball, denoted by $B(0,k_\Omega)$, which is centered at the point o_g.

In this way the global Cartesian coordinate system, which has the "vertical" axis $(0,\dots,0,1)$, is transformed into a local Cartesian coordinate system having the "vertical" axis y_g. Then the piece of $\partial\Omega$ that is defined by the local chart g can be understood as the graph of the "transported" function (again denoted by g), $g(s) = g(\tilde{s})$, over the ball $B(0,k_\Omega)$. It is customary to represent the points on $\partial\Omega$ in the local coordinate system as $(s, g(s))$ with $s \in B(0,k_\Omega)$.

In order to be able to "compare" (later) the local charts, we have postulated in the above definition that all of the balls have the same radius k_Ω, which can always be achieved in view of the compactness of $\partial\Omega$. We also notice that it is possible to find some $r_\Omega \in]0,k_\Omega[$ such that the "restricted" closed local charts defined on $\overline{B(0,r_\Omega)}$ still yield a covering of $\partial\Omega$.

Open sets of class C have the interior and the exterior *segment property* (cf. Maz'ya [1985] and Adams [1975]): for any local chart $g \in \mathcal{F}_\Omega$, there is some $a_\Omega > 0$ such that

$$R_g(\tilde{s},0) + o_g + (g(\tilde{s}) - t)y_g \in \Omega, \qquad \forall t \in]0,a_\Omega[\,, \quad \forall \tilde{s} \in \tilde{B}(0,k_\Omega),$$

$$R_g(\tilde{s},0) + o_g + (g(\tilde{s}) + t)y_g \in \mathbf{R}^d \setminus \overline{\Omega}, \qquad \forall t \in]0,a_\Omega[\,, \quad \forall \tilde{s} \in \tilde{B}(0,k_\Omega).$$

Owing to the compactness of $\partial\Omega$, we may choose the same $a_\Omega > 0$ for all local charts.

Remark. Domains of class C may have cusps on the boundary, which, for instance, may be represented by Hölder continuous local charts g. However, Definition A3.1 does not allow for domains with "cuts" or "cracks," and we have $\partial\Omega = \partial\overline{\Omega}$, i.e., domains of class C are *Carathéodory*. Indeed, the inclusion $\partial\overline{\Omega} \subset \partial\Omega$ holds for any open set Ω, while the opposite inclusion can be easily shown using the exterior segment property.

If all the mappings g are Lipschitz continuous or of class C^m, $m \in \mathbf{N}$, then the domain Ω is called *Lipschitz* or of *class* C^m, respectively. Owing to the compactness of $\partial\Omega$, the Lipschitz constant may be assumed to be the same for all local charts. We shall use the term *smooth domain* or *smooth boundary* whenever all the mappings g belong to C^∞. In the text the term *smooth* or *regular* domain will also be used in the sense that the boundary is of class C^m with sufficiently large $m \in \mathbf{N}$ such that certain properties are fulfilled. Other approaches based on local diffeomorphisms may be used alternatively, see Adams [1975].

Lipschitz domains may have corners and enjoy the so-called *cone property*: there is a d-dimensional cone whose intersection with a small ball centered at its vertex is contained in the domain after an appropriate translation and/or rotation such that its vertex lies on the boundary of the domain. Moreover, an exterior unit normal vector exists at almost every point of $\partial\Omega$ (see Necas [1967]). If the domain is even regular, then an exterior unit normal vector and a tangent hyperplane exist at each boundary point.

The *Hausdorff–Pompeiu distance* was introduced by Pompeiu [1905], according to Hausdorff [1914, p. 463], [1927, p. 280], and Kuratowski [1952, p. 106], who studied it further. Let A and B be compact subsets of $\mathbf{R}^d$. The Hausdorff–Pompeiu distance $d_H(A, B)$ between A and B is defined by

$$d_H(A,B) = \max\left\{\max_{x\in A} d(x,B), \max_{y\in B} d(y,A)\right\},$$

where

$$d(x,B) = \inf_{y\in B} |x-y| \quad (= \min_{y\in B} |x-y|, \text{ in this case}).$$

We say that a sequence of compact sets $A_n \subset \mathbf{R}^d$ converges to the compact set A in the sense of the Hausdorff–Pompeiu distance if $\lim_{n\to\infty} d_H(A_n, A) = 0$.

Apparently, the Hausdorff–Pompeiu convergence $d_H(A_n, A) \to 0$ of compact sets is equivalent to the statement that $A = \{x \in \mathbf{R}^d : \exists\, x_n \in A_n : x_n \to x\}$. Moreover, it is also equivalent to the property that $d(\cdot, A_n) \to d(\cdot, A)$ uniformly in some bounded set containing the sets A and A_n, $n \in \mathbf{N}$ (cf. Delfour and Zolésio [2001, p. 158]).

The family of compact subsets of $\mathbf{R}^d$ equipped with the distance d_H forms a complete metric space. The first statement in the following proposition shows that it is also sequentially compact (see Kuratowski [1962]):

Proposition A3.2 *Let $\{A_n\}$ denote a bounded sequence of compact subsets in $\mathbf{R}^d$. Then we have the following:*

(i) *There exist a compact set $A \subset \mathbf{R}^d$ and a subsequence (which is again indexed by n) such that $\lim_{n\to\infty} d_H(A_n, A) = 0$.*

(ii) *If A_n is connected for every $n \in \mathbf{N}$, and $\lim_{n\to\infty} d_H(A_n, A) = 0$ for some nonempty and compact set $A \subset \mathbf{R}^d$, then A is connected.*

Proof. We give only a proof of (ii), which we could not find in the literature. Assume that A is not connected. Then there exist two nonempty and compact subsets $\widetilde{A}, \widehat{A}$ such that $A = \widetilde{A} \cup \widehat{A}$ and $\widetilde{A} \cap \widehat{A} = \emptyset$. Consequently,

$$\operatorname*{Min}_{x \in \widetilde{A},\, y \in \widehat{A}} |x - y|_{\mathbf{R}^d} = c > 0.$$

By the definition of the Hausdorff–Pompeiu convergence $d_H(A_n, A) \to 0$, there exists for any $\varepsilon > 0$ some $n_\varepsilon \in \mathbf{N}$ such that $A_n \subset A_\varepsilon$ for $n \geq n_\varepsilon$, where

$$A_\varepsilon = \left\{ x \in \mathbf{R}^d : \, d(x, A) < \varepsilon \right\}.$$

If $\varepsilon < \frac{c}{4}$, then A_ε has at least two disjoint connected open components $\widetilde{A}_\varepsilon$, $\widehat{A}_\varepsilon$, that is, $A_\varepsilon = \widetilde{A}_\varepsilon \cup \widehat{A}_\varepsilon$ and $\widetilde{A}_\varepsilon \cap \widehat{A}_\varepsilon = \emptyset$. Since A_n is connected, it should be contained in one of the sets $\widetilde{A}_\varepsilon$, $\widehat{A}_\varepsilon$. This contradicts the convergence $d_H(A_n, A) \to 0$. □

The notion of distance can be extended in a natural way to bounded and open sets in $\mathbf{R}^d$: if O and Ω are two bounded open subsets in $\mathbf{R}^d$, then we define their distance as the Hausdorff–Pompeiu distance between their complements with respect to any compact set $\overline{E}$ that contains O and Ω as subsets. This notion of distance is independent of the choice of $\overline{E}$ and thus well-defined. We use the notation

$$\tilde{d}_H(O, \Omega) = d_H(\overline{E} \setminus O, \overline{E} \setminus \Omega).$$

Let us note one important fact concerning the convergences with respect to the distances d_H and $\tilde{d}_H$: if Ω_n, $n \in \mathbf{N}$, and Ω are open and bounded sets that are all contained in the same open and bounded set E and satisfy $\lim_{n\to\infty} \tilde{d}_H(\Omega_n, \Omega) = 0$, then it may happen that $\overline{\Omega}$ is not a cluster point of $\{\overline{\Omega}_n\}$ with respect to d_H (construct an example similar to Example A3.7 below). However, this situation cannot occur for domains of class C (see Proposition A3.10 below).

Next, we introduce the so-called *signed distance function* $d_\Omega : \overline{E} \to \mathbf{R}$ by

$$d_\Omega(x) = \begin{cases} d(x, \overline{E} \setminus \Omega), & \text{if } x \in \Omega, \\ 0, & \text{if } x \in \partial\Omega, \\ -d(x, \overline{\Omega}), & \text{if } x \in \overline{E} \setminus \overline{\Omega}. \end{cases}$$

Now recall (see Clarke [1983, p. 66]) that for any open set $\Omega \subset E$ the mappings $x \mapsto d(x, \overline{\Omega})$ and $x \mapsto d(x, \overline{E} \setminus \Omega)$ are Lipschitz continuous in $\overline{E}$ with Lipschitz constant $L = 1$. From this it is easily deduced that also d_Ω is Lipschitz continuous on $\overline{\Omega}$ with Lipschitz constant $L = 1$. Moreover, d_Ω has at least one nonzero directional derivative in any point belonging to $\overline{E} \setminus \partial\Omega$. Since d_Ω provides a complete geometrical description of Ω, we call it a *parametrization* of Ω; compare Definition A3.3 below. Of course, other parametrizations are also possible.

Definition A3.3 *Let $E \subset \mathbf{R}^d$ be open and bounded. We say that the sequence of open sets $\Omega_n \subset E$ is* parametrically convergent *to the open set $\Omega \subset E$ and write $\Omega = p\text{-}lim_{n\to\infty}\Omega_n$ if there are a sequence of continuous mappings $p_n : \overline{E} \to \mathbf{R}$ and a continuous function $p : \overline{E} \to \mathbf{R}$ (called parametrization functions) such that $p_n \to p$ uniformly in $\overline{E}$ and*

$$\Omega_n = \{x \in E : \; p_n(x) > 0\}, \quad E \setminus \overline{\Omega}_n = \{x \in E : \; p_n(x) < 0\},$$
$$\Omega = \{x \in E : \; p(x) > 0\}, \qquad E \setminus \overline{\Omega} = \{x \in E : \; p(x) < 0\}.$$

The notions of convergence in the Hausdorff–Pompeiu sense and of parametric convergence differ, in general, as the following example shows.

Example A3.4 Consider the open and bounded sets

$$\begin{aligned} \Omega_n &= \left\{x \in \mathbf{R}^2 : \; |x|_{\mathbf{R}^2} < 1 + \sqrt{\frac{n+2}{2n}}\right\}, \qquad n \in \mathbf{N}, \\ \Omega &= \left\{x \in \mathbf{R}^2 : \; 0 < |x|_{\mathbf{R}^2} < 1 + \frac{1}{\sqrt{2}}\right\}, \\ E &= \left\{x \in \mathbf{R}^2 : \; |x|_{\mathbf{R}^2} < 3\right\}, \end{aligned}$$

where $|x|_{\mathbf{R}^2} = (x_1^2 + x_2^2)^{1/2}$ denotes the Euclidean length of $x = (x_1, x_2) \in \mathbf{R}^2$. Obviously, $\tilde{d}_H(\Omega_n, \Omega) \not\to 0$, owing to the "hole" in the center of Ω. However, $\Omega = \text{p-lim}_{n\to\infty}\Omega_n$. To verify this, we define the continuous function $p : \mathbf{R}^2 \to \mathbf{R}$ by

$$p(x) = \begin{cases} -\left(|x|_{\mathbf{R}^2} - 1\right)^2 + \frac{1}{2}, & |x|_{\mathbf{R}^2} \geq \frac{1}{2}, \\ |x|^2_{\mathbf{R}^2}, & |x|_{\mathbf{R}^2} \leq \frac{1}{2}, \end{cases}$$

and we set $p_n(x) = p(x) + \frac{1}{n}$, $n \in \mathbf{N}$. It is easily seen that we then have the situation of Definition A3.3. Moreover, we see that the limit defined in Definition A3.3 depends on the parametrization, as shown in Proposition A3.5 below, in many cases not in an "essential manner."

One advantage of this parametrization concept is that standard set operations like finite union and intersection can be easily parametrized by taking the supremum, respectively the infimum, of the given parametrizing mappings. We also have the following result:

Proposition A3.5 *If* $\Omega = p\text{-}lim_{n\to\infty}\Omega_n$*, and if the closed set* $C = \{x \in \overline{E} : p(x) = 0\}$ *has zero measure, then the corresponding characteristic functions satisfy* $\chi_{\Omega_n} \to \chi_\Omega$ *a.e. in* E.

Proof. If $x \in \Omega$ then $p(x) > 0$, and $p_n(x) > 0$ for $n > n_1(x)$. Thus, for $x \in \Omega$ we have $x \in \Omega_n$, and $\chi_{\Omega_n}(x) = \chi_\Omega(x) = 1$ for $n > n_1(x)$. Similarly, if $x \in E\backslash\overline{\Omega}$, then $p(x) < 0$, and $p_n(x) < 0$ for $n > n_2(x)$. Hence, $\chi_{\Omega_n}(x) = \chi_\Omega(x) = 0$ for $n > n_2(x)$, in this case. Since meas$(C) = 0$, we have one of the two above situations a.e. in E, and the assertion is proved. □

Remark. In Proposition 2.3.13 in Chapter 2 it is shown that under specific conditions the convergence of the characteristic functions is necessary in order to recover certain continuity properties of the solutions to partial differential equations with respect to the underlying domain of definition.

Next, we introduce a metric on the set of measurable subsets of $\mathbf{R}^d$. To this end, let A, B be measurable. We then define

$$\rho(A, B) = \text{meas}(A \setminus B) + \text{meas}(B \setminus A).$$

The set of measurable subsets of $\mathbf{R}^d$, equipped with the distance ρ, forms a complete metric space. The distance ρ coincides with the Ekeland distance (compare Appendix 1), applied to the corresponding characteristic functions; that is, $\rho(A, B) = d_E(\chi_A, \chi_B)$.

When all the sets are contained in some bounded domain in $\mathbf{R}^d$, the corresponding convergence is equivalent to the L^1 convergence of the associated characteristic functions. We also have (cf. Pironneau [1984]) the following:

Proposition A3.6 *Let* Ω_n*,* $n \in \mathbf{N}$*, and* Ω *denote open Lipschitz domains that are all contained in some bounded subset of* $\mathbf{R}^d$*. If* $\lim_{n\to\infty} \tilde{d}_H(\Omega_n, \Omega) = 0$*, then* $\lim_{n\to\infty} \rho(\Omega_n, \Omega) = 0$.

The converse of Proposition A3.6 is only partially true. To see this, let $\rho(\overline{\Omega}_n, \overline{\Omega}) \to 0$, and suppose that all the sets Ω_n are contained in some bounded subset of $\mathbf{R}^d$. By Proposition A3.2, there is some compact set $\tilde{\Omega}$ such that on a subsequence again indexed by n, $d_H(\overline{\Omega}_n, \tilde{\Omega}) \to 0$. We claim that then meas$(\overline{\Omega} \setminus \tilde{\Omega}) = 0$.

Suppose the contrary. Then there is some $\omega \subset \overline{\Omega}$ with meas$(\omega) > 0$ such that $\omega \cap \tilde{\Omega} = \emptyset$. For any $x \in \omega$, we have

$$c_x = \inf\{d(x, \overline{\Omega}_n) : n \in \mathbf{N}\} > 0,$$

due to the definition of $\tilde{\Omega}$. Now let B_x denote the ball of radius c_x centered at x. Clearly, $\omega \subset \bigcup_{x\in\omega} B_x$. Owing to the Lindelöf property (cf. Kelley [1975]), we may select countably many sets B_x that still cover ω. Then it follows from the σ-additivity of the Lebesgue measure that there is some $\hat{x} \in \omega$ such that $\text{meas}(\omega \cap B_{\hat{x}}) > 0$. Consequently, $\text{meas}(\omega \setminus \overline{\Omega}_n) \geq \text{meas}(\omega \cap B_{\hat{x}}) > 0$ for any $n \in \mathbf{N}$, which contradicts $\rho(\overline{\Omega}_n, \overline{\Omega}) \to 0$, finishing the proof of the claim.

The opposite inclusion is not always true; that is, we may have $\text{meas}(\tilde{\Omega} \setminus \overline{\Omega}) > 0$, as the following example demonstrates.

Example A3.7 In $\mathbf{R}^2$, we take $\overline{\Omega}_n = \overline{B(0,1)} \cup K_n$, where K_n is the union of n closed "rays" of length 2 that emanate from the origin and divide the plane into sectors of equal angles. It is elementary to check that then

$$\rho(\overline{\Omega}_n, \overline{B(0,1)}) \to 0, \quad d_H(\overline{\Omega}_n, \overline{B(0,2)}) \to 0, \quad \text{meas}(\overline{B(0,2)} \setminus \overline{B(0,1)}) > 0.$$

The Hausdorff–Pompeiu convergence enjoys another property that is useful in many applications, namely, the so-called Γ *property* (see Henrot [1994]):

Proposition A3.8 *Let Ω_n, $n \in \mathbf{N}$, and Ω denote open sets that are contained in some bounded open set E and satisfy $\tilde{d}_H(\Omega_n, \Omega) \to 0$. If $K \subset \Omega$ is compact, then there is some $n_0 \in \mathbf{N}$ such $K \subset \Omega_n$ for all $n \geq n_0$.*

Proof. From the definition of the Hausdorff–Pompeiu convergence, and from the remarks made thereafter, we know that

$$d(\cdot\,, \overline{E} \setminus \Omega_n) \to d(\cdot\,, \overline{E} \setminus \Omega) \quad \text{uniformly in } \overline{E}.$$

If $K \subset \Omega$ is compact, then $d(x, \overline{E} \setminus \Omega) \geq c_K > 0$ for all $x \in K$, since $d(x, \overline{E} \setminus \Omega) > 0$ for all $x \in \Omega$, and by the Weierstrass theorem. Then it follows from the uniform convergence that for sufficiently large $n \in \mathbf{N}$,

$$d(x, \overline{E} \setminus \Omega_n) \geq \frac{1}{2} c_K > 0 \quad \forall x \in K.$$

Hence, $K \subset \Omega_n$ for sufficiently large n. □

Remark. An analogous result holds for parametric convergence. In the monograph of Delfour and Zolésio [2001], the term "compactivorous property" is used. This property was already put into evidence in Necas [1967].

If the same Γ property is also imposed for the sequence of the (open) complementary sets $\{\overline{E} \setminus \overline{\Omega}_n\}_{n \in \mathbf{N}}$ and for $\overline{E} \setminus \overline{\Omega}$, we obtain yet another concept of set convergence, which we simply denote by $\Omega_n \xrightarrow{c} \Omega$. We have this type of convergence, for instance, whenever both $d_H(\overline{\Omega}_n, \overline{\Omega}) \to 0$ and $d_H(\overline{E} \setminus \Omega_n, \overline{E} \setminus \Omega) \to 0$, as $n \to \infty$.

Let us also note that the Hausdorff–Pompeiu convergence can directly be applied to the boundaries of the considered sets, that is, $d_H(\partial\Omega_n, \partial\Omega) \to 0$ (convergence of the boundaries).

From the above considerations we see that only the Hausdorff–Pompeiu convergence enjoys a compactness property (stated in Proposition A3.2). Moreover, if the various constants appearing in Definition A3.1 are uniformly bounded, then also the corresponding families of domains turn out to be compact:

Theorem A3.9 *Suppose that* $\Omega_n \subset \mathbf{R}^d$, $n \in \mathbf{N}$, *is a sequence of domains of class* C *such that* $\Omega_n \subset E$, $n \in \mathbf{N}$, *for some open and bounded set* E, *and such that* $\lim_{n\to\infty} \tilde{d}_H(\Omega_n, \Omega_0) = 0$ *for some open set* $\Omega_0 \subset E$. *If, in addition,*

$$k_{\Omega_n} \geq k > 0, \quad r_{\Omega_n} \leq r < k, \quad a_{\Omega_n} \geq a > 0, \quad \forall n \in \mathbf{N},$$

and if the family $\mathcal{F} = \bigcup_{n\in\mathbf{N}} \mathcal{F}_{\Omega_n}$ *of all the corresponding families of local charts is equicontinuous and equibounded on* $\overline{\tilde{B}(0,k)}$, *then* Ω_0 *is of class* C *with*

$$k_{\Omega_0} \geq k, \quad r_{\Omega_0} \leq r, \quad a_{\Omega_0} \geq a.$$

Moreover, the associated characteristic functions satisfy $\chi_{\Omega_n} \to \chi_{\Omega_0}$ *a.e. in* E, *for a subsequence again indexed by* n.

Proof. Let $g \in \mathcal{F}$ be any local chart associated with Ω_n. We may assume that it is defined on $\overline{\tilde{B}(0,k)}$, in global coordinates. We recall that the local coordinates are obtained from the global ones by employing the translation by the vector o_g (giving the origin of the local system of coordinates) and some rotation R_g such that $y_g = R_g(0,0,\ldots,1)$, with y_g being the "vertical" unit vector in the local systems of coordinates. Apparently, all these rotations and translations are uniformly bounded transformations with respect to the corresponding matrix or vector norms.

The full representation of the points $(s, g(s))$ on the corresponding piece of $\partial\Omega_n$ in Cartesian coordinates is given by $R_g(\tilde{s},0)+o_g+g(\tilde{s})y_g$. The vectors y_g also give the segment property enjoyed by domains of class C (see Definition A3.1), and we may assume, without loss of generality, that E is so "large" that $\overline{E}$ contains the domains Ω_n, $n \geq 1$, together with all the associated exterior segments.

We use the abbreviation $d_n = d_{\Omega_n}$. We may extract a suitable subsequence, again indexed by n, such that $d_n \to \hat{d}$ strongly in $C(\overline{E})$ for some $\hat{d} \in C(\overline{\Omega})$.

Let

$$\Lambda = \{x \in \overline{E} : \hat{d}(x) \geq 0\},$$

which is a nonempty and closed set. To see this, we choose arbitrary boundary points $x_n \in \partial\Omega_n$, $n \in \mathbf{N}$. Then $d_n(x_n) = 0$ for all $n \in \mathbf{N}$. Since $\{x_n\}$ is bounded, we may without loss of generality assume that $x_n \to \hat{x}$ for some $\hat{x} \in \overline{E}$. Moreover, it follows from the Lipschitz continuity and from the uniform convergence of d_n that $\hat{d}(\hat{x}) = 0$ and $d_n(\hat{x}) \to 0$. Notice that such sequences may be constructed for any $\hat{x} \in \Lambda$ with $\hat{d}(\hat{x}) = 0$, by the definition of d_n.

By Definition A3.1, there is some $g_n \in \mathcal{F}_{\Omega_n}$ such that $x_n = \overline{(s_n, g_n(s_n))}$ is the representation of x_n in the local chart with some $s_n \in \overline{B(0, r_{\Omega_n})}$. By taking the interpretation in global coordinates and our assumptions into account (and using the same notation as in Definition A3.1 and its comments), we have $\tilde{s}_n \to \hat{s}$ for some $\hat{s} \in \overline{\tilde{B}(0,r)}$, as well as $g_n \to \hat{g}$ uniformly in $\overline{\tilde{B}(0,k)}$, with some bounded and continuous function $\hat{g}$ having the same modulus of continuity as the family $\mathcal{F}$. From the uniform convergence of $\{d_n\}$ and $\{g_n\}$, we therefore have

$$\hat{x} = \lim_{n\to\infty} x_n = \lim_{n\to\infty} (s_n, g_n(s_n)) = (\hat{s}, \hat{g}(\hat{s})),$$
$$d_n(s, g_n(s)) \to \hat{d}(s, \hat{g}(s)) = 0 \quad \forall s \in B(0,k).$$

We now make use of the interior segment property. To this end, take any $\varepsilon \in]0, a[$ and consider the points $(s, \hat{g}(s) - \varepsilon) \in \mathbf{R}^d$, where $s \in B(0,k)$. By Definition A3.1, we have that

$$(s, g_n(s) - \varepsilon) \to (s, \hat{g}(s) - \varepsilon), \quad (s, g_n(s) - \varepsilon) \in \Omega_n.$$

Then $d_n(s, g_n(s) - \varepsilon) > 0$ and, consequently, $\hat{d}(s, \hat{g}(s) - \varepsilon) \geq 0$ for all $s \in B(0,k)$ and $\varepsilon \in]0, a[$, that is, we have $(s, \hat{g}(s) - \varepsilon) \in \Lambda$ for such values of s, ε.

For the exterior direction sharper estimates are needed. By virtue of the equicontinuity of g_n, there is some $\delta > 0$ (depending only on $\varepsilon > 0$, and neither on $s \in \tilde{B}(0,k)$ nor on $n \in \mathbf{N}$) such that

$$|g_n(t) - g_n(s)| < \frac{\varepsilon}{2}, \quad \forall n \in \mathbf{N}, \quad \forall t \in B(s,\delta) \cap \tilde{B}(0,k).$$

Then it holds for $\varepsilon < \frac{2}{3} a$ that (in local representation)

$$d[(s, g_n(s) + \varepsilon), \partial\Omega_n] \geq \operatorname{Min}\left\{\frac{\varepsilon}{2}, \delta, a - \frac{3\varepsilon}{2}, d(s, \partial B(0,k))\right\}. \qquad (*)$$

Here, we have used the uniform outside segment property and the equicontinuity of $\{g_n\}$, which can be rephrased by saying that the "cylinder"

$$[B(0,k) \cap B(s,\delta)] \times \left[g_n(s) + \frac{\varepsilon}{2}, g_n(s) + a - \frac{\varepsilon}{2}\right]$$

cannot intersect $\partial\Omega_n$, for any $n \in \mathbf{N}$ and any $s \in B(0,k)$. And the right-hand side in $(*)$ is exactly an estimate from below of the distance between the point $(s, g_n(s) + \varepsilon)$ and the boundary of this cylinder. Notice that this point lies inside the cylinder if $\varepsilon < \frac{2}{3} a$. We conclude that

$$d_n(s, g_n(s) + \varepsilon) \leq -\operatorname{Min}\left\{\frac{\varepsilon}{2}, \delta, a - \frac{3\varepsilon}{2}, d(s, \partial B(0,k))\right\}.$$

The right-hand side in this inequality is independent of n, and we can pass to the limit as $n \to \infty$ to obtain that

$$\hat{d}(s, \hat{g}(s) + \varepsilon) \leq -\operatorname{Min}\left\{\frac{\varepsilon}{2}, \delta, a - \frac{3\varepsilon}{2}, d(s, \partial B(0,k))\right\} < 0,$$

for any $s \in B(0,k)$ and any $\varepsilon \in]0, \frac{2a}{3}[$.

Consequently, $(s, \hat{g}(s)+\varepsilon) \notin \Lambda$ for these values of s, ε. By choosing a smaller $\delta > 0$, if necessary, we can replace

$$\frac{\varepsilon}{2} \text{ by } \frac{\varepsilon}{l}, \qquad \frac{2a}{3} \text{ by } \frac{la}{l+1}, \qquad \frac{3\varepsilon}{2} \text{ by } \frac{(l+1)\varepsilon}{l}, \qquad l \in \mathbf{N},$$

in the above inequalities. Eventually, we get that

$$(s, \hat{g}(s) + \varepsilon) \notin \Lambda \quad \forall s \in B(0,k), \quad \forall \varepsilon \in]0, a[,$$

which ensures the outside segment property.

Clearly, estimates like $(*)$ are also valid for $\hat{d}(s, \hat{g}(s) - \varepsilon)$, $s \in B(0,k)$, and $\varepsilon \in]0, a[$, with the sign reversed. Consequently,

$$\widehat{\Omega} = \{x \in E : \hat{d}(x) > 0\}$$

is a nonempty and open subset of Λ. Clearly,

$$\partial\widehat{\Omega} \subset \{x \in E : \hat{d}(x) = 0\},$$

and we have already shown that this set admits a local representation of the type $(s, \hat{g}(s))$ with mappings $\hat{g}$ that have the same modulus of continuity as the family $\mathcal{F}$. The approximation of the points $(s, \hat{g}(s))$ by the points $(s, g_n(s))$ immediately yields that $\partial\widehat{\Omega} = \{x \in E : \hat{d}(x) = 0\}$ and that the constants $k_{\widehat{\Omega}}$, $r_{\widehat{\Omega}}$, $a_{\widehat{\Omega}}$ satisfy the same inequalities as k_{Ω_n}, r_{Ω_n}, a_{Ω_n}.

Next, we show that $\overline{E} \setminus \widehat{\Omega}$ is the Hausdorff–Pompeiu limit of $\overline{E} \setminus \Omega_n$, i.e., that $\widehat{\Omega} = \Omega_0$. Let $x \in \Omega_0$. Then $\lim_{n\to\infty} d(x, \overline{E} \setminus \Omega_n) > 0$, by the definition of the set convergence, and $\lim_{n\to\infty} d_n(x) = \hat{d}(x) > 0$, that is, $x \in \widehat{\Omega}$.

Conversely, if we assume that $x \in \widehat{\Omega}$ and $x \in \overline{E} \setminus \Omega_0$, then we have $\hat{d}(x) > 0$, and there exist $x_n \in \overline{E} \setminus \Omega_n$, $n \in \mathbf{N}$, such that $x_n \to x$. But then $d_n(x_n) \leq 0$ for all $n \in \mathbf{N}$, and hence $\hat{d}(x) \leq 0$, by the uniform convergence of $\{d_n\}$ to $\hat{d}$ in $\overline{E}$, which contradicts $\hat{d}(x) > 0$. We thus can conclude that $\widehat{\Omega} \cap (\overline{E} \setminus \Omega_0) = \emptyset$, which ends the argument for the claim that $\widehat{\Omega} = \Omega_0$.

Finally, we show the convergence of the characteristic functions. We remark that $\text{meas}(\partial\Omega_n) = 0$, for any $n \geq 0$, since $\partial\Omega_n$ can be represented as a finite union of graphs of continuous functions, by the hypothesis, and in view of what has been already proved.

Now, if $\hat{d}(x) > 0$, then $d_n(x) > 0$ for $n \geq n_1(x)$, and thus $\chi_{\Omega_n}(x) = \chi_{\Omega_0}(x) = 1$. By the same token, if $\hat{d}(x) < 0$, then $\chi_{\Omega_n}(x) = \chi_{\Omega_0}(x) = 0$ for

$n \geq n_2(x)$. Since one of these two cases occurs almost everywhere in E, the assertion is proved at least for the subsequence satisfying $d_n \to \hat{d}$. □

Remark. Theorem A3.9 is due to Tiba [1999]; a more accessible reference is Tiba [2003]. The theorem applies to domains with cusps or with infinitely many oscillations on the boundary having a vanishing amplitude (to preserve the equicontinuity). However, cracks or oscillations that are dense in a set of positive measure are not permitted under the assumptions of Theorem A3.9.

In the two-dimensional case, for boundaries with cusps, the segment direction is uniquely determined by the "axis" of the cusps, in each local chart. Consequently, the hypothesis $r_\Omega \leq r$ means that cusps of different type cannot cluster. The conditions $k_\Omega \geq k$, $a_\Omega \geq a$, guarantee that the local charts cannot shrink, while their number may remain undetermined.

Remark. The counterexample presented in Delfour and Zolésio [2001, p. 256], in connection with the compactness of domains having the segment property, does not fulfill the equicontinuity condition. For the "uniform cusp condition" introduced by the same authors, the compactness proof essentially depends on the fact that the "conical cusps" used there have a nonempty interior, as in the usual uniform cone property; see Chenais [1975] and Pironneau [1984].

Proposition A3.10 *Under the assumptions of Theorem A3.9, we have* $\lim_{n\to\infty} d_H(\overline{\Omega}_n, \overline{\Omega}_0) = 0$.

Proof. We have $d_H(\overline{E} \setminus \Omega_n, \overline{E} \setminus \Omega_0) \to 0$, and in view of Proposition A3.2, we may assume that $d_H(\overline{\Omega}_n, P) \to 0$ for some compact set $P \subset E$, at least for a subsequence again indexed by n. It is also the case that

$$\overline{E} = (\overline{E} \setminus \Omega_n) \cup \overline{\Omega}_n = \lim_{n\to\infty} [(\overline{E} \setminus \Omega_n) \cup \overline{\Omega}_n] = (\overline{E} \setminus \Omega_0) \cup P,$$

which is a simple property of the Hausdorff–Pompeiu convergence. This shows that $\overline{\Omega}_0 \subset P$.

Suppose now that there exists some $\hat{x} \in P \setminus \overline{\Omega}_0$. Then there is some $\lambda > 0$ such that $B(\hat{x}, \lambda) \subset E \setminus \overline{\Omega}_0$. An argument similar to that used in the proof of Theorem A3.9 yields that $B(\hat{x}, \frac{\lambda}{2}) \subset E \setminus \overline{\Omega}_n$ for sufficiently large n, due to the convergence.

But $\hat{x} \in P$ ensures that there are $x_n \in \overline{\Omega}_n$, $n \in \mathbf{N}$, such that $x_n \to \hat{x}$. This contradicts the above statement, and we can conclude that $P \setminus \overline{\Omega}_0 = \emptyset$. Hence the assertion holds at least for the selected subsequence, and since the limit set is uniquely determined, also for the entire sequence. □

Remark. In this case, $\{\Omega_n\}$ is also parametrically convergent to Ω_0 in the sense of Definition A3.3. Indeed, we may choose the signed distance functions $p_n = d_{\Omega_n}$ and $p = d_\Omega$ as parametrizations.

Unfortunately, none of the convergence concepts listed above ensures that the solutions to boundary value problems defined in the open sets Ω_n (for instance, the Dirichlet problem) will converge to the solution of the corresponding boundary value problem defined in Ω, which is the key point in the existence theory of associated optimal design problems. It is necessary to impose some supplementary assumptions (uniform regularity conditions like Lipschitz or certain continuity properties as in Theorem A3.9, or a limitation of the number of the connected components of the complement, or the boundedness of the generalized perimeter; see below and Chapter 2) in order to derive such a conclusion.

One idea, which goes to the essence of the continuity question, is to define the convergence of the domains directly in terms of the convergence of the solutions to the partial differential equations defined on them (compare Sverak [1992], [1993]): a sequence of domains $\Omega_n \subset E$, $n \in \mathbf{N}$, is said to converge to $\Omega \subset E$ if the solutions $y(\Omega_n) \in H_0^1(\Omega_n)$ and $y(\Omega) \in H_0^1(\Omega)$ to the Poisson equation with right-hand side $f \in H^{-1}(E)$ corresponding to these domains satisfy

$$\tilde{y}(\Omega_n) \to \tilde{y}(\Omega) \text{ weakly in } H^1(E) \quad \forall f \in H^{-1}(E).$$

Here, the superimposed "~" denotes the extension by zero to the whole domain E.

This condition has a minimal character, since the weak convergence in $H^1(E)$ is indispensable in order to guarantee the weak lower semicontinuity of certain cost functionals that may be taken into account. Similar ideas may be used in the case of other differential operators or boundary conditions together with appropriate extension techniques (for instance, the Calderon extension, Adams [1975]).

However, such a definition is very implicit, and only in the case of the Dirichlet boundary value problem in dimension two and of the zero extension technique has Sverak [1993] clarified its relationship with effective convergence conditions. We redenote by $y_n(f)$, $\tilde{y}_n(f)$, $y(f)$, and $\tilde{y}(f)$ the already introduced solutions to Poisson's equation, in order to put into evidence their dependence on f. The first observation in this direction is the following:

Proposition A3.11 *Let* $\mathbf{1}$ *denote the mapping defined by* $\mathbf{1}(x) = 1$ *for all* $x \in \overline{E}$. *If* $\tilde{y}_n(\mathbf{1}) \to \tilde{y}(\mathbf{1})$ *strongly in* $H_0^1(E)$, *then* $\tilde{y}_n(f) \to \tilde{y}(f)$ *strongly in* $H_0^1(E)$, *for any* $f \in H^{-1}(E)$.

Moreover, $H_0^1(\Omega)$ can be canonically identified, via the extension by zero, with a closed subspace of $H_0^1(E)$, and we may consider the associated orthogonal projection operators $P_\Omega : H_0^1(E) \to H_0^1(\Omega)$. Clearly,

$$y(\Omega) = P_\Omega \, \tilde{y}(\Omega).$$

The convergence $\Omega_n \to \Omega$ as defined above is thus nothing but the convergence $P_{\Omega_n} \to P_\Omega$ in the weak operator topology of $H_0^1(E)$ or, equivalently, in the strong operator topology (since P_{Ω_n} and P_Ω are orthogonal projections).

Although the projection operators are bounded, the compactness question related to this notion of convergence remains difficult, since in general, the weak limit of a sequence of orthogonal projections may not be a projection again.

Another related concept of set distance is obtained by assigning to each open $\Omega \subset E$ the mapping (which is different from the signed distance function)

$$\tilde{d}_\Omega(x) = d(x, \overline{E} \setminus \Omega), \quad x \in \overline{E}.$$

The family of open subsets of $\overline{E}$ is a compact metric space equipped with the distance

$$\mu(\Omega_1, \Omega_2) = \left|\tilde{d}_{\Omega_1} - \tilde{d}_{\Omega_2}\right|_{C(\overline{E})}.$$

The topology generated by μ is the same as the one generated by the Hausdorff–Pompeiu distance, via the mapping $\Omega \mapsto \overline{E} \setminus \Omega$. Notice that this concept does not enter the setting of parametric convergence as defined in Definition A3.3, and Proposition A3.5 does not apply.

In the one-dimensional case in which $E \subset \mathbf{R}$, the convergence $\tilde{d}_{\Omega_n} \to \tilde{d}_\Omega$ in $C(\overline{E})$ implies that $P_{\Omega_n} \to P_\Omega$. This follows easily since the sequence $\{P_{\Omega_n} y\}$ is bounded in $H_0^1(E)$ for every $y \in H_0^1(E)$, and we can assume that $P_{\Omega_n} y \to v$ weakly in $H_0^1(E)$. Since $E \subset \mathbf{R}$, we get that $P_{\Omega_n} y \to v$ in $C(\overline{E})$. Then, clearly, $v \in H_0^1(\Omega)$, and $v = P_\Omega y$. That is, $P_{\Omega_n} \to P_\Omega$ in the weak operator topology of $H_0^1(E)$, or equivalently, in the strong operator topology, as claimed. The result remains valid in the two-dimensional case, but the argument is no longer elementary (cf. Sverak [1993]).

Let us now consider the Dirichlet problem defined in the domains Ω_n,

$$\Delta y_n = f \quad \text{in } \Omega_n, \qquad y_n = 0 \quad \text{on } \partial\Omega_n,$$

for some given $f \in H^{-1}(E)$, where E is some open domain containing all the sets Ω_n, $n \in \mathbf{N}$, and Ω. Clearly, the above Dirichlet problem has a unique weak solution $y_n \in H_0^1(\Omega_n)$ (compare Appendix 2, Example A2.6). The corresponding weak solution in the domain Ω is denoted by $y \in H_0^1(\Omega)$. We also denote by $\tilde{y}_n$ and $\tilde{y}$ the extensions by zero of y_n and y, respectively, to the domain E.

We list some continuity-type results from the literature for the Dirichlet problem. A detailed discussion including other boundary conditions is provided in Chapter 2.

Theorem A3.12 *Assume that $\Omega_n \xrightarrow{c} \Omega$ and that Ω is Lipschitz. Then $\tilde{y}_n \to \tilde{y}$ strongly in $H_0^1(E)$.*

Remark. A minimal regularity assumption for the previous result to hold is that Ω is *stable*, i.e., it is Carathéodory, and that any function $v \in H_0^1(E)$

vanishing quasi-almost everywhere in $E \setminus \overline{\Omega}$ satisfies $v \in H_0^1(\Omega)$. It should be noted that it follows directly from the variational formulation of the Dirichlet problem that $\{\tilde{y}_n\}$ is bounded in $H_0^1(E)$, and that $\tilde{y}_n \to \hat{y}$ weakly in $H_0^1(E)$, where $\hat{y}$ satisfies a similar variational equation. The difficult point, which necessitates the introduction of various regularity hypotheses, is to show that $\hat{y}|_\Omega \in H_0^1(\Omega)$, that is, $\hat{y} = \tilde{y}$.

Theorem A3.13 (Sverak [1993]) *Let $\Omega_n \subset \mathbf{R}^2$, and assume that $\lim_{n\to\infty} \tilde{d}_H(\Omega_n, \Omega) = 0$, and that $E \setminus \Omega_n$ has at most l connected components, for all $n \in \mathbf{N}$. Then $E \setminus \Omega$ has at most l connected components, and $\tilde{y}_n \to \tilde{y}$ in $H_0^1(E)$.*

Theorem A3.14 (Henrot [1994]) *Assume that $\lim_{n\to\infty} \tilde{d}_H(\Omega_n, \Omega) = 0$, and suppose that the gradients $\{\nabla y_n\}$ are uniformly bounded in a neighborhood of $\partial\Omega$. Then $\tilde{y}_n \to \tilde{y}$ in $H_0^1(E)$.*

Another essential point in the discussion of optimization problems in a variable domains setting is given by the lower semicontinuity properties of the corresponding cost functionals. We have the following result.

Theorem A3.15 *Let $l : \mathbf{R}^d \times \mathbf{R} \times \mathbf{R}^d \to \mathbf{R}$ be nonnegative and measurable, let $l(x,\cdot,\cdot)$ be continuous on $\mathbf{R} \times \mathbf{R}^d$, and let $l(x,s,\cdot)$ be convex. If $\Omega =$p-$\lim_{n\to\infty}\Omega_n$, and if $y_n \in H^1(\Omega_n)$ and $y \in H^1(\Omega)$ satisfy that $\{|y_n|_{H^1(\Omega_n)}\}$ is bounded and $y_n|_K \to y|_K$ weakly in $H^1(K)$, for any domain K such that $\overline{K} \subset \Omega$, then*

$$\int_\Omega l(x, y(x), \nabla y(x))\, dx \le \liminf_{n\to\infty} \int_{\Omega_n} l(x, y_n(x), \nabla y_n(x))\, dx.$$

Proof. Take an increasing sequence of open sets G_j such that $\overline{G}_j \subset \Omega$ for all $j \in \mathbf{N}$ and $\cup_{j=1}^\infty \overline{G}_j = \Omega$. Then $\chi_{G_j} \to \chi_\Omega$ almost everywhere in E. For any fixed G_j, we have $y_n \to y$ weakly in $H^1(G_j)$, and

$$\int_{G_j} l(x, y(x), \nabla y(x))\, dx \le \liminf_{n\to\infty} \int_{G_j} l(x, y_n(x), \nabla y_n(x))\, dx,$$

since weak lower semicontinuity is a well-known property of convex integrands in fixed domains (cf. Proposition A1.1 in Appendix 1). Next, it follows from Fatou's lemma that

$$\begin{aligned}
\int_\Omega l(x, y(x), \nabla y(x))\, dx &= \int_\Omega \lim_{j\to\infty} \chi_{G_j}(x)\, l(x, y(x), \nabla y(x))\, dx \\
&\le \liminf_{j\to\infty} \int_{G_j} l(x, y(x), \nabla y(x))\, dx \\
&\le \liminf_{j\to\infty} \liminf_{n\to\infty} \int_{\Omega_n} l(x, y_n(x), \nabla y_n(x))\, dx
\end{aligned}$$

$$= \liminf_{n\to\infty} \int_{\Omega_n} l(x, y_n(x), \nabla y_n(x))\, dx.$$

Here, the nonnegativity of l is essential. The assertion is proved. □

In Chapter 2 a detailed treatment of continuity properties and of the existence of solutions in shape optimization problems under weak regularity assumptions may be found.

Bibliography

1. Adams, R. (1975): *Sobolev Spaces.* New York: Academic Press.

2. Ahmed, N. U., Teo, K. L. (1981): *Optimal Control of Distributed Parameter Systems.* Amsterdam: North-Holland.

3. Allaire, G. (2001): *Shape Optimization by the Homogenization Method.* New York: Springer-Verlag.

4. Alt, H. W. (1985): *Lineare Funktionalanalysis.* Berlin: Springer-Verlag.

5. Alt, H. W., Caffarelli, L. A. (1981): Existence and regularity for a minimum problem with free boundary. *J. Reine Angew. Math.* **325**, 105–144.

6. Antman, St. S. (1995): *Nonlinear Problems of Elasticity.* Berlin: Springer-Verlag.

7. Arnăutu, V. (1981/82): Approximation of optimal distributed control problems governed by variational inequalities. *Numer. Math.* **38 (3)**, 393–416.

8. Arnăutu, V. (2001): *Numerical Methods for Variational Problems.* Lecture Notes **8**, Dept. of Math. Inform. Tech., Univ. of Jyväskylä, Jyväskylä, Finland.

9. Arnăutu, V., Langmach, H., Sprekels, J., Tiba, D. (2000): On the approximation and the optimization of plates. *Numer. Funct. Anal. Optim.* **21 (3–4)**, 337–354.

10. Arnăutu, V., Neittaanmäki, P. (1998): Discretization estimates for an elliptic control problem. *Numer. Funct. Anal. Optim.* **19**, 431–464.

11. Arnăutu, V., Neittaanmäki, P. (2003): *Optimal Control from Theory to Computer Programs.* Solid Mechanics and Its Applications Vol. **III**. Dordrecht, Boston, London: Kluwer Academic Publishers.

12. Arnăutu, V., Sprekels, J., Tiba, D. (2003): Optimization problems for curved mechanical structures. *Preprint Series of the Weierstrass Institute for Applied Analysis and Stochastics*, No. **812**, Berlin, Germany. *SIAM J. Control Optim.*, to appear.

13. Arunakirinathar, K., Reddy, B. D. (1993): Mixed finite element methods for elastic rods of arbitrary geometry. *Numer. Math.* **64**, 13–43.

14. Astrakmantsev, G. P. (1978): Methods of fictitious domains of a second order elliptic equation with natural boundary conditions. *U.S.S.R. Comp. Math. and Math. Phys.* **18**, 114–121.

15. Atamian, C. (1991): *Thèse, Université de Paris VI*, Paris, France.

16. Atamian, C., Joly, P. (1993): Une analyse de la méthode des domaines fictifs pour le problème de Helmholtz extérieur. *RAIRO Modél. Math. Anal. Numér.* **27 (3)**, 251–288.

17. Attouch, H. (1984): *Variational Convergence for Functions and Operators.* Appl. Math. Ser. Boston, MA: Pitman

18. Azé, D. (1997): *Éléments d'Analyse Convexe et Variationnelle.* Paris: Ellipses.

19. Badea, L., Daripa, P. (2001): On a boundary control approach to domain embedding methods. *SIAM J. Control Optim.* **40 (2)**, 421–449.

20. Bahri, A. (1980): Résolution générique d'une équation semi-linéaire. *C. R. Acad. Sci. Paris Sér. A–B* **291 (4)**, A251–A254.

21. Bahri, A. (1981): Topological results on a certain class of functionals and application. *J. Funct. Anal.* **41 (3)**, 397–427.

22. Baiocchi, C. (1972): Su un problema a frontiera libera conesso a questioni di idraulica. *Ann. Mat. Pura Appl.* **92**, 107–127.

23. Banichuk, N. V. (1983): *Problems and Methods of Optimal Structural Design.* Mathematical Concepts and Methods in Science and Engineering Vol. **26**. New York: Plenum Press.

24. Barbu, V. (1984): *Optimal Control of Variational Inequalities.* Research Notes in Mathematics Vol. **100**. London: Pitman

25. Barbu, V. (1993): *Analysis and Control of Nonlinear Infinite Dimensional Systems.* Mathematics in Science and Engineering. Boston, MA: Academic Press.

26. Barbu, V., Friedman, A. (1991): Optimal design of domains with free-boundary problems. *SIAM J. Control Optim.* **29 (3)**, 623–637.

27. Barbu, V., Precupanu, Th. (1986): *Convexity and Optimization in Banach Spaces.* 2nd ed. Dordrecht: D. Reidel.

28. Barbu, V., Tiba, D. (1990): Optimal control of abstract variational inequalities. In: *Control of Distributed Parameter Systems* (M. Amouroux, A. El Jai, eds.), 274–277. Oxford: Pergamon Press.

29. Becker, R., Kapp, H., Rannacher, R. (2000): Adaptive finite element methods for optimal control of partial differential equations: Basic concept. *SIAM J. Control Optim.* **39 (1)**, 113–132.

30. Beckert, H. (1960): Eine bemerkswerte Eigenschaft der Lösungen des Dirichletschen Problems bei linearen elliptischen Differentialgleichungen. *Math. Ann.* **139**, 255–264.

31. Bedivan, D. M. (1996): A boundary functional for optimal shape design problems. *Appl. Math. Lett.* **9 (1)**, 9–14.

32. Bedivan, D. M. (1997): Existence of a solution for complete least squares optimal shape problems. *Numer. Funct. Anal. Optim.* **18 (5–6)**, 495–505.

33. Bedivan, D. M., Fix, G. J. (1995): Least squares methods for optimal shape design problems. *Comput. Math. Appl.* **30 (2)**, 17–25.

34. Bellman, R. (1957): *Dynamic Programming.* Princeton, NJ: Princeton University Press.

35. Bendsøe, M. (1984): Existence proofs for a class of plate optimization problems. In: *System Modelling and Optimization* (Thoft-Christensen, P., ed.), 773–779. Lecture Notes in Control and Inform. Sci. Vol. **59**. Berlin: Springer-Verlag.

36. Bendsøe, M. (1995): *Optimization of Structural Topology, Shape, and Material.* Berlin: Springer-Verlag.

37. Bendsøe, M., Kikuchi, N. (1988): Generating optimal topologies in structural design using a homogenization method. *Comput. Methods Appl. Mech. Engrg.* **71**, 197–224.

38. Benedict, B., Sokołowski, J., Zolésio, J.-P. (1984): Shape optimization for contact problems. In: *System Modelling and Optimization* (Thoft-Christensen, P., ed.), 790–799. Lecture Notes in Control and Inform. Sci. Vol. **59**. Berlin: Springer-Verlag.

39. Bénilan, Ph. (1972): Solutions intégrales d'équations d'évolution dans un espace de Banach. *C. R. Acad. Sci. Paris Sér. A–B* **274**, A47–A50.

40. Beretta, E., Vogelius, M. (1991): An inverse problem originating from magnetohydrodynamics. *Arch. Ration. Mech. Anal.* **115 (2)**, 137–152.

41. Beretta, E., Vogelius, M. (1992): An inverse problem originating from magnetohydrodynamics II. The case of the Grad–Shafranov equation. *Indiana Univ. Math. J.* **41 (4)**, 1081–1118.

42. Bergounioux, M., Tiba, D. (1996): General optimality conditions for constrained convex control problems. *SIAM J. Control Optim.* **34 (2)**, 698–711.

43. Blouza, A. (1997): Existence et unicité pour le modèle de Naghdi pour une coque peu régulière. *C. R. Acad. Sci. Paris Sér. I Math.* **324**, 839–844.

44. Blum, J. (1989): *Numerical Simulation and Optimal Control in Plasma Physics.* Chichester: J. Wiley & Sons and Montrouge: Gauthier-Villars.

45. Bonnans, J. F. (1982): *Thèse, Université de Technologie de Compiègne*, Compiègne, France.

46. Bonnans, J. F. (1998): Second order analysis for control constrained optimal control problems of semilinear elliptic systems. *Appl. Math. Optim.* **38 (3)**, 303–325.

47. Bonnans, J. F., Casas, E. (1989): Optimal control of semilinear multistate system with state constraints. *SIAM J. Control Optim.* **27 (2)**, 446–455.

48. Bonnans, J. F., Casas, E. (1991): Un principe de Pontryagine pour le contrôle des systèmes semilinéaires elliptiques. *J. Differential Equations* **90 (2)**, 288–303.

49. Bonnans, J. F., Casas, E. (1992): Some stability concepts and their applications in optimal control problems. *IMA Preprint Series, University of Minnesota*, No. **1081**, Minneapolis, USA.

50. Bonnans, J. F., Tiba, D. (1991): Pontryagin's principle in the control of semilinear elliptic variational inequalities. *Appl. Math. Optim.* **23 (3)**, 299–312.

51. Borrvall, T., Petersson, J. (2001): Topology optimization using regularized intermediate density control. *Comput. Methods Appl. Mech. Engrg.* **190**, 4911–4928.

52. Brézis, H. (1973): *Opérateurs Maximaux Monotones et Semi-Groupes de Contractions dans les Espaces de Hilbert.* North-Holland Mathematics Studies No. **5**. Amsterdam: North-Holland.

53. Brézis, H. (1983): *Analyse Fonctionnelle. Théorie et Applications.* Collection Mathématiques Appliquées pour la Maîtrise. Paris: Masson.

54. Brézis, H., Sibony, M. (1971): Équivalence de deux inéquations variationnelles et applications. *Arch. Ration. Mech. Anal.* **41**, 254–265.

55. Brézis, H., Strauss, W. A. (1973): Semilinear second order elliptic equations in L^1. *J. Math. Soc. Japan* **25**, 565–590.

56. Bucur, D., Varchon, N. (2000): Stabilité de la solution d'un problème de Neumann pour des variations de frontière. *C. R. Acad. Sci. Paris Sér. I Math.* **331 (5)**, 371–374.

57. Bucur, D., Zolésio, J. P. (1994a): Continuité par rapport au domaine dans le problème de Neumann. *C. R. Acad. Sci. Paris Sér. I Math.* **319 (1)**, 57–60.

58. Bucur, D., Zolésio, J. P. (1994b): Optimisation de forme sous contrainte capacitaire. *C. R. Acad. Sci. Paris Sér. I Math.* **318 (9)**, 795–800.

59. Bucur, D., Zolésio, J. P. (1996): Free boundary problems and density perimeter. *J. Differential Equations* **126**, 224–243.

60. Buttazzo, G. (1989): *Semicontinuity, Relaxation and Integral Representation in the Calculus of Variations.* Pitman Research Notes in Mathematics Series Vol. **207**. Harlow: Longman and New York: J. Wiley & Sons.

61. Buttazzo, G., Dal Maso, G. (1991): Shape optimization for Dirichlet problems: Relaxed formulation and optimality conditions. *Appl. Math. Optim.* **23**, 17–49.

62. Cârjă, O. (1988): On constraint controllability for linear systems in Banach spaces. *J. Optim. Theory Appl.* **56 (2)**, 215–225.

63. Cârjă, O. (1991): Constraint controllability for linear control systems. *Ann. Mat. Pura Appl. (4)* **158**, 13–32.

64. Cartan, H. (1967): *Formes Différentielles.* Paris: Hermann.

65. Casas, E. (1990): Optimality conditions and numerical approximations for some optimal design problems. *Control Cybernet.* **19 (3–4)**, 73–91.

66. Casas, E. (1992): Optimal control in coefficients of elliptic equations with state constraints. *Appl. Math. Optim.* **26 (1)**, 21–37.

67. Casas, E. (2002): Error estimates for the numerical approximation of semilinear elliptic control problems with finitely many state constraints. A tribute to J. L. Lions. *ESAIM Control Optim. Calc. Var.* **8**, 345–374.

68. Casas, E., Fernandez, L. (1991): Optimal control of quasilinear elliptic equations with nondifferentiable coefficients at the origin. *Rev. Mat. Complut.* **4 (2–3)**, 227–250.

69. Casas, E., Fernandez, L. (1993): Optimal control of semilinear elliptic equations with pointwise constraints on the gradient of the state. *Appl. Math. Optim.* **27 (1)**, 35–56.

70. Casas, E., Mateos, M. (2002a): Second order optimality conditions for semilinear elliptic control problems with finitely many state constraints. *SIAM J. Control Optim.* **40**, 1431–1454.

71. Casas, E., Mateos, M. (2002b): Uniform convergence of the FEM applications to state constrained control problems. *Comput. Appl. Math.* **21**

72. Casas, E., Tröltzsch, F. (1999): Second-order necessary optimality conditions for some state-constrained control problems of semilinear elliptic equations. *Appl. Math. Optim.* **39**, 211–227.

73. Casas, E., Tröltzsch, F. (2002): Error estimates for the finite element approximation of a semilinear elliptic control problem. *Control Cybernet.* **31 (3)**, 695–712.

74. Casas, E., Tröltzsch, F. (2003): Error estimates for linear-quadratic elliptic control problems. In: *Analysis and Optimization of Differential Systems* (Barbu, V., Lasiecka, I., Tiba, D., Varsan C., eds.), 89–100. Norwell, MA: Kluwer.

75. Castaing, C., Valadier, M. (1977): *Convex Analysis and Measurable Multifunctions.* Lecture Notes in Math. Vol. **580**. Berlin: Springer-Verlag.

76. Céa, J. (1971): *Optimisation. Théorie et Algorithmes.* Paris: Dunod.

77. Céa, J. (1986): Conception optimale ou identification de formes. Calcul rapide de la dérivée directionnelle de la fonction coût. *M^2AN* **20 (3)**, 371–402.

78. Céa, J., Garreau, St., Guillaume, Ph., Masmoudi, M. (2000): The shape and topological optimizations connection. *Comput. Methods Appl. Mech. Engrg.* **188**, 713–726.

79. Céa, J., Gioan, A., Michel, J. (1973): Quelques résultats sur l'identification de domaines. *Calcolo*, 207–232.

80. Céa, J., Malanowski, K. (1970): An example of a max-min problem in partial differential equations. *SIAM J. Control Optim.* **8 (3)**, 305–316.

81. Cesari, L. (1983): *Optimization – Theory and Applications.* New York: Springer-Verlag.

82. Cesari, L. (1990): Existence of BV absolute discontinuous minima for modified multiple integrals of the calculus of variations. *Rend. Circ. Mat. Palermo (2)* **39**, 169–209.

83. Chambolle, A., Doveri, F. (1997): Continuity of Neumann linear elliptic problems in varying two-dimensional bounded open sets. *Comm. Partial Differential Equations* **22 (5&6)**, 811–840.

84. Chan, T. F., Vese, L. A. (1997): Variational image restoration and segmentation model and approximation. *UCLA CAM Report* **97-47**.

85. Chan, T. F., Vese, L. A. (2001): Active contours without edges. *IEEE Trans. Image Process.* **10**, 266–276.

86. Chapelle, D. (1997): A locking-free approximation of curved rods by straight beam elements. *Numer. Math.* **77**, 299–322.

87. Chenais, D. (1975): On the existence of a solution in a domain identification problem. *J. Math. Anal. Appl.* **52 (2)**, 189–219.

88. Chenais, D., Paumier, J. C. (1994): On the locking phenomenon for a class of elliptic problems. *Numer. Math.* **67**, 427–440.

89. Ciarlet, Ph. G. (1978): *The Finite Element Method for Elliptic Problems.* Studies in Mathematics and Its Applications Vol. **4**. Amsterdam: North-Holland.

90. Ciarlet, Ph. G. (1982): *Introduction à l'Analyse Numérique Matricielle et à l'Optimisation.* Collection of Applied Mathematics for the Master's Degree. Paris: Masson.

91. Ciarlet, Ph. G. (2000): *Mathematical Elasticity.* Vol. III. Theory of Shells. Amsterdam: North-Holland.

92. Ciarlet, Ph. G., Raviart, P. A. (1972): Interpolation theory over curved elements, with applications to finite element methods. *Comput. Methods Appl. Mech. Engrg.* **1**, 217–249.

93. Cioranescu, D., Donato, P. (1999): *An Introduction to Homogenization.* Oxford Lecture Series in Mathematics and Its Applications Vol. **17**. New York: The Clarendon Press, Oxford University Press.

94. Cioranescu, D., Murat, F. (1982): Un terme étrange venu d'ailleurs. II. In: *Nonlinear Partial Differential Equations and Their Applications* (Brézis, H., Lions J. L., eds.), 154–178. Collège de France Seminar Vol. **III** (Paris, 1980/1981), Res. Notes in Math. Vol. **70**. Boston, MA: Pitman.

95. Clarke, F. H. (1983): *Optimization and Nonsmooth Analysis.* Canadian Mathematical Society Series of Monographs and Advanced Texts. New York: J. Wiley & Sons.

96. Cooper, J. (1998): *Introduction to Partial Differential Equations with MATLAB.* Boston, MA: Birkhäuser Boston, Inc.

97. Courant, R. (1950): *Dirichlet's Principle, Conformal Mappings and Minimal Surfaces.* New York: Interscience.

98. Courant, R., Hilbert, D. (1962): *Methods of Mathematical Physics*, Vol. **1**. New York: Interscience.

99. Crouzeix, M., Rappaz, J. (1990): *On Numerical Approximation in Bifurcation Theory.* Paris: Masson and Berlin: Springer-Verlag.

100. Daniljuk, I. I. (1971): Sur une classe de fonctionnelles intégrales à domaine variable d'intégration. In: *Actes du Congrès International des Mathématiciens (Nice, 1970)*, Tome **2**, 703–715. Paris: Gauthier-Villars.

101. Dautray, R., Lions, J.-L. (1988): *Mathematical Analysis and Numerical Methods for Science and Technology*, Vol. **2**, *Functional and Variational Methods.* Berlin: Springer-Verlag.

102. Davideanu, C. (1989): A complete solving of a control problem governed by a scalar variational inequality. In: *Proceedings of the Itinerant Seminar on Functional Equations, Approximation and Convexity (Cluj-Napoca, 1989).* University of Cluj-Napoca, 125–130.

103. Delfour, M., Payre, G., Zolésio, J.-P. (1982/83): Optimal design of a minimum weight thermal diffuser with constraint on the output thermal power flux. *Appl. Math. Optim.* **9 (3)**, 225–262.

104. Delfour, M. C., Zolésio, J.-P. (1991a): Anatomy of the shape Hessian. *Ann. Mat. Pura Appl. (4)* **159**, 315–339.

105. Delfour, M. C., Zolésio, J.-P. (1991b): Velocity method and Lagrangian formulation for the computation of the shape Hessian. *SIAM J. Control Optim.* **29 (6)**, 1414–1442.

106. Delfour, M. C., Zolésio, J.-P. (2001): *Shapes and Geometries: Analysis, Differential Calculus and Optimization.* SIAM Series on Advances in Design and Control. Philadelphia, PA

107. Duff, I. S., Erisman, A. M., Reid, J. K. (1987): *Direct Methods for Sparse Matrices.* Oxford: Clarendon Press.

108. Ekeland, I. (1979): Nonconvex minimization problems. *Bull. Amer. Math. Soc. (N.S.)* **1 (3)**, 443–474.

109. Ekeland, I., Temam, R. (1974): *Analyse Convexe et Problèmes Variationnelles.* Paris: Dunod and Gauthier-Villars.

110. Elliott, C. M., Ockendon, J. R. (1982): *Weak and Variational Methods for Moving Boundary Problems.* Research Notes in Mathematics Vol. **59**. London: Pitman.

111. Engels, H. (1980): *Numerical Quadrature and Cubature.* London: Academic Press.

112. Eppler, K. (2000): Second derivatives and sufficient optimality conditions for shape functionals. *Control Cybernet.* **29 (2)**, 485–511.

113. Falk, R. S. (1990): Approximation of inverse problems. In: *Inverse Problems in Partial Differential Equations* (Colton, D., Ewing, R., Rundell, W., eds.), 7–16. Philadelphia, PA: SIAM.

114. French, D. A., King, J. T. (1991): Approximation of an elliptic control problem by the finite element method. *Numer. Funct. Anal. Optim.* **12 (3–4)**, 299–314.

115. Friedman, A. (1982): *Foundations of Modern Analysis.* Reprint of the 1970 original. New York: Dover Publications.

116. Fujii, N. (1986): Necessary conditions for a domain optimization problem in elliptic boundary value problems. *SIAM J. Control Optim.* **24 (3)**, 346–360.

117. Gamkrelidze, R. (1975): *Fundamentals of Optimal Control* (in Russian). Tbilisi, Georgia: Tbilisi University Press.

118. Garabedian, P. R., Spencer, D. (1952): Extremal methods in cavitational flows. *J. Rational Mech. Anal.* **1**, 359–409.

119. Garreau, S., Guillaume, Ph., Masmoudi, M. (2001): The topological asymptotic for PDE systems: The elasticity case. *SIAM J. Control Optim.* **39 (6)**, 1756–1778.

120. George, P. L. (1991): *Automatic Mesh Generation. Application to Finite Element Methods.* Chichester: J. Wiley & Sons and Paris: Masson.

121. Geymonat, G., Gilardi, G. (1998): Contre-exemples à l'inégalité de Korn et au lemme de Lions dans les domaines irréguliers. In: *Equations aux dérivées partielles. Articles dédiés à J. L. Lions*, 541–548. Paris: Gauthier-Villars.

122. Geymonat, G., Sanchez-Palencia, E. (1995): On the rigidity of certain surfaces with folds and applications to shell theory. *Arch. Ration. Mech. Anal.* **129**, 11–45.

123. Gilbarg, O., Trudinger, N. (1983): *Elliptic Partial Differential Equations of Second Order*, 2nd ed. Grundlehren der Mathematischen Wissenschaften Vol. **224**. Berlin: Springer-Verlag.

124. Gonzalez de Paz, R. B. (1982): Sur un problème d'optimisation de domaine. *Numer. Funct. Anal. Optim.* **5** (**2**), 173–197.

125. Gonzalez de Paz, R. B. (1989): On the optimal design of elastic shafts. *RAIRO Modél. Math. Anal. Numér.* **23** (**4**), 615–625.

126. Gonzalez de Paz, R. B. (1994): A relaxation approach applied to domain optimization. *SIAM J. Control Optim.* **32** (**1**), 154–169.

127. Göpfert, A. (1971): Eine Anwendung des Unitätssatzes von Itô–Yamabe. *Beiträge zur Analysis* **1**, 29–41.

128. Grisvard, P. (1985): *Elliptic Problems in Nonsmooth Domains.* Monographs and Studies in Mathematics Vol. **24**. London: Pitman.

129. Gröger, K., Rehberg, J. (1989): Resolvent estimates in $W^{-1,p}$ for second order elliptic differential operators in case of mixed boundary conditions. *Math. Ann.* **285** (**1**), 105–113.

130. Gruver, W. A., Sachs, E. (1981): *Algorithmic Methods in Optimal Control.* Res. Notes in Math. Vol. **47**. London: Pitman.

131. Guillaume, Ph. (1996): Intrinsic expression of the derivatives in domain optimization problems. *Numer. Funct. Anal. Optim.* **17** (**1–2**), 93–112.

132. Guillaume, Ph., Idris, K. S. (2002): The topological asymptotic expansion for the Dirichlet problem. *SIAM J. Control Optim.* **41** (**4**), 1042–1072.

133. Guillaume, Ph., Masmoudi, M. (1994): Computation of high order derivatives in optimal shape design. *Numer. Math.* **67** (**2**), 231–250.

134. Hadamard, J. (1968): Mémoire sur un problème d'analyse relatif à l'équilibre des plaques élastiques encastrées. In: *Oeuvres de J. Hadamard*, 515–641. Paris: CNRS.

135. Haftka, R., Kamat, P. M., Gürdal, Z. (1990): *Elements of Structural Optimization.* Boston: Kluwer Academic Press.

136. Haslinger, J., Hoffmann, K.-H., Kocvara, M. (1993): Control/fictitious domain method for solving optimal shape design problems. *RAIRO Modél. Math. Anal. Numér.* **27** (**2**), 157–182.

137. Haslinger, J., Makinen, R. A. E. (2003): *Introduction to Shape Optimization. Theory, Approximation and Computation.* Philadelphia, PA: SIAM.

138. Haslinger, J., Neittaanmäki, P. (1988): *Finite Element Approximation of Optimal Shape Design.* New York: J. Wiley & Sons.

139. Haslinger, J., Neittaanmäki, P. (1996): *Finite Element Approximation for Optimal Shape, Material and Topology Design*, 2nd ed. Chichester: J. Wiley & Sons.

140. Haslinger, J., Neittaanmäki, P., Tiba, D. (1987): On state constrained optimal shape design problems. In: *Optimal Control of Partial Differential Equations II* (Hoffman, K. H., Krabs, W., eds.), 109–122. ISNM Vol. **78**. Basel: Birkhäuser.

141. Hausdorff, F. (1914): *Grundzüge der Mengenlehre.* Leipzig: Veit.

142. Hausdorff, F. (1927): *Mengenlehre.* Berlin: Walter de Gruyter.

143. He, Z. X. (1987): State constrained control problems governed by variational inequalities. *SIAM J. Control Optim.* **25 (5)**, 1119–1144.

144. Heinonen, J., Kilpeläinen, T., Martio, O. (1993): *Nonlinear Potential Theory of Degenerate Elliptic Equations.* Oxford: Oxford University Press.

145. Henrot, A. (1994): Continuity with respect to the domain for the Laplacian: a survey. Shape design and optimization. *Control Cybernet.* **23**, 427–443.

146. Henry, J. (1978): *Thèse, Université de Paris VI,* Paris, France.

147. Herwig, A. (1989): Elliptische Randwertprobleme zweiter Ordnung in Gebieten mit einer Fehlstelle. *Z. Anal. Anwendungen* **8**, 153–161.

148. Hewitt, E., Stromberg, K. (1965): *Real and Abstract Analysis.* New York: Springer-Verlag.

149. Hiriart–Urruty, J.-B., Lemaréchal, Cl. (1993): *Convex Analysis and Minimization Algorithms.* Berlin: Springer-Verlag.

150. Hlaváček, I., Bock, I., Lovíšek, J. (1984): Optimal control of a variational inequality with applications to structural analysis. I. Optimal design of a beam with unilateral supports. *Appl. Math. Optim.* **11**, 111–143.

151. Hlaváček, I., Bock, I., Lovíšek, J. (1985): Optimal control of a variational inequality with applications to structural analysis. II. Local optimization of the stress in a beam. III. Optimal design of an elastic plate. *Appl. Math. Optim.* **13 (2)**, 117–136.

152. Hoffmann, K.-H., Tiba, D. (1995): Fixed domain methods in variable domain problems. In: *Free Boundary Problems: Theory and Applications* (Diaz, J. I., ed.), 123–146. Pitman Res. Notes Math. Ser. Vol. **323**. Harlow: Longman Sci. Tech.

153. Hörmander, L. (1963): *Linear Partial Differential Operators.* Die Grundlehren der Mathematischen Wissenschaften Vol. **116**. New York: Academic Press and Berlin: Springer-Verlag.

154. Ignat, A., Sprekels, J., Tiba, D. (2001): Analysis and optimization of nonsmooth arches. *SIAM J. Control Optim.* **40**, 1107–1135.

155. Ignat, A., Sprekels, J., Tiba, D. (2002): A model of a general elastic curved rod. *Math. Methods Appl. Sci.* **25 (10)**, 835–854.

156. Ioffe, A. D., Tikhomirov, W. M. (1974): *Theory of Extremal Problems* (in Russian). Moscow: Nauka.

157. Kardestuncer, H., Norrie, D. H. (eds.) (1987): *Finite Element Handbook.* New York: McGraw-Hill.

158. Kärkkäinen, K., Tiihonen, T. (1999): Free surfaces: shape sensitivity analysis and numerical methods. *Internat. J. Numer. Methods Engrg.* **44 (8)**, 1079–1098.

159. Kawarada, H. (1979): *Numerical methods for free surface problems by means of penalty.* Lecture Notes in Math. Vol. **704**, 282–291. New York: Springer-Verlag.

160. Kawohl, B., Lang, J. (1997): Are some optimal shape problems convex? *J. Convex Anal.* **4 (2)**, 353–361.

161. Kelley, J. L. (1975): *General Topology.* Reprint of the 1955 edition [Van Nostrand, Toronto, Ont.], Graduate Texts in Mathematics, No. **27**. Berlin: Springer-Verlag.

162. Kinderlehrer, D., Stampacchia, G. (1980): *An Introduction to Variational Inequalities and Their Applications.* Pure and Applied Mathematics Vol. **88**. New York: Academic Press.

163. Kirjner-Neto, C., Polak, E. (1996): On the use of consistent approximations for the optimal design of beams. *SIAM J. Control Optim.* **34 (6)**, 1891–1913.

164. Kofler, M., Gräbe, H.-G. (2002): *Mathematica. Einführung, Anwendung, Referenz.* Addison-Wesley.

165. Komornik, V., Tiba, D. (1985): Contrôles de systèmes fortement non linéaires. *C. R. Acad. Sci. Paris Sér. I Math.* **300 (12)**, 393–396.

166. Křížek, M., Neittaanmäki, P. (1990): *Finite Element Approximation of Variational Problems and Applications.* Pitman Monographs and Surveys in Pure and Applied Mathematics Vol. **50**. Harlow: Longman and New York: J. Wiley & Sons.

167. Kuratowski, C. (1952): *Topologie*, Vol. **I** (French). 3ème ed. Monogr. Mat. Vol. XX. Warszawa: Polskie Towarzystwo Matematyczne.

168. Kuratowski, K. (1962): *Introduction to Set Theory and Topology.* Oxford: Pergamon Press.

169. Ladyzenskaya, O. A., Uraltseva, N. (1968): *Équations aux Dérivées Partielles de Type Elliptique.* Monographies Universitaires de Mathématiques, No. **31**. Paris: Dunod.

170. Lemaire, B. (1985): Application of a subdifferential of a convex composite functional to optimal control in variational inequalities. In: *LNEMS 255*, 103–118. Berlin: Springer-Verlag.

171. Levin, V. L. (1985): *Convex Analysis in Spaces of Measurable Functions and Its Applications in Mathematics and Economics* (in Russian). Moscow: Nauka.

172. Li, R., Liu, W., Ma, H., Tang, T. (2002): Adaptive finite element approximation for distributed elliptic optimal control problems. *SIAM J. Control Optim.* **41** **(5)**, 1321–1349.

173. Lions, J.-L. (1968): *Contrôle Optimale des Systèmes Gouvernés par des Équations aux Dérivées Partielles.* Paris: Dunod.

174. Lions, J.-L. (1969): *Quelques Méthodes de Résolution des Problèmes aux Limites Non Linéaires.* Paris: Dunod.

175. Lions, J.-L. (1972): *Some Aspects of the Optimal Control of Distributed Parameter Systems.* Philadelphia, PA: SIAM.

176. Lions, J.-L. (1983): *Contrôle des Systèmes Distribués Singuliers.* Méthodes Mathématiques de l'Informatique, No. **13**. Paris: Gauthier-Villars.

177. Lions, J.-L. (1988): *Controllabilité Exacte, Perturbations et Stabilisation des Systèmes Distribués.* Paris: Masson.

178. Lions, J.-L., Magenes, E. (1968): *Problèmes aux Limites Non Homogènes et Applications*, Vols. **1**, **2**. Travaux et Recherches Mathématiques, Nos. **17**, **18**. Paris: Dunod.

179. Lions, J.-L., Stampacchia, G. (1967): Variational inequalities. *Comm. Pure Appl. Math.* **20**, 493–519.

180. Litvinov, W. G. (2000): *Optimization in Elliptic Problems with Applications to Mechanics of Deformable Bodies and Fluid Mechanics.* Basel: Birkhäuser.

181. Liu, W., Neittaanmäki, P., Tiba, D. (2000): Sur les problèmes d'optimisation structurelle. *C. R. Acad. Sci. Paris Sér. I Math.* **331** **(1)**, 101–106.

182. Liu, W., Neittaanmäki, P., Tiba, D. (2003): Existence for shape optimization problems in arbitrary dimension. *SIAM J. Control Optim.* **41** **(5)**, 1440–1454.

183. Liu, W., Tiba, D. (2001): Error estimates in the approximation of optimization problems governed by nonlinear operators. *Numer. Funct. Anal. Optim.* **22** **(7&8)**, 953–972.

184. Liu, W., Yan, N. (2002): A posteriori error estimates for control problems governed by Stokes equations. *SIAM J. Numer. Anal.* **40** **(5)**, 1850–1869.

185. Liu, W., Yan, N. (2003): A posteriori error estimates for optimal control problems governed by parabolic equations. *Numer. Math.* **93** **(3)**, 497–521.

186. Lou, H. W. (2005): Existence of optimal controls in the absence of Cesari-type conditions for semilinear elliptic and parabolic systems. *J. Optim. Theory Appl.* **125** **(2)**, 367–391.

187. Lurie, K. A. (1975): *Optimal Control in Problems of Mathematical Physics* (in Russian). Moscow: Nauka.

188. Lurie, K. A. (1990): The extension of optimisation problems containing controls in the coefficients. *Proc. Roy. Soc. Edinburgh Sect. A* **114** **(1–2)**, 81–97.

189. Mäkelä, M., Neittaanmäki, P. (1992): *Nonsmooth Optimization. Analysis and Algorithms with Applications to Optimal Control.* Singapore: World Scientific.

190. Mäkinen, R., Neittaanmäki, P., Tiba, D. (1992): On a fixed domain approach for a shape optimization problem. In: *Computational and Applied Mathematics II: Differential Equations* (Ames, W. F., van der Houwen, P. J., eds.), 317–326. Amsterdam: North-Holland.

191. Mäkinen, R., Neittaanmäki, P., Tiba, D., Tiihonen, T. (1990): A boundary control approach to an optimal shape design problem. In: *Control of Distributed Parameter Systems* (Amouroux, M., El Jai, A., eds.), 415–418. Oxford: Pergamon Press.

192. Malanowski, K. (1982): Convergence of approximations vs. regularity of solutions for convex, control-constrained, optimal-control problems. *Appl. Math. Optim.* **8** **(1)**, 69–95.

193. Männikkö, T., Neittaanmäki, P., Tiba, D. (1994): A rapid method for the identification of the free boundary in two-phase Stefan problems. *IMA J. Numer. Anal.* **14**, 411–420.

194. Männikkö, T., Tiba, D. (1995): An optimal control approximation method in shape optimization problems. *Report Series of the Laboratory of Scientific Computing of the University of Jyväskylä*, No. **6**, Jyväskylä, Finland.

195. Masmoudi, M. (1987): Outil pour la conception optimale de formes. *Thèse d'Etat, Université de Nice*, Nice, France.

196. Masmoudi, M. (2001): The topological asymptotic expansion. In: *Computational Methods for Control Applications* (Glowinski, R., Kawarada, H., Periaux, J., eds.), 53–72. Gakuto Intern. Ser. Math. Sci. Appl. **16**. Tokyo: Gakkotosho.

197. Maz'ya, V. (1985): *Sobolev Spaces.* Berlin: Springer-Verlag.

198. Mignot, F., Murat, F., Puel, J. P. (1979): Variation d'un point de retournement par rapport au domaine. *Comm. Partial Differential Equations* **4** **(11)**, 1263–1297.

199. Mignot, F., Puel, J. P. (1984): Optimal control in some variational inequalities. *SIAM J. Control Optim.* **22** **(3)**, 466–476.

200. Mohammadi, B., Pironneau, O. (2001): *Applied Shape Optimization for Fluids.* New York: The Clarendon Press, Oxford University Press.

201. Mosco, U. (1971): An introduction to the approximate solutions of variational inequalities. In: *C.I.M.E.* (Geymonat, G., ed.), 497–682.

202. Murat, F. (1971): Un contre-exemple pour le problème du contrôle dans les coefficients. *C. R. Acad. Sci. Paris Sér. A-B* **273**, A708–A711.

203. Murat, F., Sili, A. (2000): Effects non locaux dans le passage 3d–1d en élasticité linéarisée anisotrope hétérogène. *C. R. Acad. Sci. Paris Sér. I Math.* **330**, 745–750.

204. Murat, F., Simon, J. (1976): Etude de problèmes d'optimal design. In: *Optimization Techniques. Modeling and Optimization in the Service of Man. Part 2* (Céa, J., ed.), Lecture Notes in Computer Science Vol. **41**, 54–62. Berlin: Springer-Verlag.

205. Nazarov, S. A., Sokołowski, J. (2003): Asymptotic analysis of shape functionals. *J. Math. Pures Appl.* **82 (2)**, 125–196.

206. Natori, M., Kawarada, H. (1981): An application of the integrated penalty method to free boundary problems of Laplace equation. *Numer. Funct. Anal. Optim.* **3 (1)**, 1–17.

207. Necas, J. (1967): *Les Méthodes Directes en Théorie des Équations Elliptiques.* Prague: Academia.

208. Neittaanmäki, P. (1991): Computer aided optimal structural design. *Surv. Math. Indust.* **1**, 173–215.

209. Neittaanmäki, P., Räisänen, T., Tiba, D. (1995): On the approximation of some ill-posed problems. In: *Qualitative Problems for Differential Equations and Control Theory* (Corduneanu, C., ed.), 251–262. River Edge, NJ: World Sci. Publishing.

210. Neittaanmäki, P., Tiba, D. (1988): A variational inequality approach to constrained control problems for parabolic equations. *Appl. Math. Optim.* **17 (3)**, 185–201.

211. Neittaanmäki, P., Tiba, D. (1994): *Optimal Control of Nonlinear Parabolic Systems. Theory, Algorithms, and Applications.* Monographs and Textbooks in Pure and Applied Mathematics Vol. **179**. New York: Marcel Dekker.

212. Neittaanmäki, P., Tiba, D. (1995): An embedding of domains approach in free boundary problems and optimal design. *SIAM J. Control Optim.* **33 (5)**, 1587–1602.

213. Neittaanmäki, P., Tiba, D. (2000): Shape optimization in free boundary systems. In: *Free Boundary Problems: Theory and Applications, II (Chiba, 1999)*, 334–343. GAKUTO Internat. Ser. Math. Sci. Appl. Vol. **14**. Tokyo: Gakkotosho.

214. Oden, J. T., Reddy, J. N. (1976): *An Introduction to the Mathematical Theory of Finite Elements.* New York: J. Wiley & Sons.

215. Okhezin, S. P. (1992): Optimal shape design for parabolic system and two-phase Stefan problem. In: *Free Boundary Problems in Continuum Mechanics*, 239–244. ISNM Vol. **106**. Basel: Birkhäuser.

216. Pascali, D., Sburlan, S. (1978): *Nonlinear Mappings of Monotone Type.* Alphen aan den Rijn: Sijthoff & Noordhoff.

217. Pavel, N. H. (1984): *Differential Equations, Flow Invariance and Applications.* Research Notes in Math. Vol. **113**. London: Pitman.

218. Pironneau, O. (1984): *Optimal Shape Design for Elliptic Systems.* Berlin: Springer-Verlag.

219. Pitkäranta, J., Leino, Y. (1994): On the membrane locking of the *h-p* finite elements in a cylindrical shell problem. *Int. J. Numer. Methods Engrg.* **37 (4)**, 415–432.

220. Pompeiu, D. (1905): Sur la continuité des fonctions de variable complexe. *Ann. Sci. Univ. Toulouse* **7 (2)**, 265–315.

221. Pontryagin, L. S. (1965): *Gewöhnliche Differentialgleichungen.* Berlin: VEB Deutscher Verlag der Wissenschaften.

222. Proskurowski, W., Widlund, O. (1976): On the numerical solution of Helmholtz equation by the capacitance matrix method. *Math. Comp.* **30 (135)**, 433–468.

223. Protter, M. H., Weinberger, H. E. (1967): *Maximum Principles in Differential Equations.* Englewood Cliffs, NJ: Prentice-Hall.

224. Raitums, U. (1989): *Optimal Control Problems for Elliptic Equations* (in Russian). Riga: Zinatne.

225. Raitums, U. (1997): *Lecture Notes on G-convergence, Convexification and Optimal Control Problems for Elliptic Equations.* Lecture Notes **39**, Univ. of Jyväskylä, Jyväskylä, Finland.

226. Renardy, M., Rogers, R. C. (1993): *An Introduction to Partial Differential Equations.* New York: Springer-Verlag.

227. Richards, D. (2002): *Advanced Mathematical Methods with Maple.* Cambridge, UK.

228. Rockafellar, R. T. (1970): *Convex Analysis.* Princeton, NJ: Princeton University Press.

229. Rousselet, B. (1983): Shape design sensitivity of a membrane. *J. Optim. Theory Appl.* **40 (4)**, 595–623.

230. Rudin, W. (1987): *Real and Complex Analysis.* New York: McGraw-Hill.

231. Saguez, C., Bermudez, A. (1985): Optimal control of variational inequalities. *Control Cybernet.* **14 (1–3)**, 9–30.

232. Schumacher, A. (1995): Topologieoptimierung von Bauteilstrukturen unter Verwendung von Lochpositionierungskriterien. *Doctoral Dissertation, Universität Gesamthochschule Siegen*, Siegen, Germany.

233. Simon, J. (1980): Differentiation with respect to the domain in boundary value problems. *Numer. Funct. Anal. Optim.* **2 (7–8)**, 649–687.

234. Simon, J. (1989): Second variations for domain optimization problems. In: *Control and Estimation of Distributed Parameter Systems*, 361–378. ISNM Vol. **91**. Basel: Birkhäuser.

235. Smirnov, W. I. (1955): *Lehrgang der Höheren Mathematik*, Teil **II**. Berlin: VEB Deutscher Verlag der Wissenschaften.

236. Sokołowski, J., Zochowski, A. (1999): On the topological derivative in shape optimization. *SIAM J. Control Optim.* **37 (4)**, 1251–1272.

237. Sokołowski, J., Zochowski, A. (2003): Optimality conditions for simultaneous topology and shape optimization. *SIAM J. Control Optim.* **42 (4)**, 1148–1221.

238. Sokołowski, J., Zolésio, J.-P. (1992): *Introduction to Shape Optimization. Shape Sensitivity Analysis.* Berlin: Springer-Verlag.

239. Soni, B. K., Thompson, J. F., Weatherill, N. P. (1999): *Handbook of Grid Generation.* New York: CRC Press.

240. Sprekels, J., Tiba, D. (1998): Propriétés de bang-bang généralisées dans l'optimisation des plaques. *C. R. Acad. Sci. Paris Sér. I Math.* **327**, 705–710.

241. Sprekels, J., Tiba, D. (1998/1999): A duality approach in the optimization of beams and plates. *SIAM J. Control Optim.* **37 (2)**, 486–501.

242. Sprekels, J., Tiba, D. (1999a): On the approximation and optimization of fourth order elliptic systems. In: *Optimal Control of Partial Differential Equations* (Hoffmann, K.-H., Leugering, G., Tröltzsch, F., eds.), 277–286. ISNM Vol. **133**. Basel: Birkhäuser.

243. Sprekels, J., Tiba, D. (1999b): A duality-type method for the design of beams. *Adv. Math. Sci. Appl.* **9 (1)**, 89–102.

244. Sprekels, J., Tiba, D. (2002): Control variational methods for differential equations. In: *Optimal Control of Complex Structures* (Hoffmann, K.-H., Lasiecka, I., Leugering, G., Sprekels, J., Tröltzsch, F., eds.), 245–257. ISNM Vol. **139**. Basel: Birkhäuser.

245. Sprekels, J., Tiba, D. (2002): An analytic approach to a generalized Naghdi shell model. *Adv. Math. Sci. Appl.* **12 (1)**, 175–190.

246. Sprekels, J., Tiba, D. (2003): Optimization of clamped plates with discontinuous thickness. *Systems Control Lett.* **48**, 289–295.

247. Stampacchia, G. (1965): Le problème de Dirichlet pour les équations elliptiques du second ordre à coefficients discontinus. *Ann. Inst. Fourier (Grenoble)* **15 (1)**, 189–258.

248. Sverak, V. (1992): On optimal shape design. *C. R. Acad. Sci. Paris Sér. I Math.* **315 (5)**, 545–549.

249. Sverak, V. (1993): On optimal shape design. *J. Math. Pures Appl.* **72 (9)**, 537–551.

250. Tahraoui, R. (1986): *Thèse de doctorat d'état, Université Paris VI*, Paris, France.

251. Tahraoui, R. (1992): Contrôle optimal dans les équations elliptiques. *SIAM J. Control Optim.* **30 (3)**, 495–521.

252. Tartar, L. (1975): Problèmes de contrôle des coefficients dans des équations aux dérivées partielles. In: *Control Theory, Numerical Methods and Computer Systems Modelling (Internat. Sympos., IRIA LABORIA, Rocquencourt, 1974)*, 420–426. Lecture Notes in Econom. and Math. Systems Vol. **107**. Berlin: Springer-Verlag.

253. Tiba, D. (1977): Subdifferentials of composed functions and applications in optimal control. *An. Ştiinţ. Univ. Al. I. Cuza Iaşi Mat. (N.S.)* **23 (2)**, 381–386.

254. Tiba, D. (1986): Une approche par inéquations variationnelles pour les problèmes de contrôle avec contraîntes. *C. R. Acad. Sci. Paris Sér. I Math.* **302 (1)**, 29–31.

255. Tiba, D. (1990a): *Optimal Control of Nonsmooth Distributed Parameter Systems.* Lecture Notes in Math. Vol. **1459**. Berlin: Springer-Verlag.

256. Tiba, D. (1990b): Une approche par contrôlabilité frontière dans les problèmes de design optimal. *C. R. Acad. Sci. Paris Sér. I Math.* **310 (4)**, 175–177.

257. Tiba, D. (1992): Controllability properties for elliptic systems, the fictitious domain method and optimal shape design problems. In: *Optimization, Optimal Control and Partial Differential Equations* (Barbu, V., Bonnans, J. F., Tiba, D., eds.), 251–261. ISNM Vol. **107**. Basel: Birkhäuser.

258. Tiba, D. (1995a): On the convex programming problem. *Preprint Series of the Institute of Mathematics of the Romanian Academy*, No. **2**, Bucharest, Romania.

259. Tiba, D. (1995b): *Lectures on The Optimal Control of Elliptic Problems.* Lecture Notes **32**, Department of Mathematics, University of Jyväskylä, Jyväskylä, Finland, 147 pages.

260. Tiba, D. (1996): Sur l'existence de multiplicateurs de Lagrange. *Preprint Series of the Institute of Mathematics of the Romanian Academy*, No. **2**, Bucharest, Romania.

261. Tiba, D. (1999): A property of Sobolev spaces and existence in optimal shape design. *Preprint Series of the Institute of Mathematics of the Romanian Academy*, No. **5**, Bucharest, Romania.

262. Tiba, D. (2003): A property of Sobolev spaces and existence in optimal design. *Appl. Math. Optim.* **47 (1)**, 45–58.

263. Tiba, D., Tröltzsch, F. (1996): Error estimates for the discretization of state constrained convex control problems. *Numer. Funct. Anal. Optim.* **17 (9–10)**, 1005–1028.

264. Tiba, D., Vodák, R. (2004): A general asymptotic model for Lipschitzian curved rods. *Preprint Series of the Weierstrass Institute for Applied Analysis and Stochastics*, No. **942**, Berlin, Germany. *Adv. Math. Sci. Appl.*, to appear.

265. Tiba, D., Xanthis, L. (1994): The method of arbitrary lines in the control of elliptic problems – error estimates. In: *Collected Papers from HERMIS '94* (Lipitakis, E. A., ed.), 205–217. Athens: Hellenic Math. Soc.

266. Tiba, D., Zălinescu, C. (2004): On the necessity of some constraint qualification conditions in convex programming. *J. Convex Anal.* **11 (1)**, 95–110.

267. Tiihonen, T. (1997): Shape optimization and trial methods for free boundary problems. *RAIRO Modél. Math. Anal. Numér.* **31 (7)**, 805–825.

268. Tiihonen, T., Gonzalez de Paz, R. (1994): A relaxation method in shape optimization. In: *Finite Element Methods* (Křížek, M., Neittaanmäki, P., Stenberg, R., eds.), 443–450. Lecture Notes in Pure and Appl. Math. Vol. **164**. New York: Marcel Dekker.

269. Trabucho, L., Viaño, J. M. (1996): Mathematical modelling of rods. In: *Handbook of Numerical Analysis*, Vol. **IV** (Ciarlet, P. G., Lions, J.-L., eds.). Amsterdam: Elsevier.

270. Tröltzsch, F. (1984): *Optimality Conditions for Parabolic Control Problems and Applications.* Teubner-Texte zur Mathematik Vol. **62**. Leipzig: Teubner.

271. Varga, R. S. (1962): *Matrix Iterative Analysis.* New Jersey: Prentice-Hall.

272. Vogelius, M. (1994): An inverse problem for the equation $\Delta u = -cu - d$. *Ann. Inst. Fourier (Grenoble)* **44 (4)**, 1181–1209.

273. Voisei, M. D. (2004): *First-order necessary optimality conditions for nonlinear optimal control problems.* Panamer. Math. J. **14 (1)**, 1–44.

274. Vulikh, B. Z. (1976): *A Brief Course in the Theory of Functions of a Real Variable.* Moscow: Mir Publishers.

275. Wang, G. (2002): Optimal control problems governed by non-well-posed semilinear elliptic equation. *Nonlinear Anal.* **49 (3)**, 315–333.

276. Wang, G., Wang, L. (2003): Maximum principle for optimal control of non-well-posed elliptic differential equations. *Nonlinear Anal.* **52 (1)**, 41–67.

277. Wang, G., Zhao, Y., Li, W. (2000): Some optimal control problems governed by elliptic variational inequalities with control and state constraint on the boundary. *J. Optim. Theory Appl.* **106 (3)**, 627–655.

278. Warga, J. (1972): *Optimal Control of Differential and Functional Equations.* New York: Academic Press.

279. Wolfram, S. (2003): *The Mathematica Book*, 5th edition. Champaign, IL: Wolfram Media, Inc.

280. Xanthis, L. S., Schwab, C. (1991): The method of arbitrary lines. *C. R. Acad. Sci. Paris Sér. I Math.* **312 (1)**, 181–187.

281. Xanthis, L. S., Schwab, C. (1992): The method of arbitrary lines – an *h-p* error analysis for singular problems. *C. R. Acad. Sci. Paris Sér. I Math.* **315 (13)**, 1421–1426.

282. Yosida, K. (1980): *Functional Analysis*, Sixth edition. Grundlehren der Mathematischen Wissenschaften Vol. **123**. Berlin: Springer-Verlag.

283. Young, L. C. (1969): *Lectures on the Calculus of Variations and Optimal Control Theory.* Philadelphia, PA: W. B. Saunders.

284. Zălinescu, C. (1980): On an abstract control problem. *Numer. Funct. Anal. Optim.* **2 (6)**, 531–542.

285. Zeidler, E. (1990): *Nonlinear Functional Analysis and its Applications*, Vol. **II**. New York: Springer-Verlag.

286. Zhikov, V. V., Kozlov, S. M., Oleinik, O. A. (1994): *Homogenization of Differential Operators and Integral Functionals.* Berlin: Springer-Verlag.

287. Zhong, X. (1998): On nonhomogeneous quasilinear elliptic equations. *Ann. Acad. Sci. Fenn. Math. Diss.* **117**, University of Jyväskylä, Jyväskylä, Finland.

288. Zolésio, J.-P. (1979): Identification de domaine par déformation. *Thèse, Université de Nice*, Nice, France.

289. Zolésio, J.-P. (1981): The material derivative (or speed) method for shape optimization. In: *Optimization of Distributed Parameter Structures*, Vol. **2** (Hang, E. J., Céa, J., eds.), 1089–1151. Alphen aan den Rijn, Netherlands: Sijthoff & Noordhoff.

290. Zolésio, J.-P. (1992): Introduction to shape optimization problems and free boundary problems. In: *Shape Optimization and Free Boundaries* (Delfour, M. C., Sabidussi, G., eds.), 397–457. Dordrecht: Kluwer.

291. Zolésio, J.-P. (1994): Weak shape formulation of free boundary problems. *Ann. Scuola Norm. Sup. Pisa Cl. Sci.* **21 (1)**, 11–44.

Index

admissibility *3, 27, 28, 39, 87, 97, 106, 112, 136, 159, 160, 168, 220, 258, 367*
algorithm
- Céa, Gioan, Michel *277 ff.*
- conjugate gradient *291, 292*
- gradient *282, 314, 361, 440*
- Rosen *361*
- Uzawa *361, 400*

approach
- complete least squares *255*
- controllability *296 ff.*
- direct *277 ff.*
- embedding *296 ff.*
- indirect *27*
- least squares *16, 267*

approximation
- ε-uniform *149, 442, 445*
- external *196*
- internal *196*
- second-order *326*
- smooth *187*
- Yosida *438*

arch *339 ff.*
- clamped *340*
- closed *347*
- curvature *340, 351*
- Gothic *347*
- Jordan *340*
- Kirchhoff–Love *340 ff.*
- smooth *349 ff.*
- Roman *362, 363, 365*
- thickness *340*

asymptotics
- topological *333 ff.*

bang-bang property *174, 183, 293 ff., 365, 403*
- generalized *124, 133, 174*

basis
- finite element *221, 222*
- local *340, 370*

beam *362*
- cantilevered *19, 42*
- unilaterally supported *19*

boundary
- free *12, 14, 100, 248 ff., 266*
- Lipschitz *46, 66, 76, 284, 404, 405, 424*
- regular *444*
- smooth *6, 126, 148, 158, 167, 296, 297, 457*

bounded
- essentially *447*
- locally *247, 441*

Calderon extension *49,326, 424, 466*
calmness *130 ff.*
capacity *43, 44, 57, 293*
coercive *27, 28, 31, 36, 112, 129, 134, 151, 206, 207, 220, 223, 226, 227, 233, 242, 252, 310, 377, 411 ff., 440, 445, 449*
compatibility *212*
- geometric *67*

condition
- Carathéodory *5, 10, 61*
- Cauchy boundary *100, 190, 191, 273, 297, 298*
- coercivity *9, 21, 87, 242, 243, 342, 439, 440, 441, 450*
- complementarity *91*
- Dirichlet boundary *8, 109, 129, 168, 222, 311, 313, 326, 334, 451, 452*
- ellipticity *6, 148, 241, 311, 451, 453*
- first-order *82, 94, 114, 116, 140, 143, 156, 163, 187, 208, 216, 223, 325, 399, 433*

- Fritz–John *142, 143, 158*
- growth *22, 55, 65, 137, 138, 140*
- interiority *87, 88, 102, 114, 121, 163, 164*
- Karush–Kuhn–Tucker *142*
- metric regularity *88*
- mixed boundary *65 ff., 252*
- Navier *374*
- Neumann boundary *8, 23, 41, 63 ff., 255, 288, 311, 313, 334*
- nonqualified *143*
- optimality *85, 91, 92, 103, 108, 109, 114, 116, 122, 140, 156, 159, 161, 168, 170, 171, 187, 208, 211, 221, 332*
- overdetermined boundary *255*
- periodicity *397*
- Robin boundary *311*
- second-order *142, 237*
- Signorini boundary *13*
- Slater *91, 102, 103, 107, 113, 114, 147, 206, 207, 210, 215*
- Stefan *256*
- Timoshenko *374*
- transmission *76, 284, 291*
- unilateral *13, 184, 369*

conductivity
- thermal *16, 256*

cone
- negative *159*
- normal *436*
- polar *144, 159*
- positive *159*
- tangent *144, 146, 399, 433*

constant
- coercivity *417*
- Lamé *375, 383, 400*
- Poincaré *33*

constraints *2, 3, 367, 396*
- active *2, 122*
- control *2, 20, 184, 220, 223, 343*
- global *7*
- integral *7, 123, 126, 421*
- local *7*
- mixed *2, 87, 112, 421*
- mixed pointwise *7*
- pointwise *7, 183*
- set *2, 207, 221, 365*
- state *2, 5, 20, 25, 30, 34, 122, 131, 158 ff., 176, 183, 258, 274, 343, 367*

control
- admissible *2, 3, 52, 77, 176, 227, 305*
- bang-bang *293 ff.*
- boundary *8, 15, 109, 126, 222*
- curves *46*
- distributed *6, 126, 175, 222, 257, 258, 260, 276, 309 ff.*
- global optimal *4*
- in the/into coefficients *10, 18, 18, 94, 167, 241, 288, 324, 339, 350, 422*
- local optimal *237*
- optimal *4, 177, 268*
- pointwise *9*
- suboptimal *31, 32*

controllability *266 ff., 300, 301*
- approximate *274, 296, 303*
- boundary *300*
- distributed *309 ff.*
- exact *259, 276, 302, 309*
- geometric *259, 296*

convergence
- domain *455 ff.*
- epi–convergence *38*
- G-convergence *38*
- Hausdorff–Pompeiu *69, 317, 457 ff.*
- H-convergence *38*
- Mosco *232*
- parametric *79, 459, 461*
- variational *38*

crack *79, 456*

cross section *14, 249, 374, 381 ff., 400*

cryogenics *253*

curve
- Jordan *369*
- polygonal *331*

- smooth *297*
- unit speed *370, 373*

cusp *80, 456*
cut *456*

Darcy's law *249*
decomposition *194*
- Cholesky *308*
- of forces *340*
- orthogonal *345*

deflection *240, 258*
δ-admissible *25*
deformation *340, 347*
- normal *349, 362*
- of curved rods *369 ff.*
- small *374*

demiclosedness *29, 31, 79, 118, 140, 286, 439*
demicontinuous *438, 450*
derivative
- conormal *7, 110, 451*
- directional *179, 180, 288, 289, 316, 325, 333, 355, 357, 396, 397, 432, 433, 435, 458*
- distributional *9*
- Fréchet *89 ff., 143, 179, 180, 397, 400, 435*
- Gâteaux *92, 104, 168, 171, 288, 353 ff., 393, 432, 435*
- local *326*
- material *325*
- normal *258, 297, 449*
- topological *338*

design
- optimal shape *16, 48, 52, 74, 182, 253ff., 277, 278, 301, 323, 349, 396*
- parameter
- shell optimal *420 ff.*

difference
- finite *196*

differential *278*
- directional *435*
- partial *432*

discretization *192 ff., 290*
displacement *373, 375, 405, 406*
- lateral *401*
- of midsurface *421*
- vertical *401*

distance
- Ekeland *460*
- Hausdorff–Pompeiu *53, 68, 259, 317, 338, 457*
- Hausdorff–Pompeiu complementary *305, 312, 314, 327, 458*

distribution *122, 125, 185, 252, 272, 341, 448*
- heat *44*
- material *76, 77, 292*
- optimal *17*

division *194, 280, 361*
domain
- admissible *52, 63, 67, 260, 261, 326, 331, 337*
- Carathéodory *280, 456, 468*
- δ-suboptimal *261, 262, 312*
- effective *2, 434*
- fictitious *15, 296 ff.*
- Lipschitz *65, 73, 76, 252, 254, 268, 448, 449, 456, 467*
- of class C *52, 55, 455, 462*
- of class C^m *455*
- of class $C^{1,1}$ *264, 310 ff.*
- optimal *58, 64, 265*
- polygonal *193, 196*
- regular *457*
- smooth *33, 109, 240, 302, 309, 310, 317, 331, 332, 457*
- stable *468*
- star-shaped *331*
- unknown *248 ff., 257, 280, 301, 313*
- variable *20 ff., 43 ff., 257*

elasticity
- linear *369*

element
- conforming *196*
- finite *192 ff., 280, 319, 361*
- Lagrange *194, 290*
- nonconforming *196*
- quadrilateral *290*
- simplex *193*
- triangular *193*

energy *36, 133, 184, 293*
- norm *198*

equation
- adjoint 85, 105, 111, 122, 167, 214, 217, 289, 399
- elliptic 14, 447 ff.
- Euler 135, 271
- fourth-order 19, 38 ff., 80, 167, 175, 178, 181, 240, 369
- in variations 330, 395, 432
- Laplace 294, 303, 334
- nonlinear 237 ff.
- of linear elasticity 375, 376
- of motion 373
- Poisson 168, 305, 334, 466
- second-order 14, 32 ff., 130, 167, 178, 451, 453
- semilinear 10, 33, 99, 126 ff., 239, 241
- state 2, 19, 22, 28, 85, 87, 110, 143, 342, 399
- variational 197, 381, 387, 392, 396, 400, 409

error
- interpolation 214, 219, 247

estimate
- a posteriori 192
- a priori 192, 202
- duality 172
- elliptic 169
- error 197 ff., 205 ff., 224, 247

family
- convergent 200, 201, 202
- of triangulations 194
- uniformly bounded 228, 231, 232
- uniformly coercive 227
- uniformly Lipschitz 243 ff.

Fenchel–Moreau conjugate 445

flux
- heat 44
- poloidal 14, 99, 297
- thermal power 323

force 400, 401
- body 373, 405, 420
- distributed 382
- external 340
- horizontal 340
- internal 340
- normal 340, 347, 362, 382
- tangential 362, 382
- torsional 382
- vertical 347
- volume 13, 380

formula
- Green 8, 13, 268, 453
- Poisson 304
- quadrature 204, 361
- variation of constants 341

frame (coordinate)
- Frenet 370, 372, 381
- Darboux 372, 381, 382

function
- characteristic 6, 16, 58, 63, 68, 77, 250, 252, 275, 283 ff., 292 ff., 306, 309, 317, 318, 361, 460, 462
- closed 434
- conjugate 445
- convex 434
- Green 40
- harmonic 40, 41, 42, 305, 327, 330, 331
- Heaviside 290
- indicator 2, 11, 25, 30, 87, 158, 165, 183, 245, 435, 450
- lower semicontinuous 434
- proper 434
- signed distance 77, 458
- state 277, 279
- strictly convex 434
- support 445
- target 46, 48, 77, 109, 158, 297
- test 60, 72, 122, 250, 330, 376, 408, 448
- value 130 ff.

functional
- compliance 37
- cost 2, 4, 7, 17, 22, 25, 31, 34, 39, 82, 102, 143, 159, 169, 184, 215, 220, 222, 243, 277, 278, 307, 318, 342, 359, 400, 404
- Dirac 9
- energy 175, 197

– quadratic 3, 255, 452, 396, 421

generalized perimeter 81
gradient 359, 360
– shape 333
– topological 333 ff.
graph
– Heaviside 77, 284, 286
– maximal monotone 77, 130, 147, 159, 240, 257, 276, 284, 454
grid 194

Hamiltonian
– pseudo- 132, 150
Hermite splines 195
hydraulic pressure 249

inclusion
– elliptic 12
inequality
– Clarkson 55
– Friedrichs 58, 129, 139
– Hölder 42, 72, 447
– Korn 339, 341, 344, 411, 424
– Poincaré 34, 44, 58, 151, 262, 306
– Schwarz 139, 181, 371
– variational 11, 13, 26, 66, 74, 142 ff., 183 ff., 225 ff., 251, 251, 257, 260, 262, 264, 276, 367 ff., 450 ff.
– Young 58, 262, 415, 419, 445
integrand
– convex 5, 129, 168, 436, 468
– normal convex 5
integration
– numerical 200, 204
interpolate 371
interpolation 224, 383
– operator 195, 196, 202, 218, 231
isotropic material 375

Jacobian 324, 328, 374, 406, 408, 421
junction 254, 255

Kronecker symbol 222

layout
– optimal 16, 284, 292
lemma
– Bramble–Hilbert 204
– Céa 202
– Fatou 62, 468
– Gronwall 353
– Hopf 269, 317
– Lax–Milgram 134, 151, 197, 207, 318, 341, 380, 396, 411, 449
– Poincaré 306
– Weyl 42
limitator 14
line
– of centroids 373, 374, 381, 387
– search 361, 400
load vector 198
local
– chart 53, 56, 67, 69, 449, 456
– coordinate frame/system 56, 315, 350, 362, 370 ff., 462
– differentiability 326

machining
– electrochemical 17, 66, 263
mapping
– Carathéodory 33, 34
– control 223
– demicontinuous 438, 450
– Dirichlet 9
– Dirichlet-to-Neumann 111, 335
– duality 116, 436
– hemicontinuous 437
– translated 57
matrix
– coefficients 6
– diagonally dominant 200
– fundamental 341 ff.
– mass 291
– rigidity 383
– sparse 193, 383
– stiffness 197, 290
– tridiagonal 199
membrane

– clamped 258
– elastic 258
mesh 194, 309, 319
method
– bundle 440
– conjugate gradient 319
– control variational 175 ff., 341 ff., 368
– direct 27
– Galerkin 197
– gradient 82
– Hilbert uniqueness 310
– integrated penalty 252, 294
– integrated topology 293, 322, 333
– least squares 15, 16, 18, 100, 255, 300
– mapping 10, 18, 283, 322 ff.
– moving grid 309, 330
– extension by reflections 49
– of lines 219 ff.
– Ritz 197
– Ritz–Galerkin 227 ff.
– speed 283, 322, 325
– steepest descent 440
– transposition 9, 453
– truncation 334
– variational inequality 26, 30
metric
– Ekeland 150, 446
– Hausdorff–Pompeiu 53
midsurface 404, 406, 421
minimal section 442
minimizing sequence 24, 27 ff., 41, 43, 45, 48, 58, 74, 78, 94, 96, 97, 104, 114, 160, 168, 269, 351, 365
minimum
– global 5
– local 4, 10, 92, 279
Minty characterization 89, 438
model
– polynomial 339
– Naghdi (generalized) 339, 372 ff., 404 ff.
– Kirchhoff–Love 339 ff.
mollifier
– Friedrichs 71, 152, 276, 288, 352, 359
multiplier
– Lagrange 122, 125, 147.

node 204, 290, 307, 308, 319, 331

observation
– boundary 7, 15
– distributed 7
operator
– bounded 107, 206, 209, 211, 213, 242, 246, 440
– canonical injection 6, 114, 213, 215
– demiclosed 171
– extension 49
– generalized divergence 21, 54, 444
– hemicontinuous 195, 196, 437, 445
– Laplace 19, 22, 62, 324
– Leray–Lions 21, 54, 444
– maximal monotone 11, 21, 31, 54, 55, 88, 89, 115, 118, 140, 148, 171, 211, 226, 286, 437, 445, 450
– monotone 21, 89, 437, 445, 450
– multivalued 88, 437
– Nemytskii 10
– observation 4, 86
– solution 392
– strictly monotone 437, 450
– strongly monotone 206, 242
– trace 51, 139
– translation 9
optimization
– domain 257, 284
– shape 46, 48, 55, 58, 66, 74, 248, 253 ff., 291, 293, 322, 333, 350, 360, 387 ff., 396, 421 ff.
– size 333
– structural 77, 284 ff., 333
– thickness 167 ff., 182
– topology 292, 333, 336
overdetermined boundary conditions 255

pair

- admissible *2, 34, 41, 79, 97, 102, 104, 144, 161, 169, 172, 176, 206, 207, 269, 270, 274, 286, 367*
- δ-admissible *25, 26*
- extremal *83, 84*
- feasible *113, 114, 119, 120*
- optimal *4, 26, 27, 34, 39, 78, 79, 84, 86, 97, 103, 110, 113, 114, 118, 120, 122, 140, 145, 166, 168, 171, 173, 176, 206, 207, 212, 215, 216, 221, 226, 227, 230, 234ff., 244, 275ff., 285, 297, 312, 346, 367*

parameter
- discretization *194, 206, 243*
- line search *361*
- minimization *1, 100, 182, 182, 255, 350*
- optimization *255, 260*
- regularization *15*
- triangulation *194*

parametrization *458*
- arc length *340, 387*
- polar coordinates *253, 303, 331, 400*

partition
- of domain *194, 280*
- of unity *56, 80*

penalization *102ff., 159, 183ff., 208, 215, 274, 294*
- adapted *96, 98, 104, 109, 114, 158*
- exact *102, 126ff.*
- exterior *102*
- external *102*
- internal *102*
- state equation *112ff., 158ff.*
- weighted *255*

piecewise
- constant *196, 217, 361*
- linear *194, 217, 218, 319, 383, 400*
- polynomial *192*

plasma *14, 99, 248, 253, 297*

plate *167ff.*
- clamped *19, 40, 175ff., 183*
- partially clamped *184, 191*
- simply supported *19, 175, 240*
- thickness *240*

point
- critical *279*
- ε-critical *280, 281*
- Lebesgue *141, 155, 279*

polar subspace *442*

polygon *193, 202, 217*

polyhedron *194*

potential
- electric *255*
- theory *81*

principle
- Dirichlet *10, 175, 341, 344*
- Ekeland's variational *148, 155, 446*
- maximum *34, 35, 62, 121, 124, 151, 173, 174, 240, 251, 302, 305, 337, 338, 453*
- Pontryagin maximum *85, 95, 112, 117, 122, 167, 182, 183, 187, 223, 271, 272, 298*
- shape variational *253*
- uniform boundedness *108*

problem
- adjoint *187*
- bottleneck *123*
- boundary control *7, 177, 219, 265, 307*
- boundary observation *111*
- Cauchy *15, 100, 186, 341*
- constrained *176*
- control *2, 25, 26, 82, 85, 109, 120, 166, 176, 183, 220, 225ff., 242, 274, 313, 341, 343, 367*
- dual *346*
- elliptic *2, 13, 96, 133, 151, 266*
- dam *249ff.*
- Dirichlet *6, 52ff., 302, 451, 452, 453, 467*
- discretized *192*
- discretized control *207, 229*
- fourth-order *38ff.*
- identification *10, 16, 94,*

100, 172, 297
- ill-posed *15*
- inverse *16*
- linear–quadratic *313*
- locking *383*
- minimization *11, 19, 26*
- mixed boundary *43, 452*
- Neumann *23, 43, 79, 451*
- nonlinear elliptic *10*
- nonlinear control *3*
- nontrivial *3, 121*
- obstacle *12, 17, 158, 239, 248, 251, 257, 263, 266, 369*
- optimal packaging *258, 262*
- quadratic control *18*
- second-order *32 ff., 43*
- Signorini *13*
- singular control *95 ff.*
- Stefan *255*
- third boundary value *451*
- transmission *318*
- unconstrained *7, 177, 345*
- unilateral *13*

programming
- convex *87, 145*
- differentiable *92, 144*
- mathematical *87 ff., 176, 307*
- sequential quadratic *307*

projection *177, 243, 466, 345, 376, 410*

property
- attractor-type *237*
- coercivity *243*
- compactivorous *461*
- cone *457*
- controllability *26*
- decomposition *442*
- descent *440*
- Γ property *63, 69, 461*
- identifiability *100, 101*
- lattice *217, 259*
- Lindelöf *460*
- minimality *106*
- orthogonality *196, 213, 214*
- segment *56, 57, 64, 68, 327, 456, 463*
- solid mean *42, 305*

qualification
- constraint *88, 113, 130*

realization *54, 445*

region
- elastic *12*
- plastic *12*
- saturated *249*
- unsaturated *249*

regularity *125, 184, 187, 190, 221, 305, 318, 339, 370, 372, 404*
- boundary *273, 326 ff.*
- global *452*
- local *78, 173, 273*
- maximal *111, 123, 173, 183*
- theory *35, 95, 317, 454*

regularization *358, 359*
- Chan–Vese *290*
- Friedrichs *328, 442*
- Tikhonov *15, 16, 297*
- Yosida–Moreau *26, 89, 103, 104, 114, 115, 208, 241, 242, 285, 439*

relaxation *38, 252, 285, 370*

resolvent *106, 115, 438*

rod
- curvature *373, 374*
- curved *369 ff.*

rule
- additivity *90, 166, 245*
- chain *93, 96, 213, 246, 315, 325, 375, 432*
- multiplier *146*
- resizing *179*

Schmidt orthogonalization *371*

semiconductor *254, 255*

semidiscretization *219 ff.*

sensitivity analysis *322, 352*

sequence
- minimizing *351, 365*

set
- admissible *25, 226, 284, 326, 365*
- coincidence *12, 66, 248, 258, 276*
- polar *442, 445*

shear *374*

shell
- cylindrical 340
- generalized Naghdi 404 ff.
- partially clamped 409
- thickness 420

solution
- classical 451
- δ-solution 24, 25, 132, 150, 152
- discrete 197
- generalized 352
- global 291, 338, 366
- local 338
- mild 341, 343
- optimal 28, 396, 401, 429
- strong 10, 14, 302, 312, 451
- transposition 86, 98, 109, 111, 177, 178, 184ff., 296, 298, 309, 453
- variational 129 ff.
- weak 2, 10, 37, 54, 93, 129, 215, 216, 220, 238, 301, 344, 363, 451, 452, 467

space
- finite element 194, 195
- fractional order 448
- observation 16
- orthogonal 342
- polar 84
- radiator 323
- Sobolev 195, 297, 448
- trace 448

spiral 381

stability
- Hedberg–Keldys 58, 252

state 2
- adjoint 110, 112, 122, 132, 140, 271, 275, 308, 433
- optimal 268, 331, 343, 367, 401

Stefan condition 256

strain tensor 375, 405

subdifferential 12, 89 ff., 108, 128, 129, 133, 158, 172, 183, 206, 209, 212, 223, 241, 245, 246, 435
- partial 83

submanifold 272

summation convention 375, 376,

surface 372, 407
- curvature 407
- cylindrical 381, 382
- lateral 373
- traction 373, 380, 405, 420

system
- adjoint 117, 178, 209, 216, 223, 298, 309, 318, 332, 336
- semidiscretized 222
- state 3, 30, 97, 109, 159, 176, 220, 294, 360, 368
- optimality 179, 188
- linear elasticity 380, 405, 410

theorem
- Arzèla–Ascoli 46, 48, 75, 141
- divergence 251
- Egorov 42, 61, 97, 287
- Fermat 446
- Havin–Bagby 57
- Holmgren 273, 276, 298
- implicit function 78, 316
- Karush–Kuhn–Tucker 88
- Lebesgue 34, 79, 136, 137, 139, 152, 169, 286
- Lions–Stampacchia 11, 13, 66, 74, 238, 252, 257, 449
- Mazur 276, 299
- open mapping 415
- Rellich 449
- Riesz 95, 329, 453
- Schauder fixed point 33, 34
- Sobolev embedding 9, 97, 123, 124, 151, 168, 171, 203, 258, 295, 301, 449
- trace 47, 55, 73, 129, 134, 176, 184, 187, 252, 258, 310, 312, 375, 449
- Vitali 61
- Weierstrass 24, 295, 461

tokamak 14, 99, 248

torsion
- elastoplastic 12, 74, 248

trace 66, 178, 269, 275, 297, 449

transformation
- Baiocchi 250, 257

- domain *326*
- geometric *373, 404*
- smooth *324*
- tangent *326*

triangulation *193 ff., 217*
- regular *194, 202*
- strongly regular *194, 196*

trick
- Aubin–Nitsche *203*

triple
- admissible *186*
- optimal *160, 161, 185 ff.*

value
- extreme *123*
- nodal *290*
- optimal *5*

variation *92, 314, 358*
- admissible *278, 325, 342, 367, 392*
- boundary *293, 322, 330, 332, 338*
- domain *277, 322 ff.*
- geometry *387*
- interior *322, 325*
- partial *163*
- spike *136, 138, 140, 152, 153, 155*

water reservoir *249*

weight
- coefficients *4*
- minimization of *20*

Young modulus *347*

Notation

$\emptyset$	the empty set
$\mathbf{N}$	set of positive integers
$\mathbf{N}_0$	$\mathbf{N} \cup \{0\}$
$\overline{1,n}$	set of integers from 1 to n
$\mathbf{Z}$	set of integers
$\mathbf{R}$	set of real numbers
$\mathbf{R}^d$	Euclidean space of dimension $d \in \mathbf{N}$
$\cdot$	scalar product in $\mathbf{R}^d, d \geq 1$
$\wedge$	vector product in $\mathbf{R}^3$
$\lvert \cdot \rvert$	Euclidean norm or modulus
$\lvert\alpha\rvert$	length of multi-index $\alpha = (\alpha_1, \ldots, \alpha_d) \in \mathbf{N}_0^d$
x_+	$\max(x, 0)$, the positive part of $x \in \mathbf{R}$
x_-	$-\min(x, 0)$, the negative part of $x \in \mathbf{R}$
Ω	domain (open connected set) in $\mathbf{R}^d$
$\dim \Omega$	dimension of the Euclidean space containing the domain Ω
$\text{meas}(\Omega)$	Lebesgue measure of the measurable set $\Omega \subset \mathbf{R}^d$
χ_Ω	characteristic function of $\Omega \subset \mathbf{R}^d$
$d(x, \Omega)$	distance of the point x from the set $\Omega \subset \mathbf{R}^d$
d_Ω	signed distance function of an open set $\Omega \subset \mathbf{R}^d$
$\partial\Omega$, Γ	boundary (or part of the boundary) of $\Omega \subset \mathbf{R}^d$
n, ν	normal vectors to $\partial\Omega$
$\mathcal{O}$	family of open sets in $\mathbf{R}^d$
$d_H(A_1, A_2)$	Hausdorff–Pompeiu distance of two compact sets A_1, A_2 in $\mathbf{R}^d$
$\tilde{d}_H(\Omega_1, \Omega_2)$	Hausdorff–Pompeiu complementary distance of two open and bounded sets $\Omega_1, \Omega_2 \subset \mathbf{R}^d$
$\dfrac{\partial y}{\partial x_i}$	partial derivative of the function y
∇, grad	gradient operator

div	divergence operator
Δ	Laplace operator
∇T	Gâteaux, Fréchet derivative of a mapping T, Jacobian of a transformation $T : \mathbf{R}^d \to \mathbf{R}^d$
$\frac{\partial y}{\partial n}, \frac{\partial y}{\partial n_A}$	normal, conormal derivatives of y to $\partial\Omega$
D^α	partial differential operator corresponding to the multi-index $\alpha \in \mathbf{N}_0^d$
$\partial\varphi$	subdifferential of the mapping φ
$\vert \cdot \vert_Z$	norm in the normed space Z
Z^*	the dual space of Z
$(\cdot,\cdot)_H$	inner product in the Hilbert space H
$(\cdot,\cdot)_{V^* \times V}$	the pairing between the space V and its dual
$L(V,W)$	space of linear and bounded operators between two normed spaces V and W
$B_r(x),\ B(x,r)$	balls of radius $r > 0$ centered at x, in metric spaces
V_x	neighborhood of the point x
$\mathrm{int}(A)$	interior of a set A
$\overline{A}$	closure of a set A
$\mathrm{conv}(A)$	convex hull of a set A
$A^\perp$	orthogonal complement of a set A in a Hilbert space
$A \times B$	Cartesian product of two sets A, B
$A \oplus B$	the direct sum of two linear spaces A, B
$[\cdot,\cdot]$	ordered pair in the Cartesian product of two sets
I_K	indicator function of the set K
I_K^*	support function of the set K
δ_{x_0}	Dirac distribution concentrated in $x_0 \in \Omega$
$C^m(\Omega),\ C^m(\overline{\Omega})$	spaces of continuously differentiable functions up to order m in Ω, respectively in $\overline{\Omega}$
$C^\infty(\Omega)$	space of infinitely differentiable functions in Ω
$C_0^\infty(\Omega) = \mathcal{D}(\Omega)$	space of test functions: infinitely differentiable and with compact support in Ω
$\mathcal{D}'(\Omega)$	space of generalized functions (distributions), the dual of $\mathcal{D}(\Omega)$
$L^p(\Omega),\ L^\infty(\Omega)$	space of p-integrable functions in Ω, $p \geq 1$
$H^s(\Omega),\ H^{-s}(\Omega),$ $W^{l,p}(\Omega),\ W^{l,\infty}(\Omega)$	various Sobolev spaces in Ω, $s \in \mathbf{R}$, $l \in \mathbf{Z}$, $p \geq 1$

$L^p(\partial\Omega)$, $H^s(\partial\Omega)$	Lebesgue and Sobolev spaces on $\partial\Omega$
φ_λ	Yosida–Moreau regularization of a convex function φ
$\operatorname{supp}\varphi$	support of the function φ
$\varphi\vert_M$	restriction of the function or operator φ to the set M
$\operatorname{dom}(\varphi)$	domain of the function or operator φ
$R(\varphi)$	range of the function or operator φ
$\operatorname{Epi}\varphi$	epigraph of the function φ
$\operatorname{cl}\varphi$	closure of the function φ
A_λ	Yosida approximation of a maximal monotone operator $A : X \to X^*$ for $\lambda > 0$
$\mathcal{F}$	family of local charts or functions
$\mathcal{T}_h$	triangulation of Ω
$\mathcal{T}$	family of triangulations
$\det A$	determinant of the matrix A
δ_{ij}	Kronecker's symbol: $\delta_{ii} = 1$, $\delta_{ij} = 0$ for $i \neq j$
$d_E(u, v)$	Ekeland distance of two functions u, v

Springer Monographs in Mathematics

This series publishes advanced monographs giving well-written presentations of the "state-of-the-art" in fields of mathematical research that have acquired the maturity needed for such a treatment. They are sufficiently self-contained to be accessible to more than just the intimate specialists of the subject, and sufficiently comprehensive to remain valuable references for many years. Besides the current state of knowledge in its field, an SMM volume should also describe its relevance to and interaction with neighbouring fields of mathematics, and give pointers to future directions of research.

Abhyankar, S.S. **Resolution of Singularities of Embedded Algebraic Surfaces** 2nd enlarged ed. 1998
Alexandrov, A.D. **Convex Polyhedra** 2005
Andrievskii, V.V.; Blatt, H.-P. **Discrepancy of Signed Measures and Polynomial Approximation** 2002
Angell, T.S.: Kirsch, A. **Optimization Methods in Electromagnetic Radiation** 2004
Ara, P.; Mathieu, M. **Local Multipliers of C*-Algebras** 2003
Armitage, D.H.; Gardiner, S.J. **Classical Potential Theory** 2001
Arnold, L. **Random Dynamical Systems** corr. 2nd printing 2003 (1st ed. 1998)
Arveson, W. **Noncommutative Dynamics and E-Semigroups** 2003
Aubin, T. **Some Nonlinear Problems in Riemannian Geometry** 1998
Auslender, A.; Teboulle, M. **Asymptotic Cones and Functions in Optimization and Variational Inequalities** 2003
Bang-Jensen, J.; Gutin, G. **Digraphs** 2001
Baues, H.-J. **Combinatorial Foundation of Homology and Homotopy** 1999
Brown, K.S. **Buildings** 3rd printing 2000 (1st ed. 1998)
Cherry, W.; Ye, Z. **Nevanlinna's Theory of Value Distribution** 2001
Ching, W.K. **Iterative Methods for Queuing and Manufacturing Systems** 2001
Crabb, M.C.; James, I.M. **Fibrewise Homotopy Theory** 1998
Chudinovich, I. **Variational and Potential Methods for a Class of Linear Hyperbolic Evolutionary Processes** 2005
Dineen, S. **Complex Analysis on Infinite Dimensional Spaces** 1999
Dugundji, J.; Granas, A. **Fixed Point Theory** 2003
Elstrodt, J.; Grunewald, F.; Mennicke, J. **Groups Acting on Hyperbolic Space** 1998
Edmunds, D.E.; Evans, W.D. **Hardy Operators, Function Spaces and Embeddings** 2004
Fadell, E.R.; Husseini, S.Y. **Geometry and Topology of Configuration Spaces** 2001
Fedorov, Y.N.; Kozlov, V.V. **A Memoir on Integrable Systems** 2001
Flenner, H.; O'Carroll, L.; Vogel, W. **Joins and Intersections** 1999
Gelfand, S.I.; Manin, Y.I. **Methods of Homological Algebra** 2nd ed. 2003
Griess, R.L. Jr. **Twelve Sporadic Groups** 1998
Gras, G. **Class Field Theory** corr. 2nd printing 2005
Hida, H. ***p*-Adic Automorphic Forms on Shimura Varieties** 2004
Ischebeck, F.; Rao, R.A. **Ideals and Reality** 2005
Ivrii, V. **Microlocal Analysis and Precise Spectral Asymptotics** 1998
Jech, T. **Set Theory** (3rd revised edition 2002)
Jorgenson, J.; Lang, S. **Spherical Inversion on SLn (R)** 2001
Kanamori, A. **The Higher Infinite** corr. 2nd printing 2005 (2nd ed. 2003)
Kanovei, V. **Nonstandard Analysis, Axiomatically** 2005
Khoshnevisan, D. **Multiparameter Processes** 2002
Koch, H. **Galois Theory of *p*-Extensions** 2002
Komornik, V. **Fourier Series in Control Theory** 2005
Kozlov, V.; Maz'ya, V. **Differential Equations with Operator Coefficients** 1999
Landsman, N.P. **Mathematical Topics between Classical & Quantum Mechanics** 1998
Leach, J.A.; Needham, D.J. **Matched Asymptotic Expansions in Reaction-Diffusion Theory** 2004
Lebedev, L.P.; Vorovich, I.I. **Functional Analysis in Mechanics** 2002
Lemmermeyer, F. **Reciprocity Laws: From Euler to Eisenstein** 2000
Malle, G.; Matzat, B.H. **Inverse Galois Theory** 1999

Mardesic, S. **Strong Shape and Homology** 2000
Margulis, G.A. **On Some Aspects of the Theory of Anosov Systems** 2004
Murdock. J. **Normal Forms and Unfoldings for Local Dynamical Systems** 2002
Narkiewicz, W. **Elementary and Analytic Theory of Algebraic Numbers** 3rd ed. 2004
Narkiewicz, W. **The Development of Prime Number Theory** 2000
Neittaanmaki, P.; Sprekels, J.; Tiba, D. **Optimization of Elliptic Systems: Theory and Applications** 2006
Parker, C.; Rowley, P. **Symplectic Amalgams** 2002
Peller, V. (Ed.) **Hankel Operators and Their Applications** 2003
Prestel, A.; Delzell, C.N. **Positive Polynomials** 2001
Puig, L. **Blocks of Finite Groups** 2002
Ranicki, A. **High-dimensional Knot Theory** 1998
Ribenboim, P. **The Theory of Classical Valuations** 1999
Rowe, E.G.P. **Geometrical Physics in Minkowski Spacetime** 2001
Rudyak, Y.B. **On Thom Spectra, Orientability and Cobordism** 1998
Ryan, R.A. **Introduction to Tensor Products of Banach Spaces** 2002
Saranen, J.; Vainikko, G. **Periodic Integral and Pseudodifferential Equations with Numerical Approximation** 2002
Schneider, P. **Nonarchimedean Functional Analysis** 2002
Serre, J-P. **Complex Semisimple Lie Algebras** 2001 (reprint of first ed. 1987)
Serre, J-P. **Galois Cohomology** corr. 2nd printing 2002 (1st ed. 1997)
Serre, J-P. **Local Algebra** 2000
Serre, J-P. **Trees** corr. 2nd printing 2003 (1st ed. 1980)
Smirnov, E. **Hausdorff Spectra in Functional Analysis** 2002
Springer, T.A.; Veldkamp, F.D. **Octonions, Jordan Algebras, and Exceptional Groups** 2000
Sznitman, S.-S. **Brownian Motion, Obstacles and Random Media** 1998
Taira, K. **Semigroups, Boundary Value Problems and Markov Processes** 2003
Talagrand, M. **The Generic Chaining** 2005
Tauvel, P.; Yu, R.W.T. **Lie Algebras and Algebraic Groups** 2005
Tits, J.; Weiss, R.M. **Moufang Polygons** 2002
Uchiyama, A. **Hardy Spaces on the Euclidean Space** 2001
Üstünel, A.-S.; Zakai, M. **Transformation of Measure on Wiener Space** 2000
Vasconcelos, W. **Integral Closure. Rees Algebras, Multiplicities, Algorithms** 2005
Yang, Y. **Solitons in Field Theory and Nonlinear Analysis** 2001